Land-Atmosphere Interactions and Effects on the Climate of the Tibetan Plateau and Surrounding Regions

Land-Atmosphere Interactions and Effects on the Climate of the Tibetan Plateau and Surrounding Regions

Editors

Yaoming Ma
Li Jia
Massimo Menenti
Lei Zhong

MDPI • Basel • Beijing • Wuhan • Barcelona • Belgrade • Manchester • Tokyo • Cluj • Tianjin

Editors

Yaoming Ma
Institute of Tibetan Plateau Research, Chinese Academy of Sciences
China

Li Jia
Aerospace Information Research Institute, Chinese Academy of Sciences
China

Massimo Menenti
Department of Geosciennce and Remote Sensing, Technische Universiteit Delft
The Netherlands

Lei Zhong
School of Earth and Space Sciences, University of Science and Technology of China
China

Editorial Office
MDPI
St. Alban-Anlage 66
4052 Basel, Switzerland

This is a reprint of articles from the Special Issue published online in the open access journal *Remote Sensing* (ISSN 2072-4292) (available at: https://www.mdpi.com/journal/remotesensing/special_issues/land_atmosphere_interactions_climate_Tibetan_Plateau).

For citation purposes, cite each article independently as indicated on the article page online and as indicated below:

LastName, A.A.; LastName, B.B.; LastName, C.C. Article Title. *Journal Name* **Year**, *Volume Number*, Page Range.

ISBN 978-3-0365-6515-6 (Hbk)
ISBN 978-3-0365-6516-3 (PDF)

Cover image courtesy of Yaoming Ma

Contents

About the Editors

Yaoming Ma

Prof. Dr. Yaoming Ma is a Professor at the Institute of Tibetan Plateau Research, Chinese Academy of Sciences. He received his PhD degree in atmospheric physics from Okayama University in Japan and PhD degree in environmental science from Wageningen University in The Netherlands. He is a distinguished research fellow of the Chinese Academy of Sciences and Post Professor of the University of the Chinese Academy of Sciences. He is Academic Director of National Observation and Research Station for Qomolongma Special Atmospheric Processes and Environmental Changes. His research interests are atmospheric boundary layer observation and satellite remote sensing application. He has been responsible for nearly 20 national major scientific research programs, including key programs of the National Natural Science Foundation of China, the National Science Fund for Distinguished Young Scholars, major international cooperation programs, etc. At the same time, he is one of the two chief coordinators of the major international cooperation research programs 'GAME/Tibet' and 'CAMP/Tibet'.

Li Jia

Prof. Dr. Li Jia is a Full Professor and Director of Laboratory for Earth Observation for Terrestrial Water Cycle and Climate Change in the Aerospace Information Research Institute (AIR), Chinese Academy of Sciences (CAS). She received her PhD degree in Environmental Science from Wageningen University of The Netherlands in 2004. Her main research disciplines are quantitative remote sensing and remote sensing applications to hydrometeorology, eco-hydrology, water resources, carbon assimilation, drought process and monitoring, and climate change. She has more than 300 publications, more than 90 of which are peer-reviewed ISI journal papers. She is a member of WCRP/GEWEX Hydroclimatology Panel (GHP) (since 2019) and IEEE. She is the co-chair of the working group on Drought Monitoring and Evaluation of the AOGEO (Asia-Oceania Group on Earth Observations) (since 2017). She serves as Editor/Associate Editor of numerous academic journals.

Massimo Menenti

Prof. Dr. Massimo Menenti is an internationally renowned scientist in the fields of earth observation and global terrestrial water cycle. He has held senior research positions in The Netherlands, France, USA, China, and Italy, received research awards in The Netherlands, France, USA, and China, and led large projects with participants from Europe, Asia, America, and Africa. His best known achievements have been attained in the aspects of surface parameter retrievals from remote sensing, remote sensing-based evapotranspiration (ET) estimation, time series analysis of remote sensing data, and the application of remote sensing technology in hydrology and climate models. Prof. Menenti initiated the use of RS to assess and monitor crop water requirements and irrigation performance in the late 1980s. He is one of the earliest researchers in using laser technology to measure surface aerodynamic roughness. He initiated the use of time series analysis of satellite data. He presented the surface energy balance index (SEBI) theory for ET estimation, which is the prototype of the following S-SEBI, SEBS, and SEBAL models.

Lei Zhong

Prof. Dr. Lei Zhong is a Professor at the School of Earth and Space Sciences, University of Science and Technology of China. He received his PhD degree in Natural Geography in 2008 at Institute of

Tibetan Plateau Research, Chinese Academy of Sciences. His main research interests are focused on land–atmosphere interaction, energy and water cycle, application of remote sensing, and monsoon climate. He won the National Natural Science Foundation for Distinguished Young Scholars in 2015. He received first prize for the Tibet Autonomous Region Science and Technology Award in 2022. He has been actively involved in major international cooperation programs, such as 'CAMP/Tibet' and the ESA-MOST Dragon program. He is a director of China Association for Scientific Exploration and Executive Youth Member of Editorial Board of Plateau Meteorology. He is also member of the IAMAS Youth Working Group.

Preface to "Land-Atmosphere Interactions and Effects on the Climate of the Tibetan Plateau and Surrounding Regions"

This book focused on recent advances in land-atmosphere interactions and their effects on the climate change over the Tibetan Plateau and surrounding regions using multisource remote sensing data and in situ measurements.

Retrieval of land surface variables and surface heat fluxes, as well as change monitoring in snow, glaciers, lakes, and other land-surface covers are of particular interest. Special attention is given to retrieval of land-surface key properties, variations of land-surface heat fluxes, estimation of precipitation and evapotranspiration, change monitoring of glacier and lakes, the responses of lakes to climate change, carbon, water and heat exchange in terrestrial ecosystems, risk assessment of snow disasters, estimation of turbulence characteristics, vegetation dynamics and its response to weather and climate.

This book was funded by the the Second Tibetan Plateau Scientific Expedition and Research (STEP) Program, Ministry of Science and Technology of the People's Republic of China (Grant No. 2019QZKK0103); the Strategic Priority Research Program of Chinese Academy of Sciences (Grant No. XDA20060101); National Natural Science Foundation of China (Grant No. 42230610, 91837208, 41875031).

Yaoming Ma, Li Jia, Massimo Menenti, and Lei Zhong
Editors

remote sensing

Editorial

Land-Atmosphere Interactions and Effects on the Climate of the Tibetan Plateau and Surrounding Regions

Yaoming Ma [1,2,3,4,5,6], Lei Zhong [7,8,9,10,*], Li Jia [11] and Massimo Menenti [11,12]

1 State Key Laboratory of Tibetan Plateau Earth System, Resources and Environment (TPESRE), Institute of Tibetan Plateau Research, Chinese Academy of Sciences, Beijing 100101, China
2 College of Earth and Planetary Sciences, University of Chinese Academy of Sciences, Beijing 100049, China
3 College of Atmospheric Science, Lanzhou University, Lanzhou 730000, China
4 National Observation and Research Station for Qomolongma Special Atmospheric Processes and Environmental Changes, Dingri 858200, China
5 Kathmandu Center of Research and Education, Chinese Academy of Sciences, Beijing 100101, China
6 China-Pakistan Joint Research Center on Earth Sciences, Chinese Academy of Sciences, Islamabad 45320, Pakistan
7 School of Earth and Space Sciences, University of Science and Technology of China, Hefei 230026, China
8 CAS Center for Excellence in Comparative Planetology, USTC, Hefei 230026, China
9 Frontiers Science Center for Planetary Exploration and Emerging Technologies, University of Science and Technology of China, Hefei 230026, China
10 Jiangsu Collaborative Innovation Center for Climate Change, Nanjing 210023, China
11 Aerospace Information Research Institute, Chinese Academy of Sciences, Beijing 100101, China
12 Department of Geoscience and Remote Sensing, Technische Universiteit Delft, 2600 GA Delft, The Netherlands
* Correspondence: zhonglei@ustc.edu.cn

Citation: Ma, Y.; Zhong, L.; Jia, L.; Menenti, M. Land-Atmosphere Interactions and Effects on the Climate of the Tibetan Plateau and Surrounding Regions. *Remote Sens.* **2023**, *15*, 286. https://doi.org/10.3390/rs15010286

Received: 27 December 2022
Accepted: 28 December 2022
Published: 3 January 2023

1. Introduction

The global climate has undergone unequivocal warming. According to the sixth assessment report of the Intergovernmental Panel on Climate Change, the global surface temperature in the first two decades of the 21st century (2001–2020) was 0.99 °C (0.84 to 1.10 °C) higher than that from 1850–1900. Moreover, the global surface temperature from 2011–2020 was 1.09 °C (0.95 to 1.20 °C) higher than that from 1850–1900 [1]. Meanwhile, accelerating climate change exerts more influence on polar regions, high-altitude zones, and ecologically fragile areas. Often referred to as 'the Third Pole' and 'the Roof of the World', the Tibetan Plateau (TP) conserves vast areas of mountain glaciers, permafrost, and seasonally frozen soil and is the largest high-elevation portion of the cryosphere sensitive to global climate change [2]. In this context, quantitative assessment of the land–atmosphere interaction processes, as well as their effects on the TP and its surrounding regions, is not only essential for understanding the energy and water cycles in the cryosphere and hydrosphere but also crucial for understanding the Asian monsoon system and predicting the climate of Asia and the Northern Hemisphere [3].

To this end, this Special Issue aimed to present recent scientific advances on (1) the estimation of key land surface properties, (2) processes in the atmospheric boundary layer, (3) monitoring of glaciers and glacial lakes, (4) hydrometeorological processes, (5) vegetation dynamics and their response to weather and climate, etc.

Twenty-three articles are published in this Special Issue, covering progress on the following: land surface energy budget, glacier elevation change and snow phenology, the spatiotemporal distribution of precipitation, properties of land surface characteristic parameters, and evaluations of current models and products. The effects of climate change on high-altitude lakes were also included in this Special Issue.

2. Highlights of Research Articles

The land surface energy budget over the TP was analyzed via both remote sensing datasets and in situ observations. Based on empirical orthogonal function analysis of the summer surface sensible heating, a decadal decreasing trend was reported as the first dominant mode over the TP with an explained variance of 20.1% [4]. An enhanced water vapor supply and convergence over the TP were revealed, which led to an increase in the total cloud cover and a decrease in surface downwelling shortwave radiation. The reduction in downwelling shortwave radiation was found to dominate the shrinkage of surface sensible heating. Meanwhile, a zonally asymmetric pattern with positive (negative) sensible heating anomalies in the western (eastern) TP represented the second dominant mode (with an explained variance of 14.2%) [4]. At the station level, the sensible (latent) heat flux was the dominant energy balance component at the Ngari and Qomolangma (Linzhi, Muztagh, Nagqu, and Nam Co) stations. The radiation/energy balance components and surface characteristic parameters exhibited distinct diurnal cycles (e.g., a 'U' shaped curve of the surface broadband albedo was reported) [5]. Based on measurements at eight plateau stations in TIPEX III (the third Tibetan Plateau experiment for atmospheric sciences), the daily mean surface heating field varied from 70.2 to 101.2 W/m^2, with sensible (latent) heat flux from 18.8 to 60.1 W/m^2 (10.1 to 74.7 W/m^2) [6]. A negative correlation between surface heating field density and the intensity of the South Asian summer monsoon was also verified at Baingoin, Nagqu, Nam Co, and Lhari stations [6].

Glaciers and snow phenology are sensitive indicators of climate change, both being involved in and affecting energy/water transfer processes. Global warming has led to significant changes in high-altitude glaciers. Shen et al. [7] developed an 'elevation-aspect bin analysis method' to estimate the intra- and interannual elevation changes of glaciers in the High Mountain Asia (HMA) region. An accelerating decreasing trend of glacier elevation was reported in most regions of the HMA, with mean change rates of -0.21 ± 0.12 m/year during 2003–2008 and -0.26 ± 0.11 m/year during 2003–2020. The variation in glacier elevation showed distinct spatial differences. The decreasing rate gradually decreased from the marginal region to the inner area of HMA, which indicates that the marginal areas of the TP may be zones facing significant risk [7]. Meanwhile, the spatiotemporal variation characteristics of snow disasters over the TP were also evaluated [8]. The frequency, duration, average snow depth, and grade of snow disasters over the TP all depicted a declining trend in the long run. Using the farmer and pastoralist well-being (FPWB) index, which has a negative relation with snow disaster risk, the whole TP area was divided into five distinct regions: Kashgar (I), Shigatse (II), Nagqu (III), Qamdo (IV), and Yushu (V), with gradually decreasing risks of snow disasters [8]. According to Wu et al. [9], factors such as precipitation, solar radiation, and air temperature significantly affect snow phenology. Precipitation was positively correlated with snow accumulation and maintenance, while solar radiation and air temperature functioned negatively. Comparatively, the quantity of snow was more sensitive to solar radiation, while its persistence was more sensitive to air temperature. The mean change rates of snow depth and snow cover maintenance days were estimated to be -0.06 cm/year and -0.37 day/year, respectively [9].

The spatiotemporal variability of precipitation was also focused on in this Special Issue. Meng et al. [10] reported that precipitation in most areas of the three rivers' source regions decreased in spring, autumn, and winter, while summer contributed the most increases. In contrast with the 2000s, the afternoon precipitation slightly decreased in the 2010s, while the nighttime precipitation increased significantly. Cao et al. [11] derived similar results, i.e., the diurnal maximum precipitation was found to be concentrated in the early evening, showing a distinct diurnal cycle. In addition, the raindrop size exhibited significant seasonal variability in the premonsoon, monsoon, and postmonsoon periods. The highest (lowest) concentration of small raindrops was observed in monsoon (winter) precipitation, while large raindrops dominated the premonsoon precipitation [12]. Shen et al. [13] found that the precipitation particles above high mountains have distinct characteristics, such as lower droplet number concentrations and larger diameters, compared with those over plains. In addition, both Wang et al. [14] and Yang et al. [15] concluded that soil moisture plays a more dominant role in precipitation-use efficiency and evapotranspiration-precipitation coupling. Moreover, Shikhovtsev et al. [16] utilized the fifth generation ECMWF (European Centre for Medium-range Weather Forecasts) atmospheric reanalysis (ERA5) precipitable water vapor (PWV) data within 2010–2020 to establish a functional relation between the PWV and the elevation, exhibiting that the decrease of PWV with the elevation was exponential with a height scale of 1000 m. The ERA5 product was also reported to overestimate the PWV values by 1–2 mm in the Big Telescope Alt-azimuthal region (40°N–50°N, 35°E–55°E) [16].

Surface characteristic parameters are significant factors affecting the accuracy of surface process assessments. Li et al. [17] developed an algorithm to generate high-spatial-resolution soil moisture during the thawing season in the permafrost environment using Sentinel-1 and Sentinel-2 data. The comparison with ERA5-Land, Global Land Data Assimilation System (GLDAS), and European Space Agency Climate Change Initiative (ESA CCI) products indicated that this proposed method is able to provide more spatial details and achieve better performance in permafrost areas over the TP. The typical land cover types, alpine desert, alpine steppe, alpine meadow, and alpine swamp meadow, displayed distinct differences in soil moisture, with mean values of 0.16, 0.20, 0.23, and 0.26 m^3/m^3, respectively [17]. The soil's apparent thermal diffusivity (k) is also vital for investigating soil surface heat transfer. Tong et al. [18] determined the magnitude of k at hourly, daily, and monthly scales via a conduction–convection method. The monthly k varied from 0.4×10^{-6} to 1.1×10^{-6} m^2/s at the wet site, with values from 1.7×10^{-7} to 3.3×10^{-7} m^2/s at two dry sites, displaying magnitude differences under different soil moisture conditions. In addition, Ren et al. [19] reported that the Bowen ratio decreased significantly with an increase in soil moisture or effective precipitation. The Bowen ratio in the semiarid region was 1.5 times higher than that in the semihumid region during the growing season. Moreover, Zhang et al. [20] analyzed the profiles of the atmospheric refractive index structure constant (C_n^2), reporting unique optical turbulence characteristics compared with plain areas.

Several articles have evaluated the current models or products of near-surface hydrometeorological variables. Huang et al. [21] conducted a systematic assessment of three widely used air temperature products, namely, ERA5L, GLDAS, and China Meteorological Administration Land Data Assimilation System (CLDAS). Among these three products, CLDAS is more consistent with observations and can better describe temperature distribution and variation details than ERA5L and GLDAS for the Asian region. CLDAS is 0.53 K higher than the in situ observation, while ERA5L and GLDAS are lower than in situ measurements by −3.45 K and −1.40 K, respectively [21]. Liu et al. [22] reported that WRF applying the default glacial albedo scheme overestimates the albedo with a mean error of 0.18, while WRF applying a modified glacial albedo scheme slightly underestimates the albedo with a mean error of only −0.08. The default glacial albedo scheme gives a relatively high albedo value of 0.68, causing an underestimation of the net shortwave and net radiation. In contrast, the modified glacial albedo scheme provides a mean albedo value of 0.35, which is close to in situ measurements. In addition, Liu et al. [23] compared the leaf area index (LAI) estimations from 35 Earth system models that participated in the sixth Coupled Model Intercomparison Project (CMIP6) and found that these models overestimated the LAI trend over alpine vegetation, grassland, and forest but underestimated it over meadow and shrub. More than 70% of the models overestimated the LAI during the 1981–2014 growing seasons, indicating that the greening of grassland in the TP was greatly overestimated. Moreover, five complementary relationship-based models, requiring only routine meteorological variables to estimate actual terrestrial evapotranspiration, were evaluated by Shang et al. [24], with biases ranging from −94.2 to 28.3 mm/year. These models provide a simple and convenient approach for evapotranspiration estimates.

Finally, the responses of TP lakes to climate change were also quantitively investigated. The lake surface water temperature of the largest freshwater lake in the TP, Ngoring Lake, was estimated to have a warming rate of 0.6 K/decade. However, comparison with its nearby small saline lake, the Hajiang Salt Pond, depicted distinct differences due to the salinity effect. The high salinity of the small saline lake made the annual mean lake surface water temperature 2.6 K higher and resulted in a 0.02 K/decade more significant warming trend than freshwater lakes at the same depth [25]. Moreover, the lakes in the Qaidam Basin were estimated to have undergone accelerated expansion. In the two study periods of 2003–2011 and 2011-present, the air temperature, precipitation, and runoff increased steadily, while the expansion rate of Tuosu Lake in the Qaidam Basin increased from 1.22 km^2/year to 3.38 km^2/year. This significant increase in the lake expansion rate reflects groundwater's substantial contribution to lake expansion [26].

3. Conclusions

This Special Issue compiles the up-to-date progress on the following: land surface energy budget, glacier/snow phenology, the spatiotemporal distribution of precipitation, properties of land surface characteristic parameters, evaluations of current models and products, and high-altitude lake processes over the TP and its surrounding regions. These selected papers are novel and timely in informing the land–atmosphere interactions driven by climate change. The collation of these papers will provide quantitative references for better assessment and prediction of the land–atmosphere interactions and their effects on "the Third Pole" and its surrounding regions.

Author Contributions: Conceptualization, Y.M. and L.Z.; formal analysis, Y.M., L.Z., L.J. and M.M.; investigation, L.Z.; writing—original draft preparation, L.Z.; writing—review and editing, Y.M., L.J. and M.M.; supervision, Y.M.; funding acquisition, Y.M. and L.Z. All authors have read and agreed to the published version of the manuscript.

Funding: This research was jointly funded by the Second Tibetan Plateau Scientific Expedition and Research (STEP) Program, Ministry of Science and Technology of the People's Republic of China (Grant No. 2019QZKK0103), the Strategic Priority Research Program of the Chinese Academy of Sciences (Grant No. XDA20060101), the National Natural Science Foundation of China (Grant Nos. 41875031, 91837208, and 4223061), and CLIMATE-Pan-TPE (ID 58516) in the framework of the ESA-MOST Dragon 5 program.

Conflicts of Interest: The authors declare no conflict of interest.

References

1. IPCC. *Climate Change 2021: The Physical Science Basis. Contribution of Working Group I to the Sixth Assessment Report of the Intergovernmental Panel on Climate Change*; Cambridge University Press: Cambridge, UK; New York, NY, USA, 2021; p. 2391.
2. Ma, Y.; Hu, Z.; Xie, Z.; Ma, W.; Wang, B.; Chen, X.; Li, M.; Zhong, L.; Sun, F.; Gu, L.; et al. A Long-Term (2005–2016) Dataset of Hourly Integrated Land-Atmosphere Interaction Observations on the Tibetan Plateau. *Earth Syst. Sci. Data* **2020**, *12*, 2937–2957. [CrossRef]
3. Su, Z.; Ma, Y.; Chen, X.; Dong, X.; Du, J.; Han, C.; He, Y.; Hofste, J.; Li, M.; Li, M.; et al. Monitoring Water and Energy Cycles at Climate Scale in the Third Pole Environment (CLIMATE-TPE). *Remote Sens.* **2021**, *13*, 3661. [CrossRef]
4. Fan, W.; Hu, Z.; Ma, W.; Ma, Y.; Han, C.; Han, X.; Yang, Y.; Yu, H.; Fu, C.; Wu, D. Dominant Modes of Tibetan Plateau Summer Surface Sensible Heating and Associated Atmospheric Circulation Anomalies. *Remote Sens.* **2022**, *14*, 956. [CrossRef]
5. Ma, J.; Wen, X.; Li, M.; Luo, S.; Zhu, X.; Yang, X.; Chen, M. Analysis of Surface Energy Changes over Different Underlying Surfaces Based on MODIS Land-Use Data and Green Vegetation Fraction over the Tibetan Plateau. *Remote Sens.* **2022**, *14*, 2751. [CrossRef]
6. Li, H.; Zhou, L.; Wang, G. The Observed Impact of the South Asian Summer Monsoon on Land-Atmosphere Heat Transfers and Its Inhomogeneity over the Tibetan Plateau. *Remote Sens.* **2022**, *14*, 3236. [CrossRef]
7. Shen, C.; Jia, L.; Ren, S. Inter- and Intra-Annual Glacier Elevation Change in High Mountain Asia Region Based on ICESat-1&2 Data Using Elevation-Aspect Bin Analysis Method. *Remote Sens.* **2022**, *14*, 1630.
8. Li, J.; Zou, Y.; Zhang, Y.; Sun, S.; Dong, X. Risk Assessment of Snow Disasters for Animal Husbandry on the Qinghai–Tibetan Plateau and Influences of Snow Disasters on the Well-Being of Farmers and Pastoralists. *Remote Sens.* **2022**, *14*, 3358. [CrossRef]
9. Wu, L.; Li, C.; Xie, X.; Lv, J.; Zou, S.; Zhou, X.; Shen, N. Land Surface Snow Phenology Based on an Improved Downscaling Method in the Southern Gansu Plateau, China. *Remote Sens.* **2022**, *14*, 2848. [CrossRef]
10. Meng, X.; Deng, M.; Liu, Y.; Li, Z.; Zhao, L. Remote Sensing-Detected Changes in Precipitation over the Source Region of Three Rivers in the Recent Two Decades. *Remote Sens.* **2022**, *14*, 2216. [CrossRef]
11. Cao, B.; Yang, X.; Li, B.; Lu, Y.; Wen, J. Diurnal Variation in Cloud and Precipitation Characteristics in Summer over the Tibetan Plateau and Sichuan Basin. *Remote Sens.* **2022**, *14*, 2711. [CrossRef]

12. Li, R.; Wang, G.; Zhou, R.; Zhang, J.; Liu, L. Seasonal Variation in Microphysical Characteristics of Precipitation at the Entrance of Water Vapor Channel in Yarlung Zangbo Grand Canyon. *Remote Sens.* **2022**, *14*, 3149. [CrossRef]
13. Shen, C.; Li, G.; Dong, Y. Vertical Structures Associated with Orographic Precipitation during Warm Season in the Sichuan Basin and Its Surrounding Areas at Different Altitudes from 8-Year GPM DPR Observations. *Remote Sens.* **2022**, *14*, 4222. [CrossRef]
14. Wang, S.; Zhang, Q.; Yue, P.; Wang, J.; Yang, J.; Wang, W.; Zhang, H.; Ren, X. Precipitation-Use Efficiency and Its Conversion with Climate Types in Mainland China. *Remote Sens.* **2022**, *14*, 2467. [CrossRef]
15. Yang, Z.; Zhang, Q.; Zhang, Y.; Yue, P.; Zhang, L.; Zeng, J.; Qi, Y. Hydrothermal Factors Influence on Spatial-Temporal Variation of Evapotranspiration-Precipitation Coupling over Climate Transition Zone of North China. *Remote Sens.* **2022**, *14*, 1448. [CrossRef]
16. Shikhovtsev, A.Y.; Kovadlo, P.G.; Khaikin, V.B.; Kiselev, A.V. Precipitable Water Vapor and Fractional Clear Sky Statistics within the Big Telescope Alt-Azimuthal Region. *Remote Sens.* **2022**, *14*, 6221. [CrossRef]
17. Li, Z.; Zhao, L.; Wang, L.; Zou, D.; Liu, G.; Hu, G.; Du, E.; Xiao, Y.; Liu, S.; Zhou, H.; et al. Retrieving Soil Moisture in the Permafrost Environment by Sentinel-1/2 Temporal Data on the Qinghai–Tibet Plateau. *Remote Sens.* **2022**, *14*, 5966. [CrossRef]
18. Tong, B.; Xu, H.; Horton, R.; Bian, L.; Guo, J. Determination of Long-Term Soil Apparent Thermal Diffusivity Using Near-Surface Soil Temperature on the Tibetan Plateau. *Remote Sens.* **2022**, *14*, 4238. [CrossRef]
19. Ren, X.; Zhang, Q.; Yue, P.; Yang, J.; Wang, S. Environmental and Biophysical Effects of the Bowen Ratio over Typical Farmland Ecosystems in the Loess Plateau. *Remote Sens.* **2022**, *14*, 1897. [CrossRef]
20. Zhang, K.; Wang, F.; Weng, N.; Wu, X.; Li, X.; Luo, T. Optical Turbulence Characteristics in the Upper Troposphere–Lower Stratosphere over the Lhasa within the Asian Summer Monsoon Anticyclone. *Remote Sens.* **2022**, *14*, 4104. [CrossRef]
21. Huang, X.; Han, S.; Shi, C. Evaluation of Three Air Temperature Reanalysis Datasets in the Alpine Region of the Qinghai–Tibet Plateau. *Remote Sens.* **2022**, *14*, 4447. [CrossRef]
22. Liu, L.; Menenti, M.; Ma, Y. Evaluation of Albedo Schemes in WRF Coupled with Noah-MP on the Parlung No. 4 Glacier. *Remote Sens.* **2022**, *14*, 3934. [CrossRef]
23. Liu, J.; Lu, Y. How Well Do CMIP6 Models Simulate the Greening of the Tibetan Plateau? *Remote Sens.* **2022**, *14*, 4633. [CrossRef]
24. Shang, C.; Wu, T.; Ma, N.; Wang, J.; Li, X.; Zhu, X.; Wang, T.; Hu, G.; Li, R.; Yang, S.; et al. Assessment of Different Complementary-Relationship-Based Models for Estimating Actual Terrestrial Evapotranspiration in the Frozen Ground Regions of the Qinghai-Tibet Plateau. *Remote Sens.* **2022**, *14*, 2047. [CrossRef]
25. Wen, L.; Wang, C.; Li, Z.; Zhao, L.; Lyu, S.; Leppäranta, M.; Kirillin, G.; Chen, S. Thermal Responses of the Largest Freshwater Lake in the Tibetan Plateau and Its Nearby Saline Lake to Climate Change. *Remote Sens.* **2022**, *14*, 1774. [CrossRef]
26. Zhang, X.; Chen, J.; Chen, J.; Ma, F.; Wang, T. Lake Expansion under the Groundwater Contribution in Qaidam Basin, China. *Remote Sens.* **2022**, *14*, 1756. [CrossRef]

remote sensing

Article

Dominant Modes of Tibetan Plateau Summer Surface Sensible Heating and Associated Atmospheric Circulation Anomalies

Weiwei Fan [1,2,3], Zeyong Hu [1,2,*], Weiqiang Ma [4,5], Yaoming Ma [3,4,5,6,7,8], Cunbo Han [4], Xiang Han [9], Yaoxian Yang [1,2], Haipeng Yu [1,2], Chunwei Fu [1,2,3] and Di Wu [1,2,3]

1 Key Laboratory of Land Surface Process and Climate Change in Cold and Arid Regions, Northwest Institute of Eco-Environment and Resources, Chinese Academy of Sciences, Lanzhou 730000, China; fanweiwei19@mails.ucas.ac.cn (W.F.); yangyaoxian@nieer.ac.cn (Y.Y.); yuhp@lzb.ac.cn (H.Y.); fuchunwei@lzb.ac.cn (C.F.); wudi@lzb.ac.cn (D.W.)
2 Nagqu Station of Plateau Climate and Environment, Northwest Institute of Eco-Environment and Resources, Chinese Academy of Sciences, Nagqu 851107, China
3 College of Earth and Planetary Sciences, University of Chinese Academy of Sciences, Beijing 100049, China; ymma@itpcas.ac.cn
4 Land-Atmosphere Interaction and Its Climatic Effects Group, State Key Laboratory of Tibetan Plateau Earth System, Resources and Environment (TPESRE), Institute of Tibetan Plateau Research, Chinese Academy of Sciences, Beijing 100101, China; wqma@itpcas.ac.cn (W.M.); cunbo.han@itpcas.ac.cn (C.H.)
5 National Observation and Research Station for Qomolongma Special Atmospheric Processes and Environmental Changes, Dingri 858200, China
6 College of Atmospheric Science, Lanzhou University, Lanzhou 730000, China
7 Kathmandu Center of Research and Education, Chinese Academy of Sciences, Beijing 100101, China
8 China-Pakistan Joint Research Center on Earth Sciences, Chinese Academy of Sciences, Islamabad 45320, Pakistan
9 Ocean College, Zhejiang University, Zhoushan 316021, China; than@zju.edu.cn
* Correspondence: zyhu@lzb.ac.cn

Citation: Fan, W.; Hu, Z.; Ma, W.; Ma, Y.; Han, C.; Han, X.; Yang, Y.; Yu, H.; Fu, C.; Wu, D. Dominant Modes of Tibetan Plateau Summer Surface Sensible Heating and Associated Atmospheric Circulation Anomalies. *Remote Sens.* **2022**, *14*, 956. https://doi.org/10.3390/rs14040956

Academic Editor: Costas Varotsos

Received: 5 January 2022
Accepted: 12 February 2022
Published: 16 February 2022

Abstract: Based on empirical orthogonal function (EOF) analysis, the dominant modes of variations in summer surface sensible heating (SH) over the Tibetan Plateau (TP), as well as the associated atmospheric circulation anomalies, were investigated in this study. The results show that the first dominant mode of summer SH presents a feature of decadal reduction over the whole TP on an interdecadal time scale, and the second dominant mode is characterized by a zonally asymmetric pattern with positive (negative) SH anomalies in the western (eastern) TP on an interannual time scale. The variations of summer SH are dominated by anomalies in downwelling surface shortwave radiation (DSWR), which are associated with atmospheric circulation changes. The first dominant mode of variation in SH is connected to the interdecadal variation of the Silk Road Pattern (SRP). Further analysis reveals that the interdecadal phase shift of the SRP induces anticyclone circulation to the northeast of the TP, leading to enhanced water vapor supply and convergence over the TP. This can lead to an increase in the total cloud cover, and a reduction in DSWR, contributing to the decadal reduction in SH over the TP. The second dominant mode of variation in SH is related to a stationary teleconnection pattern over the Eurasian continent named the North Atlantic-East and North Asia pattern (NAENA). Corresponding to the positive phase of the NAENA, there is a cyclone anomaly to the west TP, leading to anomalous water vapor convergence (divergence) over the eastern (western) TP. This can result in enhanced (decreased) cloud cover, reduced (increased) DSWR, and therefore, an anomalous decrease (enhancement) in SH over the east (west) of the TP. Furthermore, the southwesterly wind anomaly, which is accompanied by the anomalous cyclone to the west TP, leads to positive SH in the western TP.

Keywords: surface sensible heating; Tibetan Plateau; teleconnection wave trains; dominant modes; interannual and interdecadal variations

1. Introduction

As one of the highest and largest highlands in the world, the Tibetan Plateau (TP) is referred to as "the Third Pole" [1]. It contains abundant water resources with a large number of lakes, rivers, glaciers, frozen soils, and wetlands [2–4]. The TP is the source of several major rivers in Asia, which provide water for the surrounding areas, and has crucial impacts on the development of the Asian economy and civilization [1,5]. Thermodynamic forcing is closely connected to the water cycle of the TP through the "CISK-like mechanism" [6]. It also plays a critical role in Asian atmospheric circulations and thus the weather and climatic systems, such as the plateau vortices, Asian monsoon rainfall, and even the tropical signal [7–10]. As an important component of the atmospheric heat source in warm seasons, surface sensible heating (SH) can regulate the onset and maintenance of Asian summer monsoon systems confirmed by observations and numerical experiments [11–13]. In the preceding spring, the variability of SH over the TP is well connected to the onset of the East Asian summer monsoon, as well as the precipitation anomalies of East China [14,15]. In addition, the interannual variation of SH can regulate surface dust concentrations over the East Asian dust source region and the northwestern Pacific through increasing the westerly winds [16].

In consideration of the critical role of SH over the TP on surrounding weather and climate systems, research into variations of SH over the TP has important impacts for improving the understanding of the mechanism of the TP and the variability in Asian climate systems. Trend analysis indicated that SH over the TP presented significant weakening during the 1980s–2000s, which is due to the reduced surface wind speed in connection with the East Asian subtropical westerly jet under global warming [17]. Recent studies have indicated that SH over the TP has been dominated by a slightly increasing trend since the late 1990s as a result of the restored surface wind speed and difference in ground-air temperature [18]. The CMIP6 models demonstrated that SH will continue to increase in the future [19]. Moreover, the long-term trend of SH features elevation dependence with a greater variation trend at a higher elevation [20]. Observational analysis and numerical experimentation indicated that the early spring sea surface temperature anomalies over the North Atlantic can significantly impact the interannual variation of spring SH of the TP by triggering eastward propagating wave trains and intensifying the subtropical westerly jet [21,22]. Based on satellite data and observations, a recent study found that SH over the TP has increased slightly since 2001 [23].

Summer SH over the TP plays a key role in the surrounding weather and climate systems. Numerical simulations indicated that summer SH can enhance both the lower-level convergence and upper-level divergence in the TP, intensify the rising motion, and thus enhance the South Asia High [24]. Chen et al. [25] found that the disappearances of the TP vortices in the sloping terrain of the eastern TP might be attributed to the weakening of SH. Studies also showed that summer SH has crucial impacts on Sichuan-Chongqing areas [26]. However, although the trends of summer SH as a result of climatic change were investigated by previous research [17–19], few studies have investigated the interannual and interdecadal variations in summer SH of the TP and their possible causes, which is necessary for us to understand the changes in the Asian weather and climate. Therefore, the present study aims to investigate the interannual and interdecadal variations in summer SH over the TP and their associated mechanisms. This helps us to gain a deep understanding of the land and atmosphere interaction over the TP during warm seasons.

2. Data and Methods

2.1. Data

To study the variations of summer SH over the TP, ERA-interim surface sensible heat flux datasets were used in this paper, which were available at a horizontal resolution of $0.5^\circ \times 0.5^\circ$ from 1981 to 2018. To improve the handling of data bias and background error constraints, ERA-interim uses a 12-h 4D-Varassimilation system. Compared to ERA-40, ERA-Interim has additional input remote data, including Meteosat-2 clear-sky radiances,

Global Ozone Monitoring Experiment (GOME) ozone profiles, and radio occultation measurements from the Challenging Mini Satellite Payload (CHAMP), Constellation Observing System for Meteorology, Ionosphere, and Climate (COSMIC), and Gravity Recovery and Climate Experiment (GRACE). Reprocessed ocean wave height data from ERS-1 and ERS-2, as well as upper-level winds from Meteosat-2, were also included. Previous investigations have shown that ERA-interim SH data can accurately reflect the surface heat flux [27]. Element fields such as the surface wind speed, total cloud cover, surface latent heating, and surface radiation data are also provided by ERA-interim.

A remote-sensing product derived from Han et al. [28,29] was also used to analyze summer SH on the TP. A detailed retrieval algorithm of SH can be found in Han et al. [28,29].

To quantify the association between SH and atmospheric circulation, we used the 38-year (1981–2018) monthly mean geopotential height, zonal wind, and meridional wind use from the National Centers for Environmental Prediction-National Center for Atmospheric Research (NCEP/NCAR), with a horizontal grid spacing of 2.5° × 2.5°. The data we used were for the summer season (JJA). Table 1 shows the details of each variables used in this study.

Table 1. Summary of the datasets used in the study.

Variables	Data Source	Availability	Temporal Resolution	Spatial Resolution
Sensible heat flux (Reanalysis)	ERA-Interim	1981–2018	Daily	0.5° × 0.5°
Sensible heat flux (Remote Sensing)	Han et al. [28,29]	2001–2018	Monthly	0.1° × 0.1°
Surface wind speed Total cloud cover Latent heat flux Surface radiation	ERA-Interim	1981–2018	Monthly	0.5° × 0.5°
Geopotential height Zonal wind Meridional wind	NCEP/NCAR	1981–2018	Monthly	2.5° × 2.5°

2.2. Methods

2.2.1. Surface Energy Balance Analysis

Surface energy balance analysis was employed to identify the cause of change in SH. It is expressed as follows [30]:

$$SW_{\downarrow} - SW_{\uparrow} + LW_{\downarrow} - LW_{\uparrow} = SH + LH + G, \quad (1)$$

where $SW_{\downarrow}$ and $SW_{\uparrow}$ indicate the downwelling and upwelling shortwave solar radiation, respectively; $LW_{\downarrow}$ and $LW_{\uparrow}$ indicate the downwelling and upwelling longwave thermal infrared radiation, respectively; SH and LH indicate the upward surface sensible and latent heating; and G is the ground heating. The data were obtained from ERA-Interim. Therefore, the change in SH was determined by the change in the other terms in the equation, which can be expressed as follows:

$$\Delta SH = \Delta SW_{\downarrow} - \Delta SW_{\uparrow} + \Delta LW_{\downarrow} - \Delta LW_{\uparrow} - \Delta LH - \Delta G, \quad (2)$$

2.2.2. Wave Activity Flux

The wave activity flux formulated by Takaya and Nakamura [31] was used to determine the features of stationary wave propagation, which is given as follows:

$$W = \frac{1}{2|\overline{U}|}\left\{ \begin{array}{c} \overline{u}(\psi'^2_x - \psi'\psi'_{xx}) + \overline{v}\left(\psi'_x\psi'_y - \psi'\psi'_{xy}\right) \\ \overline{u}\left(\psi'_x\psi'_y - \psi'\psi'_{xy}\right) + \overline{v}(\psi'^2_y - \psi'\psi'_{yy}) \end{array} \right\} \quad (3)$$

Here, ψ is the stre-m function, u is the zonal wind v-locity, and v denotes the meridio-al wind velocity. The data were obtained from NCEP/NCAR. The overbars represent the basic states and primes represent perturbations.

2.2.3. Empirical Orthogonal Functions (EOF) Analysis

EOF analysis was used to study the features of interannual and interdecadal variations of summer SH. In addition, the EOF analysis was also used to extract the teleconnection wave trains related to variations of SH. The EOF method is frequently used to investigate potential spatial patterns of climatic variability and how they develop over time. In EOF analysis, the original climatic data were also projected on an orthogonal basis [32]. Moreover, the orthogonal basis was determined by calculating the eigenvector of the spatially weighted anomaly covariance matrix, with the corresponding eigenvalues indicating the percentage variance explained by each pattern. Therefore, the EOFs of spatiotemporal physical processes could represent mutually orthogonal spatial patterns in the data change set, in which the first pattern accounts for most of the variance, the second pattern accounts for most of the residual variance, etc. As the principal component (PC) of an EOF mode shows how the spatial pattern of this mode oscillates over time, we used the corresponding PC of the dominant mode as the reference time series for each summer of SH.

2.2.4. Linear Regression Analysis and Composite Analysis

To investigate the anomalies in surface energy flux and atmospheric circulation associated with SH, linear regression and composite analysis were employed in this study. Using the statistical analysis method of linear regression analysis, the linear relationship between two or more variables can be determined quantitatively. A composite analysis is frequently used in climate change research to explore the salient characteristics of different periods. The positive phase period of SH and the negative phase period were selected first. Second, we investigated the atmospheric circulation differences between the two periods. The significance of the regression coefficient and difference in the composite analysis was evaluated using Student's t-test.

3. Results

3.1. Dominant Modes of Variation in Summer SH over the TP

Figure 1a,b shows the spatial pattern of summer SH over the TP during the period 2001–2018 obtained from ERA-Interim reanalysis (Figure 1a) and the remote sensing product (Figure 1b). In general, ERA-Interim and remote sensing-based SH presented similar spatial patterns, with a pattern correlation of 0.8, exceeding the 99% confidence test. The values of SH were positive over the whole TP, acting as a gigantic SH air pump and having crucial climatic effects [9]. SH increased from south to north, with the maximum value in the Qaidam Basin over the northern TP. Figure 1c,d shows the spatial distributions of the standard deviation in SH derived from ERA-Interim reanalysis (Figure 1c) and the remote sensing product (Figure 1d). The spatial pattern of ERA-Interim SH standard deviation was close to the remote sensing-based SH standard deviation. In the two datasets, values of the standard deviation in SH over the northern and southern TP were also larger than those in the eastern TP, which were similar to the spatial distributions of summer seasonal-mean SH.

To identify the dominant modes of summer SH anomalies, EOF analysis was performed. The spatial distribution of the first dominant mode (EOF1) was marked by the consistent variations of summer SH over the whole TP, with an explained variance of 20.1% (Figure 2a). EOF1 was set apart from the other modes based on the North test. This means that EOF1 is considered statistically distinguishable and significant. Figure 2b shows the spatial distribution of EOF2. EOF2 accounted for 14.2% of the total variance and passed the North test [32]. A clear zonal seesaw pattern could be observed over the TP. The same analysis based on the remote sensing product derived from Han et al. [28,29] also indicated that the first leading mode of summer SH showed significant consistent variations, and the second leading mode of summer SH showed a zonal dipolar pattern (Figure S1).

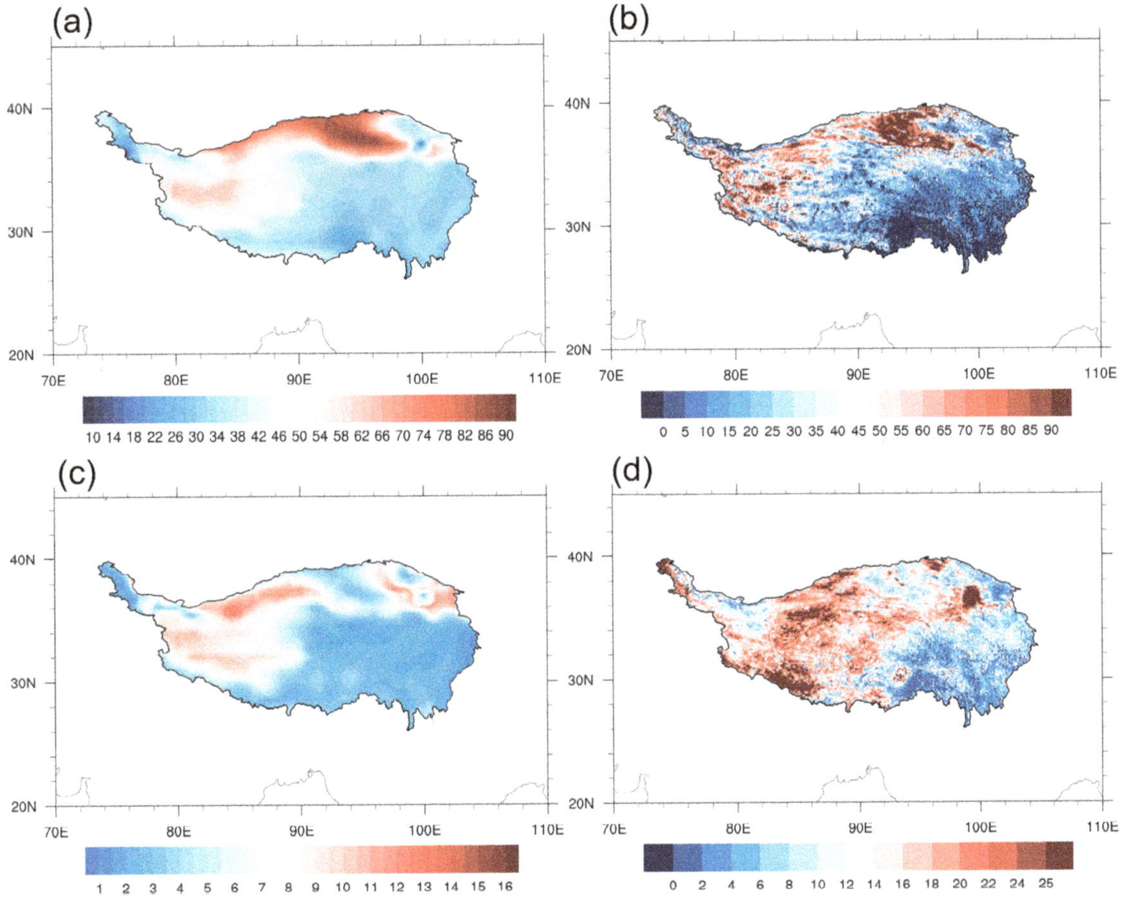

Figure 1. The spatial pattern of the TP summer SH climatology (W m^{-2}) from ERA-Interim (**a**) and remote sensing product derived from Han et al. [28,29] (**b**), standard deviation (W m^{-2}) from ERA-Interim reanalysis (**c**), and remote sensing (**d**) data.

Figure 2c shows the PC1 and its interdecadal component (PC1-ID). PC1-ID presented an interdecadal shift from the positive phase to the negative phase from around 1996. From the perspective of long-term trends, SH demonstrated reductions from the 1980s and a slight recovery from the 2000s, which is in agreement with previous studies [24,25]. The spatial distribution of the summer SH interdecadal component anomalies onto PC1-ID is shown in Figure 2e. Significant positive anomalies appeared in almost the entire TP, while weak negative anomalies occurred in the southeastern TP, which was similar to the first leading mode of SH driving from the EOF method, with a pattern correlation of up to 0.96 (Figure 2a). This indicates that the PC1-ID can effectively represent the interdecadal variation of summer SH over the TP.

Figure 2d shows the PC2 and its interannual component (PC2-IA). A remarkable feature of this was that the relation between PC2-IA and PC2 had a correlation coefficient of up to 0.95, which indicates that EOF2 presents the interannual variation of summer SH. The spatial distribution of SH anomalies associated with PC2-IA is shown in Figure 2f.

A clear zonal seesaw pattern could be observed over the TP, with negative SH in the east of the TP and positive SH over the western TP. On the interannual time scale, when the PC2-IA was in a positive phase, SH over the TP exhibited a zonal asymmetric spatial mode with anomalous strengthening in SH over the western TP and anomalous weakening in SH over the eastern TP. Anomalies in the interannual component of SH associated with PC2-IA were the same as the second leading mode of SH derived from the EOF method (Figure 2b), which indicates that the EOF2 pattern can effectively represent the dominant mode of interannual variation in summer SH over the TP.

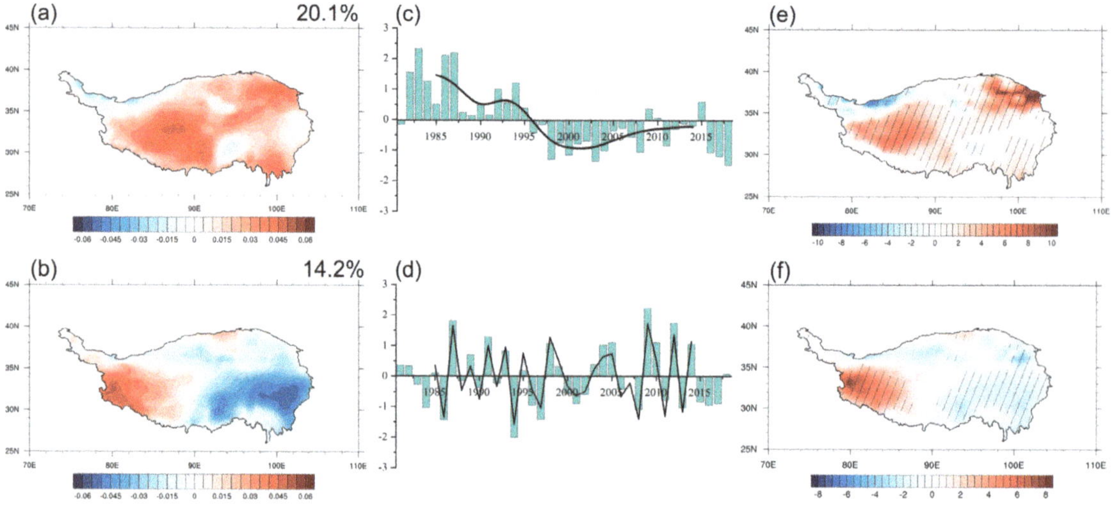

Figure 2. The first two EOFs of summer SH over the TP for the period 1981–2018 (**a**,**b**). The explained variance of EOF is in the upper right of (**a**,**b**). Normalized time series of PC1 (column) and its interdecadal component (PC1-ID) (line) (**c**). Normalized time series of PC2 (column) and its interannual component (PC2-IA) (line) (**d**). Regression of the interdecadal component of summer SH (shading, W m^{-2}) on PC1-ID (**e**). Regression of the interannual component of summer SH (shading, W m^{-2}) on PC2-IA (**f**). Grid points with statistically significant anomalies passing the 95% confidence level are denoted by an oblique line.

3.2. Physical Mechanisms of Variations in Summer SH

3.2.1. First Dominant Mode of Variation in Summer SH

The first dominant mode of variation in summer SH presented an interdecadal weakening. As SH was determined by the local surface energy budget, the local surface energy budget related to anomalies in SH on the interdecadal time scale is investigated below. Considering SH experienced a decadal phase shift around 1996, we used the two periods, 1981–1995 and 1997–2018, to represent the decadal change in SH accordingly. Figure 3 presents the difference in surface fluxes between the two periods. Anomalies in SWR bore a close resemblance to the spatial pattern of SH with negative anomalies over the majority of the TP (Figure 3a,c). The anomalous decrease in SWR corresponded to significantly reduced DSWR over the TP (Figure 3e). This indicates that the DSWR has a positive contribution to the interdecadal variation of SH. The spatial distribution of anomalies in LWR presented strengthening over the TP (Figure 3d), which was dominated by the change in DLWR (Figure 3g). Moreover, the magnitude of anomalies in LH, USWR, ULWR, and G was too small to impact SH (Figure 3b,f,h,i). The contributions of the components to changes in SH are depicted in Figure 4. It was found that the first dominant mode of SH was mainly due to anomalous SWR induced by DSWR, while the LWR anomalies induced by DLWR helped to offset the decadal weakening of SH. The contribution of the other components was insignificant.

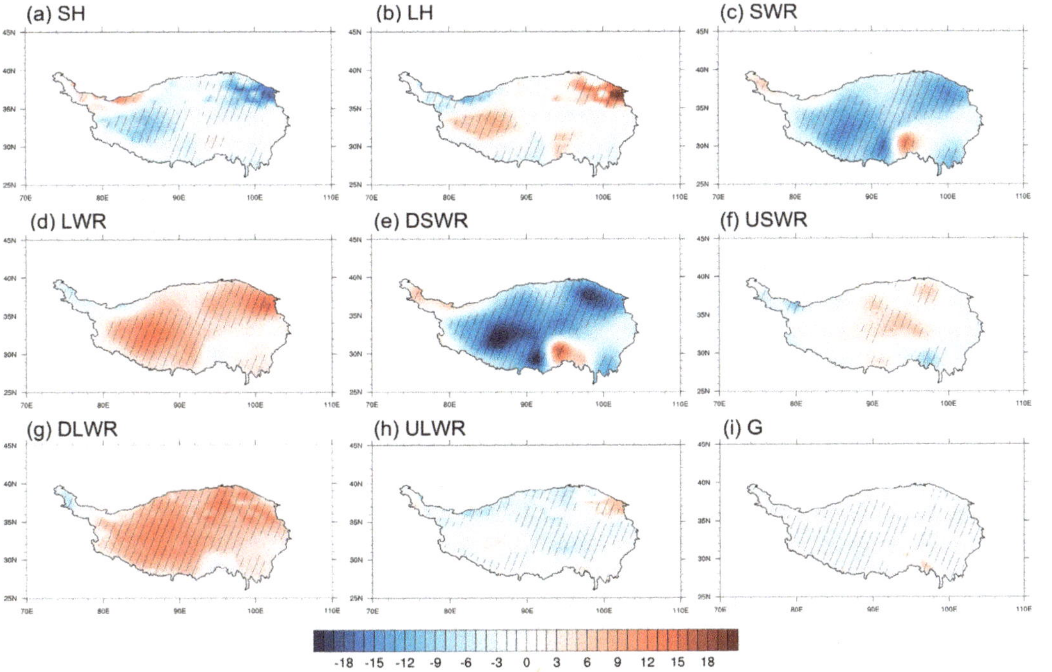

Figure 3. Spatial distribution of difference in sensible heating (SH) (**a**), latent heating (LH) (**b**), surface net shortwave radiation (SWR) (**c**), surface net longwave radiation (LWR) (**d**), downwelling surface shortwave radiation (DSWR) (**e**), upwelling surface shortwave radiation (USWR) (**f**), downwelling surface longwave radiation (DLWR) (**g**), upwelling surface longwave radiation (ULWR) (**h**), and ground heating (G) (**i**) between 1997–2018 and 1981–1995 (shading, W m^{-2}). Grid points with statistically significant anomalies passing the 95% confidence level are denoted by an oblique line.

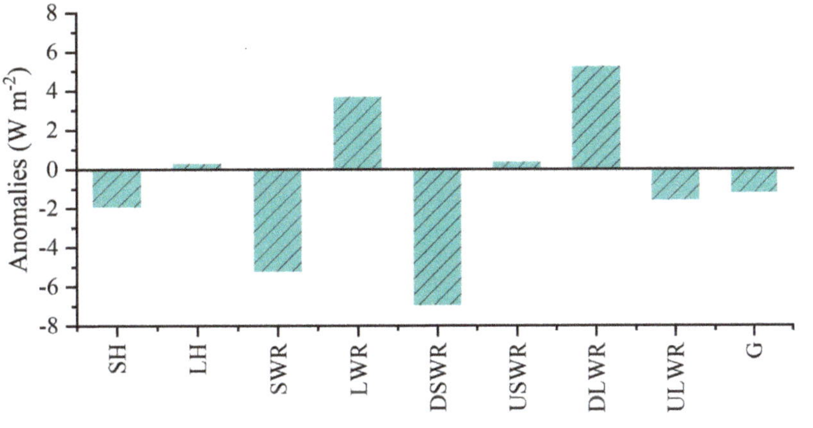

Figure 4. The difference in SH, LH, SWR, LWR, DSWR, LSWR, DLWR, ULWR, and G between 1997–2018 and 1981–1995 over the TP (W m^{-2}).

According to previous studies, cloud cover can regulate the DSWR and DLWR [33]. Therefore, the opposite variations of DSWR and DLWR may have been induced by the cloud cover. The total cloud cover variations associated with the interdecadal variation of SH were investigated next. Figure 5a displays the difference in the total cloud cover between 1997–2018 and 1981–1995. The TP was dominated by enhanced cloud cover, which induced more reflection of downwelling solar radiation and resulted in reduced DSWR. A decrease in DSWR can enhance SH owing to the surface energy budget. Figure 5b depicts the difference in water vapor flux and divergence between 1997–2018 and 1981–1995. A large westward water vapor flux emerged over the TP. This can enhance the convergence of water vapor flux over the TP by means of decreasing the water vapor exported from the eastern boundary of the TP. Associated with this increase in water vapor convergence, more summer cloud cover appeared over the TP (Figure 5a), inducing a decadal weakening of SH (Figure 3a).

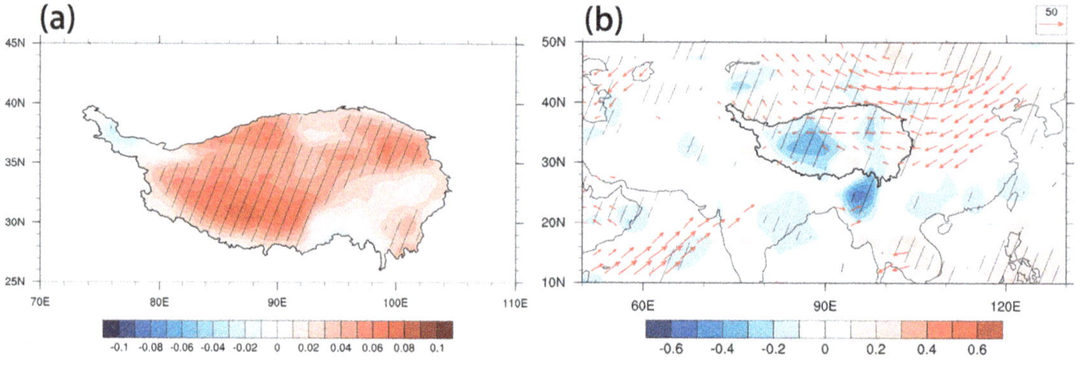

Figure 5. Spatial distribution of difference in total cloud cover (**a**), water vapor flux exceeding the 95% confidence level (vector, kg m^{-1} s^{-1}), and divergence (shading, 10^{-4} kg m^{-2} s^{-1}) (**b**) between 1997–2018 and 1981–1995. Grid points with statistically significant anomalies passing the 95% confidence level are denoted by oblique lines.

Large-scale circulation anomalies can usually contribute to regional climatic change. Studying the impact of atmospheric circulation on TP SH can improve our understanding of climatic change over the TP, which is helpful for climate prediction. The atmospheric circulation anomalies linked to the interdecadal weakening of summer SH over the TP were examined further. Figure 6a displays the 500 hPa geopotential height and horizontal wind anomalies between 1997–2018 and 1981–1995. There was a wave train in the Eurasian continent. A considerably positive geopotential height center with an anticyclonic anomaly was located to the northeast of the TP, corresponding to this anomalous wave-like train. This led to a weakened subtropical westerly jet over the TP and less water vapor export from the eastern boundary of the TP. Figure 6b shows the 200 hPa geopotential height anomalies during the two periods. A zonal wave train occurred from the North Atlantic to the Eurasian continent, with two negative-height anomalies over the North Atlantic and West Asia and two positive-height anomalies over eastern Europe and Lake Baikal. The wave activity flux indicated that the wave train's propagation direction was expected to be from the North Atlantic to the Eurasian continent. Such a circulation anomaly pattern is favorable for a reduced westerly jet and led to anomalous water vapor convergence of the TP (Figure 5b). This conclusion coincides with Zhou et al. [34], who proposed that the interdecadal variation in summer water vapor over the TP is related to a similar teleconnection pattern. Furthermore, the atmospheric circulation anomalies linked to PC1-ID were similar to those related to the interdecadal variation of SRP [35], implying that the SRP pattern may be crucial in the interdecadal variability of summer SH over the TP.

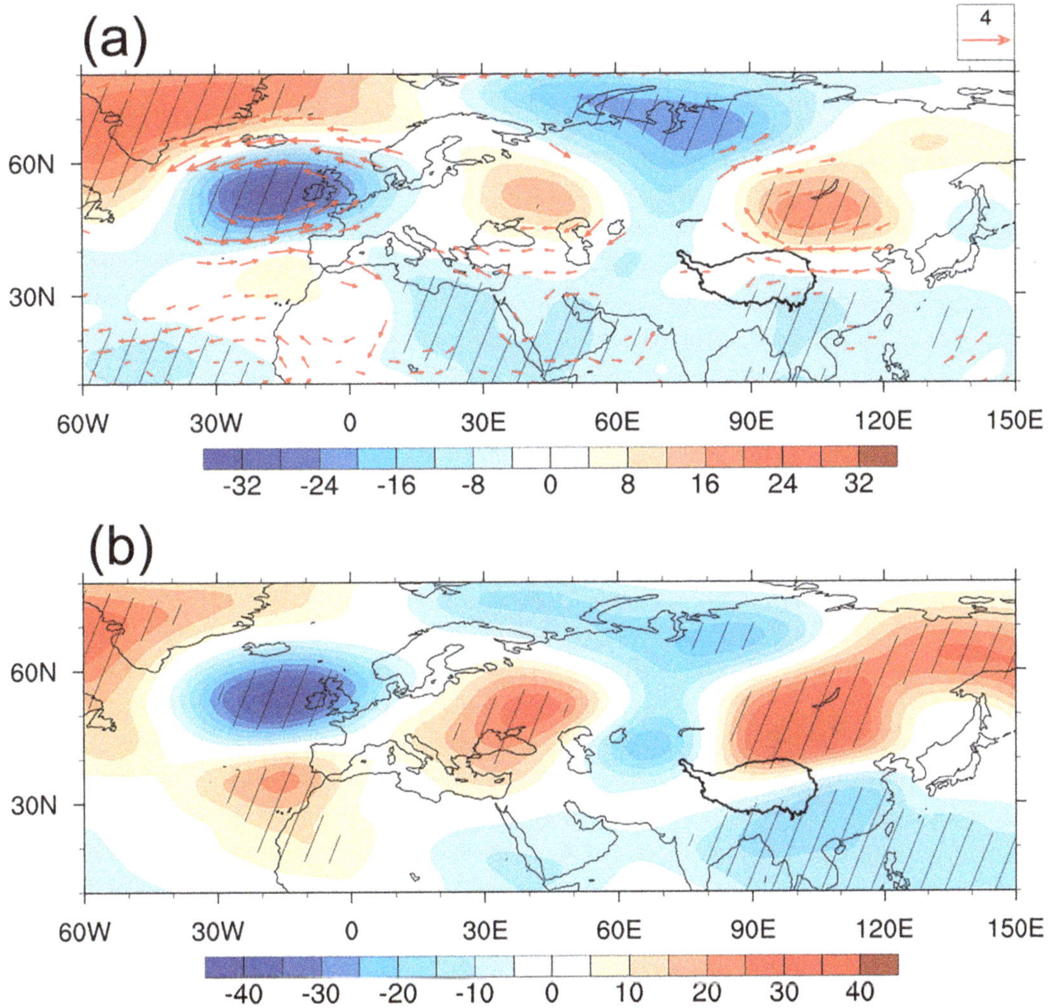

Figure 6. Spatial distribution of difference in 500 hPa geopotential height anomalies (shading, gpm) and the wind exceeding the 95% confidence level (vector, m s^{-1}) (**a**), the 200 hPa geopotential height anomalies (shading, gpm), and the wave activity flux (vector, m^2 s^{-2}) (**b**) between 1997–2018 and 1981–1995. Grid points with statistically significant anomalies passing the 95% confidence level are denoted by oblique lines.

3.2.2. Second Dominant Mode of Variation in Summer SH

We next investigated the surface flux anomalies related to the second dominant mode of variation in TP summer SH. Figure 7 displays the surface flux anomalies related to the PC2-IA of summer SH. The upwelling surface longwave and shortwave radiations ranged from the surface to the atmosphere. In general, a large decrease in surface latent heating (LH) mainly appeared in the southern TP, and an insignificant increase occurred in the northern TP. Corresponding to the EOF2 of SH, a significant increase in the surface net shortwave radiation (SWR) was observed over the western TP and a remarkable reduction appeared in the eastern TP (Figure 7c). Anomalies in SWR were dominated by the zonally asymmetric pattern of downwelling surface shortwave radiation (DSWR) (Figure 7e). However, the contribution of upwelling surface shortwave radiation (USWR) to anomalies in SWR was insignificant (Figure 7f). This means that the DSWR has a positive contribution to anomalies in SH over the TP. On the contrary, anomalies of surface longwave radiation (LWR) (Figure 7d) mainly induced by anomalous downwelling surface longwave radiation (DLWR) (Figure 7g) negatively affected SH due to the opposite spatial pattern anomalies. Moreover, anomalies in upwelling surface longwave radiation (USWR) and the ground heat flux (G) were insignificant and had little impact on the change in SH.

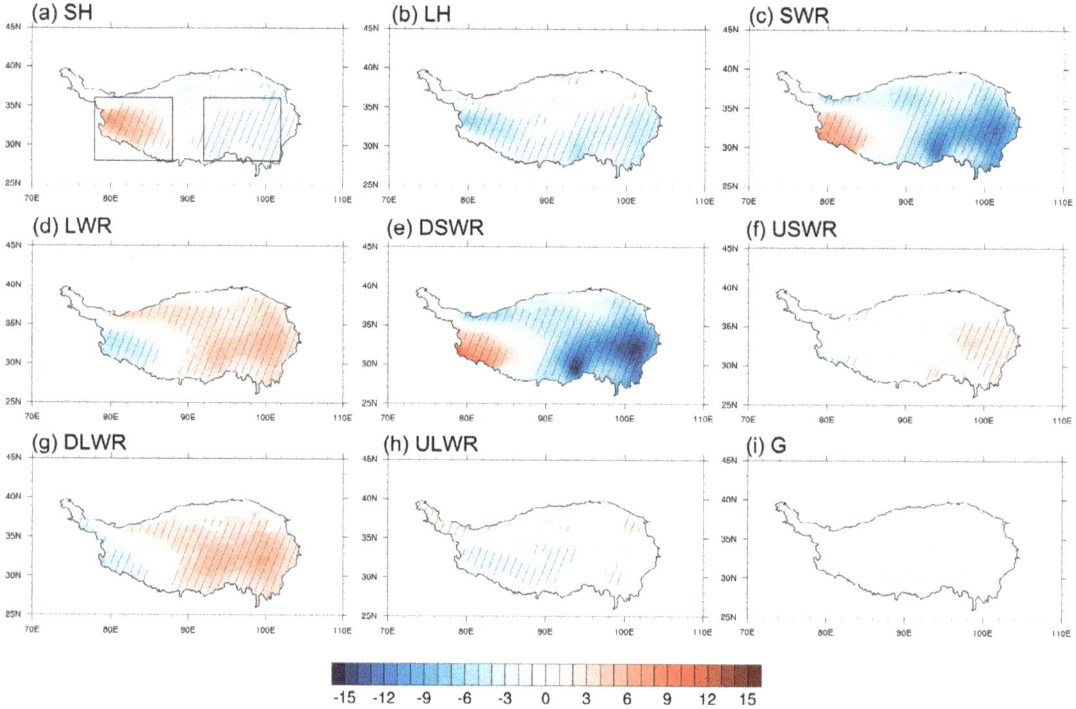

Figure 7. Anomalies in the interannual component of SH (**a**), LH (**b**), SWR (**c**), LWR (**d**), DSWR (**e**), LSWR (**f**), DLWR (**g**), ULWR (**h**), and G (**i**) regressed on PC2-IA (shading, W m^{-2}). Grid points with statistically significant anomalies passing the 95% confidence level are denoted by oblique lines. The two black rectangles denote the western TP (28–36° N, 78–88° E) and eastern TP (28–36° N, 92–102° E).

To further examine the contribution of each surface flux to SH anomalies, the average regional anomalies of SH, LH, SWR, LWR, DSWR, USWR, DLWR, ULWR, and G were calculated for the eastern TP (28–36° N, 78–88° E) (Figure 8a) and western TP (28–36° N, 92–102° E) (Figure 8b), respectively. For the eastern TP, a large decrease in SWR of approximately −9 W m^{-2} contributed to the decrease in SH and LH. The decrease in SWR is mainly attributed to the weakening of DSWR. For the western TP, enhanced SH was consistent with the increase in DSWR and decrease in LH, which means that the increase in SH was mainly due to DSWR and LH. The LWR presented a significant reduction of approximately −2.2 W m^{-2}. Among the contributions to the SH change, the role of DSWR was predominant. Therefore, we investigated the role of DSWR in changing SH on the interannual time scale.

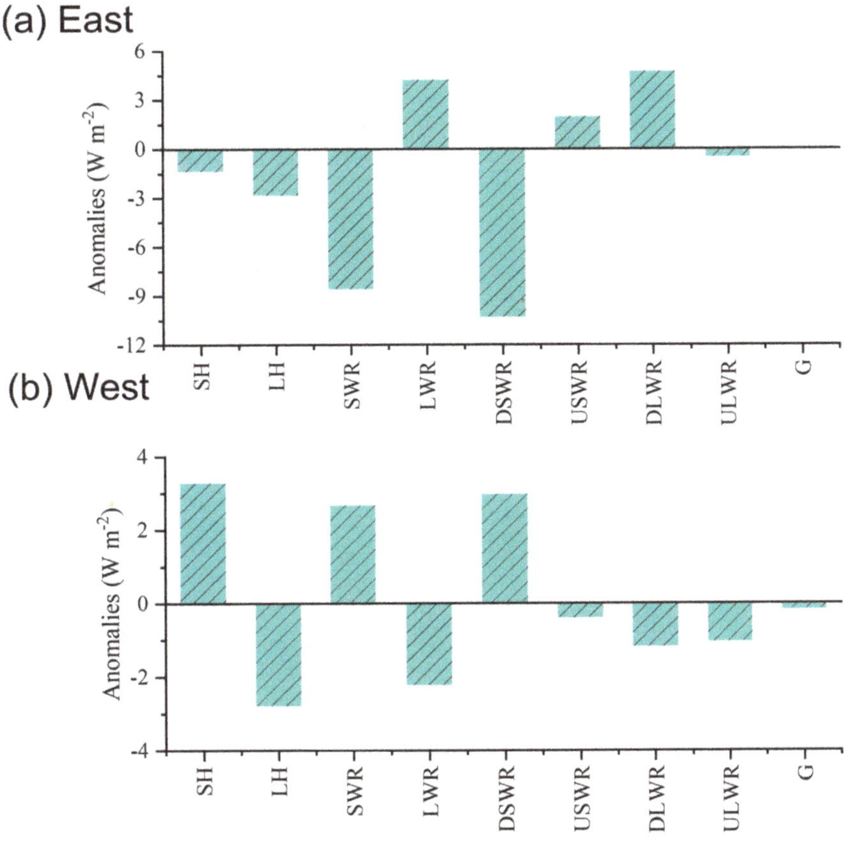

Figure 8. Anomalies in the interannual component of SH, LH, SWR, LWR, DSWR, LSWR, DLWR, ULWR, and G regressed on PC2-IA (W m^{-2}) in the region of eastern TP (**a**) and western TP (**b**).

As investigated by previous studies [33], cloud cover has an important influence on the shortwave radiation reaching the surface. Figure 9a shows anomalies in total cloud cover related to the interannual variation of the TP summer SH. In contrast, less cloud cover appeared in the western TP, and more cloud cover appeared in the eastern TP. This indicates that the anomalous DSWR leading to the zonally asymmetric pattern of SH may be partly related to the variations of total cloud cover over the TP. Figure 9b presents the regression of water vapor flux and divergence onto PC2-IA. There was a significant southwesterly vapor supply from the Indian monsoon region to the main body of the TP, which was induced by the southwesterly anomalies. The anomalous water vapor flux divergence also presented a dipole mode with divergence in the west of the TP and convergence in the east of the TP. Sufficient moisture convergence (divergence) is favorable for the increase (decrease) in total cloud cover, and therefore induced weakened (enhanced) SH over the eastern (western) TP. According to previous studies, the variations of SH significantly respond to the changes in surface wind speed [17]. Figure 9c also displays the regression of the surface wind speed on PC2-IA. Remarkable southwesterly anomalies and positive surface wind speed anomalies were located in the western TP, in accordance with the anomalous enhancement in SH of the western TP (Figure 7a). This means that the positive SH of the western TP was also partly due to the strengthened surface wind speed, owing to the southwesterly wind anomalies.

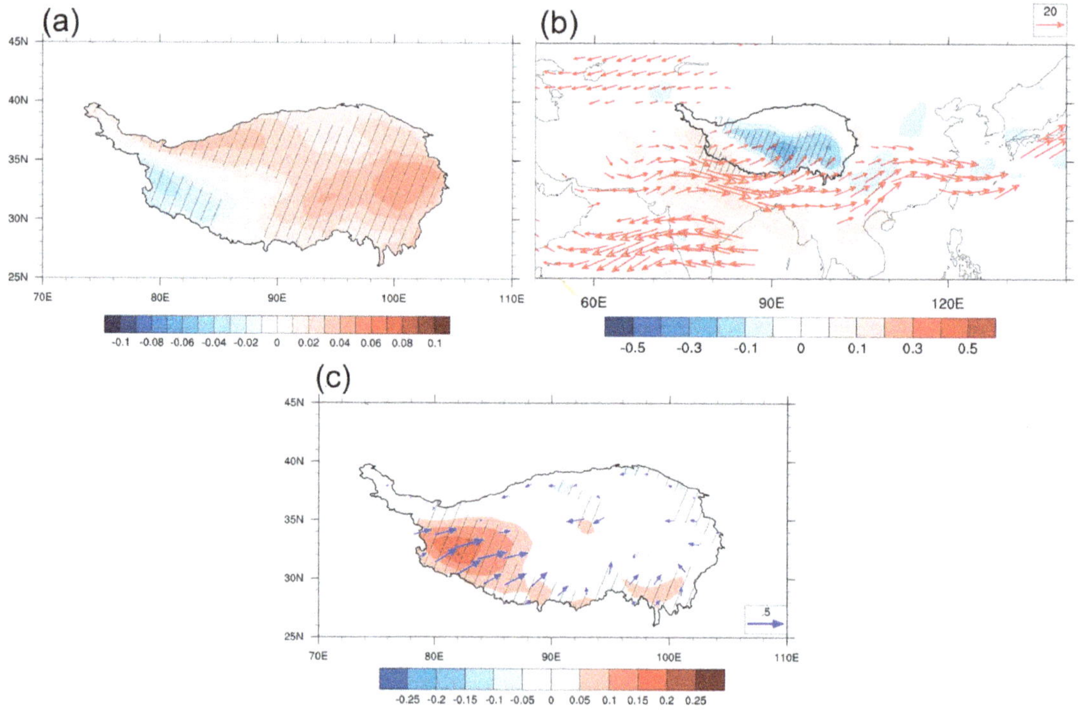

Figure 9. Anomalies in the interannual component of total cloud cover (**a**), water vapor flux exceeding the 95% confidence level (vector, kg m^{-1} s^{-1}) and divergence (shading, 10^{-4} kg m^{-2} s^{-1}) (**b**), surface wind vectors passing the 95% confidence level (vector, m s^{-1}), and wind speed (shading, m s^{-1}) (**c**) regressed on PC2-IA. Grid points with statistically significant anomalies passing the 95% confidence level are denoted by an oblique line.

The above result indicates that interannual variation in SH can be mainly explained by the DSWR change, which is associated with moisture convergence and divergence. In addition, the southwesterly wind anomaly can also partly explain the SH anomalies in the western TP. These analyses suggest that atmospheric circulation changes may have an important impact on the SH anomalies by impacting the water vapor convergence and divergence and wind anomalies over the TP. Atmosphere circulation anomalies related to the anomalies in SH were further examined. Figure 10a displays 500 hPa geopotential height and horizontal wind anomalies related to the interannual variation of SH. An evident feature is that there was a significant, negative, high-pressure center to the west of the TP corresponding to the cyclonic anomaly, which led to enhanced southwest wind anomalies at the southwestern boundary of the TP and water vapor convergence (divergence) over the eastern (western) TP. Furthermore, 200 hPa geopotential height regressed against normalized PC2-IA presented a similar spatial pattern of 500 hPa, showing an equivalent barotropic structure in the vertical direction (Figure 10b). According to the horizontal wave activity flux, the teleconnection wave associated with SH anomalies originated in the North Atlantic and traveled eastward across the Eurasian continent. This suggests that atmospheric wave-like patterns originating in the North Atlantic may be important in the interannual variation of SH.

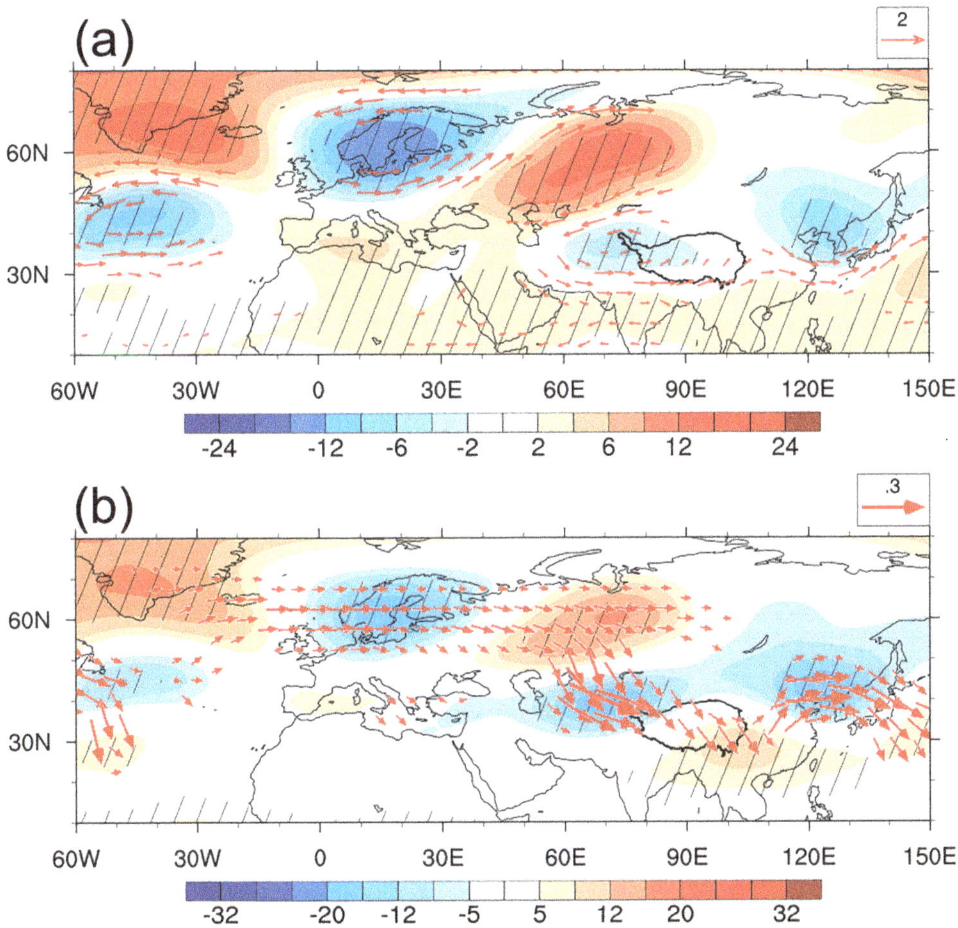

Figure 10. Anomalies in the interannual component of 500 hPa geopotential height (shading, gpm) and wind passing the 95% confidence level (vector, m/s) (**a**), the 200 hPa geopotential height (shading, gpm), and the wave activity flux (vector, $m^2 s^{-2}$) (**b**) regressed on PC2-IA. Grid points with statistically significant anomalies passing the 95% confidence level are denoted by oblique lines.

3.2.3. Association between Variations in Summer SH and Atmospheric Wave Trains

The above analysis indicated that the dominant modes of variations in summer SH over the TP had a close connection to atmospheric wave trains. Associations between variations in summer SH and atmospheric wave trains are investigated below.

We first investigated the impact of interdecadal variation in SRP on summer SH. According to previous studies [35–37], the SRP pattern is defined as the EOF1 of the summer mean 200 hPa meridional wind within the region 20–60° N, 0–150° E (Figure 11a), and the normalized PC1 is referred to as the SRP index (SRPI) (Figure 11b). A 9-year Lanczos low-pass filter was used to calculate the interdecadal component of SRPI (SRPI-ID), which can be seen in Figure 11c. The temporal correlation coefficient between PC1-ID and the SRPI-ID was up to 0.84, with a significance level of 99% (Figure 11c). Figure 11d shows the spatial distribution of SH anomalies related to the interdecadal variation of SRP. Significantly decreased anomalies occurred in almost the whole TP, while weak negative anomalies appeared in the southeastern TP. The spatial pattern was in good agreement with the distribution of interdecadal weakening of SH (Figure 3a). This suggests that the SRP is closely related to the interdecadal variability of SH. Figure 11e,f depict atmospheric circulation anomalies linked to SRP interdecadal variation. The 500 hPa geopotential height and horizontal wind anomalies regressed upon the SRPI-ID multiplied by −1 are presented in Figure 11e. To the northeast of the TP, significant positive height anomalies arose, which were related to the decadal weakening of SH. The anomalous positive height accompanied the anticyclonic anomaly to the northeast of the TP. In relation to the anticyclonic circulation at 500 hPa, easterly wind anomalies to the south of the anticyclonic center flowed into the TP from the eastern boundary of the TP, resulting in a diminished subtropical westerly jet. Corresponding to the easterly anomalies there was significant water vapor convergence and enhanced cloud cover, which could reduce DSWR and SH in the TP. The above analysis indicated that the interdecadal variation of SRP can significantly induce anomalous weakening in SH over the TP by impacting the circulation anomalies and, therefore, cloud cover and DSWR anomalies.

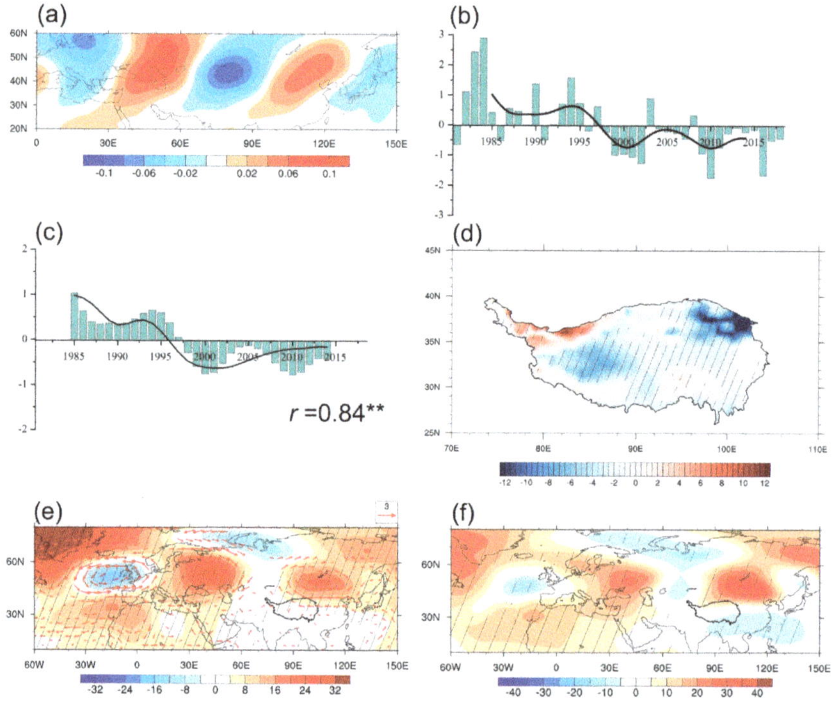

Figure 11. (**a**) The first EOF of summer means 200 hPa meridional wind over the region 20–60° N, 0–150° E for the period 1981–2018. (**b**) Normalized time series of SRPindex (column) and its interdecadal component (line). (**c**) Normalized time series of interdecadal component of SRP index (column) and PC1-ID (line). (**: significant at 99% confidence level). Anomalies in the interdecadal component of sensible heating (**d**), 500 hPa geopotential height (shading, gpm) and the wind exceeding the 95% confidence level (vector, m/s) (**e**), the 200 hPa geopotential height (shading, gpm) (**f**) regressed on SRPI-ID multiplied by −1. Grid points with statistically significant anomalies passing the 95% confidence level are denoted by oblique lines.

We also found that the second leading mode of the upper-tropospheric 200 hPa meridional wind anomalies in 20–60° N, 0–150° E during summer, which is defined as the North Atlantic-East and North Asia pattern (NAENA), had significant impacts on the interannual variation of SH (Figure 12a). The NAENA index (NAENAI) is defined as the PC2 corresponding to EOF2 of the summer 200 hPa meridional wind over the

Eurasian continent (Figure 12b). Figure 12c displays the normalized time series of the interannual component of NAENAI (NAENAI-IA) (column) and PC2-IA (black line). The temporal correlation coefficient between PC2-IA and NAENAI-IA during the period of 1981–2018 was up to 0.63, which indicated that the interannual variation in NAENA could explain approximately 40% of the anomalies in summer SH on the interannual time scale. Figure 12d shows anomalies in SH obtained by regression upon NAENAI. Positive SH anomalies predominated in the west of the TP, whereas weakened SH anomalies arose in the east of the TP, resembling the spatial pattern of EOF2 (Figure 2b). Moreover, the spatial pattern of NAENA showed close similarity to the atmospheric circulation anomalies associated with the interannual variation of SH (Figure 12e,f). The above analysis suggested that NAENA can impact the interannual variation of summer SH over the TP by means of changing the wind speed and water vapor convergence and divergence over the TP.

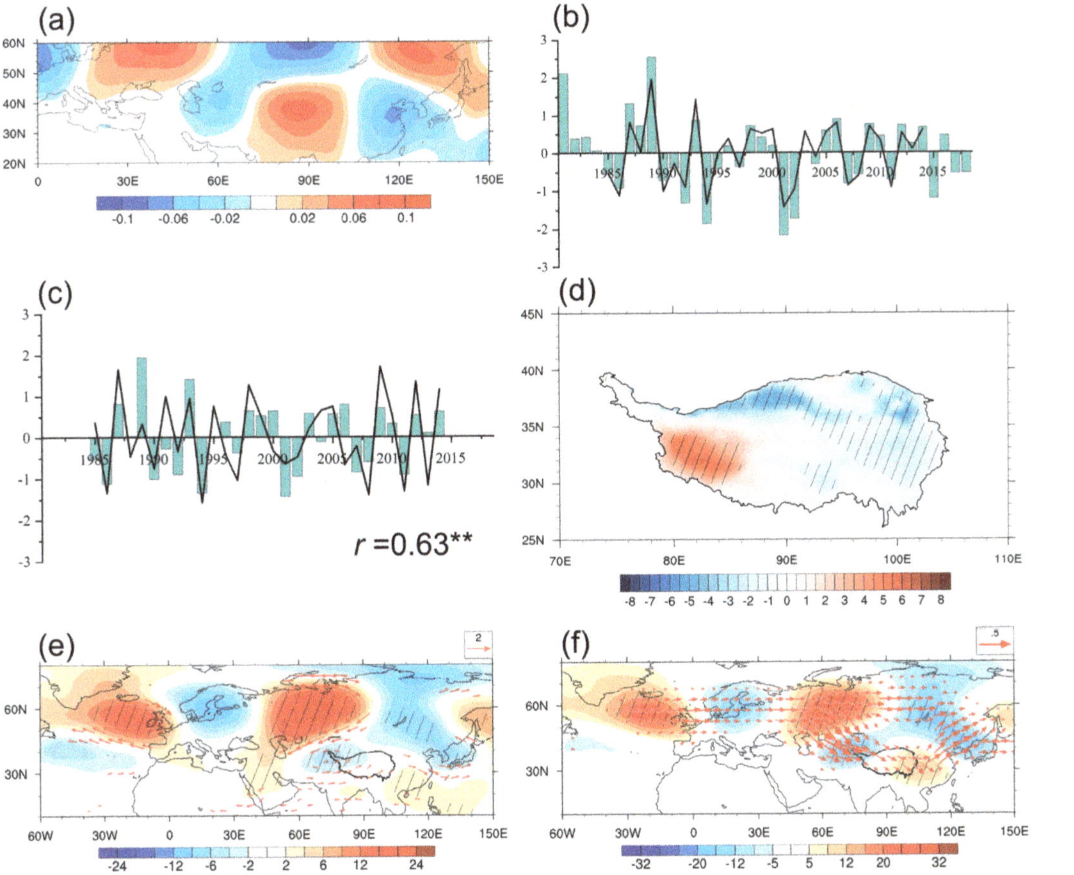

Figure 12. (**a**) The second EOF of summer mean 200 hPa meridional wind in 20–60° N, 0–150° E for the period 1981–2018. (**b**) Normalized time series of NAENA index (column) and its interannual component (line). (**c**) Normalized time series of interannual components of NAENAI (column) and PC2-IA (line), (**: significant at 99% confidience level). Anomalies in the interannual component of sensible heating (**d**), 500 hPa geopotential height (shading, gpm) and the wind exceeding the 95% confidence level (vector, m/s) (**e**), the 200 hPa geopotential height (shading, gpm), and the wave activity flux (vector, m^2 s^{-2}) (**f**) regressed on PC2-IA. Grid points with statistically significant anomalies passing the 95% confidence level are denoted by oblique lines.

4. Discussion

Summer SH over the TP had a significant influence on the regional and Asian climate. Therefore, comprehending and predicting the variability of summer SH is of great importance. Using the monthly remote sensing and reanalysis data with a period of 2001–2018, the features of dominant modes of SH were investigated in this study. EOF analysis showed that an interdecadal weakening of SH appeared over almost the whole TP from 1981–2018 in the leading mode. This is in great agreement with previous studies. Duan et al. [17] indicated that SH over the TP exhibited a decreasing trend since the 1980s, which was mainly induced by the weakening of the westerly jet. In recent years, SH has presented an increasing trend as the result of enhanced surface wind speed [18,38]. However, it was thought that the variations of SH were long-term trends due to global warming and its hiatus, while this study considered that the different trends of SH might be a manifestation of the interdecadal variation. Based on satellite data and observations, recent studies [23] have found that SH has increased slightly over the TP from around 2001. To reach more accurate SH over the TP, the introduction of satellite data will be carried out in the future, which would overcome the shortage of accuracy in reanalysis data. The EOF2 pattern of summer SH over the TP was characterized by a zonally asymmetric pattern with positive (negative) SH anomalies in the western (eastern) TP. This indicated that positive SH anomalies in the western TP are usually accompanied by negative SH anomalies in the eastern TP on an interannual time scale. Based on meteorological data and satellite products, a recent study found that the long-term variations of summer evapotranspiration (ET) over the TP present a dipole pattern, with an increasing trend in the eastern TP and a decreasing trend in the western TP [29]. It is worth studying the cause of opposite spatial patterns between SH and ET over the TP.

Further analysis showed that the interannual variation of SH had a tight connection to the stationary teleconnection pattern NAENA, which was the second leading mode of the upper-tropospheric meridional wind anomalies over the Eurasian continent in summer. The interdecadal weakening of summer SH over the TP was mainly due to the decreased westerly wind anomalies over the TP, which were induced by the anticyclonic circulation to the northeast of the TP. The atmospheric circulation anomalies linked to the interdecadal weakening of SH were characterized by a teleconnection wave train, which was similar to that linked to the interdecadal phase shift of SRP. This indicated that the decadal shift of the phase in SRP plays a crucial role in the interdecadal weakening of SH. Han et al. [38] suggested that SRP is a key factor influencing summer atmospheric heat over the inner TP on interdecadal time scales, which is consistent with our conclusion. A recent study proposed that the predictability of the TP rainfall also originates from SRP [39]. Therefore, the interdecadal variation of SRP can significantly impact the climate variations of the TP on interdecadal and multidecadal time scales and acts as a potential predictor for the climate variations for TP. Previous studies have demonstrated that the phase of the AMO is likely to provide some prediction potential for the interdecadal variations of SRP [35]. The impacts of mid-high atmospheric circulation systems on the TP call for further studies.

5. Conclusions

This study examined the dominant modes of variability in summer SH over the TP based on reanalysis and remote sensing data. The role of atmospheric teleconnection patterns in SH was also investigated. The main conclusions are as follows:

(1) The large value area in SH and its standard deviation was concentrated in the north and west of the TP. The first leading mode of the TP summer SH during the period 1981–2018 presented a decadal shift from a positive phase to a negative phase after around 1996. The second leading mode was characterized by a zonal dipole pattern with enhanced (weakened) SH anomalies in the western (eastern) TP on the interannual time scale.

(2) The interannual variation of summer SH was dominated by anomalies in DSWR, which was associated with the anomalous cloud cover over the TP. An atmospheric

pattern referred to as NAENA induced an anticyclone anomaly to the west of the TP, leading to anomalous water vapor convergence (divergence) and more (less) cloud cover in the eastern (western) TP. Corresponding to the increase (decrease) in cloud cover, DSWR presented anomalous enhancement (reduction) and resulted in a zonal dipole pattern with strengthened (weakened) SH in the western (eastern) TP.

(3) Interdecadal weakening of summer SH was associated with the interdecadal variation of DSWR induced by the enhancement of cloud cover. The decadal change in cloud cover over the TP was mainly due to the variation of water vapor transport as a result of the decadal phase shift of SRP. An anticyclone circulation to the northeast of the TP associated with SRP led to enhanced water vapor supply and convergence of the TP, which resulted in an increase in cloud cover and a reduction in DSWR, contributing to the interdecadal decrease in SH over the TP.

By investigating the relationship between large-scale circulation and the TP thermodynamic forcing, this study is crucial for us to understand the land-atmosphere interaction and its climatic effects of the TP.

Supplementary Materials: The following supporting information can be downloaded at: https://www.mdpi.com/article/10.3390/rs14040956/s1, Figure S1: The first two EOFs of summer SH over the TP for the period 2001–2018 derived from ERA-Interim (a,d) and remote sensing data (b,e). Normalized time series of PC1 (c) and PC2 (f) derived from ERA-Interim (black line) and remote sensing data (red line).

Author Contributions: Conceptualization, Z.H.; methodology, W.F. and C.H.; software, W.F.; investigation, W.F. and Z.H.; writing—original draft preparation, W.F., X.H., Y.Y. and H.Y.; writing—review and editing, W.F., Z.H., W.M., Y.M., C.H., D.W. and C.F. All authors have read and agreed to the published version of the manuscript.

Funding: This research was supported by the Second Tibetan Plateau Scientific Expedition and Research (STEP) program (grant no. 2019QZKK0103), the National Natural Science Foundation of China (grant no. 91837208, 41830650), the Strategic Priority Research Program of Chinese Academy of Sciences (grant no. XDA20060101) and the National Key Research and Development Program of China (grant no. 2018YFC1505701).

Institutional Review Board Statement: Not applicable.

Informed Consent Statement: Not applicable.

Data Availability Statement: The ERA-Interim reanalysis from the ECMWF used in this study can be accessed online (https://www.ecmwf.int/en/forecasts/datasets/reanalysis-datasets/ (accessed on 1 January 2022)). The NCEP/NCAR reanalysis used in this study can be accessed at https://psl.noaa.gov/data/reanalysis/reanalysis.shtml (accessed on 1 January 2022). The CRU dataset can be accessed at https://data.ceda.ac.uk/ (accessed on 5 January 2022).

Conflicts of Interest: The authors declare no conflict of interest.

References

1. Xu, X.; Lu, C.; Shi, X.; Gao, S. World water tower: An atmospheric perspective. *Geophys. Res. Lett.* **2008**, *35*, L20815. [CrossRef]
2. Kang, S.; Xu, Y.; You, Q.; Flügel, W.A.; Pepin, N.; Yao, T. Review of climate and cryospheric change in the Tibetan Plateau. *Environ. Res. Lett.* **2010**, *5*, 015101. [CrossRef]
3. Wang, B.; Ma, Y.; Su, Z.; Wang, Y.; Ma, W. Quantifying the evaporation amounts of 75 high-elevation large dimictic lakes on the Tibetan Plateau. *Sci. Adv.* **2020**, *6*, eaay8558. [CrossRef] [PubMed]
4. Zhao, Y.; Zhou, T. Asian water tower evinced in total column water vapor: A comparison among multiple satellite and reanalysis data sets. *Clim. Dyn.* **2020**, *54*, 231–245. [CrossRef]
5. Cuo, L.; Zhang, Y.; Zhu, F.; Liang, L. Characteristics and changes of streamflow on the Tibetan Plateau: A review. *J. Hydrol. Reg. Stud.* **2014**, *2*, 49–68. [CrossRef]
6. Xu, X.; Zhao, T.; Lu, C.; Guo, Y.; Chen, B.; Liu, R.; Li, Y.; Shi, X. An important mechanism sustaining the atmospheric "water tower" over the Tibetan Plateau. *Atmos. Chem. Phys.* **2014**, *14*, 11287–11295. [CrossRef]
7. Yanai, M.; Li, C. Mechanism of heating and the boundary layer over the Tibetan Plateau. *Mon. Weather Rev.* **1994**, *122*, 305–323. [CrossRef]

8. Wu, G.; Zhang, Y. Tibetan Plateau forcing and the timing of the monsoon onset over South Asia and the South China Sea. *Mon. Weather Rev.* **1998**, *126*, 913–927. [CrossRef]
9. Wu, G.; Liu, Y.; Zhang, Q.; Duan, A.; Wang, T.; Wan, R.; Liu, X.; Li, W.; Zhang, Z.; Wang, Z.; et al. The influence of mechanical and thermal forcing by the Tibetan Plateau on Asian climate. *J. Hydrometeorol.* **2007**, *8*, 770–789. [CrossRef]
10. Li, L.; Zhang, R.; Wen, M.; Liu, L. Effect of the atmospheric heat source on the development and eastward movement of the Tibetan Plateau vortices. *Tellus A Dyn. Meteorol. Oceanogr.* **2014**, *66*, 24451. [CrossRef]
11. Yanai, M.; Li, C.; Song, Z. Seasonal heating of the Tibetan Plateau and its effects on the evolution of the Asian summer monsoon. *J. Meteorol. Soc. Japan. Ser. II* **1992**, *70*, 319–351. [CrossRef]
12. Duan, A.; Li, F.; Wang, M.; Wu, G. Persistent weakening trend in the spring sensible heat source over the Tibetan Plateau and its impact on the Asian summer monsoon. *J. Clim.* **2011**, *24*, 5671–5682. [CrossRef]
13. Ye, D.; Wu, G. The role of the heat source of the Tibetan plateau in the general circulation. *Meteorol. Atmos. Phys.* **1998**, *67*, 181–198. [CrossRef]
14. Hsu, H.; Liu, X. Relationship between the Tibetan Plateau heating and East Asian summer monsoon rainfall. *Geophys. Res. Lett.* **2013**, *30*, 2066. [CrossRef]
15. Wang, Z.; Duan, A.; Wu, G. Time-lagged impact of spring sensible heat over the Tibetan Plateau on the summer rainfall anomaly in East China: Case studies using the WRF model. *Clim. Dyn.* **2014**, *42*, 2885–2898. [CrossRef]
16. Xie, X.; Duan, A.; Shi, Z.; Li, X.; Sun, H.; Liu, X.; Cheng, X.; Zhao, T.; Che, H.; Liu, Y. Modulation of springtime surface sensible heating over the Tibetan Plateau on the interannual variability of East Asian dust cycle. *Atmospheric Chemistry and Physics.* **2020**, *20*, 11143–11159. [CrossRef]
17. Duan, A.; Wu, G. Weakening trend in the atmospheric heat source over the Tibetan Plateau during recent decades. Part I: Observations. *J. Clim.* **2008**, *21*, 3149–3164. [CrossRef]
18. Zhu, L.; Huang, G.; Fan, G.; Qu, X.; Zhao, G.; Hua, W. Evolution of surface sensible heat over the Tibetan Plateau under the recent global warming hiatus. *Adv. Atmos. Sci.* **2017**, *34*, 1249–1262. [CrossRef]
19. Wang, M.; Wang, J.; Chen, D.; Duan, A.; Liu, Y.; Zhou, S.; Guo, D.; Wang, H.; Ju, W. Recent recovery of the boreal spring sensible heating over the Tibetan Plateau will continue in CMIP6 future projections. *Environ. Res. Lett.* **2019**, *14*, 124066. [CrossRef]
20. Zhu, L.; Huang, G.; Fan, G.; Qü, X.; Wang, Z.; Hua, W. Elevation-dependent sensible heat flux trend over the Tibetan Plateau and its possible causes. *Clim. Dyn.* **2019**, *52*, 3997–4009. [CrossRef]
21. Cui, Y.; Duan, A.; Liu, Y.; Wu, G. Interannual variability of the spring atmospheric heat source over the Tibetan Plateau forced by the North Atlantic SSTA. *Clim. Dyn.* **2015**, *45*, 1617–1634. [CrossRef]
22. Yu, W.; Liu, Y.; Yang, X.; Wu, G.; He, B.; Li, J.; Bao, Q. Impact of North Atlantic SST and Tibetan Plateau forcing on the seasonal transition of springtime South Asian monsoon circulation. *Clim. Dyn.* **2021**, *56*, 559–579. [CrossRef]
23. Ma, Y.; Wang, Y.; Han, C. Regionalization of land surface heat fluxes over the heterogeneous landscape: From the Tibetan Plateau to the Third Pole region. *Int. J. Remote Sens.* **2018**, *39*, 5872–5890. [CrossRef]
24. Qian, Y.; Zhang, Y.; Huang, Y.; Huang, Y.; Yao, Y. The effects of the thermal anomalies over the Tibetan Plateau and its vicinities on climate variability in China. *Adv. Atmos. Sci.* **2004**, *21*, 369–381.
25. Chen, B.; Qian, Z.; Zhang, L. Numerical simulation of formation and development of vortices over the Qinghai-Xizang Plateau in summer. *Chin. J. Atmos. Sci.* **1996**, *20*, 491–502.
26. Liang, L.; Li, Y.; Hu, H.; Jiang, X.; Zhang, E. Numerical study of the influence of sensible heat anomalies in summer over Qinghai-Xizang Plateau on rainfall in Sichuan-Chongqing regions. *Plateau Meteorol.* **2013**, *32*, 1538–1545. (In Chinese)
27. Yang, Y.; Liu, Y.; Li, M.; Hu, Z.; Ding, Z. Assessment of reanalysis flux products based on eddy covariance observations over the Tibetan Plateau. *Theor. Appl. Climatol.* **2019**, *138*, 275–292. [CrossRef]
28. Han, C.; Ma, Y.; Chen, X.; Su, Z. Trends of land surface heat fluxes on the Tibetan Plateau from 2001 to 2012. *Int. J. Climatol.* **2017**, *37*, 4757–4767. [CrossRef]
29. Han, C.; Ma, Y.; Wang, B.; Zhong, L.; Ma, W.; Chen, X.; Su, Z. Long-term variations in actual evapotranspiration over the Tibetan Plateau. *Earth Syst. Sci. Data* **2021**, *13*, 3513–3524. [CrossRef]
30. Su, Z. The Surface Energy Balance System (SEBS) for estimation of turbulent heat fluxes. *Hydrol. Earth Syst. Sci.* **2002**, *6*, 85–100. [CrossRef]
31. Takaya, K.; Nakamura, H. A formulation of a phase-independent wave-activity flux for stationary and migratory quasigeostrophic eddies on a zonally varying basic flow. *J. Atmos. Sci.* **2001**, *58*, 608–627. [CrossRef]
32. North, G.; Bell, T.; Cahalan, R.; Moeng, F.J. Sampling errors in the estimation of empirical orthogonal functions. *Mon. Weather Rev.* **1982**, *110*, 699–706. [CrossRef]
33. Chen, S.; Wu, R.; Liu, Y. Dominant modes of interannual variability in Eurasian surface air temperature during boreal spring. *J. Clim.* **2016**, *29*, 1109–1125. [CrossRef]
34. Zhou, C.; Zhao, P.; Chen, J. The interdecadal change of summer water vapor over the Tibetan Plateau and associated mechanisms. *J. Clim.* **2019**, *32*, 4103–4119. [CrossRef]
35. Wang, L.; Xu, P.; Chen, W.; Liu, Y. Interdecadal variations of the Silk Road pattern. *J. Clim.* **2017**, *30*, 9915–9932. [CrossRef]
36. Li, X.; Lu, R.; Ahn, J.B. Combined effects of the British-Baikal Corridor pattern and the Silk Road pattern on Eurasian surface air temperatures in summer. *J. Clim.* **2021**, *34*, 3707–3720. [CrossRef]

37. Han, Y.; Ma, W.; Yang, Y.; Ma, Y.; Xie, Z.; Sun, G.; Menenti, M.; Su, B. Impacts of the Silk Road pattern on the interdecadal variations of the atmospheric heat source over the Tibetan Plateau. *Atmos. Res.* **2021**, *260*, 105696. [CrossRef]
38. Fan, W.; Ma, W.; Hu, Z.; Ma, Y. Recovery of sensible heating and its elevation amplification over and around the Tibetan Plateau since the 2000s. *Theor. Appl. Climatol.* **2021**, *146*, 617–630. [CrossRef]
39. Hu, S.; Zhou, T. Skillful prediction of summer rainfall in the Tibetan Plateau on multiyear time scales. *Sci. Adv.* **2021**, *7*, eabf9395. [CrossRef]

remote sensing

Article

Hydrothermal Factors Influence on Spatial-Temporal Variation of Evapotranspiration-Precipitation Coupling over Climate Transition Zone of North China

Zesu Yang [1], Qiang Zhang [2,*], Yu Zhang [1], Ping Yue [2], Liang Zhang [2], Jian Zeng [1] and Yulei Qi [1]

1 Plateau Atmospheric and Environment Key Laboratory of Sichuan Province, College of Atmospheric Sciences, Chengdu University of Information Technology, Chengdu 610225, China; cqhcyzx@cuit.edu.cn (Z.Y.); yuzhang@cuit.edu.cn (Y.Z.); zengj17@lzu.edu.cn (J.Z.); qyl@cuit.edu.cn (Y.Q.)

2 Institute of Arid Meteorology, CMA, Key Laboratory of Arid Climatic Change and Reducing Disaster of Gansu Province, Key Open Laboratory of Arid Climatic Change and Disaster Reduction of CMA, Lanzhou 730020, China; yuep@iamcma.cn (P.Y.); lzhangmet@hotmail.com (L.Z.)

* Correspondence: zhangqiang@cma.gov.cn

Abstract: As a land–atmosphere coupling "hot spot", the northern China climate transition zone has a sharp spatial gradient of hydrothermal conditions, which plays an essential role in shaping the spatial and temporal pattern of evapotranspiration-precipitation coupling, but whose mechanisms still remain unclear. This study analyzes the spatial and temporal variation in land–atmosphere coupling strength (CS) in the climate transitional zone of northern China and its relationship with soil moisture and air temperature. Results show that CS gradually transitions from strong positive in the northwest to negative in the southeast and northeast corners. The spatial distribution of CS is closely related to climatic hydrothermal conditions, where soil moisture plays a more dominant role: CS increases first, and then decreases with increasing soil moisture, with the threshold of soil moisture at 0.2; CS gradually transitions from positive to negative at soil moisture between 0.25 and 0.35; CS shows an exponential decreasing trend with increasing temperature. In terms of temporal variation, CS is strongest in spring and weakens sequentially in summer, autumn, and winter, and has significant interdecadal fluctuations. The trend in CS shifts gradually from significantly negative in the west to a non-significant positive in the east. Soil moisture variability dominates the intra-annual variability of CS in the study regions, and determines the interannual variation of CS in arid and semi-arid areas. Moreover, the main reason for the positive and negative spatial differences in CS in the study area is the different driving regime of evapotranspiration (ET). ET is energy-limited in the southern part of the study area, leading to a positive correlation between ET and lifting condensation level (LCL), while in most of the northern part, ET is water-limited and is negatively correlated with LCL; LCL has a negative correlation with P across the study area, thus leading to a negative ET-P coupling in the south and a positive coupling in the north.

Keywords: land–atmosphere interaction; evapotranspiration; soil moisture; temperature; climate transition areas

Citation: Yang, Z.; Zhang, Q.; Zhang, Y.; Yue, P.; Zhang, L.; Zeng, J.; Qi, Y. Hydrothermal Factors Influence on Spatial-Temporal Variation of Evapotranspiration-Precipitation Coupling over Climate Transition Zone of North China. *Remote Sens.* **2022**, *14*, 1448. https://doi.org/10.3390/rs14061448

Academic Editor: Nicola Montaldo

Received: 7 February 2022
Accepted: 15 March 2022
Published: 17 March 2022

1. Introduction

Among the interactions amid various spheres of the climate system, the land–atmosphere interaction plays an important role in influencing the evolution of weather and climate [1]. The land surface is closely linked to the atmosphere through energy and water cycles, causing increases in the temperature variability [2] and the frequency of high-temperature heat waves [3], and exacerbating compound soil and atmospheric drought intensity [4]. The land–atmosphere coupling strength (CS) is a key indicator to characterize the land–atmosphere interaction. Areas with a stronger CS imply a greater influence of land surface on regional weather and climate. The global strong land–atmosphere coupling zone is mostly located in the arid-humid climate transition zone [5], including west-central North America, parts of Eurasia, Australia, Argentina, the Sahel region of North Africa, and South Africa [6,7]. Accurate acquisition of land–atmosphere forcing signals in these land–atmosphere coupling "hotspots" is important for improving the forecasting capabilities of the weather and climate [5,8,9].

Land–atmosphere coupling includes a series of complex processes: land surface state anomalies first cause changes in surface fluxes, which in turn affect precipitation through feedbacks from the land surface to the atmosphere [7,10]. Regulated by the surface energy balance, sensible heat fluxes change synergistically with evapotranspiration. Thus, evapotranspiration can regulate sensible heat fluxes via the Bowen ratio. Therefore, evapotranspiration is often considered as a key process in land–atmosphere coupling processes [11,12]. However, studies have shown that the influence of evapotranspiration on precipitation has the greatest uncertainty in land–atmosphere coupling processes [13]. Generally, evapotranspiration can affect precipitation in three ways. First, evapotranspiration directly affects atmospheric precipitation through water recycling. Evapotranspiration can return approximately 70% of precipitation to the atmosphere [14], and atmospheric precipitable water directly influences precipitation. This mechanism is more prominent in water-scarce areas [15,16]. Second, changes in evapotranspiration can also alter the regional pressure field, which can cause adjustments in atmospheric circulation and lead to large-scale precipitation changes [17]. Moreover, evapotranspiration and the sensible heat fluxes regulated by it affect the atmospheric stability state by altering the atmospheric temperature and humidity profiles, thus affecting convective precipitation [18]. The last of these pathways, evapotranspiration–precipitation local coupling, is the most important method for evapotranspiration to influence precipitation [13,19,20]. However, due to the complexity of the influence of evapotranspiration on atmospheric stability, evapotranspiration-precipitation local coupling has significant uncertainty and becomes a challenging problem in the current land–atmosphere coupling research.

Local evapotranspiration–precipitation coupling is controlled by many factors and the influence mechanism is very complex. Water and energy cycles are the key physical processes throughout the coupled land–atmosphere interaction [21]. Moisture and thermal factors directly affect the evapotranspiration process: in dry areas, evapotranspiration is controlled mainly by moisture factors, while in humid areas evapotranspiration is controlled mainly by thermal factors [22,23]. In turn, evapotranspiration affects the structural characteristics of the boundary layer through the transport of water and heat, and sufficient surface moisture can lead to lowered boundary layer height (BLH) and lifting condensation level (LCL) and increased moist static energy (MSE); in contrast, limited surface water and adequate thermal conditions raise the BLH and LCL. In general, lower BLH and LCL and larger MSE can lead to an increase in convective available potential energy (CAPE) and a higher probability of convective precipitation [20]. However, observation studies have found that arid conditions can also promote physical mechanisms that favor the generation of convective precipitation [24–27]. This is attributed to the large sensible heat flux caused by strong thermal factors in arid regions that weakens the convective inhibition energy (CIN) [28], thereby increasing the probability of convective precipitation; there is also a negative feedback of evapotranspiration on precipitation. Therefore, both positive and negative feedbacks between evapotranspiration and precipitation are closely related to moisture and thermal properties, i.e., moisture and thermal factors are the most critical forcing factors affecting the land–atmosphere coupling strength.

In the past 20 years, a large number of studies have paid attention to the spatial and temporal distribution of land–atmosphere coupling and its intrinsic mechanism. The Global Land-Atmosphere Coupling Experiment (GLACE) found that the strong land–atmosphere coupling regions are mostly located in semi-arid and semi-humid climate transition zones [5], and other diagnostic studies based on observational data have also verified this conclusion [6,7,29]. Due to the apparent spatial variability of land–atmosphere CS, some typical regions have attracted wide attention. In North America, the spatial distribution of land–atmosphere coupling was correlated with the multi-year average soil moisture, and the strong coupling area was mainly distributed in areas with soil moisture ranging from 0.4 to 0.55 [30]. Due to interannual variations in soil moisture, land–atmosphere coupling shows significant interannual fluctuations [31,32]. In southeastern South America, the spatial and temporal distribution of land–atmosphere coupling is correlated not only with soil moisture but also with wet static energy and its vertical gradient that is controlled by soil moisture [33]. Moreover, under future climate warming and humidification, the intensity of land–atmosphere coupling will be significantly weakened due to the gradual shift of evapotranspiration from moisture limitation to energy limitation [34]. In Europe, the northward expansion of the Hadley circulation has caused a northward shift of the climate transition zone, leading to a northward shift of the strong land–atmosphere coupling zone [2]. In Africa, land–atmosphere coupling was negatively correlated with the spatial and temporal distribution of soil moisture, with areas of lower soil moisture and periods of dry moisture exhibiting stronger land–atmosphere coupling. Spatial differences in the soil moisture lead to enhanced sensible heat fluxes in the dry zone and reduced sensible heat fluxes in the wet zone, which in turn trigger mesoscale circulation, and the upward branch of this circulation in the dry zone is an important factor in triggering deep convection [26]. In East Asia, land–atmosphere coupling is strong in north China, where soil moisture is low [35]; land–atmosphere coupling is strong in southwest China in spring, when soil moisture is low [36]; land–atmosphere coupling is strongly influenced by the snow cover in the dry season and by the soil moisture in the rainy season over Tibetan Plateau [37]; the land–atmosphere coupling degree is closely related to the state of surface vegetation in northwest China, and the improvement of vegetation state can improve the surface moisture condition, reduce the land surface evapotranspiration, and decrease the strength of land–atmosphere coupling [38].

The above studies mainly focused on the spatial distribution and temporal variation in the land–atmosphere CS and its relationship with soil moisture, and there is lack of research on the role of thermal factors on the land–atmosphere CS. Theoretically, the thermal properties also play a substantial role in the land–atmosphere coupling. The role of moisture and thermal factors in regulating the land–atmosphere coupling is similar to that of evapotranspiration, as it is regulated by moisture in water-scarce areas and by thermal and energy in water-sufficient areas [39].

Most of northern China makes up a dry-wet climate transitional zone, with dramatic spatial and temporal variations in water and heat characteristics. From the northwest to the southeast, moisture availability decreases, and energy availability increases, while the evapotranspiration control factor gradually changes from moisture to energy limitation [23]. This inevitably affects the spatial and temporal variation of regional land–atmosphere coupling. However, the spatial and temporal distribution of land–atmosphere coupling in north China's climate transition zone remains unclear, and it is also unknown how hydrothermal factors affect the spatial-temporal variation in land–atmosphere coupling.

In this study, an evapotranspiration–precipitation coupling index proposed by Zeng et al. [6] was used to diagnose the land–atmosphere CS in the climate transitional zone of north China, and the main objectives were to (i) analyze the spatial distribution and temporal evolution characteristics of the land–atmosphere coupling and (ii) explain the impacts of moisture and thermal factors on the land–atmosphere CS. The results of the study are expected to enhance the understanding of land–atmosphere coupling mechanisms in the climate transitional zone of northern China.

2. Data and Methods

2.1. Study Area and Sites

The climate transitional zone of northern China (CTZNC) is selected as the study area in this work, with the spatial extent of the region between 33°N and 45°N and 100°E and 125°E. The geographical area and the climatic background are shown in the black box of Figure 1. The climate background is classified using the dryness index defined by the United Nations Environment Programme:

$$AI = P/PET, \tag{1}$$

where, AI is the dryness index, P is the average annual precipitation in mm, and PET is the average annual potential evapotranspiration in mm. AI < 0.05 is classified as a hyper-arid zone, 0.05 < AI < 0.2 as an arid zone, 0.2 < AI < 0.5 as semi-arid zone, 0.5 < AI < 0.65 as a sub-humid zone, and AI > 0.65 as a humid zone. As shown in Figure 1, the study region mainly includes arid, semi-arid, sub-humid, and humid climate. It is not only a climate transition region but also a typical ecological transition zone, as well as a major activity area of the northern edge of the East Asian summer monsoon.

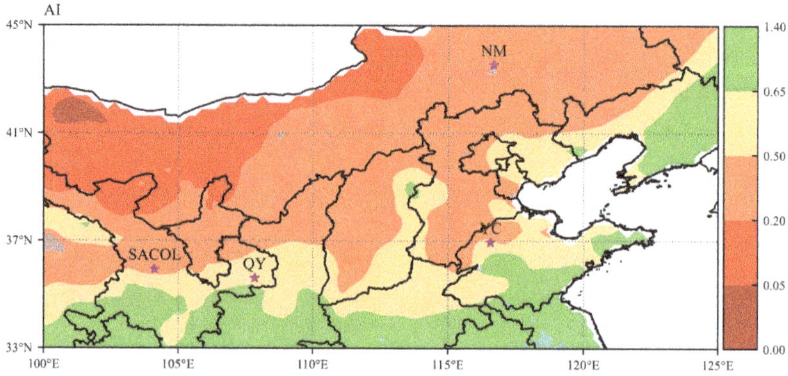

Figure 1. Climatic background of climate transitional zone of northern China (hyper-arid: AI < 0.05, arid: 0.05 < AI < 0.2, semi-arid: 0.2 < AI < 0.5, sub-humid: 0.5 < AI < 0.65, and humid: AI > 0.65). The soil moisture observation sites are indicated by stars.

Four sites are selected for study, the underlying surfaces of which are either grassland or cropland. The locations of the sites are shown in Figure 1, and the climate and environmental background are given in Table 1.

Table 1. Brief description of the soil moisture observatories.

Station	Location	Elevation (m)	Land Cover	Precipitation (mm)	Mean Air Temperature (°C)	Climate
SACOL	35°57′N, 104°08′E	1966	Grassland	381.8	6.7	Semi-arid
QY	35°41′N, 107°51′E	1280	Cropland	562	8.8	Sub-humid
NM	43°33′N, 116°40′E	1250	Grassland	336	0.8	Semi-arid
YC	36°50′N, 116°34′E	28	Cropland	582	13.1	Sub-humid

2.2. Data

We selected a gridded evapotranspiration dataset—Derived Optimal Linear Combination Evapotranspiration (DOLCE) [40]. DOLCE is a hybrid evapotranspiration dataset under observational constraints, which merges four available global ET datasets: ERA5-land, FLUXCOM METEO+RS, GLEAM v3.5a, and GLEAM v3.5b. The contribution of each dataset to DOLCE ET is based on its ability to match field observations as well as its dependence to the other parent datasets. The latest version of DOLCE-ET V3.0 also provides time-varying estimates of its uncertainty errors. DOLCE-ET V3.0 has a temporal resolution of months and a spatial resolution of 0.25°. It spans the years 1980–2018 and covers the global land surface.

We use the China Meteorological Forcing Dataset (CMFD) for precipitation and near-surface air temperature data [41]. CMFD uses Princeton reanalysis data, GLDAS (global land data assimilation system) data, GEWEX-SRB (the global energy and water exchanges) radiation data, and tropical rainfall measuring mission precipitation data as background fields, and merges the conventional meteorological observation data of the China Meteorological Administration to produce a regional high spatial and temporal resolution dataset. The dataset has a temporal resolution of months and a spatial resolution of 0.1°, with a spatial range of 60–140°E and 15–55°N.

For soil moisture, we selected the Climate Change Initiative (CCI) data set of the European Space Agency (ESA) [42]. The CCI SM v04.7 product consists of three sets of surface soil moisture data: active products, passive products, and combined products. The "active product" and "passive product" are generated by merging the soil moisture outputs of the scatterometer and radiometer, respectively; the "combined product" is a hybrid product based on the first two data sets, and used in this study. The dataset has a temporal resolution of a day and a spatial resolution of 0.25°, and spans the period from November 1978 to December 2019. Site observations was used to validate CCI data.

Boundary layer heights used ECMWF monthly averaged ERA5 data with a spatial resolution of 0.25°. All data were time-scaled to months and interpolated to 0.25° × 0.25° spatial resolution.

2.3. Methods

2.3.1. Evapotranspiration-Precipitation Coupling Strength

An evapotranspiration–precipitation coupling index proposed by Zeng et al. [6] was employed to diagnose the strength of land–atmosphere coupling. Considering that surface state variables always affect the atmospheric state via near-surface fluxes, evapotranspiration was selected as the surface impact factor. In the index, the covariance of evapotranspiration and precipitation reflects the synchronous change in evapotranspiration and precipitation, and the ratio of the covariance to the precipitation variance reflects the contribution of evapotranspiration changes to total precipitation changes. This method has a solid physical mechanism and is widely used in diagnostic studies of land–atmosphere CS. The equation for its calculation is as follows:

$$\Gamma = \frac{\sum_{i=1}^{N} P_i{}' E_i{}'}{\sum_i^N P_i{}'^2} \tag{2}$$

This can also be rewritten as:

$$\Gamma = r_{P,E}\frac{\sigma_E}{\sigma_P}, \tag{3}$$

where Γ is the land–atmosphere CS, $P_i{}'$ and $E_i{}'$ are anomalies of precipitation and evapotranspiration, respectively, N is total number of months or years, $r_{P,E}$ is correlation coefficient of precipitation and evapotranspiration, σ_E and σ_P are the standard deviation of evapotranspiration and precipitation, respectively. This index reflects the proportion of precipitation changes caused by evapotranspiration in total precipitation using the relative magnitude of the covariance between precipitation and evapotranspiration and the variance of precipitation. The more consistent the pace of change between the two variables, and the larger the magnitude of change, and the stronger the land–atmosphere coupling. The positive and negative values can also reflect the respective coupling between evapotranspiration and precipitation.

This method examines the ET–P relationship statistically and does not reflect the specific physical processes and influence mechanisms. To identify the significant strong CS areas, a correlation-coefficient significance test can be used to test the significance of the CS [7]. i.e., the ET–P coupling is deemed to be significant if the p value of the correlation coefficient is less than 0.05.

The evapotranspiration–precipitation CS was calculated using monthly or yearly ET and P data. The bulk CS in Figure 4 was calculated using data of all months during the study period. The seasonal CS in Figure 5 was calculated using monthly data in each of the four seasons. The decadal CS in Figure 6 was calculated by using yearly data in each decade. The yearly CS was calculated using monthly data of each year, and subsequently the yearly CS was used to calculate the linear trend of CS in Figure 7. The warm season CS was calculated using monthly data of April-September.

2.3.2. Lifting Condensation Level

The lifting Condensation Level (*LCL*) can be calculated by:

$$LCL \approx 125(T_{2m} - D_{2m}) \tag{4}$$

where T_{2m} and D_{2m} are 2-m air temperature and dew point temperature, respectively.

2.4. Validation of CCI Soil Moisture

ESA CCI soil moisture data were validated using observations at four sites in the study area in 2007. Figure 2 shows that the correlation coefficients between CCI and observations are 0.71, 0.69, 0.66, and 0.65 for NM, QY, SACOL, and YC, respectively. The RMSE values for the four sites are 0.033, 0.041, 0.043, and 0.072 m^3/m^3, respectively. YC station, in a semi-humid region with higher soil moisture, has a larger root mean squared error (RMSE) compared to other sites. This is consistent with the results of other studies [43,44], and CCI has a relatively high accuracy in the climate transitional zone of northern China with a correlation coefficient of about 0.7. Therefore, CCI soil moisture is applicable in the climate transitional zone of northern China.

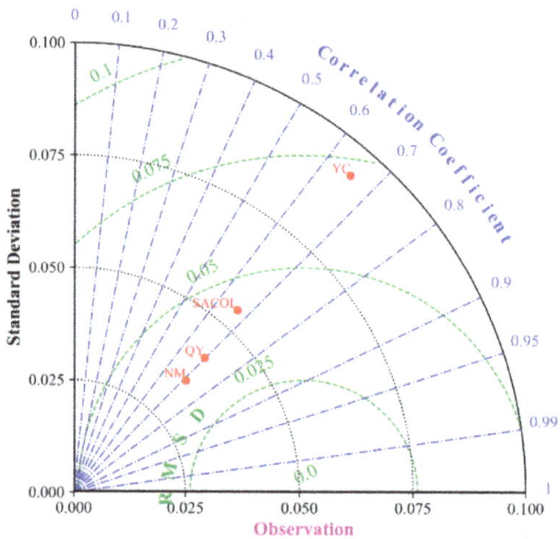

Figure 2. Taylor diagram of ESA CCI soil moisture data.

3. Results

3.1. Spatial and Temporal Variation of Evapotranspiration-Precipitation Coupling Strength

To evaluate the spatial pattern of evapotranspiration-precipitation coupling over the climate transitional zone of northern China, the spatial distribution of evapotranspiration, precipitation, and their variations are shown in Figure 3. Precipitation gradually transitions from more than 800 mm in the southeast to less than 100 mm in the northwest. The standard deviation of precipitation has similar spatial distribution to annual precipitation, decreasing from southeast to northwest. This spatial distribution of precipitation is consistent with the situation of the study area of north China in the transition zone from the East Asian summer monsoon-influenced zone to the non-monsoon zone, where the monsoon precipitation gradually increases from the non-monsoon zone to the monsoon zone. Similarly, evapotranspiration likewise decreases from the southeast to northwest, with a maximum of about 600 mm in the southeast and a minimum of a few tens of millimeters in the northwest. The spatial distribution of evapotranspiration is similar to that of precipitation, indicating a close relationship between evapotranspiration and precipitation. Furthermore, the distribution of standard deviation of evapotranspiration is more complicated, which is larger in the middle region of transition area and smaller in wet and dry areas.

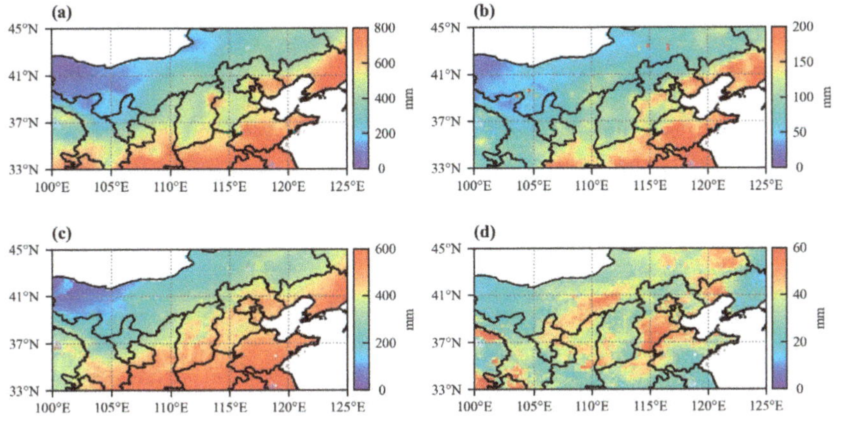

Figure 3. Spatial distribution of (**a**) climatology of annual precipitation, (**b**) standard deviation of annual precipitation, (**c**) climatology of evapotranspiration, and (**d**) standard deviation of evapotranspiration in the climate transitional zone of northern China.

The results from the above analysis highlight a general consistency of the spatial distribution of mean precipitation and evapotranspiration. This suggests a close relationship between precipitation and evapotranspiration in the study region. Furthermore, an index of evapotranspiration–precipitation CS was applied to assessing the spatial and temporal variation in CS in the climate transitional zone of northern China. The spatial pattern of CS shows sharp transition features in the climate transitional zone of northern China (Figure 4), decreasing from the northwest to southeast. The northwest half of the study region is a strong positive coupling area, with CS between 0.2 and 0.6 (passing 0.05 significance test), while the southeast and northeast horn depict negative coupling zone, with CS ranging from −0.2 to −0.5 (passing 0.05 significance test). The middle region of the two regions marks the transition zone from positive to negative coupling, and CS is relatively small.

Due to the temporal changes in climate variables, land–atmosphere couplings also vary with time. Despite having a similar spatial pattern in all seasons, the CS is strongest in spring, when it is significantly larger than the other seasons, followed by summer, and the smallest coupling in autumn and winter (Figure 5). This indicates that the contribution of surface evapotranspiration to precipitation occurs mainly in spring.

On an interdecadal scale (Figure 6), the CS was highest in the 1980s and lowest in the 1990s. The spatial distribution of CS with stronger coupling in the 1980s and 2000s is similar to the distribution of annual coupling; however, it shows a more heterogeneous spatial pattern in the 2010s and 1990s.

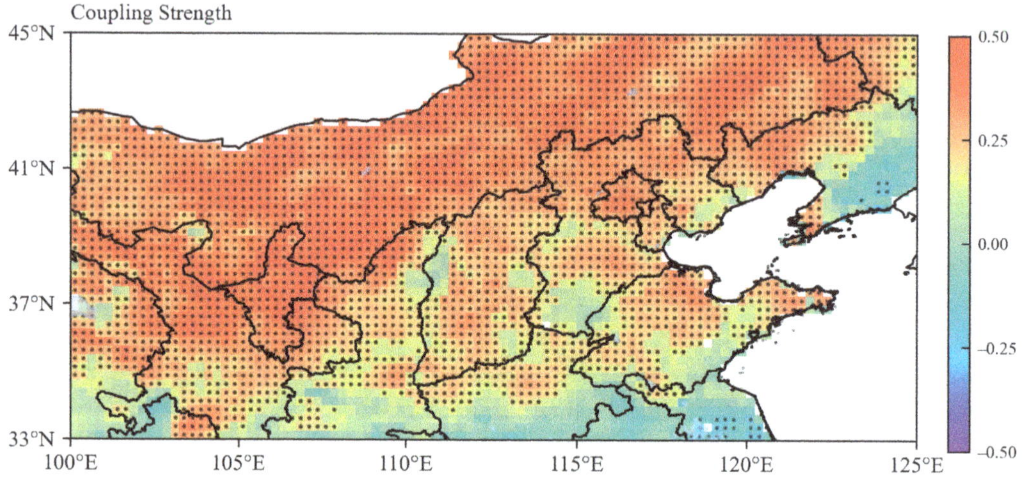

Figure 4. Spatial distribution of evapotranspiration-precipitation CS in the climate transitional zone of northern China (dot denotes CS passing 0.05 significance test).

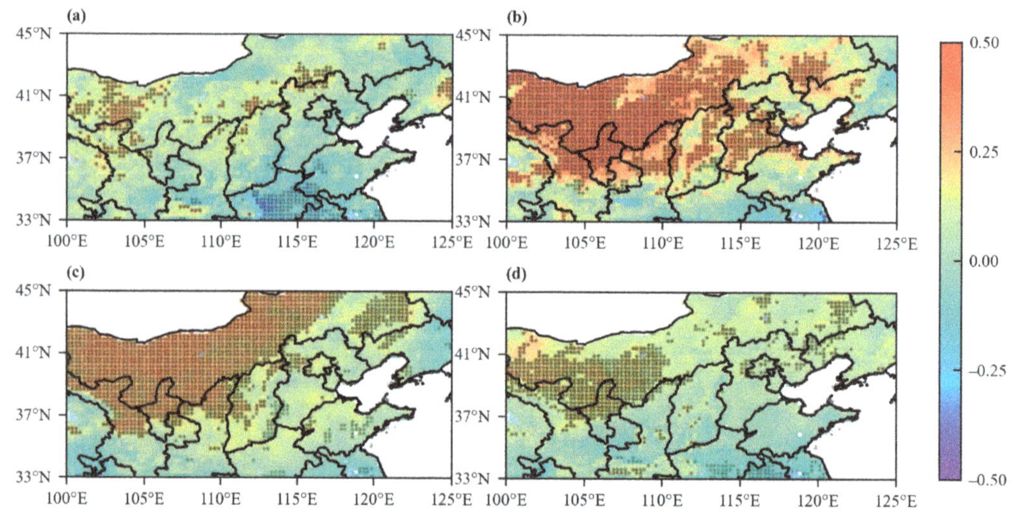

Figure 5. Seasonal changes in spatial distribution of evapotranspiration–precipitation CS in the climate transitional zone of northern China, (**a**) for winter, (**b**) for spring, (**c**) for summer, and (**d**) for autumn (dot denotes CS passing 0.05 significance test).

The trend of annual CS was examined for the climate transitional zone of northern China for the period 1980–2018 (Figure 7). The CS showed a significant decreasing trend in the central and western parts. Except for a small area showing an increase trend in the northwest corner, the overall trend of CS gradually shifts from significant negative in the west to insignificant positive in the east. The northwest region has the strongest decreasing trend of CS, about −0.003/year, whereas the southeast region has a weak increasing trend of CS, with a rate of about 0.001/year.

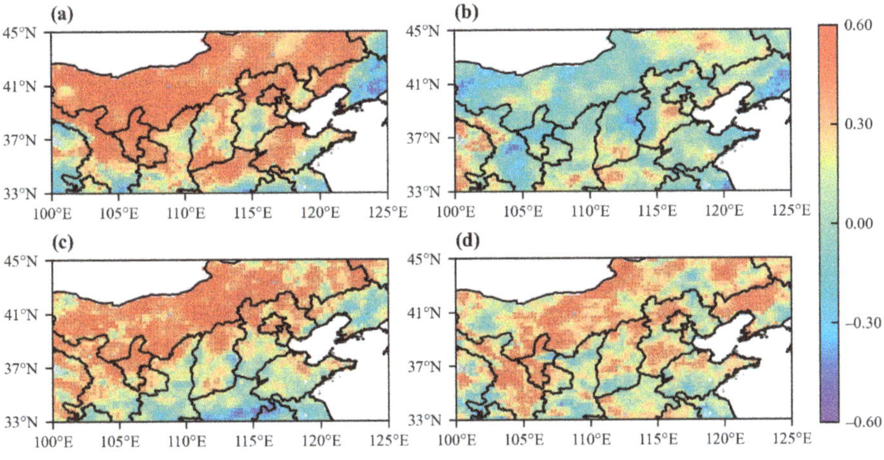

Figure 6. Decadal changes in spatial distribution of evapotranspiration–precipitation coupling in the climate transitional zone of northern China, (**a**) for 1980–1989, (**b**) for 1990–2000, (**c**) for 2000–2009, and (**d**) for 2010–2019 (dot denotes CS passing 0.05 significance test).

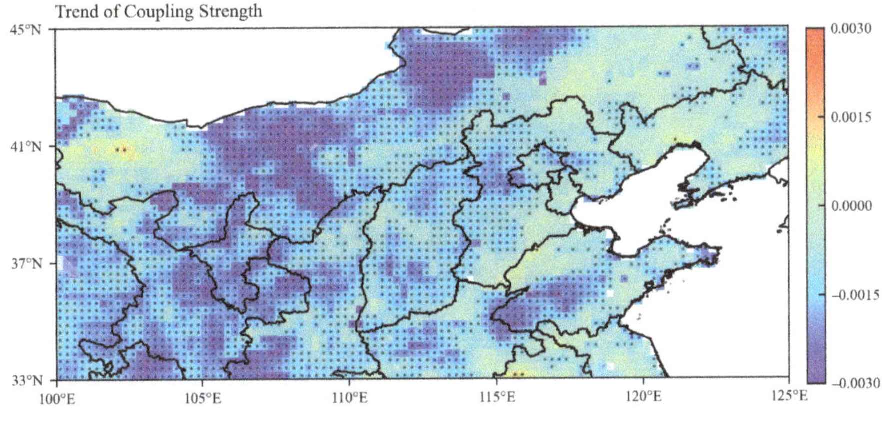

Figure 7. Spatial distribution of evapotranspiration–precipitation coupling trends in the climate transitional zone of northern China (dot denotes trend passing the significance test).

3.2. *Spatial and Temporal Variation of Evapotranspiration-Precipitation Coupling in Relation to Moisture and Thermal Conditions*

3.2.1. Spatial Variation of CS in Relation to Spatial Moisture and Thermal Conditions

The CS has large spatial differences and displays transitional characteristics in the climate transitional zone of northern China, which is closely related to the fact that the region is in a climatic transition zone with large spatial gradients of hydrothermal conditions in the region. Soil moisture and air temperature can aptly reflect hydrothermal conditions in the study region. Therefore, this section analyzes the dependences of CS on soil moisture and air temperature.

First, the spatial patterns of climatological mean soil moisture and air temperature were analyzed for the study region. Soil moisture has large spatial variability in the climate transitional zone of northern China, gradually increasing from 0.1 in the northwest to 0.4 in the southeast (Figure 8a). From the northwest to southeast, the climate is arid, semi-arid, sub-humid, and humid. Most of the study area belongs to a semi-arid or sub-humid climate, with only the northwest and southeast corners being arid and humid zones. Moreover, temperature has a similar spatial pattern as soil moisture (Figure 8b), increasing from the northwest of −8 °C to southeast of 16 °C in the study area. Notably, the southeast area was generally warmer with a lower gradient. In total, the soil moisture and air temperature have a general reversed spatial pattern compared to CS, and the spatial pattern of soil moisture is closer to that of CS in the study area.

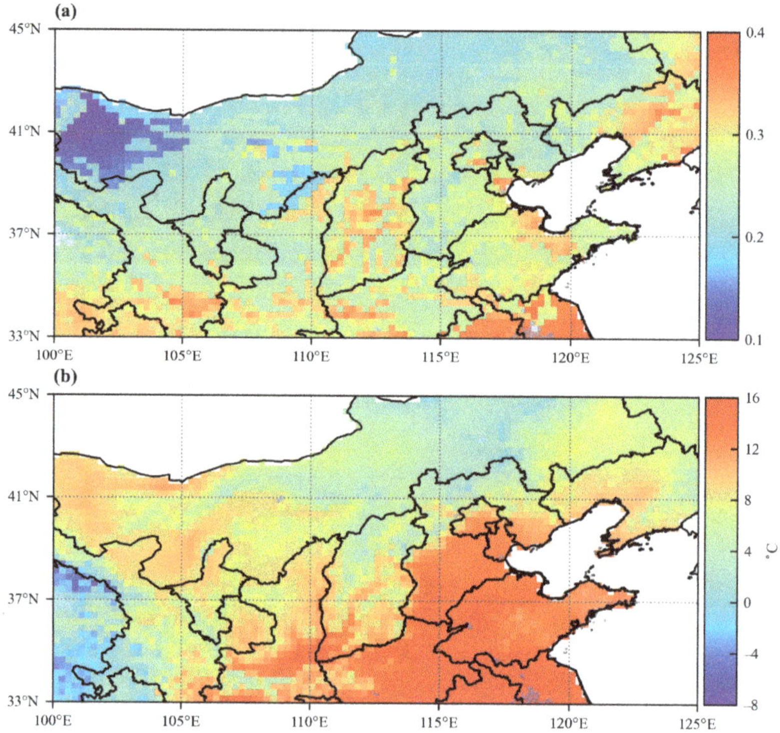

Figure 8. Distribution of climatology of (**a**) soil moisture and (**b**) air temperature in northern China.

To examine the influence of soil moisture on spatial CS, Figure 9a displays the relationship between CS and climatological soil moisture. Generally, CS increases slightly and is maintained at a strong level when the soil moisture is below 0.2, and CS decreases gradually with increase in soil moisture when the soil moisture is larger than 0.2 (Figure 8a). In areas where the soil moisture is greater than 0.35, the CS is negative; in areas where the soil moisture is below 0.25, the CS is positive; in areas where the soil moisture is between 0.25 and 0.35, CS gradually transits from negative to positive. The determination coefficient R^2 of 0.4 indicates that variation in soil moisture explains 40% of the variation in CS. Figure 9b illustrates the relationship between the CS trend and soil moisture. The relationship between the CS trend and soil moisture is roughly opposite to that of between soil moisture and CS. In the range of soil moisture below 0.2, the CS trend decreases with increasing soil moisture; while in the range of soil moisture larger than 0.2, the CS trend increases with increasing soil moisture. Further, a negative CS trend occurs at moderate soil moisture, while a positive CS trend occurs at very dry or wet soil moisture.

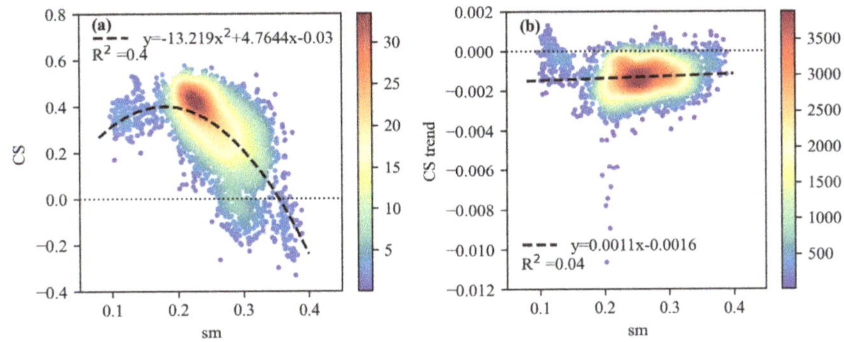

Figure 9. (**a**) Variation in evapotranspiration–precipitation CS and (**b**) its trend with soil moisture (shades of color indicate the density of the points).

Similarly, the relationships between CS, CS trend, and air temperature were analyzed to examine the influence of air temperature on the spatial distribution of CS and the CS trend. Figure 10a illustrates that CS logarithmically decreases with increasing temperature. CS is mainly positive below zero degrees, and both positive and negative coupling exist above zero degree. The percentage of negative coupling increases as the temperature rises. The determination coefficient R^2 of 0.4 indicates that Ta only explains 26% of the variation in CS. In the contrary, the CS trend increases with increasing temperature (Figure 10b). The CS trend is negative below zero degrees, and the proportion of positive trend increases with increasing temperature. Clearly, the relationship between CS and temperature shows a significantly wider spread than that between CS and soil moisture. The spatial variation in soil moisture explains more of the spatial variation in CS compared to the spatial variation in TA. Therefore, the climatological soil moisture plays a more dominant role in determining the spatial pattern of CS compared to the temperature.

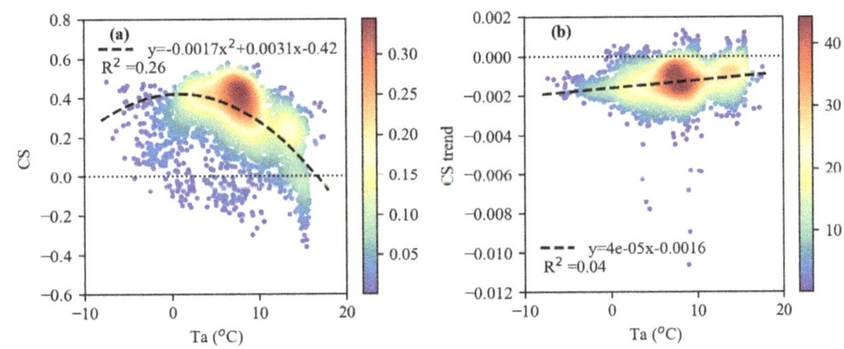

Figure 10. (**a**) Variation in evapotranspiration-precipitation CS and (**b**) its trend with air temperature (shades of color indicate the density of the points).

3.2.2. Temporal Variation of CS in Relation to Hydrothermal Conditions

Inner-Annual Variability

The intra-annual variations of soil moisture and temperature were analyzed first (Figure 11). The soil moisture exhibits an evident intra-annual cycle reaching its minimum in the winter, followed by a rise in spring and autumn, and reaching its maximum in the summer (Figure 11a). Precipitation mainly concentrates in the summer over the water-scared northern areas, which serves as the primary method to replenish the soil moisture. Figure 10b shows the intra-annual variation in soil moisture variability. The soil moisture variability was small in winter and relatively larger in spring, summer, and autumn in all areas. The soil moisture variability is small in arid areas due to low soil moisture. Moreover, the air temperature in all regions shows a unimodal pattern of a low winter and high summer, peaking in July (Figure 11c). Temperature variability is U-shaped, with large variations in the winter and small in the summer (Figure 11d). Furthermore, the temperature variability is larger in semi-arid regions than in others.

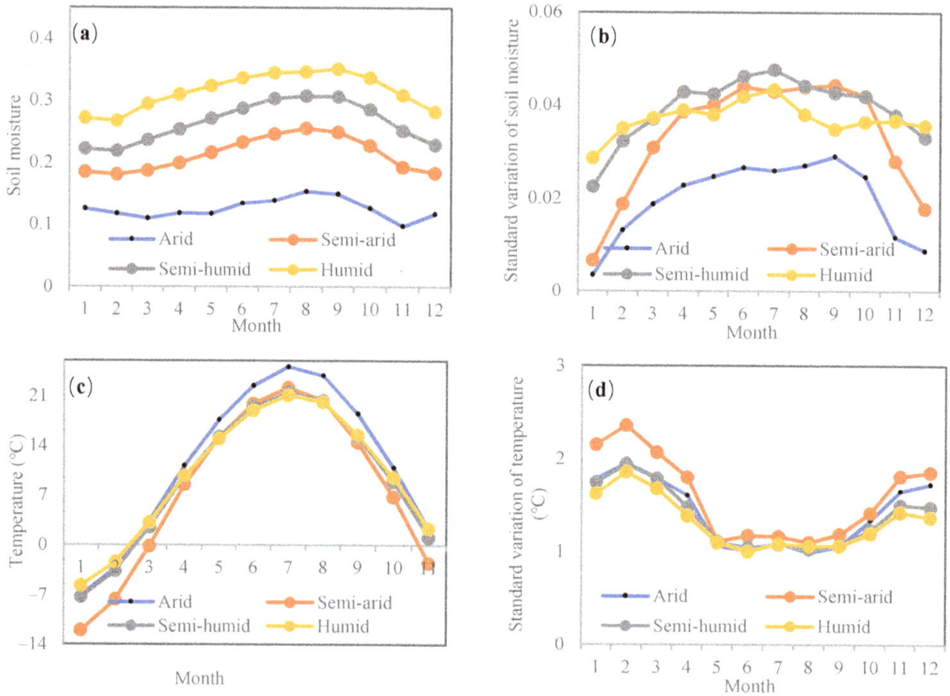

Figure 11. Intra-annual variation in (**a**) soil moisture, (**b**) standard deviation of soil moisture, (**c**) temperature, and (**d**) standard deviation of temperature (standard deviation of each month data for the 39 years) under different dry-wet climatic backgrounds.

Figure 12a further illustrates the intra-annual variation in CS, and shows that CS is smallest in winter months, reaching the maximum in spring months, and then decreasing again in summer and autumn months across all areas. The semi-arid region has the largest CS, followed by arid and semi-humid areas, and it has the smallest CS in humid areas. The CS is weak in the humid region, with small negative or positive values fluctuating around zero.

To determine this intra-annual variability of CS in relation to moisture and thermal factors, the intra-annual pattern of CS was compared to that of moisture (i.e., soil moisture and its variability) and thermal factors (i.e., temperature and its variability) for each dry and wet climate background. The intra-annual variation in CS is similar to the intra-annual variation in soil moisture variability, and temperature, and has roughly opposite characteristics to the intra-annual variation in temperature variability. Notably, soil moisture peaks in March–October, temperature peaks in July–August, temperature variability is at its minimum in May–August, whereas CS peaks in March–May. Generally, the coupling is most similar to the intra-annual variation of soil moisture variability.

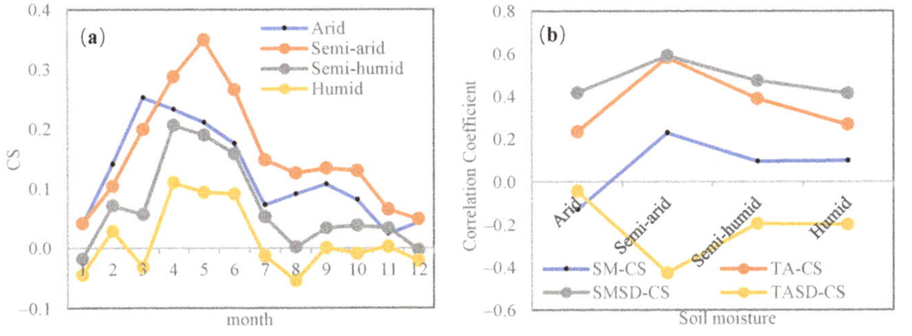

Figure 12. (**a**) Intra-annual variation in ET-P CS and (**b**) its correlation coefficients with soil moisture (SM), standard deviation of soil moisture (SMCD), air temperature (TA) and standard deviation of temperature (TASD) under different dry-wet climatic backgrounds.

From the correlation analysis of the CS with moisture and thermal factors (Figure 12b), soil moisture variability was found to have the highest correlation coefficient with CS, with the correlation coefficients larger than 0.4 in all regions. This suggests that a large soil moisture variability causes a large ET variability, and subsequently a large P variability, leading to a stronger ET-P coupling. The correlation coefficients of CS with temperature and temperature variability are large in semi-arid regions, but small in other regions, indicating that thermal factors have an important influence in semi-arid regions. Moreover, a higher temperature and temperature variability supplies more energy for the land–atmosphere interaction. The correlation coefficient between CS and soil moisture is low. Therefore, soil moisture variability is the main factor dominating the intra-annual variation of CS in the climate transitional zone of northern China.

Inter-Annual Variability

Because the CS is the highest in the warm season (April–September) in the climate transitional zone of northern China, warm season CS was chosen as a representative to analyze the inter-annual variation of the CS with moisture and thermal factors. First, the inter-annual evolution of soil moisture and air temperature and their variability were analyzed. Soil moisture changed little during the study period, with a weak increase in all areas (Figure 13a). Instead, soil moisture variability fluctuated dramatically and declined during the study period (Figure 13b). Soil moisture variability was larger in the arid zone than other regions. In addition, temperature showed a significant increasing trend (Figure 13c), while the temperature variability displayed a fluctuating decreasing trend during the study period (Figure 13d).

Responding to changes in climatic conditions, the CS fluctuated strongly during the study period, and showed a slight decreasing trend in all subregions (Figure 14a). To find the dominant factors of inter-annual variation in CS, first, the time evolution of CS was compared with that of moisture and thermal factors. Soil moisture and temperature fluctuations were small, while soil moisture variability and temperature variability fluctuations were large and more similar to the inter-annual variation of CS. Figure 14b further presents the correlation of CS with soil moisture and temperature related variables for different soil moisture backgrounds in the warm season. The CS was significantly positive and correlated with soil moisture variability in arid and semi-arid regions, suggesting that inter-annual variation in soil moisture variability has a significant impact on the variation in ET and subsequently the variation in P. CS was negatively correlated with the soil moisture, demonstrating that soil moisture experiencing a relative dry state could cause a stronger CS. In the humid and sub-humid region, soil moisture and temperature related variables were weakly correlated with CS. This may be due to the joint influence of soil moisture and air temperature on CS giving rise to a more complex influence mechanism.

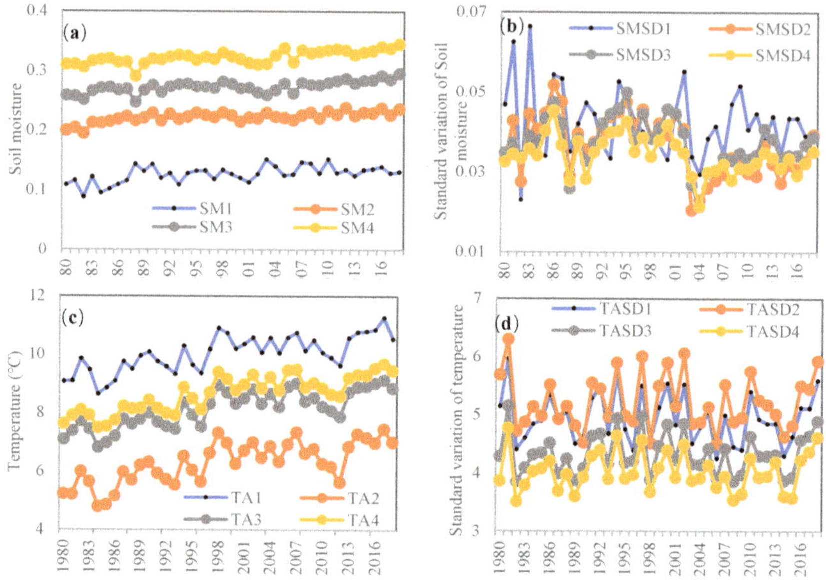

Figure 13. Inter-annual variation in warm season (**a**) soil moisture, (**b**) standard deviation (standard deviation of monthly data within warm season of a year) of soil moisture, (**c**) temperature, and (**d**) standard deviation of temperature under different dry-wet climatic backgrounds.

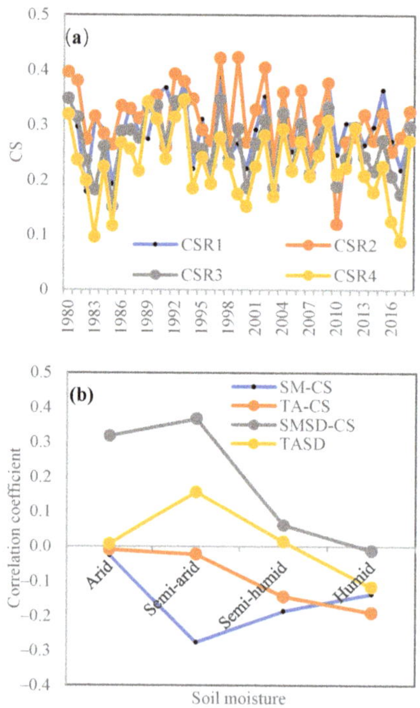

Figure 14. (**a**) Inter-annual variation of warm season ET-P CS and (**b**) its correlation coefficients with soil moisture (SM), standard deviation of soil moisture (SMCD), air temperature (TA) and standard deviation of temperature (TASD) under different dry-wet climatic conditions.

3.3. Reasons of Spatial Differences in Coupling Strength

To investigate the mechanism for the opposite signs of CS over different regions of the climate transitional zone of northern China, we analyzed boundary layer characteristics as an intermediate process of ET-P coupling. The distribution of correlation coefficients between the LCL and precipitation is shown in Figure 15a. The precipitation and LCL are negatively correlated in the majority of study areas, demonstrating that a lower LCL is more likely to trigger precipitation. This implies that the role of water vapor in precipitation is very prominent in the study area. The more saturated the atmosphere is, the lower the LCL, and the easier it is to trigger precipitation. This also shows that the effect of moisture recycling from evapotranspiration on precipitation is the main pathway of ET-P coupling.

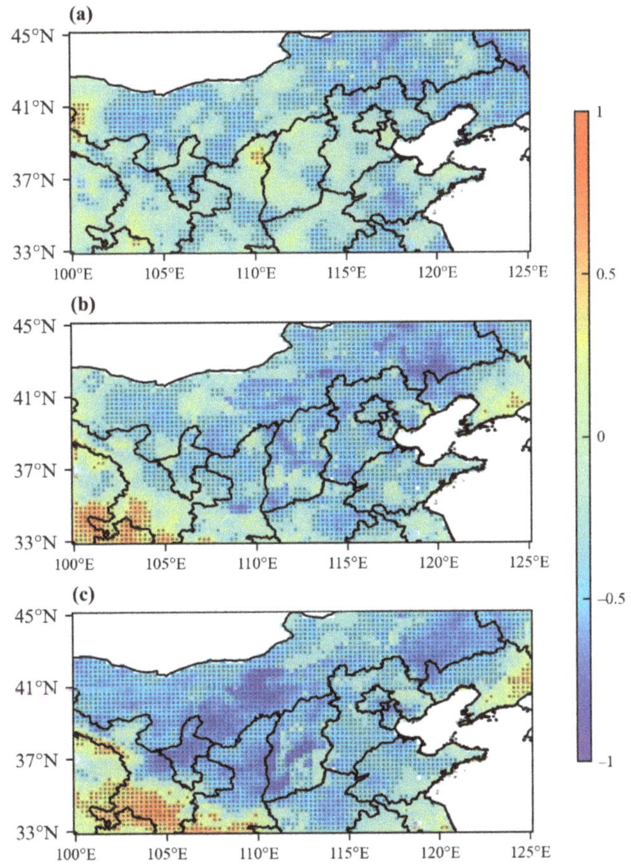

Figure 15. Distribution of correlation coefficients between (**a**) P and LCL, (**b**) ET and LCL, (**c**) ET and BLH (dot denotes correlation coefficient passing 0.05 significance test).

The spatial distribution of correlation coefficients between ET and LCL is shown in Figure 15b. ET and LCL are negatively correlated in the majority of the study area, where CS are mainly positive; ET and LCL are positively correlated in the southern humid zone and eastern northeast, where CS is mainly negative. In fact, in the negative ET-LCL correlation zones in the climate transitional zone of northern China, the ET type is water-limited, and the increase in soil moisture causes an increase in ET, which increases air humidity and thus decreases LCL. Meanwhile, the increase in ET reduces the energy partitioning available to sensible heat, and the decrease in sensible heat inhibits the boundary layer development and decreases the boundary layer height (BLH), resulting in a negative correlation between ET and BLH in these regions (Figure 15c). In contrast, in the positive ET-LCL correlation zone in the south and northeast, the ET type is energy-limited, and an increase in the available energy leads to an increase in both sensible and latent heat. Hence, the boundary layer is developed, resulting in an increase in LCL and BLH.

The main reason behind positive and negative ET-P coupling is the different driving regimes of evapotranspiration in the study area: ET is energy-limited in the southern and northeast corner of the study area, leading to a positive correlation between ET and LCL, while ET is water-limited, and ET is negatively correlated with LCL in most of the northern part. Meanwhile, LCL has a negative correlation with P in the whole study area, it therefore leads to a negative ET-P coupling in the south and northeast corner and a positive coupling in the most northern region. Combined with the scatter plot of CS and soil moisture in Figure 8a, CS is positive in areas with soil moisture below 0.25, corresponding to moisture-limited evapotranspiration; both positive and negative CSs exist in areas with soil moisture in the range of 0.25–0.35, corresponding to the transition zone of evapotranspiration from moisture-limited to energy-limited; CS is negative in areas with soil moisture greater than 0.35, corresponding to an energy-limited evapotranspiration.

4. Discussion

4.1. Determination of Water and Temperature Factors via Spatial Pattern of CS

Both observation and simulation studies showed that the strong land–atmosphere coupling zone is mainly located in the semi-arid and sub-humid climate transition zone [5–7,29]. Because coupling is influenced by the evapotranspiration variability, the sensitivity of evapotranspiration to soil moisture, and sufficient water vapor conditions, which are optimal in the transition zone following a compromise, it is strongest in the transition zone. In these land–atmosphere coupling "hot spots", the CS is further influenced by hydrothermal factors.

In semi-arid regions of southeastern South America and Africa, land–atmosphere coupling (soil moisture–precipitation coupling) is negatively correlated with soil moisture, with stronger coupling occurring in areas with lower soil moisture [26,33]. Wei et al. [30] found that the spatial distribution of soil moisture -precipitation CS is linked to the mean soil moisture, and the strong coupling area is mainly distributed in the areas with a soil moisture of 0.4–0.5.

Land–atmosphere coupling can be separated into two components: the terrestrial leg and the atmospheric leg [7]. For instance, soil moisture–precipitation coupling can be separated into soil moisture–evapotranspiration coupling (terrestrial leg) and evapotranspiration coupling (atmospheric leg). The current study focused on ET-P coupling, which is the atmospheric leg of land-precipitation coupling. The CS is found to be positively related to climatological soil moisture, and the data reveal the strong coupling in the climate transition zone with soil moisture in the range of 0.15–0.25, with relatively weak coupling in the arid and humid areas. The soil moisture range of the strongest coupling differs from the strong coupling zone of 0.4–0.55 in Wei's study, which may be related to the different soil moisture data, where they used MERRA-LAND reanalysis soil moisture data (top 1 m), while we used CCI remotely sensed soil moisture (surface 5–10 cm).

Studies have shown that the areas where land–atmosphere coupling is controlled by thermal energy factors are mostly located in moist areas with sufficient moisture [35,39]. In the climate transitional zone of northern China, where the climate is non-humid, the influence of temperature is weak, and the relationship between the spatial distribution of CS and temperature is considerably weaker than that with soil moisture. Therefore, the moisture factor is the main factor dominating the spatial distribution of ET-P coupling.

4.2. Determination of Water and Temperature Factors on Temporal Variation of CS

The studies on the temporal variation of CS are fewer than those on the spatial pattern of CS. The findings based on GLACE and MERRA-LAND both indicate that interannual variation in land–atmosphere coupling is mainly caused by soil moisture variation, and suggest a phenomenon of "see-saw" that the CS is stronger in the wet period in the dry area and in the dry period in the wet area [30,31]. This is explained by the fact that where the CS is strongest in the transition zone, either the dry zone becomes wet or the wet zone becomes dry, and the coupling is thus enhanced. Recently, Lo et al. showed that hydrological events have a significant effect on temporal evolution of CS by changing the surface state [45]. After large-scale intensive precipitation events, the soil moisture increases significantly, causing evapotranspiration to change from moisture to transitional limitation. Thus, the dependence of evapotranspiration on soil moisture decreases, resulting in a decrease in CS.

In the current study, responding to the intra- and inter-annual fluctuations of environmental conditions, land–atmosphere coupling exhibits distinct intra-annual cycles and inter-annual fluctuations. Soil moisture variability (standard deviation) is the most important influencing factor in determining the CS in the northern China climate transition zone. This is in relation to evapotranspiration being moisture-limited across most of the climate transition zone, and a larger soil moisture variability causes a larger evapotranspiration variability, and subsequently a precipitation variability. This effect is more significant in arid and semi-arid regions.

4.3. Positive and Negative Coupling Mechanisms

Land–atmosphere couplings could be positive or negative. Drylands tend to show positive coupling, i.e., the lager the soil moisture, the higher the evapotranspiration, and the more likely to trigger convective precipitation [5,16]. The mechanism responsible for positive coupling involves dominant moisture recycling in land–atmosphere coupling. Negative coupling was also found in some studies, i.e., negative coupling exists in north Africa [26]. Negative coupling implies that a lower soil moisture is more likely to trigger precipitation. The mechanism responsible for negative CS is that in areas where the boundary layer is wet with a dry surface with strong heating, the convective available potential energy (CAPE) is large and convective inhibition (CIN) is small, causing the boundary layer to be more likely to develop deeper. Although a dry and hot boundary layer causes LCL lift, the well-developed BLH would exceed LCL and trigger convective precipitation. This mechanism is similar to the land–atmosphere coupling mechanism in the southern region of the study area in the current study.

LCL is a key variable in the linkage between surface and precipitation, and the development of LCL is closely related to the type of evapotranspiration [46]. The ET in the southern region of the study area is energy-limited, and the increase in available energy causes both sensible and latent heat to increase, and the increased sensible heat heats the boundary layer and increases the LCL, leading to a positive correlation between ET and LCL. In most northern regions, ET is moisture-limited, and increasing soil moisture results in an increase in ET and a decrease in sensible heat, causing LCL to decrease and the boundary layer to become wet and cold, leading negative correlation of ET with LCL. In contrast, over the whole study area, a lower LCL is more likely to trigger precipitation, and hence, LCL has a negative correlation with P. Thus, it leads to negative ET-P coupling in the part of south region and positive coupling in the north region. Therefore, the main reason for the positive and negative differences in ET-P coupling in the study area is the different driving regimes of evapotranspiration.

5. Conclusions

Employing an evapotranspiration–precipitation coupling index (CS) proposed by Zeng (2010), this study found that CS decreases gradually from the northwest to southeast in the north China climate transition zone, with strong positive coupling in the northwest and negative coupling in the southeast and northeast corners. The CS decreases sequentially in the spring, summer, autumn and winter, and is considerably stronger in spring than in other seasons. On the interdecadal scale, coupling is highest in the 1980s, lowest in the 1990s, and moderate in the 2000s and 2010s. The trend of CS gradually shifts from a significant declining trend in the west to an increasing trend (not significant) in the east.

The spatial distribution of CS is closely related to the distribution of climatology of moisture and temperature. The CS remained at a strong level and increased slightly by increasing soil moisture when it was below 0.2, and decreased with increasing soil moisture when the soil moisture was above 0.2. In the zone of study, areas with soil moisture below 0.25 have positive CS, areas with soil moisture between 0.25 and 0.35 experience a transitional coupling from positive to negative, and areas with soil moisture greater than 0.35 exhibit negative CS. The relationship between soil moisture and the CS trend is roughly opposite to that between the soil moisture and CS. The CS shows an exponential decreasing trend with the increase in temperature, while the CS trend gradually increases with increasing temperature. Climatological soil moisture plays a more dominant role in determining the distribution of CS.

The CS exhibits evident intra- and inter-annual variability in the climate transitional zone of northern China. Soil moisture variability has the highest correlation coefficient with the intra-annual CS, dominating the intra-annual variation in ET-P coupling in the northern region. At the interannual scale, soil moisture variability is significantly and positively correlated with CS in arid and semi-arid regions, determining the interannual variability in CS in these regions. In humid and semi-humid areas, the CS is more complex in relation to the hydrothermal factors, and subject to the combined effect of hydrothermal factors.

The boundary layer thermodynamic analysis revealed that the main reason for positive and negative differences in CS across the study area is the different driving regimes of evapotranspiration. ET is energy-limited in the southern part of the study area, leading to a positive correlation between ET and LCL, while in most of the northern part, ET is moisture-limited, and ET is negatively correlated with LCL. The effect of moisture recycling from evapotranspiration on precipitation represents the main pathway of ET-P coupling, and LCL has a negative correlation with P across the study area, therefore leading to a negative ET-P coupling in part of the south and a positive coupling in the north.

Author Contributions: Conceptualization, Z.Y. and Q.Z.; Methodology, Z.Y. and Q.Z.; Investigation and Data Acquisition, Z.Y.; Formal Analysis, Z.Y., Q.Z. and P.Y.; Writing—Original Draft Preparation, Z.Y.; Writing—Review & Editing, Z.Y., Q.Z., Y.Z., P.Y., L.Z. and J.Z.; Visualization, Z.Y. and Y.Q.; Funding Acquisition, Z.Y., Q.Z. and Y.Z. All authors have read and agreed to the published version of the manuscript.

Funding: This work was funded by the National Natural Science Foundation of China (Grant No. 42005071 and 41630426) supporting Z.Y. and Q.Z. The Second Tibetan Plateau Scientific Expedition and Research (STEP) program (grant no. 2019QZKK0102) supporting Y.Z. and Z.Y.

Institutional Review Board Statement: Not applicable.

Informed Consent Statement: Not applicable.

Data Availability Statement: The data presented in this study are available online: Evapotranspiration product can be download from https://dapds00.nci.org.au/thredds/catalog/ks32/CLEX_Data/DOLCE/v3-0/catalog.html (accessed on 30 July 2021); Precipitation and near-surface air temperature data from China Meteorological Forcing Dataset can be obtained from http://data.tpdc.ac.cn/zh-hans/ (accessed on 10 May 2021); CCI soil moisture data are available at http://www.esa-soilmoisture-cci.org/node/202 (accessed on 19 November 2020); Boundary layer heights of ERA5 are available at https://cds.climate.copernicus.eu/cdsapp#!/dataset/reanalysis-era5-single-levels-monthly-means-preliminary-back-extension?tab=form (accessed on 18 November 2021).

Conflicts of Interest: The authors declare no conflict of interest.

References

1. Huang, R. Progresses in Research on the Formation Mechanism and Prediction Theory of Severe Climatic Disasters in China. *Adv. Earth Sci.* **2006**, *21*, 564–575.
2. Seneviratne, S.I.; Lüthi, D.; Litschi, M.; Schär, C. Land–atmosphere coupling and climate change in Europe. *Nature* **2006**, *443*, 205–209. [CrossRef] [PubMed]

3. Zhang, J.; Wu, L. Land-atmosphere coupling amplifies hot extremes over China. *Chin. Sci. Bull.* **2011**, *56*, 1905–1909. [CrossRef]
4. Zhou, S.; Williams, A.P.; Berg, A.M.; Cook, B.I.; Zhang, Y.; Hagemann, S.; Lorenz, R.; Seneviratne, S.I.; Gentine, P. Land-atmosphere feedbacks exacerbate concurrent soil drought and atmospheric aridity. *Proc. Natl. Acad. Sci. USA* **2019**, *116*, 18848–18853. [CrossRef] [PubMed]
5. Koster, R.D. Regions of Strong Coupling Between Soil Moisture and Precipitation. *Science* **2004**, *305*, 1138–1140. [CrossRef] [PubMed]
6. Zeng, X.; Barlage, M.; Castro, C.; Fling, K. Comparison of Land–Precipitation Coupling Strength Using Observations and Models. *J. Hydrometeorol.* **2010**, *11*, 979–994. [CrossRef]
7. Dirmeyer, P.A. The terrestrial segment of soil moisture-climate coupling. *Geophys. Res. Lett.* **2011**, *38*, L16702. [CrossRef]
8. Koster, R.D.; Sud, Y.C.; Guo, Z.; Dirmeyer, P.A.; Bonan, G.; Oleson, K.W.; Chan, E.; Verseghy, D.; Cox, P.; Davies, H.; et al. GLACE: The Global Land–Atmosphere Coupling Experiment. Part I: Overview. *J. Hydrometeorol.* **2006**, *7*, 590–610. [CrossRef]
9. Chen, H.; Zhou, J. Impact of Interannual Soil Moisture Anomaly on Simulation of Extreme Climate Events in China.Part II: Sensitivity Experiment Analysis. *Chin. J. Atmos. Sci.* **2013**, *37*, 1–13.
10. Guo, W.; Ma, Z.; Yao, Y. Regional Characteristics of Soil Moisture Evolution in Northern China over Recent 50 Years. *Acta. Geogr. Sin.* **2003**, *58*, 83–90.
11. Williams, I.N.; Torn, M.S. Vegetation controls on surface heat flux partitioning, and land-atmosphere coupling. *Geophys. Res. Lett.* **2015**, *42*, 9416–9424. [CrossRef]
12. Weiqiang, M.; Yaoming, M.; Hirohiko, I. Evaluation of the SEBS for upscaling the evapotranspiration based on in-situ observations over the Tibetan Plateau. *Atmos. Res.* **2014**, *138*, 91–97.
13. Seneviratne, S.I.; Corti, T.; Davin, E.L.; Hirschi, M.; Jaeger, E.B.; Lehner, I.; Orlowsky, B.; Teuling, A.J. Investigating soil moisture–climate interactions in a changing climate: A review. *Earth-Sci. Rev.* **2010**, *99*, 125–161. [CrossRef]
14. Eltahir, E.A.B.; Bras, R.L. Precipitation recycling. *Rev. Geophys.* **1996**, *34*, 367–378. [CrossRef]
15. Wei, J.; Dirmeyer, P.A.; Guo, Z. How Much Do Different Land Models Matter for Climate Simulation? Part II: A Decomposed View of the Land–Atmosphere Coupling Strength. *J. Clim.* **2010**, *23*, 3135–3145. [CrossRef]
16. Goessling, H.F.; Reick, C.H. What do moisture recycling estimates tell us? Exploring the extreme case of non-evaporating continents. *Hydrol. Earth Syst. Sci.* **2011**, *15*, 3217–3235. [CrossRef]
17. Gao, C.; Chen, H.; Li, G.; Ma, H.; Li, X.; Long, S.; Xu, B.; Li, X.; Zeng, X.; Yan, H.; et al. Land–atmosphere interaction over the Indo-China Peninsula during spring and its effect on the following summer climate over the Yangtze River basin. *Clim. Dyn.* **2019**, *53*, 6181–6198. [CrossRef]
18. Findell, K.L.; Eltahir, E.A.B. Atmospheric Controls on Soil Moisture–Boundary Layer Interactions. Part II: Feedbacks within the Continental United States. *J. Hydrometeorol.* **2003**, *4*, 570–583. [CrossRef]
19. Berg, A.; Findell, K.; Lintner, B.R.; Gentine, P.; Kerr, C. Precipitation Sensitivity to Surface Heat Fluxes over North America in Reanalysis and Model Data. *J. Hydrometeorol.* **2013**, *14*, 722–743. [CrossRef]
20. Santanello, J.A.; Dirmeyer, P.A.; Ferguson, C.R.; Findell, K.L.; Tawfik, A.B.; Berg, A.; Ek, M.; Gentine, P.; Guillod, B.P.; van Heerwaarden, C.; et al. Land–Atmosphere Interactions: The LoCo Perspective. *Bull. Am. Meteorol. Soc.* **2018**, *99*, 1253–1272. [CrossRef]
21. Ma, Y.; Yao, T.; Hu, Z.; Wang, J. The Cooperative Study on Energy and Water Cycle over the Tibetan Plateau. *Adv. Earth Sci.* **2009**, *24*, 1280.
22. Yang, Z.; Zhang, Q.; Hao, X.; Yue, P. Changes in Evapotranspiration Over Global Semiarid Regions 1984–2013. *J. Geophys. Res. Atmos.* **2019**, *124*, 2946–2963. [CrossRef]
23. Zhang, Q.; Yang, Z.; Hao, X.; Yue, P. Conversion features of evapotranspiration responding to climate warming in transitional climate regions in northern China. *Clim. Dyn.* **2019**, *52*, 3891–3903. [CrossRef]
24. Findell, K.L.; Eltahir, E.A.B. Atmospheric Controls on Soil Moisture–Boundary Layer Interactions. Part I: Framework Development. *J. Hydrometeorol.* **2003**, *4*, 552–569. [CrossRef]
25. Guillod, B.P.; Orlowsky, B.; Miralles, D.G.; Teuling, A.J.; Seneviratne, S.I. Reconciling spatial and temporal soil moisture effects on afternoon rainfall. *Nat. Commun.* **2015**, *6*, 6443. [CrossRef]
26. Petrova, I.Y.; van Heerwaarden, C.C.; Hohenegger, C.; Guichard, F. Regional co-variability of spatial and temporal soil moisture–precipitation coupling in North Africa: An observational perspective. *Hydrol. Earth Syst. Sci.* **2018**, *22*, 3275–3294. [CrossRef]
27. Luan, L.; Meng, X.; Lv, S.; Han, B.; Li, Z.; Zhao, L.; Li, R. Simulation on Afternoon Convective Precipitation Triggered by Soil Moisture over the Qinghai-Tibetan Plateau. *Plateau Meteorol.* **2018**, *37*, 873–885.
28. Ford, T.W.; Rapp, A.D.; Quiring, S.M.; Blake, J. Soil moisture–precipitation coupling: Observations from the Oklahoma Mesonet and underlying physical mechanisms. *Hydrol. Earth Syst. Sci.* **2015**, *19*, 3617–3631. [CrossRef]
29. Dirmeyer, P.A.; Chen, L.; Wu, J.; Shin, C.S.; Huang, B.; Cash, B.A.; Bosilovich, M.G.; Mahanama, S.; Koster, R.D.; Santanello, J.A.; et al. Verification of land-atmosphere coupling in forecast models, reanalyses and land surface models using flux site observations. *J. Hydrometeorol.* **2018**, *19*, 375–392. [CrossRef]
30. Wei, J.; Dirmeyer, P.A. Dissecting soil moisture-precipitation coupling. *Geophys. Res. Lett.* **2012**, *39*, L19711. [CrossRef]
31. Guo, Z.; Dirmeyer, P.A. Interannual Variability of Land–Atmosphere Coupling Strength. *J. Hydrometeorol.* **2013**, *14*, 1636–1646. [CrossRef]

32. Phillips, T.J.; Klein, S.A. Land-atmosphere coupling manifested in warm-season observations on the U.S. southern great plains. *J. Geophys. Res. Atmos.* **2014**, *119*, 509–528. [CrossRef]
33. Ruscica, R.C.; Sörensson, A.A.; Menéndez, C.G. Pathways between soil moisture and precipitation in southeastern South America. *Atmos. Sci. Lett.* **2015**, *16*, 267–272. [CrossRef]
34. Ruscica, R.C.; Menéndez, C.G.; Sörensson, A.A. Land surface–atmosphere interaction in future South American climate using a multi-model ensemble. *Atmos. Sci. Lett.* **2016**, *17*, 141–147. [CrossRef]
35. Li, M.; Ma, Z.; Gu, H.; Yang, Q.; Zheng, Z. Production of a combined land surface data set and its use to assess land-atmosphere coupling in China. *J. Geophys. Res. Atmos.* **2017**, *122*, 948–965. [CrossRef]
36. Gao, C.; Chen, H.; Sun, S.; Xu, B.; Ongoma, V.; Zhu, S.; Ma, H.; Li, X. Regional Features and Seasonality of Land–Atmosphere Coupling over Eastern China. *Adv. Atmos. Sci.* **2018**, *35*, 689–701. [CrossRef]
37. Genhou, S.; Zeyong, H.; Yaoming, M.; Zhipeng, X.; Fanglin, S.; Jiemin, W.; Song, Y. Analysis of local land atmosphere coupling characteristics over Tibetan Plateau in the dry and rainy seasons using observational data and ERA5. *Sci. Total Environ.* **2021**, *774*, 145138.
38. Zhao, J.; Liu, S. Research on the impact of vegetation change on land-atmosphere coupling strength in northwest China. *Chin. J. Geophys.* **2015**, *58*, 47–62.
39. Zscheischler, J.; Orth, R.; Seneviratne, S.I. A submonthly database for detecting changes in vegetation-atmosphere coupling. *Geophys. Res. Lett.* **2015**, *42*, 9816–9824. [CrossRef]
40. Hobeichi, S.; Abramowitz, G.; Evans, J.; Ukkola, A. Derived Optimal Linear Combination Evapotranspiration (DOLCE): A global gridded synthesis ET estimate. *Hydrol. Earth Syst. Sci.* **2018**, *22*, 1317–1336. [CrossRef]
41. Yang, K.; He, J. China meteorological forcing dataset (1979–2018). In *National Tibetan Plateau Data*; National Tibetan Plateau Data Center: Beijing, China, 2019.
42. Gruber, A.; Scanlon, T.; van der Schalie, R.; Wagner, W.; Dorigo, W. Evolution of the ESA CCI Soil Moisture climate data records and their underlying merging methodology. *Earth Syst. Sci. Data* **2019**, *11*, 717–739. [CrossRef]
43. Dorigo, W.A.; Gruber, A.; De Jeu, R.A.M.; Wagner, W.; Stacke, T.; Loew, A.; Albergel, C.; Brocca, L.; Chung, D.; Parinussa, R.M.; et al. Evaluation of the ESA CCI soil moisture product using ground-based observations. *Remote Sens. Environ. Interdiscip. J.* **2015**, *162*, 380–395. [CrossRef]
44. Ma, S.; Zhu, K.; Li, M.; Ma, Z. A Comparative Study of Multi-source Soil Moisture Data for China's Regions. *Clim. Environ. Res.* **2016**, *21*, 121–133.
45. Lo, M.-H.; Wu, W.-Y.; Tang, L.I.; Ryu, D.; Rashid, M.; Wu, R.-J. Temporal Changes in Land Surface Coupling Strength: An Example in a Semi-Arid Region of Australia. *J. Clim.* **2021**, *34*, 1503–1513. [CrossRef]
46. Wei, J.; Zhao, J.; Chen, H.; Liang, X.-Z. Coupling Between Land Surface Fluxes and Lifting Condensation Level: Mechanisms and Sensitivity to Model Physics Parameterizations. *J. Geophys. Res. Atmos.* **2021**, *126*, e2020JD034313. [CrossRef]

remote sensing

MDPI

Article

Inter- and Intra-Annual Glacier Elevation Change in High Mountain Asia Region Based on ICESat-1&2 Data Using Elevation-Aspect Bin Analysis Method

Cong Shen [1,2], Li Jia [1,*] and Shaoting Ren [3]

1 State Key Laboratory of Remote Sensing Science, Aerospace Information Research Institute, Chinese Academy of Sciences, Beijing 100101, China; shencong192@mails.ucas.ac.cn
2 University of Chinese Academy of Sciences, Beijing 100049, China
3 Key Laboratory of Tibetan Environment Changes and Land Surface Processes, Institute of Tibetan Plateau Research, Chinese Academy of Sciences, Beijing 100101, China; renst@itpcas.ac.cn
* Correspondence: jiali@aircas.ac.cn; Tel.: +86-10-6480-7982

Abstract: Glaciers are sensitive indicators of climate change and have a significant influence on regional water cycle, human survival and social development. Global warming has led to great changes in glaciers over the High Mountain Asia (HMA) region. Glacier elevation change is a measure of glacier mass balance driven by the processes of energy and mass exchange between the glacier surface and the atmosphere which are influenced by climatic factors and glacier surface properties. In this study, we estimated the inter-annual and intra-annual elevation changes of glaciers in the HMA region in 2003–2020 using Ice, Cloud and land Elevation Satellite (ICESat) data and Shuttle Radar Terrain Mission (SRTM) digital elevation model (DEM) data by developing an "elevation-aspect bin analysis method" that considered the difference of glacier elevation changes in different elevations and aspects of glacier topography. The results showed that: (1) The inter-annual change of glacier elevation in 2003–2020 had large spatial heterogeneity. Glacier elevation reduction mainly occurred in the marginal region of the HMA with the maximum decline in the Nyainqentanglha region, while glacier elevation showed increase in the West Kunlun of inner HMA regions in 2003–2020. The glacier elevation change rate showed an accelerating reduction trend in most of the HMA regions, except in the west HMA where the glacier elevation reduction rate showed slowdown tendency. Specifically, the glacier elevation change rate in the entire HMA was -0.21 ± 0.12 m/year in 2003–2008 and -0.26 ± 0.11 m/year in 2003–2020, respectively. (2) The intra-annual change of HMA glacier elevation in 2019 and 2020 showed obvious spatiotemporal heterogeneity, and the glacier thickening period was gradually delayed from the marginal area to the inner area of the HMA. The glaciers in the western marginal part of the HMA (the Tienshan Mountains, Pamir and Hindu Kush and Spiti Lahaul) and Karakoram thickened in winter or spring, the glaciers in the Nyainqentanglha Mountains exhibited spring accumulation. The glaciers in West Kunlun accumulated in two time periods, i.e., from March to June and from July to September. The glaciers in the Inner Tibetan Plateau and Bhutan and Nepal areas experienced spring or summer accumulation, especially in June or July. Moreover, we found that the inter-annual and intra-annual change of glacier elevation could be explained by the changes in temperature and precipitation. A similar analysis can be extended to mountain glaciers in other regions of the world, and glacier change trends could be further explored over a longer time span with the continuous operation of ICESat-2.

Keywords: ICESat-1; ICESat-2; glacier elevation change; inter-annual and intra-annual changes; elevation-aspect bin analysis method; HMA

Citation: Shen, C.; Jia, L.; Ren, S. Inter- and Intra-Annual Glacier Elevation Change in High Mountain Asia Region Based on ICESat-1&2 Data Using Elevation-Aspect Bin Analysis Method. *Remote Sens.* **2022**, *14*, 1630. https://doi.org/10.3390/rs14071630

Academic Editors: Gareth Rees and Peter Romanov

Received: 24 January 2022
Accepted: 16 March 2022
Published: 29 March 2022

1. Introduction

In the past two decades, with global warming, great changes have taken place in global glaciers [1–3]. A recent study based on satellite data and modeling has revealed that the potential contribution of the world's melting glaciers to sea level rise is 257 ± 85 mm in 2017–2018 [4]. High Mountain Asia (HMA) is the region with the largest distribution of glaciers and diverse glacier types in the middle latitudes of the earth. It is the birthplace of major rivers in Asia, such as the Amu Darya River, the Syr Darya River, the Yangtze River, the Yellow River, the Salween River, the Brahmaputra River and the Indus River [5–7]. To date, the glaciers in the HMA region have undergone significant changes, triggering a series of disasters and having a major impact on people's production and lives. The glacier changes not only directly drive changes in the natural environment in the HMA region, but also have feedbacks to climate change in the whole northern hemisphere and even the whole world. Therefore, accurate monitoring of glacier changes in the HMA region is crucial for studying global glacier and climate change, as well as for understanding the potential impact of glacier retreat [8–11].

The method based on satellite stereoscopic imagery has good spatial coverage and is a common method to monitor glacier elevation changes [12,13]. Existing studies have shown that the glaciers in the HMA are generally in a melting state, but the rate of change varies in different regions and different time spans [9,14,15]. For example, in the Nyanqentanglha Mountains, Ren et al. [16] analyzed the ZY-3 satellite stereo image pair data and found that the glacier thinning rate in 2013–2017 was faster than that in 2000–2013. The results of Brun et al. [17] showed that the glacier thickness across the HMA region decreased by 0.21 m per year from 2000 to 2016. However, glaciers in East Pamir, Karakoram and West Kunlun are in a state of equilibrium or slightly rising, which is called the "Karakoram Anomaly" [18–21].

Lidar altimetry data, such as ICESat-1&2 data, have higher vertical accuracy than satellite-based stereo image pair data. Using lidar data to monitor glacier elevation change is a hot research topic [22]. At present, ICESat-1 equipped with the Geoscience Laser Altimeter System (GLAS), as the first generation of space-borne laser point cloud satellite, can provide global laser point cloud data from 2003 to 2009. The data have a high point frequency (40 Hz) and are widely used in ice sheet monitoring in the Arctic and Antarctic regions [23]. ICESat-1 GLAS data were also used to explore the mass balance changes of glaciers in the Tibetan Plateau, and found that the annual mass loss of glaciers from 2003 to 2009 was −26 ± 12 Gt [24]. However, ICESat-1 data points in the mid-latitude region are sparse, which limits its use in the HMA. In addition, ICESat-1 stopped operation after 2009. ICESat-2 is a new generation of spaceborne lidar satellite equipped with an Advanced Topographic Laser Altimeter System (ATLAS). Since 2018, the ATLAS has provided abundant laser point data every year. Compared with ICESat-1 data, ICESat-2 data not only improves the observation accuracy but can also observe more glaciers in the same time span, thus providing more complete and accurate glacier elevation information [25–27]. Therefore, ICESat-2 provides a new perspective for monitoring inter-annual glacier elevation changes.

At the same time, it would be interesting to explore the intra-annual changes of glacier elevation. Seasonal meltwater from glaciers is a guarantee of water resources in the surrounding and downstream regions of the HMA, in particular in arid and semi-arid regions [28]. Many studies on glacier elevation changes in the HMA region have focused on the inter-annual variability, and studies on monthly/seasonal changes in glacier elevation are still lacking. Although many studies have estimated the intra-annual loss of glaciers, most studies infer monthly/seasonal glacier elevation changes based on relationships with precipitation rather than directly extracting glacier elevation information from satellites observations. Existing studies have shown that the timing of glacier accumulation or melting varies widely in different regions. For example, Ageta [29] found that most of the glaciers in the inner Tibetan Plateau thickened in summer based on continuous ground observations of some specific glaciers, but the spatial patterns still need to be understood to a large extent. Maussion et al. [30] discovered that most glaciers in Pamir and Spiti Lahaul thickened in winter, but they used a clustering algorithm based on monthly precipitation data to calculate monthly changes in glacier elevation, rather than using direct observations of monthly glacier elevation. Wang et al. [31] utilized ICESat-1 and Gravity

Recovery and Climate Experiment (GRACE) data to explore seasonal changes of glaciers in the HMA. However, due to sensor failure, the ICESat-1 data could only be obtained two to three times every year [32], and most of them were concentrated in March, June and October, and could not provide observations for each month. However, studying monthly/seasonal changes of glaciers is of great significance for disaster warning and an in-depth understanding of the mechanism of glacier changes [31]. In short, none of the existing studies have analyzed glacier elevation changes on a monthly/seasonal scale [33]. The lack of existing observations and effective monitoring methods has limited researchers to accurately analyze the intra-annual variation of glacier elevation. The rich data volume of ICESat-2 data, along with the increase of laser point emission rate (10 KHz) and the number of laser point beams (six beams), provides a new perspective for exploring the intra-annual changes of glacier elevation.

Terrain factors, i.e., elevation, slope and aspect, have impacts on the glacier mass balance and changes [34]. In general, the lower the altitude, the more glacier mass is lost, and vice versa. Existing studies show that precipitation affects glacier accumulation [35]. It is found that the glaciers on the windward side of the mountain retreated less, which was attributed to the effective recharge produced by more precipitation than that on the leeward side of the mountain [36–38]. For example, Wang et al. [39] found that due to the influence of climatic conditions, the glaciers on the northern slope of the Tienshan Mountains decreased more than that on the southern slope. This may be because the precipitation brought by the westerly wind supplemented the mass loss of the glacier on the southern slope, resulting in less glacier mass loss. Therefore, it is important to consider terrain aspect and elevation when monitoring glacier elevation changes using laser point data. Previous studies have rarely considered topographical aspect effects when extracting changes in glacier elevation using ICESat-1&2 data.

The objectives of this study are:

(1) To improve the analysis method for glacier elevation change by comprehensively considering the glacier terrain elevation and slope aspect.
(2) To analyze the inter-annual differences in the rate of glacier elevation change.
(3) To analyze the intra-annual differences in glacier elevation changes on a monthly/seasonal scale.

2. Study Area and Datasets

2.1. Study Area

The HMA (26°N~47°N, 65°E~104°E), located in the central Asia (Figure 1), with an average altitude of more than 4000 m. It is the birthplace of many major rivers in central Asia, East Asia, Southeast Asia and South Asia [40]. The HMA is located in the intersection of various climates. The south of HMA is mainly affected by the Indian monsoon [41], the west and northwest of HMA are mainly affected by the westerly wind, the east is mainly affected by the East Asian monsoon, and the central region is mainly affected by the continental climate. The HMA is sensitive to climate change. Previous studies have shown that a 1.5 °C increase in global temperature will lead to a 2.1 °C increase in the temperature in the Tibetan Plateau [11]. In addition, various types of glaciers are formed in the HMA region due to the dense mountains and complex topography. According to the Randolph Glacier Inventory (RGI) 6.0 glacier catalog data, there are about 100,000 glaciers in the HMA region, covering an area of nearly 100,000 square kilometers [42]. As these glaciers are less affected by human activities, their changes are largely driven by natural factors. Therefore, glacier change is an important indicator of global warming, and it is of great significance to explore the changes in the HMA glacier and its relationship with the climate variations.

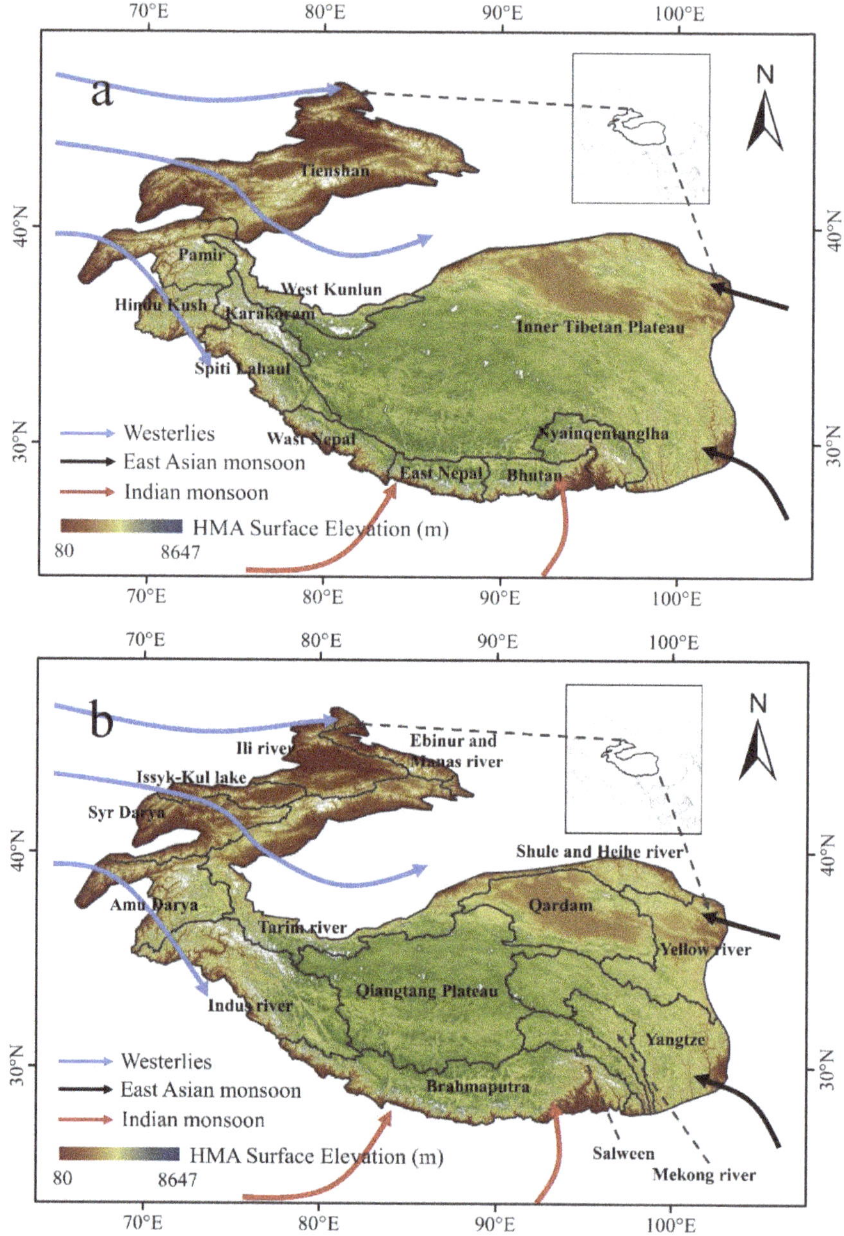

Figure 1. Overview of the High Mountain Asia with background image as elevation above sea level. The white regions in the HMA are the glacier areas. The blue, black and red arrows show the climate-influencing sphere of the Westerly, East Asian monsoon and Indian monsoon, respectively. (**a**) The HMA sub-regions according to the regional division of Brun et al. [17]; (**b**) Major river basins in the HMA according to [43].

2.2. *Datasets*

2.2.1. ICESat-1 Data

ICESat-1 is part of NASA's earth observation system and the first satellite with a lidar sensor to monitor the earth's elevation information. The GLAS onboard ICESat-1 satellite emits 40 Hz laser pulses vertically along the orbit and estimates the surface elevation from the round-trip time of the laser pulses. The vertical detection accuracy of the GLAS is less than 34 cm [44]. Its main scientific purpose is to observe changes in the glacier mass balance, cloud layer and vegetation features from 2003–2009. GLAS14 data is GLAS Level 2 altimetry data that includes the geographical location of the laser beam footprint and the correction parameters for elevation measurement. In this study, GLAS14 data was used to extract glacier elevation information at laser points, including elevation, longitude, latitude, ellipsoid correction parameters, etc. The data were downloaded from the National Snow & Ice Data Center (NSIDC) (https://nsidc.org/data/icesat, accessed on 1 1 September 2020).

2.2.2. ICESat-2 Data

ICESat-2 uses ATLAS to monitor the elevation information of the earth's surface [45]. Compared to the ICESat-1 GLAS, the working technology of the ICESat-2 ATLAS is greatly improved. The ATLAS emits six laser beams at a time, divided into three groups. Each group consists of a strong laser beam and a weak laser beam. The energy ratio of the strong laser beam to the weak laser beam is about 4:1. The smaller spatial sampling interval (~0.7 m) and higher frequency (10 kHz) [27] of the ATLAS allow it to collect denser data points for more effective monitoring of glacier elevation changes. For example, the data points observed by ATLAS in 2020 can occupy about 30,000 glaciers in the HMA region, which is about nine times as many as all data points observed by GLAS in 2003–2008 (Figure 2). In addition, the ATLAS has higher detection accuracy; Zhang et al. [46] found that the ICESat-2 data can extract glacier elevation in the Qilian Mountains with an accuracy of 0.08 m compared with the data of Unmanned Aerial Vehicle (UAV). In the Antarctic area, the accuracy can reach 1–2 cm [47]. Brunt et al. [48] used the Global Navigation Satellite Systems (GNSS) to verify that the accuracy of ICESat-2 data in extracting the elevation change of Antarctic ice sheet was better than 3 cm. In this study, the ICESat-2 ATL06 data were used, including elevation, time, latitude, longitude, confidence parameters, etc. The data were downloaded from the NSIDC (https://nsidc.org/data/icesat-2, accessed on 1 June 2021).

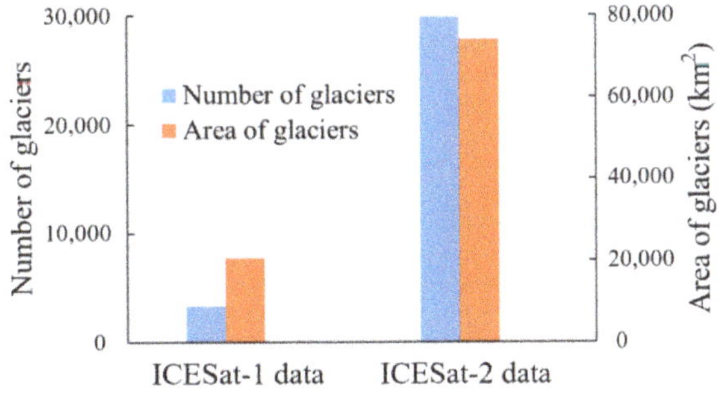

Figure 2. Comparison of observation capabilities by ICESat-1 and ICESat-2: the number of glaciers and the total area of glaciers in the HMA region observed by ICESat-1 data in 2003–2008 and by ICESat-2 data in 2020.

2.2.3. SRTM DEM

SRTM uses Synthetic Aperture Radar (SAR) technology to collect the earth's surface elevation data. It uses C-band Synthetic Aperture Radar (C-SAR) and X-band Synthetic Aperture Radar (X-SAR) to collect data of the earth's environment. The earth's surface elevation information from these data is converted to the height information specified in the DEM and used to create an accurate earth map. The revisit period of the SRTM mission is 11 days. Since February 2000, the SRTM has used the phase difference between two SAR images to calculate the DEM information from 60°N to 56°S [23]. It obtains the topographic information of 80% of the earth's surface (except the Arctic and Antarctic) and 95% of the residential areas. In this study, the SRTM DEM from C-SAR data were used as the reference elevation to obtain the glacier surface elevation information of the HMA in 2000. The data were downloaded from the EARTHDATA platform (https://search.earthdata.nasa.gov/, accessed on 1 September 2020).

2.2.4. ERA5

The fifth-generation European Centre for Medium-Range Weather Forecasts (ECMWF) global climate and weather reanalysis v5 (ERA5) data were used to analyze the changes in temperature and precipitation in the HMA region over the past 20 years. The reanalysis process combined model data with observations from the world into a globally complete and consistent dataset. We extracted the temperature and precipitation data for the HMA region from 2003 to 2020. The temperature data is the air temperature at 2 m height above the surface of land, ocean or inland water; the precipitation data is the accumulated liquid and frozen water that falls on the earth's surface, including rain and snow, but does not include fog, dew or moisture that evaporates in the atmosphere before reaching the earth's surface. The data were downloaded from the ECMWF (https://www.ecmwf.int/en/forecasts/datasets/reanalysis-datasets/era5, accessed on 1 September 2021).

2.2.5. Auxiliary Data

This study used the Randolph Glacier Inventory (RGI 6.0) data to determine glacier boundaries. The RGI 6.0 data provide a global list of glaciers, with Landsat TM/ETM+ images as the primary data source, and interpreted with high-resolution imagery and topographic maps such as ASTER, IKONOS and Systeme Probatoire d'Observation de la Terre (SPOT) as supplementary data.

According to the regional division of HMA by Brun et al. [17], the HMA is divided into Bhutan, Nepal, Nyainqentanglha, Spiti Lahaul, Karakoram, West Kunlun, Tienshan, Inner TP (Tibetan Plateau), Pamir and Hindu Kush (Figure 1a). Additionally, using the "One Belt, One Road" boundary map of the major Asian river basins [43], the HMA is divided into 14 basins, namely Amu Darya, Brahmaputra, Ebinur and Manas river, Ili river, Shule and Heihe river, Indus river, Qiangtang Plateau, Mekong river, Qardam, Salween river, Issyk-Kul lake, Syr Darya, Tarim river and Yangtze, for analysis of changes in glacier elevation in basin scale (Figure 1b).

3. Methods

Since the ICESat-1&2 data are laser point data and the spatial distribution of the data points is not uniform in space, the ICESat-1&2 data therefore cannot provide pixel-wised information covering the entire surface of glacier like the ZY-3 stereo image pair data. The spatial distribution of ICESat-1&2 data points on each glacier differs due to different satellite overpassing tracks and different sensor laser beam trajectories. In addition, due to the different atmospheric conditions and different radiation energy received by glaciers in different elevation and orientations, the glacier mass balance will differ in different aspects at the same altitude. Based on the above reasons, this paper proposed an analysis method, namely "elevation-aspect bin analysis method", to investigate glacier elevation changes by considering the glacier terrain elevation, slope aspect and the corresponding glacier areas.

Our research workflow consists of: (1) extraction of the elevation differences between the ICESat-1&2 data and the SRTM DEM data in the ICESat-1&2 footprint scale; (2) extraction of glacier elevation difference in $1^\circ \times 1^\circ$ grids; (3) estimation of the change of glacier elevation in sub-regions or river basins; and (4) uncertainty analysis for the quantitative assessment of the influences of data errors, processing errors and other factors on the results. Figure 3 shows the specific workflow, and the method details are described in the following sections.

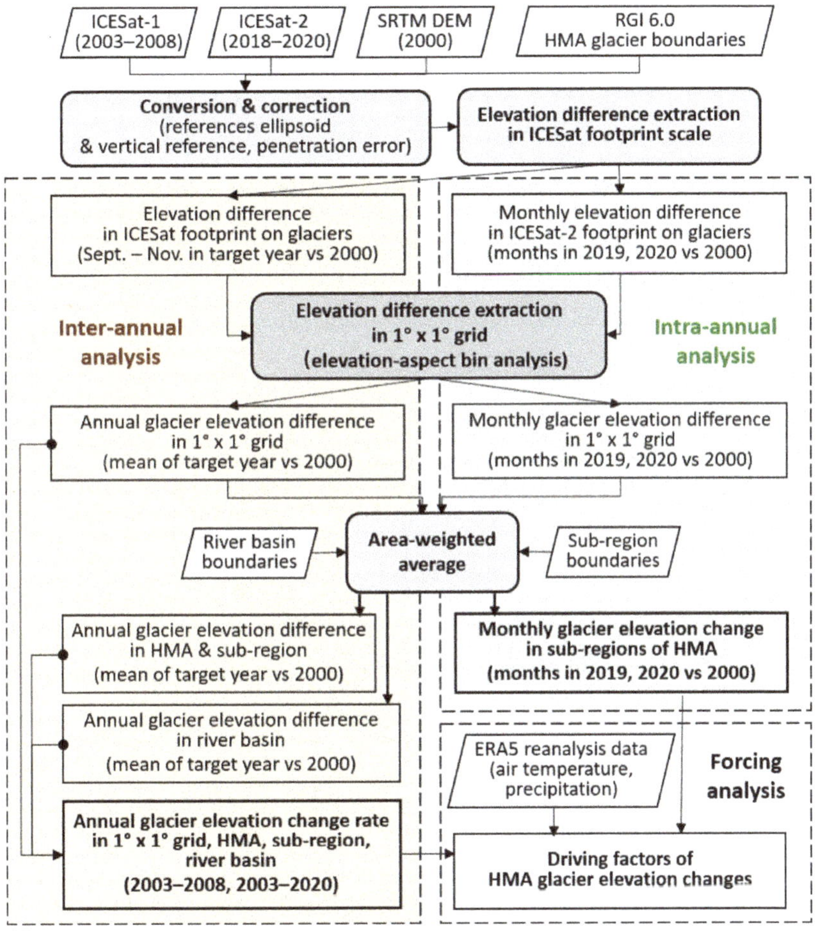

Figure 3. Flowchart of glacier elevation change extraction and analysis using ICESat-1&2 data and SRTM DEM data.

3.1. *Extraction of Elevation Difference in the ICESat-1&2 Data Footprints*

Due to the differences in reference ellipsoid for projection and differences in vertical reference system between ICESat-1&2 data and SRTM DEM data, we first need to convert the elevation datum to extract elevation change information and eliminate outliers. In this study, bilinear interpolation method was used to extract the elevation information of SRTM DEM data in the footprint of the ICESat-1&2 data. Two corrections were applied before the difference between the elevation from ICESat-1&2 data and from SRTM DEM data can be calculated: (1) The Topex/Poseidon ellipsoid used by ICESat-1 data was converted to the World Geodetic System 1984 (WGS84) ellipsoid used by ICESat-2 data and SRTM DEM data [49,50]. (2) The Earth Gravity Model (EGM) 1996 used by the SRTM DEM geoid elevation was converted to EGM2008 used by ICESat-1&2 data [35].

The glacier elevation change in each year, ΔH_{yr} (in meter), at each footprint point can then be calculated by:

$$\Delta H_{yr} = H_{ICESat,\ yr} - H_{SRTM} + P_{SRTM} \tag{1}$$

where $H_{ICESat,yr}$ (m) is the elevation from ICESt-1&2 data in a year "yr" after ellipsoid correction; H_{SRTM} (m) is the elevation from SRTM DEM data in 2000 after leveling correction and bilinear interpolation within the footprint of ICESat-1&2 data point; P_{SRTM} (m) is the penetration depth of the SRTM DEM data in the glacier region, we set P_{SRTM} as 2.4 m in this study according to [51].

To reduce the influence of slope, ICESat-1&2 data points in slopes greater than 30° were excluded according to [35,52]. To remove outliers affected by detector saturation, only ICESat-1 GLAS14 data flagged with class "0" or "1" in the data quality layer "satCorrFlg" were selected according to [53]. For ICESat-2 data, we only selected the data flagged with category "0" in the data quality layer "atl06_quality_summary" to ensure the data quality according to [48]. Additionally, we excluded the data points with elevation differences greater than 100 m between ICESat-1&2 data and the SRTM DEM data to eliminate the influence of outliers that were possibly introduced by cloud interference [35].

3.2. Extraction of Glacier Elevation Difference in 1° × 1° Grids

To reduce the uncertainty caused by the uneven spatial distribution of the ICESat-1&2 data points in each elevation and slope aspect on the results, the core idea of the "elevation-aspect bin analysis method" proposed in this paper is to calculate the glacier elevation change in each 1° × 1° grid according to the distribution of ICESat-1&2 data points in different elevations and slope aspects and respective glacier areas. The whole HMA is divided into 1° × 1° grids. In each 1° × 1° grid, the elevation is divided into elevation bins (denoted by i), each elevation bin is divided into 8 aspect bins (denoted by j) with an interval of 45°.

The elevation difference for each elevation bin in a 1° × 1° grid is calculated by:

$$\Delta H_{yr,\ grid}(i) = \sum_{j=1}^{8}\left(F_{yr,grid}(i,j) \cdot \Delta H_{yr,\ grid}(i,j)\right),\ (j = 1,2,\ldots,8, \textit{number of aspect bins}) \tag{2}$$

where, $F_{yr,grid}(i,j)$ and $\Delta H_{yr,\ grid}(i,j)$ are the fraction of glacier area and the median of glacier elevation difference (between a target year and 2000) in aspect bin j of the elevation bin i in a 1° × 1° grid.

The elevation difference between a target year and the reference year 2000 in a 1° × 1° grid, $\Delta H_{yr,grid}$, was the area-weighted average of elevation difference for all elevation bins and calculated as:

$$\Delta H_{yr,\ grid} = \sum_{i=1}^{N}\left(F_{yr,grid}(i) \cdot \Delta H_{yr,\ grid}(i)\right),\ (i = 1,2,\ldots,N,\ \textit{number of elevation bins}) \tag{3}$$

where $F_{yr,grid}(i)$ is the fraction of glacier area of the elevation bin i in a 1° × 1° grid. Following the method in [35], we repeated the calculation using Equations (2) and (3) by setting 6 different intervals of elevation bin, i.e., 200 m, 300 m, 400 m, 500 m, 600 m and 700 m, and used the average of the six results as the final glacier elevation difference in each 1° × 1° grid. ICESat-1 and ICESat-2 data from September to November of each year in 2003–2008 and 2018–2020, respectively, were used for annual change calculation.

Monthly glacier elevation difference between all months in 2019–2020 and the reference year 2000 in each 1° × 1° grid was calculated in the same way using the "elevation-aspect bin" area-weighting method.

3.3. Estimation of Annual and Monthly Change of Glacier Elevation in Sub-Regions

The annual glacier elevation changes in a sub-region or a river basin, $\Delta H_{yr,\ subregion}$, was calculated by the area-weighted average of glacier elevation differences in all 1° × 1° grids in a sub-region or river basin as:

$$\Delta H_{yr,\ subregion} = \sum_{k=1}^{M}\left(F_{yr}(k) \cdot \Delta H_{yr,\ grid}(k)\right),\ (k = 1,2,\ldots,M,\ \textit{number of grids in a region}) \tag{4}$$

where $F_{yr}(k)$ is the ratio of glacier area of each $1^\circ \times 1^\circ$ grid over the total glacier area in either a sub-region or a river basin.

Following the method used in Kääb et al. [54], the rate of change of glacier elevation over multiple years was calculated by applying robust linear regression to the time series of $\Delta H_{yr,\ subregion}$. This way, the rate of change in glacier elevation between 2003–2008 and between 2003–2020 were obtained using ICESat-1 and ICESat-2 data, respectively.

3.4. Error Analysis

It is difficult to directly assess the uncertainty of ICESat-1&2 and SRTM DEM data using ground data due to insufficient in situ measurements of glacier elevations. Factors affecting the accuracy of the results include the vertical height error of SRTM DEM data, the penetration depth error of SRTM DRM data in the glacier area and the error of ICESat-1&2 data itself. In this study, the uncertainty assessment method of glacier monitoring by Wang et al. [35] was adopted, as shown below:

$$\sigma_{DH} = \sqrt{\sigma_{std}^2 + \sigma_{dh}^2} \tag{5}$$

where σ_{DH} is the glacier elevation change error (m), σ_{std} is the standard deviation of the result in different elevation bins (m). The σ_{dh} includes the SRTM DEM vertical deviation (<16 m), ICESat-1&2 error (centimeter and decimeter scale) and radar penetration error (meter scale). In this study, the absolute error of each ICSat-1/2 footprint point is set to 20 m.

The uncertainty of glacier elevation change rate is obtained by

$$\sigma_{DH/dt} = \sqrt{\sigma_{fit}^2 + \sigma_{spat}^2 + \sigma_{temp}^2 + \sigma_{bias}^2} \tag{6}$$

where $\sigma_{DH/dt}$ is the glacier elevation change rate error, σ_{fit} is the linear fitting error, σ_{spat} and σ_{temp} are spatial and temporal sampling errors, σ_{bias} is the comparison bias and unknown system uncertainty of crustal uplift [55]. We followed Wang et al. [22] and set σ_{spat}, σ_{temp} and σ_{bias} to 0.06 m/year in this study.

4. Results

This study analyzed the change in glacier elevation from two perspectives, inter-annual and intra-annual. In the first part, we investigated the change in glacier elevation at different altitudes and slope aspects (Section 4.1) and analyzed the spatial differences in the inter-annual rate of change in glacier elevation (Section 4.2). In the third part, we conducted monthly/seasonal study by analyzing the 24-month pattern of glacier elevation changes from January 2019 to December 2020 (Section 4.3).

4.1. Glacier Elevation Change in Different Elevations and Slope Aspects

Firstly, we analyzed the glacier elevation changes between 2000 (from SRTM DEM data) and 2020 (from ICESat-2 data) in different elevations and slope aspects in the HMA sub-regions according to the regional division of Brun et al. [17], as shown in Figure 1a. Overall, the glacier elevation changes (i.e., the glacier thickness change) between 2000 and 2020 decreased with increasing altitudes in all sub-regions of the HMA (Figure 4a), which is consistent with the findings of Ragettli et al. [56]. In addition, the accumulation and melting characteristics of glaciers also varied in different slope aspects (Figure 4b). In general, in most sub-regions, the glacier elevation changes on the southern slope decreased more than that on the northern slopes from 2000 to 2020, because the southern slopes received more solar radiation, resulting in more melting of the glaciers. However, different features were found in the Nyainqentangla Mountains, Bhutan, Nepal and the Tienshan Mountains. In the Nyainqentangla Mountains, the glaciers thinning on the northern slopes thinned more than that in the southwestern slopes. Additionally, in Bhutan, the glacier on the northern slopes thinned more than that on the eastern and southwestern slopes. The reason might be that these regions were strongly influenced by monsoon which brings abundant precipitation and leads to less reduction in glacier thickness on the windward sides [57]. In Nepal, for example, the southern slopes of the mountains are windward, and the glaciers on the southern slopes were replenished by precipitation from the Indian monsoon, resulting in less reduction in glacier thickness. Similarly, in the Tienshan Mountains, under the influence of the westerly winds, the glacier thickness on the southwestern slopes (windward) decreased less than that on the northeastern slopes. To sum up, the water vapor carried by monsoon will affect the melting rate of the glaciers in the HMA region, resulting in a large difference in the glacier elevation changes on different slope aspects, especially in the areas greatly affected by the monsoon.

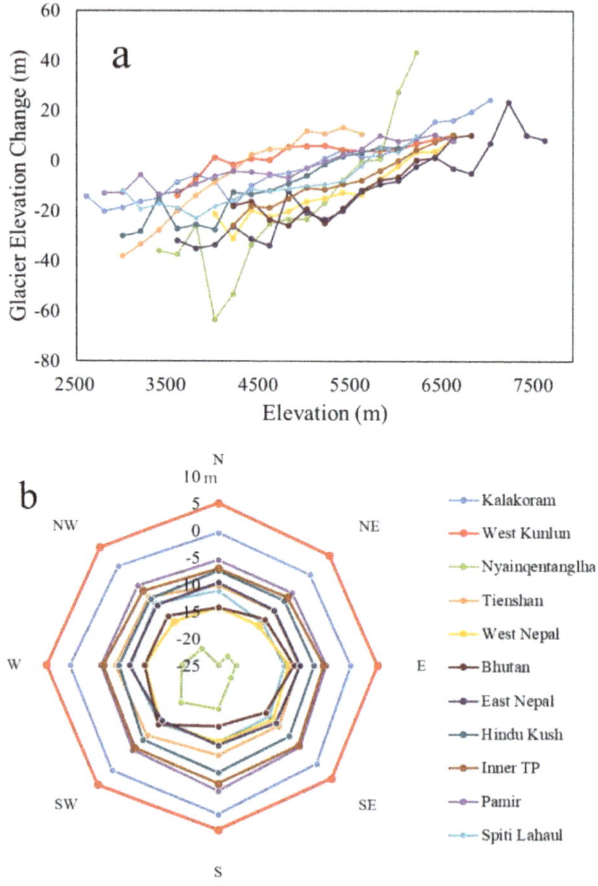

Figure 4. Glacier elevation change between 2000 and 2020 in each sub-region of the HMA in: (**a**) different elevations, and (**b**) different slope aspects. (The sub-regions were defined in Figure 1a).

4.2. Inter-Annual Change Rate of Glacier Elevation

4.2.1. Spatial Pattern of Annual Glacier Elevation Change Rate

To calculate inter-annual rate of change in glacier elevation, ICESat-1&2 data in autumn months (September–November) from 2003 to 2008 and from 2018 to 2020 were selected (procedure described in Section 3). Autumn was chosen for the annual trend analysis because the ICESat-1&2 and SRTM observations are less influenced by snow cover in this season [35,54]. The year 2009 was not included in the trend analysis due to too little data available in autumn to ensure reliable results (ICESat-1 eventually ceased work in 2010). Figure 5 showed the spatial distribution of the annual rate of change in glacier elevation in the HMA region from 2003 to 2008 and from 2003 to 2020, respectively. The detailed annual changes in glacier elevation and the respective error information were shown in Table A1. The results showed that the spatial variation of glacier elevation change rate in the HMA was very significant, and the change rate was very different in some areas between 2003–2008 and 2003–2020. The glacier elevations have thinned across much of the HMA region. For the entire HMA, the annual rate of change in glacier elevation from 2003 to 2008 and from 2003 to 2020 were -0.21 ± 0.12 m/year and -0.26 ± 0.11 m/year, respectively, indicating a faster thinning rate in recent years. In Nyainqentanglha, the annual glacier elevation change rate increased from -0.81 ± 0.15 m/year in 2003–2008 to -1.12 ± 0.13 m/year in 2003–2020, which is the region with the fastest glacier loss. The annual glacier change rate in the Tienshan Mountains increased from -0.27 ± 0.13 m/year in 2003–2008 to -0.33 ± 0.11 m/year in 2003–2020. In contrast, the rate of glacier thinning decreased significantly in the west HMA. For example, in the Pamir, the glacier elevations decreased by 0.22 m/year in 2003–2008 and 0.13 m/year in 2003–2020, indicating a slowing down trend in glacier thinning rate. Although the elevation of most glaciers generally decreased between 2003–2020, some glaciers in the inner region of HMA showed an increase in elevation. Previous studies found that glaciers in the West Kunlun was thickening [17]. Our results showed that in West Kunlun, the glacier elevation change rate was 0.18 ± 0.11 m/year in 2003–2020 and 0.10 ± 0.13 m/year in 2003–2008. As a summary of our results, the glacier thinning rate across the entire HMA is gradually accelerating, except for the West Kunlun, Karakoram, Pamir and Hindu Kush.

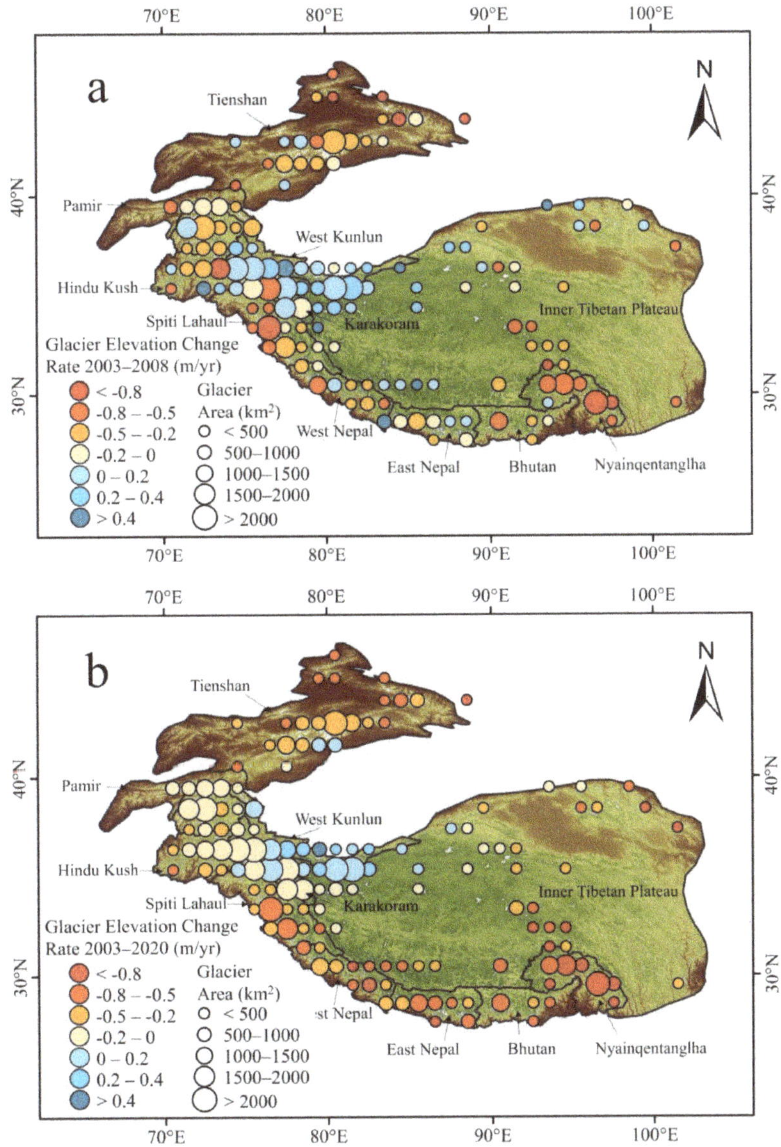

Figure 5. Glacier elevation change rate in the HMA in 2003–2008 (**a**) and in 2003–2020 (**b**) (Statistics in 1° × 1° grids).

4.2.2. Glacier Elevation Change Rate in Basin Scale

The melting of the HMA glaciers has an important impact on the changes in river flow and water resources in the HMA and its downstream regions. Therefore, obtaining information on glaciers changes in different basins of the HMA is of great significance for understanding the changes of water resources in the HMA and its surrounding areas. We estimated the glacier elevation change rate in each basin of the HMA during 2003–2008 and 2003–2020 (Table 1) (method described in Section 3.3).

Table 1. The rate of change in glacier elevation in the basins of the HMA in 2003–2008 and 2003–2020. The errors are given at the 1σ level calculated by Equation (6).

Basin Name	**Glacier Elevation Change Rate (m/year)**	
	2003–2008	**2003–2020**
Amu Darya	−0.21 ± 0.13	−0.12 ± 0.11
Brahmaputra	−0.43 ± 0.18	−0.78 ± 0.13
Ili river	−0.46 ± 0.13	−0.40 ± 0.12
Indus river	−0.25 ± 0.15	−0.20 ± 0.12
Qiangtang Plateau	0.07 ± 0.12	−0.04 ± 0.11
Shule and Heihe river	−0.11 ± 0.18	−0.44 ± 0.15
Ebinur and Manas river	−0.10 ± 0.30	−0.42 ± 0.18
Mekong river	−0.34 ± 0.43	−1.2 ± 0.33
Qardam	−0.21 ± 0.20	−0.37 ± 0.13
Issyk-Kul lake	0.10 ± 0.2	−0.39 ± 0.14
Salween	−0.69 ± 0.20	−1.02 ± 0.14
Syr Darya	−0.67 ± 0.16	−0.34 ± 0.13
Tarim river	0.08 ± 0.14	0.08 ± 0.11
Yangtze	−0.43 ± 0.19	−0.48 ± 0.13

Due to the differences in temperature, precipitation, evapotranspiration and other factors in different basins, rates of changes in glacier elevations varied greatly in time and space (Table 1). The results in Table 1 showed that from 2003 to 2020, the glacier elevations in almost all basins of the HMA showed a decreasing trend, except for Tarim. From 2003 to 2020, the Mekong River basin experienced the fastest thinning of glaciers, followed by the Salween and Brahmaputra basins. However, we found that the rate of thinning of glaciers in the Indus, Syr Darya and Amu Darya river basins slowed down. We will analyze the reasons for this phenomenon in Section 5.2.1.

4.3. *Intra-Annual Change of Glacier Elevation*

The period of thickening or thinning of glaciers varied in different regions of the HMA due to climate and topography. In this section, the HMA was divided into eight sub-regions to explore the monthly/seasonal characteristics of the glacier elevation changes in the HMA caused by differences in glacier accumulation and melting over time. The eight sub-regions are Spiti Lahaul, Bhutan and Nepal, Nyainqentanglha mountain, West Kunlun, Pamir and Hindu Kush, Tienshan mountain, Karakoram and Inner TP (Tibetan Plateau). The ICESat-2 monthly data in 2019–2020 were used for monthly/seasonal analysis.

In this study, we found that the period of glacier thickening from the marginal regions of the HMA to the interior HMA was gradually delayed from autumn to summer. The results in Figure 6 revealed that in Pamir and Hindu Kush and Karakoram, glaciers showed similar seasonal changes in thickening in autumn (SON: September–October–November) or winter (DJF: December–January–February). In Pamir and Hindu Kush, although the glaciers were at their highest elevations in spring, significant glacier thickening occurred mainly in October–March, followed by a significant thinning period. The glaciers in Karakoram had almost the same seasonal variation as in Pamir and Hindu Kush, except for a sudden drop in elevation in July 2019, which might be attributed to relatively low precipitation and warmer temperatures in July 2019 (see the discussion in Section 5.2.2). In West Kunlun and Bhutan and Nepal, the glaciers exhibited similar seasonal thickening variation in spring (MAM: March–April–May) or summer (JJA: June–July–August). The glaciers in West Kunlun Mountains showed two accumulation periods with obvious thickening, i.e., in March–June and in July–September. In the Bhutan and Nepal regions, the month with the least glacier elevation decline in 2019–2020 occurred in July. In the inner TP, the glaciers did not show an obvious consistency of accumulation (or melting) in the two years, although the months with the highest glacier elevations appeared in the summer of the two years. In the Tienshan Mountains, the glacier elevations fluctuated greatly, the glacier thickening time was mainly concentrated from December to April, and there was a weak thickening trend from September to December. In Spiti Lahual, the southwest of the HMA, the glacier thickening period was mainly concentrated in winter and spring, with thinning in summer. The glacier elevations in the Nyainqentanglha Mountains reached their highest points in spring.

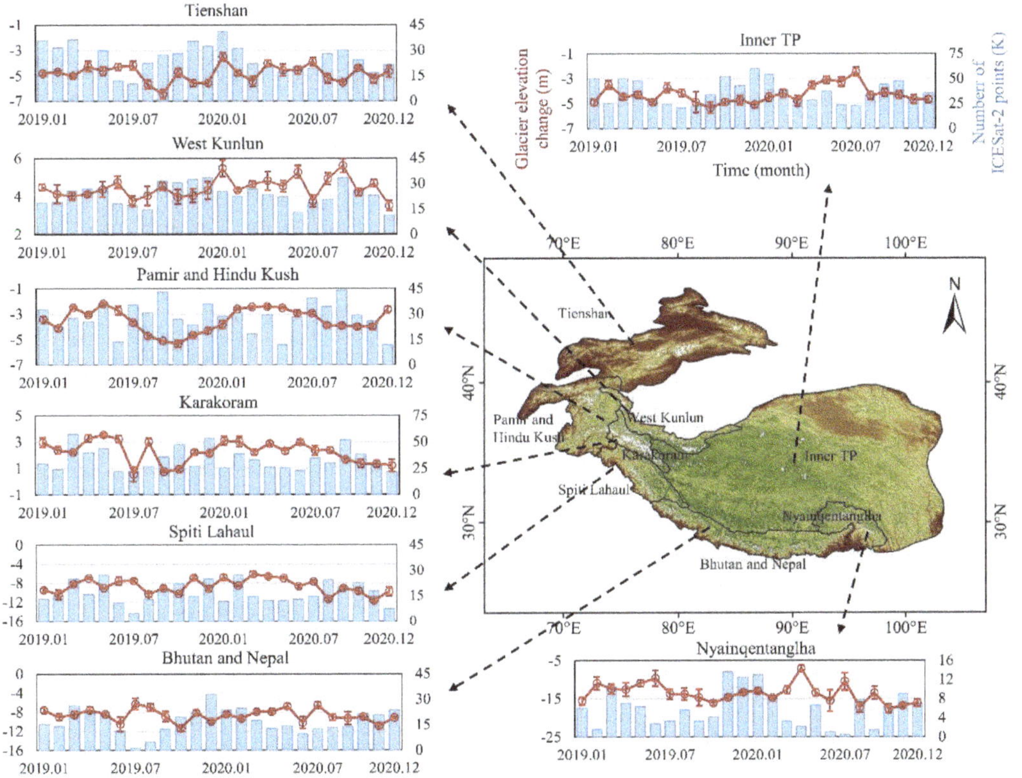

Figure 6. The monthly change of glacier elevation in different regions of the HMA from January of 2019 to December of 2020. The red line (left *Y*-axis) represents the glacier elevation difference between the ICESat-2 data in 2019–2020 and the SRTM DEM data in 2000, the blue column (right *Y*-axis) represents the number of ICESat-2 data points in each month. The error bars are the values of standard deviation of spatial glacier elevation difference in the corresponding months calculated by Equation (5) at 1σ level.

5. Discussion

5.1. Comparison with Existing Studies

5.1.1. Comparison of Inter-Annual Glacier Elevation Change Rate

We compared our results with those by other studies in each sub-region of the HMA (Figure 7). Overall, our results are consistent with most previous studies.

However, there are some discrepancies between our results and existing studies, which may be due to the use of different methods and data. As shown in Section 4.1, glacier elevation changes varied with elevations and slope aspects. As shown in Figure 4b, the glacier elevation changes on the east and west slopes of the Nyainqentanglha Mountains were different. At the same time, the distribution of ICESat-1&2 data points in different slope aspects will also affect the changes of glacier elevations. The ICESat-1&2 laser point data cannot be guaranteed to be evenly distributed in space. In general, a large area may have more data points. However, due to factors such as satellite orbit or cloud cover, the spatiotemporal distribution of available laser points is irregular. We selected the computing unit (i.e., one 1° × 1° grid) around the Yalong glacier (96°E~97°E, 29°N~30°N) to investigate the distribution of ICESat-2 data and found that the data volume of ICESat-2 data in 2020 in the eastern slopes is about 1.6 times of that in the western slopes, while the area of glaciers in the eastern slopes is only 1.2 times of that in the western slope. In brief, the distribution of ICESat-1&2 data points will be different in various slope aspects due to the influence of topography, climate and satellite operation mode, which in turn affects the accuracy of the results of glacier elevation changes. Therefore, in this study, elevation and aspect are both factors that must be considered simultaneously when calculating the glacier elevation change.

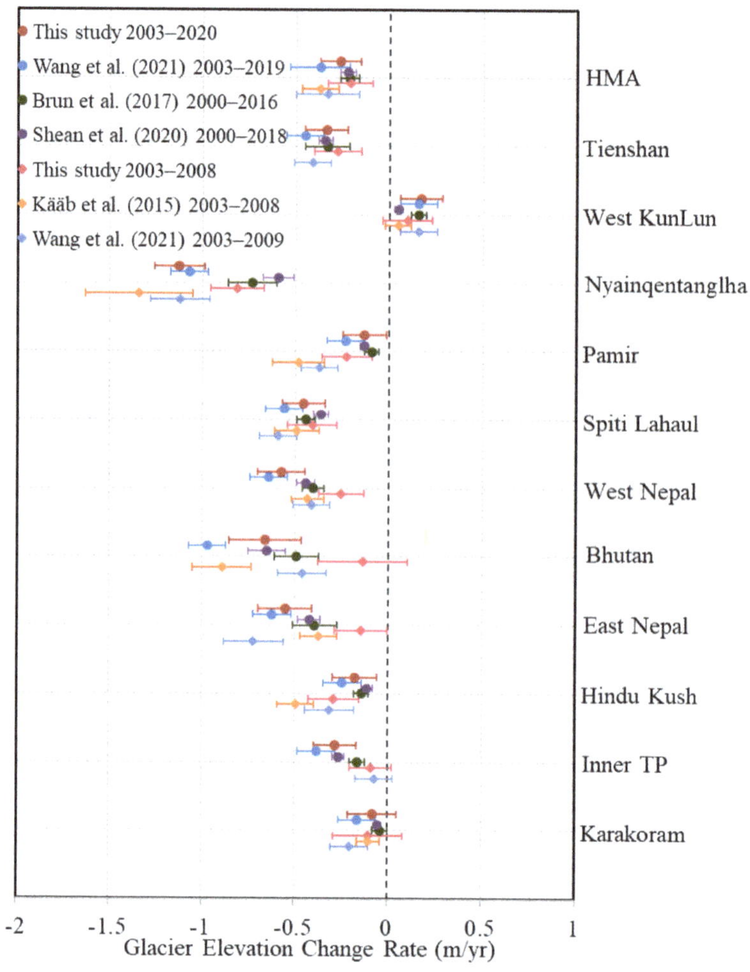

Figure 7. Comparison of results on glacier elevation change rate in the HMA in 2003–2020 and 2003–2008 between this study with existing studies. Wang et al. [22] calculated the glacier elevation change rate in 2003–2019 using the ICESat-1&2 data. Brun et al. [17] and Shean et al. [10] calculated the glacier elevation change rates in 2000–2016 and 2000–2018 using ASTER data in 2016/2018 and SRTM DEM data in 2000, respectively. Kääb et al. [58] calculated the glacier elevation change rate in 2003–2008 using ICESat-1 data. The error bars are the standard deviation of spatial glacier elevation change rate calculated by Equation (6) at 1σ level.

Existing studies have shown that the glacier thinning rate based on ICESat-1&2 data is greater than that based on stereo pair data in the Nyainqentanglha region [22,59]. However, with the exception of Nyainqentanglha region, the difference between our results in 2003–2020 and the results of Brun et al. [17] and Shean et al. [10], both using stereo pair data, is very small (within two standard deviations), indicating that our results of glacier elevation change rate are within an acceptable range.

5.1.2. Comparison of Intra-Annual Glacier Elevation Change

The results of this study are basically consistent with the existing studies (Table 2) on intra-annual variation of glacier elevation. In general, glaciers in the marginal regions of the HMA (the Tienshan Mountains, Spiti Lahaul, etc.) thicken in winter. However, similar to existing studies, the glaciers in Bhutan and Nepal have an obvious thickening trend in summer. This may be due to the accumulation of water vapor in the whole Tibet Plateau mainly in summer. The southern slopes of Bhutan and Nepal are affected by the strong monsoon from the Indian Ocean and the Bay of Bengal, with the most precipitation in summer [60]. The impacts of precipitation outweigh the impacts of temperature, leading to summer replenishment of glaciers in Himalayas. However, there are also large fluctuations in glacier elevation changes in Bhutan and Nepal. Our findings show that in 2019 the glaciers generally accumulated between January and June, while in 2020 the opposite was true, although the months with the highest glacier elevation were all in July. Kansakar et al. [61] found large variation in precipitation patterns in Nepal suggesting that there may be large differences in glacier elevation changes in the Himalayas Mountain, which is consistent with studies by Maussion et al. [30] and Wang et al. [31]. Similarly, the greater variability in the glacier elevation change in the Nyainqentanglha Mountains may be partly due to the influence of hydrothermal conditions, because the inter-annual variability of precipitation due to abnormal anticyclones in the northern Indian subcontinent and the Bay of Bengal varies greatly [62,63]. Different from the existing research, in the Tienshan Mountains, our results showed a clear trend of glacier thickening in autumn, although the glacier elevation in autumn was the lowest. This is mainly due to the large glacier loss from July to September. However, looking at the monthly distribution, we found a slight increase in autumn during the two-year period (September to December). The same phenomenon occurred in Nyainqentanglha, Pamir and Hindu Kush. This is also an advantage of our study based on monthly data from ICESat-2, as we could highlight the details between seasons. In short, using the ICESat-2 data, we can more precisely monitor the intra-annual changes in glacier elevation. However, when the analysis is carried out on a monthly/seasonal scale, the amount of data is reduced, which also affects the accuracy of regional results. For example, some glaciers may have only a few hundred points in a month, which may be the reason for the large fluctuations in glacier elevation change in some areas. Nevertheless, our method is a direct calculation of glacier elevation change that can represent the intra-annual variation pattern of glacier elevation. With the continuous observation by ICESat-2 satellite, we will obtain more accurate monthly/seasonal variation characteristics of glacier in the future.

Table 2. The period of glacier accumulation in the sub-regions of the HMA from this study and the comparison with previous studies.

Sub-Regions	This Study	Wang et al. [31]	Maussion et al. [30]
Spiti Lahaul	December–March	December–March	December–February
Bhutan and Nepal	February–July	/	March–August
Nyainqentanglha	October–April (June)	December–June	December–August
West Kunlun	March–June, July–September	March–June (August)	March–May
Pamir and Hindu Kush	October–March	November–April	December–February
Tienshan	September–April	December–June	/
Karakoram	October–January	/	December–February
Inner TP	summer	April–August	June–August

5.2. *Factors Controlling Glacier Elevation Change*

5.2.1. Factors Controlling Inter-Annual Glacier Elevation Change Rate

Previous studies have shown that the sensitivity of glaciers to climate is the main controlling factor of the HMA glacier change [64]. The rate at which a glacier melts is related to the energy gain or loss on the glacier surface. The main factors affecting glacier changes include temperature and precipitation. Warmer temperatures will lead to accelerated glacier loss [65], while increased precipitation will compensate for glacier mass loss. To explore the factors affecting the changes of glacier thickness, we calculated the change of glacier elevation change rate from 2003 to 2020 and from 2003 to 2008 (Figure 8a), and used the ERA5 data to investigate the average temperature and precipitation changes in autumn in the HMA from 2003 to 2020 and from 2003 to 2008 (Figure 8b,c).

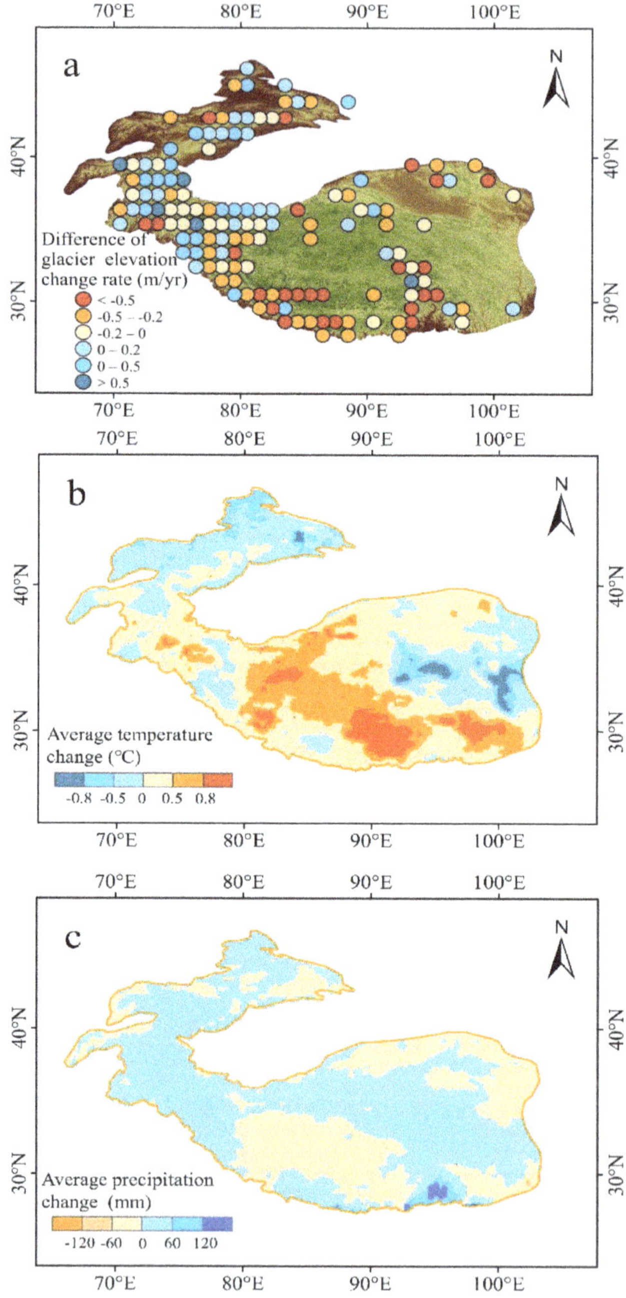

Figure 8. Difference between 2003–2020 and 2003–2008 (in autumn) in: (**a**) glacier elevation change rate, (**b**) mean temperature, and (**c**) mean precipitation.

The result showed strong warming in the central and southeastern regions of the HMA, while the temperature decreased in the western, northwestern and southwestern regions. The precipitation in the HMA generally showed an overall increasing trend, but the increased precipitation was mainly concentrated in the Tienshan Mountains in the northwest of the HMA, the western part of the HMA, parts of the central HMA and southern edge of the HMA. Our study revealed that in the context of global warming, most glaciers in the HMA experienced accelerated mass loss. Spatially, the change of glacier elevation change rate in the Tienshan Mountains is consistent with the change in precipitation, indicating that the change of glaciers in the Tienshan Mountains may be more affected by precipitation. However, the rate of glacier thinning in the southwest of HMA has slowed down. Guo et al. [66] found that, compared with the warming trend (+0.18 °C/decade) of the Tibetan Plateau since 2001, the temperature in the southwest of the Tibetan Plateau decreased by 0.15 °C/decade, and we found that Pamir and Hindu Kush also appeared to have a similar pattern. Overall, a decrease in temperature (Figure 8b) and an increase in precipitation (Figure 8c) explained the slowing of glacier mass loss in the west HMA region. The analysis of climatic factors of precipitation and temperature showed that the regional pattern of glacier elevation changes was consistent with the patterns of precipitation and temperature. Existing studies showed that for every 1 degree increase in temperature, precipitation needs to increase by 25–35% to compensate for the impact of temperature on glaciers [67]. Both temperature and precipitation within the HMA are increasing overall, and while more precipitation can compensate for the mass loss of the glaciers, it is still far from being able to compensate for the overall melting of glaciers caused by the increase in temperature, which contributes to the accelerated glacier mass loss in most parts of the HMA.

5.2.2. Factors Controlling Intra-Annual Glacier Elevation Variation

The glaciers in the HMA region have multiple types of accumulation and ablation patterns, which are largely related to precipitation [68,69]. This study analyzed the monthly/seasonal variations in temperature and precipitation from January 2019 to December 2020. In all regions of the HMA, the temperature throughout the year showed a trend of first rising and then falling. The changes in precipitation can explain the characteristics of intra-annual glacier elevation variations (Figure 9).

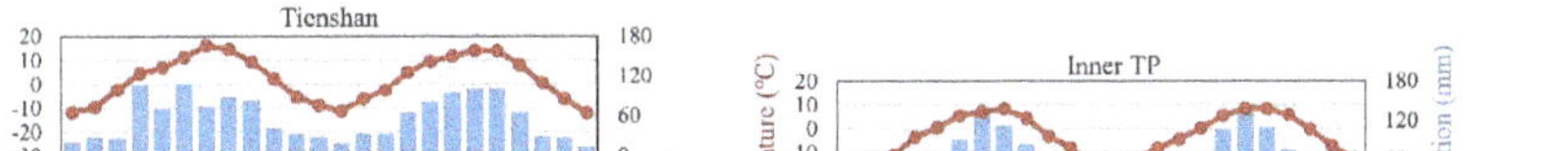

Figure 9. Monthly temperature and precipitation from January 2019 to December 2020 in each sub-region of the HMA. The red line (left Y-axis) represents the monthly temperature, and the blue column (right Y-axis) represents the monthly precipitation.

For example, precipitation in the Himalayas has a clear upward trend in summer, which explains the accumulation of its glaciers in summer. In the Nyainqentanglha Mountains, the thickening of the glaciers in spring is attributed to that the concentration of about 20−40% of the precipitation in spring [70]. The strength and nature of the coupling between the monsoon system and the westerly system are important factors that cause precipitation changes. Glaciers in Pamir, Hindu Kush and Spiti Lahaul, located at the intersection zones of westerly and monsoon air flows, may be more sensitive to changes in weather and atmospheric circulation [30]. Compared with other regions, Spiti Lahaul and Pamir have significantly more precipitation in winter and spring, which is the reason for the thickening of glaciers here. April in particular is the rainiest month in Pamirs and in Spiti Lahaul from November to March, while the rest of the HMA is around July. As shown in Section 4.3, the overall trend of glacier elevation changes in Karakoram is similar to that in Pamir and Hindu Kush, except in July 2019, which may be due to less precipitation in July 2019. Existing studies have shown that in areas affected by the westerly climate, the precipitation in winter is more than that in summer [71]. At the same time, affected by the atmospheric circulation and the deviation of the earth's rotation, the Tienshan Mountains, Pamir and Hindu Kush are mainly affected by the Atlantic southwesterly wind. Coupled with the barriers of mountains, the westerly wind in spring will bring sufficient precipitation here and reduce the mass loss of glaciers. For example, in Pamir and Hindu Kush, the glacier elevation is highest in spring, due to the continuous thickening of the glaciers from October to March. There is less glacier mass loss in spring as precipitation remains high from March to May. In conclusion, the difference of precipitation can explain the difference in the glacier accumulation in different regions of the HMA.

5.3. Advantages and Disadvantages of ICESat-1&2 Data in Estimating Glacier Elevation Change

The ICESat-1&2 data have high vertical detection accuracy, and their applications in the cryosphere will be worth exploring. The emergence of ICESat-1&2 data allows us to obtain a large amount of data every year or even every month, which provides us with an opportunity to understand the glacier changes at higher temporal resolutions (monthly/seasonal), which is of great significance to disaster prevention and rational use of water resources [24,72]. For example, the ICESat-2 data are observed more frequently and therefore have more data points in space, providing a new perspective for the monitoring changes in glacier elevation. Its main disadvantage is that the large orbital spacing in the middle latitudes [73]. As shown in Figure 2, the ICESat-1 GLAS data have less data density. Although the amount of ICESat-2 ATLAS data and glacier coverage have been greatly improved, it still cannot cover all glaciers completely, which is an unavoidable problem when using ICESat-1&2 data in mid latitudes. The number and spatial distribution of data points will directly affect the accuracy of the results. In some regions, the reduction in data volume can increase the uncertainty of the results. For example, in the monthly analysis in Section 4.3, there were only a few hundred points per month in some regions, resulting in large fluctuations in annual performance. The spatiotemporal sampling of the data will affect the accuracy of the results, especially the ICESat-1 data with a small amount of data. There is no doubt that the more complete the data points, the more reliable the results will be. In future study, systematic analysis should be carried out to explore the impact of the spatial distribution of the ICESat-1&2 data points on the accuracy of the results. The second disadvantage is that the orbital revisit positions of ICESat-1&2 are not constant, so the observation points of each orbit do not repeat at the same position. While the track crossing method or plane fitting method can help reduce uncertainty, these methods are mostly applicable to polar regions [74] and are not reliable in mid-latitudes. The method proposed in this paper can better calculate the glacier elevation changes, but the uncertainty of the ICESat-1&2 data itself cannot be eliminated compared with studies based on stereo image pairs.

It should be noted that the data record of ICESat-1&2 is short (2003–2008 & 2018-onwards), making it is difficult to draw ultimate and firm conclusions about the trend in glacier elevation changes. With more ICESat-2 data becomes available over time, the glacier thickness change and seasonal dynamics can be monitored with longer record and data of better quality.

6. Conclusions

This study applied the "elevation-aspect bin analysis method" to ICESat-1&2 data to estimate glacier elevation changes in the HMA region and explored the inter-annual and intra-annual changes of glacier elevation in the HMA. The main conclusions of this study are as follows: (1) The "elevation-aspect bin analysis method" can efficiently capture the glacier elevation change and reduce the uncertainty caused by uneven spatial distribution of data points of ICESat-1&2 observations. (2) The result of the inter-annual rate of change in glacier elevation in the HMA showed spatial heterogeneity.

The glacier elevation in the marginal regions of HMA declined more (i.e., thinned faster), while the elevation of the glaciers in West Kunlun rose. The declined rate of glacier elevation in the HMA in 2003–2020 (-0.26 ± 0.11 m/year) was faster than that in 2003–2008 (-0.21 ± 0.12 m/year). Glacier retreat is accelerating in all regions of the HMA except in the western part of the HMA. The regional variability of the glacier elevation change rate from 2003 to 2020 was large, ranging from -1.12 ± 0.13 m/year in the Nyaingentanglha Mountains to $+0.18 \pm 0.11$ m/year in the West Kunlun Mountains. (3) For the intra-annual variation of glacier elevation, the results show that glacier elevation change has spatial heterogeneity, and the glacier thickening period is gradually delayed from the marginal regions to inner regions of the HMA. The glaciers in the Spiti Lahaul (December to March) and the Tienshan Mountains (September to April) tend to thicken during winter to spring, while glacier elevation in the Tienshan Mountains tends to rise slightly in autumn. The glaciers in Pamir and Hindu Kush (October to March) and Karakoram (October to January) thicken during winter. The glaciers in Nyainqentanglha thicken during spring (October to April or June). The glaciers in West Kunlun, Inner TP and Bhutan and Nepal thicken in spring and summer. West Kunlun has two accumulation periods (March–June and July–September). The glaciers in the Bhutan and Nepal (February to July) thicken in spring and summer, with elevation peaking in July. The glacier elevation in Inner TP reaches the highest level in June or July, but the accumulation trend is not obvious. In addition, the factors affecting glacier elevation changes are analyzed and the results indicate that the inter-annual and intra-annual changes in glacier elevation are consistent with the changes in air temperature or precipitation patterns.

Author Contributions: Conceptualization, C.S. and L.J.; methodology, C.S., L.J. and S.R.; software, C.S.; validation, C.S. formal analysis, C.S. and L.J.; investigation, C.S. and L.J.; resources, L.J. and C.S; data curation, C.S., L.J.; writing—original draft preparation, C.S.; writing—review and editing, C.S., L.J. and S.R.; visualization, C.S.; supervision, L.J.; project administration, L.J.; funding acquisition, L.J. All authors have read and agreed to the published version of the manuscript.

Funding: This research was funded jointly by the Second Tibetan Plateau Scientific Expedition and Research Program (STEP) (grant no. 2019QZKK0103); the projects of National Natural Science Foundation of China (grant no. 91737205, 42171039); the Strategic Priority Research Program of the Chinese Academy of Sciences (grant no. XDA19070102); the MOST High-Level Foreign Expert program (grant no. GL20200161002).

Conflicts of Interest: The authors declare no conflict of interest.

Appendix A

Table A1. The annual elevation difference and error of glaciers in each region of HMA based on ICESat-1&2 data and SRTM DEM data (m). The errors are the values of standard deviation of spatial glacier elevation difference in the corresponding years calculated by Equation (5) at 1σ level.

Region	2003	2004	2005	2006	2007	2008	2018	2019	2020
Bhutan	$-2.1 \pm$ 1.63	$0.69 \pm$ 1.3	$0.52 \pm$ 1.53	$0.11 \pm$ 1.51	$-1.33 \pm$ 1.74	$-1.21 \pm$ 1.08	$-8.46 \pm$ 0.89	$-8.92 \pm$ 0.46	$-9.19 \pm$ 0.61
East Nepal	$-1.67 \pm$ 1.06	$-1.64 \pm$ 1.07	$-2.86 \pm$ 1.13	$-2.70 \pm$ 1.17	$-2.70 \pm$ 1.18	$-1.99 \pm$ 1.04	$-9.34 \pm$ 0.3	$-10.84 \pm$ 0.36	$-9.86 \pm$ 0.22
Hindu Kush	$-2.51 \pm$ 0.69	$-2.27 \pm$ 1.01	$-1.85 \pm$ 0.91	$-2.57 \pm$ 1.1	$-3.62 \pm$ 0.93	$-3.55 \pm$ 1.32	$-4.87 \pm$ 0.21	$-5.37 \pm$ 0.16	$-4.83 \pm$ 0.35
Inner TP	$-0.36 \pm$ 0.41	$-0.53 \pm$ 0.55	$-0.35 \pm$ 0.53	-1 ± 0.59	$-0.9 \pm$ 0.55	$-0.62 \pm$ 0.48	$-4.22 \pm$ 0.19	$-4.59 \pm$ 0.15	$-4.78 \pm$ 0.12
Karakoram	$2.33 \pm$ 0.46	$1.98 \pm$ 0.57	$3.37 \pm$ 0.56	3.65 ± 0.6	$2.23 \pm$ 0.55	$1.38 \pm$ 0.47	$-0.15 \pm$ 0.17	$1.76 \pm$ 0.15	$2.02 \pm$ 0.11
West Kunlun	$2.28 \pm$ 0.47	$2.02 \pm$ 0.63	$3.19 \pm$ 0.637	$2.16 \pm$ 0.63	$2.23 \pm$ 0.63	$3.04 \pm$ 0.62	$4.60 \pm$ 0.62	$5.10 \pm$ 0.26	$4.92 \pm$ 0.13
Nyainqentanglha	$-1.12 \pm$ 0.76	$-0.66 \pm$ 0.94	$-2.13 \pm$ 0.89	$-3.38 \pm$ 0..99	$-4.48 \pm$ 1.36	$-4.28 \pm$ 0.92	$-16.13 \pm$ 0.8	$-18.18 \pm$ 0.38	$-19.52 \pm$ 0.49
Pamir	$-1.12 \pm$ 0.61	$-1.09 \pm$ 0.75	$-0.43 \pm$ 0.72	$-1.68 \pm$ 0.97	$-1.90 \pm$ 0.65	$-2.14 \pm$ 0.78	$-2.68 \pm$ 0.3	$-3.35 \pm$ 0.13	$-3.37 \pm$ 0.19
Tienshan	$0.25 \pm$ 0.5	$-0.18 \pm$ 0.58	$0.48 \pm$ 0.55	$-0.47 \pm$ 0.54	$-0.39 \pm$ 0.57	$-1.39 \pm$ 0.54	$-4.99 \pm$ 0.2	$-4.94 \pm$ 0.15	$-4.68 \pm$ 0.16

Table A1. *Cont.*

Region	2003	2004	2005	2006	2007	2008	2018	2019	2020
West Nepal	−2.08 ± 0.79	−3.15 ± 0.95	−2.58 ± 0.97	−3.17 ± 1.04	−3.35 ± 0.99	−3.54 ± 0.9	−10.96 ± 0.21	−10.47 ± 0.19	−11.35 ± 0.23
Spiti Lahaul	−1.10 ± 0.6	−1.55 ± 0.76	−0.93 ± 0.81	−1.76 ± 0.73	−2.94 ± 0.76	−2.98 ± 0.75	−7.82 ± 0.26	−7.53 ± 0.15	−8.74 ± 0.27
HMA	0.11 ± 0.17	−0.07 ± 0.22	0.48 ± 0.2	−0.15 ± 0.24	−0.71 ± 0.23	−0.87 ± 0.25	−3.63 ± 0.11	−3.85 ± 0.04	−4.06 ± 0.04

References

1. Ai, B.-B.; Qin, C.-Z.; Ye, Q.; Zhu, A.-X.; Cogley, G. An approach to extracting surface supply relationships between glaciers and lakes on the Tibetan Plateau. *Int. J. Digit. Earth* **2017**, *11*, 1151–1165. [CrossRef]
2. Immerzeel, W.W.; Van Beek, L.P.H.; Bierkens, M.F.P. Climate change will affect the Asian Water Towers. *Science* **2010**, *328*, 1382–1385. [CrossRef] [PubMed]
3. Raper, S.C.B.; Braithwaite, R. Low sea level rise projections from mountain glaciers and icecaps under global warming. *Nature* **2006**, *439*, 311–313. [CrossRef]
4. Millan, R.; Mouginot, J.; Rabatel, A.; Morlighem, M. Ice velocity and thickness of the world's glaciers. *Nat. Geosci.* **2022**, *15*, 124–129. [CrossRef]
5. Jacob, T.; Wahr, J.M.; Pfeffer, W.T.; Swenson, S.C. Recent contributions of glaciers and ice caps to sea level rise. *Nature* **2012**, *482*, 514–518. [CrossRef] [PubMed]
6. Kaser, G.; Groβhauser, M.; Marzeion, B. Contribution potential of glaciers to water availability in different climate regimes. *Proc. Natl. Acad. Sci. USA* **2010**, *107*, 20223–20227. [CrossRef] [PubMed]
7. Liang, L.; Cuo, L.; Liu, Q. Long-Term Temporal Scale-Dependent Warming Effects on the Mass Balance in the Dongkemadi Glacier, Tibetan Plateau. *J. Geophys. Res. Atmos.* **2020**, *125*. [CrossRef]
8. Xiao, C.-D.; Wang, S.-J.; Qin, D.-H. A preliminary study of cryosphere service function and value evaluation. *Adv. Clim. Chang. Res.* **2015**, *6*, 181–187. [CrossRef]
9. Huss, M.; Hock, R. Global-scale hydrological response to future glacier mass loss. *Nat. Clim. Chang.* **2018**, *8*, 135–140. [CrossRef]
10. Shean, D.E.; Bhushan, S.; Montesano, P.; Rounce, D.R.; Arendt, A.; Osmanoglu, B. A Systematic, Regional Assessment of High Mountain Asia Glacier Mass Balance. *Front. Earth Sci.* **2020**, *7*, 363. [CrossRef]
11. Kraaijenbrink, P.; Bierkens, M.F.; Lutz, A.; Immerzeel, W. Impact of a global temperature rise of 1.5 degrees Celsius on Asia's glaciers. *Nature* **2017**, *549*, 257–260. [CrossRef]
12. Gardelle, J.; Berthier, E.; Arnaud, Y.; Kääb, A. Region-wide glacier mass balances over the Pamir-Karakoram-Himalaya during 1999–2011. *Cryosphere* **2013**, *7*, 1263–1286. [CrossRef]
13. Bajracharya, S.R.; Maharjan, S.B.; Shrestha, F. The status and decadal change of glaciers in Bhutan from the 1980s to 2010 based on satellite data. *Ann. Glaciol.* **2014**, *55*, 159–166. [CrossRef]
14. Wu, K.; Liu, S.; Jiang, Z.; Xu, J.; Wei, J.; Guo, W. Recent glacier mass balance and area changes in the Kangri Karpo Mountains from DEMs and glacier inventories. *Cryosphere* **2018**, *12*, 103–121. [CrossRef]
15. Paul, F.; Bolch, T.; Kääb, A.; Nagler, T.; Nuth, C.; Scharrer, K.; Shepherd, A.; Strozzi, T.; Ticconi, F.; Bhambri, R.; et al. The glaciers climate change initiative: Methods for creating glacier area, elevation change and velocity products. *Remote Sens. Environ.* **2015**, *162*, 408–426. [CrossRef]
16. Ren, S.; Menenti, M.; Jia, L.; Zhang, J.; Zhang, J.; Li, X. Glacier Mass Balance in the Nyainqentanglha Mountains between 2000 and 2017 Retrieved from ZiYuan-3 Stereo Images and the SRTM DEM. *Remote Sens.* **2020**, *12*, 864. [CrossRef]
17. Brun, F.; Berthier, E.; Wagnon, P.; Kääb, A.; Treichler, D. A spatially resolved estimate of High Mountain Asia glacier mass balances from 2000 to 2016. *Nat. Geosci.* **2017**, *10*, 668–673. [CrossRef]
18. Dehecq, A.; Gourmelen, N.; Trouve, E. Deriving large-scale glacier velocities from a complete satellite archive: Application to the Pamir–Karakoram–Himalaya. *Remote Sens. Environ.* **2015**, *162*, 55–66. [CrossRef]
19. Yin, B.; Zeng, J.; Zhang, Y.; Huai, B.; Wang, Y. Recent Kyagar glacier lake outburst flood frequency in Chinese Karakoram unprecedented over the last two centuries. *Nat. Hazards* **2018**, *95*, 877–881. [CrossRef]
20. Bhambri, R.; Hewitt, K.; Kawishwar, P.; Pratap, B. Surge-type and surge-modified glaciers in the Karakoram. *Sci. Rep.* **2017**, *7*, 15391. [CrossRef]
21. Rankl, M.; Kienholz, C.; Braun, M. Glacier changes in the Karakoram region mapped by multimission satellite imagery. *Cryosphere* **2014**, *8*, 977–989. [CrossRef]
22. Wang, Q.; Yi, S.; Sun, W. Continuous Estimates of Glacier Mass Balance in High Mountain Asia Based on ICESat-1,2 and GRACE/GRACE Follow-On Data. *Geophys. Res. Lett.* **2021**, *48*, e2020GL09095. [CrossRef]
23. Nuth, C.; Kääb, A. Co-registration and bias corrections of satellite elevation data sets for quantifying glacier thickness change. *Cryosphere* **2011**, *5*, 271–290. [CrossRef]

24. Gardner, A.S.; Moholdt, G.; Cogley, J.G.; Wouters, B.; Arendt, A.A.; Wahr, J.; Berthier, E.; Hock, R.; Pfeffer, W.T.; Kaser, G.; et al. A Reconciled Estimate of Glacier Contributions to Sea Level Rise: 2003 to 2009. *Science* **2013**, *340*, 852–857. [CrossRef] [PubMed]
25. Yuan, C.; Gong, P.; Bai, Y. Performance Assessment of ICESat-2 Laser Altimeter Data for Water-Level Measurement over Lakes and Reservoirs in China. *Remote Sens.* **2020**, *12*, 770. [CrossRef]
26. Ma, Y.; Xu, N.; Liu, Z.; Yang, B.; Yang, F.; Wang, X.H.; Li, S. Satellite-derived bathymetry using the ICESat-2 lidar and Sentinel-2 imagery datasets. *Remote Sens. Environ.* **2020**, *250*, 112047. [CrossRef]
27. Neumann, T.A.; Martino, A.J.; Markus, T.; Bae, S.; Bock, M.R.; Brenner, A.C.; Brunt, K.M.; Cavanaugh, J.; Fernandes, S.T.; Hancock, D.W.; et al. The Ice, Cloud, and Land Elevation Satellite—2 mission: A global geolocated photon product derived from the Advanced Topographic Laser Altimeter System. *Remote Sens. Environ.* **2019**, *233*, 111325. [CrossRef]
28. Pritchard, H.D. Asia's glaciers are a regionally important buffer against drought. *Nature* **2017**, *545*, 169–174. [CrossRef]
29. Ageta, Y. Characteristics of mass balance of summer-accumulation type glaciers in the Himalayas and Tibetan Plateau. *Z. Fur Gletsch. Und Glazialgeol.* **1996**, *32*, 61–65.
30. Maussion, F.; Scherer, D.; Mölg, T.; Collier, E.; Curio, J.; Finkelnburg, R. Precipitation Seasonality and Variability over the Tibetan Plateau as Resolved by the High Asia Reanalysis. *J. Clim.* **2013**, *27*, 1910–1927. [CrossRef]
31. Wang, Q.; Yi, S.; Chang, L.; Sun, W. Large-Scale Seasonal Changes in Glacier Thickness across High Mountain Asia. *Geophys. Res. Lett.* **2017**, *44*, 10427–10435. [CrossRef]
32. Schutz, B.E.; Zwally, H.J.; Shuman, C.A.; Hancock, D.; DiMarzio, J.P. Overview of the ICESat Mission. *Geophys. Res. Lett.* **2005**, *32*, L21S01. [CrossRef]
33. Zhang, G.; Yao, T.; Shum, C.K.; Yi, S.; Yang, K.; Xie, H.; Feng, W.; Bolch, T.; Wang, L.; Behrangi, A.; et al. Lake volume and groundwater storage variations in Tibetan Plateau's endorheic basin. *Geophys. Res. Lett.* **2017**, *44*, 5550–5560. [CrossRef]
34. Kaldybayev, A.; Chen, Y.; Vilesov, E. Glacier change in the Karatal river basin, Zhetysu (Dzhungar) Alatau, Kazakhstan. *Ann. Glaciol.* **2016**, *57*, 11–19. [CrossRef]
35. Wang, Q.; Yi, S.; Sun, W. Precipitation-driven glacier changes in the Pamir and Hindu Kush mountains. *Geophys. Res. Lett.* **2017**, *44*, 2817–2824. [CrossRef]
36. Yao, T.D.; Thompson, L.; Yang, W.; Yu, W.S.; Gao, Y.; Guo, X.J.; Yang, X.X.; Duan, K.Q.; Zhao, H.B.; Xu, B.Q.; et al. Different glacier status with atmospheric circulations in Tibetan Plateau and surroundings. *Nat. Clim. Chang.* **2012**, *2*, 663–667. [CrossRef]
37. Lin, H.; Li, G.; Cuo, L.; Hooper, A.; Ye, Q. A decreasing glacier mass balance gradient from the edge of the Upper Tarim Basin to the Karakoram during 2000–2014. *Sci. Rep.* **2017**, *7*, 6712. [CrossRef]
38. Zhang, Z.; Liu, S.Y. Area Changes and Mass Balance of Glaciers in KangzhagRi of the Tibetan Plateau from 1970 to 2016 Derived from Remote Sensing Data. *J. Geo-Inf. Sci.* **2018**, *20*, 1338–1349. [CrossRef]
39. Wang, S.; Zhang, M.; Li, Z.; Wang, F.; Li, H.; Li, Y.; Huang, X. Glacier area variation and climate change in the Chinese Tianshan Mountains since 1960. *J. Geogr. Sci.* **2011**, *21*, 263–273. [CrossRef]
40. Nuimura, T.; Sakai, A.; Taniguchi, K.; Nagai, H.; Lamsal, D.; Tsutaki, S.; Kozawa, A.; Hoshina, Y.; Takenaka, S.; Omiya, S.; et al. The GAMDAM glacier inventory: A quality-controlled inventory of Asian glaciers. *Cryosphere* **2015**, *9*, 849–864. [CrossRef]
41. Webster, P.J.; Magaña, V.O.; Palmer, T.; Shukla, J.; Tomas, R.A.; Yanai, M.; Yasunari, T. Monsoons: Processes, predictability, and the prospects for prediction. *J. Geophys. Res. Earth Surf.* **1998**, *103*, 14451–14510. [CrossRef]
42. Mu, J. The global glacierized area: Current situation and recent change, based on the Randolph Glacier Inventory(RGI6.0) published in 2017. *J. Glaciol. Geocryol.* **2018**, *440*, 238–248. [CrossRef]
43. Ran, Y.; Wang, L.; Zeng, T.; Ge, C.; Li, H. "One belt, one road" boundary map of key basins in Asia. National Tibetan Plateau Data Center 2020. In Proceedings of the 22nd EGU General Assembly, held online, 4–8 May 2020. [CrossRef]
44. Magruder, L.A.; Webb, C.E.; Urban, T.J.; Silverberg, E.C.; Schutz, B.E. ICESat Altimetry Data Product Verification at White Sands Space Harbor. *IEEE Trans. Geosci. Remote Sens.* **2007**, *45*, 147–155. [CrossRef]
45. Markus, T.; Neumann, T.; Martino, A.; Abdalati, W.; Brunt, K.; Csatho, B.; Farrell, S.; Fricker, H.; Gardner, A.; Harding, D.; et al. The Ice, Cloud, and land Elevation Satellite-2 (ICESat-2): Science requirements, concept, and implementation. *Remote Sens. Environ.* **2017**, *190*, 260–273. [CrossRef]
46. Zhang, Y.; Pang, Y.; Cui, D.; Ma, Y.; Chen, L. Accuracy Assessment of the ICESat-2/ATL06 Product in the Qilian Mountains Based on CORS and UAV Data. *IEEE J. Sel. Top. Appl. Earth Obs. Remote Sens.* **2020**, *14*, 1558–1571. [CrossRef]
47. Li, R.; Li, H.; Hao, T.; Qiao, G.; Cui, H.; He, Y.; Hai, G.; Xie, H.; Cheng, Y.; Li, B. Assessment of ICESat-2 ice surface elevations over the Chinese Antarctic Research Expedition (CHINARE) route, East Antarctica, based on coordinated multi-sensor observations. *Cryosphere* **2021**, *15*, 3083–3099. [CrossRef]
48. Brunt, K.M.; Neumann, T.A.; Smith, B.E. Assessment of ICESat-2 Ice Sheet Surface Heights, Based on Comparisons Over the Interior of the Antarctic Ice Sheet. *Geophys. Res. Lett.* **2019**, *46*, 13072–13078. [CrossRef]
49. Wang, Q.; Yi, S.; Sun, W. The changing pattern of lake and its contribution to increased mass in the Tibetan Plateau derived from GRACE and ICESat data. *Geophys. J. Int.* **2016**, *207*, 528–541. [CrossRef]
50. Zhang, G.; Chen, W.; Xie, H. Tibetan Plateau's Lake Level and Volume Changes from NASA's ICESat/ICESat-2 and Landsat Missions. *Geophys. Res. Lett.* **2019**, *46*, 13107–13118. [CrossRef]
51. Li, C.; Jiang, L.; Liu, L.; Wang, H. Regional and Altitude-Dependent Estimate of the SRTM C/X-Band Radar Penetration Difference on High Mountain Asia Glaciers. *IEEE J. Sel. Top. Appl. Earth Obs. Remote Sens.* **2021**, *14*, 4244–4253. [CrossRef]

52. Wu, H.; Guo, Z. Characteristics of ICESat-GLAS full waveforms and the errors of GLAS footprints elevation change influenced by the surface slope and roughness. *Geogr. Geo-Inf. Sci.* **2016**, *32*, 02S03. [CrossRef]
53. Huang, T.; Jia, L.; Menenti, M.; Lu, J.; Zhou, J.; Hu, G. A New Method to Estimate Changes in Glacier Surface Elevation Based on Polynomial Fitting of Sparse ICESat—GLAS Footprints. *Sensors* **2017**, *17*, 1803. [CrossRef] [PubMed]
54. Kääb, A.; Berthier, E.; Nuth, C.; Gardelle, J.; Arnaud, Y. Contrasting patterns of early twenty-first-century glacier mass change in the Himalayas. *Nature* **2012**, *488*, 495–498. [CrossRef] [PubMed]
55. Moholdt, G.; Wouters, B.; Gardner, A. Recent mass changes of glaciers in the Russian High Arctic. *Geophys. Res. Lett.* **2012**, *39*. [CrossRef]
56. Ragettli, S.; Bolch, T.; Pellicciotti, F. Heterogeneous glacier thinning patterns over the last 40 years in Langtang Himal, Nepal. *Cryosphere* **2016**, *10*, 2075–2097. [CrossRef]
57. Romatschke, U.; Houze, R.A. Characteristics of Precipitating Convective Systems in the South Asian Monsoon. *J. Hydrometeorol.* **2011**, *12*, 3–26. [CrossRef]
58. Kääb, A.; Treichler, D.; Nuth, C.; Berthier, E. Brief Communication: Contending estimates of 2003–2008 glacier mass balance over the Pamir–Karakoram–Himalaya. *Cryosphere* **2015**, *9*, 557–564. [CrossRef]
59. Zhao, F.; Long, D.; Li, X.; Huang, Q.; Han, P. Rapid glacier mass loss in the Southeastern Tibetan Plateau since the year 2000 from satellite observations. *Remote Sens. Environ.* **2022**, *270*, 112853. [CrossRef]
60. Wu, G.; Liu, Y.; He, B.; Bao, Q.; Duan, A.; Jin, F.-F. Thermal Controls on the Asian Summer Monsoon. *Sci. Rep.* **2012**, *2*, 404. [CrossRef]
61. Kansakar, S.R.; Hannah, D.M.; Gerrard, J.; Rees, G. Spatial pattern in the precipitation regime of Nepal. *Int. J. Clim.* **2004**, *24*, 1645–1659. [CrossRef]
62. Chen, B.; Xu, X.-D.; Yang, S.; Zhang, W. On the origin and destination of atmospheric moisture and air mass over the Tibetan Plateau. *Theor. Appl. Climatol.* **2012**, *110*, 423–435. [CrossRef]
63. Feng, L.; Zhou, T. Water vapor transport for summer precipitation over the Tibetan Plateau: Multidata set analysis. *J. Geophys. Res. Earth Surf.* **2012**, *117*, D20. [CrossRef]
64. Fujita, K.; Ohta, T.; Ageta, Y. Characteristics and climatic sensitivities of runoff from a cold-type glacier on the Tibetan Plateau. *Hydrol. Process.* **2007**, *21*, 2882–2891. [CrossRef]
65. Wang, J.; Ye, B.; Cui, Y.; He, X.; Yang, G. Spatial and temporal variations of albedo on nine glaciers in western China from 2000 to 2011. *Hydrol. Process.* **2014**, *28*, 3454–3465. [CrossRef]
66. Guo, D.; Sun, J.; Yang, K.; Pepin, N.; Xu, Y.; Xu, Z.; Wang, H. Satellite data reveal southwestern Tibetan plateau cooling since 2001 due to snow-albedo feedback. *Int. J. Clim.* **2019**, *40*, 1644–1655. [CrossRef]
67. Oerlemans, J. Extracting a Climate Signal from 169 Glacier Records. *Science* **2005**, *308*, 675–677. [CrossRef]
68. Fujita, K. Effect of precipitation seasonality on climatic sensitivity of glacier mass balance. *Earth Planet. Sci. Lett.* **2008**, *276*, 14–19. [CrossRef]
69. Shi, Y.; Liu, S. Estimation on the response of glaciers in China to the global warming in the 21st century. *Chin. Sci. Bull.* **2000**, *45*, 668–672. [CrossRef]
70. Ouyang, L.; Yang, K.; Lu, H.; Chen, Y.; La, Z.; Zhou, X.; Wang, Y. Ground-Based Observations Reveal Unique Valley Precipitation Patterns in the Central Himalaya. *J. Geophys. Res. Atmos.* **2020**, *125*. [CrossRef]
71. Huang, W.; Chen, F.; Feng, S.; Chen, J.; Zhang, X. Interannual precipitation variations in the mid-latitude Asia and their association with large-scale atmospheric circulation. *Chin. Sci. Bull.* **2013**, *58*, 3962–3968. [CrossRef]
72. Bolch, T. Asian glaciers are a reliable water source. *Nature* **2017**, *545*, 161–162. [CrossRef] [PubMed]
73. Treichler, D.; Kääb, A. ICESat laser altimetry over small mountain glaciers. *Cryosphere* **2016**, *10*, 2129–2146. [CrossRef]
74. Moholdt, G.; Nuth, C.; Hagen, J.O.; Kohler, J. Recent elevation changes of Svalbard glaciers derived from ICESat laser altimetry. *Remote Sens. Environ.* **2010**, *114*, 2756–2767. [CrossRef]

remote sensing

MDPI

Article

Lake Expansion under the Groundwater Contribution in Qaidam Basin, China

Xi Zhang [1], Jiaqi Chen [2,*], Jiansheng Chen [3], Fenyan Ma [3] and Tao Wang [4]

[1] School of Earth Sciences and Engineering, Hohai University, Nanjing 211100, China; zhangxi19@hhu.edu.cn
[2] College of Computer and Information, Hohai University, Nanjing 211100, China
[3] College of Civil and Transportation Engineering, Hohai University, Nanjing 210098, China; jschen@hhu.edu.cn (J.C.); fyma@hhu.edu.cn (F.M.)
[4] College of Water Conservancy and Hydropower Engineering, Hohai University, Nanjing 210098, China; wangtao77@hhu.edu.cn
* Correspondence: jiaqichen@hhu.edu.cn

Abstract: The relationship between groundwater and lakes in Qaidam Basin is often overlooked. Therefore, we employed Landsat satellite images and meteorological data to investigate the causes of lake expansion through model calculation and statistical analysis and then determine groundwater sources through isotope analysis (^{2}H, ^{3}H, and ^{18}O). In the two study periods of 2003–2011 and 2011–present, temperature, precipitation, and runoff increased at a steady rate, whereas the expansion rate of Tuosu Lake increased from 1.22 km^2/year to 3.38 km^2/year. This significant increase in the rate of lake expansion reflects the substantial contribution of groundwater to lake expansion. The groundwater contribution to the lake includes not only the glacial meltwater that infiltrates the piedmont plain but also other, more isotopically deleted water sources from other basins. It is speculated that the 2003 M_s 6.4 earthquake in the northwest of the Delingha region was a possible mechanism for lake expansion. Earthquakes can enhance crustal permeability and keep fractures open, which promotes groundwater contribution to lakes and in turn causes rapid lake expansion and an increased groundwater level. This study is important for understanding the sources, circulation, and evolution of groundwater in Qaidam Basin.

Keywords: Qaidam Basin; lake expansion; groundwater contribution; oxygen and hydrogen isotopes; climate change

Citation: Zhang, X.; Chen, J.; Chen, J.; Ma, F.; Wang, T. Lake Expansion under the Groundwater Contribution in Qaidam Basin, China. *Remote Sens.* **2022**, *14*, 1756. https://doi.org/10.3390/rs14071756

Academic Editors: Massimo Menenti, Yaoming Ma, Li Jia and Lei Zhong

Received: 11 February 2022
Accepted: 1 April 2022
Published: 6 April 2022

1. Introduction

In recent decades, many lakes in the Tibetan Plateau have exhibited continued and rapid expansion [1–3], in contrast to a general trend of lake shrinkage in other regions and basins around the world. Qaidam Basin, located in the northeastern margin of the Tibetan Plateau, is an arid alpine region with scarce precipitation and intense evaporation. Thus, the regional hydrological cycle is sensitive to both climate change and human activity. As an important part of the hydrological cycle in arid areas, lakes are possibly affected by climate change processes such as increased precipitation and temperature [3,4]. For example, higher temperatures accelerate glacier melting and increase runoff into the lake, leading to lake expansion. Simultaneously, increased precipitation may partly contribute to lake expansion [5–7].

In addition to climate change, the impact of groundwater contribution to lakes should also be considered. The role of groundwater in lakes has previously been observed in Qinghai Lake [8] (China), Nalengele River [9] (China), Pyhajarvi Lake (Finland) [10], and other basins [11], although the sources of groundwater have not been thoroughly studied. A recent study [12] revealed that the endorheic Qiangtang Basin has a large amount of missing water (up to 540 × 10^8 m^3/year), which leaks through six rifts in the south of the basin and may subsequently upwell in surrounding areas. This observation invalidates the traditional water budget theory of watersheds and likely also affects the water budget of surrounding areas. Qaidam Basin, which is adjacent to Qiangtang Basin, is a key area for water discharge. Studies of radon (^{222}Rn) isotopes have revealed that lakes in the Qaidam Basin have an extensive groundwater contribution [9,11]. Therefore, in this study, we investigated lake surface area changes in response to climate change and groundwater contributions in the northeast of Qaidam Basin, which is important to understand the sources, circulation, and evolution patterns of regional groundwater.

The expansion or shrinkage of a lake directly reflects the lake water budget. In inland basins of alpine regions, mountain precipitation and/or glacial meltwater converge into rivers. When these rivers flow through the piedmont plain, some of the water infiltrates as groundwater, and some continues to flow downstream, flowing through alluvial plains to eventually form endorheic lakes. Endorheic lakes are generally located at the lowest elevation in the basin, where they form a confluence of surface water [9,13]. The regional distribution of groundwater heads determines whether the lake discharges to groundwater or groundwater contributes to the lake, which controls groundwater inflow and outflow in the water balance of the endorheic lake. The groundwater level in Qaidam Basin gradually decreases from the mountains to the plains; thus, groundwater flows from the mountains to the lakes in the plains, where it eventually contributes to the lakes [8,9,11]. Therefore, the water input component of these endorheic lakes predominantly includes surface runoff, lake precipitation, and groundwater inflow, and the water output component is principally lake evaporation; groundwater outflow can typically be ignored [14].

Thus, it is important to accurately quantify lake evaporation before exploring the causes of lake expansion. By comparing the eddy covariance system with several combined evaporation models, McJannet [15,16] concluded that the Penman–Monteith model was most suitable for estimating lake evaporation because it considers the effects of both vapor pressure gradient and wind speed. However, the meteorological monitoring network in alpine areas is typically sparse, with stations often located far from lakes; therefore, it is difficult to obtain long time-series of meteorological monitoring data near lakes. The improved Penman–Monteith model proposed a general application of wind functions and thus can be used to calculate lake evaporation from remote overland meteorological measurements [15,16], which has been applied in several subsequent studies. Lake surface area changes can also reflect changes in lake water storage [17–19]. However, because of the remote environment and lack of monitoring stations, lakes in the Tibetan Plateau lack long-term monitoring data related to water levels and lake surface area. Thus, satellite remote sensing technology is used in this study to observe long-term changes in lake surface area and water level [2,4,20,21]. Moreover, ^{2}H and ^{18}O isotopes in water bodies are ideal natural tracers for identifying groundwater sources and tracing hydrologic cycles [11,13]. Furthermore, tritium (^{3}H) can determine the rate of groundwater circulation and the groundwater age [20].

We investigated three lakes in this study (Tuosu Lake, Keluk lake, and Gahai Lake), all of which are located in the northeast of the extremely arid Qaidam Basin. Remote sensing techniques, model calculations, and statistical analyses were used to analyze lake surface area changes in response to climate change and groundwater contribution in the study area, and potential groundwater sources were identified using stable isotopes. The aims of this study were to (1) provide scientific support for the utilization and management of water resources in Qaidam Basin and (2) propose measures for mitigating future environmental and geological problems related to continued lake expansion, thereby protecting inhabitants and production in the study region.

2. Study Region

The study area (96°34′–97°54′E, 36°58′–37°40′N) is located in the northeast of Qaidam Basin, China, which is surrounded by the Buhete Mountains to the east, the Delingha uplift to the west, the Denan hills to the south, and the Zongwulong Mountains to the north (Figure 1). The landscape is predominantly mountainous, alluvial–proluvial plain, and alluvial lacustrine plain. The study area has a typical plateau continental climate, with annual average temperature, precipitation, and pan-evaporation values of 4.7 °C, 211 mm/year, and up to 1845 mm/year, respectively.

The main river (Bayin River) in the study area originates from the Zongwulong Mountains and has a total length of 188 km and a catchment area of 7281 km^2. The Bayin River flows east to west between Zongwulong and Buhete Mountains and then north to south after flowing through the Heishishan reservoir [11]. The Bayin River seeps underground in the middle of the alluvial–proluvial plain and then upwells as springs at the front edge of the alluvial plain. Because of the influence of the Denan hills, the Bayin River flows westward downstream across the alluvial lacustrine plain and eventually flows into Keluke Lake. Keluke Lake is connected to Tuosu Lake by the Lianshui River, forming a terminal endorheic lake. Gahai Lake is another terminal endorheic lake located in the southeast of the study area, which has a weak hydraulic connection with the Bayin River through southeast groundwater runoff.

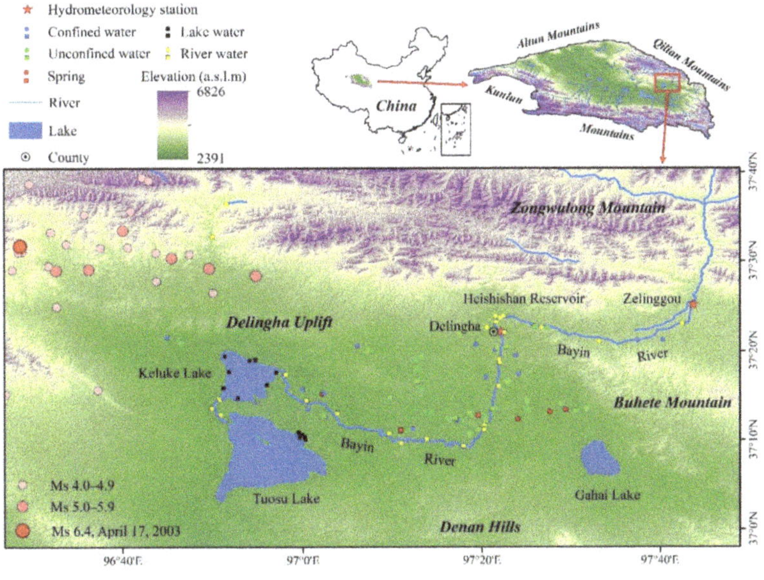

Figure 1. Spatial map representing the distribution of samples, including river, lake, spring, confined groundwater, and phreatic groundwater in the northeast of the Qaidam Basin, China. Squares represent samples from the literature [11].

Groundwater in the Delingha area mainly occurs in porous quaternary loose sediments in the plain areas. At the top and middle of the alluvial–proluvial plain, the single-layer alluvial aquifer is more than 300 m thick and mainly composed of sand and gravel; the depth of the groundwater level is 80–120 m and 10–30 m, respectively. At the end of the alluvial–proluvial plain, the aquifer system changes from a single-layer structure to a multi-layer structure, sediments are mostly fine-grained, and the groundwater level is shallow (<10 m) [13]. In the lacustrine plain downstream of Bayin River, the aquifer system consists of a multi-layer aquifer with interbedded clay and fine sand. In the study area, groundwater flows from south to north in the upper part of the alluvial–proluvial plain and then flows westward again because of the effect of the Denan hills [21], which are predominantly tertiary clastic rocks interspersed with mudstone and gypsum with limited infiltration capacity. In addition, the scarce precipitation and low precipitation intensity (a single precipitation event is less than 10 mm) hinders groundwater formation. In the western part of the alluvial–proluvial plain, groundwater flows from north to south and southeast because of the Delingha uplift, before finally flowing westward [11]. In the eastern part of the alluvial–proluvial plain, a small amount of groundwater flows southeast along the ancient river channel and converges at Gahai Lake.

Since 2000, Tuosu and Gahai Lakes have exhibited rapid expansion. The continuous expansion of Gahai Lake has caused groundwater levels to rise in the vicinity, which threatens the safety of inhabitants and their livelihoods. Therefore, the causes of lake expansion have become a key area of research for local governments. Delingha has a permanent population of 88,200 and a population density of only 2.88 people/km^2 (the seventh National Census). Because of their remote environment and minimal human impact, the lakes in the study area are suitable for studying the effects of climate change and groundwater on lake expansion.

The Bayin River has two hydrologic stations (Figure 1), the Delingha (96°16′E, 37°22′N) and Zelinggou stations (97°48′E, 37°25′N). Zelinggou station is located 7 km upstream of Delingha station, near the exit of the mountain. The runoff measured at this station is mainly glacial meltwater from the mountainous area; hence, Zelinggou station runoff represents the amount of glacier melt in the lakes. The catchment between Delingha and Zelinggou station is defined as the upstream area. The runoff measured at Delingha station includes glacial meltwater and runoff generated by precipitation in the upstream area. The monitoring period for Zelinggou station was 1959–1984, as the station was withdrawn after 1984. Data from these hydrological stations were used for subsequent statistical analysis of runoff trends.

3. Methods

3.1. Landsat Data and Lake Surface Area Extraction

Long-term changes in the surface area of Tuosu, Keluke, and Gahai Lakes were extracted from remote sensing images. Since 1984, Landsat satellites have acquired high-resolution Earth observation images, which are widely used for feature identification. In this study, the Landsat 5 Thematic Mapper (TM) and Landsat 8 Operational Land Imager (OLI), which are the longest time-series data currently available, provided observation data for different periods. The Universal Transverse Mercator (UTM) and World Geodetic System 1984 were used as geocoordinate references to construct the temporal and spatial sequence of lake changes. The spatial resolution of the extracted lake surface area was 30 m, which is sufficient for studying lakes measuring several tens to hundreds of square kilometers. Although the spatial resolution is moderate, its effect on the results is very limited [22–25]. All Landsat data were downloaded from the United States Geological Survey (http://glovis.usgs.gov/, accessed on 10 January 2021), Geospatial Data Cloud (https://geocloud.cgs.gov.cn/, accessed on 10 January 2021), and Chinese Academy of Sciences (http://www.gscloud.cn/, accessed on 10 January 2021). The necessary image pre-processing steps, such as radiation calibration and atmospheric correction, were performed using ENVI 5.3 software.

The repeat coverage of Landsat 5 and Landsat 8 is 18 and 16 days, respectively, providing 1–2 images per month. Considering the influence of seasonal variations, monthly variations of the Tuosu Lake surface area in 2003 and 2020 were also extracted (Figure 2), and it showed that the seasonal variation had only a slight disturbance to the annual trend of lake surface area (see Section 4.1 for details). To observe the lake surface area in more detail, this study selected 20 remote sensing images of the Delingha region from 2000 to 2020 and extracted changes in the surface areas in Tuosu, Keluke, and Gahai Lakes. As optical images are significantly affected by weather conditions, particularly clouds, satellite images were typically only selected from July to August. This is because the summer in Qaidam Basin is dry and sunny with minimal clouds. If no images were available, we substituted images from adjacent months. The data sources used in the study are listed in Table 1. It should be noted that there was a lack of images available in 2012.

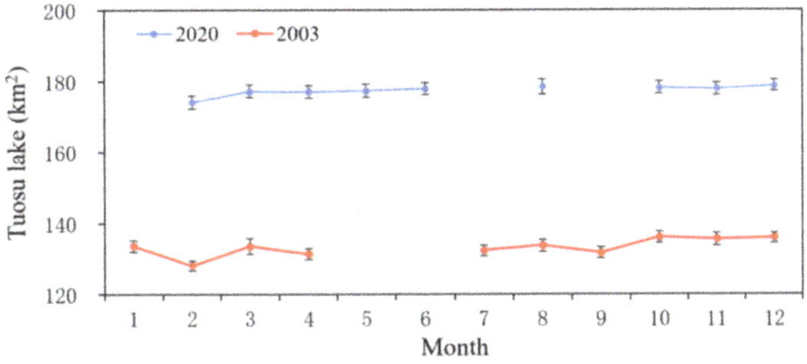

Figure 2. Changes in the surface area of Tuosu Lake in 2003 and 2020. Black error bar represents the uncertainty of the extracted lake surface area.

Table 1. Sources of remote sensing data used in this study.

Index	Time	Data Source	Index	Time	Data Source
1	26 May 2000	Landsat 5(TM)	11	2010/7/27	Landsat 5(TM)
2	2 July 2001	Landsat 5(TM)	12	2011/7/14	Landsat 5(TM)
3	5 July 2002	Landsat 5(TM)	13	2013/6/1	Landsat 8(OLI)
4	10 September 2003	Landsat 5(TM)	14	2014/8/23	Landsat 8(OLI)
5	27 August 2004	Landsat 5(TM)	15	2015/8/10	Landsat 8(OLI)
6	13 July 2005	Landsat 5(TM)	16	2016/9/29	Landsat 8(OLI)
7	1 August 2006	Landsat 5(TM)	17	2017/7/14	Landsat 8(OLI)
8	21 September 2007	Landsat 5(TM)	18	2018/10/5	Landsat 8(OLI)
9	5 July 2008	Landsat 5(TM)	19	2019/5/1	Landsat 8(OLI)
10	9 August 2009	Landsat 5(TM)	20	2020/10/10	Landsat 8(OLI)

Uncertainty in the extracted lake surface area mainly originates from the positioning accuracy and indistinguishable mixed pixels in the image [26]. According to previous studies, the registration error is 6 m for TM images [27] and 5 m for OLI images [28]. Considering the registration error, the positioning accuracy in the image, and the clarity of the lake boundary, the uncertainty in the lake surface area was estimated by Equation (1) [29] and is shown as error bars in Figure 3a. Here, E_A is the uncertainty in the extracted lake surface area; l is the length of the lake boundary; LRE_{year} is the resolution error of Landsat images in different years, which should be half the resolution of the image pixel; and E_{co} is the registration error in the image:

$$E_A = l \times \sqrt{LRE_{year}{}^2 + E_{co}{}^2} \tag{1}$$

Because of a lack of detailed bathymetric maps of the lakes, the change in lake storage was estimated based on the lake surface area and the slope of the lakeshore zone. The change in water storage between the two stages can be approximated as a frustum, and its volume can be estimated by the following equation [30]:

$$\Delta V = \frac{1}{3} \times \left(S_1 + S_2 + \sqrt{S_1 \times S_2}\right) \times H \tag{2}$$

where S_1 and S_2 are the lake surface area in the two stages; ΔV is the change in lake storage; and H is the lake water level interval. The lakes in the study area are located at the end of the alluvial lacustrine plain, with gentle topography and a lakeshore slope of less than 5%, where the slope $I = H/L$, and L is the horizontal distance. For H, the shapes of Tuosu,

Keluk, and Gahai Lakes are approximately equilateral triangles, whose side lengths can be calculated according to the lake surface area in different periods:

$$H = \frac{2}{3} \times I \times \left(\sqrt{S_2} - \sqrt{S_1}\right) \tag{3}$$

here, slope values of 4, 3, and 2 were used to calculate the change in lake water storage, and the calculation results are shown in Figure 3b.

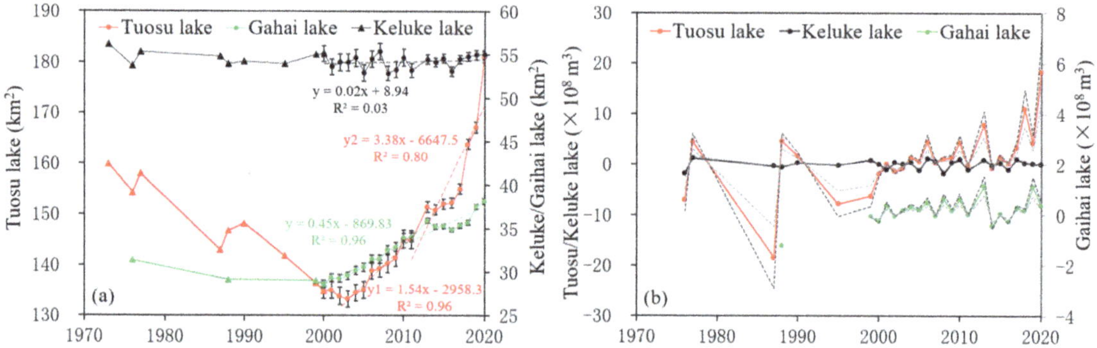

Figure 3. (**a**) Plot depicting the change of lake surface area in the study region extracted from Landsat satellite images. Triangles (before 2000) represent data in the literature [31,32]. y, y1, and y2 represent the least square fitting line of the surface area in Gahai Lake after 2000, Tuosu Lake from 2003 to 2011, and after 2011, respectively. Black error bar represents the uncertainty of the extracted lake surface area. (**b**) Calculated change in lake storage (ΔV). The solid line, dashed lines in dark gray and light gray are the calculation results when the slope (I) is 3, 2, and 4, respectively.

3.2. Hydrological and Meteorological Monitoring Data

The meteorological monitoring network in the alpine region is sparse; however, the single weather station is generally considered to represent the climatic conditions of the basin. The distance between Delingha station (97°22′E, 37°22′N) and Tuosu, Keluke, and Gahai Lakes is only 42, 41, and 27 km, respectively. The terrain between the weather station and the lakes is flat, with no obstruction from mountains, and the climate conditions exhibit minimal spatial variability. Therefore, data from Delingha station were used to represent the climatic conditions near the lakes. Meteorological data at Delingha station were downloaded from the National Meteorological Data Center (http://data.cma.cn/, accessed on 9 August 2021); this included daily data of relative humidity, minimum relative humidity, mean temperature, maximum temperature, minimum temperature, mean wind speed, minimum wind speed, and annual precipitation. Delingha station lacks solar radiation data; therefore, these data were replaced by monitoring data from the nearest station in Golmud (94°54′E, 36°25′N).

Runoff data from Delingha station (1957–2018) and Zelinggou station (1957–1983) were compiled from the *Hydrological Yearbook*. Lake water temperature data for Tuosu, Keluke, and Gahai Lakes were obtained from a dataset of daily lake surface temperature over the Tibetan Plateau (1978–2017) compiled by the National Tibetan Plateau Data Center (http://data.tpdc.ac.cn, accessed on 11 August 2021) [33], which uses the improved lake water temperature model (air2water) to simulate the annual surface temperature every day. Considering the different available periods of sequences, the calculation interval for analyzing lake evaporation was 1984–2016 in this study.

3.3. Penman–Monteith Model

Accurate quantification of lake evaporation is essential for determining the water budget of a lake, especially for endorheic lakes, where evaporation is the most important output term of the lake water budget [34–36]. Based on the energy conservation formula, Penman [37] first proposed the Penman formula for calculating evapotranspiration, from which many water surface evaporation models have been developed [38,39]. After comparing 14 evaporation models with baseline Bowen ratio energy budget measurements, Rosenberry [40] identified that the De BruinKeijman, Priestley Taylor, and Penman models provided the best estimates of water surface evaporation. McJannet then compared eddy covariance measurements with the De BruinKeijman, Priestley Taylor, and Penman–Monteith models, [15,16] and found that the Penman–Monteith model was most suitable for water surface evaporation because it considers both the vapor pressure gradient and wind speed, and produced estimates of total evaporation that varied from the actual measurements by less than 1%. In addition, the Penman–Monteith model has been improved by proposing the general application of wind functions, making it applicable for calculating evaporation for water bodies ranging from tens to hundreds of kilometers. This improved Penman–Monteith model has been applied in several subsequent studies [17–19].

In this study, we used the improved Penman–Monteith model to calculate lake surface evaporation (see McJannet [15,16] for details). Uncertainty in the evaporation value mainly derived from data measurement and parameter calculation. The measurement error comes from the measurement of wind speed, temperature, humidity, and accumulated solar radiation. These errors are inevitable but have little impact on the final calculation results.

3.4. Sampling and Isotope Measurements

Hydrogen (^{2}H) and oxygen (^{18}O) isotopes are widely used to study hydrological cycles [41,42] and qualitatively identify water sources and trace groundwater runoff processes [11,13]. In this study, 47 water samples were collected from rivers, phreatic groundwater, confined groundwater, springs, and lakes (Figure 1). River water samples (R07, R09, and R12) were collected from the upper reaches of Bayin River, close to where the river exits the mountains, mainly from glacial meltwater; thus, they are minimally influenced by precipitation and groundwater. These three samples represent the hydrogen (^{2}H) and oxygen (^{18}O) isotope characteristics of glacial meltwater. In this study, only one lake water sample was collected from Keluke Lake, which was analyzed together with 12 lake water samples collected from Yang Lake [11], which are discussed in Section 4.4.

Stable isotopes of hydrogen and oxygen (^{2}H, ^{3}H, ^{18}O) were measured at the State Key Laboratory of Hydrology, Water Resources, and Hydraulic Engineering, Hohai University. ^{18}O/^{16}O and ^{2}H/^{1}H ratios were measured using a MAT253 mass spectrometer. The isotope ratio 'δ' was expressed as follows:

$$\delta_{sample}(‰) = \left(R_{sample}/R_{standard} - 1\right) \times 1000 \quad (4)$$

where R_{sample} and $R_{standard}$ are the isotope ratios (^{18}O/^{16}O, ^{2}H/^{1}H) of the sample and standard, respectively, and the international standard is the δ^2H and δ^{18}O of Vienna mean seawater. The measurement errors of δ^{18}O and δ^2H were $\pm$0.1‰ and $\pm$1‰, respectively. The tritium content of samples was measured by a liquid scintillation meter (TRI-CARB 3170 TR/SL) with a detection limit of 0.2 TU and precision of >0.8 TU. The measurement results for hydrogen (^{2}H, ^{3}H) and oxygen (^{18}O) isotopes are shown in Table 2.

Table 2. Isotope measurement results (^2H, ^{18}O, and tritium) for different water samples in the study area.

Sam. No.	Type	Latitude (N)	Longitude (E)	Depth (m)	δ^2H (‰)	$\delta^{18}O$ (‰)	^3H (TU)
R01	RW	37°23′15″	97°21′39″		−58.6	−8.81	14.0
R02	RW	37°23′43″	97°22′04″		−58.4	−8.76	
R03	RW	37°23′55″	97°21′35″		−52.1	−7.34	
R07	RW	37°22′38″	97°26′40″		−57.1	−6.60	
R09	RW	37°21′06″	97°33′12″		−56.9	−8.82	17.8
R10	RW	37°21′11″	97°37′31″		−58.4	−8.97	11.6
R12	RW	37°23′03″	97°42′34″		−56.7	−8.63	16.7
R22	RW	37°19′06″	97°21′59″		−57.4	−8.77	
R25	RW	37°11′44″	97°20′21″		−56.3	−8.80	
R31	RW	37°09′21″	97°17′56″		−56.1	−8.39	
R33	RW	37°9′39″	97°10′58″		−56.3	−8.49	
R36	RW	37°15′07″	97°02′14″		−50.8	−7.20	
R39	RW	37°32′47″	96°50′03″		−59.6	−9.23	
R40	RW	37°32′47″	96°50′03″		−59.8	−9.26	
R41	RW	37°41′14″	96°31′03″		−53.6	−7.59	
R43	RW	37°36′31″	96°51′35″		−59.8	−9.44	
R47	RW	37°22′38″	97°20′37″		−60.1	−9.34	
L37	LW	37°18′56″	96°54′07″		−46.1	−6.42	
L42	LW	37°43′32″	96°28′43″		−22.1	−1.05	
G17	SW	37°12′20″	97°24′02″		−58.3	−8.82	11.2
G18	SW	37°13′10″	97°27′36″		−60.0	−9.16	11.9
G20	SW	37°13′24″	97°29′24″		−62.4	−9.38	5.0
G35	SW	37°15′08″	97°02′12″		−66.6	−10.08	3.7
G26	PGW	37°12′07″	97°17′54″	0.26	−58.2	−8.89	9.0
G19	PGW	37°13′07″	97°27′35″	0.6	−57.8	−8.68	
G27	PGW	37°12′18″	97°16′14″	1.2	−60.0	−9.08	
G24	PGW	37°12′38″	97°20′20″	1.4	−62.5	−9.36	
G16	PGW	37°13′24″	97°25′08″	1.9	−60.5	−8.82	9.1
G08	PGW	37°22′33″	97°27′16″	2.0	−42.0		
G28	PGW	37°13′05″	97°14′11″	3.5	−58.7	−9.19	
G05	PGW	37°22′40″	97°26′00″	3.8	−62.2	−9.39	15.1
G45	PGW	37°20′59″	96°46′32″	26.4	−53.6	−8.31	
G44	PGW	37°21′00″	96°46′25″	27.6	−50.8	−8.26	11.8
G29	PGW	37°15′46″	97°13′12″	12.0	−64.8	−9.92	5.5
G30	PGW	37°17′55″	97°12′47″	14.9	−62.5	−9.56	
G34	PGW	37°19′05″	97°12′36″	18.2	−58.8	−8.88	9.6
G06	PGW	37°22′44″	97°25′54″	20.0	−65.4	−9.28	
G23	CGW	37°17′02″	97°20′28″	31.8	−58.9	−8.84	10.9
G15	CGW	37°17′23″	97°24′07″	40.4	−60.3	−9.08	
G11	CGW	37°21′14″	97°40′20″	50.0	−71.1	−10.60	1.3
G14	CGW	37°19′31″	97°23′49″	63.3	−58.2	−8.86	
G04	CGW	37°22′45″	97°25′51″	65.0	−60.4	−9.08	15.6

Note: PGW, CGW, RW, SW, and LW represent phreatic water, confined water, river water, spring water, and lake water, respectively.

4. Results

4.1. Long-Term Changes in Lake Surface Area

Lakes generally have seasonal variations, which possibly affect the interannual trend analysis of change in the lake surface area. Considering this effect of seasonal variations, the monthly variations of the Tuosu Lake surface area in 2003 and 2020 were extracted (Figure 2). The results show that the lake surface areas of Tuosu Lake in 2003 and 2020 were 128.2–136.1 km^2 and 174.0–178.7 km^2 with uncertainties within 1.6% and 1.2%, respectively. This indicates that the extracted lake surface areas are reliable. The maximum differences of lake surface area caused by seasonal variation are 7.9 km^2 and 4.7 km^2 in 2003 and 2020, respectively, and the relative deviation is within 4% and 2%, respectively. Therefore, the seasonal variation has only a slight effect on the interannual trend of the lake surface area.

Changes in the surface areas of Tuosu, Keluke, and Gahai Lakes from 2000 to 2020, as well as changes in lake water storage calculated from the lake surface area and slope (I = 2, 3, and 4), are shown in Figure 3a,b, respectively. The average surface areas of Keluke, Tuosu, and Gahai Lakes were 54.2, 146.5, and 33.1 km^2, with uncertainties of within 1.3%, 1.7%, and 1.6%, respectively. Keluke Lake remained stable because it is connected to Tuosu Lake through the Lianshui River. When the water level of Keluke Lake rises, excess water flows into Tuosu Lake through the river. Therefore, changes in lake surface area mainly occurred in Tuosu Lake, with three distinct trends: continuous shrinkage (before 2003), slow expansion (2003–2011), and rapid expansion (after 2011). Before 2003, Tuosu Lake shrank at a rate of 2.50 km^2/year and then gradually increased at a rate of 1.54 km^2/year from 2003 to 2011, with an average increase in lake water storage of 1.45 $\times$ 10^8 m^3/year (Figure 3b). After 2011, Tuosu Lake expanded rapidly at a rate of up to 3.38 km^2/year, which was much higher than the overall increase of 2.19 km^2/year (R^2 = 0.88) after 2003. At this time, the average increase in lake water storage was as high as 5.75 $\times$ 10^8 m^3/year (I = 3). Gahai Lake shrank gradually prior to 2000 and then expanded steadily after 2000 at a rate of 0.45 km^2/year (R^2 = 0.96), with an average increase in lake water storage of 0.29 $\times$ 10^8 m^3/year. Although detailed lakeshore slopes were not obtained in the calculation of lake water storage changes, the gentle lakeshore zone allowed changes in water storage to be largely reflected in the lake surface area rather than the lake height, which supports our subsequent analysis of the water budget based on surface area changes.

Similar lake expansion occurred at the edge of the Tibetan Plateau. Taitema Lake in the north of the Altun Mountains reappeared in 2003 after prolonged drying over many years, and then expanded rapidly [43]. Similarly, Qinghai Lake in the northeast margin of the Tibetan Plateau expanded rapidly at a rate of 8.67 km^2/year after 2003 (https://hydroweb.theia-land.fr/hydroweb, accessed on 27 November 2021), with a simultaneous increase in groundwater level in the Hexi corridor in the north of the Qilian Mountains [44], which may imply similar lake response patterns.

4.2. Lake Evaporation Calculated by the Improved Penman–Monteith Model

Lake expansion is a direct reflection of the water budget of a lake. Regional groundwater head distribution determines groundwater inflow and groundwater outflow in the water balance of the endorheic lake. The groundwater level in the study area gradually decreases from the mountains to the plains, and the groundwater flows from the mountains to the lakes and eventually contributes to the lakes. For endorheic lakes, the input components of the lake water budget mainly include precipitation, runoff, and groundwater inflow, while the output component is mainly lake evaporation, and groundwater outflow can be neglected. In this study, lake evaporation was quantified by using the improved Penman–Monteith model and used to analyze the influence of the major output component (lake evaporation) on lake expansion in the study area. Based on the monitoring data of Delingha station, evaporation values for Tuosu, Keluke, and Gahai Lakes were 1233–1476 mm/year, 1164–1379 mm/year, and 1407–1700 mm/year, with average values of 1342, 1274, and 1542 mm/year, respectively (Figure 4). These results show that evaporation varied significantly between lakes with different lake surface areas and depths. All three lakes exhibited stable interannual evaporation values, with insignificant variation trends and relative deviations of 9.9%, 9.6%, and 10.2%, respectively. Therefore, evaporation did not cause significant changes in lake surface area and cannot explain the rapid expansion of the lakes since 2003. This suggests that climate change processes such as increased temperature and precipitation did not significantly affect lake evaporation and were not the main factors affecting the lake water budget during the study period.

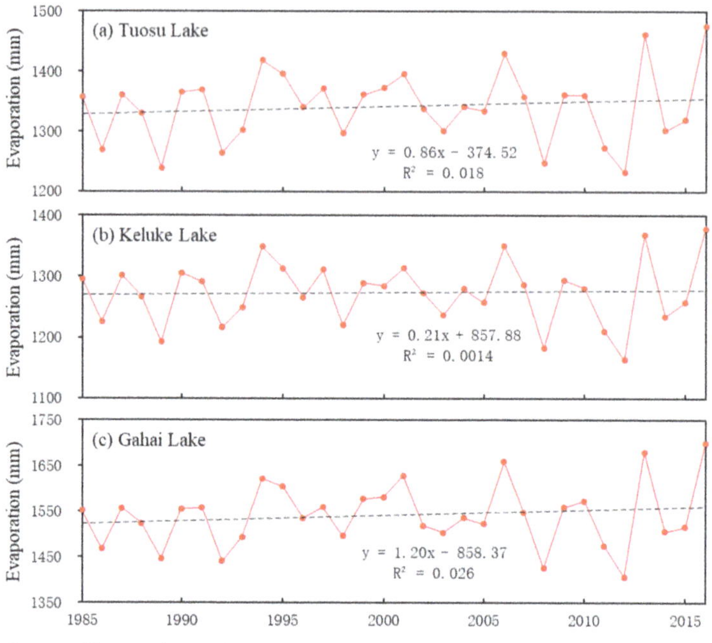

Figure 4. Plots exhibiting the evaporation of Tuosu (**a**), Keluke (**b**), and Gahai Lakes (**c**) calculated by the improved Penman–Monteith model.

4.3. Annual Hydrometeorological Trends

Considering the lack of significant changes in the main output component of the lake water budget (lake evaporation), rapid lake expansion may instead have been caused by the input components, which include precipitation, runoff, and groundwater inflow. Rainfall and runoff data were derived from long-term monitoring at Delingha meteorological and hydrological stations. Because of the complexity and hidden nature of the groundwater runoff process, the groundwater contribution to lakes is difficult to directly quantify and observe long-term. Therefore, we first analyzed the long-term variation trends of rainfall and runoff.

The Mann–Kendall test was used to inspect the long-term trends of hydrometeorological data and identify abrupt changes in the time-series data. An abrupt change in the precipitation time-series occurred in 1989 ($\alpha = 0.05$), with average precipitation before and after this change equal to 156 mm/year and 215 mm/year, respectively, representing an increase of more than 37.7%. However, this abrupt change in precipitation occurred much earlier than the beginning of lake expansion in 2003. This indicates that, although precipitation increased annually, it was not the dominant factor influencing lake expansion. Additionally, precipitation in the plain area is only 50 mm/year; thus, its contribution to the lake water budget can be neglected (Comprehensive Investigation Committee of Chinese Academy of Sciences, 1984).

Conversely, there was no obvious abrupt change in the runoff measured at Delingha station, although there was a significant difference in average runoff before and after 2002 (3.19×10^8 m^3/year and 4.55×10^8 m^3/year, respectively), representing a difference of 42.6% (Figure 5b). Runoff from Zelinggou station represents glacier meltwater, whereas Delingha station is located at the boundary between the upper and middle reaches of the Bayin River; thus, runoff is derived from both glacier meltwater and precipitation in mountainous areas. A comparison of the runoff values between the two stations during 1957–1983 revealed strong similarity, with a correlation coefficient of up to 0.88 (Figure 6). Runoff at Zelinggou station accounts for 90% of that at Delingha station, which indicates that glacial meltwater is the main source of the Bayin River. The rate of glacial melt is controlled by the average air temperature, which changed significantly ($\alpha = 0.05$) in 1997, from 4.1 °C to 5.1 °C, representing an increase of 23.3% (Figure 5d). This increase in air temperature likely accelerated glacier meltwater.

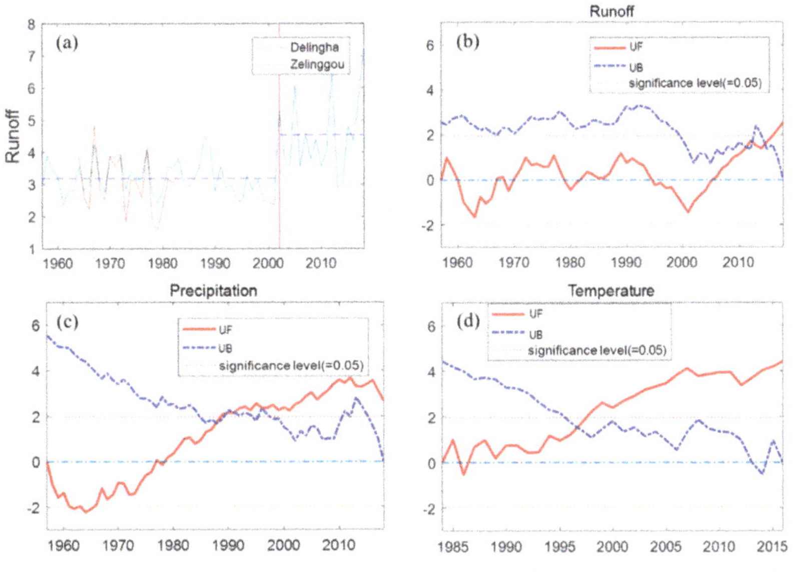

Figure 5. Plots depicting the change of runoff at Delingha and Zelinggou stations (**a**), results of MK test for runoff (**b**), precipitation (**c**), and temperature (**d**) at Delingha station.

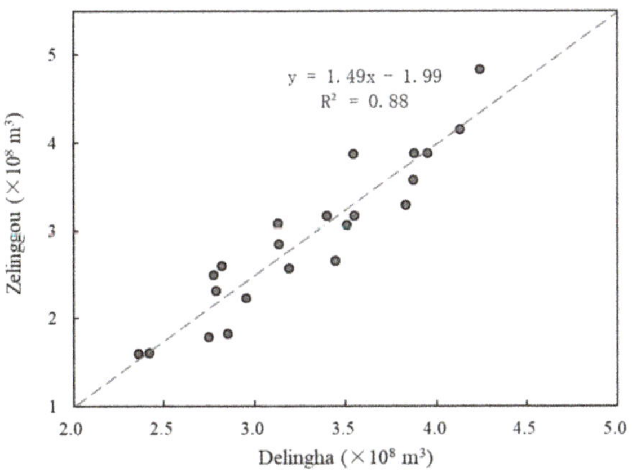

Figure 6. Graph representing the comparison between runoff at Delingha and Zelinggou stations from 1957 to 1983.

4.4. Isotopic Characteristics of Surface Water, Groundwater, and Spring Water

To trace groundwater processes and sources, the stable isotopes (^2H, ^{18}O) of collected water samples were analyzed (Figure 7 and Table 2). Samples R12, R09, and R07 represent glacial meltwater, as discussed earlier. The δ^2H and $\delta^{18}O$ of glacial meltwater, phreatic groundwater, and river water lie in the range of −57.1 to −56.7‰ and −8.8 to −6.6‰, −65.4 to −42.0‰ and −9.9 to −8.3‰, and −60.1 to −50.8‰ and −9.4 to −6.6‰, respectively. The points of river water were distributed along with the least square fitting line in Figure 7, i.e., $\delta^2H = 3.43\ \delta^{18}O - 27.63$ ($R^2 = 0.77$), showing the evaporation characteristics of river water. Most of the river points fell within the range of phreatic groundwater. This implies a strong interaction between river water and phreatic groundwater and that the river was recharged by both phreatic groundwater and glacial meltwater. The fitting line of lake water was $\delta^2H = 5.1883\ \delta^{18}O - 12.599$ ($R^2 = 0.81$), and its intersection (−54.7‰, −8.1‰) with the global meteoric water line (GWML) fell within the range of river water, suggesting that phreatic groundwater contributed to the lake after flowing into the river.

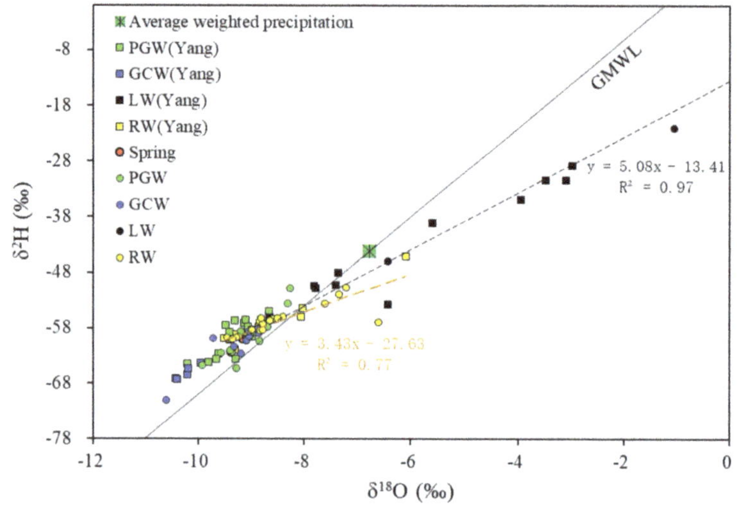

Figure 7. Plot illustrating $\delta^{18}O$ vs. δ^2H diagram of groundwater, lake water, river water, and spring water in the study area of the Qaidam Basin, China. Squares represent samples from the literature [11].

Zhang [45] also collected precipitation samples from Delingha from September 1991 to December 1992 and derived a weighted average of δ^2H and $\delta^{18}O$ in precipitation of −44.2‰ and −6.8‰, respectively. The δ^2H and $\delta^{18}O$ values of confined groundwater ranged from −71.1‰ to −58.2‰ and from −10.6‰ to −8.8‰, respectively. Thus, the confined groundwater was more depleted in deuterium and oxygen-18 than phreatic groundwater and local meteoric precipitation. The concentrations of tritium in river water, spring water, and confined water were 11.6–17.8 TU, 3.7–11.9 TU, and 1.3–15.6 TU, respectively.

5. Discussion

5.1. Potential Causes of Lake Expansion

Long-term changes in the surface areas of Tuosu, Keluke, and Gahai Lakes during the study period were the result of a combination of climate change and groundwater contributions, with climate change processes mainly including increased precipitation and temperature. Expansion of Tuosu Lake was particularly obvious; hence, we used this lake as an example to analyze the reasons for lake expansion. Long-term lake evaporation was stable at 1342 mm/year, indicating that increased temperatures affect the lake surface area by accelerating glacier melting rather than promoting lake evaporation. According to long-term meteorological data, lake evaporation and runoff remained stable prior to 2003, whereas precipitation slowly increased (2.279 mm/year). However, the surface area of Tuosu Lake decreased from 159.9 km^2 in 1973 to 132.3 km^2 in 2003; thus, lake shrinkage was likely caused by the decrease in the groundwater contribution to the lake.

From 2003 to 2011, temperature and precipitation increased by 18.6% and 34.1%, respectively (Table 3), which caused an increase in runoff of 1.21×10^8 m^3/year and 0.15×10^8 m^3/year at Delingha station, respectively. Temperature increases during this period mainly promoted lake expansion by accelerating glacial melting, which resulted in more surface runoff into the lake. This reflects the contribution of glacial meltwater to lake expansion. The rates of temperature, precipitation, and runoff increase were similar between the two periods of 2003–2011 and 2011–present; however, the expansion rate of Tuosu Lake increased rapidly between these periods, from 1.22 km^2/year to 3.38 km^2/year. Thus, the increase in runoff caused by accelerated glacial melting cannot fully explain the observed lake expansion, which implies that groundwater was an important reason for rapid lake expansion.

Table 3. Variation of factors affecting the Tuosu Lake surface area before 2003 and after 2003.

	Time	Runoff	Precipitation	Temperature	Evaporation	Lake Surface Area
Average	Before 2003	3.2	171	4.3	1341.8	148.6
	After 2003	4.5	229	5.1	1343.0	146.5
Variation	Whole time	1.3	58	0.8	1.1	−2.1
Variation ratio	Whole time	40.6%	34.1%	18.6%	0.09%	−1.4%
Variation rate	Before 2003	−0.002	2.279	0.068	3.087	−0.867
	After 2003	0.092	−0.777	0.034	3.080	2.190
	Whole time	0.024	2.083	0.051	0.858	0.131

Note: units of runoff, precipitation, evaporation, and lake surface area are 10^8 m^3/year, mm/year, mm/year, and km^2, respectively.

When glacial meltwater flows through the piedmont plain, some infiltrates the ground as subsurface runoff, which is an important contributor of water to the lake. However, not all of the groundwater contribution to the lakes is derived from glacial meltwater, as groundwater collected near the lakes showed greater isotopic depletion than glacial meltwater in the basin. This indicates the existence of other water sources with more depleted isotopes, which may be related to the mechanism of rapid lake expansion. Several earlier studies have revealed the important role of groundwater in lake expansion and shrinkage [8–11]. For example, the groundwater contribution flux estimated by the radon isotope (^{222}Rn) is $0.55 - 2.49 \times 10^{-4}$ $m^3/(s \times m)$; however, this flux was only measured at a certain time [11]. In future studies, long-term observations of groundwater contributions are required to determine and predict the effects of groundwater on lakes.

5.2. Sources of Groundwater Contribution to Lakes

Deuterium and oxygen-18 isotopes revealed that confined groundwater is characterized by significant isotopic depletion. The origin of depleted confined groundwater in alpine arid basins is controversial. According to ^{14}C-dating of groundwater, it is generally believed that glacial meltwater generated a large amount of recharge after the last glacial period [46,47], or some studies have suggested that meteoric precipitation during glacial and interglacial periods recharged the confined groundwater [48–50]. ^{14}C-dating of groundwater age strictly requires that endogenous CO_2 from no other sources is dissolved in the groundwater system, which can dilute the ^{14}C concentration in the groundwater, resulting in overestimation of the significant groundwater age. However, Qaidam Basin is an active geological environment containing multiple crisscrossing fractures. Thus, mantle-derived endogenous CO_2 with low ^{14}C activity can migrate upward through active structures such as fault zones and dissolve into the groundwater, leading to significant overestimation of groundwater age [51,52] by tens of thousands of years. As such, tritium was used to identify the groundwater renewal cycle in this study. The half-life of tritium is only 12.3 a, and the background value of tritium in natural groundwater systems is generally less than 1 TU. After 1952, global nuclear explosion tests caused a peak in the atmospheric tritium concentration. Therefore, groundwater systems with tritium values >5 TU are considered to have a groundwater renewal cycle of several decades. The confined groundwater samples in this study exhibited high tritium concentrations of between 8.5 TU and 15.6 TU, which demonstrates that confined groundwater is rapidly circulated and recharged by modern water since the global nuclear explosion tests. Therefore, confined groundwater in the study area may be recharged by water sources with more depleted isotopic signatures from other areas.

Previous research [12] has revealed an enormous amount of missing water in Qiangtang Basin, with a leakage water volume of up to 540 × 10^8 m^3/year, which is related to tectonic activity such as earthquakes. Leakage occurs in six major rift valleys in the southern part of Qiangtang Basin and is transported to other basins by underground runoff. Although the drainage area was not identified in previous literature, this leakage was likely discharged to surrounding areas at lower elevations, causing groundwater levels to rise and lakes to expand. Delingha, Qinghai Lake, Taklamakan Desert, and Hexi Corridor, located on the northern edge of the Tibetan Plateau, have a relatively low altitude and an active geological environment with frequent earthquakes, providing suitable conditions for the remote discharge of groundwater. Moreover, earthquakes are known to increase groundwater discharge [53–55]. In 2003, groundwater release induced by earthquakes was observed in Qinghai Lake (https://hydroweb.theia-land.fr/hydroweb, accessed on 27 November 2021), Taklimakan Desert [43], and the Hexi corridor [44], resulting in the emergence of new lakes, the expansion of existing lakes, and an increase in groundwater levels. An earthquake with a magnitude of 6.1 can affect areas as far as 80 km from the earthquake source [56]. The 2003 M_s 6.4 earthquake that occurred in the northwestern part of the study area (Figure 1) was only 57 km away from Tuosu Lake; thus, it very likely led to an increase in groundwater contribution to the lake. The endorheic lakes in the study area represent places of convergence for surface water and groundwater, and stable isotope analysis showed that groundwater is an important contributor to the lakes. Outflow from the lake to groundwater is limited, which in turn supports that lake evaporation is the main output of lake water balance. The distinctly depleted H and O isotopes in the confined groundwater indicate remote discharge from a high-altitude water source with a more depleted isotopic signature. Therefore, it is speculated that either the 2003 M_s 6.4 earthquake in the northwest of Delingha or the 2001 M_s 8.1 earthquake in the Kunlun Mountains were possible mechanisms for expansion of the lakes in the study area. Earthquakes enhance crustal permeability and keep fractures open [43,44], which promotes the groundwater contribution to lakes and in turn causes rapid lake expansion.

5.3. Uncertainty in Lake Evaporation Calculations

Uncertainty in the calculated evaporation values is derived from the lack of solar radiation monitoring data at the Delingha station, which was replaced by data from the nearby Golmud station in this study. Because the intensity of solar radiation is mainly related to latitude, and the latitude difference between Delingha and Golmud stations is only approximately 1°, the uncertainty caused by solar radiation data can be ignored. Uncertainty also originates from the lake surface temperature simulated by the improved lake water temperature model (air2water), with a deviation of ±0.55 °C, which also has a limited impact on the calculation results. Combined with remote sensing and meteorological data, evaporation from Tuosu Lake is 1333 mm/year [57], which only differs by 0.7% from the value calculated in this study. Earlier studies have reported substantial variation in evaporation between lakes, even if the lake surface areas are similar. For example, Laguo and Yang Lakes have only a 1.3% difference in surface area but a 24.4% difference in lake evaporation [57]. The improved Penman–Monteith model considers the effect of lake surface area and depth on evaporation, leading to more accurate calculation results.

5.4. Mitigating the Environmental Effects of Lake Expansion

Lake expansion likely affects the groundwater runoff process, leading to geological and environmental problems. For example, Gahai Lake is located at the southeastern edge of the alluvial fan and has no inflow from surface runoff. Water from the Bayin River seeps into groundwater in the middle reaches, with weak groundwater runoff in the southeast direction being one of the sources of Gahai Lake. The Gahai irrigation area (D) is located in this groundwater flow path (Figure 8). The geological formation in front of the alluvial fan is mainly composed of fine-grained sediment, which has a strong water-blocking effect. The water level of Gahai Lake has been rising continuously since 2000, which slows down the groundwater discharge rate and enhances groundwater level rises in the surrounding area. After 2006, the groundwater level increased at a rate of 0.5 m/year. In 2012, the groundwater level came close to the surface and overflowed, rising as high as 11 m in some areas [58]. This directly led to problems such as foundation collapses, and soil salinization occurred in the vicinity of Gahai Lake, which threatened the lives and livelihoods of nearby residents. Therefore, measures should be taken to mitigate the effects of lake expansion in the study area.

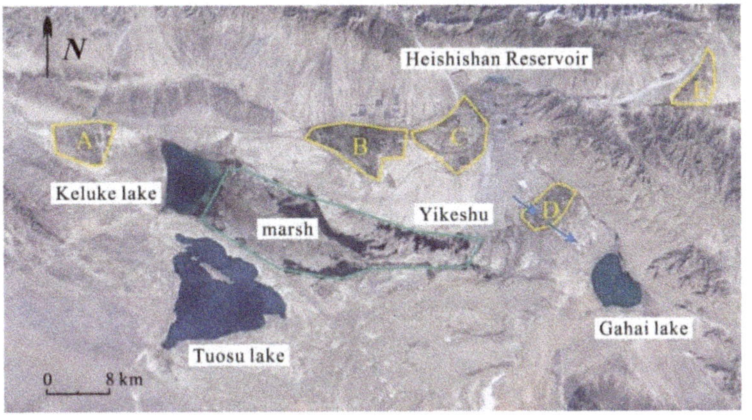

Figure 8. The distribution of irrigation area in Delingha. A, B, C, D, F represents Huaitoutala, Gebi, Delingha, Gahai, and Zelinggou irrigation area, respectively. The blue arrow represents the direction of weak groundwater runoff after the Bayan River leakage.

Furthermore, when the water level of Keluke Lake rises, the water can be discharged to Tuosu Lake through the Lianshui River. The Huaitoutala irrigation area (A) and nearby villages are located northwest of Keluke Lake (Figure 8), at higher elevation than both Keluke Lake (by 36 m) and Tuosu Lake (by 48 m). Therefore, the future expansion of Tuosu Lake will have little impact on the nearby irrigation areas and villages but is likely to promote expansion of the marsh in the northeast of Tuosu Lake. Considering the long-term trend of climate warming, a rise in groundwater levels in the vicinity of Gahai Lake is inevitable. Therefore, to ensure the safety of residents and continued operation of the irrigation area, drainage channels can be excavated at the end of the alluvial fan to divert groundwater to the Bayin River. Despite the potential for further increases in the water level of Tuosu Lake after channel excavation, the lack of villages and farmland around the lake makes this an appropriate management solution.

6. Conclusions

Remote sensing techniques, model calculations, and statistical analyses were used to analyze lake surface area changes in response to climate change and groundwater in Qaidam Basin, and stale isotopes were used to identify potential sources of groundwater. Our analysis suggests that long-term increases in temperature and precipitation had a certain promotion effect on lake expansion, with higher temperatures accelerating glacier melting rather than promoting lake evaporation. However, the significant increase in the rate of lake expansion also indicated the important contribution of groundwater to lake expansion, which includes not only glacial meltwater that infiltrates the piedmont plain but also other sources of water. Isotope data revealed that confined groundwater can circulate rapidly and is recharged by modern water since the global nuclear explosion tests. The more depleted isotopic signature of the confined groundwater compared to that of local meteoric precipitation and glacial meltwater suggested recharge by water sources with more isotopic depletion from other basins. Therefore, it is speculated that the 2003 M_s 6.4 earthquake in the northwest of Delingha may be a possible mechanism for the expansion of the lakes in the study area by enhancing crustal permeability and keeping fractures open, which promotes the groundwater contribution to lakes and in turn causes rapid lake expansion and increased groundwater levels.

The expansion of Gahai Lake has caused an increase in surrounding groundwater levels, which threatens the lives and livelihoods of residents. Under the long-term trend of climate warming, Gahai Lake will inevitably continue to expand in the future. Therefore, to ensure the safety of residents and continued operation of the irrigation area, it is suggested that drainage channels can be excavated at the end of the alluvial fan to divert groundwater to the Bayin River and eventually Tuosu Lake. This study emphasizes the important role of groundwater in lake expansion and improves our understanding of groundwater sources, circulation, and evolution patterns in Qaidam Basin and the arid area of northwest China. Currently, the future contribution of groundwater to lake expansion cannot be predicted because of a lack of groundwater monitoring data in the study area; therefore, future work should include long-term monitoring of the groundwater contribution to lakes in Qaidam Basin.

Author Contributions: Conceptualization, J.C. (Jiaqi Chen) and X.Z.; methodology, X.Z.; investigation, X.Z. and F.M.; writing—original draft preparation, X.Z.; writing—review and editing, J.C. (Jiaqi Chen), J.C. (Jiansheng Chen), and T.W.; supervision, J.C. (Jiansheng Chen). All authors have read and agreed to the published version of the manuscript.

Funding: This work was funded by the National Key Research and Development Program of China (Grant/Award Number: 2018YFC0406601), National Natural Science Foundation of China (Grant/Award Number: 61771183, and 42101021), and Fundamental Research Funds for the Central Universities (Grant/Award Number: B200202184).

Institutional Review Board Statement: Not applicable.

Informed Consent Statement: Not applicable.

Data Availability Statement: Not applicable.

Acknowledgments: The authors are grateful for the help of the State Key Laboratory of Hydrology–Water Resources and Hydraulic Engineering at Hohai University, where all analyses were performed.

Conflicts of Interest: The authors declare no conflict of interest.

References

1. Yang, K.; Wang, J.; Lei, Y.; Chen, Y.; Zhu, L.; Ding, B.; Qin, J. Quantifying evaporation and its decadal change for Lake Nam Co, central Tibetan Plateau. *J. Geophys. Res. Atmos.* **2016**, *121*, 7578–7591. [CrossRef]
2. Wang, X.; Gong, P.; Zhao, Y.; Xu, Y.; Cheng, X.; Niu, Z.; Luo, Z.; Huang, H.; Sun, F.; Li, X. Water-level changes in China's large lakes determined from ICESat/GLAS data. *Remote Sens. Environ.* **2013**, *132*, 131–144. [CrossRef]
3. Zhang, G.; Yao, T.; Xie, H.; Zhang, K.; Zhu, F. Lakes' state and abundance across the Tibetan Plateau. *Chin. Sci. Bull.* **2014**, *59*, 3010–3021. [CrossRef]
4. Zhang, G.; Xie, H.; Kang, S.; Yi, D.; Ackley, S.F. Monitoring lake level changes on the Tibetan Plateau using ICESat altimetry data (2003–2009). *Remote Sens. Environ.* **2011**, *115*, 1733–1742. [CrossRef]
5. Lei, Y.; Yao, T.; Bird, B.W.; Yang, K.; Zhai, J.; Sheng, Y. Coherent lake growth on the central Tibetan Plateau since the 1970s: Characterization and attribution. *J. Hydrol.* **2013**, *483*, 61–67. [CrossRef]
6. Lei, Y.; Yang, K.; Wang, B.; Sheng, Y.; Bird, B.W.; Zhang, G.; Tian, L. Response of inland lake dynamics over the Tibetan Plateau to climate change. *Clim. Change* **2014**, *125*, 281–290. [CrossRef]
7. Song, C.; Huang, B.; Richards, K.; Ke, L.; Phan, V.H. Accelerated lake expansion on the Tibetan Plateau in the 2000s: Induced by glacial melting or other processes? *Water Resour. Res.* **2014**, *50*, 3170–3186. [CrossRef]
8. Li, X.Y.; Xu, H.Y.; Sun, Y.L.; Zhang, D.S.; Yang, Z.P. Lake-level change and water balance analysis at lake Qinghai, West China during recent decades. *Water Resour. Manag.* **2007**, *21*, 1505–1516. [CrossRef]
9. Tan, H.; Zhang, Y.; Rao, W.; Guo, H.; Ta, W.; Lu, S.; Cong, P. Rapid groundwater circulation inferred from temporal water dynamics and isotopes in an arid system. *Hydrol. Process.* **2021**, *35*, e14225. [CrossRef]
10. Wiebe, A.J.; Conant, B.; Rudolph, D.L.; Korkka-Niemi, K. An approach to improve direct runoff estimates and reduce uncertainty in the calculated groundwater component in water balances of large lakes. *J. Hydrol.* **2015**, *531*, 655–670. [CrossRef]

11. Yang, N.; Zhou, P.; Wang, G.; Zhang, B.; Shi, Z.; Liao, F.; Li, B.; Chen, X.; Guo, L.; Dang, X.; et al. Hydrochemical and isotopic interpretation of interactions between surface water and groundwater in Delingha, Northwest China. *J. Hydrol.* **2021**, *598*, 126243. [CrossRef]
12. Yong, B.; Wang, C.Y.; Chen, J.; Chen, J.; Barry, D.A.; Wang, T.; Li, L. Missing water from the Qiangtang Basin on the Tibetan Plateau. *Geology* **2021**, *49*, 867–872. [CrossRef]
13. Wen, G.; Wang, W.; Duan, L.; Gu, X.; Li, Y.; Zhao, J. Quantitatively evaluating exchanging relationship between river water and groundwater in Bayin River Basin of northwest China using hydrochemistry and stable isotope. *Arid Land Geogr.* **2018**, *41*, 734–743.
14. Qin, B.; Qun, H. Evaluation of the climatic change impacts on the inland lake—A case study of Lake Qinghai, China. *Clim. Change* **1998**, *39*, 695–714. [CrossRef]
15. McJannet, D.L.; Cook, F.J.; Burn, S. Comparison of techniques for estimating evaporation from an irrigation water storage. *Water Resour. Res.* **2013**, *49*, 1415–1428. [CrossRef]
16. McJannet, D.L.; Webster, I.T.; Cook, F.J. An area-dependent wind function for estimating open water evaporation using land-based meteorological data. *Environ. Model. Softw.* **2012**, *31*, 76–83. [CrossRef]
17. Rodrigues, I.S.; Costa, C.A.G.; Lima Neto, I.E.; Hopkinson, C. Trends of evaporation in Brazilian tropical reservoirs using remote sensing. *J. Hydrol.* **2021**, *598*, 126473. [CrossRef]
18. Li, W.; Brunner, P.; Hendricks Franssen, H.J.; Li, Z.; Wang, Z.; Zhang, Z.; Wang, W. Potential evaporation dynamics over saturated bare soil and an open water surface. *J. Hydrol.* **2020**, *590*, 125140. [CrossRef]
19. Altho, D.; Neiva, L.; David, D.; Couto, H. Improving methods for estimating small reservoir evaporation in the Brazilian Savanna. *Agric. Water Manag.* **2019**, *216*, 105–112. [CrossRef]
20. Zhang, G.; Yao, T.; Shum, C.K.; Yi, S.; Yang, K.; Xie, H.; Feng, W.; Bolch, T.; Wang, L.; Behrangi, A.; et al. Lake volume and groundwater storage variations in Tibetan Plateau's endorheic basin. *Geophys. Res. Lett.* **2017**, *44*, 5550–5560. [CrossRef]
21. Yang, K.; Yao, F.; Wang, J.; Luo, J.; Shen, Z.; Wang, C.; Song, C. Recent dynamics of alpine lakes on the endorheic Changtang Plateau from multi-mission satellite data. *J. Hydrol.* **2017**, *552*, 633–645. [CrossRef]
22. Xiao, Y.; Shao, J.; Frape, S.K.; Cui, Y.; Dang, X.; Wang, S.; Ji, Y. Groundwater origin, flow regime and geochemical evolution in arid endorheic watersheds: A case study from the Qaidam Basin, northwestern China. *Hydrol. Earth Syst. Sci.* **2018**, *22*, 4381–4400. [CrossRef]
23. Zhang, B.; Zhao, D.; Zhou, P.; Qu, S.; Liao, F.; Guangcai, W. Hydrochemical Characteristics of Groundwater and Dominant Water—Rock Interactions in the Delingha. *Water* **2020**, *12*, 836. [CrossRef]
24. Yao, F.; Wang, J.; Yang, K.; Wang, C.; Walter, B.A.; Crétaux, J.F. Lake storage variation on the endorheic Tibetan Plateau and its attribution to climate change since the new millennium. *Environ. Res. Lett.* **2018**, *13*, 064011. [CrossRef]
25. Ma, Y.; Xu, N.; Sun, J.; Wang, X.H.; Yang, F.; Li, S. Estimating water levels and volumes of lakes dated back to the 1980s using Landsat imagery and photon-counting lidar datasets. *Remote Sens. Environ.* **2019**, *232*, 111287. [CrossRef]
26. Xu, N.; Ma, Y.; Zhang, W.; Wang, X.H. Surface-Water-Level Changes during 2003-2019 in Australia Revealed by ICESat/ICESat-2 Altimetry and Landsat Imagery. *IEEE Geosci. Remote Sens. Lett.* **2021**, *18*, 1129–1133. [CrossRef]
27. Paul, F.; Barrand, N.E.; Baumann, S.; Berthier, E.; Bolch, T.; Casey, K.; Frey, H.; Joshi, S.P.; Konovalov, V.; Le Bris, R.; et al. On the accuracy of glacier outlines derived from remote-sensing data. *Ann. Glaciol.* **2013**, *54*, 171–182. [CrossRef]
28. Storey, J.C.; Choate, M.J. Landsat-5 bumper-mode geometric correction. *IEEE Trans. Geosci. Remote Sens.* **2004**, *42*, 2695–2703. [CrossRef]
29. Guan, W.; Cao, B.; Pan, B.; Chen, R.; Shi, M.; Li, K.; Zhao, X.; Sun, X. Updated Surge-Type Glacier Inventory in the West Kunlun Mountains, Tibetan Plateau, and Implications for Glacier Change. *J. Geophys. Res. Earth Surf.* **2022**, *127*, e2021JF006369. [CrossRef]
30. Minora, U.; Bocchiola, D.; D'Agata, C.; Maragno, D.; Mayer, C.; Lambrecht, A.; Vuillermoz, E.; Senese, A.; Compostella, C.; Smiraglia, C.; et al. Glacier Area Stability in the Central Karakoram National Park (Pakistan) in 2001–2010: The "Karakoram Anomaly" in the Spotlight. *Prog. Phys. Geogr.* **2016**, *40*, 629–660. [CrossRef]
31. Du, Y.; Liu, B.; He, W.; Zhou, J.; Duan, S. Analysis on the variation and cause of the lake area in Qaidam Basin from 1976 to 2017. *J. Glaciol. Geocryol.* **2018**, *40*, 1275–1284. [CrossRef]
32. Liu, X.; Wen, Z.; Shu, L.; Lu, C.; Liu, B.; He, H. Analysis of surface area changes of Keluke and Tuosu lakes over past 40 years and influencing factors. *Water Resour. Prot.* **2014**, *30*, 28–33+63.
33. Guo, L.; Wu, Y.; Zheng, H.; Zhang, B.; Wen, M. *An Integrated Dataset of Daily Lake Surface Temperature Over Tibetan Plateau (LSWT_TPv1) (1978~2017)*; National Tibetan Plateau Data Center: Beijing, China, 2021.
34. Haginoya, S.; Fujii, H.; Kuwagata, T.; Xu, J.; Ishigooka, Y.; Kang, S.; Zhang, Y. Air-Lake interaction features found in heat and water exchanges over Nam Co on the Tibetan Plateau. *Sci. Online Lett. Atmos.* **2009**, *5*, 172–175. [CrossRef]
35. Zhu, L.P.; Xie, M.P.; Wu, Y.H. Quantitative analysis of lake area variations and the influence factors from 1971 to 2004 in the Nam Co basin of the Tibetan Plateau. *Chin. Sci. Bull.* **2010**, *55*, 1294–1303. [CrossRef]
36. Monteith, J.L. Evaporation and Environment. The Stage and Movement of Water in Living Organisms. In *Proceedings of the Symposia of the Society for Experimental Biology*; The Company of Biologists: Cambridge, UK, 1965.
37. Penman, H.L. Natural Evaporation from Open Water, Bare Soil and Grass. *Proc. R. Soc. Lond. Ser. A* **1948**, *193*, 120–145.
38. Priestley, C.; Taylor, R.J. On the Assessment of Surface Heat Flux and Evaporation Using Large Scale Parameters. *Mon. Weather. Rev.* **1972**, *100*, 81–92. [CrossRef]

39. De Bruin, H.A.R.; Keijman, J.Q. The Priestley-Taylor Evaporation Model Applied to a Large, Shallow Lake in the Netherlands. *J. Appl. Meteorol. Climatol.* **1979**, *18*, 898–903. [CrossRef]
40. Rosenberry, D.O.; Winter, T.C.; Buso, D.C.; Likens, G.E. Comparison of 15 evaporation methods applied to a small mountain lake in the northeastern USA. *J. Hydrol.* **2007**, *340*, 149–166. [CrossRef]
41. Tian, L.; Yao, T.; MacClune, K.; White, J.W.C.; Schilla, A.; Vaughn, B.; Vachon, R.; Ichiyanagi, K. Stable isotopic variations in west China: A consideration of moisture sources. *J. Geophys. Res. Atmos.* **2007**, *112*, 1–12. [CrossRef]
42. Clarke, I.; Fritz, P. *Environmental Isotope in Hydrogeology*; Springer: Berlin/Heidelberg, Germany, 1997. [CrossRef]
43. Chen, J.; Wang, C.-Y.; Tan, H.; Rao, W.; Liu, X.; Sun, X. New lakes in the Taklamakan Desert. *Geophys. Res. Lett.* **2012**, *39*, 22402. [CrossRef]
44. Chen, J.S.; Wang, C.Y. Rising springs along the Silk Road. *Geology* **2009**, *37*, 243–246. [CrossRef]
45. Zhang, X.; Shi, Y.; Yao, T. Variation of δ 18O in precipitation over the Northeastern Tibetan Plateau. *Sci. China Ser. B* **1995**, *38*, 540–547.
46. Salamon, T. Subglacial conditions and Scandinavian Ice Sheet dynamics at the coarse-grained substratum of the fore-mountain area of southern Poland. *Quat. Sci. Rev.* **2016**, *151*, 72–87. [CrossRef]
47. Post, V.E.A.; Groen, J.; Kooi, H.; Person, M.; Ge, S.; Edmunds, W.M. Offshore fresh groundwater reserves as a global phenomenon. *Nature* **2013**, *504*, 71–78. [CrossRef]
48. Zongyu, C.; Jixiang, Q.; Jianming, X.; Jiaming, X.; Hao, Y.; Yunju, N. Paleoclimatic interpretation of the past 30 ka from isotopic studies of the deep confined aquifer of the North China plain. *Appl. Geochem.* **2003**, *18*, 997–1009. [CrossRef]
49. Currell, M.J.; Cartwright, I.; Bradley, D.C.; Han, D. Recharge history and controls on groundwater quality in the Yuncheng Basin, north China. *J. Hydrol.* **2010**, *385*, 216–229. [CrossRef]
50. Gates, J.B.; Edmunds, W.M.; Darling, W.G.; Ma, J.; Pang, Z.; Young, A.A. Conceptual model of recharge to southeastern Badain Jaran Desert groundwater and lakes from environmental tracers. *Appl. Geochem.* **2008**, *23*, 3519–3534. [CrossRef]
51. Wang, T.; Chen, J.; Zhang, C.; Zhan, L.; Li, L. 14C-Dating Model for Groundwater Affected by CO_2 Inputs From Deep Underground Formations. *Water Resour. Res.* **2020**, *56*, 2–12. [CrossRef]
52. Wang, T.; Chen, J.; Zhang, C. Estimation of fossil groundwater mass fraction accounting for endogenic carbon input across California. *J. Hydrol.* **2021**, *595*, 126034. [CrossRef]
53. Wang, C.; Wang, C.; Manga, M. Coseismic release of water from mountains Evidence from the 1999 (Mw = 7.5) Chi-Chi, Taiwan, earthquake. *Geology* **2004**, *32*, 769–772. [CrossRef]
54. Muir-Wood, R.; King, G. Hydrological signatures of earthquake strain. *J. Geophys. Res. Solid Earth* **1993**, *98*, 22035. [CrossRef]
55. Rojstaczer, S.; Michel, R.; Wolf, S. Permeability enhancement in the shallow crust as a cause of earthquake-induced hydrological changes. *Nature* **1995**, *373*, 237–239. [CrossRef]
56. Manga, M.; Wang, C.-Y. Earthquake hydrology. In *Treatise on Geophysics*; Schubert, G., Ed.; California Institute of Technology: Pasadena, CA, USA, 2007; pp. 293–320.
57. Wang, B.; Ma, Y.; Su, Z.; Wang, Y.; Ma, W. Quantifying the evaporation amounts of 75 high-elevation large dimictic lakes on the Tibetan Plateau. *Sci. Adv.* **2020**, *6*, eaay8558. [CrossRef] [PubMed]
58. Zhao, Z.; Chen, H. Study on Control and Exploration Schemes of Groundwater Level Rising in Gahai Lake Delingha City of Qinghai Province. *Site Investig. Sci. Technol.* **2014**, 45–48.

remote sensing

Article

Thermal Responses of the Largest Freshwater Lake in the Tibetan Plateau and Its Nearby Saline Lake to Climate Change

Lijuan Wen [1,*], Chan Wang [1], Zhaoguo Li [1], Lin Zhao [1], Shihua Lyu [2], Matti Leppäranta [3], Georgiy Kirillin [4] and Shiqiang Chen [1]

1 Key Laboratory of Land Surface Process and Climate Change in Cold and Arid Regions, Northwest Institute of Eco-Environment and Resources, Chinese Academy of Sciences, Lanzhou 730000, China; wangchan@lzb.ac.cn (C.W.); zgli@lzb.ac.cn (Z.L.); zhaolin_110@lzb.ac.cn (L.Z.); csq@lzb.ac.cn (S.C.)
2 School of Atmospheric Sciences, Chengdu University of Information Technology, Chengdu 610225, China; slu@cuit.edu.cn
3 Institute of Atmospheric and Earth Sciences, University of Helsinki, 00014 Helsinki, Finland; matti.lepparanta@helsinki.fi
4 Department of Ecohydrology, Leibniz-Institute of Freshwater Ecology and Inland Fisheries, 12587 Berlin, Germany; georgiy.kirillin@igb-berlin.de
* Correspondence: wlj@lzb.ac.cn

Abstract: There are thousands of lakes in the Tibetan Plateau (TP), and most are saline. However, little is known about the responses of TP lakes to climate change, especially saline ones. We investigated the thermal responses of the largest freshwater lake (Ngoring Lake) in the TP and its nearby small saline lake (Hajiang Salt Pond) to climate change using the improved lake scheme in the Community Land model (CLM4-LISSS), in which we primarily developed the salinity parameterizations previously evaluated in the Great Salt Lake in USA and further considered the effect of salinity on the temperature of the maximum density of saline water in the present study. The improved lake model with salinity parameterizations was first applied to a saline lake in the TP, where saline lakes make up the majority of water bodies. The CLM4-LISSS model could effectively simulate lake surface water temperature (LSWT), lake water temperature (LT) and ice thickness in Ngoring Lake. Additionally, the model including our salinity parameterizations significantly improved simulations of LSWT and LT in Hajiang Salt Pond, especially in winter. The LSWT of the two completely opposite lakes were warming in the simulations at a rate above 0.6 °C/decade. Meteorological forces were the main driving factor, with increasing downward longwave radiation, air temperature and air humidity, as well as weakening winds contributing to LSWT increase. Compared to a hypothetical shallow freshwater lake, the greater depth of Ngoring Lake made its surface warm faster, and salinity slightly accelerated the warming of Hajiang Salt Pond. Monthly mean LSWT differences between the two lakes were induced by salinity effects in cold periods and lake depth in the unfrozen period. In response to a warming climate, the LSWT in the ice-free Hajiang Salt Pond rapidly increased from January to April due to the warming climate, whereas the LSWT of Ngoring Lake increased faster in the first and last month of the ice-cover period due to later ice-on and earlier ice-off. This study will provide a useful tool for saline lakes in the TP and help deepen our knowledge about the responses of TP lakes, especially the saline lakes, to climate change, as well as response differences between freshwater and saline lakes and the reasons for these differences.

Keywords: Ngoring Lake; Tibetan Plateau; saline lake; lake temperature; climate change; CLM4-LISSS; salinity parameterizations

Citation: Wen, L.; Wang, C.; Li, Z.; Zhao, L.; Lyu, S.; Leppäranta, M.; Kirillin, G.; Chen, S. Thermal Responses of the Largest Freshwater Lake in the Tibetan Plateau and Its Nearby Saline Lake to Climate Change. *Remote Sens.* **2022**, *14*, 1774. https://doi.org/10.3390/rs14081774

Academic Editor: Monica Rivas Casado

Received: 25 January 2022
Accepted: 4 April 2022
Published: 7 April 2022

1. Introduction

The Tibetan Plateau (abbreviated as TP, see Abbreviation List at the end of the text) is known as "the Third Pole", with an average altitude of 4000 m above sea level. The TP, with a total area of about 50,000 km^2, contains 1424 lakes ($\geq$1 km^2 each) [1], most of which are saline, accounting for more than half of the total lake coverage in China. The specific TP climatic environment (low air density, pressure, and temperature, all-year intensive solar radiation) creates unique lake–atmosphere interactions [2–8]. TP lakes significantly influence the local and regional climate by heat and mass exchanges between lakes and the atmosphere, and resonate with the adjacent and remote regions [9,10].

Lakes are sentinels of large-scale climate variability which interact strongly with the atmosphere and respond fast and widely to climate change, especially in the TP. The TP is influenced by elevation-dependent intensive warming, at up to three times the global warming rate [8,11]. The lake surface water temperature (LSWT) has rapidly increased globally, with a mean increasing trend of 0.34 °C/decade in summer averaged over 235 lakes worldwide between 1985 and 2009 [12]. However, TP lakes have shown an overall warming trend of 0.37 °C/decade, based on data from 374 inland lakes [13]. This rate was slightly higher than the global mean, because the TP climate and the warming of TP lakes are highly heterogeneous [14]. The majority of the TP lakes are warming at a higher rate of 0.76 °C/decade primarily due to the increasing air temperature, downward longwave radiation, and decreasing wind speeds, while some lakes are cooling due to glacier meltwater inflow or reduced salinity [3,13–17]. Changes in thermal conditions profoundly influence a lake's biological and chemical processes [18–20]. These processes may undergo substantial alterations, even with relatively small changes in lake temperature [21]. Moreover, the changing thermal characteristics of lakes further modulate local air–lake interactions, with significant impacts on the local climate. Therefore, a comprehensive investigation of the response of the thermal structure in TP lakes to climate change is needed to predict changes in lake ecosystems and the regional climate.

Most previous studies about the thermal responses of TP lakes were based on remote sensing data, which only reflected LSWT changes and the correlation between LSWT and possible driving factors. Additionally, results were mainly derived from statistical methods. However, this approach does not reveal the changes of internal phenomena in lakes, the quantitative contribution of driving factors, and the detailed mechanisms in lake processes. Numerical simulations appear to be the efficient method to reveal these key processes. A series of lake models, such as the Lake model, Flake (Freshwater lake) model, WRF (Weather Research and Forecast)-Lake, CLM (Community Land model)-Lake, CLM4-LISSS (the Lake, Ice, Snow, and Sediment Simulator), and the General Lake Model have been applied to studies of the TP lakes [2,4,7,15,16], with results showing that the vertically integrated mean lake water temperature (MLT) has been consistently changing corresponding to the increasing LSWT, while the bottom lake temperature (BLT) has varied in different ways depending on the lake depth [16,18]. However, with a scarcity of data, the development of TP lake models and numerical studies about the thermal responses of lakes to climate change have been focused solely on several large lakes, such as Nam Co Lake, Qinghai Lake, Ngoring Lake (NL) and Gyraing Lake [3,4,15], etc. In the absence of sufficient observational data and accurate forcing datasets, a previous long-term NL study employed the NCEP (National Centers for Environmental Prediction) and ERA-Interim (European Centre for Medium-Range Weather Forecasts Re-Analysis) data, in which the solar radiation was too great and decreasing quickly compared to the observations [22]. This resulted in predictions of insignificant NL warming in the simulations, whereas NL was actually warming, as shown by remote sensing data [14,23]. As such, the response of NL to climate warming should be restudied based on more accurate forcing data.

Most TP lakes are saline, but their responses to climate warming have been less studied than those of large lakes in the TP, mostly because of their small areas and the scarcity of observational data. Studies of saline lakes should be strengthened. The Hajiang Salt Pond (HSP, rich in soluble salts) is a paleo-saline lake which was formed by the joint action of traceability development of the Yellow River and climate change due to evaporation and condensation resulting from strong wind and sun. The pond is only about 11.2 km from NL and provides an ideal contrast as a saline lake to a freshwater lake NL with a similar climate.

The effects of salinity on responses of saline lakes to climate warming were poorly understood because of the lack of salinity parameterizations in commonly used models. Salinity could affect lake temperature, evaporation and ice appearance, etc. As such, a lake model considering salinity parameterizations is necessary for numerical studies of the majority of TP lakes.

Therefore, the CLM4-LISSS lake model, parameterized with salinity effects on several lake water characteristics and developed by ourselves and applied to the Great Salt Lake in USA, was introduced. Our previous saline lake model ignored the temperature of the maximum density (Tmaxd) of saline water that decreases with increased salinity and could affect the vertical thermal structure during the cold season in the TP [2,24]. To further improve our lake model, Tmaxd was further parameterized.

In the present study, we applied the CLM4-LISSS lake model developed with salinity parameterizations, in situ lake data, remote sensing data, and an assimilated meteorological dataset to study the thermal response of the largest freshwater lake NL in the TP and its nearby saline lake to climate change. The aim of the study is:

- to further improve our developed lake model with salinity parameterizations, to apply it to a TP saline lake, and to evaluate the model performance in the TP;
- to study the long-term warming trends in a deep freshwater lake and a saline lake in the TP; and
- to quantify the contribution of meteorological factors and salinity effects on the thermal regime changes of a deep freshwater lake and a saline lake.

2. Study Area, Data and Methods

2.1. Study Area

2.1.1. Freshwater Ngoring Lake

Ngoring Lake (NL), with a surface area of 610 km^2 and mean depth of 17 m, is the largest freshwater lake in the TP (Figure 1, 97.5~97.92°E, 34.75~35.08°N, 4274 m a.s.l.). Mineralization is low. A cold, semi-arid continental climate prevails in the NL basin. The monthly mean air temperature varies from 11.6 °C (August 2016) to −26.6 °C (January 1978), the annual average air temperature is −3.5 °C (1953–2016), and the annual precipitation is 322.4 mm (Data from China's National Climate Center) at Maduo meteorological station (Figure 1, 34.91°N, 98.22°E, 4272 m a.s.l.). The lake is covered with ice from early December to early April.

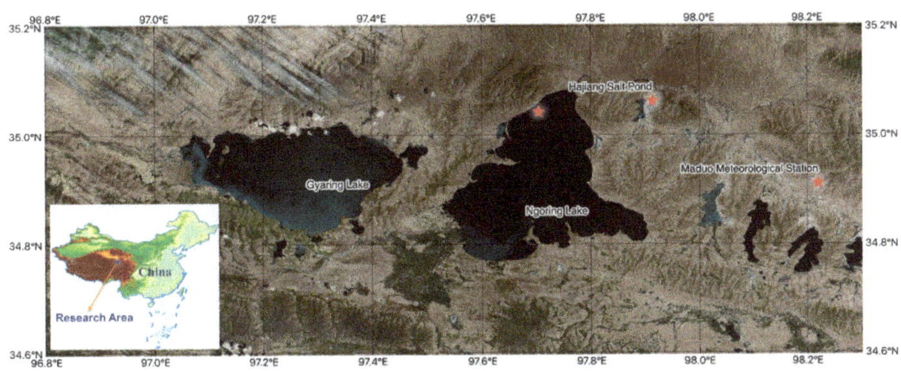

Figure 1. Map of the research area, locations of the Ngoring Lake (NL) and Hajiang Salt Pond (HSP), and three observation sites (marked by red stars).

2.1.2. Hajiang Salt Pond

Hajiang Salt Pond (HSP, 97.88–97.92°E, 35.02–35.05°N) is a small and shallow saline lake with less than 1 m depth and about 220 g L^{-1} salinity [25]. The freezing point caused by the salinity is low enough to prevent the lake from freezing normally. It is located approximately 11.2 km east from NL at an altitude of 4240 m. The lake developed from the large Hajiang paleo-lake and currently covers an area of about 10 km^2.

2.2. Data

2.2.1. Observations Data in NL and HSP

The NL lake temperature below the surface was observed (at a distance of ~2 km from the shore) using a Campbell 109 L logger sensor from June to October 2012, and from May to September 2013, and using RBR SOLO sensor from September 2015 to September 2016. The lake ice thickness was manually measured at irregular intervals from December 2012 to March 2013 and from December 2015 to March 2016 near the shore. Water temperature in HSP was measured (~1 km from the shore) under the lake surface from September 2015 to April 2016 using HOBO Water Temperature Pro v2 Data Loggers U22-001. These data were used to evaluate the performance of the lake model in the TP.

2.2.2. ITPCAS Data and Its Correction

Data for the lake model forcing were obtained from the China meteorological forcing dataset (1979–2018) (ITPCAS), developed by the Institute of Tibetan Plateau Research, Chinese Academy of Sciences [26]. The data include air temperature and specific humidity at 2 m above the surface, wind speed, surface air pressure, precipitation as well as downward shortwave and longwave radiation. The spatial resolution is 0.1° and the temporal resolution is 3 h [27]. The dataset was produced by merging a variety of data sources, including Princeton reanalysis data, Global Land Data Assimilation System data, the Global Energy and Water Cycle Experiment-surface radiation budget shortwave radiation dataset, Tropical Rainfall Measuring Mission satellite precipitation analysis data and China Meteorological Administration (CMA) station data.

The trend of annual mean ITPCAS air temperature (Ta) in the period of 1979–2016 was 0.77 °C/decade ($p < 0.01$), i.e., higher than that measured at the Maduo meteorological station (0.55 °C/decade ($p < 0.01$)). A closer inspection of the data revealed that the warming rate, i.e., 0.38 °C/decade, was identical in the two datasets for 1979–1997, with a systematic cold bias in the ITPCAS data compared to observations (Figure 2). The difference reversed in 1998–2006 but returned in 2007. Consequently, the forcing data from ITPCAS was bias-corrected based on the monthly observational data from the Maduo Station from 1979 to 2016 before being used to drive the lake model.

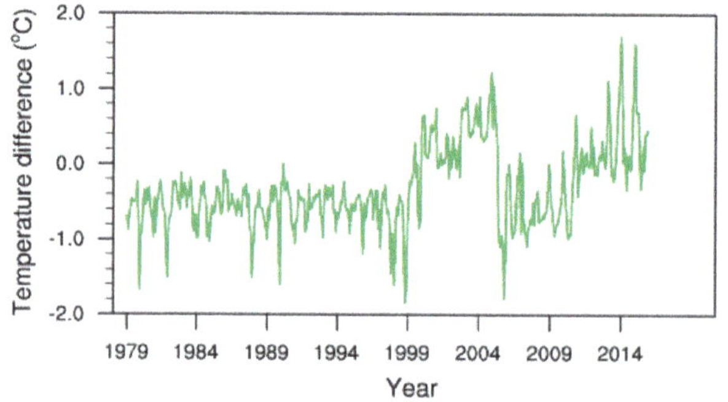

Figure 2. Monthly air temperature difference between ITPCAS and observations at the Maduo station (ITPCAS-Observation).

2.2.3. MODIS Data

The global 8-day composite daytime and nighttime LSWT from MODIS AQUA product data MYD11C2 V006 (0.05° resolution) and MYD11A2 V006 (1 km resolution) from

2003–2016 were used to evaluate the results of the long-term simulations for NL and HSP, respectively. Albedo from MODIS MCD43B3 V006 with 1-km resolution was used to set up the lake model.

2.3. Lake Model and Setup

In this study, we used an enhanced version of the Hostetler 1-D lake thermal model code, known as CLM4-LISSS. It was originally the lake scheme coupled in CLM, which is the land component of the Community Earth System Model (Details may be found in Subin et al., 2012). The lake scheme includes 0–5 snow layers, 10 lake liquid water and ice layers, and 10 sediment layers, which were also applied in WRF and the RegCM (Regional Climate Model) [28,29], etc.

CLM4-LISSS has been shown to provide reasonable performance in simulations of lake temperature, surface energy fluxes, and ice and snow thicknesses for several different-sized lakes around the world [30–33]. The model has also been improved in terms of ice albedo and mixing process, etc., for the TP [34,35]. It has been shown to effectively simulate the amplitude and pattern of temperature variability at all depths [2,7,36]. However, there is little progress for the saline lake model in the TP. Therefore, we first applied the developed CLM4-LISSS model with salinity parameterizations (evaluated in the Great Salt Lake in USA) to a simulation study of a saline lake (HSP) in the TP [17,30,37]. Further, the change of Tmaxd caused by dissolved salt in the water was parameterized in the current salinity scheme because it could significantly change the thermal processes of the lake water [2,24].

2.3.1. Lake Model with Salinity Parameterizations

The following equations, except for Tmaxd, were incorporated into the developed CLM4-LISSS lake model to evaluate for the salinity [17,37].

The dependence on salinity s (‰) of the freezing point (Tf, °C) was approximated by the seawater formula [38]:

$$\mathrm{Tf} = -0.0575\, s \tag{1}$$

The effect of dissolved salts on evaporation was expressed by the ratio of the saturated vapor pressure Rsvp over saline water to that over freshwater as [39]:

$$\mathrm{Rsvp} = \exp\left(-2/55.51 \times (s/(1 - s/1000)/58.44 + 0.77)\right) \tag{2}$$

The specific heat capacity of saline water c_{psw} (kJ/kg/K) was determined as [17,37]:

$$c_{psw} = 4.188 - 4.4\, s/1000 \tag{3}$$

The thermal conductivity of saline water λ_{sw} was determined as [17,37]:

$$\lambda_{sw} = \lambda_{fw}\left[1.0 - 0.22\, s/1000 + 0.1(s/1000)^2\right] \tag{4}$$

where λ_{fw} is the freshwater thermal conductivity, equal to 0.6 W $m^{-1}\cdot K$.

In addition to the previously used salinity parametrizations [37], the effect of salinity on the temperature of the maximum density (Tmaxd) of saline water was introduced in the study as Tmaxd decrease with increased salinity [40]:

$$\mathrm{Tmaxd} = 3.98 - 0.216\, s \tag{5}$$

2.3.2. Model Parameters for the Freshwater NL

In the model settings for NL (Table 1), the fraction of the net solar radiation absorbed near the lake surface was set to $\beta = 0.5$ in the absence of snow, and derived from the snow optics sub-model when snow was present [30].

Table 1. Numerical experiments S-(Lake) and model parameters.

Parameter \ Experiment	S-NL	S-D1F	S-HSP
Lake	NL	D1F	HSP
Salinity (‰)	0	0	220
Lake depth (m)	17	1	1
Albedo	Equation (7)	Equation (7)	0.15
Parameter of the light extinction coefficient η_0	1.1925	1.1925	2 × 1.1925
Fraction of absorbed surface solar radiation β	0.5	0.5	0.6
Meteorological forcing	Bias-corrected ITPCAS	Bias-corrected ITPCAS	Bias-corrected ITPCAS

The light extinction coefficient η was modelled as

$$\eta = \eta_0 \, d^{-0.424} \tag{6}$$

where d is the lake depth (m). The η_0 parameter was set as 1.1925 in the lake scheme of the CLM model [30].

The lake albedo was fixed as 0.06 for open water conditions according to observations at the NL station without considering the diurnal change, as in CLM4-LISSS, and calculated for an ice-covered lake with the following function [30,41]:

$$\alpha = \alpha_{max} - \alpha_{max}\, x + \alpha_{min}\, x,\ x = \exp\left(-95.6\,(Tf - LSWT)/Tf\right) \tag{7}$$

where α_{max} and α_{min} are the max and min values of the lake ice albedo, respectively. In CLM4-LISSS, there are different (α_{max}, α_{min}) values for near infrared and visible radiation. Without making a distinction between the two radiation types in the study, α_{max} and α_{min} were set to 0.6 and 0.1, respectively [30,41].

In the lake model, NL depth was set to 17 m, i.e., the same as the mean lake depth. Variations of lake depth were not considered, for two reasons: (1) Some lake models, such as the General Lake Model [15], are capable of considering changes in lake depth, but they have not yet been coupled with atmospheric models. In recent lake–air coupled simulation studies, the lake model CLM4-LISSS and Flake model are the two most commonly applied lake models in the TP [42,43]. Both models use a fixed lake depth. Returning to this study, long-term lake–air interactions will be further studied in the coupled atmospheric model, so the CLM4-LISSS lake model with a fixed lake depth was employed for the sake of consistency. (2) The NL lake level varies by less than 1 m per year, and varied by less than 3 m from 1985 to 2014 [44]. Such variations only induced small effects on the simulated LSWT [7].

2.3.3. Model Parameters for the Saline HSP

The salinity in HSP was set to 220‰ (Table 1) according to observations [25], and the lake depth was set to 1 m. Salinity variations were not considered because there was always insoluble salt at the bottom of HSP and the lake was shallow and well mixed.

Due to its shallowness, high salinity (corresponding Tf around −20 °C), and turbidity, HSP rarely freezes and has a higher albedo than the ice-free NL. The albedo of HSP was set to 0.15 (Table 1) as its annual mean albedo, as shown by MODIS.

The parameters β and η_0 were set to 0.6–0.8 and three times the freshwater lake value η_0 for the shallow turbid hypereutrophic Taihu Lake, respectively [45]. HSP has less phytoplankton and more transparent water than Taihu; as such, β was set to 0.6 and η_0 was set to twice the freshwater value (Table 1).

2.3.4. Numerical Experiments Design

The bias-corrected ITPCAS data and the two lake configurations from Table 1 were used to simulate the temperature conditions in NL and HSP in two model runs henceforth referred to as S-NL and S-HSP, respectively (Table 1).

Additionally, to segregate the effects of lake depth and salinity on the lake heat budget, a sensitivity model run called S-D1F (Table 1) was performed for a hypothetical freshwater lake with a depth of 1 m under the same ITPCAS atmospheric forcing. In this way, the only difference between the S-D1F and S-NL configurations was the lake depth, and the differences between S-HSP and S-D1F runs were caused solely by the salinity.

To understand the effects of climate change on lake warming, we performed sensitivity experiments in which the monotonic trend in each meteorological forcing variable (air temperature Ta, wind speed WS, specific humidity Q, downward shortwave radiation SWD and downward longwave radiation LWD) was removed individually based on a linear regression analysis and control runs using S-NL, S-D1F and S-HSP for NL, D1F and HSP, respectively. Experiments were called S-(Lake)-d(meteorological variable), as shown in Table 2. Owing to the consistency and significant impacts on the simulated lake temperature of Ta and LWD, more experiments with detrended Ta and LWD together were run. Some of the above forcing meteorological variables could be interconnected, and the above sensitivity experiments were quite artificial in nature. However, these experiments shed light on the controlling factors of lake warming [15,19] and quantified their individual effects on lake warming rate.

Table 2. S-(Lake)-d(meteorological variable) sensitivity experiments.

Experiment	Lake	Forcing
S-NL-dTa	NL	Same as S-NL except that Ta was detrended
S-NL-dWS	NL	Same as S-NL except that WS was detrended
S-NL-dQ	NL	Same as S-NL except that Q was detrended
S-NL-dSWD	NL	Same as S-NL except that SWD was detrended
S-NL-dLWD	NL	Same as S-NL except that LWD was detrended
S-NL-dTa&LWD	NL	Same as S-NL except that Ta and LWD were detrended
S-D1F-dTa	D1F	Same as S-D1F except that Ta was detrended
S-D1F-dWS	D1F	Same as S-D1F except that WS was detrended
S-D1F-dQ	D1F	Same as S-D1F except that Q was detrended
S-D1F-dSWD	D1F	Same as S-D1F except that SWD was detrended
S-D1F-dLWD	D1F	Same as S-D1F except that LWD was detrended
S-D1F-dTa&LWD	D1F	Same as S-D1F except that Ta and LWD were detrended
S-HSP-dTa	HSP	Same as S-HSP except that Ta was detrended
S-HSP-dWS	HSP	Same as S-HSP except that WS was detrended
S-HSP-dQ	HSP	Same as S-HSP except that Q was detrended
S-HSP-dSWD	HSP	Same as S-HSP except that SWD was detrended
S-HSP-dLWD	HSP	Same as S-HSP except that LWD was detrended
S-HSP-dTa&LWD	HSP	Same as S-HSP except that Ta and LWD were detrended

To estimate the individual effect of salinity on lake heat budget, we ran an additional series of sensitivity experiments without considering the effects of salinity on each parameter (α, η_0, β, Rsvp, Tmaxd and Tf), i.e., S-HSP-(parameter) in Table 3. The salinity effects on simulated lake temperature caused by the specific heat capacity, and the thermal conductivity in saline lakes were not studied as they were negligible [17,37].

Table 3. S-HSP-(parameter of salinity effect) sensitivity experiments.

Experiment	Model Setting
S-HSP-α	Same as S-HSP except that albedo was set to that in S-NL
S-HSP-η_0	Same as S-HSP except that the parameter of the light extinction coefficient η_0 was set from 2 × 1.1925 to 1.1925 as in S-NL
S-HSP-β	Same as S-HSP except that the fraction of absorbed surface solar radiation β was set from 0.6 to 0.5 as in S-NL
S-HSP-Rsvp	Same as S-HSP except that the ratio of the saturated vapor pressure Rsvp over the saline water to that over the fresh water in Equation (2) was set to 1 as in S-NL
S-HSP-Tmaxd	Same as S-HSP except that the temperature of the maximum density Tmaxd was changed from minus to 3.98 °C as in S-NL
S-HSP-Tf	Same as S-HSP except that the freezing temperature Tf was changed from −12.65 °C to 0 °C as in S-NL

2.3.5. Model Performance Criteria

The performance of the model was tested against the observed temperature and heat fluxes using three model efficiency scores: bias, root mean square error (RMSE) and the correlation coefficient (R) [46]:

$$\text{Bias} = \sum_{i=1}^{n}(S_i - O_i)/n \tag{8}$$

$$\text{RMSE} = \sqrt{\sum_{i=1}^{n}(S_i - O_i)^2/n} \tag{9}$$

$$\text{R} = \sum_{i=1}^{n}((S_i - \overline{S})(O_i - \overline{O}))/\left(\sqrt{\sum_{i=1}^{n}(S_i - \overline{S})^2}\sqrt{\sum_{i=1}^{n}(O_i - \overline{O})^2}\right) \tag{10}$$

where O_i represents the observations, n is the total number of observations, and S_i represents the simulated results.

3. Results and Analysis

3.1. *Performance of the Lake Model*

3.1.1. Performance on the Freshwater Lake (NL)

The model showed long-term seasonal variations with maximum and minimum values consistent with the MODIS LSWT (Figure 3a). The simulated LSWT forced by ITPCAS was slightly overestimated compared to the MODIS data with bias = 0.6 °C, RMSE = 3.2 °C, and R = 0.94. The simulated LSWT precision was close to that of Qinghai Lake and Nam Co Lake [4,15,16] using long-term ITPCAS. Except for the model errors, the simulated errors were from two sources: (1) compared to in situ observations, MODIS had the bias averaged from −1.5 °C to 0.2 °C and RMSE of around 2.0 °C owing to the cool skin effect [14,47,48]; and (2) ITPCAS was closer to the observations than other reanalysis data (e.g., NCEP and ERA) but still not very accurate, especially in lake basins with strong underlying heterogeneity. When the model was driven by observations from the NL lakeshore, the bias and RMSE between the simulated and observed LSWT were only −0.21 °C and 1.44 °C, respectively, in 2012 [49]. The model produced a better simulation with in situ observed forcing, but still could not reproduce long-term lake thermal conditions which were consistent with the assimilated meteorological dataset.

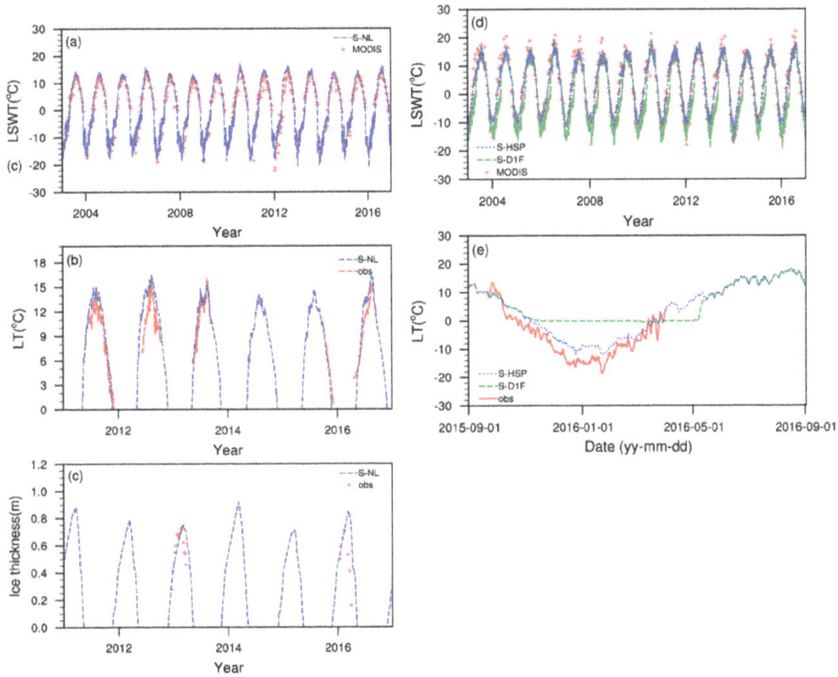

Figure 3. Observed and simulated LSWT (**a**), LT (**b**) and ice thickness (**c**) in S-NL, and LSWT (**d**) and LT (**e**) in S-HSP and S-D1F.

The model was able to accurately reproduce the lake temperature (Figure 3b), i.e., compared to the shallow-layer observations in the ice-free period with bias = 0.5 °C, RMSE = 2.8 °C, and R = 0.91. Similarly, the bias and RMSE between the simulated and observed lake temperature could be reduced to −0.25 °C and 0.41 °C, respectively, in 2016, when the model was driven by observations taken on the shore of NL [49].

The observed ice thickness remained above 0.6 m from the middle of January to early March, with the maximum value measured in late February (Figure 3c). With bias < 0.1 m and RMSE < 0.2 m, the model had a good ability to simulate the ice thickness.

3.1.2. Performance over a Saline Lake

The S-HSP numerical experiment with representative fixed salinity yielded more accurate LSWT simulations (Figure 3d) than the reference S-D1F experiment without the salinity effects, although both simulations were able to mimic the variations of LSWT. Compared to MODIS LSWT, the bias and RMSE of LSWT in S-D1F were 8.3 °C and 10.3 °C, respectively, while in S-HSP, they were reduced 3.4 °C and 7.0 °C, mainly because the application of a lower freezing point led to improved simulations in winter. The bias and RMSE from S-D1F to S-HSP in winter were reduced from 7.4 °C to 3.0 °C and from 9.3 °C to 3.1 °C, respectively. S-HSP could effectively reproduce the observed drop and increase in lake water temperature in winter, while S-D1F resulted in an ice cover and a nearly fixed water temperature below the ice. Factoring in salinity effects, the lake model was able to reflect the unfrozen state of the saline lake and the real variability of the lake temperature in winter, which is essential for studying the physics and chemistry of cold saline lakes.

3.2. Lake Temperature Variations and the Influence of Forcing Data

3.2.1. LSWT and Variation Trends

Annual LSWT

The annual mean LSWT of NL varied from −2 to 3 °C during 1979–2016, with a long-term average of 0.37 °C; the long-term mean LSWT of HSP was 2.95 °C, varying within 1–6 °C from one year to another (Figure 4a,c). The 2.6 °C lower mean LSWT in freshwater NL than in non-freezing saline HSP was conditioned by the ice cover reflecting solar radiation in winter. In the reference simulation, i.e., D1F with 1 m freshwater lake depth, the annual mean LSWT of 0.38 °C was close to that of NL and much smaller than that of HSP (Figure 4). Compared with salinity, lake depth did not have significant effects on the annual mean LSWT in the three simulations because the ice formation played the dominant role.

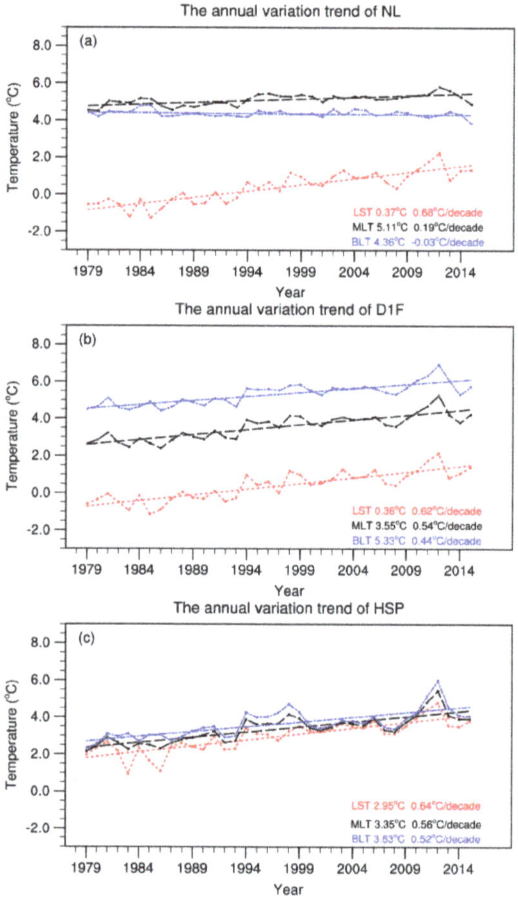

Figure 4. Annual LSWT (red), MLT (black) and BLT (blue) and their trends in S-NL (**a**), S-HSP (**b**) and S-D1F (**c**) Monthly LSWT.

The trends of annual LSWT, MLT and BLT in each of the three simulations except NL BLT surpassed 99% significance. The annual LSWT in NL increased by 0.68 °C/decade during 1979–2016, which was slightly faster than the 0.64 °C/decade simulated for HSP. The increasing rate of annual LSWT in S-D1F was 0.62 °C/decade, the lowest among the three simulations. Still, the difference between the three experiments was not significant as long as the meteorological forcing was the same and remained the main driver responsible for the annual LSWT changing rate.

In monthly means, LSWT differences between the three experiments were mainly controlled by salinity during the cold periods, while lake depth was the primary factor during open-water periods. From November to May, the monthly LSWT of NL and D1F were similar to each other and about 4 °C lower than those in HSP (Figure 5a), because the high salinity of the latter prevented the development of ice cover. In turn, during the warm period from June to October, the LSWT of D1F and HSP closely followed air temperature variations; both lake surfaces were 2 °C warmer than that of the deeper NL in mid-summer (June and July), and both cooled faster than NL in autumn, with LSWT differences increasing gradually from 1 to 3 °C from September to October.

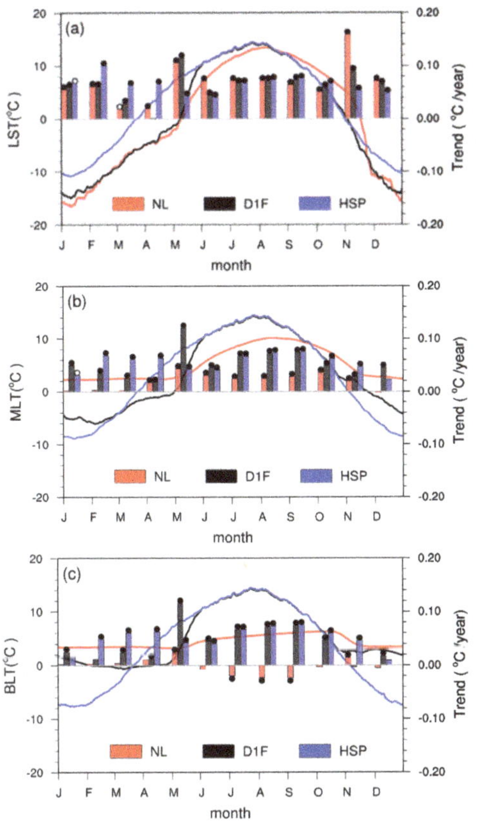

Figure 5. The monthly mean temperatures (°C, lines) and their long-term trends (°C/year, bars) in the model runs S-NL (red), S-HSP (blue) and S-D1F (black). Panels (**a**–**c**) correspond to LSWT, MLT and BLT, respectively. Solid and hollow points at the end of bars mean passing the significance test of $p < 0.01$ and $p < 0.05$, respectively.

The long-term LSWT trends of the three lakes were positive for all months. In non-freezing HSP, the temperatures between January and April increased at 0.6 °C/decade (Figure 5a), with the largest increase taking place in February (1.0 °C/decade) in response to climate change. The LSWT of NL and D1F increased significantly in May and November because of the shortened ice period and ice-albedo feedback. In May, the LSWT in D1F warmed a little faster than in NL (1.2 °C/decade vs. 1.1 °C/decade), and in November, the warming rate of NL LSWT (1.6 °C/decade) was the most acute, demonstrating that LSWT increase in months of delayed ice cover is faster in deeper lakes.

3.2.2. MLT and the Variation Trends

Annual MLT

The annual vertically averaged lake water temperature (MLT) was 5.11, 3.55, and 3.35 °C in NL, D1F, and HSP respectively (Figure 4b). The annual MLT of HSP was the lowest because of the stronger heat loss from the open surface of HSP in winter compared to the ice-covered NL and D1F.

The annual NL MLT increased by 0.19 °C/decade, i.e., much slower than the annual LSWT in NL (0.68 °C/decade) and slower than the annual MLT trends in D1F (0.54 °C/decade) and HSP (0.56 °C/decade) (Figure 4). Apparently, the MLT of the deep lake increased more slowly due to its large heat capacity and thermal inertia. Salinity differences between shallow lakes D1F and HSP did not have a strong influence on the observed MLT trend.

Monthly MLT

The monthly means of HSP MLT in winter were the lowest, reaching around −10 °C in January (Figure 5b). In turn, because of the shallow depth, the MLT of HSP and D1F were the highest in summer, reaching about 14 °C in July. Hence, lake depth was the main factor controlling the magnitude of the annual and summer monthly MLT in the three simulations. In contrast, salinity determined the winter MLT minima by preventing ice cover formation in saline lakes.

In NL, the long-term trends of monthly MLT were less than 0.6 °C/decade, and variations therein were much smaller than those of monthly mean LSWT (0.2–1.6 °C/decade) (Figure 5). In NL, the monthly MLT showed a strong positive trend from April to November and stayed almost stable during the ice period, while the monthly LSWT became higher throughout the whole year, especially in freezing and breakup months.

In shallow lakes (HSP and D1F), the trends of monthly MLT in summer coincided with monthly LSWT trends (0.4–0.8 °C/decade); the difference (<0.3 °C/decade) between MLT and LSWT in shallow lakes was only seen in winter and at times of ice-formation and melting. Still, their differences were much smaller than in NL.

Akin to the surface temperatures, the MLT of HSP in winter (February to April) increased at the highest rate among the three simulations, i.e., at about 0.6 °C/decade, compared to around 0.2 °C/decade in D1F and <0.1 °C/decade in NL.

3.2.3. Bottom Temperature and the Variation Trends

Annual BLT

An annual BLT of 3.63 °C in HSP was the lowest among the three experiments, with 4.36 °C and 5.33 °C in NL and D1F, respectively (Figure 4). The simulated BLT of the two shallow lake experiments, i.e., in S-HSP and D1F, were warming by 0.52 and 0.44 °C/decade, while that of NL tended to become slightly cooler (−0.03 °C/decade) without passing the significance test. This is consistent with results from other deep dimictic lakes [16,50,51].

Monthly BLT

In NL, the monthly BLT became obviously lower (about −0.4 °C/decade) from July to September with a 0.01 significance level (Figure 5c). Its warming surface intensified the stability in the stratification period of the deep lake and resulted in less heat being transferred to the bottom. Changes in NL BLT in other months were not insignificant.

Owing to the shallow depth of D1F and HSP, there were no big differences between monthly LSWT, MLT, and BLT in the two runs during ice-free periods (Figure 5). The shallow lake depth and high salinity made the winter BLT increase faster in HSP than in other runs.

3.3. Effects of Local Climate Drivers on the Lake Warming

3.3.1. Correlation between LSWT and Meteorological Forcing

The study region was experiencing the same rapid climate change as most of the TP during the study period [52]. All trends of ITPCAS meteorological variables (Figure 6) over the study region passed the 0.01 significance level except for solar radiation, with a 0.1 significance level. Ta increased by 0.49 °C/decade, downward longwave radiation LWD increased at a rate of 3.40 W m^{-2}/decade, and the specific humidity Q grew at 0.16 g kg^{-1}/decade.

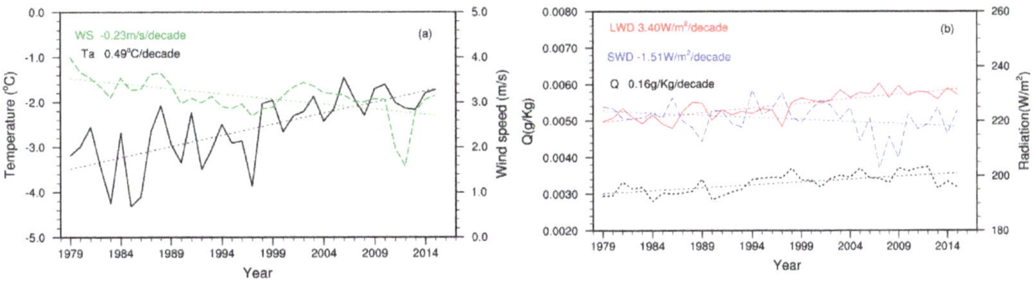

Figure 6. The ITPCAS annual mean meteorological variables and their trends: air temperature Ta and wind speed WS (**a**) and downward shortwave radiation SWD, downward longwave radiation LWD and specific humidity Q (**b**).

The increasing Ta, LWD, and Q had positive effects on lake warming, with around 0.8 correlation with LSWT in the three experiments (Figure 7a). Wind speed decreased by −0.23 m s^{-1}/decade and was negatively correlated with the LSWT (r = −0.6). The downward shortwave radiation SWD decreased at −1.51 W m^{-2}/decade, acting in the opposite direction, with a correlation coefficient of −0.15.

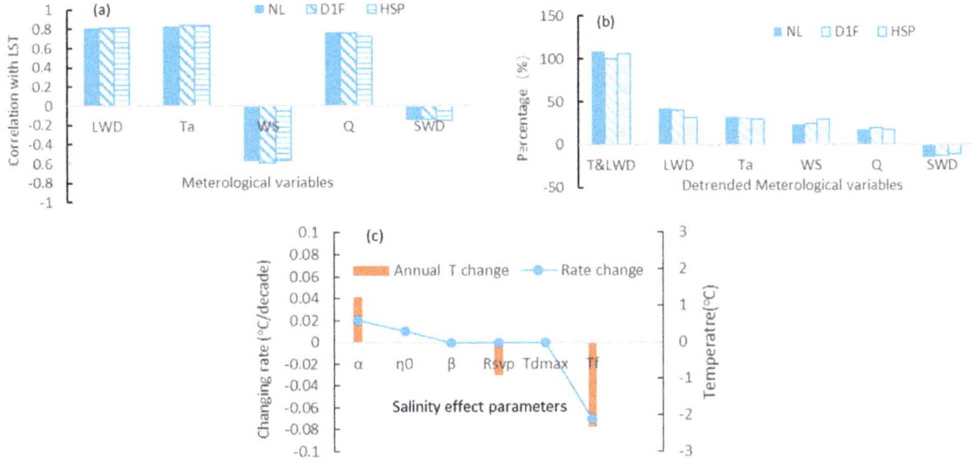

Figure 7. Correlations between the meteorological variables and LSWT (**a**), the percentage difference of LSWT warming rate between the control simulation S-(lake) and the sensitivity simulations S-(Lake)-d (variable) in NL, D1F, and HSP (**b**), the difference of LSWT and the warming rate between the S-HSP control simulation and S-HSP-(parameter of salinity effect) sensitivity simulations (**c**).

3.3.2. Quantified Contribution of Individual Meteorological Forcing to Lake Warming

The above correlation analysis showed the effects of meteorological variables on lake warming, but their quantified contributions were still unknown. Therefore, detrended sensitivity experiments, referred to as S-(Lake)-d (variable) in Table 2, were performed to attempt to answer this question, although the sensitivity experiments were a little artificial. The increase of atmospheric longwave radiation contributed the most to the warming of TP lakes, causing 30–40% of the annual LSWT change (Figure 7b). Increasing Ta induced a 30% increase of LSWT. The combined increase of atmospheric longwave radiation and Ta could explain almost the all of the observed LSWT warming. The decreasing WS caused a 20–30% increase in LSWT, especially in HSP. Air humidity increase accelerated the lake surface warming by 20% and, consequently, should not be ignored, as it was in a study of the Nam Co Lake in TP [15]. The decrease of SWD decelerated warming by about 10%.

3.4. *Effects of Salinity Parameters on the Lake Warming*

The cumulative effect of salinity on the lake water properties caused the simulated lake surface to be 2.6 °C warmer and experience 0.02 °C/decade faster warming in S-HSP than in its freshwater counterpart (Figure 4).

The salinity effect on Tf and Rsvp were simulated, making the saline lake surface 2.34 °C and 0.9 °C warmer (Figure 7c) in S-HSP than in the two sensitivity experiments of salinity effects (S-HSP-Tf and S-HSP-Rsvp), inducing about 90% and 31% annual differences between the saline lake and the freshwater one. The lower freezing point also significantly accelerated the increase of simulated LSWT in saline lake warming to 0.07 °C/decade (3.5 times the warming rate difference between the saline lake and the hypothetical freshwater lake with the same 1 m depth), while the salinity effect on the saturation water vapor pressure had no impact on the long-term temperature trend.

The differences between the S-HSP and S-HSP-η_0 simulations showed that the lower transparency of salt water with higher η_0 increased the annual LSWT by only 0.02 °C and slowed the warming by 0.01 °C/decade.

The difference between the S-HSP and S-HSP-α simulations showed that the higher α of salt water cooled the lake surface by 1.24 °C annually (about −47% of the annual LSWT difference between the saline lake and the freshwater one) and slowed the warming rate by 0.02 °C/decade (the same magnitude as the warming rate caused by all the salinity effects).

Changes in the temperature of the maximum density (Tmaxd) of saline water and the absorption of solar radiation (β) by the lake surface caused by salinity did not affect the annual mean temperature or the changing rate.

4. Discussion

4.1. *Salinity Effects and Parameterizations*

Most TP lakes are saline, but existing numerical studies have focused exclusively on several large lakes with small amounts of salt because of the lack of observations and salinity parameterizations in lake models. Based on previous salinity parameterizations coupled in CLM applied to the Great Salt Lake in USA and the significant impacts of Tmaxd on density convection and thermal stratification, the effects of Tmaxd were further parameterized in our lake model. The improved model was first applied to the TP saline lake and significantly reduced errors in the simulation of LSWT and LT in the saline lake, especially in winter. The salinity-extended lake model will be an efficient tool for studying saline lakes in the TP.

Salinity parameterization of Tmaxd had no obvious impacts on the warming of a small saline lake, mainly because of the shallow lake depth, absence of ice cover and the strong effects of wind in terms of turbulence mixing in HSP. However, Tmaxd could alter the thermal structure of a lake with a certain depth [2,24]. Therefore, considering the salinity effect on Tmaxd makes the developed saline lake model more accurate.

Salinity will play a major role in terms of the impact of climate change on TP lakes in future. Salinity decreases due to increased precipitation and inflow of glacial meltwater, but it could also increase due to increased evaporation. Since many lakes are fairly small, even small changes in the water balance can be important. Also, salinity evolution influences the vertical stratification of lakes and, consequently, the water temperature structure. Variations in salinity will be addressed in future according to the mass balance.

4.2. Simulated Warming Rates of LSWT in Different Studies

LSWTs globally have increased rapidly, with a mean trend of 0.34 °C/decade in summer between 1985 and 2009 [12]. Although the TP is warming at twice or even three times the global warming rate [8,11], TP lakes have overall been warming with a trend of 0.37 °C/decade, based on data from 374 inland lakes. They are warming slightly more rapidly than the global mean because the warming of TP lakes is highly heterogeneous [14]. Most TP lakes are warming with the higher 0.76 °C/decade rate, while some lakes are cooling due to glacier meltwater inflow or reduced salinity [3,13–17].

The simulated LSWT warming rate of NL was 0.68 °C/decade, as shown by remote sensing data that the lake was warming [14,23], and its warming rate was between the simulated rates of 0.52 ± 0.25 °C/decade in Nam Co by the GLM and 0.74 °C/decade in Qinghai Lake by Flake. The trend was not insignificant as in the previous NL simulation, in which the model was forced by the NCEP and ERA data where solar radiation was larger and decreased quickly compared to the observations [22]. Thus, insignificant warming in NL was concluded in the previous study. Although field work is hard, enough observations should be performed and accurate forcing datasets should be built for more accurate simulation studies.

Our results make it clear that the largest freshwater lake in the TP and a nearby small saline lake have indeed warmed over the last several decades and are warming faster owing to the amplification effect of their high altitude. The warming rate of LSWT in NL (0.68 °C/decade) significantly exceeded that of the regional air temperature (0.49 °C/decade) and was similar to Qinghai Lake and Lake Superior due to reduced ice cover [16,53]. While lakes in some temperate climate regions are warming in line with increased air temperatures [50,54], even the warming rates of tropical lakes are smaller than those of air [55,56]. Moreover, the bottom layers of NL were simulated to isolate from direct atmospheric influence, and tended to show long-term cooling at a rate −0.03 °C/decade on account of strengthening stratification. This result is similar to those reported from other stratified dimictic lakes, such as Qinghai Lake, Heiligensee Lake, and so on [16,50].

5. Conclusions

Compared to our observations, the salinity-extended lake model demonstrated a good ability to represent lake–air interactions and the thermal regime in both a freshwater lake and a saline lake. The newly introduced salinity parameterization significantly improved the model performance for a saline lake in winter.

The simulated long-term increasing rates of the annual LSWT in NL, HSP, and a shallower hypothetical freshwater lake amounted to more than 0.6 °C/decade, mainly due to meteorological forcing. Increasing LWD and Ta, weakening wind, and increased air humidity had positive effects on the warming trend of TP lakes in decreasing order, while solar radiation dimming counteracted the warming. The LWD and Ta contributed the most to lake warming in the sensitivity experiment; although it was overlooked in previous studies [3,15], increasing atmospheric humidity over TP should be considered a significant climatic factor.

The shallow lakes in experiments S-D1F and S-HSP lacked seasonal thermal stratification and were well-mixed vertically, revealing similar long-term warming trends across their depths. Comparing to a fresh water lake with 1 m depth, the 17-m-deep NL experienced a 0.06 °C/decade faster surface warming and a slower MLT rise of 0.35 °C/decade.

High salinity prevented ice cover formation in HSP and induced more heat release in winter and lower MLT and BLT than in freshwater lakes NL and D1F. However, the high salinity made the annual mean LSWT 2.6 °C higher and resulted in a 0.02 °C/decade stronger warming trend than in the freshwater lake with the same depth. The salinity effect on the freezing point contributed most to this difference, inducing a 90% higher LSWT compared to the freshwater D1F. The salinity effect on evaporation caused a 31% higher LSWT in HSP. The opposite salinity effect on the lake surface albedo cooled the lake surface and decelerated the warming trend.

The monthly mean LSWT differences between Ngoring Lake and the Hajiang Salt Pond were induced by salinity effects in cold periods and lake depth in the unfrozen period. The LSWT in ice-free Hajiang Salt Pond increased rapidly from January to April due to climate change, whereas the LSWT of Ngoring Lake increased faster in the first and last months of the ice-cover period due to later ice-on and earlier ice-off.

Author Contributions: L.W. initiated this work, carried out modeling experiments, analyzed the results and wrote the original draft; C.W. and S.C. collected the data, Z.L., L.Z., S.L., M.L. and G.K. gave constructive suggestions on the design and modification of the manuscript. All the authors contributed to the writing and editing of the manuscript. All authors have read and agreed to the published version of the manuscript.

Funding: This study was supported by the National Key Research and Development Program of China (2019YFE0197600), CAS "Light of West China" Program (E129030101, Y929641001), the National Natural Science Foundation of China (41975081, 41930759).

Data Availability Statement: ITPCAS dataset is available in the Third Pole Environment Database (https://data.tpdc.ac.cn/en/). Maduo data could be downloaded from http://www.tpdc.ac.cn/zh-hans/data/52c77e9c-df4a-4e27-8e97-d363fdfce10a/. MODIS data could be downloaded from https://modis-land.gsfc.nasa.gov/MODLAND_grid.html.

Conflicts of Interest: The authors declare no conflict of interest.

Abbreviation

Abbreviation	Explain	Unit
BLT	bottom lake temperature	°C
CMA	China Meteorological Administration	
CLM4-LISSS	the Lake, Ice, Snow, and Sediment Simulator in the Community Land model V4.0	
c_{psw}	specific heat capacity of saline water	kJ/kg/K
d	lake depth	m
ITPCAS	Institute of Tibetan Plateau Research, Chinese Academy of Sciences	
LWD	downward longwave radiation	W m^{-2}
HSP	Hajiang Salt Pond	
LSWT	lake surface water temperature	°C
LT	lake water temperature	°C
MLT	mean lake water temperature	°C
n	total number of observations	
NL	Ngoring Lake	
Q	specific humidity	g kg^{-1}
O_i	represents the observations	
R	correlation coefficient	
RMSE	root mean square error	
Rsvp	the ratio of the saturated vapor pressure	
s	salinity	‰
S_i	represents the simulated results	
SWD	downward shortwave radiation	W m^{-2}
Ta	air temperature	°C

Tf	freezing point	°C
Tmaxd	temperature of the maximum density of saline water	°C
TP	The Tibetan Plateau	
WS	wind speed	m s^{-1}
α	lake albedo	
α_{max}	max values of the lake ice albedo	
α_{min}	min values of the lake ice albedo	
β	fraction of absorbed surface solar radiation	
λ_{sw}	thermal conductivity of saline water	W m^{-1}·K
η	light extinction coefficient	
η_0	constant value 1.1925	

References

1. Zhang, G.; Luo, W.; Chen, W.; Zheng, G. A robust but variable lake expansion on the Tibetan Plateau. *Sci. Bull.* **2019**, *64*, 1306–1309. [CrossRef]
2. Huang, A.; Lazhu; Wang, J.; Dai, Y.; Yang, K.; Wei, N.; Wen, L.; Wu, Y.; Zhu, X.; Zhang, X.; et al. Evaluating and Improving the Performance of Three 1-D Lake Models in a Large Deep Lake of the Central Tibetan Plateau. *J. Geophys. Res.* **2019**, *124*, 3143–3167. [CrossRef] [PubMed]
3. Kirillin, G.; Wen, L.; Shatwell, T. Seasonal thermal regime and climatic trends in lakes of the Tibetan highlands. *Hydrol. Earth Syst. Sci.* **2017**, *21*, 1895–1909. [CrossRef]
4. Lazhu; Yang, K.; Wang, J.; Lei, Y.; Chen, Y.; Zhu, L.; Ding, B.; Qin, J. Quantifying evaporation and its decadal change for Lake Nam Co, central Tibetan Plateau. *J. Geophys. Res.* **2016**, *121*, 7578–7591. [CrossRef]
5. Ma, Y.; Wang, B.; Zhong, L.; Ma, W. The regional surface heating field over the heterogeneous landscape of the Tibetan Plateau using MODIS and in-situ data. *Adv. Atmos. Sci.* **2012**, *29*, 47–53. [CrossRef]
6. Wang, Y.; Gao, Y.; Qin, H.; Huang, J.; Liu, C.; Hu, C.; Wang, W.; Liu, S.; Lee, X. Spatiotemporal characteristics of lake breezes over lake Taihu, China. *J. Appl. Meteorol. Climatol.* **2017**, *56*, 2053–2065. [CrossRef]
7. Wen, L.; Lyu, S.; Kirillin, G.; Li, Z.; Zhao, L. Air-lake boundary layer and performance of a simple lake parameterization scheme over the Tibetan highlands. *Tellus A* **2016**, *68*, 31091. [CrossRef]
8. Qiu, J. China: The third pole. *Nature* **2008**, *454*, 393–396. [CrossRef]
9. Yang, Y.; Cheng, B.; Kourzeneva, E.; Semmler, T.; Rontu, L.; Lepparanta, M.; Shirasawa, K.; Li, Z. Modelling experiments on air snow ice interactions over Kilpisjarvi, a lake in northern Finland. *Boreal Environ. Res.* **2013**, *18*, 341–358.
10. Wu, G.; Duan, A.; Liu, Y.; Mao, J.; Ren, R.; Bao, Q.; He, B.; Liu, B.; Hu, W. Tibetan Plateau climate dynamics: Recent research progress and outlook. *Natl. Sci. Rev.* **2015**, *2*, 100–116. [CrossRef]
11. You, Q.; Chen, D.; Wu, F.; Pepin, N.; Cai, Z.; Ahrens, B.; Jiang, Z.; Wu, Z.; Kang, S.; AghaKouchak, A. Elevation dependent warming over the Tibetan Plateau: Patterns, mechanisms and perspectives. *Earth Sci. Rev.* **2020**, *210*, 103349. [CrossRef]
12. O'Reilly, C.M.; Sharma, S.; Gray, D.K.; Hampton, S.E.; Read, J.S.; Rowley, R.J.; Schneider, P.; Lenters, J.D.; McIntyre, P.B.; Kraemer, B.M.; et al. Rapid and highly variable warming of lake surface waters around the globe. *Geophys. Res. Lett.* **2015**, *42*, 10773–10781. [CrossRef]
13. Wan, W.; Zhao, L.; Xie, H.; Liu, B.; Li, H.; Cui, Y.; Ma, Y.; Hong, Y. Lake surface water temperature change over the Tibetan Plateau from 2001–2015: A sensitive indicator of the warming climate. *Geophys. Res. Lett.* **2018**, *45*, 11177–11186. [CrossRef]
14. Zhang, G.; Yao, T.; Xie, H.; Qin, J.; Ye, Q.; Dai, Y.; Guo, R. Estimating surface temperature changes of lakes in the Tibetan Plateau using MODIS LST data. *J. Geophys. Res.* **2014**, *119*, 8552–8567. [CrossRef]
15. Huang, L.; Wang, J.; Zhu, L.; Ju, J.; Daut, G. The Warming of Large Lakes on the Tibetan Plateau: Evidence from a Lake Model Simulation of Nam Co, China, during 1979–2012. *J. Geophys. Res.* **2017**, *122*, 13095–13107. [CrossRef]
16. Su, D.; Hu, X.; Wen, L.; Lyu, S.; Gao, X.; Zhao, L.; Li, Z.; Du, J.; Kirillin, G. Numerical study on the response of the largest lake in China to climate change. *Hydrol. Earth Syst. Sci.* **2019**, *23*, 2093–2109. [CrossRef]
17. Wen, L. Impacts of a Saline Lake and Its Salinity on Local Precipitation. *Adv. Meteorol.* **2015**, *2015*, 679634. [CrossRef]
18. Antonopoulos, V.Z.; Gianniou, S.K. Simulation of water temperature and dissolved oxygen distribution in Lake Vegoritis, Greece. *Ecol. Model.* **2003**, *160*, 39–53. [CrossRef]
19. Ito, Y.; Momii, K. Impacts of regional warming on long-term hypolimnetic anoxia and dissolved oxygen concentration in a deep lake. *Hydrol. Process.* **2015**, *29*, 2232–2242. [CrossRef]
20. Farrell, K.J.; Ward, N.K.; Krinos, A.I.; Hanson, P.C.; Daneshmand, V.; Figueiredo, R.J.; Carey, C.C. Ecosystem-scale nutrient cycling responses to increasing air temperatures vary with lake trophic state. *Ecol. Model.* **2020**, *430*, 16. [CrossRef]
21. Adrian, R.; O'Reilly, C.M.; Zagarese, H.; Baines, S.B.; Hessen, D.O.; Keller, W.; Livingstone, D.M.; Sommaruga, R.; Straile, D.; Van Donk, E.; et al. Lakes as sentinels of climate change. *Limnol. Oceanogr.* **2009**, *54*, 2283–2297. [CrossRef] [PubMed]
22. Du, J.; Wen, L.; Su, D. Reliability of three reanalysis dataset in simulation of three alpine lakes on the Qinghai-Tibetan Plateau. *Plateau Meteorol.* **2019**, *38*, 101–103. (In Chinese with English abstract)

23. Song, K.; Wang, M.; Du, J.; Yuan, Y.; Ma, J.; Wang, M.; Mu, G. Spatiotemporal Variations of Lake Surface Temperature across the Tibetan Plateau Using MODIS LST Product. *Remote Sens.* **2016**, *8*, 854. [CrossRef]
24. Lazhu; Yang, K.; Hou, J.; Wang, J.; Lei, Y.; Zhu, L.; Chen, Y.; Wang, M.; He, X. A new finding on the prevalence of rapid water warming during lake ice melting 2 on the Tibetan Plateau. *Sci. Bull.* **2021**, *66*, 2358–2361. [CrossRef]
25. Zhou, J.; Han, F.; Pang, X.; Luo, C.; Yan, J. Preliminary investigation of Hajiang Salt Pond and Kuhai Lake in Yellow River Source Area. *J. Salt Lake Res.* **2010**, *18*, 18–22. (In Chinese with English Abstract) [CrossRef]
26. Yang, K.; He, J. *China Meteorological Forcing Dataset (1979–2018)*; National Tibetan Plateau Data Center: Beijing, China, 2019; Available online: http://data.tpdc.ac.cn/en/data/8028b944-daaa-4511-8769-965612652c49/ (accessed on 24 January 2022).
27. Yang, K.; He, J.; Tang, W.; Qin, J.; Cheng, C.C.K. On downward shortwave and longwave radiations over high altitude regions: Observation and modeling in the Tibetan Plateau. *Agric. Forest. Meteorol.* **2010**, *150*, 38–46. [CrossRef]
28. Yang, X.; Wen, J.; Huang, A.; Lu, Y.; Meng, X.; Zhao, Y.; Wang, Y.; Meng, L. Short-Term Climatic Effect of Gyaring and Ngoring Lakes in the Yellow River Source Area, China. *Front. Earth Sci.* **2022**, *9*, 770757. [CrossRef]
29. Bennington, V.; Notaro, M.; Holman, K.D. Improving climate sensitivity of deep lakes within a regional climate model and its impact on simulated climate. *J. Clim.* **2014**, *27*, 2886–2911. [CrossRef]
30. Subin, Z.M.; Riley, W.J.; Mironov, D. An improved lake model for climate simulations: Model structure, evaluation, and sensitivity analyses in CESM1. *J. Adv. Mod. Earth Sys.* **2012**, *4*, M02001. [CrossRef]
31. Deng, B.; Liu, S.; Xiao, W.; Wang, W.; Jin, J.; Lee, X. Evaluation of the CLM4 lake model at a large and shallow freshwater lake. *J. Hydrometeorol.* **2012**, *14*, 636–649. [CrossRef]
32. Stepanenko, V.; Joehnk, K.D.; Machulskaya, E.; Perroud, M.; Subin, Z.; Nordbo, A.; Mammarella, I.; Mironov, D. Simulation of surface energy fluxes and stratification of a small boreal lake by a set of one-dimensional models. *Tellus A* **2014**, *66*, 21389. [CrossRef]
33. Hu, C.; Wang, Y.; Wang, W.; Liu, S.; Piao, M.; Xiao, W.; Lee, X. Trends in evaporation of a large subtropical lake. *Theor. Appl. Climatol.* **2017**, *129*, 159–170. [CrossRef]
34. Li, Z.; Ao, Y.; Lyu, S.; Lang, J.; Wen, L.; Stepanenko, V.; Meng, X.; Zhao, L. Investigation of the ice surface albedo in the Tibetan Plateau lakes based on the field observation and MODIS products. *J. Glaciol.* **2018**, *64*, 506–516. [CrossRef]
35. Zhang, Q.; Jin, J.; Wang, X.; Budy, P.; Barrett, N.; Null, S.E. Improving lake mixing process simulations in the Community Land Model by using K profile parameterization. *Hydrol. Earth Syst. Sci.* **2019**, *23*, 4969–4982. [CrossRef]
36. Xu, L.; Liu, H.; Du, Q.; Wang, L. Evaluation of the WRF-lake model over a highland freshwater lake in southwest China. *J. Geophys. Res.* **2016**, *121*, 13989–14005. [CrossRef]
37. Wen, L.; Nagabhatla, N.; Zhao, L.; Li, Z.; Chen, S. Impacts of salinity parameterizations on temperature simulation over and in a hypersaline lake. *Chin. J. Oceanol. Limnol.* **2015**, *33*, 790–801. [CrossRef]
38. Unesco. Algorithms for Computation of Fundamental Properties of Seawater. 1983. Available online: https://unesdoc.unesco.org/ark:/48223/pf0000059832 (accessed on 24 January 2022).
39. Low, R.D.H. A Generalized Equation for the Solution Effect in Droplet Growth. *J. Atmos. Sci.* **1969**, *26*, 608–611. [CrossRef]
40. Caldwell, D.R. The maximum density points of pure and saline water. *Deep Sea Res.* **1978**, *25*, 175–181. [CrossRef]
41. Mironov, D.; Heise, E.; Kourzeneva, E.; Ritter, B.; Schneider, N.; Terzhevik, A. Implementation of the lake parameterisation scheme FLake into the numerical weather prediction model COSMO. *Boreal Environ. Res.* **2010**, *15*, 218–230.
42. Su, D.; Wen, L.; Gao, X.; Leppäranta, M.; Song, X.; Shi, Q.; Kirillin, G. Effects of the largest lake of the Tibetan Plateau on the regional climate. *J. Geophys. Res.* **2020**, *125*, e2020JD033396. [CrossRef]
43. Wu, Y.; Huang, A.; Lazhu; Yang, X.; Tang, Y. Improvements of the coupled WRF-Lake model over Lake Nam Co, Central Tibetan Plateau. *Clim. Dyn.* **2020**, *55*, 2703–2724. [CrossRef]
44. Duan, S.; Fan, S.; Cao, G.; Liu, X.; Sun, Y. The changing features and cause analysis of the lakes in the source regions of the Yellow River from 1976 to 2014. *J. Glaciol. Geocryol.* **2015**, *37*, 745–756. (In Chinese with English abstract)
45. Gu, H.; Shen, X.; Jin, J.; Xiao, W.; Wang, Y. An application of a 1-D thermal diffusion lake model to Lake Taihu. *Acta Meteorol. Sin.* **2013**, *71*, 719–730. (In Chinese with English abstract)
46. Wilks, D.S. *Statistical Methods in the Atmospheric Sciences*; Academic Press: New York, NY, USA, 2011; 676p.
47. Crosman, E.T.; Horel, J.D. MODIS-derived surface temperature of the Great Salt Lake. *Remote Sens. Environ.* **2009**, *113*, 73–81. [CrossRef]
48. Liu, B.; Wan, W.; Xie, H.; Li, H.; Zhu, S.; Zhang, G.; Wen, L.; Hong, Y. A long-term dataset of lake surface water temperature over the Tibetan Plateau derived from AVHRR 1981–2015. *Sci. Data* **2019**, *6*, 48. [CrossRef]
49. Song, X.; Wen, L.; Li, M. Comparative study on applicability of different lake models to typical lakes in Qinghai-Tibetan Plateau. *Plateau Meteorol.* **2020**, *39*, 213–225. (In Chinese with English Abstract)
50. Kirillin, G. Modeling the impact of global warming on water temperature and seasonal mixing regimes in small temperate lakes. *Boreal Environ. Res.* **2010**, *15*, 279–293.
51. Pilla, R.M.; Williamson, C.E.; Adamovich, B.V.; Adrian, R.; Anneville, O.; Chandra, S.; Colom-Montero, W.; Devlin, S.P.; Dix, M.A.; Dokulil, M.T.; et al. Deeper waters are changing less consistently than surface waters in a global analysis of 102 lakes. *Sci. Rep.* **2020**, *10*, 20514. [CrossRef]
52. Yang, K.; Wu, H.; Qin, J.; Lin, C.; Tang, W.; Chen, Y. Recent climate changes over the Tibetan Plateau and their impacts on energy and water cycle: A review. *Glob. Planet. Chang.* **2014**, *112*, 79–91. [CrossRef]

53. Austin, J.A.; Colman, S.M. Lake Superior summer water temperatures are increasing more rapidly than regional air temperatures: A positive ice-albedo feedback. *Geophys. Res. Lett.* **2007**, *34*. [CrossRef]
54. Pius, B.; Marszelewski, W. Effect of climatic changes on the development of the thermal-ice regime based on the example of Lake Charzykowskie (Poland). *Bull. Geography. Phys. Geogr. Ser.* **2016**, *11*, 27–33. [CrossRef]
55. Woolway, R.I.; Kraemer, B.M.; Lenters, J.D.; Merchant, C.J.; O'Reilly, C.M.; Sharma, S. Global lake responses to climate change. *Nat. Rev. Earth Environ.* **2020**, *1*, 388–403. [CrossRef]
56. Wang, W.; Lee, X.; Xiao, W.; Liu, S.; Schultz, N.; Wang, Y.; Zhang, M.; Zhao, L. Global lake evaporation accelerated by changes in surface energy allocation in a warmer climate. *Nat. Geosci.* **2018**, *11*, 410–414. [CrossRef]

remote sensing

MDPI

Article

Environmental and Biophysical Effects of the Bowen Ratio over Typical Farmland Ecosystems in the Loess Plateau

Xueyuan Ren [1], Qiang Zhang [1,2], Ping Yue [1,2,*], Jinhu Yang [3] and Sheng Wang [1,2]

1 Institute of Arid Meteorology, China Meteorological Administration, Lanzhou 730020, China; renxy@iamcma.cn (X.R.); zhangqiang@cma.gov.cn (Q.Z.); wangs@iamcma.cn (S.W.)
2 Key Laboratory of Arid Climatic Change and Reducing Deserter of Gansu Province, China Meteorological Administration, Lanzhou 730020, China
3 Lanzhou Regional Climate Center, Lanzhou 730020, China; yangjh@iamcma.cn
* Correspondence: yuep@iamcma.cn

Abstract: The Bowen ratio (β) comprehensively reflects physical characteristics of the land-surface climate. In this study, eddy covariance systems installed at Dingxi and Qingyang were used to conduct energy distribution measurements and observations characteristic of semi-arid and semi-humid farmland ecosystems on the China Loess Plateau. We studied mechanisms by which eco-environmental factors influence β. Additionally, we investigated responses of physiological and ecological factors to water and heat exchange under seasonally dry and wet conditions within each farmland ecosystem. Our results showed that sensible heat flux in the semi-arid farmland was the main consumer of available energy. In the semi-humid area, latent heat flux in summer had the dominant role in energy distribution (mean β 0.71). The β in the semi-arid region was 1.5 times higher than that in the semi-humid region during the growing season. β increased with an increase in the vapor pressure deficit (*VPD*) and ground–air temperature difference ($Ts - Ta$), and decreased significantly with an increase in effective precipitation and soil moisture. The change in β with environmental factors was more clear-cut in semi-arid areas than in semi-humid areas. The Priestley–Taylor coefficient (α) and β satisfied a power function law in the growing season. There was a strong correlation between them, with the coefficients of determination for semi-humid and semi-arid areas being 0.62 and 0.72, respectively. β decreased with an increase in the normalized difference vegetative index (*NDVI*), with this phenomenon being more obvious in the semi-humid zone (R^2 = 0.40). β responded more rapidly to *NDVI* in the semi-arid area than in the semi-humid area. There was a negative exponential relationship between canopy stomatal conductance (*Gs*) and β, which displayed a stronger declining trend with the increase in *Gs* in the semi-arid area than in the semi-humid area. This study provides an important reference for the determination of land-surface characteristics of semi-arid and semi-humid farmland ecosystems on the Loess Plateau and for improving parameterization of land-surface processes.

Keywords: Loess Plateau; farmland ecosystem; physiological and ecological factors; Bowen ratio

Citation: Ren, X.; Zhang, Q.; Yue, P.; Yang, J.; Wang, S. Environmental and Biophysical Effects of the Bowen Ratio over Typical Farmland Ecosystems in the Loess Plateau. *Remote Sens.* **2022**, *14*, 1897. https://doi.org/10.3390/rs14081897

Academic Editors: Massimo Menenti, Yaoming Ma, Li Jia and Lei Zhong

Received: 20 March 2022
Accepted: 11 April 2022
Published: 14 April 2022

1. Introduction

The farmland ecosystem is the foundation of the existence and development of human society, and represents an orderly structure composed of organisms in an environment that can realize the conversion of energy and matter [1]. In-depth research and scientific understanding of the influencing factors and regulatory mechanisms of farmland ecosystems can provide high-quality information that may potentially guarantee the sustainable development of society. Due to the interactions between biogeochemical cycles, climate, soil available water, and plant physiology, the distributions of sensible and latent heat fluxes in farmland ecosystems differ [2–4]. Studies have shown that climate change affects the energy change between the earth and the atmosphere through the water cycle [5–8]. In addition to climate change, human activities (such as water conservancy projects and changes in land utilization) can also alter the water balance, thereby affecting the evapotranspiration process [5,9,10]. Meteorological and environmental factors, and the development of vegetation, can affect the distribution of surface energy during the growing season [11,12], in which the latent and sensible heat fluxes can change the environmental variables that affect matter and energy transfer between the atmosphere and the ecosystem. It has also been found that evapotranspiration in farmland ecosystems is mainly controlled by net radiation, but the regulation of latent heat transfers by vegetation indexes and canopy stomatal conductance (*Gs*) cannot be ignored [13–15]. In addition, vegetation phenology also affects the partitioning of net radiation to turbulent fluxes and soil heat flux [16–18].

The Bowen ratio (β) is a comprehensive physical index of the land surface climate, which comprehensively reflects the effects of microclimate and hydrological processes on ecosystem energy distribution and water use [19]. In previous studies of land–atmosphere interactions in ecosystems, β has been found to be a very important factor [2,20]. However, due to changes in regional climate conditions (such as temperature, precipitation, and soil moisture) [12,21,22], and seasonal differences in the physiological characteristics of vegetation [13–15], there are often large differences in β of ecosystems [23,24]. AmeriFlux observations have shown that the monthly average value of β in farmland ecosystems is between 0.26 and 1.3 [25–27]. Even during a relatively stable growing season, there are still significant differences in β among different farmland ecosystems [2,28]. For different ecosystems in the same climate region, there is an obvious contrast in their ability to regulate water and heat exchange, which is an internal factor leading to an apparent discrepancy in β. Precipitation is the most important driving factor in this process [2,24]. In the Loess Plateau, where precipitation fluctuates substantially, the ecological environment is fragile. The process of water and heat exchange in this region is also extremely sensitive to climate change [24], which makes β more dependent on the driving effect of environmental factors. Therefore, studying the seasonal variation in β of the typical farmland ecosystem of the Loess Plateau is of great significance for better understanding the land–atmosphere interaction mechanism in semi-arid regions.

The Loess Plateau in China is located within a typical semi-arid and semi-humid climate zone, which is not only a transitional zone for the East Asian summer monsoon, but is also positioned at the intersection of the water and heat gradient zones in China [25]. Therefore, the spatial distribution and temporal variation in land surface physical parameters in the Loess Plateau are very significant and highly sensitive to the advance and retreat of the monsoon and changes in its intensity. Due to the influence of the summer monsoon, the annual precipitation in this region is relatively concentrated, with about 65% of the total annual precipitation received from June to September. However, the interannual variability of precipitation is very large, which leads to visible spatial differences in the vegetation distribution [29,30]. The seasonal fluctuations of precipitation will undoubtedly lead to seasonal changes in β, which will cause the water and heat exchange of farmland ecosystems to display significant dry–wet conversion characteristics in turn [31,32]. The β and its influence on grassland in the Loess Plateau have been studied in depth [24], and the evapotranspiration in farmland ecosystems and its environmental impact in this region are also well understood [33–35]. However, there have been few studies of water and heat exchange in farmland ecosystems on the Loess Plateau, especially in terms of β and its influencing factors, despite it being a surface parameter that can comprehensively reflect the effects of water and heat. This has prevented the interactions between the land surface and the atmosphere in the farmland ecosystem of the Loess Plateau from being fully understood, and has prevented an in-depth understanding of water and heat exchange.

This study aimed to identify the effects of environmental factors on β of farmland ecosystems in different climate regions of the Loess Plateau using experimental land–atmosphere interaction data for two typical farmland ecosystems in Dingxi and Qingyang, which are semi-arid and semi-humid regions, respectively. The remainder of this paper is organized as follows: The study area, data and method employed are described in Section 2.

The results of physiological and ecological factors relating to water and heat exchange are investigated briefly in Section 3. The discussion of the results is provided in Section 4. In Section 5, the conclusions of this paper are presented.

2. Materials and Methods

2.1. Site Description

Dingxi Station (35.58°N, 104.62°E) is located in the elevated extension area of the Loess Plateau, with an altitude of 1896.7 m. Precipitation from June to September accounts for 66% of total annual precipitation. The mean annual temperature and precipitation are 6.7 °C and 386 mm, respectively. The average annual pan evaporation is 1400 mm, and the annual mean sunshine duration of 2344 h is typical for a semi-arid climate. Qingyang Station (35.44°N, 107.38°E) is located in Dongzhiyuan on the Longdong Loess Plateau at an altitude of 1421 m and has an average annual temperature and precipitation of 8.8 °C and 562 mm, respectively. Precipitation from June to September accounts for 67% of the annual total. The average annual pan evaporation is 1470 mm, and the annual average sunshine duration of 2250 h is typical for a semi-humid climate. During the study, the principal crops in Dingxi were potatoes and spring wheat, whereas the principal crops in Qingyang were winter wheat and spring corn. The canopy height of the crops during the vigorous growth period was approximately 50 cm [34]. Both experimental sites are rain-fed farmland. Figure 1 shows the specific geographic locations.

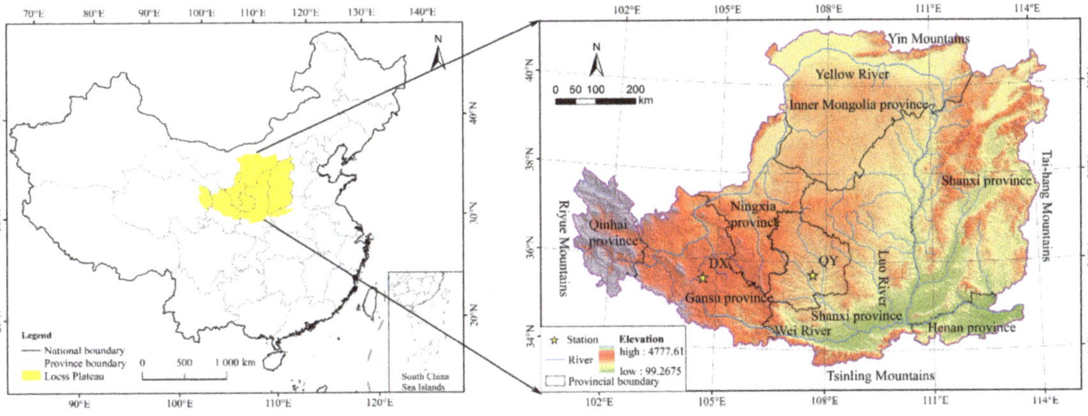

Figure 1. Geographical location of the study area. The stars show the Dingxi and Qingyang stations.

2.2. Observation Method and Data Processing

The data used in this paper include turbulence flux data observed by the eddy covariance system, temperature, humidity, and wind gradient data observed by a near-ground gradient tower, and radiation, soil temperature, and humidity gradient. Conventional observation data from meteorological stations were also used. The data period of turbulent flux in Dingxi is August 2016–May 2019; for Qingyang, multi-segment data were used, such as July 2011–July 2012, May 2013–October 2013, December 2015–May 2016, and May 2018–July 2019. In addition, the observation data of Dingxi and Xifeng meteorological stations from 1980 to 2010 were used. The installation height and details of the specific models of the measuring instruments are shown in Table 1.

Table 1. Measurement instruments and installation height.

Instrument	Type	Installation Height	
		Dingxi	Qingyang
Open Path CO_2/H_2O Gas Analyzer	Li-7500, Li-Cor (Lincoln, NE, USA)	2.5 m	3 m
Three-dimensional (3D) sonic anemometer	CSAT-3, Campbell (Logan, UT, USA)	2.5 m	3 m
Temperature and relative humidity probe	HMP45C-L, Vaisala (Vantaa, Finland)	1, 2, 4, 10, and 16 m	2, 4, 8, and 18 m
Net radiometer	CNR4, Kipp and Zoned (Delft, The Netherlands)	1.5 m	1.5 m
Self-calibrating heat flux sensor	HFP01SC-L50, Hukseflux (Delft, The Netherlands)	2, 5, and 10 cm	1, 2.5 and 5 cm
Soil temperature profile sensor	STP01-L50, Hukseflux	0, 5, 10, 20, 40, 50, and 80 cm	0, 5, 10, 20, 40, 60, and 90 cm
Water content reflectometer	CS616-L, Campbell	5, 10, 20, 40, 50, and 80 cm	5, 10, 20, 40, 60, and 90 cm

The turbulent flux data were processed by the EdiRe software (v1.5.0.32, Robert Clement, University of Edinburgh, Edinburgh, UK), which was developed by the University of Edinburgh for quality control and pre-processing. The operations included wild point removal, rotation coordinates, turbulence stationarity calculation, and water and CO_2 lag corrections. After quality control, the data were processed into 30 min average results. After excluding the outliers and precipitation period data, missing data for periods of less than 6 h were linearly interpolated, whereas missing data for periods of more than 6 h were interpolated by a look-up table method, which was based on the correlation between sensible heat, latent heat, net radiation, and water vapor pressure deficit (*VPD*) [36]. In addition, turbulent flux is greatly affected at night [37–40], and using midday data (09:00–15:00) can make the calculation results more reliable [2,41]. Beijing time was used in the study.

Due to the lack of station vegetation index observation data, the normalized difference vegetative index (*NDVI*) retrieved from the Aqua Moderate Resolution Imaging Spectroradiometer (MODIS, Phoenix, AZ, USA) data was used, with a temporal resolution of 16 days and a spatial resolution of 250 m (https://ladsweb.modaps.eosdis.nasa.gov/search/order/1/MOD13Q1--61). The *NDVI* of the experimental site was obtained from the average values of the four nearest grid points.

2.3. Energy Balance

The surface energy balance can be expressed as:

$$Rn = H + LE + G + S + Q \tag{1}$$

where Rn is the net radiation (W/m^2), H is the sensible heat flux (W/m^2), LE is the latent heat flux (W/m^2), G is the soil heat flux (W/m^2), S is canopy heat storage, and Q is the sum of all additional energy sources and sinks. Typically, Q is neglected as a small term. McCaughey [42] and Moore [43] suggested that canopy heat storage had a great effect on the degree of energy balance closure when the vegetation height was more than 8 m. Hence, the canopy heat storage term (S) was not taken into account in this study. The two principal methods for evaluating the degree of surface energy closure are the energy balance ratio (*EBR*) and the ordinary least squares (OLS) methods.

The EBR determines the degree of surface energy closure by calculating the ratio of turbulent flux to available energy during the study period:

$$EBR = \frac{\sum(H + LE)}{\sum(Rn - G)} \tag{2}$$

For EBR = 1, the surface energy is in an ideal equilibrium state. This method is ideal for evaluating the long-term energy closure state.

The OLS method is a simple regression model based on a hypothesis. It is based on the principle of the least squares method and is widely used in parameter estimation. When the sum of squares between the estimated value of the model and the experimental observation is at a minimum, the estimated value model is considered the optimal fitting model, and can describe the relationship between turbulent flux and available energy to the greatest extent. The slope of the regression model reflects the degree of surface energy closure. When the intercept of the regression curve is 0 and the slope is 1, the surface energy reaches the ideal closed state. Figure 2 shows the surface energy closure obtained by the OLS method. The black dotted line is the ideal state, and the grey shadow part (the slope ranges from 0.49 to 0.81) is the result reported by Li et al. [44] in evaluating the energy closure of flux observations of the terrestrial ecosystem in China. The slope calculated by the OLS method is closer to 1, indicating that the degree of surface energy closure is higher.

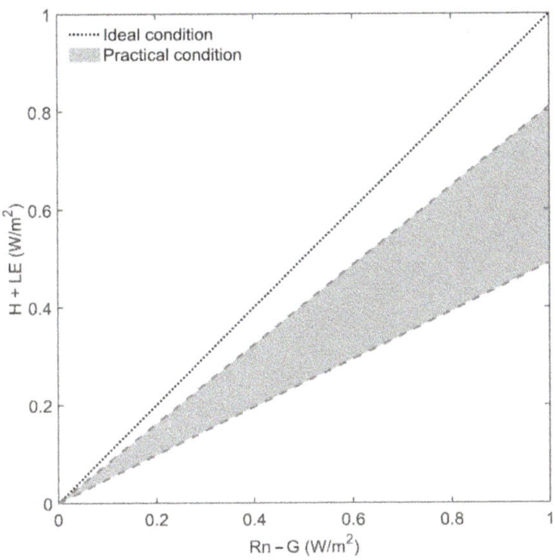

Figure 2. Surface energy closure obtained by the OLS method.

2.4. Soil Heat Flux Correction

The soil heat flux can be corrected to the surface value by the temperature integral method using the soil heat flux observed at 5 cm and soil temperatures at depths of 0 and 5 cm [45,46]:

$$G = G_5 + \frac{\rho_s c_s}{\Delta t} \sum_{z=5\text{cm}}^{z=0} [T((z_i, t) + \Delta t) - T(z_i, t)]\Delta z \tag{3}$$

where G is the soil heat flux corrected to the surface (W/m^2); G_5 is the soil heat flux at 5 cm measured by the heat flux observation board (W/m^2); $T(z_i, t)$ is the soil temperature (°C) at depths of 0 and 5 cm; and $\rho_s c_s$ is the volumetric heat capacity of the soil, which was $1.24 \times 10^6\ \text{J}/(\text{m}^3{\cdot}\text{K})$ in the calculation. The soil temperature at 0 cm can be converted from surface long-wave radiation as follows:

$$T_0 = \left(\frac{R_L^{\uparrow} - (1 - \varepsilon_g) R_L^{\downarrow}}{\varepsilon_g \sigma} \right)^{1/4} \tag{4}$$

where $R_L^{\uparrow}$ and $R_L^{\downarrow}$ are the upward and downward long-wave radiation from the surface (W/m^2), respectively; ε_g is the surface specific emissivity (0.96); and σ is the Stefan–Boltzmann constant (5.67×10^{-8} W/$\left(m^2 \cdot K^2\right)$).

2.5. Bowen Ratio

The Bowen ratio (β) comprehensively reflects the impact of climate and hydrological processes on the energy distribution of land surface ecosystems and is expressed as the ratio of the sensible and latent heat fluxes:

$$\beta = \frac{H}{LE} \tag{5}$$

For $\beta > 1$, the sensible heat flux plays a dominant role in the energy distribution; for $\beta < 1$, the latent heat flux plays the leading role.

2.6. Overall Land Surface Parameters

Canopy resistance (R_s), dynamic resistance (R_a), and climate resistance (R_i) are important parameters that affect the study of land-atmosphere interaction [14]. R_s is obtained by the Penman–Monteith equation [47]:

$$R_s = \frac{1}{Gs} = \frac{\rho c_p VPD + R_a LE(\Delta\beta - \gamma)}{\gamma LE} \tag{6}$$

where Gs is stomatal conductance (m/s); ρ is air density $\left(kg/m^3\right)$; c_p is the specific heat of air (1005 J/(kg·K)); VPD is the saturated vapor pressure deficit (kPa); LE is the latent heat flux (W/m^2); Δ is the slope of saturated vapor pressure curve (kPa/K); β is the Bowen ratio; γ is the dry and wet bulb constant (kPa/K); and R_a is the aerodynamic impedance at the height of the canopy (s/m). R_a can be calculated by the Monteith–Unsworth equation [48]:

$$R_a = \frac{u}{u_*^2} + 6.2u_*^{-0.67} \tag{7}$$

where u is the wind speed at 2 m (m/s); and u_* is the friction velocity (m/s). Δ can be calculated by the following formula:

$$\Delta = \frac{4098\left[0.6108exp\left(\frac{17.27T}{T+237.3}\right)\right]}{(T+237.3)^2} \tag{8}$$

where T is the air temperature (K).

Climatic resistance R_i reflects the degree of atmospheric demand for moisture under different surface available energy conditions [14]:

$$R_i = \frac{\rho c_p VPD}{\gamma(Rn - G)} \tag{9}$$

Using Equations (6), (7), and (9), it can be shown that R_s, R_a, and R_i satisfy the following relationship:

$$\frac{R_s}{R_a} = k_0 + k\sqrt{\frac{R_i}{R_a}} \tag{10}$$

where k_0 and k are empirical coefficients that depend on vegetation physiology and soil moisture status. For a clearer understanding of the impact of vegetation physiological processes on the water and heat exchange of the ecosystem, the normalized surface resistance

Rs^* is defined to eliminate the difference in aerodynamic resistance and climate resistance caused by local changes in the underlying surface [23]:

$$R_s^* = \frac{R_s}{\sqrt{R_i R_a}} \tag{11}$$

In addition, the Priestley–Taylor coefficient (α) can reflect the influence of environmental meteorological elements and vegetation physiological factors on ecosysterm evapotranspiration:

$$\alpha = \frac{LE}{LE_{eq}} \tag{12}$$

where LE_{eq} (W/m^2) is the latent heat flux on a wide surface that is not restricted by moisture, defined as:

$$LE_{eq} = \frac{\Delta(Rn - G)}{\Delta + \gamma} \tag{13}$$

The value of α can be used to determine whether the evapotranspiration of the ecosystem is restricted by water conditions. When $\alpha < 1$, the evaporation of the ecosystem is limited by water. When $\alpha > 1.26$, there is no water stress in the ecosystem, and the factor affecting evaporation is only surface available energy $(Rn - G)$ [49].

3. Results

3.1. Environmental Factor Variations

Figure 3 shows the seasonal variation in the characteristics of the environmental factors at Dingxi and Qingyang. The temperatures at the two stations had unimodal distributions (Figure 3(a1,b1)), reaching a maximum in midsummer and a minimum in January. The average temperature (*Ta*) at Dingxi was 0.2 °C higher than the 30 y (from 1988–2017) historical average (6.9 °C) during the experimental period. The monthly average minimum and maximum temperatures were −8.63 and 22.61 °C, respectively. *Ta* at Qingyang (10.2 °C) was 0.5 °C higher than the historical average; the average monthly minimum and maximum temperatures were −2.93 and 22.32 °C, respectively. During the experiment, maximum monthly precipitation in Dingxi and Qingyang was 148.1 (August 2017) and 246.2 mm (July 2018), respectively. Considering the monthly average precipitation over the past 30 years, the dry months at Dingxi station (April–October) accounted for 57.9% of the total; the dry months at Qingyang accounted for 35.0%. Due to the summer monsoon, more than 65% of the precipitation at the two stations was concentrated from July to September. Soil moisture greater than 40 cm in the tillage layer is very sensitive to precipitation. The monthly average *VPD* at Dingxi Station was 0.78 kPa and was largest in July 2017 (1.14 kPa) and smallest in January 2018 (0.21 kPa). The monthly average *VPD* at Qingyang Station was 0.69 kPa, with minimum and maximum values in November 2015 (0.13 kPa) and July 2019 (1.22 kPa), respectively. During the growing season, the average *VPD* values in Dingxi and Qingyang were 1.10 and 0.89 kPa, respectively. The seasonal variation patterns of the *NDVI* in Dingxi and Qingyang were basically the same. Vegetation growth and the *NDVI* increased in spring as precipitation and temperature gradually increased. The annual average values of the *NDVI* in Dingxi and Qingyang were 0.31 and 0.50, respectively; the respective growing season values were 0.40 and 0.56. The *Gs* trend was very similar to that of the *NDVI*, with an annual maximum from June to August. Precipitation was sufficient during the summer monsoon from June to August, and vegetation photosynthesis and transpiration were the strongest, which made the *Gs* reach the maximum. Vegetation physiological factors had the greatest impact on ecosystem evapotranspiration at this time, resulting in the peak value of α (Figure 3(a2,b2)).

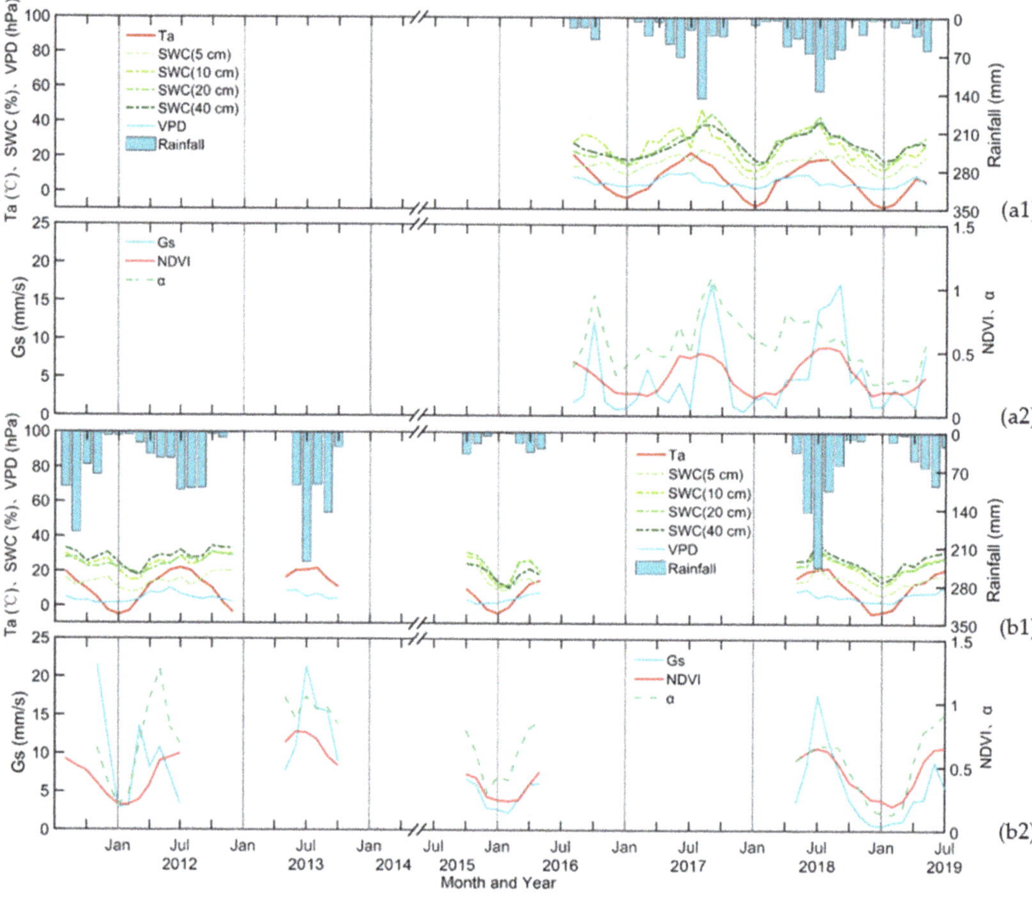

Figure 3. Seasonal and interannual variations of eco-environmental factors at Dingxi (**a1**,**a2**) and Qingyang (**b1**,**b2**) stations. *Ta* is the monthly mean temperature, *SWC* is the soil water content, *VPD* is the vapor pressure deficit, Rainfall represents monthly precipitation, *Gs* is the canopy stomatal conductance, α is the Priestley–Taylor coefficient, and *NDVI* is the normalized vegetation index.

3.2. Energy Balance Characteristics

The degree of energy closure is one of the most important criteria used to measure the quality and reliability of turbulent flux observation data [49]. The linear regression relationship between turbulent fluxes ($LE + H$) and available energy ($Rn - G$) is usually adopted to evaluate the reliability of eddy correlation system observations [50,51]. A large number of studies have shown that the energy imbalance observed by the eddy correlation method is between 10% and 30% [49,52]. Table 2 presents the energy closure of the Dingxi and Qingyang flux stations in the daytime, at night, and throughout the day. From the *OLS* results, the energy closure at both Dingxi and Qingyang was greater in daytime than that at night. This is due to the strong solar radiation during the day. The turbulent air is heated by the surface warming in the near-ground atmosphere, and the exchange is therefore greater [49,53], resulting in a high degree of energy closure. At night, due to the effect of stable atmosphere and low wind speed, the turbulent mixing is insufficient, resulting in low closing energy [12,54]. The increased uncertainty of net radiation measurement at night is also one of the reasons for the low energy closure [55]. When analyzing the degree of energy closure of flux observation systems, it has been reported that the energy balance components of the land surface are not in the same physical measurement plane, which is also the objective reason for the phenomenon of energy non-closure [56].

Table 2. Characteristics of the energy balance in different regions.

Site	Midday				Night				All Day			
	Sample Numbers	OLS		*EBR*	Sample Numbers	OLS		*EBR*	Sample Numbers	OLS		*EBR*
		Slope	R^2			Slope	R^2			Slope	R^2	
Dingxi	10,516	0.65	0.68	0.89	11,190	0.18	0.05	0.03	40,350	0.76	0.81	0.68
Qingyang	5149	0.71	0.71	0.81	7773	0.11	0.05	0.49	24,064	0.73	0.85	0.60

The *OLS* method commonly used in energy balance analyses is a simplified processing method based on the assumption of "no random error". The residual frequency distribution can be used to determine whether the model satisfies the hypothesis. Figure 4 shows the residual frequency distribution of the research site during the daytime. It can be seen that both residual density curves followed a normal distribution, indicating that the linear model obtained by the *OLS* method satisfies the assumption of "no random error". For observation data with a longer time scale, the *EBR* can balance the influence of error on energy closure. A large number of studies have shown that the energy closure calculated by this method is generally higher than that calculated by the *OLS* method [12]. It can be seen from Table 2 that the magnitude of the whole-day energy closure of Dingxi and Qingyang stations was between that of day and night, i.e., the closure of surface energy in daytime is greater than that during the whole day and at night. Compared with Li et al. [44], who used the *OLS* (0.49–0.81) and *EBR* (0.58–1.00) methods to evaluate the energy closure of ChinaFlux sites, the energy closure of the research site used in this study was slightly higher. This indicates that the accuracy of the observation data was high, and was suitable for a study of water and heat exchange in farmland ecosystems.

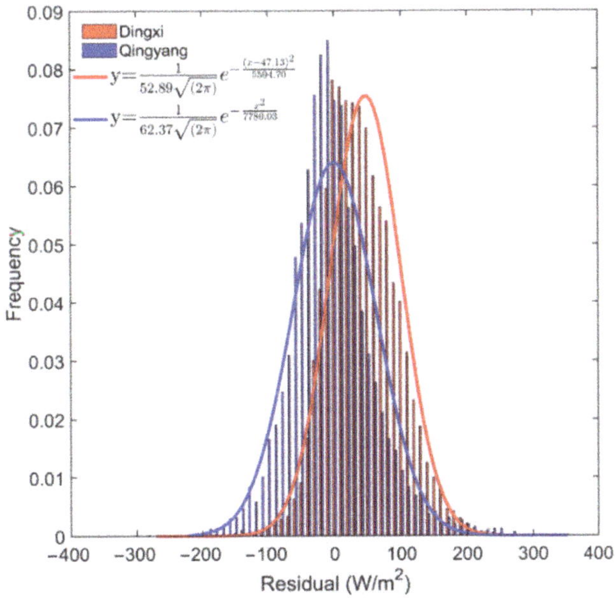

Figure 4. Frequency distribution and probability density curve of the energy balance residual.

3.3. Diurnal Cycle and Seasonal Variation in Energy Flux

When the annual average daily change in energy flux (Figure 5) was considered, the difference in the daily peak of net radiation between Dingxi and Qingyang was only 14.3 W/m^2 (Table 3). The daily peak values of the sensible heat flux were almost the same (120.3 and 122.3 W/m^2 in Dingxi and Qingyang, respectively). However, the difference in the latent heat flux between the two locations was large. The daily peak value of the latent heat flux in Qingyang in the semi-humid climate region was almost twice than that in Dingxi in the semi-arid climate region. The daily peak soil heat flux in Qingyang was approximately 2/3 of the Dingxi value. In addition, the latent heat flux (20.95 W/m^2) in the arid region of the Loess Plateau was less than the sensible heat flux (28.98 W/m^2), whereas the latent heat flux in the semi-humid region was larger than the sensible heat flux, with values of 41.41 and 28.50 W/m^2, respectively. According to the ratio of turbulent flux to net radiation (Figure 5e,f), the sensible heat flux in Dingxi was higher than the latent heat flux, with an average difference of 10.3%. In Qingyang, on the contrary, the average difference between the proportion of latent heat flux and the proportion of sensible heat flux was 7.6%. As a consequence, the sensible heat flux played a dominant role in the energy distribution in the semi-arid region of the Loess Plateau, whereas the latent heat flux was dominant in the semi-humid region.

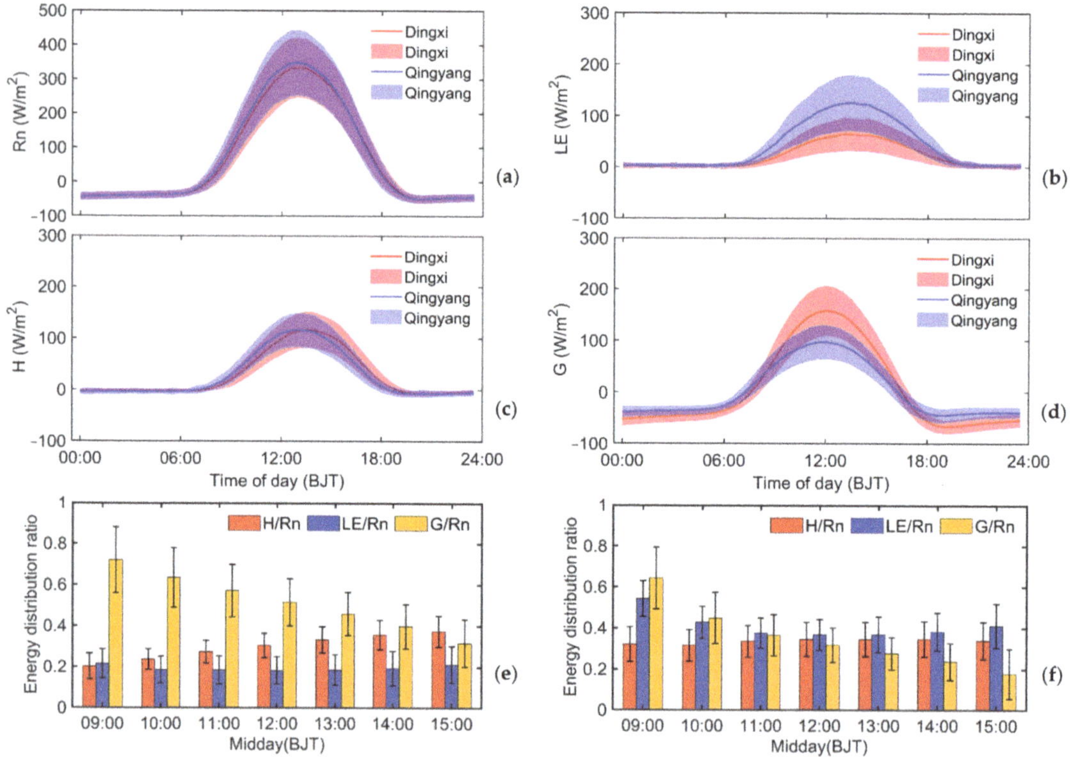

Figure 5. Annual average diurnal variation in energy flux (**a**–**d**), and the ratio of *H*, *LE*, and *G* to *Rn* at midday at Dingxi (**e**) and Qingyang (**f**). *Rn* is net radiation, *H* is the sensible heat flux, *LE* is the latent heat flux, and *G* is the soil heat flux.

Table 3. Annual average daily peak values and average daily values of the energy components.

Site	*Rn* (W/m²)		*LE* (W/m²)		*H* (W/m²)		*G* (W/m²)	
	Peak	Mean	Peak	Mean	Peak	Mean	Peak	Mean
Dingxi	334.10	67.29	64.14	20.95	120.29	28.98	157.88	6.08
Qingyang	348.37	71.74	125.83	41.41	122.32	28.50	96.93	1.62

Figure 6 shows the daily distribution characteristics of the energy flux on a seasonal scale. The net radiation in the semi-arid and semi-humid regions of the Loess Plateau (represented by Dingxi and Qingyang) reached a maximum in summer, with daily average radiation intensities of 109.9 and 119.8 W/m^2, respectively, and a minimum in winter, with average values of 25.3 and 26.2 W/m^2, respectively (Table 4). The daily peak values of net radiation in Dingxi and Qingyang occurred at 13:00, with summer values of 440.2 and 460.22 W/m^2 and winter values of 226.52 and 240.9 W/m^2, respectively. In summer and autumn, the latent heat flux at Qingyang in the semi-humid area accounted for 61.8% and 77.7% of the net radiation, respectively; whereas the sensible heat flux accounted for 25.5% and 33.8% of the net radiation, respectively. The latent heat flux at Dingxi in the semi-arid area in summer and autumn accounted for 32.9% and 41.3% of the net radiation, respectively; whereas the sensible heat flux accounted for 37.8% and 36.3%, respectively. The maximum value of soil heat flux at Dingxi and Qingyang appeared in summer, 125.82 W/m^2 and 214.49 W/m^2, respectively. Compared with other energy components, soil heat flux has a larger nighttime variability. This is related to two factors: the change in the direction of soil heat transfer caused by the process of soil freezing and thawing [57]; and the different energy intensity of soil radiation to the atmosphere caused by the diurnal variation in ground temperature difference in different seasons. Zhang et al. [58] reported a similar conclusion in the Loess Plateau.

Figure 7 shows the seasonal variation in energy flux in Dingxi and Qingyang. The net radiation in the semi-arid and semi-humid areas of the Loess Plateau had single-peak distributions, with maximum values in July (111.2 W/m^2) and June (125.0 W/m^2), respectively, and minimum values in December (21.6 and 19.5 W/m^2, respectively). Sensible heat flux and latent heat flux are not only restricted by net radiation, but are also affected by surface vegetation and soil moisture. Dingxi and Qingyang are bare land in the non-growing season, where precipitation is less than 20% of the annual total. Therefore, whether it is a semi-arid or semi-humid area of the Loess Plateau, the ratio of sensible heat flux to net radiation is relatively large (Figure 7(a1,a2)). For the semi-arid area of the Loess Plateau, with the increase in net radiation from March to May, and under the constraint of water conditions, net radiation is mainly transformed into sensible heat flux. Nevertheless, as the summer monsoon advances, the region that is located at the northern edge of the typical summer monsoon transition zone is affected by monsoon precipitation; as a result, the latent heat flux from June to September is generally equivalent to the sensible heat flux (Figure 7(a2)). However, due to the large fluctuation in monsoon precipitation, this area often experiences the phenomenon in which sensible heat and latent heat flux alternately dominate the energy distribution. In contrast, the unimodal distribution of the latent heat flux in Qingyang in the semi-humid region of the Loess Plateau was more prominent than that in Dingxi in the semi-arid region, with a peak value of 77.9 W/m^2 in July. The experimental results showed that the latent heat flux in the growing season in the semi-humid region of the Loess Plateau was 2.4 times than that of the sensible heat flux. The average latent and sensible heat fluxes were 69.4 and 29.1 W/m^2, respectively. In the same period, the latent and sensible heat fluxes in Dingxi were similar (33.1 and 39.4 W/m^2, respectively).

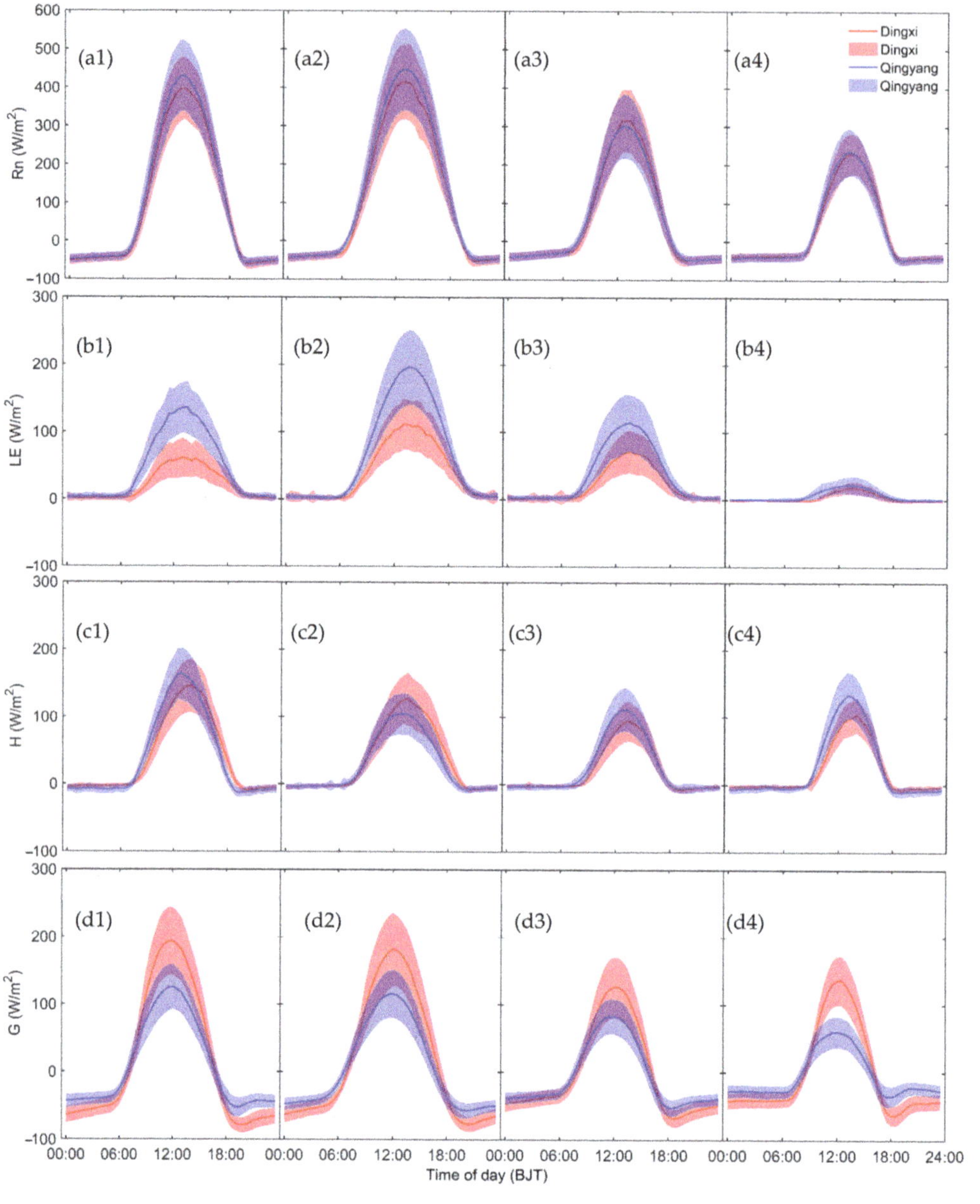

Figure 6. Seasonal average diurnal variation in the energy flux (**a**–**d**). Columns 1, 2, 3, and 4 represent spring, summer, autumn, and winter, respectively.

Table 4. Peak and daily average values of the seasonal average daily energy fluxes.

Energy Component (W/m^2)		Spring		Summer		Autumn		Winter	
		Dingxi	Qingyang	Dingxi	Qingyang	Dingxi	Qingyang	Dingxi	Qingyang
Rn	Peak	345.02	376.71	440.16	460.16	362.38	348.03	226.45	240.88
	Mean	67.73	77.14	109.87	119.84	78.76	75.02	25.34	26.20
LE	Peak	39.31	82.61	103.10	169.41	101.19	168.67	22.55	21.63
	Mean	11.77	25.70	36.00	74.11	32.54	58.31	5.23	6.64
H	Peak	148.75	162.54	143.10	112.81	107.21	110.68	93.27	118.66
	Mean	34.85	34.02	41.31	30.49	28.59	25.33	17.93	21.93
G	Peak	173.24	109.18	214.49	125.82	124.15	97.69	136.16	56.19
	Mean	10.34	6.68	20.07	8.67	1.24	−2.29	−3.29	−6.94

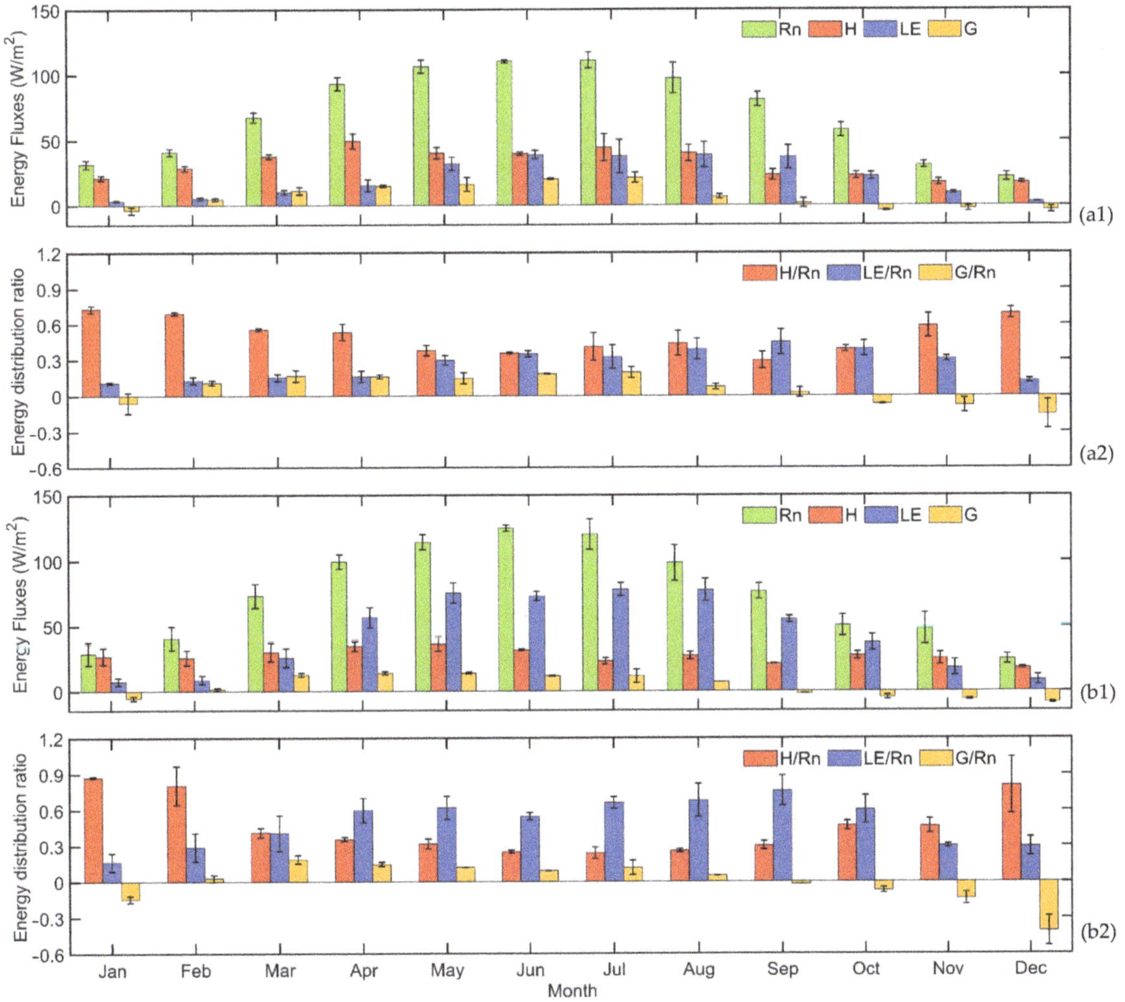

Figure 7. Seasonal variation in the energy components (**a1**,**a2**) and the ratio of *H*, *LE*, and *G* to *Rn* in Dingxi (**b1**) and Qingyang (**b2**).

In addition, the cumulative fraction curves of the sensible heat flux in Qingyang during the whole year and the growing season basically overlapped, when the cumulative fraction reached 0.6 (Figure 8a). When the cumulative fraction was about 0.4, the annual sensible heat flux was basically consistent with the latent heat flux in Dingxi. The cumulative fraction curves of the sensible and latent heat fluxes in the growing season in the two regions show that the sensible and latent heat fluxes in Dingxi in the semi-arid region were basically the same for cumulative fractions > 0.6, whereas the sensible heat flux in Qingyang in the semi-humid region was almost half of the latent heat flux. These results were consistent with the fluctuation in the sensible and latent heat fluxes in Figure 8b. The average latent heat flux (31.2 W/m^2) was less than the average sensible heat flux (34.1 W/m^2) in the semi-arid area during the growing season; contrary results were found for the semi-humid area. The interquartile ranges of sensible and latent heat fluxes in semi-arid regions are 23.27–43.42 W/m^2 and 21.00–42.74 W/m^2, respectively, and in sub-humid areas the ranges are 22.46–36.72 W/m^2 and 42.45–70.07 W/m^2, respectively. Zhang et al. [58] also showed that the summer latent heat and sensible heat flux in the semi-arid area of Northwest China are equivalent, and the summer latent heat flux in the semi-humid area is about twice the sensible heat flux.

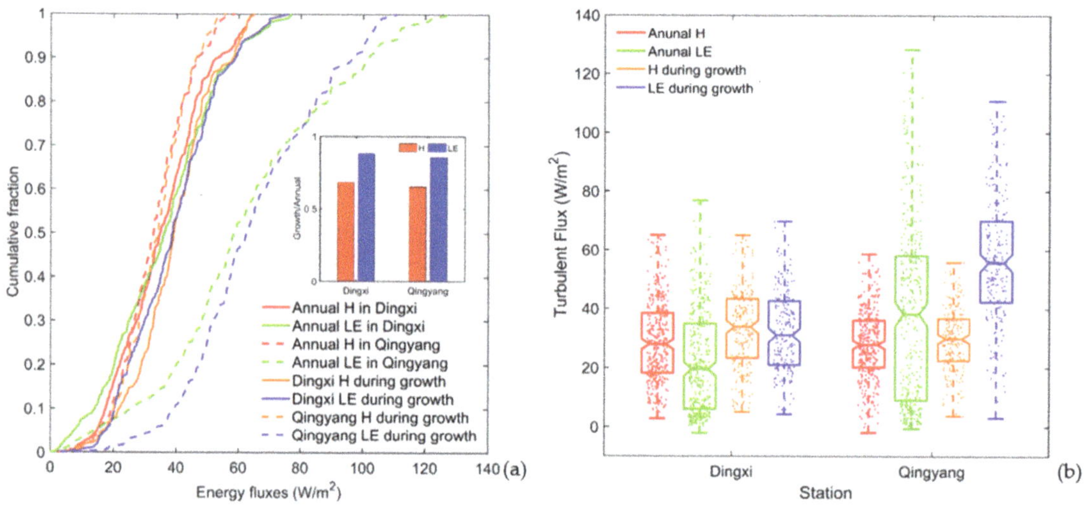

Figure 8. Cumulative fraction curve (**a**) and box plot (**b**) of the sensible and latent heat fluxes in Dingxi and Qingyang.

The soil heat flux changed from negative to positive from January to February, and from positive to negative from August to September (Figure 7(a2,b2)). This shows that the conversion of the heat source and heat sink occurred in the soil during these two periods. The soil is the conversion period of the heat sink and heat source in spring and summer. In winter, the soil heat flux is transferred from the deep layer to the shallow layer, which serves as a heat source to heat the atmosphere. Yue et al. [45] obtained consistent results in the study of semi-arid grassland in the Loess Plateau.

3.4. *Bowen Ratio Variation*

Figure 9 shows the seasonal variation in β. Overall, β of the two stations first decreased, then fluctuated slightly, and finally increased. The seasonal average β at Dingxi (6.58) and Qingyang (5.85) was highest in the winter and lowest in the summer (2.51 and 0.71, respectively). The growing season β at Dingxi fluctuated around 1, whereas that at Qingyang was < 1. Furthermore, β of both stations was low, with multi-year mean values of 2.11 and 0.77, respectively. Precipitation at Dingxi (320 mm) was 1.4 times that of Qingyang

(446 mm) during the same period, and β in the semi-arid area was 2.7 times that of the semi-humid area.

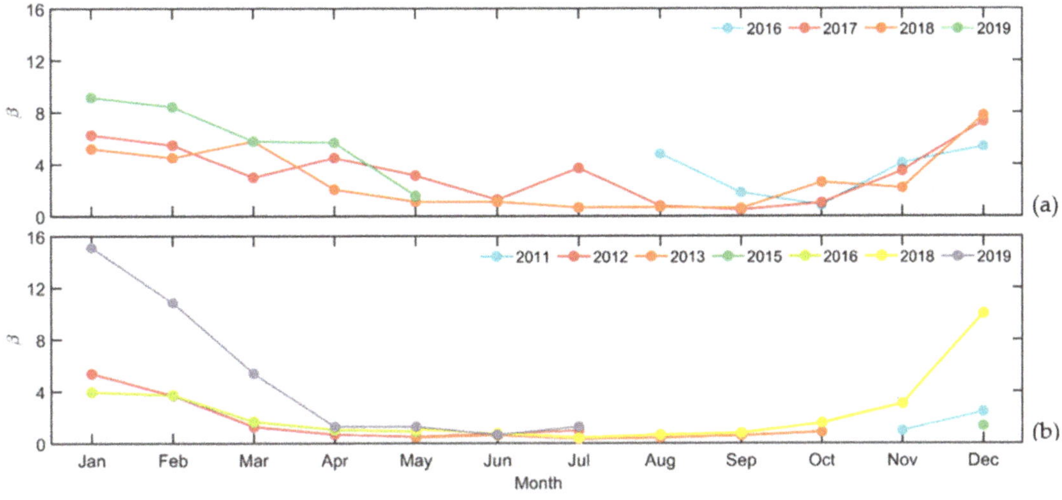

Figure 9. Seasonal variation in β in Dingxi (**a**) and Qingyang (**b**).

3.5. Environmental and Ecological Controls on Bowen Ratio

To fully understand the influence of environmental impacts on β of farmland ecosystems under dry and wet conditions on the Loess Plateau, the rainfall data from Dingxi and Qingyang Meteorological Stations in the past 30 years were used to divide the growth period into dry and wet months. Months with monthly precipitation greater than the average of the same period over many years were defined as wet months, whereas the opposite pattern indicated dry months. It was found that β was mainly affected by $Ts - Ta$, VPD, shallow SWC, and precipitation. Figure 10 and Table 5 show the relationship between β and environmental factors on a monthly scale. The regularity between β and $Ts - Ta$ in the semi-humid region ($R^2 = 0.51$) was better than that in the semi-arid region ($R^2 = 0.36$) whether under dry or wet conditions (Table 5). Under drought conditions, the correlation between VPD and β in the semi-humid region was more significant ($R^2 = 0.44$), and the coefficient of determination in the semi-arid area was only 0.29. Under humid conditions, the opposite result was observed (Figure 10b). The relationship between effective precipitation (defined as the daily precipitation amount that exceeded 0.5 mm in winter and 4.0 mm in other seasons [28]) and β was more significant, and β decreased significantly as effective precipitation increased (Figure 10c). As can be seen from the scatter points in Figure 10c, under dry conditions, β decreased more rapidly with increased precipitation in the semi-arid area than in the semi-humid area. Figure 10d shows the relationship between β and SWC; the decrease with SWC was more prominent in semi-arid areas. Under the humid condition, the goodness of fit in the semi-humid region was the highest, reaching 0.63.

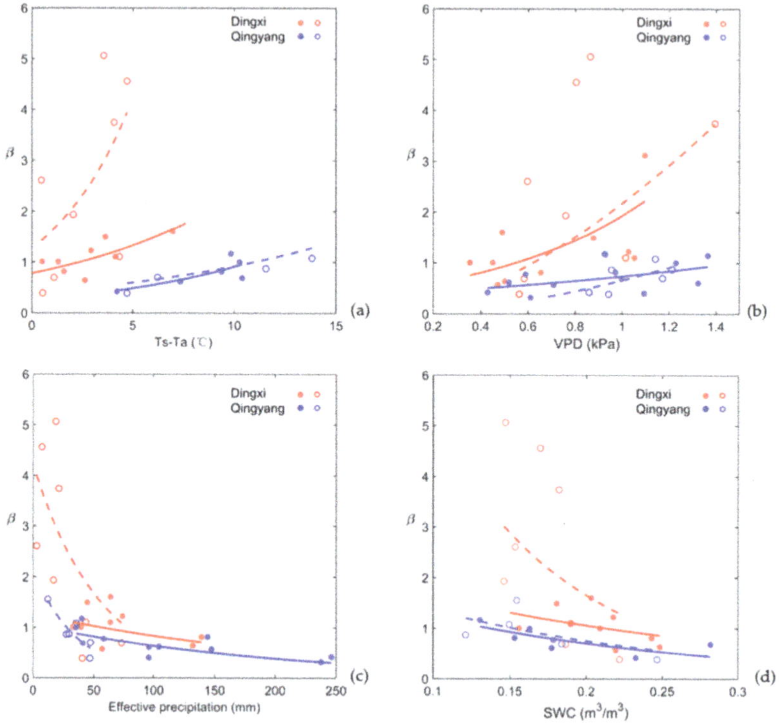

Figure 10. Relationship between environmental factors and β under different moisture conditions. The solid and hollow dots represent wet and dry conditions, respectively. The red and blue dashed line are the relationships of (**a**) $Ts - Ta$, (**b**) VPD, (**c**) effective precipitation, (**d**) SWC, and β in Dingxi and Qingyang, respectively.

Table 5. Statistical relationships between environmental factors and the Bowen ratio under dry and wet conditions.

Environmental Factors	Equation Parameters	Dingxi		Qingyang	
		Dry	Wet	Dry	Wet
$Ts - Ta$ (°C)		$y = ae^{bx}$	$y = ae^{bx}$	$y = ae^{bx}$	$y = ae^{bx}$
	a	1.27	0.78	0.39	0.26
	b	0.24	0.11	0.09	0.13
	R^2	0.36	0.59	0.51	0.58
	p	<0.01	<0.01	<0.01	<0.01
VPD (kPa)		$y = ax^b$	$y = ae^{bx}$	$y = ax^b$	$y = ae^{bx}$
	a	2.18	0.46	0.66	0.39
	b	1.62	1.45	1.76	0.65
	R^2	0.29	0.38	0.44	0.22
	p	<0.05	<0.05	<0.05	<0.05
Effective precipitation (mm)		$y = ae^{bx}$	$y = ae^{bx}$	$y = ae^{bx}$	$y = ae^{bx}$
	a	4.25	1.27	2.16	1.05
	b	−0.02	−0.004	−0.03	−0.01
	R^2	0.37	0.27	0.80	0.60
	p	<0.01	<0.01	<0.01	<0.01
SWC (m^3/m^3)		$y = ae^{bx}$	$y = ae^{bx}$	$y = ae^{bx}$	$y = ax^b$
	a	15.23	2.49	2.48	0.12
	b	−11.08	−4.24	−5.92	−1.10
	R^2	0.19	0.33	0.36	0.63
	p	<0.05	<0.01	<0.01	<0.01

The relationship between *NDVI*–*Gs*, *Gs*–α, and α–β was determined to explore the influence of ecological factors on the hydrothermal process. *Gs* increased exponentially as the *NDVI* increased. When the *NDVI* was the same in both areas, *Gs* in the semi-arid area was smaller than in the semi-humid area, and the correlation between *NDVI* and *Gs* in the semi-humid area was more significant ($R^2 = 0.57$). The regulation of transpiration by *Gs* is reflected by the Priestley–Taylor coefficient. Figure 11b shows that α increased logarithmically with *Gs*. The increasing trend of α in semi-humid areas as *Gs* increased is more significant than that in the semi-arid area, with tangent slopes of 0.19 and 0.26, respectively. In addition, β decreased exponentially as α increased; this trend was more pronounced in the semi-arid area (Figure 11c).

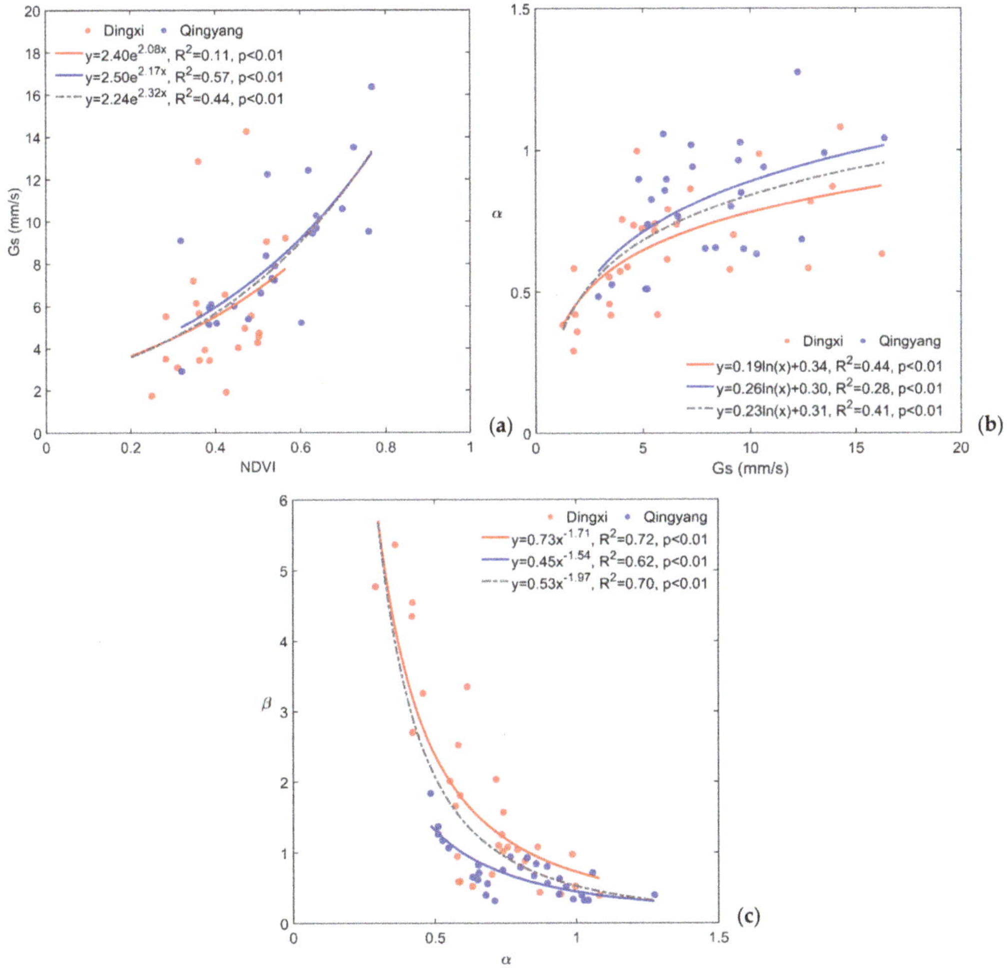

Figure 11. Relationship between (**a**) the *NDVI* and *Gs*, (**b**) *Gs* and α, (**c**) α and β on a monthly scale.

We produced path diagrams of the Dingxi and Qingyang stations to further analyze the direct and indirect effects of various influencing factors on β (Table 6). On a daily scale, eco-environmental factors at Dingxi and Qingyang explained 60% and 58% of the change in β, respectively. From an impact factor perspective, β in the semi-arid area was primarily influenced by the direct effect of *NDVI* and the indirect effect of *SWC* (path coefficients of −0.68 and −0.41, respectively). *Ta* and *NDVI* in the semi-humid area were the most important direct and indirect influencing factors, with contributions of 48% and 17%, respectively. In addition, *SWC* had a more significant overall regulatory effect on β in the semi-arid area, with a total path coefficient of −0.63. In the semi-humid area, β was primarily affected by *Ta*, with a total path coefficient of −0.53. It is worth noting that the effect of *Ta* on Bowen ratio in the semi-arid and semi-humid regions was completely opposite. This phenomenon can be explained by the influence of the mechanism found by Zhang et al. [59] in the transition zone affected by the summer monsoon, which is related to *Ta* and land surface evapotranspiration. Under the humid condition, the increase in temperature significantly increases the evapotranspiration, whereas under the drought condition, the increase in temperature decreases the land surface *SWC*, thus inhibiting the surface evapotranspiration. Yue et al. [33] found that the effects of *Ta* in dry and wet years on evapotranspiration were similar by studying the long series of observation data of the semi-arid grassland ecosystem on the Loess Plateau.

Table 6. Path coefficient between Bowen ratio and impact factor.

Site	Correlation Effect	*NDVI*	*Ta*	*VPD*	*SWC*
Dingxi	Direct	−0.68	0.42	0.11	−0.21
	Indirect	0.12	−0.36	0.39	−0.41
	Total	−0.57	0.06	0.51	−0.63
Qingyang	Direct	0.33	−0.90	0.54	−0.35
	Indirect	−0.84	0.37	−0.26	−0.01
	Total	−0.51	−0.53	0.28	−0.36

To comprehensively assess the effects of farmland ecosystem *Gs*, near-ground aerodynamic characteristics, and the local climate background on β, Cho et al. [24] defined the normalized surface impedance (Rs^*). Figure 12a shows the relationship between β and the monthly mean Rs^*. There was a significant linear relationship between the β and Rs^* in the farmland ecosystem of the Loess Plateau, with a slope of 0.49. As expected, α decreased more slowly as Rs^* increased in the semi-arid region than in the semi-humid region due to the growing season. The goodness of fit in the two regions was basically the same ($R^2 = 0.81$).

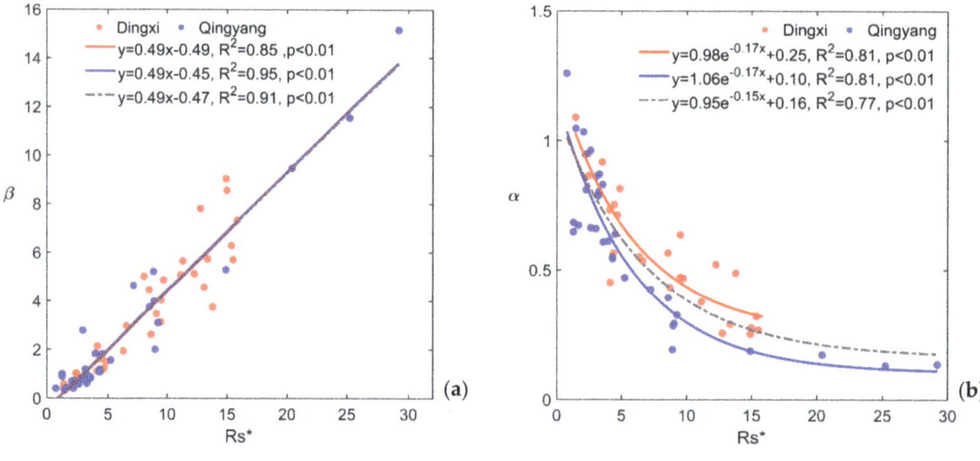

Figure 12. Relationships between the normalized surface impedance (Rs^*) and (**a**) β and (**b**) α on a monthly time scale. The gray dotted line in the figure represents the overall trend of the above relationship.

4. Discussion

4.1. Bowen Ratio Variation

β comprehensively reflects the wet and dry conditions of an ecosystem. A β of 1 is usually used as the critical value for assessing dry and wet conditions in an ecosystem [58]. Figure 9 indicates that β in the Loess Plateau was extremely sensitive to climate fluctuations; semi-arid areas represented by Dingxi were mostly dry. β still fluctuated around 1 in summer when precipitation was relatively concentrated. Thus, the land surface ecosystem in the region was largely on the border of the dry and slightly wet areas and was primarily in a state of drought in winter and spring. The fluctuation in β at Dingxi in the semi-arid area during the growing season was more significant than that of Qingyang in the semi-humid area. This was due to the low annual total precipitation in this area and the large precipitation variability during the summer monsoon, which led to prominent non-uniformity in the spatio-temporal distribution of soil moisture in the underlying surface and quickly transformed the surface energy distribution. In addition, the precipitation in the semi-humid area was 1.4 times higher than that in the semi-arid area; however, β of 2.7 was higher in the semi-arid area than in the semi-humid area. Thus, the β—which reflects the comprehensive effect of land surface water and heat—is more sensitive than relative precipitation in the transition zone from the semi-humid to semi-arid regions of the Loess Plateau.

Huang et al. [60] found that climate change aggravates drought in semi-arid regions of East Asia and humidifies semi-arid regions of North America. Our findings are comparable with β (0.26–0.36) in a farmland ecosystem observed by AmeriFlux [26–28]. The ratio in the farmland ecosystem in the Loess Plateau was approximately 2–3 times higher than that in similar climate regions of North America, indicating that the degree of water stress of the farmland ecosystem in the Loess Plateau is higher than that in North America. This also means that the impact of climate change, especially drought, on the ecosystem of the Loess Plateau in China is greater than that in North America.

4.2. Influence of Environmental and Ecological Factors on the Bowen Ratio

Solar radiation [61], rainfall dynamics, and irrigation [60] affect the water potential gradient and surface resistance, and change the soil moisture and evaporation composition of LE, thus impacting β [3,62]. For the same climatic region, β is more sensitive to the changes in $Ts - Ta$ under arid conditions. Further analysis showed that β decreased with an increase in SWC. This trend was more prominent in semi-arid areas, because the land surface parameters in semi-arid areas were more sensitive to changes in SWC [34]. Effective precipitation indirectly affects energy distribution by affecting the content of shallow soil moisture (Figure 10c) [28]. Comparing the relationship between β and effective precipitation in semi-arid and semi-humid regions of the Loess Plateau, it was found that the response speed of β to effective precipitation is faster than that of the latter. Thus, the change in the effective precipitation in this region had an important impact on the key physical parameters of the land surface ecosystem, and regulated the water and heat exchange of the ecosystem through ecological factors. In this study, the change in β with environmental factors was clearer in semi-arid areas than in semi-humid areas; further, the response of β to environmental factors was more severe under dry conditions than under humid conditions.

Ecological factors are the principal regulators of ET and affect the process of energy distribution through canopy conductance. Evapotranspiration is also closely related to vegetation growth [20,61]. Yue et al. [28] indicated that the $NDVI$ regulated transpiration in the ecosystem by affecting Gs, which in turn had an important regulatory effect on β. Differences between climatic regions with various vegetation growth statuses lead to differences in the correlation between the $NDVI$ and Gs. The ecosystem was less affected by water stress in the semi-humid area than in the semi-arid area, and the growth state of vegetation was better. Because vegetation transpiration is very sensitive to stomatal impedance, the intensity of transpiration increases as Gs increases. In theory, when Gs

reaches a critical value, α tends to reach equilibrium as *Gs* increases; thus, the magnitude of evapotranspiration is not limited by *Gs*, and vegetation transpiration reaches a maximum [63]. McHaughton and Spriggs [64] reported the theoretical critical value of ecosystem *Gs* (12–16 mm/s). Yue et al. [28] found that the critical value of *Gs* in semi-arid grasslands on the Loess Plateau was 8.2 mm/s. However, this study did not find an obvious *Gs* threshold in the farmland ecosystem in the Loess Plateau, indicating an ecosystem under water stress. As *Gs* increased, the α value in the semi-arid area increased less than in the semi-humid area, indicating that ecological factors in the semi-arid area had a more prominent inhibitory effect on evapotranspiration.

4.3. Biometeorological Controls on the Bowen Ratio

The water and heat distributions of farmland ecosystems in different climatic regions are restricted by different influencing factors. For semi-humid areas, the energy distribution is affected by the air temperature, which indicates the type of energy constraint. However, the hydrological conditions of β are more significant in semi-arid areas and show the characteristics of water restriction. Because the surface of the semi-humid area was relatively humid and the vegetation grew well, the increase in *Ta* promoted surface evaporation and vegetation transpiration, resulting in the latent heat flux being the dominant energy distribution process. The surface was dry in the semi-arid area, and therefore the lower *SWC* inhibited soil evaporation and vegetation transpiration, and energy distribution through the sensible heat flux played the leading role in the increase in β. The effect of *Ta* on β was the opposite in the semi-arid and semi-humid regions. This phenomenon is related to *Ta* and land surface evapotranspiration [32,59]. Under humid conditions, a temperature increase significantly increases evapotranspiration; under drought conditions, a temperature increase decreases the land surface *SWC*, thus inhibiting surface evapotranspiration.

In this study, Rs^* was used to show the consistencies and differences among different climatic zones of farmland ecosystems on the Loess Plateau. The consistency is reflected in the relationship between Rs^* and β, which was significantly linear (Figure 12a). Fraedrich et al. [65] found that Rs^* was larger when the surface was dry, that the physiological activity of vegetation was weak, and that the Bowen ratio was accordingly larger. Cho et al. [24] analyzed AmeriFlux observation data and found a positive correlation between β and Rs^*, which is very sensitive to the physiological vegetation processes. The slope of the regression equation between β and Rs^* was 0.21 ($R^2 = 0.65$). Yue et al. [28] found that the slope of β and Rs^* in the semi-arid grassland of the Loess Plateau was 0.34 ($R^2 = 0.95$). The slope of the regression equation for the relationship between β and Rs^* of the farmland ecosystem in the Loess Plateau was 0.49 ($R^2 = 0.91$), which indicates that the comprehensive influence of the eco-environmental factors of the farmland ecosystem on the Loess Plateau led to a larger β than that of the grassland ecosystem in this region. Therefore, the effect of water stress on the farmland ecosystem is more serious than that on the grassland ecosystem. Figure 12b shows the difference between the semi-arid and sub-humid regions and the relationship between α and Rs^*. A significant negative correlation between the two was found in the semi-arid farmland ecosystem of the Loess Plateau. For the same α value, the Rs^* of the semi-arid area was greater than that of the semi-humid area, indicating that Rs^* increased as the aridity of the regional climate increased. Additionally, Rs^* had a stronger limiting effect on evapotranspiration in semi-arid areas.

5. Conclusions

Under the background of global change, differences in precipitation caused by the duration of the dry and wet seasons in different climate regions are likely to be enhanced. The Loess Plateau in China has both semi-arid and semi-humid climate zones, and it is located in the transitional zone of the East Asian summer monsoon, in which the seasonal variation in precipitation is particularly obvious. Due to the intensity of the monsoon and its advancing northern edge, the land surface processes in the farmland ecosystem of this region display large interannual and seasonal changes. In particular, β, which characterizes the intensity of land surface water and heat exchanges, is very sensitive to environmental factors. The precipitation in Qingyang during the growing season was 1.4 times that in Dingxi, but the Bowen ratio in Dingxi was 2.7 times that in Qingyang, indicating that the land surface water and heat exchanges on the Loess Plateau were more sensitive than the precipitation variations. In addition, the farmland ecosystem of the Loess Plateau was more affected by water stress than farmland ecosystems in North America, and the impact of drought on the ecosystem in this region was also greater than in North America. According to the experimental observations, β in the Loess Plateau was extremely sensitive to climate fluctuations, and most of the time in the semi-arid region it indicated a dry state. Even during the summer monsoon, the regional land surface ecosystem was subject to water stress. The fluctuation in β around 1 indicated that the semi-arid region of the Loess Plateau was on the borderline of dry and slightly wet conditions for long periods. Compared with the semi-humid region of the Loess Plateau, there was less annual total precipitation in the semi-arid region and the precipitation variability was larger during the summer monsoon, which led to obvious fluctuations in the contributions of the latent and sensible heat fluxes to the energy distribution; this was a factor affecting the stability of β in the semi-arid region of the Loess Plateau during the summer monsoon.

The main environmental factors affecting β of the farmland ecosystem under different dry and wet conditions on the Loess Plateau are $Ts - Ta$, *VPD*, shallow *SWC*, and precipitation. The positive correlation between β and $Ts - Ta$ in the Loess Plateau was stronger in the semi-humid region than in the semi-arid region. Under drought conditions, the correlation between *VPD* and β in the semi-humid area was more significant. β of the farmland ecosystem in this region decreased with the increase in *SWC*, especially in semi-arid areas, because the land surface water and heat exchanges in semi-arid areas were more sensitive to changes in *SWC*. Ecological factors regulated evapotranspiration through canopy conductance, which then affected β. The *NDVI* controlled the transpiration process within the ecosystem by affecting *Gs*, and then played an important role in regulating β of the ecosystem. Theoretically, when *Gs* reaches a critical value, α tends to remain stable with a further increase in *Gs*, and the transpiration of vegetation reaches a maximum at this point. According to our observations, there was no obvious threshold for the farmland ecosystem on the Loess Plateau, but previous studies have found that there is a sensitivity threshold for *Gs* in the semi-arid grassland on the Loess Plateau, demonstrating that the farmland ecosystem in this region is in a state of water stress. Therefore, from the response of the land surface water and heat exchange processes to the summer monsoon, restoring farmland to grassland in the Loess Plateau may reduce the demand for water evapotranspiration and help to maintain the stability of the regional ecosystem. A path analysis showed that *NDVI* and *SWC* had obvious direct and indirect effects on β in the semi-arid area, whereas β in the semi-humid area was directly and indirectly affected by *Ta* and *NDVI*. The influence of *Ta* on β in the semi-humid and semi-arid regions had the opposite effect. An increase in temperature in the semi-humid region significantly increased evapotranspiration, whereas an increase in temperature in the semi-arid region decreased the land surface *SWC* and inhibited surface evapotranspiration.

Author Contributions: Conceptualization, Q.Z. and P.Y.; Data curation, X.R.; Formal analysis, X.R.; Funding acquisition, P.Y.; Resources, S.W.; Supervision, Q.Z., P.Y. and J.Y.; Validation, P.Y. and J.Y.; Visualization, X.R.; Writing–original draft, X.R.; Writing–review and editing, X.R. and P.Y. All authors have read and agreed to the published version of the manuscript.

Funding: This research was funded by the National Natural Science Foundation of China under grant Nos. U2142208, 41975016, 41705075, and the Basic Science Fund for Creative Research Groups of Gansu Province, grant number 20JR5RA121.

Informed Consent Statement: Not applicable.

Data Availability Statement: Data not available due to legal restrictions and observation team requirements.

Acknowledgments: We are grateful to the NASA Goddard Space Center for providing remote sensing data for this study.

Conflicts of Interest: The authors declare no conflict of interest.

References

1. Rana, G.; Katerji, N.; Mastrorilli, M.; El Moujabber, M.; Brisson, N. Validation of a model of actual evapotranspiration for water stressed soybeans. *Agric. For. Meteor.* **1997**, *86*, 215–224. [CrossRef]
2. Wilson, K.B.; Baldocchi, D.D.; Aubinet, M.; Berbigier, P.; Bernhofer, C.; Dolman, H.; Falge, E.; Field, C.; Goldstein, A.; Granier, A.; et al. Energy partitioning between latent and sensible heat flux during the warm season at FLUXNET sites. *Water Res. Res.* **2002**, *38*, 30-1–30-11. [CrossRef]
3. Chen, S.P.; Chen, J.Q.; Lin, G.H.; Zhang, W.L.; Miao, H.X.; Wei, L.; Huang, J.H.; Han, X.G. Energy balance and partition in Inner Mongolia steppe ecosystems with different land use types. *Agric. For. Meteor.* **2009**, *149*, 1800–1809. [CrossRef]
4. Biudes, M.S.; Vourlitis, G.L.; Machado, N.G.; Arruda, P.H.Z.D.; Nogueira, J.D.S. Patterns of energy exchange for tropical ecosystems across a climate gradient in Mato Grosso, Brazil. *Agricu. For. Meteor.* **2015**, *202*, 112–124. [CrossRef]
5. Milly, P.C.D.; Dunne, K.A. A Hydrologic Drying Bias in Water-Resource Impact Analyses of Anthropogenic Climate Change. *J. Am. Water Resour. Assoc.* **2017**, *53*, 822–838. [CrossRef]
6. Ashraf Vaghefi, S.; Mousavi, S.J.; Abbaspour, K.C.; Srinivasan, R.; Yang, H. Analyses of the impact of climate change on water resources components, drought and wheat yield in semiarid regions: Karkheh River Basin in Iran. *Hydrol. Process.* **2014**, *28*, 2018–2032. [CrossRef]
7. Mittal, N.; Bhave, A.G.; Mishra, A.; Singh, R. Impact of human intervention and climate change on natural flow regime. *Water Resour. Manag.* **2016**, *30*, 685–699. [CrossRef]
8. Lama, G.F.C.; Sadeghifar, T.; Azad, M.T.; Sihag, P.; Kisi, O. On the Indirect Estimation of Wind Wave Heights over the Southern Coasts of Caspian Sea: A Comparative Analysis. *Water* **2022**, *14*, 843. [CrossRef]
9. Kalinowska, M.B. Effect of water–air heat transfer on the spread of thermal pollution in rivers. *Acta Geophys.* **2019**, *67*, 597–619. [CrossRef]
10. Jodar-Abellan, A.; Valdes-Abellan, J.; Pla, C.; Gomariz-Castillo, F. Impact of land use changes on flash flood prediction using a sub-daily SWAT model in five Mediterranean ungauged watersheds (SE Spain). *Sci. Total Environ.* **2019**, *657*, 1578–1591. [CrossRef]
11. Yuan, G.H.; Zhang, L.; Liang, J.N.; Cao, X.J.; Liu, H.; Yang, Z.H. Understanding the partitioning of the available energy over the semi-arid areas of the Loess Plateau, China. *Atmosphere* **2017**, *8*, 87. [CrossRef]
12. Rahmana, M.M.; Zhang, W.; Wang, K. Assessment on surface energy imbalance and energy partitioning using ground and satellite data over a semi-arid agricultural region in north China. *Agric. Water Manag.* **2019**, *213*, 245–259. [CrossRef]
13. Lei, H.M.; Yang, D.W. Interannual and seasonal variability in evapotranspiration and energy partitioning over an irrigated cropland in the North China Plain. *Agric. For. Meteorol.* **2010**, *150*, 581–589. [CrossRef]
14. Ding, R.S.; Kang, S.Z.; Li, F.S.; Zhang, Y.Q.; Tong, L. Evapotranspiration measurement and estimation using modified Priestley–Taylor model in an irrigated maize field with mulching. *Agric. For. Meteorol.* **2013**, *168*, 140–148. [CrossRef]
15. Qiu, R.J.; Liu, C.W.; Cui, N.B.; Wu, Y.J.; Wang, Z.C.; Li, G. Evapotranspiration estimation using a modified Priestley-Taylor model in a rice-wheat rotation system. *Agric. Water Manag.* **2019**, *224*, 105755. [CrossRef]
16. Lama, G.F.C.; Crimaldi, M.; Pasquino, V.; Padulano, R.; Chirico, G.B. Bulk drag predictions of riparian arundo donax stands through UAV-acquired multispectral images. *Water* **2021**, *13*, 1333. [CrossRef]
17. Vélez-Nicolás, M.; García-López, S.; Barbero, L.; Ruiz-Ortiz, V.; Sánchez-Bellón, Á. Applications of unmanned aerial systems (UASs) in hydrology: A review. *Remote Sens.* **2021**, *13*, 1359. [CrossRef]
18. Yue, P.; Zhang, Q.; Ren, X.Y.; Yang, Z.S.; Li, H.Y.; Yang, Y. Environmental and biophysical effects of evapotranspiration in semiarid grassland and maize cropland ecosystems over the summer monsoon transition zone of China. *Agric. Water Manag.* **2022**, *264*, 107462. [CrossRef]
19. Perez, P.J.; Castellvi, F.; Martínez-Cob, A. A simple model for estimating the Bowen ratio from climatic factors for determining latent and sensible heat flux. *Agric. For. Meteorol.* **2008**, *148*, 25–37. [CrossRef]
20. Kang, M.; Zhang, Z.; Noormets, A.; Fang, X.; Zha, T.; Zhou, J.; Sun, G.; McNulty, S.G.; Chen, J. Energy partitioning and surface resistance of a poplar plantation in northern China. *Biogeoscience* **2015**, *12*, 4245–4259. [CrossRef]
21. Chen, X.; Yu, Y.; Chen, J.B.; Zhang, T.T.; Li, Z.C. Seasonal and interannual variation of radiation and energy fluxes over a rain-fed cropland in the semi-arid area of Loess Plateau, northwestern China. *Atmos. Res.* **2016**, *176*, 240–253. [CrossRef]
22. Feng, Y.; Gong, D.Z.; Mei, X.R.; Hao, W.P.; Tang, D.H.; Cui, N.B. Energy balance and partitioning in partial plastic mulched and non-mulched maize fields on the Loess Plateau of China. *Agric. Water Manag.* **2017**, *191*, 193–206. [CrossRef]
23. Yang, Z.S.; Zhang, Q.; Hao, X.C. Environmental and biological controls on monthly and annual evapotranspiration in China's Loess Plateau. *Theor. Appl. Clim.* **2019**, *137*, 1675–1692. [CrossRef]
24. Cho, J.; Oki, T.; Yeh, P.J.F.; Kim, W.; Kanae, S.; Otsuki, K. On the relationship between the Bowen ratio and the near-surface air temperature. *Theor. Appl. Clim.* **2012**, *108*, 135–145. [CrossRef]

25. Lu, Y.Q.; Kueppers, L.M. Surface energy partitioning over four dominant vegetation types across the United States in a coupled regional climate model (Weather Research and Forecasting Model 3–Community Land Model 3.5). *J. Geophys. Res.-Atmos.* **2012**, *117*, D06111. [CrossRef]
26. Wilson, K.B.; Baldocchi, D.D. Seasonal and interannual variability of energy fluxes over a broadleaved temperate deciduous forest in North American. *Agric. For. Meteorol.* **2000**, *100*, 1–18. [CrossRef]
27. Wever, L.A.; Flanagan, L.B.; Carlson, P.J. Seasonal and interannual variation in evapotranspiration, energy balance and surface conductance in a northern temperate grassland. *Agric. For. Meteorol.* **2002**, *112*, 31–49. [CrossRef]
28. Yue, P.; Zhang, Q.; Yang, Y.; Zhang, L.; Zhang, H.L.; Hao, X.C.; Sun, X.Y. Seasonal and inter-annual variability of the Bowen smith ratio over a semi-arid grassland in the Chinese Loess Plateau. *Agric. For. Meteorol.* **2018**, *252*, 99–108.
29. Wang, Y.; Wang, C.; Zhang, Q. Synergistic effects of climatic factors and drought on maize yield in the east of Northwest China against the background of climate change. *Theor. Appl. Clim.* **2020**, *143*, 1017–1033. [CrossRef]
30. Wang, X.; Wen, J.; Wei, Z.G.; Tian, H.; Wang, L.; Li, Z.Z.; Shi, X.G.; Zhang, T.T.; Liu, R.; Zhang, J.H. Study on water deficit of the topsoil over the Chinese Loess Plateau mesa region. *Plateau Meteorol.* **2009**, *28*, 530–538. (In Chinese)
31. Zhang, Q.; Zhang, L.; Huang, J.; Zhang, L.Y.; Wang, W.Y.; Sha, S. Spatial distribution of surface energy fluxes over the Loess Plateau in China and its relationship with climate and the environment. *Sci. China Earth Sci.* **2014**, *57*, 2135–2147. [CrossRef]
32. Zhang, Q.; Wang, S. On land surface processes and its experimental study in Chinese Loess Plateau. *Adv. Earth Sci.* **2008**, *23*, 167–173. (In Chinese)
33. Yue, P.; Zhang, Q.; Zhang, L.; Yang, Y.; Wei, W.; Yang, Z.S.; Li, H.Y.; Wang, S.; Sun, X.Y. Biometeorological effects on carbon dioxide and water-use efficiency within a semiarid grassland in the Chinese Loess Plateau. *J. Hydrol.* **2020**, *590*, 125520. [CrossRef]
34. Sun, M.; Dong, Q.G.; Jiao, M.Y.; Zhao, X.N.; Gao, X.R.; Wu, P.T.; Wang, A. Estimation of Actual Evapotranspiration in a Semiarid Region Based on GRACE Gravity Satellite Data—A Case Study in Loess Plateau. *Remote. Sens.* **2018**, *10*, 2032. [CrossRef]
35. Gao, X.R.; Sun, M.; Luan, Q.H.; Zhao, X.N.; Wang, J.C.; He, G.H.; Zhao, Y. The spatial and temporal evolution of the actual evapotranspiration based on the remote sensing method in the Loess Plateau. *Sci. Total Environ.* **2020**, *708*, 135111. [CrossRef]
36. Falge, E.; Baldocchi, D.; Olson, R.; Anthoni, P.; Aubinet, M.; Bernhofer, C.; Burba, G.; Ceulemans, G.; Clement, R.; Dolman, H.; et al. Gap filling strategies for long term energy flux data sets. *Agric. For. Meteorol.* **2001**, *107*, 71–77. [CrossRef]
37. Lavigne, M.B.; Ryan, M.G.; Anderson, D.E.; Baldocchi, D.D.; Crill, P.M.; Fitzjarrald, D.R.; Goulden, M.L.; Gower, S.T.; Massheder, J.M.; McCaughey, J.H.; et al. Comparing nocturnal eddy covariance measurements to estimates of ecosystem respiration made by scaling chamber measurements at six coniferous boreal sites. *J. Geophys. Res.* **1997**, *102*, 28977–29985. [CrossRef]
38. Mahrt, L. Stratified atmospheric boundary layers. *Bound.-Layer Meteorol.* **1999**, *90*, 375–396. [CrossRef]
39. Law, B.E.; Goldstein, A.H.; Anthoni, P.M.; Unsworth, M.H.; Panek, J.A.; Bauer, M.R.; Fracheboud, J.M.; Hultman, N. Carbon dioxide and water vapor exchange by young and old ponderosa pine ecosystems during a dry summer. *Tree Physiol.* **2001**, *21*, 299–308. [CrossRef]
40. Aubinet, M.; Grelle, A.; Ibrom, A.; Rannik, Ü.; Moncrieff, J.; Foken, T.; Kowalski, A.S.; Martin, P.H.; Berbigier, P.; Bernhofer, C.; et al. Estimates of the Annual Net Carbon and Water Exchange of Forests: The EUROFLUX Methodology. *Adv. Ecol. Res.* **1999**, *30*, 114–175.
41. McGloin, R.; Šigut, L.; Fischer, M.; Foltýnová, L.; Chawla, S.; Trnka, M.; Pavelka, M.; Marek, M.V. Available energy partitioning during drought at two Norway spruce forests and a European Beech forest in Central Europe. *J. Geophys. Res-Atmos.* **2019**, *124*, 3726–3742. [CrossRef]
42. McCaughey, J.H. Energy balance storage terms in a mature mixed forest at Petawawa, Ontario—A case study. *Bound.-Layer Meteorol.* **1985**, *31*, 89–101. [CrossRef]
43. Moore, C.J. Frequency response corrections for eddy correlation systems. *Bound.-Layer Meteorol.* **1986**, *37*, 17–35. [CrossRef]
44. Li, Z.Q.; Yu, G.R.; Wen, X.F.; Zhang, L.M.; Ren, C.Y.; Fu, Y.L. Energy balance closure at ChinaFLUX sites. *Sci. China Earth Sci.* **2005**, *48*, 51–62.
45. Yue, P.; Zhang, Q.; Zhao, W.; Wang, R.Y.; Zhang, L.; Wang, W.Y.; Shi, J.S.; Hao, X.C. Influence of environmental factors on land-surface water and heat exchange during dry and wet periods in the growing season of semiarid grassland on the Loess Plateau. *Sci. China Earth Sci.* **2015**, *58*, 2002–2014. [CrossRef]
46. Yang, K.; Wang, J.M. A temperature prediction-correction method for estimating surface soil heat flux from soil temperature and moisture data. *Sci. China Ser. D-Earth Sci.* **2008**, *51*, 721–729. [CrossRef]
47. Kumagai, T.; Saitoh, T.M.; Sato, Y.; Morooka, T.; Manfroi, O.J.; Kuraji, K.; Suzuki, M. Transpiration, canopy conductance and the decoupling coefficient of a lowland mixed dipterocarp forest in Sarawak, Borneo: Dry spell effects. *J. Hydrol.* **2004**, *287*, 237–251. [CrossRef]
48. Hossen, M.S.; Mano, M.; Miyata, A.; Baten, M.A.; Hiyama, T. Surface energy partitioning and evapotranspiration over a double-cropping paddy field in Bangladesh. *Hydrol. Proc.* **2012**, *26*, 1311–1320. [CrossRef]
49. Arain, M.A.; Black, T.A.; Barr, A.G.; Griffis, T.J.; Morgenstern, K.; Nesic, Z. Year-round observations of the energy and water vapour fluxes above a boreal black spruce forest. *Hydrol. Proc.* **2003**, *17*, 3581–3600. [CrossRef]
50. Rodrigues, T.R.; Vourlitis, G.L.; Lobo, F.D.; de Oliveira, R.G.; Nogueira, J.D. Seasonal variation in energy balance and canopy conductance for a tropical savanna ecosystem of south central Mato Grosso, Brazil. *J. Geophys. Res.-Biogeosci.* **2014**, *119*, 1–13. [CrossRef]

51. Tian, F.Q.; Yang, P.J.; Hu, H.C.; Liu, H. Energy balance and canopy conductance for a cotton field under film mulched drip irrigation in an arid region of northwestern China. *Agric. Water Manag.* **2017**, *179*, 110–121. [CrossRef]
52. Foken, T.; Wimmer, F.; Mauder, M.; Thomas, C.; Liebethal, C. Some aspects of the energy balance closure problem. *Atmos. Chem. Phys.* **2006**, *6*, 4395–4402. [CrossRef]
53. Twine, T.E.; Kustas, W.P.; Norman, J.M.; Cook, D.R.; Houser, P.R.; Meyers, T.P.; Prueger, J.H.; Starks, P.J.; Wesely, M.L. Correcting eddy-covariance flux underestimates over a grassland. *Agric. For. Meteor.* **2000**, *103*, 279–300. [CrossRef]
54. Sánchez, J.M.; Caselles, V.; Rubio, E.M. Analysis of the energy balance closure over a FLUXNET boreal forest in Finland. *Hydrol. Earth System Sci.* **2010**, *14*, 1487–1497. [CrossRef]
55. Majozi, N.P.; Mannaerts, C.M.; Ramoelo, A.; Mathieu, R.; Nickless, A.; Verhoef, W. Analysing surface energy balance closure and partitioning over a semi-arid savanna FLUXNET site in Skukuza, Kruger National Park, South Africa. *Hydrol. Earth Syst. Sci.* **2017**, *21*, 3401–3415. [CrossRef]
56. Yue, P.; Zhang, Q.; Niu, S.J.; Cheng, H.; Wang, X.Y. Effects of the soil heat flux estimates on surface energy balance closure over a semi-arid grassland. *Acta Meteorol. Sin.* **2011**, *25*, 774–782. [CrossRef]
57. Zhang, H.H.; Xiao, H.B.; Qi, D.L. Features of land surface process over wetland at Tibetan Plateau during soil freezing and thawing periods. *Acta. Meteor. Sin.* **2017**, *75*, 481–491. (In Chinese)
58. Zhang, Q.; Sun, Z.X.; Wang, S. Analysis of Variation Regularity of Land-Surface Physical Quantities Over the Dingxi Region of the Loess Plateau. *Chin. J. Geophys.* **2011**, *54*, 436–447. (In Chinese) [CrossRef]
59. Zhang, Q.; Yang, Z.; Hao, X.; Yue, P. Conversion features of evapotranspiration responding to climate warming in transitional climate regions in northern China. *Clim. Dyn.* **2019**, *52*, 3891–3903. [CrossRef]
60. Huang, J.P.; Yu, H.P.; Guan, X.D.; Wang, G.Y.; Guo, R.X. Accelerated dryland expansion under climate change. *Nat. Clim. Change* **2016**, *6*, 166–171. [CrossRef]
61. Yang, Z.S.; Zhang, Q.; Hao, X.C.; Yue, P. Changes in evapotranspiration over global semiarid regions 1984–2013. *J. Geophys. Res.-Atmos.* **2019**, *124*, 2946–2963. [CrossRef]
62. Ozdogan, M.; Rodell, M.; Beaudoing, H.K.; Toll, D.L. Simulating the Effects of Irrigation over the United States in a Land Surface Model Based on Satellite-Derived Agricultural Data. *J. Hydrometeorol.* **2010**, *11*, 171–184. [CrossRef]
63. Blanken, P.D.; Black, T.A.; Yang, P.C.; Neumann, H.H.; Nesic, Z.; Staebler, R.; Hartog, G.D.; Noval, M.D.; Lee, X. Energy balance and canopy conductance of a boreal aspen forest: Partitioning overstory and understory components. *J. Geophys. Res.-Atmos.* **1997**, *102*, 28915–28927. [CrossRef]
64. Mchaughton, K.G.; Spriggs, T.W. A mixed-layer model for regional evaporation. *Bound.-Layer Meteorol.* **1986**, *34*, 243–262. [CrossRef]
65. Fraedrich, K.; Kleidon, A.; Lunkeit, F. A green planet versus a desert world: Estimating the effect of vegetation extremes on the atmosphere. *J. Clim.* **1999**, *12*, 3156–3163. [CrossRef]

remote sensing

MDPI

Article

Assessment of Different Complementary-Relationship-Based Models for Estimating Actual Terrestrial Evapotranspiration in the Frozen Ground Regions of the Qinghai-Tibet Plateau

Chengpeng Shang [1,2], Tonghua Wu [1,3,*], Ning Ma [4], Jiemin Wang [1], Xiangfei Li [1], Xiaofan Zhu [1], Tianye Wang [5], Guojie Hu [1], Ren Li [1], Sizhong Yang [1], Jie Chen [1], Jimin Yao [1] and Cheng Yang [1,6]

1 Cryosphere Research Station on the Qinghai-Tibet Plateau, State Key Laboratory of Cryospheric Science, Northwest Institute of Eco-Environment and Resources, Chinese Academy of Sciences, Lanzhou 730000, China; cpshang@lzb.ac.cn (C.S.); jmwang@lzb.ac.cn (J.W.); lxf@lzb.ac.cn (X.L.); zxf_jc@lzb.ac.cn (X.Z.); huguojie123@lzb.ac.cn (G.H.); liren@lzb.ac.cn (R.L.); syang@gfz-potsdam.de (S.Y.); chenjie@lzb.ac.cn (J.C.); yjm@lzb.ac.cn (J.Y.); yangc@hebtu.edu.cn (C.Y.)
2 University of Chinese Academy of Sciences, Beijing 100049, China
3 Southern Marine Science and Engineering Guangdong Laboratory, Guangzhou 511458, China
4 Key Laboratory of Water Cycle and Related Land Surface Processes, Institute of Geographic Sciences and Natural Resources Research, Chinese Academy of Sciences, Beijing 100101, China; ningma@igsnrr.ac.cn
5 School of Water Conservancy Engineering, Zhengzhou University, Zhengzhou 450001, China; wangtianye@zzu.edu.cn
6 School of Geographical Sciences, Hebei Normal University, Shijiazhuang 050024, China
* Correspondence: thuawu@lzb.ac.cn

Abstract: Actual evapotranspiration (ET_a) is important since it is an important link to water, energy, and carbon cycles. Approximately 96% of the Qinghai-Tibet Plateau (QTP) is underlain by frozen ground, however, the ground observations of ET_a are particularly sparse–which is especially true in the permafrost regions–leading to great challenge for the accurate estimation of ET_a. Due to the impacts of freeze-thaw cycles and permafrost degradation on the regional ET process, it is therefore urgent and important to find a reasonable approach for ET_a estimation in the regions. The complementary relationship (CR) approach is a potential method since it needs only routine meteorological variables to estimate ET_a. The CR approach, including the modified advection-aridity model by Kahler (K2006), polynomial generalized complementary function by Brutsaert (B2015) and its improved versions by Szilagyi (S2017) and Crago (C2018), and sigmoid generalized complementary function by Han (H2018) in the present study, were assessed against in situ measured ET_a at four observation sites in the frozen ground regions. The results indicate that five CR-based models are generally capable of simulating variations in ET_a, whether default and calibrated parameter values are employed during the warm season compared with those of the cold season. On a daily basis, the C2018 model performed better than other CR-based models, as indicated by the highest Nash-Sutcliffe efficiency (NSE) and lowest root mean square error (RMSE) values at each site. On a monthly basis, no model uniformly performed best in a specific month. On an annual basis, CR-based models estimating ET_a with biases ranging from −94.2 to 28.3 mm year^{-1}, and the H2018 model overall performed best with the smallest bias within 15 mm year^{-1}. Parameter sensitivity analysis demonstrated the relatively small influence of each parameter varying within regular fluctuation magnitude on the accuracy of the corresponding model.

Keywords: actual evapotranspiration; complementary relationship; permafrost; seasonally frozen ground; Qinghai-Tibet Plateau

Citation: Shang, C.; Wu, T.; Ma, N.; Wang, J.; Li, X.; Zhu, X.; Wang, T.; Hu, G.; Li, R.; Yang, S.; et al. Assessment of Different Complementary-Relationship-Based Models for Estimating Actual Terrestrial Evapotranspiration in the Frozen Ground Regions of the Qinghai-Tibet Plateau. *Remote Sens.* **2022**, *14*, 2047. https://doi.org/10.3390/rs14092047

Academic Editor: Sayed M. Bateni

Received: 10 March 2022
Accepted: 21 April 2022
Published: 25 April 2022

1. Introduction

Terrestrial evapotranspiration (ET) is an important part of the water cycle and surface energy balance, which is not only the nexus of the water cycle in the soil–vegetation–atmosphere continuum, but also a crucial link in the mass exchange and energy balance of the land surface [1,2]. Accurate estimation of ET is important to deeply understand how much liquid or solid water from land transforms as vapor in the atmosphere, which has significant implications for regional weather or climate conditions, and water resource management, such as ecological conservation and agricultural management, etc. [3–5].

In practice, ET estimation is often challenging because of complex land–atmosphere interactions, which often involve relevant micrometeorological, hydrological, or ecological methods [6,7]. In recent decades, great progress has been made on the theories and technologies of actual evapotranspiration (ET_a) regardless of observations or simulations, significantly contributing to our knowledge of the ET process [8–12]. Current ET estimation approaches, such as land surface models or remote sensing technologies, have been successfully applied to various climatic or ecosystem zones [13–15]; however, the above approaches generally have relatively higher uncertainty in parameterization schemes at specific zones and require detailed information on vegetation or soil properties [16], which is especially difficult for data-scarce regions. For example, the Penman-Monteith (PM) [17,18] equation is often used to calculate potential evapotranspiration (ET_p), which is close to ET_a under ample water supply conditions, and with a higher accuracy. However, under water-limited conditions, the PM method assumes typically that ET_a is proportional to the ET_p (rescaling with soil moisture content), giving rise to practical difficulties such as soil moisture data unavailability in harsh environments where the estimation has a higher uncertainty. In addition, the assumption of the positive relationship between ET_a and ET_p may be rejected by observational evidence that supports negative correlations between the two variables [19]. Han [20] emphasized that the correlation between ET_a and ET_p depends mainly on water availability rather than always being positive.

Bouchet [21] first proposed another important hypothesis about ET_a estimation, often called the "Bouchet hypothesis". In contrast to the PM method, Bouchet [21] described the negative relationship between ET_a and ET_p caused by land-atmosphere feedback, which is also referred to as the complementary relationship (CR) approach. The greatest advantage of the CR approach is that it does not require any vegetation and soil information, and only a few model parameters and routine meteorological forcing data are needed [22,23]. Various models based on the CR approach have been proposed over the past decades [24]. The most widely used CR-based models include the AA (advection-aridity) model [25], CRAE (complementary relationship areal evapotranspiration) model [26], and GG (Granger and Gray) model [27], which are all based on the linearly symmetric CR approach. However, some later studies [28,29] have found that the complementary relationships are not symmetric, and they further proposed various asymmetric linear CR-based functions. Nevertheless, Han [30] found that both symmetric and asymmetric linear CR-based functions have poor generality and are applicable only to some mild climate regions. To extend the applicability of the CR function, Han [31] and Brutsaert [32] generalized the CR approach; subsequently, many studies have further refined the CR approach based on the above [33–37]. At present, various CR-based models have been applied to different ecosystems (shrubland, cropland, grassland, etc.) or climate (arid, semiarid, temperate) regions around the world on annual, monthly, daily, and sub-daily timescales [38–42], but very few studies focused on frozen ground regions on the Qinghai-Tibet Plateau (QTP), which may limit the development and application of the CR approach in frozen ground regions.

The QTP accounts for approximately a quarter of China's land area, with an average altitude greater than 4000 m above sea level, known as the "roof of the world" and the "third Pole" [43]. Its vast topography and high altitude have significant impacts on the weather and climate of East Asia and even the world through thermal and dynamic forcing [44,45]. Additionally, frozen ground is widely distributed on the QTP [46], accounting for 96% of the area (including approximately 40% permafrost and 56% seasonally frozen ground). In these regions, freeze-thaw seasonal cycles and permafrost degradation [47,48] have certainly affected the regional hydrothermal balance of the soil–vegetation–atmosphere system; however, one of the current obstacles is that little is known about how seasonal and long-term ET_a rates have changed in the frozen ground regions. In recent decades, several comprehensive observation sites have been established [49,50], improving our understanding of the ET process, but large-scale ET_a estimation in the frozen ground regions of the QTP is still sparse. Hence, the above CR approach using only routine meteorological observations seems to be a feasible method. Wang [51] have improved and validated one CR-based model when applied to a permafrost sites on the QTP, but it is still unclear about which CR-based models are best for ET_a estimation in this region.

This study utilized four comprehensive field sites, which are all located in the frozen ground regions of the QTP, based on five CR-based models with in situ measurement meteorological and eddy-covariance (EC) flux data, to calculate variables and optimize parameters in CR-based models. The objective of this study was to clarify the applicability of five widely used CR-based models in the frozen ground regions of the QTP and further investigate the uncertainty of the CR approach in ET_a estimation. The results of this study could provide a reference for the applicability of CR-based models in the frozen ground regions of the QTP and provide a simple and feasible option for future large-scale terrestrial ET_a in such data-scarce regions.

2. Materials and Methods

2.1. Site Description

In this study, in situ measurement data were collected from four comprehensive observation field sites (Figure 1). Tanggula (TGL, 91.86°E, 32.58°N, 5100 m asl) and Xidatan (XDT, 94.13°E, 35.72°N, 4538 m asl) lie in the south edge and northern limit of permafrost regions of QTP, covered by alpine steppe and alpine meadow, where the annual average air temperature is about −4.7 °C and −3.6 °C, and the annual accumulated precipitation is about 352 and 384.5 mm, respectively [52]. Another two field sites, Nagqu (BJ, 91.90°E, 31.37°N, 4509 m asl) and Nam Co (NAMORS, 90.96°E, 30.77°N, 4730 m asl), lie in the seasonally-frozen ground regions of QTP, covered by alpine meadow and alpine steppe, where the annual average air temperature is about 0.54 °C and −0.36 °C, and the annual accumulated precipitation is about 436 and 462 mm, respectively [53]. All the sites are near the Qinghai-Tibet highway.

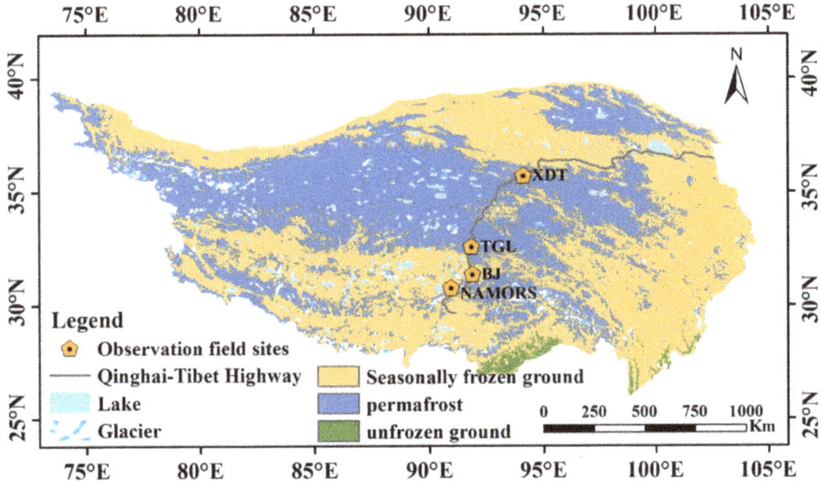

Figure 1. Locations of four observation field sites on the QTP.

2.2. In Situ Measurement Data and Data Processing

All four selected field sites include eddy covariance flux data, meteorological data, soil temperature, soil moisture content, and soil heat flux data in a shallow soil layer. Eddy covariance flux data, including latent heat flux (LE) and sensible heat flux (H), were used to identify the performance of the different CR-based models. All EC systems were installed on towers about 3 m distance from the ground, consisting of a three-dimensional sonic anemometer (CAST3, Campbell Scientific Inc., USA) that measured instantaneous horizontal (u,v), vertical (w) wind speeds and sonic air temperature fluctuations, and an open path infrared gas analyser (Li-7500, LI-cor Inc., USA) that measured the water vapor density and carbon dioxide concentrations fluctuations. The EC instruments were all sampled at a frequency of 10 Hz. The corresponding meteorological data, including roughly 2 m downwards/upwards shortwave and longwave radiation, air temperature, relative humidity, wind speed, and pressure, were employed to drive each CR-based model. Soil temperature, soil moisture content, and soil heat flux data in shallow soil layer (within 10 cm from ground surface) were used to calculate ground heat flux for variables in CR-based models. All above observations were recorded at intervals of 30 min. Data of TGL and XDT were provided from Cryosphere Research Station on the Qinghai-Tibet Plateau, Northwest Institute of Eco-Environment and Resources, CAS (http://new.crs.ac.cn/, 1 September 2021). Data of BJ and NAMORS were provided from National Tibetan Plateau Third Pole Environment Data Center, CAS (https://data.tpdc.ac.cn/en/data/b9ab35b2-81fb-4330-925f-4d9860ac47c3/, 1 September 2021). In addition, two different selected study periods of each field site were determined by data integrity and continuity; observation of first and second year was taken as calibration and validation period, respectively.

The EC flux raw data processing was conducted on EddyPro7.0.6 software, and the main processing steps included spike detection, lag corrections of H_2O/CO_2 relative to the vertical wind component, sonic virtual temperature correction, coordinate rotation (planar fit rotation), corrections for Webb-Pearman-Leuning density fluctuations, and frequency response correction. The 30 min data output by EddyPro software was screened as follows: (1) the data of instrument error was removed; (2) 30-min record data with more than 10% missing values were excluded; (3) night data with weak turbulence (friction velocity (u*) less than 0.1 m s^{-1}) were eliminated; (4) data with quality control code "2" ("2" indicate poor quality) were filtered out [54]. Then, half-hourly data during data available period accounts for filling the gaps; for gaps within seven consecutive days, they were filled using REddyProcWeb online tool (https://www.bgc-jena.mpg.de/bgi/index.php/Services/REddyProcWeb, 1 September 2021); for gaps longer than seven consecutive days, they were filled alternatively using aerodynamic method [55] or Bowen ratio energy balance method [56] for 30-min intervals. Finally, 30 min gap-filled flux data were aggregated into daily values. Ground heat flux were computed by linear method and thermal diffusion equation (TDE); detailed calculation processes are described in Yang [57] and Yao [58]. The daily energy balance closure ratios are 0.85, 0.79, 0.95, and 1.1 at TGL, XDT, BJ, and NAMORS, respectively (Figure 2).

It should be noted that on snow cover days, snow sublimation was one of the main evapotranspiration forms for four selected sites. Due to lack of observation of snow cover, this study employed the albedo instead, defined as the ratio of upward short wave radiation to downward shortwave radiation, to identify whether the observation field was snow-free or not. To exclude the influence of solar elevation angle, daily mean albedo was calculated as the average of half-hourly albedo from 10:00 to 14:00, local time. According to the variations of albedo at each observation fields, snow cover days were determined. In addition, snowfall still occurs during warm season at some observation fields, however, snow generally melts quickly to liquid water in few hours; thus, on a daily basis, only sublimation during cold season (from October to April in the next year) were considered. Then, the latent heat of vaporization—λ_v (J Kg^{-1}), for snow free days—and the latent heat of sublimation—λ_s (J Kg^{-1}), for snow cover days—were calculated using Equation (1), and the results were then used to calculate the actual evapotranspiration or sublimation [59]:

$$\begin{cases} \lambda_v = (2500 - 2.4T_a) \times 10^3 \\ \lambda_s = (2834.1 - 0.149T_a) \times 10^3 \end{cases} \quad (1)$$

where T_a (°C) is the air temperature measured at 2 m high.

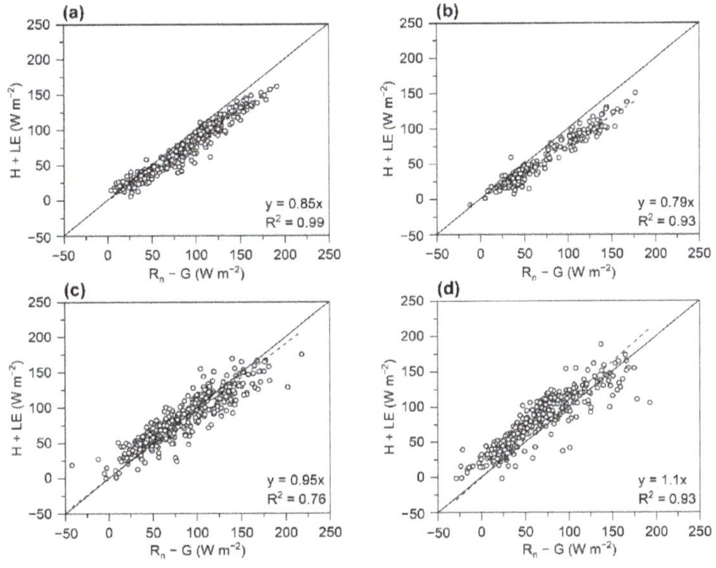

Figure 2. Comparison of available energy ($R_n - G$) and sum of turbulent fluxes (H + LE) for the daily-averaged measurements after gap-filled during study period at four selected sites: (**a**) TGL; (**b**) XDT; (**c**) BJ; (**d**) NAMORS. The black solid line is the 1:1 line, and the black dash line fitted by the linear regression.

2.3. Complementary Relationship (CR) Approach

The CR approach involves three types of evaporation, namely, the actual evaporation (E, note: followed by "E" instead), the apparent potential evaporation (E_{pa}), and the wet-environment evaporation (E_{po}). The CR approach is expressed as follows: for relatively larger and homogeneous surfaces with minimum advection, under certain net radiation inputs, E decreases with the availability of limited water over the underlying surface, and the energy that would be consumed by the latent heat flux thus becomes the sensible heat, thus increasing E_{pa}. Hence, one can predict water-limited E by gauging how much E_{pa} is raised from the hypothetical evaporation rate that should occur under the full wetness (E_{po}). Previous studies [60] hold that a unit decrease in E yields a corresponding unit increase in E_{pa}, signifying a symmetric CR. This can be expressed as follows:

$$E_{pa} - E_{po} = E_{po} - E \tag{2}$$

E_{pa} and E_{po} is calculated as follows:

$$E_{pa} = E_{rad} + E_{aero} = \frac{\Delta(R_n - G)}{\Delta + \gamma} + \frac{\gamma f(U)(e_o - e_a)}{\Delta + \gamma} \tag{3}$$

$$E_{po} = \alpha_e E_{rad} = \alpha_e \frac{\Delta(R_n - G)}{\Delta + \gamma} \tag{4}$$

where E_{rad} is the radiation term, E_{aero} is the aerodynamic term, α_e is an analog of the dimensionless Priestley-Taylor coefficient, $(R_n - G)$ is the available energy (mm d^{-1}), Δ is the slope of saturation vapor pressure curve at the air temperature T_a (kPa °C^{-1}), γ is the psychrometric constant (kPa °C^{-1}), and e_o and e_a are the saturation and actual vapor pressure of the air (kPa), respectively. $f(U)$ (mm d^{-1} kPa^{-1}) is the wind function calculated by Penman's original empirical linear equation:

$$f(U) = 2.6(1 + 0.54U_2) \tag{5}$$

where U_2 is the measured wind speed (m s^{-1}) at 2 m height.

Here, five CR-based models were selected to estimate E: the modified AA model [28] represents the linear CR version; the polynomial CR function [32] and sigmoid CR function [36] both represents the generalized nonlinear CR concept, and they represent different improved versions of the original CR approach mainly in physical boundary constraints; in addition, the calibration-free CR function and rescaled CR function are two improved versions based on the polynomial CR function [33–35].

2.3.1. Modified AA Model

The modified AA model extended the symmetric CR principle (herein referred to as K2006) and the equation becomes asymmetry as follows:

$$y_K = \frac{b+1}{b} x_K - \frac{1}{b} \tag{6}$$

where b is a coefficient that depicts the proportion of the sensible heat that increases E_{pa}, $x_K = E_{po}/E_{pa}$, $y_K = E/E_{pa}$. The only physical constraint is $E \le E_{po} \le E_{pa}$; when the water availability of the landscape is not limited, E is assumed to proceed at E_{pa} and $E = E_{po} = E_{pa}$.

2.3.2. Polynomial Generalized Complementary Function

Brutsaert [32] reformulated a new general polynomial complementary function satisfying boundary conditions based on strictly physical constraints, referred to as the B2015 model, and its boundary conditions were as follows:

$$\begin{cases} y_B = 1, x_B = 1 \\ y_B = 0, x_B = 0 \\ \frac{dy_B}{dx_B} = 1, x_B = 1 \\ \frac{dy_B}{dx_B} = 0, x_B = 0 \end{cases} \tag{7}$$

where $x_B = E_{po}/E_{pa}$, $y_B = E/E_{pa}$. The function form is as follows:

$$y_B = (2-c){x_B}^2 - (1-2c){x_B}^3 - c{x_B}^4 \tag{8}$$

c is an adjustable parameter which need to be locally calibrated.

2.3.3. Calibration-Free CR Function

Szilagyi [35] utilized the same function forms as the B2015 model (herein referred to as S2017), and the equation is as follows:

$$y_S = 2{X_S}^2 - {X_S}^3 \tag{9}$$

where y_S is the same as y_B, X_S is defined as

$$X_S = \frac{E_{p\max} - E_{pa}}{E_{p\max} - E_{po\_c}} \frac{E_{po\_c}}{E_{pa}} \tag{10}$$

Here two variables ($E_{p\,\max}$ and E_{po_c}) are introduced to improve the performance of S2017 function. $E_{p\,\max}$ is the maximum value that E_{pa} can theoretically reach, which may appear when the air loses all moisture, that is,

$$E_{p\max} = \frac{\Delta_{Tdry}(R_n - G)}{\Delta_{Tdry} + \gamma} + \frac{\gamma f(U)\left(e_{o,Tdry} - 0\right)}{\Delta_{Tdry} + \gamma} \quad (11)$$

where Δ_{Tdry} and $e_{o,Tdry}$ are the slope of the saturation vapor pressure curve and saturated vapor pressure, respectively, at extreme dry environment air temperature, T_{dry} (°C). T_{dry} can be estimated from the adiabatic line as follows:

$$T_{dry} = \frac{e_{o,Twb}(T_a - T_{wb})}{e_{o,Twb} - e_a} + T_{wb} \quad (12)$$

where T_{wb} (°C) is the wet-bulb temperature, and $e_{o,Twb}$ (kPa) is the saturated vapor pressure at T_{wb}. T_{wb} is derived by another iteration of solving for the Bowen ratio during adiabatic changes, that is,

$$\gamma \frac{T_{wb} - T_a}{e_{o,Twb} - e_a} = -1 \quad (13)$$

Another one variable E_{po_c} is defined as

$$E_{po\_c} = \alpha_e E_{rad} = \alpha_e \frac{\Delta_{wea}(R_n - G)}{\Delta_{wea} + \gamma} \quad (14)$$

where Δ_{wea} (kPa °C^{-1}) is the slope of the saturation vapor pressure curve at T_{wes}. E_{po_c} is the wet environment evapotranspiration rate calculated with Δ_{wea} instead of Δ. Note that S2017 function uses E_{po_c} instead of E_{po} because the latter variable is based on local air temperature, which is physically improperly employed to calculate wet environment evaporation. Szilagyi and Jozsa (2008) [61] suggested Equation (4) should use wet environment air temperature (T_{wea}). T_{wea} cannot be measured directly under water-limited conditions but can be approximated by the wet surface temperature (T_{wes}). According to Szilagyi [62], T_{wes} can be solved iteratively by implementing the Bowen ratio of a small wet patch, that is,

$$\beta_w = \frac{R_n - G - E_{pa}}{E_{pa}} \approx \gamma \frac{T_{wes} - T_a}{e_{o,Twes} - e_a} \quad (15)$$

in which β_w is the Bowen ratio of the wet patch (assuming that the available energy for the wet patch is close to that of the drying surface), and $e_{o,Twes}$ is the saturated vapor pressure at $T_{wes}(\approx T_{wea})$. Note that T_{wes} may be larger than T_a when air is close to saturation, and in such cases, T_{wea} should be replaced by T_a.

2.3.4. Rescaled Complementary Function

Crago [34] used a similar method to the S2017 model, with $E_{p\ \max}$ introduced to rescale x_B, and new analytical forms (herein referred to as C2018) have been proposed as follows:

$$\begin{cases} x_{\min} = \frac{E_{po\_c}}{E_{p\max}} \\ X_C = \frac{x - x_{\min}}{1 - x_{\min}} \\ y_C = X_C \end{cases} \quad (16)$$

where y_C are consistent with the corresponding variables of y_S in the S2017 model.

2.3.5. Sigmoid Generalized Complementary Function

Han [36] brought in the minimum and maximum limits of E_{rad}/E_{pen} ($x_{\min}$ and $x_{\max}$) under an assumed constant E_{rad} and rederived four new boundary conditions fitting to two widely accepted assumptions following Penman's combination theory, namely, $dE/dE_{pen} = 0$ in extremely arid environments and $E = E_{pen}$ in completely wet environments. The boundary conditions are as follows:

$$\begin{cases} y_H = 0, x_H \to x_{\min} \\ y_H = 1, x_H \to x_{\max} \\ \frac{dy_H}{dx_H} = 0, x_H \to x_{\min} \\ \frac{dy_H}{dx_H} = 0, x_H \to x_{\max} \end{cases} \tag{17}$$

The following new sigmoid function in accordance with the above boundary conditions were also proposed (herein referred to as H2018):

$$y_H = \frac{1}{1 + m\left(\frac{x_{\max} - x_H}{x_H - x_{\min}}\right)^n} \tag{18}$$

where $x_H = E_{rad}/E_{pen}$, $y_H = E/E_{pen}$, and the calculation of E_{rad} and E_{pen} (equal to E_{pa}) refers to Equation (3); m and n are two unknown parameters that need to be locally calibrated.

$$\begin{cases} n = \frac{4\alpha\left(1+b^{-1}\right)(x_{0.5} - x_{\min})(x_{\max} - x_{0.5})}{x_{\max} - x_{\min}} \\ m = \left(\frac{x_{0.5} - x_{\min}}{x_{\max} - x_{0.5}}\right)^n \end{cases} \tag{19}$$

where $x_{0.5}$ is a variable that corresponds to y_H = 0.5 (simply reflecting a temperate environment). Han [36] conducted that $x_{0.5} = \frac{0.5+b^{-1}}{\alpha_e\left(1+b^{-1}\right)}$, $x_{\max}$ and $x_{\min}$ can be simply set as 1 and 0 on a daily time scale, respectively. For longer time scales, $x_{\max}$ and $x_{\min}$ needed to be locally calibrated.

2.4. Model Parameter Calibration Methods

For the CR approach, one of the most important steps is to determine unknown parameter values. Generally, "optimal" parameter values could be either derived from transplanting from elsewhere, or are locally calibrated. First, "optimal" parameter values between two different sites are roughly equal under similar climatic or underlying surface conditions. In our study, the vegetation type of all four selected sites was grassland, so we set the "optimal" parameter values from other grassland sites reported in previous literature as the default values in each CR-based model. Note that the unknown parameters in the K2006, B2015, and H2018 models used corresponding values from the US-Fpe flux site [36] as default values, and the unknown parameters in S2017 and C2018 model used corresponding values from the Riggs Creek site at Australia [34] as default values, respectively. Second, unknown parameters of each CR-based model were all optimized by minimizing the root mean square error of the simulated and measured daily E during the calibrated period, and then the calibrated parameter values were used for the validation period.

2.5. Model Evaluation Criteria

To assess the model performance, five statistical metrics, including root mean square error (RMSE), mean absolute error (MAE), mean bias error (MBE), Nash-Sutcliffe efficiency (NSE), and correlation coefficient (CC) are used to evaluate the accuracy of simulated E with in situ measurements:

$$\text{RMSE} = \sqrt{\frac{1}{N}\sum_{i=1}^{N}(E_{sim,i} - E_{obs,i})^2} \tag{20}$$

$$\text{MAE} = \frac{1}{N}\sum_{i=1}^{N}\left|E_{sim,i} - E_{obs,i}\right| \tag{21}$$

$$\text{MBE} = \frac{1}{N}\sum_{i=1}^{N}(E_{sim,i} - E_{obs,i}) \tag{22}$$

$$\text{NSE} = 1 - \frac{\sum_{i=1}^{N} \left(E_{sim,i} - E_{obs,i}\right)^2}{\sum_{i=1}^{N} \left(E_{obs,i} - \overline{E_{obs}}\right)^2} \tag{23}$$

$$\text{CC} = \frac{\sum_{i=1}^{N} \left(E_{sim,i} - \overline{E_{sim}}\right)\left(E_{obs,i} - \overline{E_{obs}}\right)}{\sqrt{\sum_{i=1}^{N} \left(E_{sim,i} - \overline{E_{sim}}\right)^2}\sqrt{\sum_{i=1}^{N} \left(E_{obs,i} - \overline{E_{obs}}\right)^2}} \tag{24}$$

where N is the number of observation, $E_{sim,i}$ (i = 1, 2, 3, ..., N) is the simulated actual evapotranspiration value, $E_{obs,i}$ is the observed actual evapotranspiration value, $\overline{E_{sim}}$ and $\overline{E_{obs}}$ are the mean values of $E_{sim,i}$ and $E_{obs,i}$, respectively.

3. Results

3.1. Variations in ET Rates at Four Observation Sites

The daily variations in actual evapotranspiration (E), sublimation, radiation term (E_{rad}), and apparent potential evapotranspiration (E_{pa}) rate at four observation field sites are shown in Figure 3. Note that relatively larger errors of observation may exist at NAMORS during some periods, especially in the warm season when E is larger than E_{pa}, which is discrepant with Penman's evaporation theory. Visibly, among four observation field sites–whether located in permafrost regions or seasonally frozen ground regions–are both clear seasonal variations of each ET variable: higher E and E_{pa} usually occurs during the warm season, maximum E could reach approximately 4 mm d^{-1} at TGL and XDT, and 5 mm d^{-1} or more at BJ and NAMORS. For E_{pa}, which exhibited greater daily fluctuations compared with E and E_{rad}, the maximum E_{pa} ranges from 5 mm d^{-1} to 7 mm d^{-1} at four selected sites: low E occurred during cold seasons and was generally close to zero when soils were frozen, except at NAMORS, and E_{pa} and E_{rad} was relatively lower, which was mainly controlled by net radiation. Additionally, some obvious differences between permafrost regions and seasonally frozen ground regions were that greater variations in E occurred at seasonally frozen ground regions than permafrost regions, and the onset time of increasing E after the frozen period in seasonally frozen ground regions (usually in late March) was earlier than that of permafrost regions (usually in late April), which was due to the onset time of thawing being earlier in the former. As the ground ice thaws, surface soil moisture increases, and E also increases.

Some studies found that E is larger than rainfall at the annual scale at observation sites on the QTP [59,63]; the same phenomenon was also found in our study (see Table 1). Due to missing values of rainfall during the study period at BJ and NAMORS, we instead used data from the nearby Naqu and Dangxiong country meteorological station provided from China Meteorological Administration. The exception of this was when there were observation errors about precipitation or ET; this phenomenon is often explained by surface soil moisture providing available water for evaporation. Although limited by land surface water supply, soil evaporation is one of the main forms of water consumption for the four selected field sites.

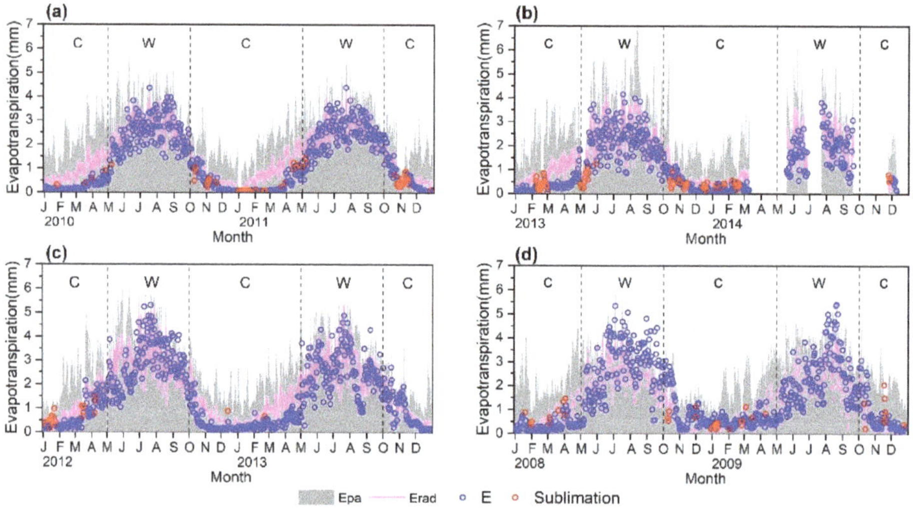

Figure 3. The daily variations of *E* (blue circles), sublimation (red circles), E_{rad} (pink line), and E_{pa} (gray area) during the study period at the observation fields: (**a**) TGL; (**b**) XDT; (**c**) BJ; (**d**) NAMORS. "W" represents warm season and "C" represents cold season.

Table 1. Sum of daily rainfall (mm), *E* plus sublimation (mm), E_{pa} (mm) for two periods of every observational year. The warm season is from May to September, the cold season is from January to April and October to December.

Sites	Variables	First Year			Second Year		
		Warm Season	Cold Season	Annual	Warm Season	Cold Season	Annual
TGL (N_1 = 365; N_2 = 360)	Rainfall	339.5	54.3	393.8	355.9	36	391.9
	E plus Sublimation	362.1	70.9	433	364.7	81.6	446.3
	E_{pa}	537.5	409.7	947.2	496	391.1	887.1
XDT (N_1 = 363; N_2 = 185)	Rainfall	299.3	65.1	364.4	200.6	7	207.6
	E plus Sublimation	312.4	82.9	395.3	178	24.4	202.4
	E_{pa}	515.9	410.1	926	309.7	141.9	451.6
BJ (N_1 = 363; N_2 = 350)	Rainfall [a]	-	-	-	386	83.2	469.2
	E plus Sublimation	432.9	118.9	551.8	385.5	117.3	502.8
	E_{pa}	580.8	421.5	1002.3	549.8	363.4	913.2
NAMORS (N_1 = 337; N_2 = 352)	Rainfall [b]	494.7	54.2	548.9	327	47.3	374.3
	E plus Sublimation	413.6	128.9	542.5	346	121.2	467.2
	E_{pa}	493.1	336.1	829.2	531.6	411.4	943

Note: N_1 represents available samples of first observational year, N_2 represents available samples of second observational year. [a] Rainfall data on second year from Naqu country meteorological station; [b] Rainfall data on second year from Dangxiong country meteorological station.

3.2. Evaluating Model Performance

3.2.1. Model Performance with Default and Calibrated Parameter Values on a Daily Timescale

This study first employed the default and calibrated parameter values to simulate the daily actual evapotranspiration (E_{sim}) and then compared with in situ measurements, respectively (Figure 4).

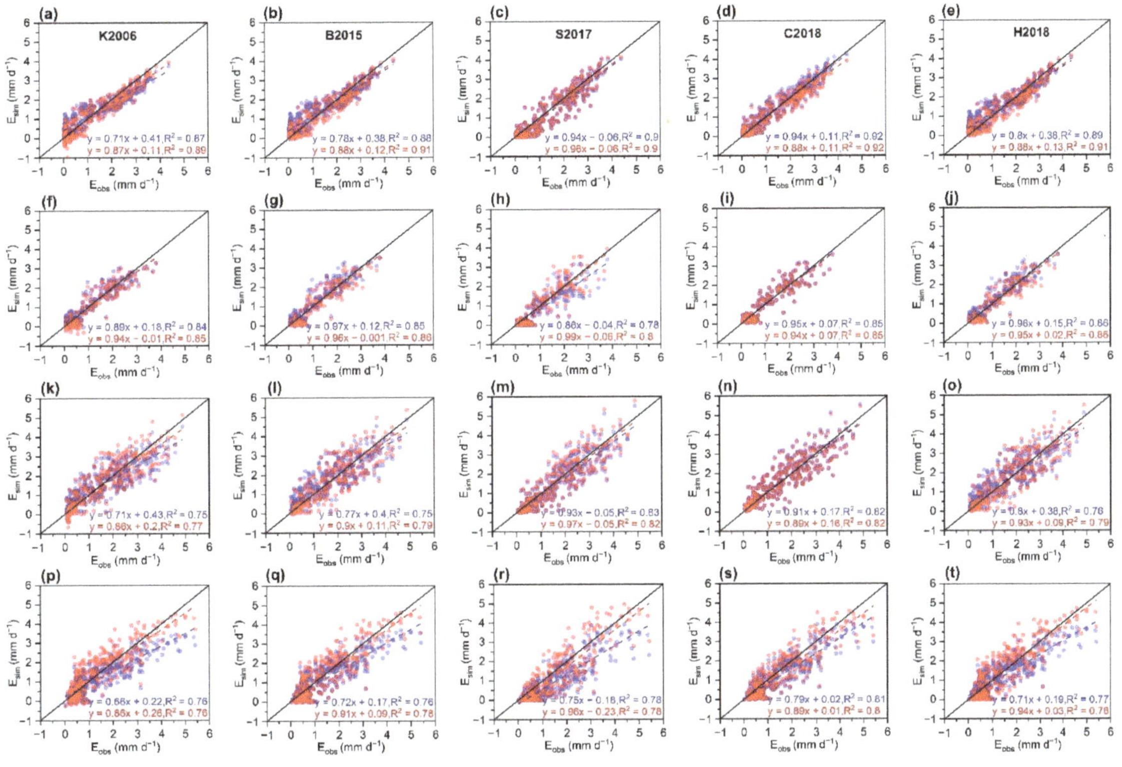

Figure 4. Comparison of simulated daily actual evapotranspiration (E_{sim}) rates by five CR-based models with default parameter against measurements made using EC system at **(a–e)** TGL; **(f–j)** XDT; **(k–o)** BJ; **(p–t)** NAMORS, respectively. The blue scattered dots and dash lines represent E estimated by default parameter values and its fitting line; the red scattered dots and dash lines represent E estimated by calibrated parameter values and its fitting line; the black solid line is the 1:1 line.

The results indicate that E_{sim} was underestimated to different degrees by each CR-based model at all four selected sites, whether the default or calibrated parameter values were employed; the highest NSE value was 0.92 at TGL and the lowest NSE value was 0.58 at NAMORS. The average RMSE value was 0.4 mm d^{-1} at TGL and at XDT, 0.58 mm d^{-1} at BJ, and 0.63 mm d^{-1} at NAMORS when default values were employed (Table 2). After being locally calibrated, the highest NSE value was 0.92 at TGL and lowest NSE value was 0.69 at NAMORS. The average RMSE value was 0.36 mm d^{-1} at TGL, 0.38 mm d^{-1} at XDT, 0.56 mm d^{-1} at BJ, and 0.6 mm d^{-1} at NAMORS (Table 2), with only a small improvement in accuracy when compared with the results simulated by default parameter values. In addition, the CC values of five CR-based models all exceeded 0.87. Therefore, a better performance of the CR approach was found for daily E estimation in the frozen regions of the QTP, and unknown parameter values from other similar climatic or underlying surfaces could also be applicable without local calibration due to a lack of in situ measured E, such as in data-scarce regions.

Table 2. Statistical results of CR-based models with default and calibrated (in parentheses) parameter values against daily actual evapotranspiration measurements from four observation sites. Bold font indicates the best model based on maximum NSE values among five CR-based models.

Sites	Model	αe	b	c	RMSE (mm d^{-1})	MAE (mm d^{-1})	MBE (mm d^{-1})	NSE	R^2	Average (mm d^{-1})
TGL	K2006	0.88 (1.01)	16.67 (2.41)	-	0.457 (0.381)	0.375 (0.31)	0.053 (−0.056)	0.84 (0.889)	0.872 (0.891)	1.24
	B2015	0.92 (1.03)	-	−1.35 (2.22)	0.428 (0.347)	0.338 (0.269)	0.109 (−0.032)	0.86 (0.908)	0.88 (0.91)	
	S2017	1.12 (1.13)	-	-	0.394 (0.388)	0.279 (0.275)	−0.136 (−0.112)	0.881 (0.885)	0.897 (0.899)	
	C2018	1.12 (1.08)	-	-	0.321 (0.324)	0.244 (0.245)	0.039 (−0.041)	**0.921 (0.919)**	0.924 (0.924)	
	H2018	0.97 (1.07)	5.56 (1.44)	-	0.415 (0.343)	0.328 (0.266)	0.128 (−0.01)	0.868 (0.91)	0.889 (0.91)	
XDT	K2006	0.88 (0.94)	16.67 (4.1)	-	0.372 (0.375)	0.282 (0.281)	0.052 (−0.07)	0.839 (0.837)	0.845 (0.852)	1.09
	B2015	0.92 (1.01)	-	−1.35 (1.48)	0.388 (0.361)	0.287 (0.268)	0.093 (−0.049)	0.826 **(0.849)**	0.852 (0.861)	
	S2017	1.12 (1.18)	-	-	0.486 (0.467)	0.354 (0.341)	−0.198 (−0.07)	0.725 (0.746)	0.778 (0.799)	
	C2018	1.12 (1.11)	-	-	0.37 (0.366)	0.277 (0.274)	0.017 (−3.4 × 10^{-4})	**0.841** (0.844)	0.852 (0.852)	
	H2018	0.97 (1.04)	5.56 (1.8)	-	0.381 (0.36)	0.289 (0.269)	0.105 (−0.042)	0.831 (0.849)	0.856 (0.859)	
BJ	K2006	0.88 (1.03)	16.67 (2.63)	-	0.609 (0.59)	0.463 (0.439)	0.001 (0.003)	0.746 (0.762)	0.748 (0.773)	1.44
	B2015	0.92 (1.05)	-	−1.35 (2.55)	0.604 (0.578)	0.453 (0.42)	0.074 (−0.031)	0.75 (0.772)	0.755 (0.789)	
	S2017	1.12 (1.14)	-	-	0.544 (0.553)	0.405 (0.41)	−0.156 (−0.092)	0.798 (0.791)	0.826 (0.824)	
	C2018	1.12 (1.11)	-	-	0.531 (0.523)	0.377 (0.371)	0.038 (0.013)	**0.808 (0.813)**	0.819 (0.821)	
	H2018	0.97 (1.11)	5.56 (1.17)	-	0.596 (0.584)	0.45 (0.421)	0.092 (−0.013)	0.757 (0.767)	0.764 (0.792)	
NAMORS	K2006	0.88 (1.13)	16.67 (10.07)	-	0.637 (0.601)	0.458 (0.464)	−0.231 (0.08)	0.704 (0.737)	0.755 (0.757)	1.33
	B2015	0.92 (1.17)	-	−1.35 (0.76)	0.607 (0.585)	0.448 (0.455)	−0.204 (−0.033)	0.731 (0.751)	0.764 (0.774)	
	S2017	1.12 (1.25)	-	-	0.756 (0.653)	0.573 (0.516)	−0.516 (−0.273)	0.583 (0.689)	0.78 (0.785)	
	C2018	1.12 (1.2)	-	-	0.579 (0.549)	0.422 (0.413)	−0.264 (−0.139)	**0.756 (0.78)**	0.806 (0.803)	
	H2018	0.97 (1.28)	5.56 (1.72)	-	0.601 (0.6)	0.431 (0.475)	−0.197 (−0.041)	0.737 (0.738)	0.769 (0.776)	

However, a relatively larger bias occurred when there were lower E values for each model at all four sites, which indicated poor applicability of the CR approach. There were even some negative E_{sim} values by the K2006 model when E was low at all four observation sites, probably because there were no strict boundary conditions constrained in the K2006 model, which led to abnormal values near the border of the boundary. In general, the CR approach may be inappropriate during the cold season in the frozen regions of the QTP because lower E values mainly occurred during the cold season. The S2017 model performed poorly among the five CR-based models, with relatively lower NSE values and higher RMSE and MBE values (Table 2), especially at NAMORS, where the MAE values reached 0.57 mm d^{-1} and 0.52 mm d^{-1} with the default and calibrated parameter values, respectively, signifying the largest bias during the study period. The C2018 model performed best at TGL, BJ, and NAMORS, and performed very well at XDT, and the NSE values of the C2018 model were all above 0.75. The overall performances of the B2015 and H2018 models were slightly inferior to that of the C2018 model, and the NSE and RMSE values of the above two models were close to those of the C2018 model.

Note that for the C2018 model, although there was only one unknown parameter to calibrate, there were more variables, such as wet and dry environment temperature, which meant a relatively more complex calculation process. For the B2015 model, previous studies have pointed out its deficiency in physical boundaries, which may be inappropriate for extremely dry or wet environments and periods. Therefore, the H2018 model, with relatively robust physical boundary constraints and a simpler calculation process, may be the most appropriate CR-based model for daily E estimation in the frozen regions of the QTP.

3.2.2. Performance of Different CR-Based Functions against Relationships among Three Evapotranspiration Variables

To determine relationships among three evapotranspiration variables (E, E_{po} and E_{pa}) during warm and cold seasons and to further elucidate the applicability and optimal form of CR-based functions in the frozen ground regions of the QTP, the performance of different function forms using calibrated parameter values combined with observation data is presented in Figure 5. Here, all data at each field site were divided into three parts: daily ET variables that fulfill the preconditions of each CR-based model, judged by "po (E_{po_c}) $\leq E_{pa}$" for the K2006, B2015, S2017, and C2018 models, "$E \leq E_{po}$ and $E \leq E_{pen}$" for the H2018 model, and the above data can be used to further divide based on the warm and cold season, respectively. The remaining data are daily evapotranspiration variables that do not fulfill the CR-based models. Note that although parameter calibration was performed, there were still some data that did not satisfy the CR-based models, which may have been due to observation errors or inapplicability of the CR approach to that time.

The results indicated that observation data were unevenly scattered around CR-fitting lines at each site, and all CR-based function forms basically captured relationships among the three types of evaporation during the warm season in the frozen ground regions of the QTP, similar to the simulated results from some other sites around the world [36,42,64]. However, when variable y converges to 0 (that is, E is low), which often occurs in cold seasons, the dimensionless variable x (or X) has a relatively wide range, even above 0.6, because strong radiation (E_{rad}) on the QTP leads to a high proportion of radiation to apparent potential evapotranspiration (E_{pa}). Usually, a high proportion of radiation indicates there is strong evaporation energy available and higher actual evaporation; however, for water-limited frozen ground regions, water availability is another important factor affecting evaporation. Figure 5a–t shows relatively small deviations during warm seasons and the CR-fitting lines cross near the center of the observation data; five different CR-based functions both captured the daily E values well without evident differences, however, differences were clear in the lower E values. Because most studies about CR approach focus on warm seasons or growing seasons when E is relatively high, it is not enough to take into account the conditions of small E values during cold seasons; difficulties lie in high uncertainty of observation data during cold seasons, which leads to a lack of reliable data to validate whether the hypothesis of the CR approach is applied to cold and high-altitude areas or cold seasons from the aspect of the mechanism.

3.2.3. Model Performance with Calibrated Parameter Values on a Monthly Timescale

We also investigated the performance of five CR-based models with calibrated parameter values on a monthly timescale. Figure 6 illustrates that the monthly E_{sim} got acceptable accuracy at TGL, XDT, and BJ, but relatively larger deviations occurred at NAMORS. The simulated performance between each CR-based model was very close, which indicated more stable performance to the CR approach on longer timescales. Compared with variations in precipitation, the peak of E was not always synchronized with the peak of precipitation on a monthly timescale. Taking TGL (Figure 6a) for an example, the peak of E occurred in August and was higher than concurrent precipitation; higher E may be contributed by water stored in the soil layer by antecedent precipitation, due to the peak of precipitation in July.

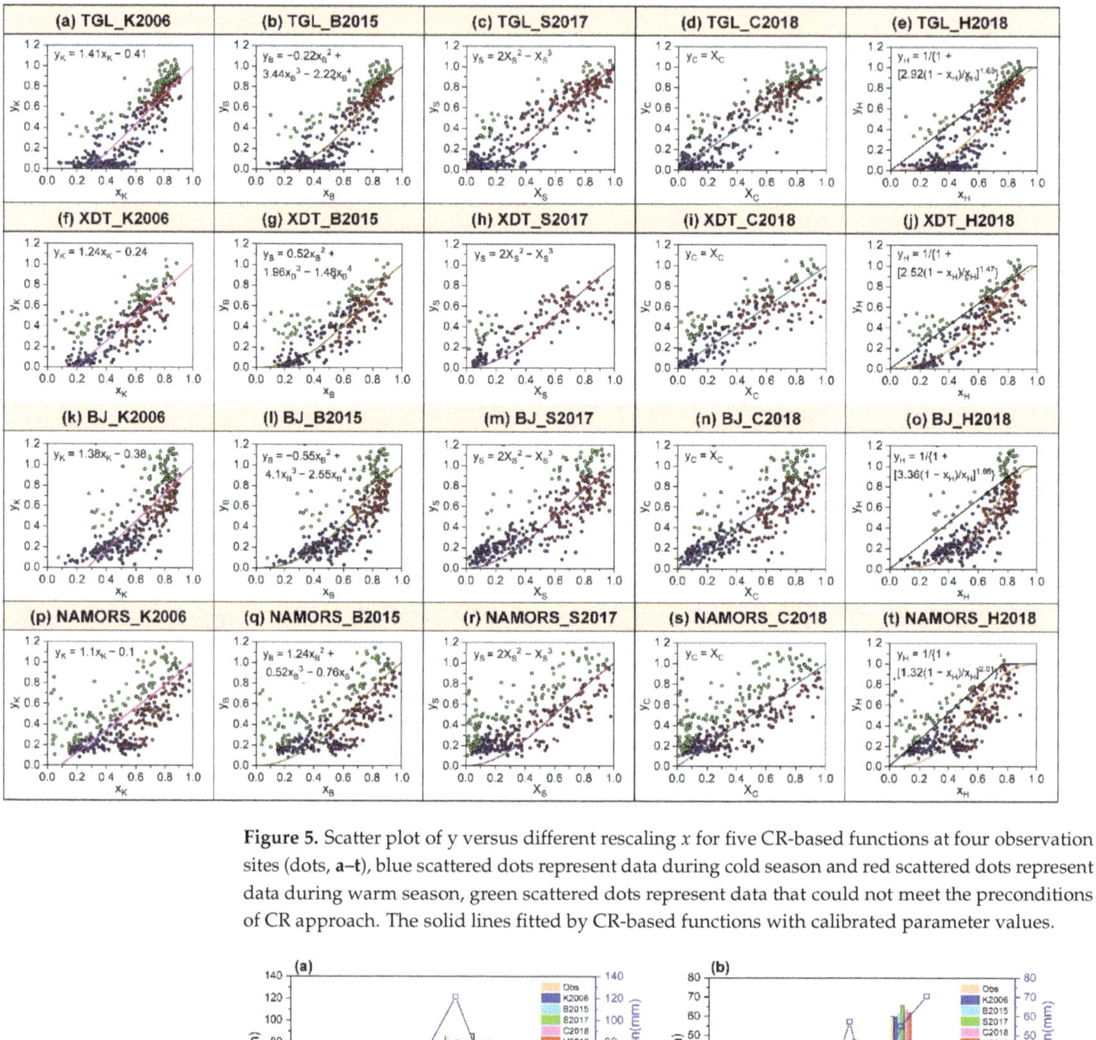

Figure 5. Scatter plot of y versus different rescaling x for five CR-based functions at four observation sites (dots, **a**–**t**), blue scattered dots represent data during cold season and red scattered dots represent data during warm season, green scattered dots represent data that could not meet the preconditions of CR approach. The solid lines fitted by CR-based functions with calibrated parameter values.

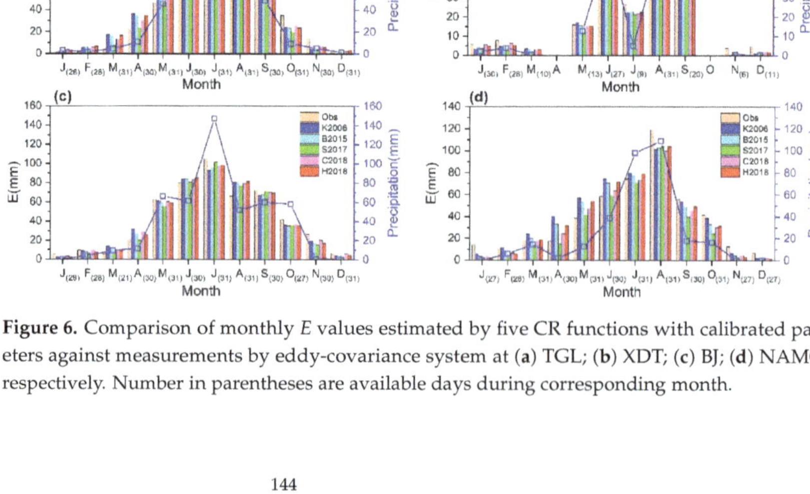

Figure 6. Comparison of monthly E values estimated by five CR functions with calibrated parameters against measurements by eddy-covariance system at (**a**) TGL; (**b**) XDT; (**c**) BJ; (**d**) NAMORS, respectively. Number in parentheses are available days during corresponding month.

To further determine the contributions of monthly simulated bias to annual total bias at each field site, the monthly absolute bias of each CR-based model was calculated (Figure 7). Although the performance of the five CR-based models varied from one site to another, overall, the H2018 model seemed to have a smaller annual bias than the other models. According to Figure 7e,i,o,t, the H2018 model underestimated annual E at each site, and the bias was within 15 mm. The K2006, B2015, and C2018 models performed slightly inferior to the H2018 model; larger biases all occurred at NAMORS, and the rest of the sites performed relatively well. The S2017 model performed worst among the five CR-based models at each site, and the largest bias occurred at NAMORS, which was close to 100 mm on an annual timescale. Considering the contribution of monthly or seasonal bias to annual total bias (see Table 3), negative annual bias was mainly contributed by the negative bias of the warm season at TGL; however, the negative bias of the cold season contributed more to the total bias at XDT, BJ, and NAMORS. The larger negative monthly bias frequently occurred in October or November, and the maximum bias was −15.3 mm (in October) at TGL, −5.42 mm (in July) at XDT, −11.72 mm (in July) at BJ, and −17.58 mm (in October) at NAMORS.

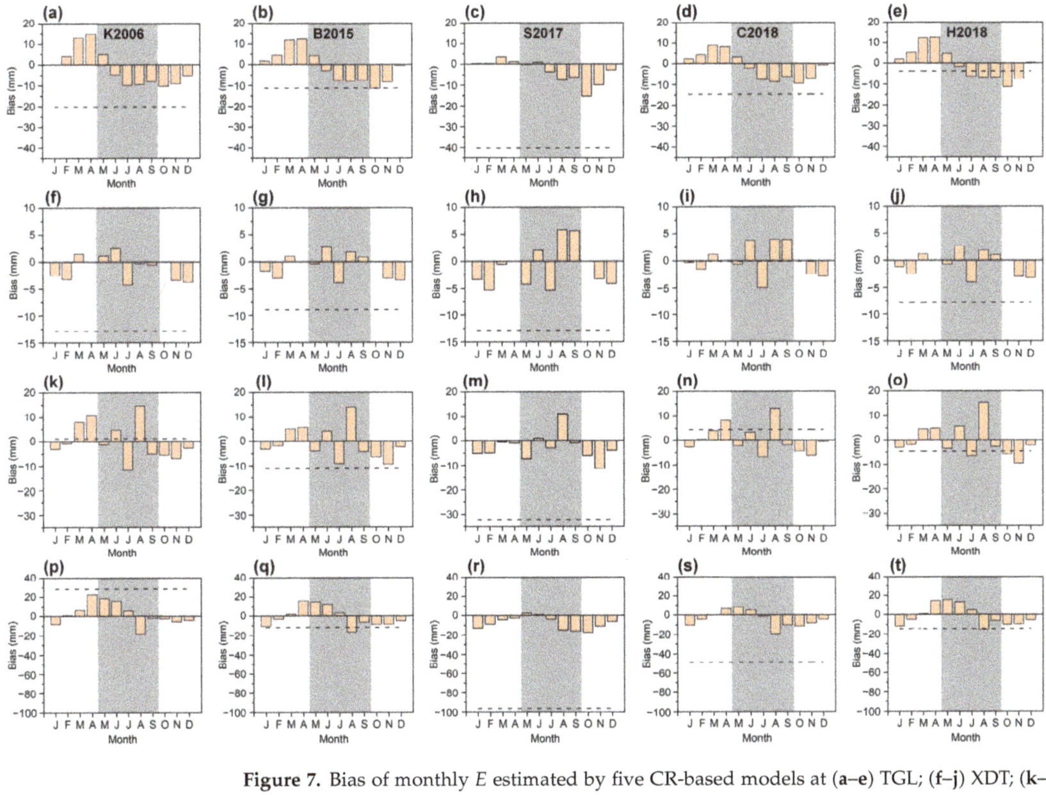

Figure 7. Bias of monthly E estimated by five CR-based models at (**a–e**) TGL; (**f–j**) XDT; (**k–o**) BJ; (**p–t**) NAMORS, respectively. Dash lines represent annual total bias. The grey filled area represents the warm season.

Table 3. Bias of simulated E by CR-based models with calibrated parameter values against measurements by EC on different timescales (monthly, seasonal, annual) at TGL, XDT, BJ and NAMORS. Grey-filled area is warm season period (May to September), bold font stands for model with minimum absolute bias during corresponding period.

Site	Model	Jan	Feb	Mar	Apr	May	Jun	Jul	Aug	Sep	Oct	Nov	Dec	Warm Season	Cold Season	Year
TGL	K2006	0.06	4	13.07	14.84	4.91	−4.75	−9.91	−9.49	−8.03	−10.39	−9.12	−5.21	−27.27	7.25	−20.02
	B2015	1.65	4.49	12.01	12.54	4.18	−2.86	−7.55	−7.9	−7.44	−11.54	−8.36	−0.59	−21.57	10.2	−11.37
	S2017	0.22	0.2	3.35	1.15	−0.32	0.69	−3.77	−7.25	−6.33	−15.3	−10.06	−2.86	−16.98	−23.3	−40.28
	C2018	2.23	4.28	9.02	8.33	3.09	−2.08	−7.31	−8.68	−6.21	−9.51	−7	−0.8	−21.19	**6.55**	−14.64
	H2018	1.94	5.12	12.38	12.61	4.51	−1.67	−6.21	−6.76	−6.72	−11.41	−7.85	0.3	**−16.85**	13.09	**−3.76**
XDT	K2006	−2.54	−3.19	1.47	-	1.12	2.48	−4.26	−0.27	−0.52	-	−3.41	−3.77	−1.45	−11.44	−12.89
	B2015	−1.68	−3.06	1.02	-	−0.37	2.82	−3.92	1.81	0.88	-	−3.07	−3.42	1.22	−10.21	−8.99
	S2017	−3.4	−5.35	−0.52	-	−4.28	2.02	−5.42	5.8	5.67	-	−3.29	−4.2	3.79	−16.76	−12.97
	C2018	−0.28	−1.55	1.19	-	−0.61	3.76	−5.01	3.93	3.9	-	−2.52	−2.87	5.97	**−6.03**	**−0.06**
	H2018	−1.15	−2.54	1.17	-	−0.65	2.66	−4.04	1.87	1.05	-	−2.99	−3.23	**0.89**	−8.74	−7.85
BJ	K2006	−3.07	−0.8	8.06	10.73	−1.25	4.63	−11.72	14.75	−5.06	−5.56	−6.92	−2.66	1.35	**−0.22**	**1.13**
	B2015	−3.13	−1.77	5.08	5.79	−3.9	4.11	−9.11	14.12	−4.22	−6.35	−9.28	−2.25	1	−11.91	−10.91
	S2017	−5.14	−4.91	−0.5	−0.89	−7.31	0.89	−2.93	10.88	−0.86	−6.06	−11.33	−3.98	**0.67**	−32.81	−32.14
	C2018	−2.66	0.25	4.02	8.24	−2.22	3.27	−6.72	13.17	−1.86	−4.37	−6.23	−0.48	5.64	−1.23	4.41
	H2018	−2.99	−1.92	4.54	4.86	−3.48	5.65	−6.55	15.5	−2.68	−5.9	−9.49	−2.21	8.44	−13.11	−4.67
NAMORS	K2006	−7.92	0.73	6.17	22.71	18.67	15.86	5.41	−18	−2.06	−2.72	−5.96	−4.59	19.88	**8.42**	28.3
	B2015	−10.5	−3.52	1.7	15.62	14.63	12.23	3.11	−16.32	−6.57	−8.49	−8.37	−5.12	**7.08**	−18.68	**−11.6**
	S2017	−13.01	−8.75	−4.44	−2.77	2.54	0.67	−3.92	−15.14	−16.21	−17.58	−11.14	−6.4	−32.06	−64.09	−96.15
	C2018	−10.07	−4.58	0.08	6.63	8.45	5.21	−1.21	−19.3	−10.15	−11.23	−8.03	−4.66	−17	−31.86	−48.86
	H2018	−11.56	−5.13	0.77	14.39	15.45	13.3	4.75	−15.05	−6.28	−9.85	−9.5	−5.7	12.17	−26.58	−14.41

4. Discussion

4.1. Uncertainty of Actual Evapotranspiration Estimation by the CR Approach

Using only routine meteorological variables (air temperature, relative humidity, wind speed, and net radiation) in the CR approach can better estimate daily or monthly E in the frozen ground regions of the QTP according to the present study; however, the uncertainty of the CR approach itself, physical variable calculations, and key parameter values are still non-negligible for estimating E on different spatiotemporal scales.

For the K2006 model, characterizing variables such as E_{pa} and E_{po} is difficult. Although using the Penman equation to calculate E_{pa} is widely accepted, determining E_{po} is controversial because a fixed analog of the Priestley-Taylor coefficient α_e usually cannot vary with climate, which means that the value of α_e cannot truly reflect the interaction between the land and atmosphere [65,66]. We also need to know atmospheric status–such as air temperature–under well-supplied water conditions at the land surface before determining E_{po}, despite some progress [61,62] that has been made in obtaining air temperature under well-supplied water conditions. However, usual iterative solving exists for unsolved or anomalous solutions, which makes it difficult to promote the above method as a universal approach.

For generalized complementary functions, in addition to the above common problems, the definition of boundary constraint conditions is another one important problem. Han [65] noted that boundary constraint conditions determine the domain of definition and the analytic formula of complementary functions. Many of the latest debates [67,68] on the CR-based functions have focused on boundary constraint conditions, which is controversial for the CR approach. The above-mentioned problems reflect a great lack of understanding of the evaporation process on different spatiotemporal scales for frozen ground regions on the QTP, where the current CR function forms seem to perform inferiorly in the specific regions described by Wang [42].

For the CR principle, an important prerequisite of the CR approach is that at large and homogeneous land surfaces, the influence of air advection could be negligible, so atmospheric evaporation demand is totally caused by feedback of the land surface. Morton [26] pointed out that the CR principle should be applied at spatial resolutions larger than 1 km and temporal resolutions longer than five days, since large-scale weather fronts may bring air masses over the land with a moisture signature decoupled from the underlying surface, which thus may temporarily disrupt the dynamic equilibrium of air humidity and surface fluxes in the land–atmosphere system. Although previous studies [28,38,42,69] have applied the CR approach to hourly or daily E estimations, the theory of the CR principle on short timescales still needs to be improved.

4.1.1. Influence of Parameter Values on Actual Evapotranspiration Estimation

Determining the parameter values of CR functions is urgent for the application of each CR-based model to estimate E, as well as for the development of a CR approach. The K2006 and H2018 both have two parameters, α_e and b; the B2015, S2017, and C2018 all have one parameter, α_e (note: parameter c in the B2015 model is an adjustable parameter, $c = 0$ often in most conditions, so, here we did not discuss parameter c). To determine the impacts of parameter values on evapotranspiration estimation, we discuss the following two problems: (i) difference in simulated results by each CR-based model with parameter calibration on different time lengths and (ii) parameter sensitivity for actual evapotranspiration estimation.

Here, we first investigated the influence of calibrated parameter values by different time lengths (whole year and warm season) on simulating daily E during the warm season. For the cold season, due to higher uncertainty of observations, we did not discuss the results here. Table 4 compares the parameter values and simulated NSE values during the warm season by the calibration period of one whole year and the corresponding warm season, respectively. The results indicated that the time length of parameter calibration did not have much of an impact on simulated E during the warm season, parameter α_e

was very close between the two calibration periods, and the differences in parameter b was slightly larger. According to our results, the simulated E was insensitive to the variations in parameter b in the K2006 and H2018 model, because even if the difference in parameter b was large, like the K2006 model at NAMORS, the simulated daily E between the two calibrated parameter values were still approximate. However, variations in parameter α_e in all five CR-based models exerted more of an influence on the simulated results. The above results indicated that parameter calibration—especially α_e—in CR-based models at specific sites is more important to simulate daily E compared with different time lengths of calibration periods in the frozen ground regions of the QTP.

Table 4. Comparison of simulated results by different CR-based models using two different calibrated parameter values during the warm season. One is calibrated by a whole year period, another is calibrated by a corresponding warm season.

Site	Model	Calibrated Period by Warm Season				Calibrated Period by Whole Year			
		α_e	b	c	NSE	α_e	b	c	NSE
TGL	K2006	1.02	2.49	-	0.759	1.01	2.41	-	0.73
	B2015	1	-	0.8	0.764	1.03	-	2.22	0.752
	S2017	1.12	-	-	0.559	1.13	-	-	0.581
	C2018	1.08	-	-	0.731	1.08	-	-	0.731
	H2018	1.01	2.58	-	0.764	1.07	1.44	-	0.762
XDT	K2006	1	2.93	-	0.657	0.94	4.1	-	0.669
	B2015	0.98	-	0.38	0.656	1.01	-	1.48	0.657
	S2017	1.18	-	-	0.376	1.18	-	-	0.376
	C2018	1.12	-	-	0.607	1.11	-	-	0.617
	H2018	0.99	3.03	-	0.657	1.04	1.8	-	0.65
BJ	K2006	1.11	1.13	-	0.323	1.03	2.63	-	0.357
	B2015	1.1	-	4.66	0.294	1.05	-	2.55	0.361
	S2017	1.14	-	-	0.429	1.14	-	-	0.429
	C2018	1.11	-	-	0.466	1.11	-	-	0.466
	H2018	1.17	0.75	-	0.27	1.11	1.17	-	0.341
NAMORS	K2006	1.18	3.81	-	0.577	1.13	10.07	-	0.58
	B2015	1.19	-	1.01	0.577	1.17	-	0.76	0.588
	S2017	1.25	-	-	0.492	1.25	-	-	0.492
	C2018	1.2	-	-	0.622	1.2	-	-	0.622
	H2018	1.35	1.03	-	0.525	1.28	1.72	-	0.573

4.1.2. Sensitivity Analysis of CR-Based Models to Parameter Values

Then, we tested the sensitivity of five CR-based models to parameter values at each site, by adding increments from −50 to 50% at an interval of 10% to optimized parameter values for each CR-based model. Figure 8 displays the RMSE of the simulated E values for the in situ measurement. It is clear that variations in b values combined with constant α_e values have little impact on evapotranspiration estimation, and a 50% variation in the α_e value combined with constant b values led to increased RMSE values within 0.5 mm d^{-1}. The K2006, B2015 and H2018 models exhibited lower sensitivity to variations in parameter values, and the H2018 model had the lowest sensitivity among all CR-based models.

We also noticed that larger RMSE values for the S2017 and C2018 models when the optimized α_e value increased more than 20% because there were many abnormal E values, which indicated that the S2017 and C2018 models were more sensitive to parameter values, especially when the α_e value increased more than 20%. Brutsaert [70] found that α_e was closely related to the aridity index (AI, AI = E_{pa}/P); accordingly, based on the global distribution of the parameter α_e, the α_e values were mainly from 0.8 to 1.2 around the world, except in extremely arid and wet regions. Assuming α_e values were within the above scope in the present study–which means α_e mainly varied from about −10% to 20% of its optimized values for all five CR models at each field site–according to Figure 8, the averaged RMSE increased by 0.14 mm d^{-1}, 0.12 mm d^{-1}, 0.15 mm d^{-1}, 0.09 mm d^{-1} from the RMSE of the E estimated when using optimized α_e values at TGL, XDT, BJ, and NAMORS, respectively.

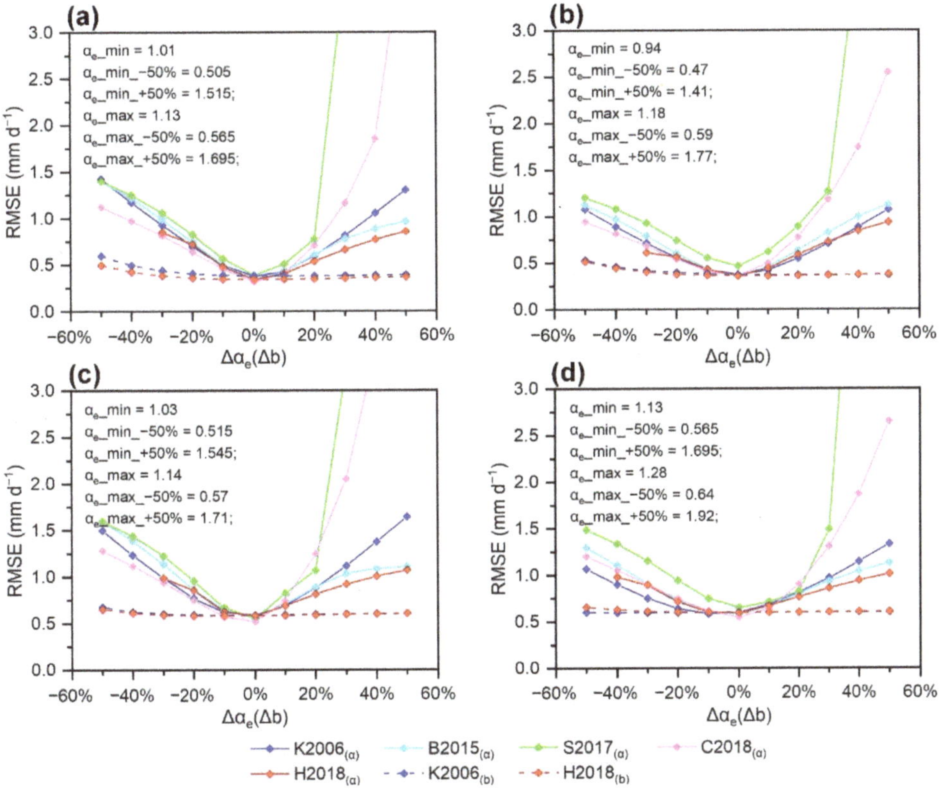

Figure 8. Sensitivity analysis of CR-based models to the parameter α_e and b at (**a**) TGL, (**b**) XDT, (**c**) BJ, and (**d**) NAMORS. $\Delta\alpha_e$ represents the increment to the optimized α_e values for all five CR-based models and Δb represents the increment to the optimized b values for K2006 and H2018 model. Suffix "(α)" and "(b)" represents simulated actual ET with perturbations of α (with constant b) and b (with constant α), respectively.

Thus, the parameter α_e value is very important to the accuracy of E estimation according to our study; certainly, the smaller deviation of optimized α_e value would exert little impact on the accuracy of E estimation. Furthermore, the K2006 and H2018 models were completely insensitive to deviations of parameter b values.

4.2. Comparison with Previous Studies on the QTP at a Single Point Scale

Here, we compared estimation results of monthly E by five CR-based models in the present study with the latest two improved remote-sensing ET models that have been validated for the QTP. One is an improved MOD16 model by Yuan [71], referred to as "MOD16_Yuan" in this study; the another one is an improved SEBS model by Han [72], referred to as "SEBS_Han" in this study. Both obtained better accuracy with in situ measurements than that in previous results. Due to shared use observation data with Yuan's work at NAMORS in 2009 and with Han's work at BJ in 2013, the above two field sites during the corresponding period were chosen for comparison. Note that there are some differences in the observed E between this study and the above two works, which may be caused by different data procedure processes, such as quality control and gap-filling approaches.

Figure 9a exhibits the estimation results of monthly E by the CR approach and MOD16_Yuan model at NAMORS. Statistical results showed that in situ measured annual E values ranged from 467.2 to 481.6 mm, and the annual E values simulated from the five CR-based models were 495.5 mm (K2006), 455.6 mm (B2015), 371.1 mm (S2017), 418.4 mm (C2018), and 452.8 mm (H2018), respectively. For the MOD16_Yuan model, the annual E values reached 539.7 mm which was overestimated by 12.1% compared with in situ measurements. The largest bias occurred in June when the positive bias could reach 36 mm. The K2006 model overestimated the annual E by 6.1% with smaller positive deviations than the MOD16_Yuan model. The B2015 and H2018 models obtained better accuracy with smaller negative bias–approximately 20 mm on an annual scale–and both two CR-based models performed better than the K2006 and MOD16_Yuan models. However, the S2017 and C2018 models both performed poorly due to large negative biases, especially the S2017 model, with a bias of approximately 100 mm on an annual scale.

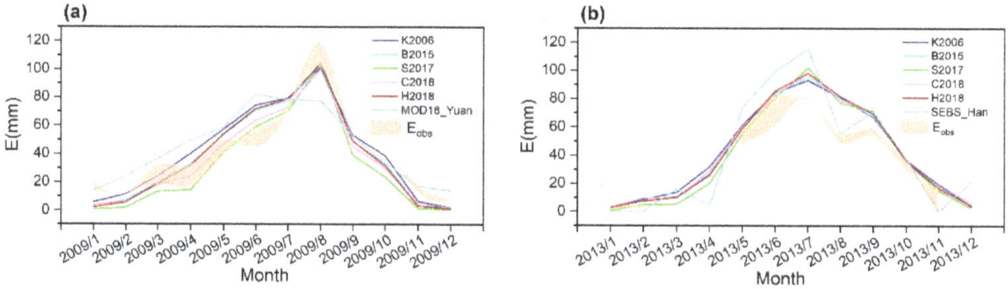

Figure 9. Comparison of monthly observed E values (filled area enclosed by calculated results from this study and another corresponding work) and simulated E values (solid lines) of CR-based models with improved MOD16 evapotranspiration model by Yuan [71] at (**a**) NAMORS; with improved SEBS model by Han [72] at (**b**) BJ.

Figure 9b is the same as Figure 9a, but for BJ. Due to the lack of observed E data in some months, only available data during the corresponding month were analyzed. The simulated monthly E values by the five CR-based models were very close to each other– which could also capture the monthly variations in observed E–but a larger positive bias occurred in August. The SEBS_Han model performed well in August; however, it significantly overestimated monthly E values in May, June, and July, with a larger positive bias than that of the five CR-based models. Thus, the overall performance of the CR approach is still better than that of the SEBS_Han model in the present study.

Both the MOD16_Yuan and the SEBS_Han models have solid physical foundations, and the MOD16_Yuan model based on the original MOD16 model takes soil moisture and soil texture into soil evaporation estimation, and simultaneously optimizes canopy transpiration estimation. The SEBS_Han model introduces the description of the form drag caused by subgrid-scale topographical obstacles, the effective roughness lengths for momentum, and sensible heat transfer into the SEBS model. The above two models both improved the description capability of remote sensing models for the physical process of ET. For the CR approach, from another aspect of feedback of atmospheric evaporation demand to land surface moisture conditions, with a few unknown parameters and routine meteorological variables, the CR approach has also been demonstrated to have comparable accuracy with current sophisticated ET models in the present study.

4.3. Perspectives from the Present CR-Based Model Evaluations

This study provided a relatively comprehensive assessment about applicability of the CR-based models to frozen ground regions on the QTP. A performance comparison of five

CR-based models was evaluated at four observation field sites. After local calibration, the E estimated by all five CR-based models captured daily variations; however, consistent with the findings of previous studies, it is difficult to capture daily variations during cold seasons (Figure 5). The reason for poor performance during cold seasons is still unclear. Theoretically, variable E_{po} (approximated by the Priestley-Taylor equation) in the CR approach assumes evaporation for an extensive saturation with minimal advection; however, during cold seasons, air advection is usually stronger than that in warm seasons. Thus, greater deviations may be derived from the conduction of E_{po} by only local air temperature and humidity in cold seasons, and the degree of land–atmosphere coupling during the cold season is usually weaker than that in warm seasons. The above conditions for cold seasons hardly satisfy the requirements of the CR approach.

Han [65] pointed out that accurate terrestrial ET estimation depends on precisely determining the land surface and atmospheric status; however, the deficiencies of CR-based models are focused only on the atmospheric status and neglect the land surface status, and the influence of land surface status on evaporation processes may not be fully captured through changes in atmospheric status alone, especially at small spatiotemporal scales. The CR principle may be an alternatively valuable approach when there is a lack of land surface information in the past, but for now, remote-sensing technologies make obtaining information on land surface status significantly easier. The CR approach also needs to take land surface information into account. Currently, some studies [70,73,74] have introduced shallow soil moisture, the vegetation index, or AI into the CR approach when applied at large-scale regions. Some other studies [75,76] also adhere to using only atmospheric status information, developing a free-calibrated CR approach when estimating large-scale ET. Further assessments about two different CR approaches need to be explored in the future.

5. Conclusions

This work assessed the actual evapotranspiration estimated by the CR approach in the frozen ground regions of the QTP. The uncertainties associated with observation and modeling of E at point scale were analyzed by investigating the parameter determined by different lengths of calibrated periods and parameter deviations on the accuracy of E estimation. Finally, two more latest works were compared with the CR-based models in this study at BJ and NAMORS.

Five CR-based models–whether with default or calibrated parameters–performed well in daily E estimation: the NSE values were both above 0.7, the accuracy of each CR-based model was close to another, the C2018 and H2018 models performed better among five CR-based models on daily and monthly timescales at four field sites, and the S2017 model performed poorest. The five CR-based models could estimate daily E during warm seasons, however, they are hardly applied during cold seasons. Nevertheless, the amount of E during cold seasons only account for small proportion of annual evaporation. Therefore, CR-based models with fewer data requirements and parameters provide a feasible approach to estimate daily and monthly E without needing detailed information on moisture and vegetation in the frozen ground regions of the QTP.

For calibrated parameter α_e and b in CR-based models, all models are more sensitive to variations of α_e than b. Specifically, within a certain range of deviations of parameter α_e (for example, deviations from -10 to 30% in this study), the RMSE values will increase about 0.1 mm d^{-1} compared with RMSE of E estimation with optimized parameter values. Parameter b has little impact on the accuracy of E estimation. Thus, reasonable parameter α_e values are of much importance to the performance of CR-based models.

Compared with previous studies about the application of CR-based models on the QTP, this study uses more in situ measurement data and more CR-based models. The present study also demonstrates better capability of CR approach on estimating E in high altitude and cold regions over QTP. At present, the CR approach has been developed as a benchmarking tool for large-scale evapotranspiration estimates, however, discrepancies in different spatiotemporal scales with other mainstream evapotranspiration products indicated that with the exception of developing physically CR-based models, hybrid modelling approaches such as combining CR approach with data-driven models are of great significance to large-scale and high-accuracy ET estimation in the future.

Author Contributions: C.S.: Formal analysis, Writing—original draft, Visualization. T.W. (Tonghua Wu): Conceptualization, Supervision, Writing—review & editing. N.M.: Conceptualization, Methodology, Writing—review & editing. J.W.: Conceptualization, Methodology, Writing—review & editing. X.L.: Writing—review & editing. X.Z.: Writing—review & editing. T.W. (Tianye Wang): Writing—review & editing. G.H.: Writing—review & editing. R.L.: Conceptualization. S.Y.: Conceptualization. J.C.: Conceptualization. J.Y.: Conceptualization. C.Y.: Writing—review & editing. All authors have read and agreed to the published version of the manuscript.

Funding: This research was funded by the National Natural Science Foundations of China (grant number 41771076, 41690142, 42001071), the West Light Foundation of Chinese Academy of Sciences (Tonghua Wu), the "Hundred Talents program" of Chinese Academy of Sciences (Sizhong Yang), the Open Research Fund Program of State Key Laboratory of Cryospheric Science, Northwest Institute of Eco-Environment and Resources, Chinese Academy of Sciences (grant number: SKLCSOP-2020-11). And The APC was funded by Tonghua Wu.

Data Availability Statement: The data that support the findings of this study are available from the corresponding author upon request (Tonghua Wu, thuawu@lzb.ac.cn).

Acknowledgments: The authors would like to thank Xuelong Chen, Cunbo Han and Ling Yuan for sharing the original and modelling data at BJ and NAMORS.

Conflicts of Interest: The authors declare that they have no known competing financial interest or personal relationships that could have appeared to influence the work reported in this paper.

References

1. Wang, K.; Dickinson, R.E. A review of global terrestrial evapotranspiration: Observation, modeling, climatology, and climatic variability. *Rev. Geophys.* **2012**, *50*, RG2005. [CrossRef]
2. Jung, M.; Reichstein, M.; Ciais, P.; Seneviratne, S.I.; Sheffield, J.; Goulden, M.L.; Bonan, G.; Cescatti, A.; Chen, J.; Jeu, R.; et al. Recent decline in the global land evapotranspiration trend due to limited moisture supply. *Nature* **2010**, *467*, 951–954. [CrossRef]
3. Fisher, J.B.; Melton, F.; Middleton, E.; Hain, C.; Anderson, M.; Allen, R.; McCabe, M.F.; Hook, S.; Baldocchi, D.; Townsend, P.A.; et al. The future of evapotranspiration: Global requirements for ecosystem functioning, carbon and climate feedbacks, agricultural management, and water resources. *Water Resour. Res.* **2017**, *53*, 2618–2626. [CrossRef]
4. Anderson, M.C.; Allen, R.G.; Morse, A.; Kustas, W.P. Use of Landsat thermal imagery in monitoring evapotranspiration and managing water resources. *Remote Sens. Environ.* **2012**, *122*, 50–65. [CrossRef]
5. Xiang, K.; Li, Y.; Horton, R.; Feng, H. Similarity and difference of potential evapotranspiration and reference crop evapotranspiration—A review. *Agric. Water Manag.* **2020**, *232*, 106043. [CrossRef]
6. Allen, R.G.; Pereira, L.S.; Howell, T.A.; Jensen, M.E. Evapotranspiration information reporting: I. Factors governing measurement accuracy. *Agric. Water Manag.* **2011**, *98*, 899–920. [CrossRef]
7. McMahon, T.A.; Peel, M.C.; Lowe, L.; Srikanthan, R.; McVicar, T.R. Estimating actual, potential, reference crop and pan evaporation using standard meteorological data: A pragmatic synthesis. *Hydrol. Earth Syst. Sci.* **2013**, *17*, 1331–1363. [CrossRef]
8. Salvucci, G.D.; Gentine, P. Emergent relation between surface vapor conductance and relative humidity profiles yields evaporation rates from weather data. *Proc. Natl. Acad. Sci. USA* **2013**, *110*, 6287–6291. [CrossRef] [PubMed]
9. Fisher, J.B.; Lee, B.; Purdy, A.J.; Halverson, G.H.; Dohlen, M.B.; Cawse-Nicholson, K.; Wang, A.; Anderson, R.G.; Aragon, B.; Arain, M.A.; et al. ECOSTRESS: NASA's Next Generation Mission to Measure Evapotranspiration From the International Space Station. *Water Resour. Res.* **2020**, *56*, e2019WR026058. [CrossRef]
10. Mueller, B.; Hirschi, M.; Jimenez, C.; Ciais, P.; Dirmeyer, P.A.; Dolman, A.J.; Fisher, J.B.; Jung, M.; Ludwig, F.; Maignan, F.; et al. Benchmark products for land evapotranspiration: LandFlux-EVAL multi-data set synthesis. *Hydrol. Earth Syst. Sci.* **2013**, *17*, 3707–3720. [CrossRef]
11. Zhang, K.; Kimball, J.S.; Running, S.W. A review of remote sensing based actual evapotranspiration estimation. *Wires. Water.* **2016**, *3*, 834–853. [CrossRef]
12. Jung, M.; Koirala, S.; Weber, U.; Ichii, K.; Gans, F.; Camps-Valls, G.; Papale, D.; Schwalm, C.; Tramontana, G.; Reichstein, M. The FLUXCOM ensemble of global land-atmosphere energy fluxes. *Sci. Data* **2019**, *6*, 74. [CrossRef]
13. Fisher, J.B.; Tu, K.P.; Baldocchi, D.D. Global estimates of the land–atmosphere water flux based on monthly AVHRR and ISLSCP-II data, validated at 16 FLUXNET sites. *Remote Sens. Environ.* **2008**, *112*, 901–919. [CrossRef]

14. Vinukollu, R.K.; Wood, E.F.; Ferguson, C.R.; Fisher, J.B. Global estimates of evapotranspiration for climate studies using multi-sensor remote sensing data: Evaluation of three process-based approaches. *Remote Sens. Environ.* **2011**, *115*, 801–823. [CrossRef]
15. Wang, K.; Ma, N.; Zhang, Y.; Qiang, Y.; Guo, Y. Evapotranspiration and energy partitioning of a typical alpine wetland in the central Tibetan Plateau. *Atmos. Res.* **2022**, *267*, 105931. [CrossRef]
16. Pan, S.; Pan, N.; Tian, H.; Friedlingstein, P.; Sitch, S.; Shi, H.; Arora, V.K.; Haverd, V.; Jain, A.K.; Kato, E.; et al. Evaluation of global terrestrial evapotranspiration using state-of-the-art approaches in remote sensing, machine learning and land surface modeling. *Hydrol. Earth Syst. Sci.* **2020**, *24*, 1485–1509. [CrossRef]
17. Penman, H.L. Natural evaporation from open water, bare soil and grass. *Proc. R. Soc. Lond. Ser. A—Math. Phys. Sci.* **1948**, *193*, 120–145. [CrossRef]
18. Monteith, J.L. *Evaporation and Environment. 19th Symposia of the Society for Experimental Biology*; University Press: Cambridge, UK, 1965; Volume 19, pp. 205–234.
19. Ramírez, J.A.; Hobbins, M.T.; Brown, T.C. Observational evidence of the complementary relationship in regional evaporation lends strong support for Bouchet's hypothesis. *Geophys. Res. Lett.* **2005**, *32*, L15401. [CrossRef]
20. Han, S.; Tian, F.; Hu, H. Positive or negative correlation between actual and potential evaporation? Evaluating using a nonlinear complementary relationship model. *Water Resour. Res.* **2014**, *50*, 1322–1336. [CrossRef]
21. Bouchet, R.J. Evapotranspiration reelle, evapotranspiration potentielle, et production agricole. *Ann. Agron.* **1963**, *14*, 743–824.
22. Ma, N.; Zhang, Y.; Xu, C.Y.; Szilagyi, J. Modeling actual evapotranspiration with routine meteorological variables in the data-scarce region of the Tibetan Plateau: Comparisons and implications. *J. Geophys. Res. Biogeosci.* **2015**, *120*, 1638–1657. [CrossRef]
23. Ma, N.; Szilagyi, J.; Zhang, Y.; Liu, W. Complementary-Relationship-Based Modeling of Terrestrial Evapotranspiration Across China During 1982–2012: Validations and Spatiotemporal Analyses. *J. Geophys. Res. Atmos.* **2019**, *124*, 4326–4351. [CrossRef]
24. Han, S.; Tian, F. A review of the complementary principle of evaporation: From the original linear relationship to generalized nonlinear functions. *Hydrol. Earth Syst. Sci.* **2020**, *24*, 2269–2285. [CrossRef]
25. Brutsaert, W.; Stricker, H. An advection-aridity approach to estimate actual regional evapotranspiration. *Water Resour. Res.* **1979**, *15*, 443–450. [CrossRef]
26. Morton, F.I. Operational estimates of areal evapotranspiration and their significance to the science and practice of hydrology. *J. Hydrol.* **1983**, *66*, 1–76. [CrossRef]
27. Granger, R.J.; Gray, D.M. Evaporation from natural nonsaturated surfaces. *J. Hydrol.* **1989**, *111*, 21–29. [CrossRef]
28. Kahler, D.M.; Brutsaert, W. Complementary relationship between daily evaporation in the environment and pan evaporation. *Water Resour. Res.* **2006**, *42*, W05413. [CrossRef]
29. Szilagyi, J. On the inherent asymmetric nature of the complementary relationship of evaporation. *Geophys. Res. Lett.* **2007**, *34*, L02405. [CrossRef]
30. Han, S.; Hu, H.; Yang, D. A complementary relationship evaporation model referring to the Granger model and the advection-aridity. *Hydrol. Process.* **2011**, *25*, 2094–2101. [CrossRef]
31. Han, S.; Hu, H.; Tian, F. A nonlinear function approach for the normalized complementary relationship evaporation model. *Hydrol. Process.* **2012**, *26*, 3973–3981. [CrossRef]
32. Brutsaert, W. A generalized complementary principle with physical constraints for land-surface evaporation. *Water Resour. Res.* **2015**, *51*, 8087–8093. [CrossRef]
33. Crago, R.; Szilagyi, J.; Qualls, R.; Huntington, J. Rescaling the complementary relationship for land surface evaporation. *Water Resour. Res.* **2016**, *52*, 8461–8471. [CrossRef]
34. Crago, R.D.; Qualls, R.J. Evaluation of the Generalized and Rescaled Complementary Evaporation Relationships. *Water Resour. Res.* **2018**, *54*, 8086–8102. [CrossRef]
35. Szilagyi, J.; Crago, R.; Qualls, R. A calibration-free formulation of the complementary relationship of evaporation for continental-scale hydrology. *J. Geophys. Res. Atmos.* **2017**, *122*, 264–278. [CrossRef]
36. Han, S.; Tian, F. Derivation of a Sigmoid Generalized Complementary Function for Evaporation With Physical Constraints. *Water Resour. Res.* **2018**, *54*, 5050–5068. [CrossRef]
37. Gao, B.; Xu, X. Derivation of an exponential complementary function with physical constraints for land surface evaporation estimation. *J. Hydrol.* **2021**, *593*, 125623. [CrossRef]
38. Crago, R.; Crowley, R. Complementary relationships for near-instantaneous evaporation. *J. Hydrol.* **2005**, *300*, 199–211. [CrossRef]
39. Huntington, J.L.; Szilagyi, J.; Tyler, S.W.; Pohll, G.M. Evaluating the complementary relationship for estimating evapotranspiration from arid shrublands. *Water Resour. Res.* **2011**, *47*, W05533. [CrossRef]
40. Xu, X.; Li, X.; Wang, X.; He, C.; Tian, W.; Tian, J.; Yang, L. Estimating daily evapotranspiration in the agricultural-pastoral ecotone in Northwest China: A comparative analysis of the Complementary Relationship, WRF-CLM4.0, and WRF-Noah methods. *Sci. Total Environ.* **2020**, *729*, 138635. [CrossRef]
41. Zhou, H.; Han, S.; Liu, W. Evaluation of two generalized complementary functions for annual evaporation estimation on the Loess Plateau, China. *J. Hydrol.* **2020**, *587*, 124980. [CrossRef]
42. Wang, L.; Han, S.; Tian, F. At which timescale does the complementary principle perform best in evaporation estimation? *Hydrol. Earth Syst. Sci.* **2021**, *25*, 375–386. [CrossRef]
43. Qiu, J. China: The third pole. *Nature* **2008**, *454*, 393–396. [CrossRef] [PubMed]

44. Duan, A.M.; Wu, G.X.; Liu, Y.M.; Ma, Y.M.; Zhao, P. Weather and climate effects of the Tibetan Plateau. *Adv. Atmos. Sci.* **2012**, *29*, 978–992. [CrossRef]
45. Wu, G.; Duan, A.; Liu, Y.; Mao, J.; Ren, R.; Bao, Q.; He, B.; Liu, B.; Hu, W. Tibetan Plateau climate dynamics: Recent research progress and outlook. *Natl. Sci. Rev.* **2015**, *2*, 100–116. [CrossRef]
46. Zou, D.; Zhao, L.; Sheng, Y.; Chen, J.; Hu, G.; Wu, T.; Wu, J.; Xie, C.; Wu, X.; Pang, Q.; et al. A new map of permafrost distribution on the tibetan plateau. *Cryosphere* **2017**, *11*, 2527–2542. [CrossRef]
47. Cheng, G.; Wu, T. Responses of permafrost to climate change and their environmental significance, Qinghai-Tibet Plateau. *J. Geophys. Res. Earth.* **2007**, *112*, F02S03. [CrossRef]
48. Zhao, L.; Zou, D.; Hu, G.; Du, E.; Pang, Q.; Xiao, Y.; Li, R.; Sheng, Y.; Wu, X.; Sun, Z.; et al. Changing climate and the permafrost environment on the Qinghai–Tibet (Xizang) plateau. *Permafrost. Periglac.* **2020**, *31*, 396–405. [CrossRef]
49. Ma, Y.; Kang, S.; Zhu, L.; Xu, B.; Tian, L.; Yao, T. ROOF OF THE WORLD: Tibetan Observation and Research Platform. *Bull. Am. Meteorol. Soc.* **2008**, *89*, 1487–1492. [CrossRef]
50. Ma, Y.; Ma, W.; Zhong, L.; Hu, Z.; Li, M.; Zhu, Z.; Han, C.; Wang, B.; Liu, X. Monitoring and Modeling the Tibetan Plateau's climate system and its impact on East Asia. *Sci. Rep.* **2017**, *7*, 44574. [CrossRef] [PubMed]
51. Wang, G.; Lin, S.; Hu, Z.; Lu, Y.; Sun, X.; Huang, K. Improving Actual Evapotranspiration Estimation Integrating Energy Consumption for Ice Phase Change Across the Tibetan Plateau. *J. Geophys. Res. Atmos.* **2020**, *125*, e2019JD031799. [CrossRef]
52. Zhao, L.; Zou, D.; Hu, G.; Wu, T.; Du, E.; Liu, G.; Xiao, Y.; Li, R.; Pang, Q.; Qiao, Y.; et al. A synthesis dataset of permafrost thermal state for the Qinghai–Tibet (Xizang) Plateau, China. *Earth Syst. Sci. Data* **2021**, *13*, 4207–4218. [CrossRef]
53. Ma, Y.; Hu, Z.; Xie, Z.; Ma, W.; Wang, B.; Chen, X.; Li, M.; Zhong, L.; Sun, F.; Gu, L.; et al. A long-term (2005–2016) dataset of hourly integrated land-atmosphere interaction observations on the Tibetan Plateau. *Earth Syst. Sci. Data* **2020**, *12*, 2937–2957. [CrossRef]
54. Liu, S.M.; Xu, Z.W.; Wang, W.Z.; Jia, Z.Z.; Zhu, M.J.; Bai, J.; Wang, J.M. A comparison of eddy-covariance and large aperture scintillometer measurements with respect to the energy balance closure problem. *Hydrol. Earth Syst. Sci.* **2011**, *15*, 1291–1306. [CrossRef]
55. You, Q.; Xue, X.; Peng, F.; Dong, S.; Gao, Y. Surface water and heat exchange comparison between alpine meadow and bare land in a permafrost region of the Tibetan Plateau. *Agric. For. Meteorol.* **2017**, *232*, 48–65. [CrossRef]
56. Gu, L.; Yao, J.; Hu, Z.; Zhao, L. Comparison of the surface energy budget between regions of seasonally frozen ground and permafrost on the Tibetan Plateau. *Atmos. Res.* **2015**, *153*, 553–564. [CrossRef]
57. Yang, K.; Wang, J. A temperature prediction-correction method for estimating surface soil heat flux from soil temperature and moisture data. *Sci. China. Ser. D Earth Sci.* **2008**, *51*, 721–729. [CrossRef]
58. Yao, J.; Zhao, L.; Gu, L.; Qiao, Y.; Jiao, K. The surface energy budget in the permafrost region of the Tibetan Plateau. *Atmos. Res.* **2011**, *102*, 394–407. [CrossRef]
59. Ma, N.; Zhang, Y.; Guo, Y.; Gao, H.; Zhang, H.; Wang, Y. Environmental and biophysical controls on the evapotranspiration over the highest alpine steppe. *J. Hydrol.* **2015**, *529*, 980–992. [CrossRef]
60. Parlange, M.B.; Katul, G.G. An Advection-Aridity evaporation model. *Water Resour. Res.* **1992**, *28*, 127–132. [CrossRef]
61. Szilagyi, J.; Jozsa, J. New findings about the complementary relationship-based evaporation estimation methods. *J. Hydrol.* **2008**, *354*, 171–186. [CrossRef]
62. Szilagyi, J. Temperature corrections in the Priestley–Taylor equation of evaporation. *J. Hydrol.* **2014**, *519*, 455–464. [CrossRef]
63. Dai, L.; Fu, R.; Guo, X.; Ke, X.; Du, Y.; Zhang, F.; Li, Y.; Qian, D.; Zhou, H.; Cao, G. Evaluation of actual evapotranspiration measured by large-scale weighing lysimeters in a humid alpine meadow, northeastern Qinghai-Tibetan Plateau. *Hydrol. Process.* **2021**, *35*, e14051. [CrossRef]
64. Wang, L.; Tian, F.; Han, S.; Wei, Z. Determinants of the Asymmetric Parameter in the Generalized Complementary Principle of Evaporation. *Water Resour. Res.* **2020**, *56*, e2019WR026570. [CrossRef]
65. Han, S.; Tian, F. Research Progress of the Generalized Nonlinear Complementary Relationships of Evaporation. *Adv. Earth Sci.* **2021**, *36*, 849–861. (In Chinese with English Abstract)
66. Gan, G.; Liu, Y.; Chen, D.; Zheng, C. Investigation of a non-linear complementary relationship model for monthly evapotranspiration estimation at global flux sites. *J. Hydrometeorol.* **2021**, *22*, 2645–2658. [CrossRef]
67. Szilagyi, J.; Crago, R.D. Comment on "Derivation of a Sigmoid Generalized Complementary Function for Evaporation With Physical Constraints" by S. Han and F. Tian. *Water Resour. Res.* **2019**, *55*, 868–869. [CrossRef]
68. Crago, R.D.; Szilagyi, J.; Qualls, R. Comment on: "A review of the complementary principle of evaporation: From the original linear relationship to generalized nonlinear functions" by Han and Tian (2020). *Hydrol. Earth Syst. Sci.* **2021**, *25*, 63–68. [CrossRef]
69. Ma, N.; Zhang, Y.; Szilagyi, J.; Guo, Y.; Zhai, J.; Gao, H. Evaluating the complementary relationship of evapotranspiration in the alpine steppe of the Tibetan Plateau. *Water Resour. Res.* **2015**, *51*, 1069–1083. [CrossRef]
70. Brutsaert, W.; Cheng, L.; Zhang, L. Spatial Distribution of Global Landscape Evaporation in the Early Twenty-First Century by Means of a Generalized Complementary Approach. *J. Hydrometeorol.* **2020**, *21*, 287–298. [CrossRef]
71. Yuan, L.; Ma, Y.; Chen, X.; Wang, Y.; Li, Z. An Enhanced MOD16 Evapotranspiration Model for the Tibetan Plateau During the Unfrozen Season. *J. Geophys. Res. Atmos.* **2021**, *126*, e2020JD032787. [CrossRef]
72. Han, C.; Ma, Y.; Wang, B.; Zhong, L.; Ma, W.; Chen, X.; Su, Z. Long-term variations in actual evapotranspiration over the Tibetan Plateau. *Earth Syst. Sci. Data* **2021**, *13*, 3513–3524. [CrossRef]

73. Zhang, L.; Brutsaert, W. Blending the Evaporation Precipitation Ratio With the Complementary Principle Function for the Prediction of Evaporation. *Water Resour. Res.* **2021**, *57*, e2021WR029729. [CrossRef]
74. Zhou, H.; Li, Z.; Liu, W. Connotation analysis of parameters in the generalized nonlinear advection aridity model. *Agric. For. Meteorol.* **2021**, *301*, 108343. [CrossRef]
75. Szilagyi, J.; Crago, R.; Ma, N. Dynamic Scaling of the Generalized Complementary Relationship Improves Long-term Tendency Estimates in Land Evaporation. *Adv. Atmos. Sci.* **2020**, *37*, 975–986. [CrossRef]
76. Ma, N.; Szilagyi, J. The CR of Evaporation: A Calibration-Free Diagnostic and Benchmarking Tool for Large-Scale Terrestrial Evapotranspiration Modeling. *Water Resour. Res.* **2019**, *55*, 7246–7274. [CrossRef]

remote sensing

MDPI

Article

Remote Sensing-Detected Changes in Precipitation over the Source Region of Three Rivers in the Recent Two Decades

Xianhong Meng [1,2,3], Mingshan Deng [1,2,4], Yumeng Liu [1,2,4], Zhaoguo Li [1,2,*] and Lin Zhao [1,2]

1 Key Laboratory of Land Surface Process and Climate Change in Cold and Arid Regions, Northwest Institute of Eco-Environment and Resources, Chinese Academy of Sciences, Lanzhou 730000, China; mxh@lzb.ac.cn (X.M.); dengmingshan@lzb.ac.cn (M.D.); liuyumeng@nieer.ac.cn (Y.L.); zhaolin_110@lzb.ac.cn (L.Z.)
2 Zoige Plateau Wetland Ecosystem Research Station, Northwest Institute of Eco-Environment and Resources, Chinese Academy of Sciences, Lanzhou 730000, China
3 Southern Marine Science and Engineering Guangdong Laboratory (Zhuhai), Zhuhai 519000, China
4 University of Chinese Academy of Sciences, Beijing 100029, China
* Correspondence: zgli@lzb.ac.cn

Abstract: The source region of three rivers (SRTR) is an important water conservation area, also known as the Water Tower of Asia. Precipitation is one of the most important factors affecting the ecological system and water resources over the SRTR. However, the characteristics and mechanism of its change at different time scales are still uncertain. Using the GSMaP remote sensing products and ERA5 reanalysis data, this study analyzes the spatial and temporal variability of precipitation and water vapor transport in the SRTR over the past two decades. The annual precipitation slightly reduces in the north and west and slightly increases in the east and south parts of the SRTR. The spring, autumn and winter dominate the decrease in precipitation in most areas of the SRTR, while the summer contributes the most increases. In contrast with the 2000s, the afternoon precipitation slightly reduced in the 2010s, while the nighttime precipitation increases significantly. The changes in nighttime precipitation, especially its intensity, associated with the water vapor transport contribute to the changes in precipitation over the SRTR.

Keywords: remote sensing; afternoon and nighttime precipitation; source region of three rivers; Tibetan Plateau

Citation: Meng, X.; Deng, M.; Liu, Y.; Li, Z.; Zhao, L. Remote Sensing-Detected Changes in Precipitation over the Source Region of Three Rivers in the Recent Two Decades. *Remote Sens.* **2022**, *14*, 2216. https://doi.org/10.3390/rs14092216

Academic Editor: Simone Lolli

Received: 12 April 2022
Accepted: 4 May 2022
Published: 5 May 2022

1. Introduction

The Tibetan Plateau (TP), due to its unique high altitude, large topography and hollow heating effect, plays an important role in the modulation of Asian and even global atmospheric circulation [1,2]. In recent decades, the TP has been experiencing rapid warming and humidification characteristics, with the warming rate almost 1.5 times the global average value [3]. Thus, the TP is known as the "initiator" and "amplifier" of climate change. The source region of Three Rivers (SRTR) is located on the eastern TP and includes the source region of the Yangtze River, Yellow River and Lantsang River and is honored as the "Asia's Water Tower" [4]. Precipitation is one of the most important climatic factors affecting the ecological system and water resources over the SRTR [5]. Because it is located in the intersection area of the Indian monsoon, East Asian monsoon and westerly belt, the SRTR has a complex variety of climate types and inter-annual variation of precipitation [6]. Generally, the precipitation in the eastern and southern parts of the SRTR is significantly more than that in the northwest [7], and there is a complex coupling relationship between land surface processes and precipitation in different regions. Therefore, it is of great value for climate change, water resources research and ecological protection to study the characteristics and mechanisms of precipitation change in the SRTR.

The precipitation in the SRTR has been widely analyzed by using the in situ observation data. Besides the dominant pattern with high and low-value centers located in the southeast and northwest SRTR, respectively, a dipole pattern with southwest–northeast reverse distribution also exists in the SRTR [8]. From 1961 to 2019, the SRTR average annual precipitation was 470.7 mm and increased by 10.31 mm·10 a^{-1} [9], and the frequency of extreme events has increased [10]. The period 1971–1980 was the driest period since the year 1961, and 2001–2015 was the wettest period [11]. The trends in precipitation variation during spring, summer and autumn decreased from northwest to southeast, but the opposite trend was observed in winter [9]. Even in cold seasons, the precipitation has discordant trends in different months, with an increasing trend in November and February and a decreasing trend in other winter months [12].

In addition to meteorological station data, remote sensing products and reanalysis data have also been used to analyze precipitation in the SRTR. The Global Precipitation Climatology Project (GPCP) data are in agreement with the in situ measured precipitation [13]. The Integrated Multisatellite Retrievals for Global Precipitation Measurement (IMERG) products are affected by the temporal scale, precipitation intensity and phase, and the performance in the wet season is superior to that in the dry season [14]. Compared to the in situ observation data, the Climatic Research Unit (CRU) dataset underrated the annual precipitation but gave a similar variation characteristic in the SRTR [15]. On the same time scale, the consistency of NOAA Climate Prediction Center (CPC) products and Tropical Rainfall Measuring Mission (TRMM) products is better than that of the NOAA PERSIANN Precipitation Climate Data Record (PERSIANN-CDR) products [16]. Overall, remote sensing products have a higher ability to detect precipitation in high-altitude areas (>3000 m) than in low-altitude areas (<3000 m), and they have a better detection performance for light rain than moderate and heavy rain events [16].

Many studies have focused on the sources of water vapor and mechanisms of precipitation variation over the SRTR under different climates, but the results remain inconclusive. A study using the GPCP data suggests that the abnormal wind convergence and the low-pressure system, combined with the effects of the western Pacific subtropical high and the Mongolian high, provide conditions for the transport of water vapor and precipitation over the SRTR [13]. Another study suggests that Niño3.4, North Atlantic oscillation and Arctic oscillation play more important roles in the variation of dryness/wetness patterns in the SRTR [12]. In the cold season, the mechanisms for the interannual variation in precipitation are significantly different in different months. The main factors modulating the interannual variability of precipitation are the anomalous westerly water vapor transport (WVT) branch in November and southwesterly WVT anomalies in January and February [13].

In general, previous studies mostly focused on the analysis of long-term interannual or seasonal variations of SRTR precipitation and rarely discusses the changes in specific precipitation types (such as afternoon convective precipitation or nocturnal precipitation). A few studies have found that precipitation in the TP occurred mostly in the afternoon and night due to the thermal processes and the longwave radiation cooling [17,18]. In this study, we used the remote sensing precipitation product to diagnose precipitation changes in the SRTR in the last two decades. As previous studies emphasized an important influence of the hydrological cycle on local precipitation [19–21], we also present variation of afternoon precipitation as it is a dominant part of local triggered precipitation and is strongly related to the local thermal and hydrological processes. The paper is organized as follows. Section 2 introduces the study area and data used in this study. Section 3 presents the results. Section 4 is the discussion. Section 5 presents the conclusions.

2. Study Area and Data

2.1. Study Area

The SRTR is located in the northeastern of the TP, with an average elevation of 3500 m; we mainly focused on the area of 30–37°N and 88–104°E in this study (Figure 1). Previous studies show that 38% of runoff in the source region of the Yellow River, 15% of runoff in the source region of the Lantsang River and a considerate amount of runoff in the Yangtze River originate from the SRTR [4,22]. The GSMaP_Gauge is densely covered with rivers, lakes, wetlands, snow-capped mountains and glaciers and thus is an important ecological shelter zone in China [23].

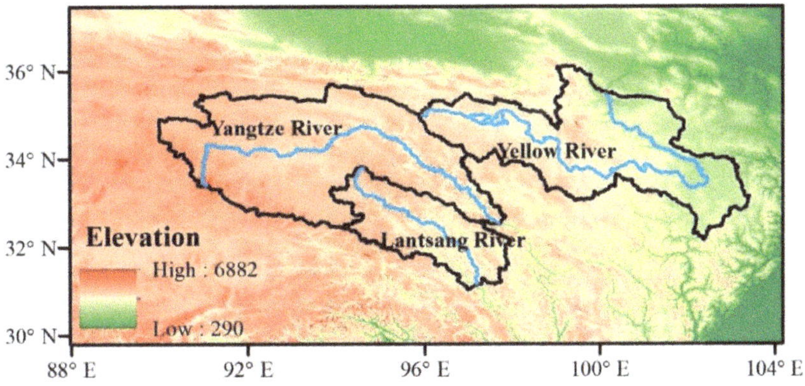

Figure 1. The overview of the Source Region of Three Rivers.

2.2. The GSMaP Precipitation Product

GSMaP (Global Satellite Mapping of Precipitation) and IMERG (Integrated Multi-satellite Retrievals for GPM) are two widely used satellite precipitation products in the GPM era, with high spatial and temporal resolutions. GSMaP (Global Satellite Mapping of Precipitation), developed by the Japan Aerospace Exploration Agency (JAXA) (https://sharaku.eorc.jaxa.jp/GSMaP_CLM/index.htm, accessed on 10 May 2020), is one of the most popular algorithms in the era of GPM [24,25]. The GSMaP_Guage product we used in this study is a gauge-calibrated product that adjusts the GSMaP_MVK estimation with CPC (Climate Precipitation Center) gauge-based analysis of global daily precipitation, whose spatial and temporal resolutions are 0.1° × 0.1° and 1 h, respectively. Kentaro et al. (2015) compared GSMaP_Gauge and GSMaP_MVK products in Japan and found that GSMaP_Gauge products have a better detection performance under different time scales and precipitation intensities [26]. Previous studies show that GSMaP gets some improvements in inversion accuracy and hydrological simulation utility compared to TRMM (Tropical Rainfall Measuring Mission) products over the Tibetan Plateau [27]. In the Yellow River basins. The latest GSMaP data is evaluated as having a relatively higher accuracy than IMERG [28].

2.3. ERA5 Reanalysis Data

ERA5 is the fifth generation of global climate atmospheric reanalysis information from the Copernicus Climate Change Service (C3S) at the European Centre for Medium-range Weather Forecasts (ECMWF) (https://www.ecmwf.int/en/forecasts/dataset/ecmwf-reanalysis-v5, accessed on 30 August 2021), which uses an advanced modeling and data assimilation system to combine model data with observations from around the world to form a globally complete and consistent dataset. Compared to its predecessor, ERA5 has a finer horizontal grid of about 30 km while also improving vertical resolution and providing hourly estimates of a large number of atmospheric, terrestrial and oceanic climate variables [29–31]. Moreover, ERA5 effectively corrects for overestimating some physical quantities of thermodynamics and can be used for general analysis of the Tibetan Plateau [32,33].

3. Results

3.1. Climatology of Precipitation

Figure 2 shows the climatology of annual and seasonal precipitation on the SRTR. The annual precipitation ranges from 500 to 1000 mm/a, presenting a pattern of gradual decrease from southeast to northwest. Among the different seasons, summer (June-July-

August, JJA) precipitation dominates the pattern of annual precipitation, accounting for about 61.3% of the annual precipitation (Figure 3). In spring (March-April-May, MAM), summer and autumn (September-October-November, SON), the precipitation distribution presents a pattern of decrease from southeast to northwest, while in winter (December-January-February, DJF), precipitation has no significant spatial distribution characteristics. In addition, the summer precipitation is mainly in July and August, suggesting influences of water vapor transport by the summer monsoon.

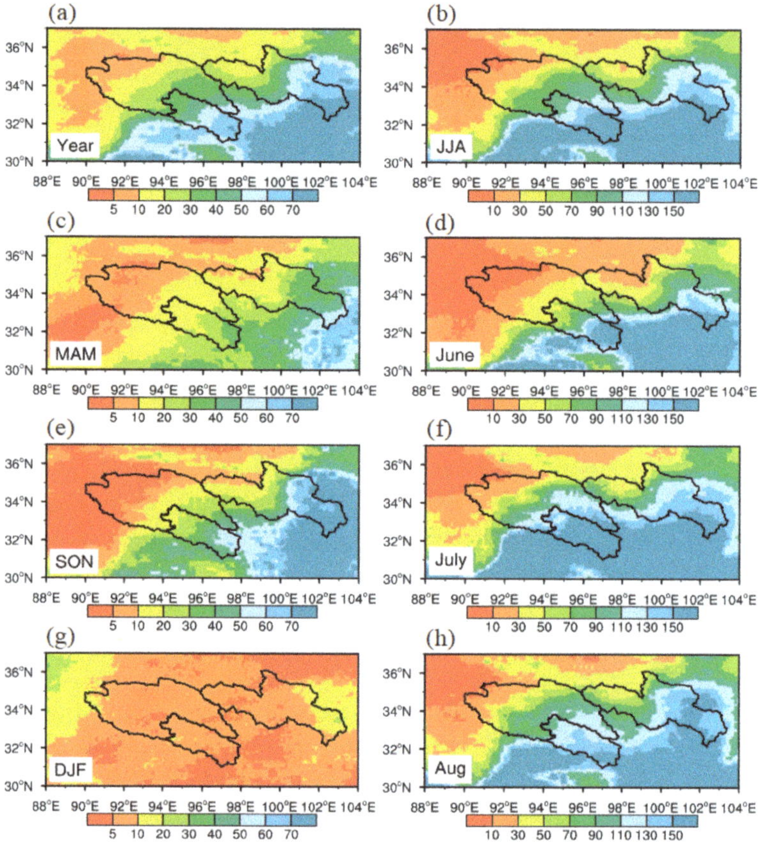

Figure 2. Precipitation (mm/month) in (**a**) annual average, (**b**) June-July-August (JJA) average, (**c**) March-April-May (MAM) average, (**d**) June average, (**e**) September-October-November (SON) average, (**f**) July average, (**g**) December-January-Februrary (DJF) average, and (**h**) August average from 2001 to 2019.

In terms of precipitation occurrence ratio (i.e., proportion of precipitation days in the total days, Figure 4), the distribution characteristics for multi-year climatology, MAM, JJA and SON consistently decline from southeast to northwest, the same as precipitation amount shown in Figure 2. There is not much distinction between MAM and SON. However, for JJA, precipitation days occupy more than 60%, particularly in June, almost 80% of the days have precipitation events, followed by July and August. In the west of the SRTR, precipitation is suppressed most of the time. However, for the east part of the source region of the Yellow River, it always presents a relatively higher frequency of precipitation in all seasons, which has also become an important water supply area for the Yellow River. In DJF, precipitation only occurs in this region, suggesting a very dry condition in other regions of the SRTR.

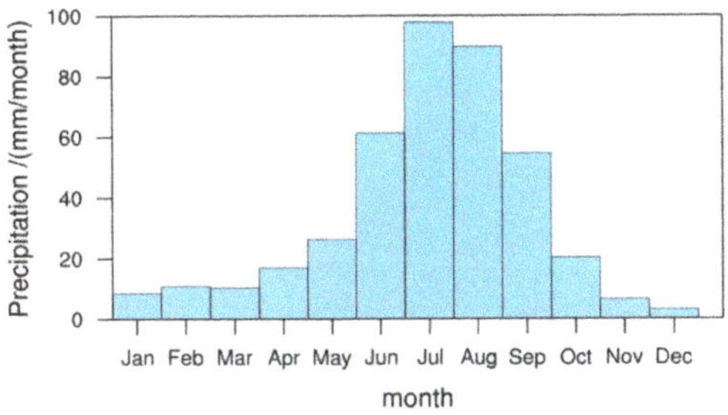

Figure 3. Monthly accumulated precipitation on the SRTR (averaged from 2001 to 2019).

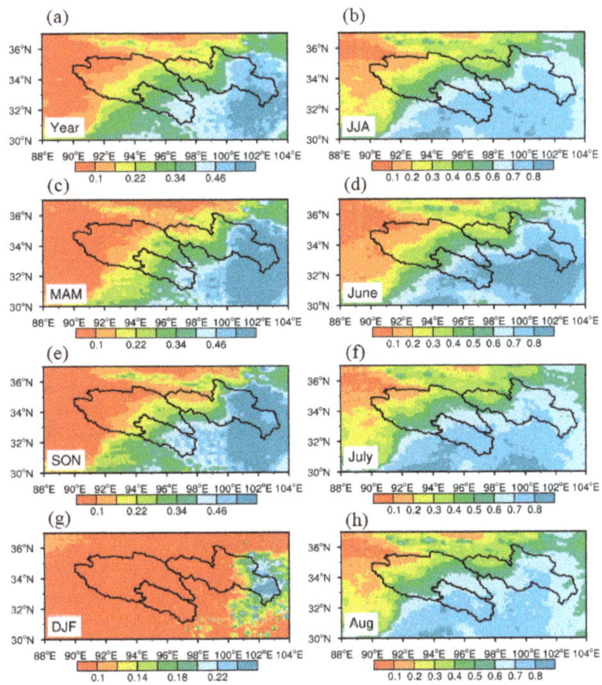

Figure 4. Precipitation occurrence ratio in (**a**) annual average, (**b**) June-July-August (JJA) average, (**c**) March-April-May (MAM) average, (**d**) June average, (**e**) September-October-November (SON) average, (**f**) July average, (**g**) December-January-Februrary (DJF) average, and (**h**) August average from 2001 to 2019.

3.2. *Changes in Precipitation*

3.2.1. Precipitation Amount

Figure 5 shows differences between the climatology of precipitation in the two decades, i.e., 2010 to 2019 and 2001 to 2010. For the annual average precipitation, it shows a slight decrease in the north and west of the SRTR and a slight increase in the eastern and southern parts. In a large area of central SRTR, precipitation shows tiny variation. The most significant changes happened in the southeast of the source region of the Yellow River and the Lantsang River. Considering the contribution of different seasons, MAM, SON and DJF dominate the decreasing of precipitation in most area of the SRTR, while JJA contributes the most increases. Precipitation in spring presents a similar pattern, with the annual average, while in SON, precipitation in most of the area shows a drying trend except the southeast of the source region of the Yellow and Lanstang Rivers. The winter presents a total drying pattern in contrast with the total wetting variation in most areas in summer. For different months in summer, precipitation shows a significant increase in June, especially in the center to the south, while it has an increase in the north in August. In July, it shows basically a drying trend in the northwest of the SRTR and a wetting trend in the south.

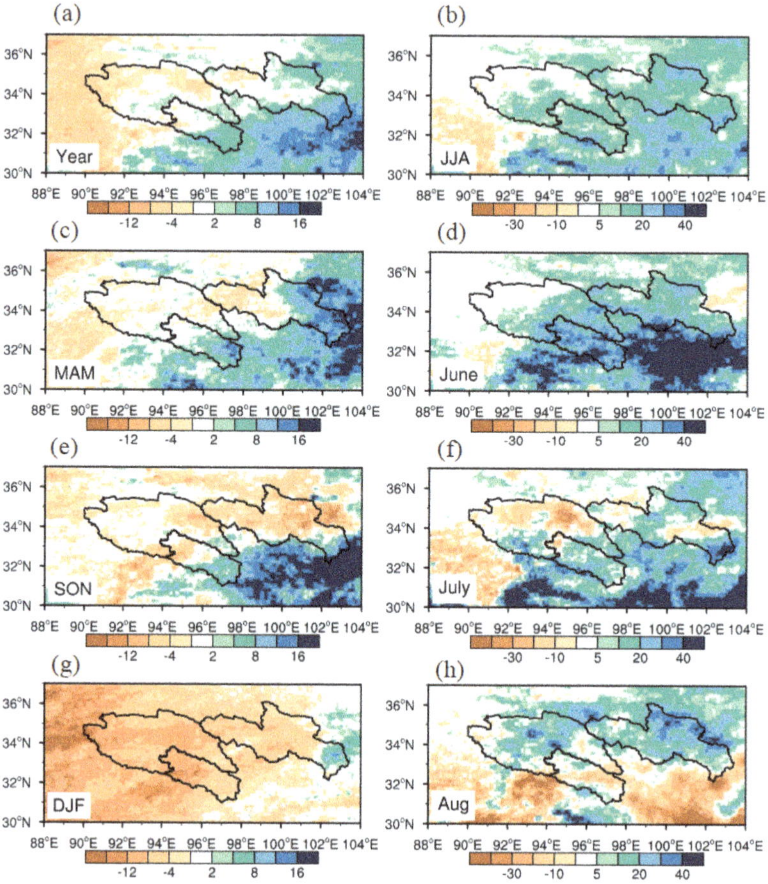

Figure 5. Difference between precipitation (**a**) annual average, (**b**) June-July-August (JJA) average, (**c**) March-April-May (MAM) average, (**d**) June average, (**e**) September-October-November (SON) average, (**f**) July average, (**g**) December-January-Februrary (DJF) average, and (**h**) August average from 2010 to 2019 and from 2001 to 2010 (mm/month).

3.2.2. Precipitation Frequency

As to the changes in precipitation frequency, for the annual average, MAM, SON and DJF, most areas show a reduction in the precipitation occurrence ratio (Figure 6). Only in June and August do the precipitation occurrence ratios increase in most areas. In the west of the Yangtze River headwater region, a significant rise in the precipitation occurrence ratio is presented, although the precipitation amount does not increase correspondingly in this region. The spatial correlation coefficients between the variations of precipitation amount and precipitation occurrences ratio over the two decades are 0.689, 0.752, 0.48, 0.697 and 0.437 for the annual average, MAM, JJA, SON and DJF, respectively. The higher spatial correlation in MAM and SON suggests the possibility of precipitation reduction caused by the decreasing precipitation frequency in these two seasons, while this is not the same in JJA and DJF. In June, July and August, the spatial correlation coefficients are 0.195, 0.604 and 0.611, indicating the inconsistency in changes in precipitation amount and frequency. '

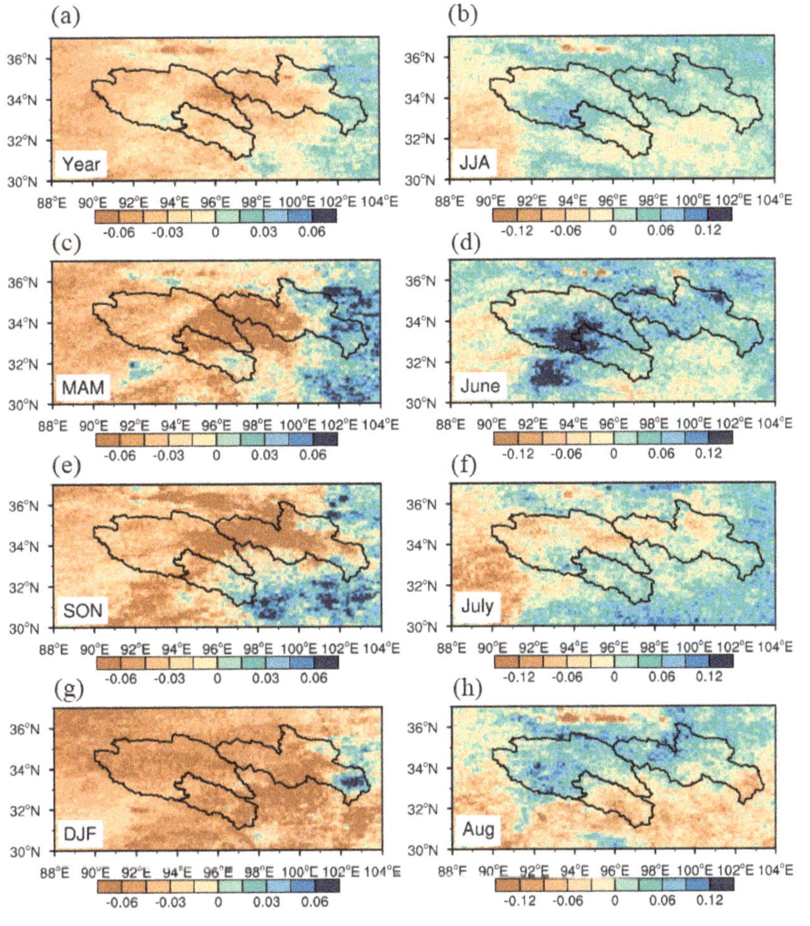

Figure 6. Difference in precipitation occurrence ratio between 2010 and 2019 and from 2001 to 2010 in (**a**) annual average, (**b**) June-July-August (JJA) average, (**c**) March-April-May (MAM) average, (**d**) June average, (**e**) September-October-November (SON) average, (**f**) July average, (**g**) December-January-Februrary (DJF) average, and (**h**) August average.

3.2.3. Afternoon and Nighttime Precipitation

Figure 7 shows the changes in afternoon precipitation in the same period as Figure 5. To address the contribution of afternoon precipitation to precipitation changes in the two decades, we calculated spatial correlation coefficients between the afternoon precipitation differences and total precipitation differences between the two decades (Table 1). Comparing Figure 5 with Figure 7, there was a similar pattern in Year, MAM, SON and JJA, but the results were quite different in the three months in summer. In terms of the spatial correlation coefficients, the maximum is from the total precipitation changes between the two decades, and the correlation coefficients are 0.552, 0.438, 0.518 and 0.805 for MAM, JJA, SON, and DJF, respectively, which are all larger than that in each month in summer.

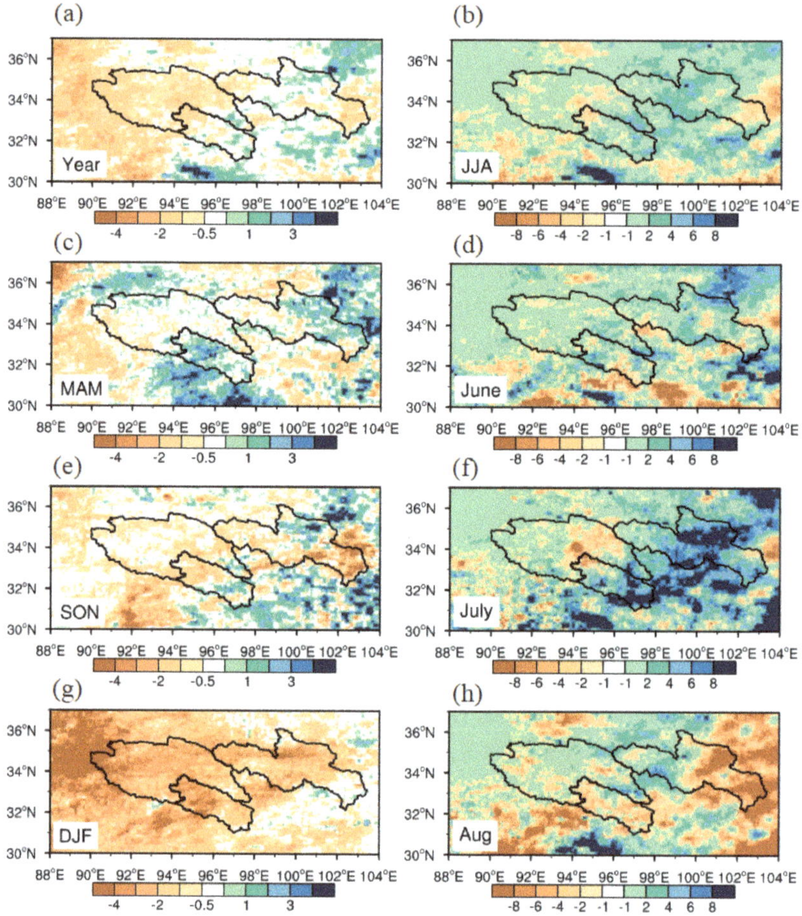

Figure 7. Difference between afternoon precipitation in (**a**) annual average, (**b**) June-July-August (JJA) average, (**c**) March-April-May (MAM) average, (**d**) June average, (**e**) September-October-November (SON) average, (**f**) July average, (**g**) December-January-Februrary (DJF) average, and (**h**) August average from 2010 to 2019 and from 2001 to 2010 (mm/month).

Table 1. Spatial correlation coefficients (Cor) between afternoon precipitation differences and total precipitation differences among the two decades in different periods.

	Year	Season				Month		
period	Year	MAM	JJA	SON	DJF	Jun	Jul	Aug
Cor	0.629	0.552	0.438	0.518	0.805	0.186	0.485	0.326

As shown in Figure 8, the nighttime precipitation changes show similar patterns as in Figure 5, except the magnitude in the night is different. The spatial correlation coefficients are around 0.9 for almost all time periods (Table 2), suggesting a dominant contribution of nighttime precipitation to the total precipitation changes. Figure 9 shows diurnal changes in precipitation rates between the two decades. In the morning in the local time (i.e., 00:00 to 04:00 UTC), there are no significant changes between the two decades. For afternoon precipitation (04:00 UTC to 12:00 UTC), there is a slight reduction in the 2010s in contrast with the 2000s, whilst there is a strong increase shown for the nighttime precipitation, emphasizing the contribution of nighttime precipitation variation to the total precipitation.

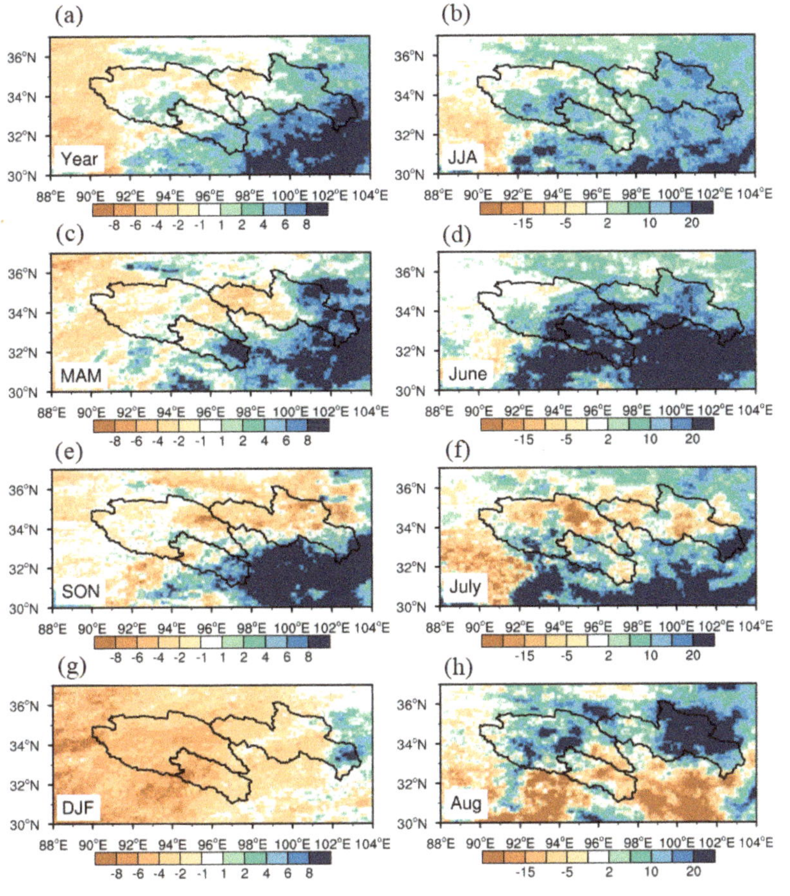

Figure 8. Difference between nighttime precipitation in (**a**) annual average, (**b**) June-July August (JJA) average, (**c**) March-April-May (MAM) average, (**d**) June average, (**e**) September-October-November (SON) average, (**f**) July average, (**g**) December-January-Februrary (DJF) average, and (**h**) August average from 2010 to 2019 and from 2001 to 2010 (mm/month).

Table 2. Spatial correlation coefficients between nighttime precipitation differences and precipitation differences among the two decades in different time periods.

	Year	**Season**				**Month**		
period	Year	MAM	JJA	SON	DJF	Jun	Jul	Aug
Cor	0.964	0.943	0.908	0.965	0.916	0.934	0.917	0.884

In order to quantify the causes of precipitation changes, we calculated the correlation coefficients of precipitation difference with precipitation probability, afternoon precipitation difference, nighttime precipitation difference, afternoon precipitation frequency difference and nighttime precipitation frequency difference (Table 3). It is very clear that the maximums of the correlation coefficients occur between the precipitation difference and the nighttime precipitation difference, suggesting the domination of nighttime precipitation changes in the total variations. When comparing the precipitation intensity and frequency in the nighttime (Table 4), the nighttime precipitation intensity difference shows a higher correlation with the night precipitation changes, implying the domination of nighttime precipitation.

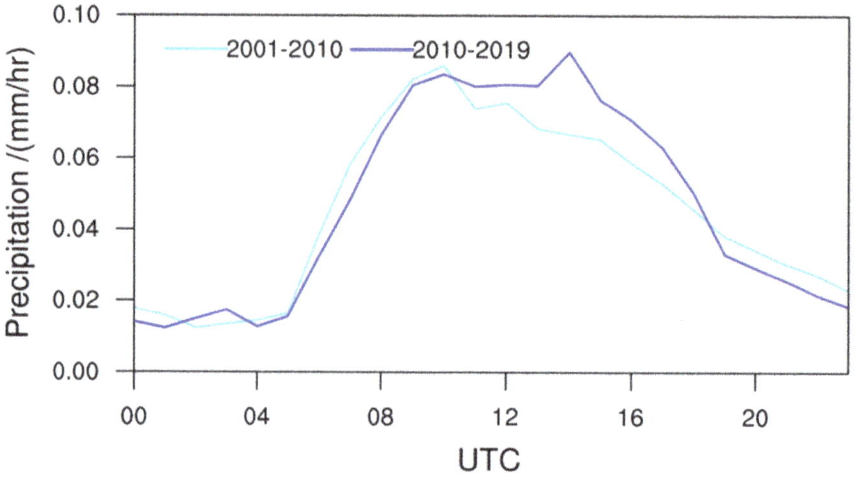

Figure 9. Diurnal changes in precipitation rate (mm/h) between the two decades.

Table 3. Correlation coefficients of precipitation difference with precipitation probability (Pp), afternoon precipitation difference (Pa), nighttime precipitation difference (Pe), afternoon precipitation frequency difference (Ppa) and nighttime precipitation frequency difference (Ppe).

Cor	Year	MAM	JJA	SON	DJF	Jun	Jul	Aug
Pp	0.689	0.752	0.480	0.697	0.437	0.195	0.604	0.611
Pa	0.629	0.552	0.438	0.518	0.805	0.186	0.485	0.326
Pe	0.964	0.943	0.908	0.965	0.916	0.934	0.917	0.884
Ppa	0.835	0.834	0.638	0.743	0.620	0.586	0.629	0.646
Ppe	0.818	0.844	0.562	0.699	0.589	0.465	0.599	0.595

Table 4. Correlation coefficients of nighttime precipitation difference with nighttime precipitation intensity difference (Se) and nighttime precipitation frequency difference (Pa).

Cor	Year	MAM	JJA	SON	DJF	Jun	Jul	Aug
Se	0.963	0.946	0.890	0.959	0.912	0.916	0.909	0.929
Pa	0.795	0.823	0.455	0.705	0.528	0.393	0.538	0.469

4. Discussion

As was shown above, the nighttime precipitation intensity dominates the variations of night precipitation and then precipitation amount in the SRTR. Previous studies have investigated the interdecadal variability of regional precipitation in the SRTR. For example, Shang et al. (2021) found that the increased precipitation in the cold season over the SRTR is associated with the strengthened easterly anomalies over the TP and water vapor meridional transport enhancement from tropical oceans and the South China Sea [34]. Zhao et al. (2021) found the heavy precipitation events, which mainly contribute to the interannual increasing trend of summer precipitation over the SRTR, could be largely attributed to the enhanced moisture sources from the neighboring northeastern areas of the SRTR [35]. It is necessary to diagnose the changes in water vapor between the two decades.

Figure 10 shows the difference between the climatology of water vapor and its divergence between the two decades. For the changes in annual average and different seasons, the variation in atmospheric water vapor convergence and dispersion is in good agreement with the variation in the nighttime precipitation (Figure 8). For example, in SON and July, the increase in water vapor transported from the south side and the decrease in water vapor on the north side result in a distribution characteristic of a dry north and a wet south, despite the spatial gradients being different. In MAM and DJF, the water vapor variation is not evident, and with the strong decrease of vertical velocity (figure not shown), the nighttime precipitation of the SRTR generally presents a reduction trend (Figure 8). For JJA, a clearly drying trend is shown, although the pattern varies from month to month. Basically, the water vapor transport increases from the south ocean in the monsoon period, while it shows a decreasing trend in the monsoon retreat period. It is worth noting that the substantial drying of the atmospheric water vapor content and decrease of vertical upward motion in July in the past two decades over the whole SRTR has led to a decrease in summer precipitation, which are different from other studies focusing on longer time periods [12,35–37].

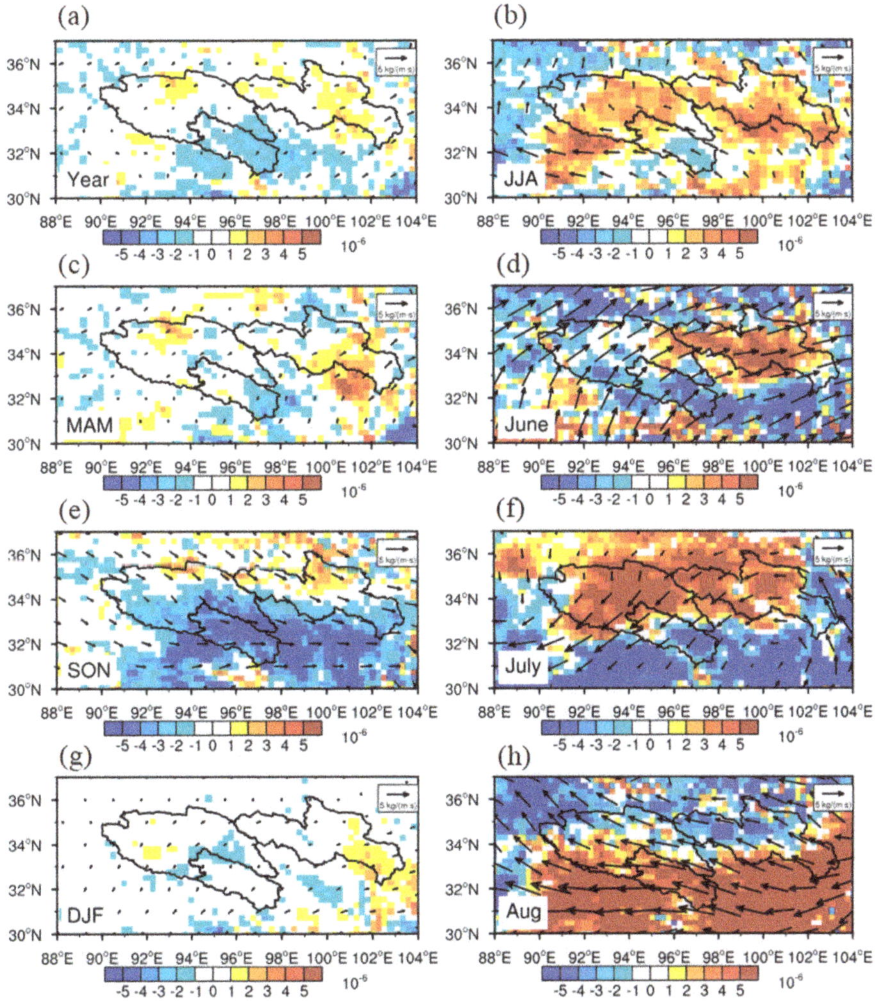

Figure 10. Differences between climatology of vertical water vapor (Vectors, kg/(m.s)) and its divergence (Shaded, kg/(m^2.s)) in (**a**) annual average, (**b**) June-July-August (JJA) average, (**c**) March-

April-May (MAM) average, (**d**) June average, (**e**) September-October-November (SON) average, (**f**) July average, (**g**) December-January-Februrary (DJF) average, and (**h**) August average from 2010 to 2019 and 2001 to 2010.

5. Conclusions

This study focuses on the spatial and temporal variability of precipitation in the SRTR over the past two decades based on satellite observations and reanalysis data. The annual precipitation shows a slight reduction in the north and west of the SRTR and a slight increase in the east and south parts. In most regions of central SRTR, there is little variation in precipitation. For the annual variation, JJA contributes the most increases, while the remaining three seasons dominate the decrease in precipitation in most areas of the SRTR. In the west of the Yangtze River headwater region, although the precipitation amount does not increase, the precipitation occurrences ratio significantly raises. The higher spatial correlation between the precipitation amount and precipitation occurrences ratio in MAM and SON suggests the possibility of precipitation reduction caused by the decreasing precipitation frequency in these two seasons, while this is not the same in JJA and DJF. In contrast with the 2000s, the afternoon precipitation slightly reduced in the 2010s, while there is a strong increase shown for the nighttime precipitation. The spatial correlation coefficients between nighttime and total precipitation differences between the two decades are around 0.9 for almost all time periods, suggesting a dominant contribution of nighttime precipitation to the total precipitation changes. The water vapor transport changes dominate the precipitation changes in different seasons and months in summer, while the decreasing vertical velocity depresses precipitation to some extent. The changes in nighttime precipitation, especially its intensity, associated with the water vapor transport, contribute to the changes in precipitation over the SRTR region.

Author Contributions: Data curation, M.D. and Y.L.; Formal analysis, Z.L.; Project administration, X.M.; Software, M.D. and Y.L.; Writing—original draft, X.M.; Writing—review and editing, Z.L. and L.Z. All authors have read and agreed to the published version of the manuscript.

Funding: This study was conducted under the auspices of the National Natural Science Foundation of China (41930759, 91837209, 42075089, 41975012), the Science and Technology Research Plan of Gansu Province (20JR10RA070).

Data Availability Statement: Not applicable.

Acknowledgments: We would like to that the GSMaP science team for making the satellite precipitation data available and thanks to the ECMWF for providing ERA5 reanalysis data.

Conflicts of Interest: The authors declare no conflict of interest.

References

1. Liu, Y.M.; Lu, M.M.; Yang, H.J.; Duan, A.M.; He, B.; Yang, S.; Wu, G. Land–atmosphere–ocean coupling associated with the Tibetan Plateau and its climate impacts. *Natl. Sci. Rev.* **2020**, *7*, 534–552. [CrossRef] [PubMed]
2. You, Q.L.; Chen, D.L.; Wu, F.Y.; Pepin, N.; Cai, Z.Y.; Ahrens, B.; Jiang, Z.; Wu, Z.; Kang, S.; AghaKouchak, A. Elevation dependent warming over the Tibetan Plateau: Patterns, mechanisms and perspectives. *Earth-Sci. Rev.* **2020**, *210*, 103349. [CrossRef]
3. Yang, K.; Wu, H.; Qin, J.; Lin, C.G.; Tang, W.J.; Chen, Y.Y. Recent climate changes over the Tibetan Plateau and their impacts on energy and water cycle: A review. *Glob. Planet. Chang.* **2014**, *112*, 79–91. [CrossRef]
4. Mao, T.X.; Wang, G.X.; Zhang, T. Impacts of climatic change on hydrological regime in the Three-River headwaters region, China, 1960–2009. *Water Resour. Manag.* **2016**, *30*, 115–131. [CrossRef]
5. Li, Y.; Xiao, J.S.; Yan, D.L. Precipitation pattern of Three River source area in Qinghai from 1964 to 2014. *Agric. Res. Arid Areas* **2016**, *34*, 282–288.
6. Meng, X.; Chen, H.; Li, Z.; Zhao, L.; Zhou, B.; Lü, S.; Deng, M.; Liu, Y.; Li, G. Review of Climate Change and Its Environmental Influence on the Three-River Regions. *Plateau Meteorol.* **2020**, *39*, 1133–1143.
7. Wei, Y.L.; Han, F.X.; Xie, W.X. Analysis on change characteristics of precipitation in the Three-River Headwaters Region in the past 53 years. *Sci. Technol. Qinghai Agric. For.* **2015**, *2*, 45–48.

8. Dong, Y.Y.; Zhai, J.Q.; Zhao, Y.; Li, H.H.; Wang, Q.M.; Jiang, S.; Chang, H.; Ding, Z. Teleconnection patterns of precipitation in the Three-River Headwaters region, China. *Environ. Res. Lett.* **2020**, *15*, 104050. [CrossRef]
9. Cai, Y.X.; Luo, S.H.; Wang, J.; Qi, D.L.; Hu, X.Y. Spatiotemporal variations in precipitation in the Three-River Headwater region from 1961 to 2019. *Pratacultural Sci.* **2022**, *39*, 10–20.
10. Xi, Y.; Miao, C.Y.; Wu, J.W.; Duan, Q.Y.; Lei, X.H.; Li, H. Spatiotemporal Changes in Extreme Temperature and Precipitation Events in the Three-Rivers Headwater Region, China. *J. Geophys. Res. Atmos.* **2018**, *123*, 5827–5844. [CrossRef]
11. Li, S.S.; Yao, Z.J.; Wang, R.; Liu, Z. Dryness/wetness pattern over the Three-River Headwater Region: Variation characteristic, causes, and drought risks. *Int. J. Climatol.* **2020**, *40*, 3550–3566. [CrossRef]
12. Wang, T.; Sun, B.; Wang, H.J. Interannual variations of monthly precipitation and associated mechanisms over the Three River Source region in China in winter months. *Int. J. Climatol.* **2020**, *41*, 2209–2225. [CrossRef]
13. Chen, Y.; Wen, J.; Liu, R.; Zhou, J.; Liu, W. The characteristics of water vapor transport and its linkage with summer precipitation over the Source Region of the Three-River. *J. Hydrometeorol.* **2022**, 441–455. [CrossRef]
14. Meng, C.C.; Mo, X.G.; Liu, S.X.; Hu, S. Extensive evaluation of IMERG precipitation for both liquid and solid in Yellow River source region. *Atmos. Res.* **2021**, *256*, 105570. [CrossRef]
15. Shi, H.Y.; Li, T.J.; Wei, J.H. Evaluation of the gridded CRU TS precipitation dataset with the point rain gauge records over the Three-River Headwaters Region. *J. Hydrol.* **2017**, *548*, 322–332. [CrossRef]
16. Wang, X.Y.; Lu, F.; Zhu, K.; Zhou, Y.Y.; Wu, Z. Comparison of different remote sensing precipitation products at multiple time scales: A case study in the source region of Three Rivers. *China Rural Water Hydropower* **2022**, 1–14. Available online: http://kns.cnki.net/kcms/detail/42.1419.TV.20220218.1747.022.html (accessed on 10 May 2020).
17. Guo, J.P.; Zhai, P.M.; Wu, L.; Cribb, M.; Li, Z.Q.; Ma, Z.Y.; Wang, F.; Chu, D.; Wang, P.; Zhang, J. Diurnal variation and the influential factors of precipitation from surface and satellite measurements in Tibet. *Int. J. Climatol.* **2014**, *34*, 2940–2956. [CrossRef]
18. Li, J. Hourly station-based precipitation characteristics over the Tibetan Plateau. *Int. J. Climatol.* **2018**, *38*, 1560–1570. [CrossRef]
19. Dirmeyer, P.A.; Schlosser, C.A.; Brubaker, K.L. Precipitation, recycling, and land memory: An integrated analysis. *J. Hydrometeorol.* **2009**, *10*, 278–288. [CrossRef]
20. Findell, K.L.; Eltahir, E.A. Atmospheric controls on soil moisture–boundary layer interactions. Part II: Feedbacks within the continental United States. *J. Hydrometeorol.* **2003**, *4*, 570–583. [CrossRef]
21. Zhao, C.L.; Meng, X.H.; Li, Y.H.; Lyu, S.H.; Guo, J.; Liu, H. Impact of soil moisture on afternoon convection triggering over the Tibetan Plateau based on 1-D boundary layer model. *J. Geophys. Res. Atmos.* **2022**, *127*, e2021JD035591. [CrossRef]
22. Zhang, Y.; Fu, Y.F.; Hou, B.J. Analysis of the causes for runoff evaluation in the Yellow River source region. *Yellow River* **2013**, *35*, 22–24.
23. Sheng, W.P.; Zhen, L.; Xiao, Y.; Hu, Y.F. Ecological and socioeconomic effects of ecological restoration in China's Three Rivers Source Region. *Sci. Total Environ.* **2019**, *650*, 2307–2313. [CrossRef] [PubMed]
24. Kachi, M.; Kubota, T.; Aonashi, K.; Ushio, T.; Oki, R. Recent improvements in the global satellite mapping of precipitation (GSMaP). In Proceedings of the 2014 IEEE Geoscience and Remote Sensing Symposium, Quebec City, QC, Canada, 13–18 July 2014.
25. Okamoto, K.I.; Ushio, T.; Iguchi, T.; Takahashi, N.; Iwanami, K. The global satellite mapping of precipitation (GSMaP) project. In Proceedings of the IEEE International Geoscience & Remote Sensing Symposium, Seoul, Korea, 25–29 July 2005; Volume 5, pp. 3414–3416.
26. Kentaro, T.; Saavedra, V.; Masahiro, R.; Kazuki, T.; Tomoo, U.; Takuji, K. Spatiotemporal evaluation of the gauge-adjusted global satellite mapping of precipitation at the basin scale. *J. Meteorol. Soc. Jpn.* **2016**, *94*, 185–195.
27. Lu, D.K.; Yong, B. Evaluation and hydrological utility of the latest GPM IMERG v5 and GsMAP v7 precipitation products over the Tibetan Plateau. *Remote Sens.* **2018**, *10*, 2022. [CrossRef]
28. Ning, S.W.; Song, F.; Parmeshwar, U.; Jin, J.L.; Thapa, B.R.; Ishidaira, H. Error analysis and evaluation of the latest GSMap and IMERG precipitation products over Eastern China. *Adv. Meteorol.* **2017**, *2017*, 1803492. [CrossRef]
29. Kalverla, P.C.; Duncan, J.B., Jr.; Steeneveld, G.J.; Holtslag, A. Low-level jets over the North Sea based on ERA5 and observations: Together they do better. *Wind Energy Sci. Discuss.* **2019**, *4*, 193–209. [CrossRef]
30. Hersbach, H.; Bell, B.; Berrisford, P.; Hirahara, S.; Horányi, A.; Muñoz-Sabater, J.; Nicolas, J.; Peubey, C.; Radu, R.; Schepers, D.; et al. The ERA5 global reanalysis. *Q. J. R. Meteorol. Soc.* **2020**, *146*, 1999–2049. [CrossRef]
31. Lei, Y.H.; Letu, H.S.; Shang, H.S.; Shi, J.C. Cloud cover over the Tibetan Plateau and eastern China: A comparison of ERA5 and ERA-Interim with satellite observations. *Clim. Dyn.* **2020**, *54*, 2941–2957. [CrossRef]
32. Zhao, J.; Li, T.J.; Shi, K.F.; Qiao, Z.; Xia, Z.Y. Evaluation of ERA-5 precipitable water vapor data in plateau areas: A case study of the Northern Qinghai-Tibet Plateau. *Atmosphere* **2021**, *12*, 1367. [CrossRef]
33. Xin, Y.F.; Liu, J.B.; Liu, X.W.; Liu, G.; Cheng, X.H.; Chen, Y.L. Reduction of uncertainties in surface heat flux over the Tibetan Plateau from ERA-Interim to ERA5. *Int. J. Climatol.* **2022**. *early view*. [CrossRef]
34. Shang, S.S.; Zhu, G.F.; Wei, J.H.; Li, Y.; Zhang, K.; Li, R.; Arnault, J.; Zhang, Z.; Laux, P.; Yang, Q.; et al. Associated atmospheric mechanisms for the increased cold season precipitation over the Three-River Headwaters Region from the Late 1980s. *J. Clim.* **2021**, *34*, 8033–8046. [CrossRef]
35. Zhao, R.Y.; Chen, B.; Xu, X.D. Intensified moisture sources of heavy precipitation events contributed to interannual trend in precipitation over the Three-Rivers-Headwater Region in China. *Front. Earth Sci.* **2021**, *9*, 674037. [CrossRef]

36. Zhang, Y.; Li, T.J.; Li, J.Y.; Zhong, D.Y. Influence of the westerlies and the South Asia monsoon on water vapor transport and precipitation in the Three-River Headwaters Region during the rainy season. *Adv. Water Sci.* **2019**, *30*, 348–358.
37. Zhao, D.; Zhang, L.X.; Zhou, T.J. Detectable anthropogenic forcing on the long-term changes of summer precipitation over the Tibetan Plateau. *Clim. Dyn.* **2022**, 1–14. [CrossRef]

remote sensing

MDPI

Article

Precipitation-Use Efficiency and Its Conversion with Climate Types in Mainland China

Suping Wang [1,2], Qiang Zhang [1,2], Ping Yue [1,*], Jianshun Wang [2], Jinhu Yang [3], Wei Wang [1], Hongli Zhang [4] and Xueyuan Ren [1]

1 Institute of Arid Meteorology, CMA, Key Laboratory of Arid Climatic Change and Reducing Disaster of Gansu Province, Key Laboratory of Arid Climatic Change and Disaster Reduction of CMA, Lanzhou 730020, China; wangsp@iamcma.cn (S.W.); zhangqiang@cma.gov.cn (Q.Z.); wangw@iamcma.cn (W.W.); renxy@iamcma.cn (X.R.)
2 College of Atmospheric Sciences, Lanzhou University, Lanzhou 730000, China; wangjsh17@lzu.edu.cn
3 Lanzhou Regional Climate Center, Lanzhou 730020, China; yangjh@iamcma.cn
4 College of Resources and Environmental Engineering, Tianshui Normal University, Tianshui 741000, China; zhanghl13@lzu.edu.cn
* Correspondence: yuep@iamcma.cn

Abstract: The impacts of climate change on ecosystem productivity and water resources over a long term in China are not well quantified. Precipitation-use efficiency (PUE) is a key parameter that describes carbon and water exchange in terrestrial ecosystems. Research on the response of regional PUE to climate change and its driving forces is of great significance to climate-change mitigation and the sustainable development of regional ecology. Based on an improved actual evapotranspiration (ET_a) model, the responses of ET_a, net primary productivity (NPP), and PUE to climate change in different climatic regions of China were analyzed; the contributions of various environmental factors to PUE changes were quantified; and the conversion characteristics and regulatory mechanisms of the PUE regime in different climatic regions were identified. The results indicate that the improved ET_a model, after considering the limiting effect of energy on ET_a in humid regions, can simulate the ET_a distribution in China well. Over the past 58 years (1960–2017), ET_a and NPP have increased in the western regions and decreased in the eastern regions, with the boundary at 103° E. PUE presents a "low-high-low" spatial distribution from northwest to southeast in China. It is noteworthy that there was a zonal distribution for a high value area of PUE, which coincided with the summer monsoon transition zone. The soil moisture (SM) increase in arid regions is the main driving force of the PUE increase, whereas the annual net radiation (Rn) change in humid regions is the main driving force of the PUE change. The transition zone is the conversion zone, where the prevailing factor limiting vegetation growth transitions from water to energy.

Keywords: typical climatic zones; precipitation-use efficiency; net primary productivity; contribution of environmental factors; transformation characteristics

Citation: Wang, S.; Zhang, Q.; Yue, P.; Wang, J.; Yang, J.; Wang, W.; Zhang, H.; Ren, X. Precipitation-Use Efficiency and Its Conversion with Climate Types in Mainland China. *Remote Sens.* **2022**, *14*, 2467. https://doi.org/10.3390/rs14102467

Academic Editors: Massimo Menenti, Yaoming Ma, Li Jia and Lei Zhong

Received: 23 March 2022
Accepted: 19 May 2022
Published: 20 May 2022

1. Introduction

Global warming and the increase of extreme weather events are having a serious impact on the structure, function, and processes of global ecosystems [1], and have become a focal issue of common concern to governments, the public, and the scientific community. In IPCC AR6 [2], it is reported that the frequency and intensity of some extreme weather and climate events will continue to increase under medium and high emission scenarios, and the increased extreme events (e.g., droughts, heat waves, and heavy rainfall) will affect 25–40% of global ecosystem structure and function.

The terrestrial ecosystem carbon cycle is a key process driving ecosystem change, and changes in the ecosystem carbon cycle are sensitive to climate change. China is one of the most sensitive and vulnerable regions to climate change. Climate change decreased the capacity of carbon storage [3], and extreme climate events such as drought, extreme heat, and extreme precipitation all have serious impacts on the carbon cycle of terrestrial ecosystems. Droughts have weakened vegetation growth [4], and prolonged and persistent droughts have reduced carbon accumulation, causing grassland ecosystems in Inner Mongolia to change from a carbon sink to a carbon source in a typical year [5]. Heat waves and droughts significantly reduced regional GPPs and crop yields in summer 2013 [6]. The ice storm in early 2008 also resulted in a decrease in annual evapotranspiration and GPP in southeastern China [7].

Precipitation-use efficiency (PUE) describes the response of net primary productivity (NPP) to the temporal and spatial distribution of precipitation. PUE is a comprehensive physiological and ecological index for evaluating the appropriate degree of vegetation growth while reflecting the carbon and water cycles and the carbon–water relationship in the ecosystem [8]. Research on the characteristics of PUE and its control mechanism can help evaluate and predict the impact of global changes on the carbon–water cycle of ecosystems and provide a theoretical basis for regional plant protection and restoration. PUE distribution and changes are affected by several factors, such as topography, soil conditions, climate change, and human activities. Climate change is the most important and active factor [9,10]. Earlier studies addressed the response of the spatiotemporal PUE pattern to climate change but did not reach a consistent conclusion due to different temporal and spatial scales. Some studies concluded that the PUE spatially decreases with increasing drought and potential evapotranspiration and increases with increasing precipitation [11]. However, other studies concluded that the PUE spatially decreases with increasing precipitation [12,13]. It has also been demonstrated that there is no obvious relationship between the spatial PUE distribution and precipitation [14]. In addition, some studies have reported that the PUE exhibits a unimodal distribution that first increases and then decreases with increasing precipitation and reaches its peak at a specific precipitation value [13,15,16]. The feedback mechanisms between the water-carbon cycle of ecosystems and climate are relatively complex, and the response of NPP and PUE to climate change has a large spatial and temporal heterogeneity. China's vast land area, complex topography, diverse climate and vegetation types, and high spatial and temporal variability in precipitation and ecosystem carbon fluxes hinder the accurate assessment of carbon fluxes [17–20]. Earlier studies focused mostly on small local areas with precipitation or temperature limits or specific vegetation types [11,16,21,22], and little attention has been paid to variations in PUE along a climatic gradient. At the same time, prior studies mostly focused on the effects of single climate factors on PUE, lacking a comprehensive understanding of the specific contributions of each climate factor and regional differences [16,20,21], which cannot fully reveal the difference and transformation characteristics of PUE with climate and vegetation-gradient distribution, thereby limiting the in-depth understanding of PUE characteristics and driving forces in different regions.

NPP is a key for calculating the PUE. There are several methods for obtaining the NPP. In situ measurements have high data accuracy but are limited by the amount of data and are time-consuming and labor-intensive. Thus, they can only be used during surveys of small areas. Model estimation is an effective means of obtaining NPP on a regional or global scale. NPP estimation models can be roughly divided into three categories: ecophysiological process models, light-use efficiency models, and climate statistical models [23]. Ecophysiological process models simulate NPP based on the ecophysiological characteristics and growth mechanisms of plants [24]. Representative models include the Biome-BGC, CEVSA, and BEPS models. This type of model has strong mechanisms and is systematic. However, they are complex, and the required parameters are many and difficult to obtain. Light-use efficiency models use photosynthesis from vegetation and a resource balance view as the theoretical basis. They apply remote sensing data to drive ecological models for NPP simulation on regional or global scales and have been used worldwide. Representative models are the CASA and Glo-PEM models. Although this type of model has clear mechanisms and complete structures, the assignment and correction of model parameters are complicated, with large uncertainties. There are three main types of parameters based on the input of the light-use efficiency model: solar radiation data, maximum light-use efficiency, and environmental factors. The algorithms and required data for each parameter type are diverse. Different calculation methods have substantial differences in simulating vegetation NPP, especially in simulating environmental factors and maximum light use efficiency, because it is difficult to assign a large number of soil parameters to certain regions or special geomorphic types [25]. Additionally, these models are completely dependent on the availability and quality of remote sensing data [26]; in regions with strong spatial heterogeneity and complex terrain, the accuracy of the model is highly uncertain.

Climate statistical models estimate NPP based on the correlation between plant growth and environmental factors. Representative models include the Miami, Thornthwaite memorial, Chicago, and comprehensive models. Although these models lack a mechanism, they are simple, intuitive, and highly applicable. Hence, they constitute the easiest and most convenient method for estimating NPP. Since their development in the 1970s, they have been applied in vegetation NPP research worldwide [20], particularly in relation to large-scale research. Various land-atmosphere mutual observation experiments also provide the possibility for model calibration, thereby continuously improving the accuracy of these models.

In the context of global climate change, the characteristics and changes of China's regional PUE are not yet fully understood, as well as the regional differences and driving forces of the PUE response to climate change. What are the spatial distribution characteristics of the carbon flux in mainland China? How do environmental factors relate to PUE? Does PUE spatial conversion occur with climate type? What is the possible mechanism of action? Therefore, our primary objectives were to: (1) improve the ET_a calculation model by comprehensively considering the water and energy conditions in different regions; (2) analyze the characteristics of ET_a, NPP, and PUE in different climate regions of China; (3) reveal the driving forces, transformation characteristics, and control mechanism of the PUE distribution.

2. Materials and Methods

2.1. Study Area

This study focused on mainland China. There are great climatic differences between east and west and south and north of China. Precipitation decreases from southeast to northwest, and climate transitions from humid to arid, presenting a basic pattern of humid in the east and south, but arid in the west and north, and there is a narrow zonal climate transition zone between humid and arid climate regions. We divided the study region into three sub-regions according to the distribution of average annual precipitation (P): the arid region (P < 200 mm), transition zone (600 > P ≥ 200 mm), and humid region (P ≥ 600 mm) [27]. The distribution of meteorological and flux stations and sub-regions is shown in Figure 1.

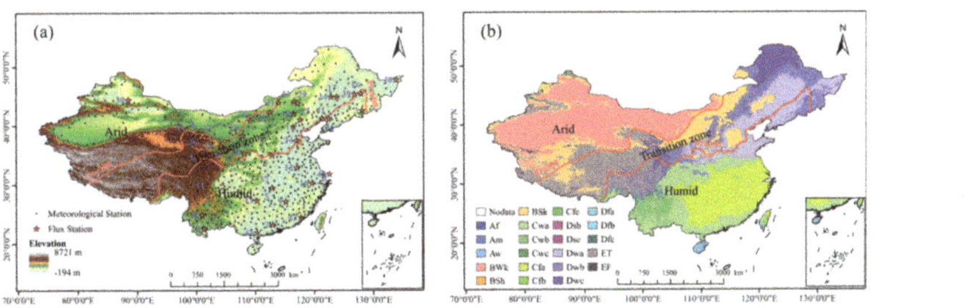

Figure 1. Locations of 693 meteorological stations (black solid dots), 44 flux stations (red stars), and the three sub-regions (i.e., arid region, transition zone, and humid region) (**a**), and the Köppen-Geiger climatic zones (**b**).

2.2. Data

Daily meteorological data from 693 stations in China between 1960 and 2017 were used, including maximum, mean, and minimum temperatures (Tmax, Tmean, and Tmin, respectively); sunshine hours (H); wind velocity (U); relative humidity (Rh); and P. These data were obtained through the meteorological data sharing network of the China Meteorological Administration and were checked for homogenization and quality, including controls for time and space consistency, extreme values, and climate-limit or allowable values [28].

Observations from 44 flux stations were used to validate the ET_a model. Nine of these stations are part of China FLUX (The Chinese Terrestrial Ecosystem Flux Research Network). Data from the remaining stations were obtained from published articles. These flux stations are widely distributed in space, including 16 forest, 17 grassland, 6 farmland, and 5 wetland stations, covering the main climatic regions and typical ecosystem types in China. Details on the data are given in Table 1.

Table 1. Metadata for each flux station including station name, location, altitude, years of available data, and references.

No.	Stations	Latitude	Longitude	Observational Periods	Reference
1	Daxing	39.53° N	116.25° E	2006	Zhang [29]
2	Dinghushan	23.17° N	112.53° E	2003–2010	China FLUX
3	Changbaishan	42.40° N	128.10° E	2003–2010	China FLUX
4	Guantan	38.53° N	100.25° E	2011	Zhu et al. [30]
5	Yueyang	29.31° N	112.51° E	2006	Wang [31]
6	Kubiqi	40.54° N	108.69° E	2006	Wilske et al. [32]
7	Xiaolangdi	35.02° N	112.47° E	2007–2009	Guo [33]
8	Xishuangbannan	21.96° N	101.20° E	2003–2010	China FLUX
9	Qianyanzhou	26.74° N	115.06° E	2003–2010	China FLUX
10	Ailaoshan	24.53° N	101.02° E	2010	China FLUX
11	**Gonggashan**	29.58° N	102.00° E	**2009**	Lin et al. [34]
12	**Huaining**	33.00° N	117.00° E	**2005–2006**	Han et al. [35]
13	Huitong	26.83° N	109.75° E	2009	Wang et al. [36]
14	Laoshan	45.33° N	127.57° E	2004–2006	Cui [37]
15	Miyun	40.63° N	117.32° E	2008–2010	Liu et al. [38]
16	Taihuyuan	30.18° N	119.34° E	2011	Lin et al. [39]
17	Changling	44.58° N	123.50° E	2007–2008	Dong et al. [40]
18	Duolun County	42.05° N	116.28° E	2006	Chen et al. [41]
19	Haibei [a]	37.66° N	101.29° E	2003–2010	China FLUX
20	**Haibei [b]**	37.60° N	101.30° E	**2002–2004**	Gu et al. [42]
21	Siziwang Banner, grazed	41.79° N	111.90° E	2010	Shao et al. [43]
22	Xilinhot, grazed	43.55° N	116.67° E	2006	Chen et al. [41]
23	Xilinhot, typical fenced	44.13° N	116.33° E	2004–2006	Wang et al. [44]
24	Tongyu	44.59° N	122.52° E	2003–2008	Liu and Feng [45]
25	**Yuzhong**	35.95° N	104.14° E	**2007–2012**	Yue et al. [46]
26	Aro	38.04° N	100.46° E	2008	Wang et al. [47]
27	Dangxiong	30.85° N	91.08° E	2003–2010	China FLUX
28	Yanchi	37.81° N	107.48° E	2016	Liu [48]
29	Xinlin Gol	43.55° N	116.68° E	2003–2010	China FLUX
30	Fukang	44.28° N	87.93° E	2004	Liu et al. [49]
31	**Sunitezuo**	44.00° N	113.57° E	**2008–2009**	Zhang et al. [50]
32	Sanjiangyuan	34.35° N	100.50° E	2006–2008	Li et al. [51]
33	Tianjun	38.42° N	98.32° E	2011	Wu et al. [52]
34	Yucheng	36.95° N	116.60° E	2003–2010	China FLUX
35	Dingxi	35.55° N	104.58° E	2010	Yang et al. [53]
36	Jinzhou	41.15° N	121.20° E	2006	Zhou et al. [54]
37	**Yueyang**	29.31° N	112.51° E	**2015**	Chen and Huang [55]
38	**Changwu**	35.23° N	107.67° E	**2008**	Wu et al. [56]

Table 1. *Cont.*

No.	Stations	Latitude	Longitude	Observational Periods	Reference
39	**Yushu**	44.85° N	126.52° E	**2009**	Guo et al. [57]
40	**Sanjiang Plains**	47.51° N	133.51° E	**2017**	Guo et al. [58]
41	**Panjin**	41.14° N	121.90° E	**2005, 2006–2007**	Zhou et al. [59]
42	**Dongtan**	31.52° N	121.96° E	**2005**	Guo et al. [60]
43	**Yunxiao**	23.92° N	117.42° E	**2009–2010**	Yan [61]
44	Gaoqiao	21.57° N	109.76° E	2010	Chen et al. [62]

Nos.1 to 16 are forest stations, 17 to 33 are grassland stations, 34 to 39 are cropland stations, and 40 to 44 are wetland stations. Periods in **bold** are the **Validation Data**. Haibei [a] and Haibei [b] are two sites in different geographical locations.

Actual evapotranspiration data (unit: mm) and the 0–10 cm depth, monthly average soil water-content data (unit: kg·m^{-2}) in GLDAS_Noah025_M 2.0 and 2.1 datasets were used, with 0.25 × 0.25° spatial resolution and a time range from January 1960 to December 2017. GLDAS data are global land-surface characteristics and flux data generated by advanced land–surface models and data assimilation technology [63].

The validation NPP data are the MOD17A3 surface vegetation NPP data provided by the EOS/MODIS (TERRA satellite) of NASA. MOD17A3 has been verified and widely applied in research regarding vegetation growth, biomass estimation, environmental monitoring, and global change in different regions of China and the world. In this study, we used NPP data with a 0.5 × 0.5 km resolution from 2000 to 2017. Data corresponding to the 693 meteorological stations were extracted using the neighboring grid method.

2.3. Methods

PUE was identified as the ratio of NPP to P [11]:

$$PUE = \frac{NPP}{P} \tag{1}$$

where NPP is the annual net primary productivity (unit: g·m^{-2}) and P is the annual precipitation (unit: mm). NPP is obtained through the Thornthwaite memorial model [64]:

$$NPP = 3000\left(1 - e^{-0.0009695(ET_a - 20)}\right) \tag{2}$$

where ET_a is the actual evapotranspiration. It can be obtained as Zhou and Zhang [65]. Although there are many methods to estimate ET_a, such as the water balance method, surface energy balance method, remote sensing analysis method, etc., they all have limitations, such as complicated parameters and difficulty to determine [66,67]. Zhou and Zhang's method fully reflects the limiting effect of energy and water on evapotranspiration, with its few parameters and clear physical significance making it high practicability. The method is as follows:

$$ET_a = \frac{P \times R \times \left(P^2 + R^2 + P \times R\right)}{(P + R) \times \left(P^2 + R^2\right)} \tag{3}$$

Here, P is the annual precipitation, and R is the annual net radiation correction factor, which can be obtained by Equation (4):

$$R = (ET_0 \times P)^{0.5} \times (0.369 + 0.598 \times (ET_0/P)^{0.5} \tag{4}$$

where ET_0 is the potential evapotranspiration calculated using the FAO Penman-Monteith method. See Allen et al. [68] for details.

The sensitivity coefficient and the relative change were used to measure the contribution of environmental factors to the PUE change:

$$Con_X = RC_X \times S_X \tag{5}$$

$$RC_X = \frac{58 \times Trend_X}{|ave_X|} \times 100\% \tag{6}$$

where Con_X is the contribution of a factor to the PUE change, RC_X is the relative change rate of the factor, $Trend_X$ and ave_X are the change rate and average of the factor, respectively, and Sx is the sensitivity coefficient of the PUE with respect to environmental factor X. Sx can be obtained as McCuen [69]:

$$S_x = \lim_{\Delta X \to 0} \left(\frac{\Delta PUE/PUE}{\Delta X/X} \right) = \frac{\partial PUE}{\partial X} \times \frac{X}{PUE} \tag{7}$$

3. Results

3.1. ET_a Estimation and Its Change Characteristics

ET_a was calculated using Equation (3). Compared with the observations from the flux stations (Figure 2), both the observed and estimated ET_a were found to increase significantly with increased precipitation in the region where P < 600 mm. Furthermore, when $P \geq 600$ mm, the observed ET_a first increased and then decreased with increasing P, although the estimated ET_a still increased significantly with increased P. Many factors affect ET_a, the most important of which are water and energy conditions. In arid and semiarid regions, the amount of precipitation is relatively small, but the energy is sufficient. The main factor affecting ET_a is water, and a change in P largely determines the change in ET_a. In the semi humid region, increased P makes energy the controlling factor of ET_a instead of water, and the rate at which ET_a increases with P slows down. In humid areas with abundant rainfall, ET_a no longer increased with increasing P. Due to more precipitation, energy conditions limit ET_a, and higher P results in smaller ET_a. Earlier studies confirmed the switch of the ET_a controlling factor from water to energy with increased precipitation [70]. Equation (3) does not consider this switch when estimating the ET_a in the humid region and does not reflect its energy constraint. Estimates are largely dependent on precipitation, which exaggerates the results. We further used P = 600 and P = 1400 mm as the boundaries and conducted regression corrections on the estimated ET_a. The specific regression equation is given by Equation (8).

$$ET_{a\_new} = \begin{cases} \mathrm{ET_a} \text{ if Pre} \leq 600\text{mm} \\ 0.815 * \mathrm{ET_a} + 26.2 \text{ if } 600 \text{ mm} < \text{Pre} \leq 1400\text{mm } (\mathrm{r} = 0.82, \mathrm{a} < 0.01) \\ -0.615 * \mathrm{ET_a} + 1249.8 \text{ if Pre} > 1400\text{mm } (\mathrm{r} = 0.53, \mathrm{a} < 0.01) \end{cases} \tag{8}$$

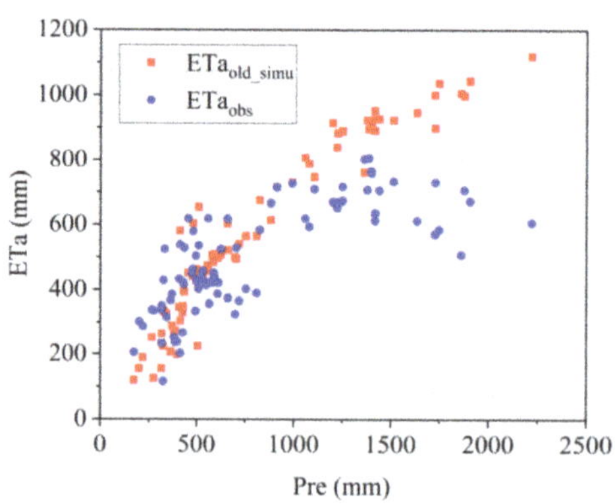

Figure 2. Distribution of observed and estimated ET_a with P.

The improved ET_a in the southern humid region is significantly smaller than the ET_a before improvement (Figure 3). Regions with the highest ET_a move north from the southeastern coastal area to the middle and lower reaches of the Yangtze River. To further verify the accuracy of the improved ET_a model, we compared the estimated values with the ET_a observations from the flux stations during the validation period and the ET_a from GLDAS during the same period. This comparison showed that the improved ET_a is highly consistent with the observed and GLDAS ET_a (Figure 4). The correlation coefficients were 0.95 and 0.85, respectively, and the standard deviation between the estimated and observed values was much smaller than that between the GLDAS assimilation and observed values. GLDAS data are sparser in the high evapotranspiration regions, and the values are too large. Wang et al. [71] also identified an overestimation in the high ET_a regions of southern China in the GLDAS data. Hence, our improved ET_a model can simulate the ET_a in the study region more accurately.

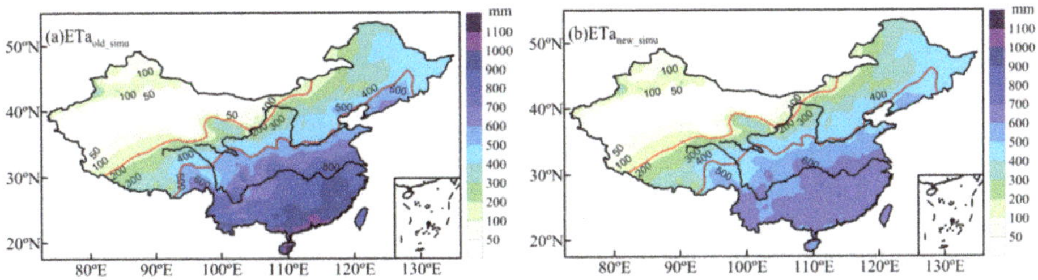

Figure 3. ET_a distribution (1981–2010 average) before (**a**) and after (**b**) improvement.

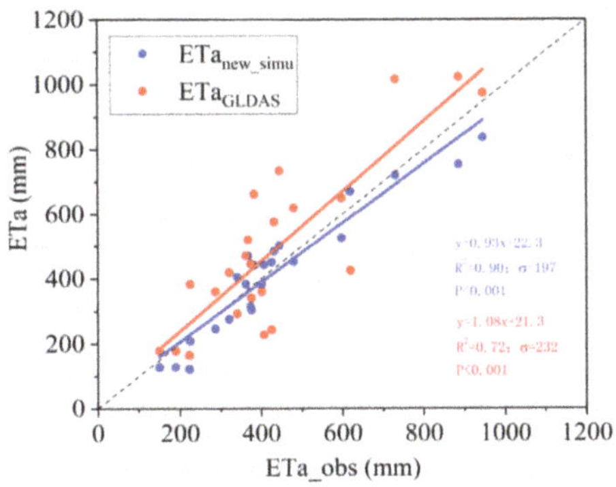

Figure 4. Comparison of the improved ET_a values (ET_{anew_simu}) with the observed and GLDAS ET_a values.

Over the past 58 years (1960–2017), the ET_a trend was roughly bounded by 103° E, with ET_a increasing to the west and decreasing to the east (Figure 5a). ET_a in the arid region is the smallest, with an average of 68.8 mm and a fluctuation range of 21.3–142.7 mm. ET_a in the transition zone is 350.2 mm with a 79.2–485.0 mm fluctuation range. The ET_a fluctuation range in the humid region is between 439.8 and 745.6 mm, with an average of 602.5 mm. Over the past 58 years, ET_a in the arid region increased at an average rate of 2.9 mm·10 a^{-1}. The amount of water expenditure in the region increased, and this increase was significant in some areas. The ET_a in the humid area decreased at a rate of −1.7 mm· 10 a^{-1}. Moreover, the ET_a in the transition zone generally shows a slight overall decrease, and the regional differences are obvious (Figure 5b).

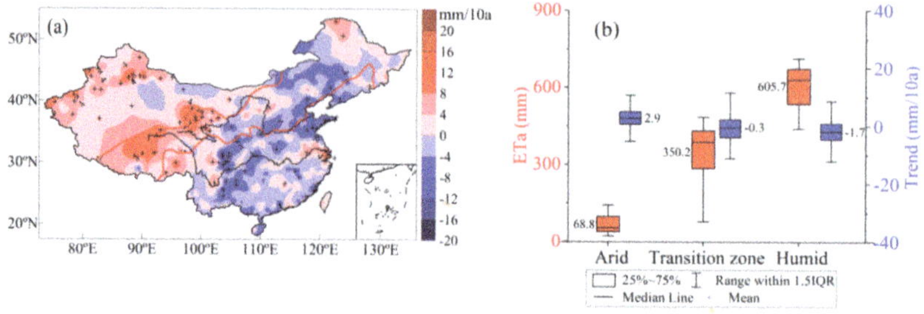

Figure 5. ET_a trend (1960–2017) and (**a**) distribution of average values and trends in sub regions and (**b**) +indicates that the trend was significant at the 0.05 level.

3.2. NPP Estimation and Change Characteristics

Figure 6 shows the estimated annual NPP before and after ET_a improvement. As depicted, the NPP in China is small in the northwest and large in the southeast, which is consistent with previous study [72]. The NPP of the southern humid region obtained before the improvement of the ET_a model was between 1400 and 1700 $g \cdot m^{-2} \cdot a^{-1}$. Precipitation is proportional to the NPP. In some regions of the southeast coast, the NPP is above 1700 $g \cdot m^{-2} \cdot a^{-1}$. The improved NPP range was 22–1510 $g \cdot m^{-2} \cdot a^{-1}$. The region of maximum NPP shifts northward and is located south of the Yangtze River. The annual average NPP was above 1400 $g \cdot m^{-2} \cdot a^{-1}$. In arid regions, water has a positive effect on vegetation productivity, which means that productivity increases with improved water conditions. However, in humid regions with sufficient water supply, NPP tends to be saturated and is no longer controlled by water [73]. In addition, this region is the cloudiest region in China [74], which is generally proportional to the light stress on vegetation growth. In coastal regions with abundant rainfall, vegetation growth is more likely to be regulated by radiation. Before the improvement, the ET_a model did not consider the energy constraint in the south, the estimated NPP was overly dependent on precipitation, and the estimated values were too large. The improved NPP better reflects the response of vegetation growth to the regional climate.

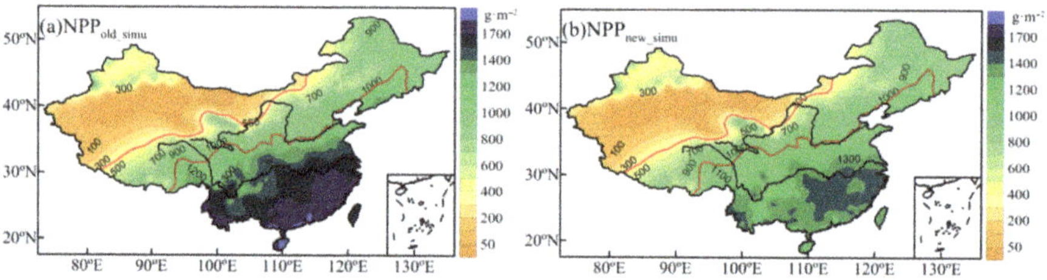

Figure 6. NPP distribution before (**a**) and after(**b**) improvement.

A comparison between the estimated NPP values (average values from 2000 to 2017) and the MOD17A3 multiyear average data shows that the estimated NPP after improvement is significantly correlated with the MOD17A3 NPP ($R = 0.67$, $p < 0.001$) (Figure 7). In the arid region, the MOD17A3 NPP was larger than the estimated NPP. However, in the humid region, the MOD17A3 NPP was relatively small. Due to estimation errors regarding reflectance, maximum light-use efficiency, and radiation, MODIS NPP products are overestimated in low-productivity regions and underestimated in high-productivity regions [75]. In addition, different methods and scale conversions can also lead to different comparisons. Therefore, the estimated NPP in this study has a certain rationality and superiority.

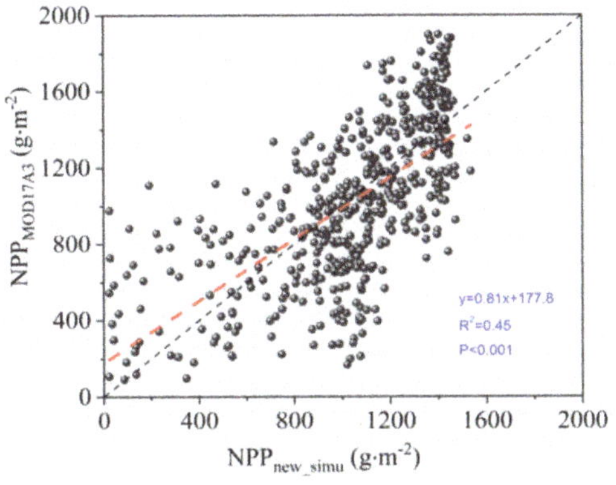

Figure 7. Comparison of estimated NPP and MOD17A3 NPP (average values from 2000 to 2017).

The distribution of the average values and trends of the NPP in the sub regions in China from 1960 to 2017 are shown in Figure 8. Over the past 58 years, the NPP increased to the west of 103° E, while it decreased to the east. NPP in arid regions typically increases, with a rate of 4.3 $g{\cdot}m^{-2}{\cdot}10\ a^{-1}$. NPP in the transition zone generally exhibits a slight decrease. However, the regional differences are large. NPP in the humid region typically displays a decreasing trend of −3.3 $g{\cdot}m^{-2}{\cdot}10\ a^{-1}$. The distribution of the average NPP in different climatic regions shows that the NPP gradually increases from arid to humid regions. The main vegetation types in arid regions are desert grassland and lowland meadows. The soil is severely desertified and salinized in regions with low vegetation coverage. The annual average NPP is 140.4 $g{\cdot}m^{-2}{\cdot}a^{-1}$. The NPP fluctuation range in the humid region is the smallest, with an average of 1287.9 $g{\cdot}m^{-2}{\cdot}a^{-1}$. The average NPP in the transition zone is 807.2 $g{\cdot}m^{-2}{\cdot}a^{-1}$, ranging from 226.6 to 1084.8 $g{\cdot}m^{-2}{\cdot}a^{-1}$, thereby exhibiting the largest regional difference, which is related to the complex and diverse climate types and vegetation types in the area.

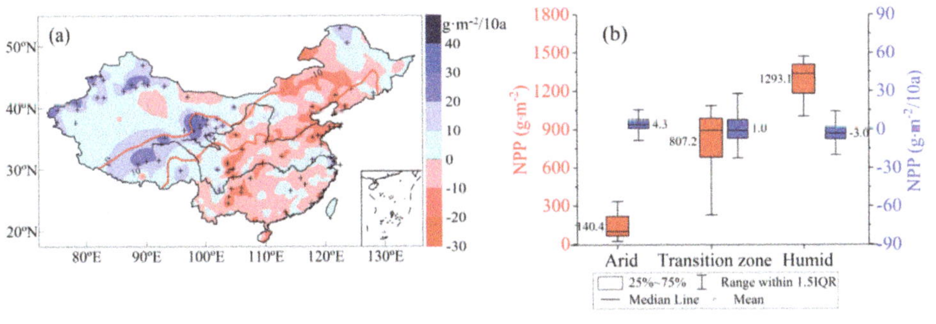

Figure 8. NPP trend (1960–2017) and (**a**) distribution of average values and trends in sub regions and (**b**) +indicates that the trend was significant at the 0.05 level.

3.3. Spatial Distribution and Temporal Variation in PUE

The distribution of the multiyear average PUE presents a "low-high-low" band from northwest to southeast (Figure 9). PUE is relatively low in arid and humid regions and is the lowest in extremely arid and extremely humid regions. The transition zone exhibited the highest PUE. PUE reached its highest value of 2.2 $g{\cdot}m^{-2}{\cdot}mm^{-1}$ in the area where the annual precipitation was 414 mm. The regional differences in PUE distribution are closely related to the regional topography, landform, and water expenditure modes. Arid regions have sparse precipitation, sufficient energy, and low and sparse vegetation, and water is mostly spent in the form of soil evaporation. Therefore, vegetation PUE is low. Humid regions have abundant precipitation, but there are many rainstorms of large intensity [76]. Precipitation is dissipated in the form of runoff, canopy interception, and soil evaporation, which may produce more ineffective water. In addition, in the humid region, there are mostly mountainous and hilly land forms with large surface runoff. Hence, PUE is also low there.

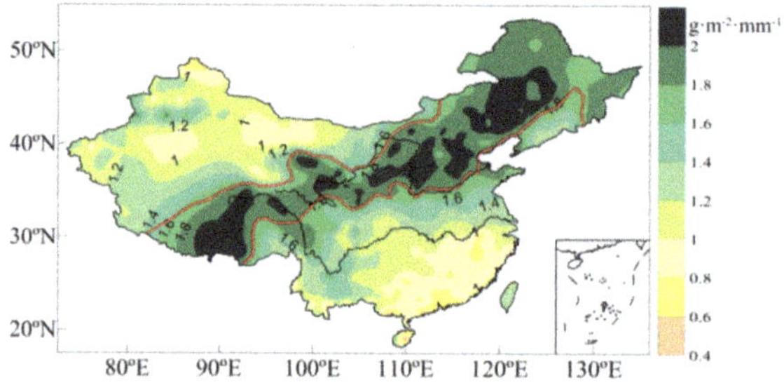

Figure 9. Spatial distribution of mean precipitation-use efficiency (PUE) in China during 1981–2010.

Figure 10 shows the PUE trend in China from 1960 to 2017 and the distribution of average values and trends in the sub regions. Over the past 58 years, PUE in the arid region increased at a rate of 0.014 $g{\cdot}m^{-2}{\cdot}mm^{-1}{\cdot}10a^{-1}$, indicating that the ability of vegetation in those regions to convert water and nutrients into biomass has increased. PUE in the transition zone was the highest, with an average of 1.92 $g{\cdot}m^{-2}{\cdot}mm^{-1}$, and generally showed a slight decreasing trend. In the western part of the transition zone (i.e., west of 103° E), PUE decreases and ecology deteriorates, which is consistent with current grassland degradation in the upper reaches of the Yellow River [77]. In the middle of the transition zone (i.e., between 103 and 120° E), PUE increases. PUE in the eastern part (i.e., east of 120° E) decreases. Most of the PUE in the humid region exhibits a decreasing trend of -0.003 $g{\cdot}$ $m^{-2}{\cdot}mm^{-1}{\cdot}10$ a^{-1}. In recent years, rainstorm intensity in humid regions has significantly increased, as has the proportion of rainstorms in annual precipitation [76]. Rainstorms are more likely to form runoff. Hence, this change in precipitation intensity is one of the reasons for the PUE decrease in the region.

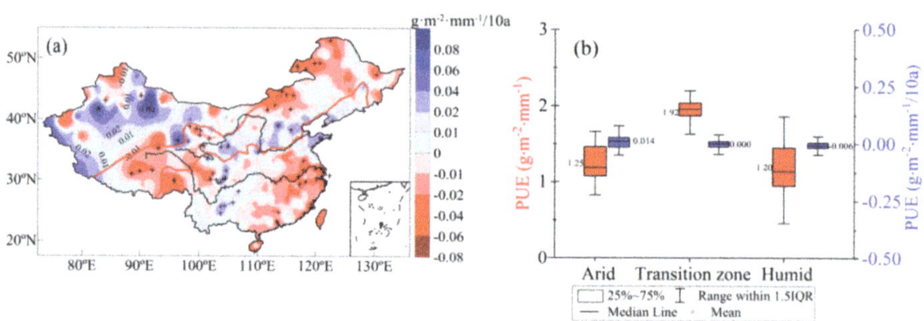

Figure 10. Precipitation-use efficiency (PUE) trend (1960–2017) and (**a**) distribution of average values and trends in the sub regions and (**b**) +indicates that the trend was significant at the 0.05 level.

3.4. Driving Force of PUE Changes and Its Corresponding Conversion Characteristics

Fluctuations in environmental factors have a significant effect on the PUE. Here, we selected Tmean, Rn, U, Rh, and soil moisture (SM) as influencing factors to characterize the energy, dynamic, and water statuses, respectively. Regression models between regional environmental factors and PUE were established (Table S1). Based on the trends and sensitivity analysis, the contribution of each factor to the PUE change was obtained.

Figure 11 shows the trends of environmental factors from 1960 to 2017. Over the past 58 years, Tmean in China exhibited a significant increasing trend. Air-temperature increases in the arid region and transition zone are particularly evident. Rn, U, and Rh exhibited decreasing trends. Among them, Rn decreased most significantly in the eastern transition zone and the humid region, especially in Beijing-Tianjin-Hebei and the lower reaches of the Yangtze River, which is closely related to increased aerosols in these areas [78]. U also decreased most significantly in the eastern transition zone and east of the humid region. In addition, U exhibited significant decreasing trends in most of the arid regions. Rh change is more complicated. Rh in the western Tianshan Mountains, which are located in an arid region, exhibited an increasing trend because the climate in the region tends to be warm and humid [50]. Rh decreased in most of the rest of the country. SM changes differed from east to west. The arid region and western transition zone exhibited significant SM increases, whereas the Middle Eastern transition zone and humid region exhibited SM decreases. Overall, soil tended to become arid.

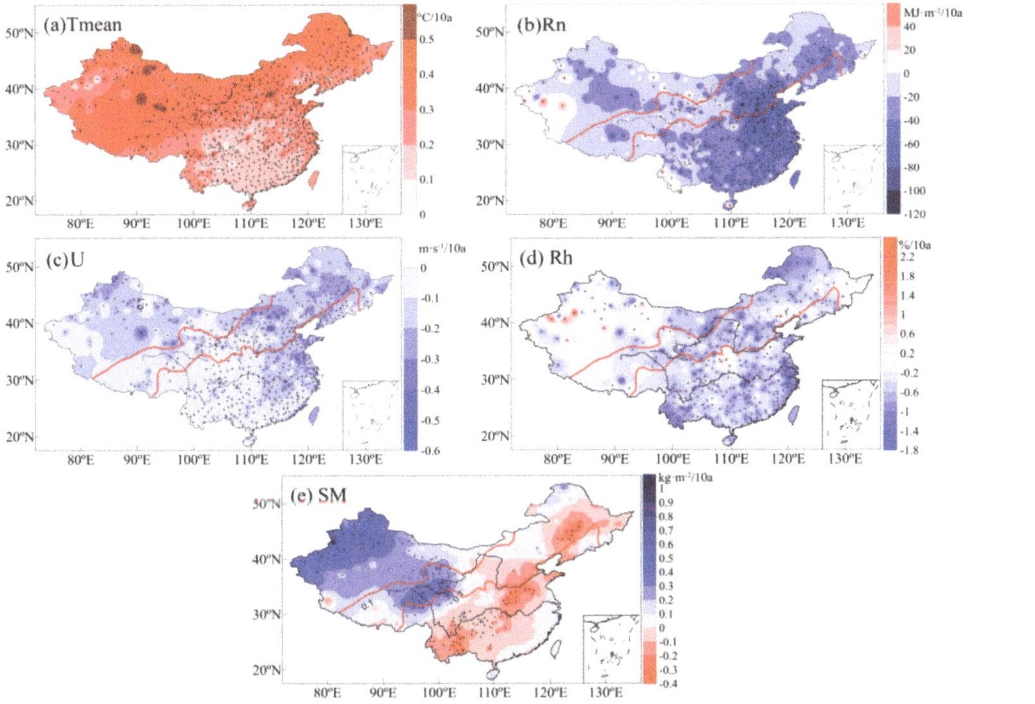

Figure 11. Trends of each environmental factor from 1960 to 2017, (**a**) Tmean, (**b**) Rn, (**c**) U, (**d**) Rh, (**e**) SM, +indicates that the trend was significant at the 0.05 level.

PUE changes in different regions caused by changes in environmental factors are shown in Figure 12. In the arid region, PUE increased by 6.5% over the past 58 years. The Tmean increase reduced the PUE by 9.3%. The SM increase and U decrease increased the PUE by 8.4% and 6.1%, respectively. Rn and Rh increased the PUE by 1.3 and 1.9%, respectively. SM was the main driving force of regional PUE increases. In the humid region, PUE decreased by 3.5% over the past 58 years. The relative change rates of PUE caused by changes in Rn, U, Tmean, Rh, and SM were −15.3, 13.0, −3.2, −1.3, and 1.6%, respectively. Rn changes were the main driving force of regional PUE decreases. Over the past 58 years, atmospheric aerosols in humid regions increased significantly, whereas net radiation decreased significantly [78]. As a result, vegetation photosynthesis is inhibited, vegetation productivity decreases, and PUE decreases accordingly. The impact of climate factors on PUE in the transition zone is more complicated. From the regional average sequence, the positive contribution of the Tmean increase and the U decrease to PUE offset the adverse effects of the Rn and SM decreases, which makes the PUE change insignificant. However, the PUE change in the transition zone exhibited obvious regional differences. The PUE decrease in the western plateau area was dominated by a significant SM increase. The PUE increase in the central regions was mainly due to the positive effect of a U decrease. PUE in the eastern and northeastern regions of the transition zone was dominated by Rn, which means that a significant Rn decrease can reduce the PUE. Overall, the PUE trend was dominated by water in the northern and arid regions and by energy in the southern and humid regions of China.

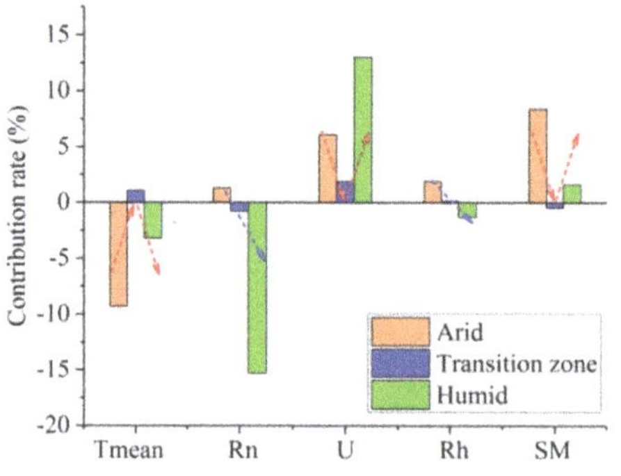

Figure 12. Contribution of environmental factors to precipitation-use efficiency (PUE). Tmean is air temperature, Rn is the net radiation, U is the wind speed at 10 m, Rh is the relative humidity, and SM is soil moisture.

The contribution of the factors is consistent or opposite depending on the factor and region. Rn exhibits a negative effect in southern humid regions and a positive effect in northern arid regions. Rh exhibits a positive effect in northern arid regions and a negative effect in southern humid regions. Furthermore, the effects of Tmean, SM, and U follow apparent "V" shapes, with positive and negative directions or turns from large to small and subsequently to large in the transition zone. To further reveal the turning characteristics of the PUE responses to water, energy, and dynamics and to clarify the precipitation climate zone where turning occurs, the multiyear average precipitation at each station was taken as a spatial climate type at every 200 mm. For example, the P100 climate type represents a spatial climate type with average annual precipitation between 0 and 200 mm. The NPP and PUE responses to changes in Tmean, SM, and U in different precipitation climate types were further analyzed.

Figure 13 shows the distribution patterns of annual NPP and PUE with ΔT, ΔSM, and ΔU changes in different precipitation climate types. Therefore, ΔT, ΔSM, and ΔU are the increments of Tmean, U, and SM, expressed as $\Delta X = X - X_{min}$, where X is T, SM, or U, and Xmin is the minimum value of each factor. In the arid region and transition zone (i.e., P100, P300, and P500), NPP and PUE were the most sensitive to various factors, especially ΔSM (Figure 13b,e). NPP and PUE increased significantly with positive ΔSM. In the humid region, NPP and PUE changed slightly with ΔSM. However, in the extremely humid area (i.e., $P \geq 1400$), NPP and PUE decreased with positive ΔSM. Soil moisture in the arid region and transition zone is close to the withering humidity. Vegetation growth is affected mainly by water factors. Soil moisture in the humid region always maintains a relatively high value. However, vegetation is less sensitive to soil moisture. Hence, extremely humid soil can restrict the oxygen supply to vegetation roots and soil microorganisms due to excessive moisture. The NPP and PUE responses to a positive ΔSM range from significantly increased to decreased between the arid and extremely humid regions, respectively. The NPP and PUE responses to ΔT and ΔU also have conversion characteristics (Figure 13a,c,d,f). Water available for evapotranspiration in arid regions is limited. Changes in NPP and PUE with ΔT were not obvious. However, in the transition zone, more obvious air temperature increases resulted in stronger water restriction for vegetation growth and smaller PUE. In the humid region, NPP and PUE were less sensitive to ΔT, but they increased slightly with positive ΔT in the extremely humid region, which reflects the promotion of vegetation growth in extremely humid regions by improved energy conditions. Wind velocity had the greatest impact on vegetation growth in the arid region and transition zone. Increased wind velocity was more conducive to evapotranspiration, thereby causing faster water loss and decreased NPP and PUE. In the humid region, a positive ΔU resulted in more favorable evapotranspiration of super humid water vapor, which indirectly promoted NPP and PUE.

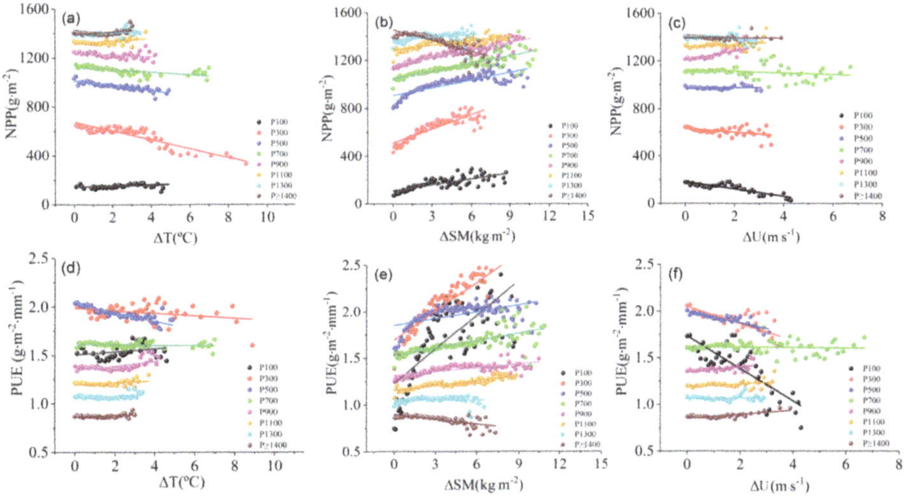

Figure 13. Trends of annual NPP and PUE with ΔT (**a**,**d**), ΔSM (**b**,**e**), and ΔU (**c**,**f**) in different precipitation climate types.

4. Discussion

4.1. Effect of Energy and Water on ET_a in Different Climatic Regions of China

The Budyko curve shows the relationship between ET_a-Pre-ET_0 and can reveal the limiting relationship between energy and water on ET_a [79–81]. Figure 14 shows the Long-term mean values of annual ET_a, Pre, and ET_0 together with Fu's curves with the regional average values of parameter ω, where ω is the plant-available water coefficient, representing the relative difference in the way plants use soil water for transpiration, and larger values of ω tend to promote evapotranspiration. It can be seen that the relationship in most areas conforms to the Budyko curve, and the data are within the boundary conditions of the hydrothermal coupling assumption, with ω in the humid region greater than in the arid region. In humid regions, ET_a is limited by available energy. As it asymptotically approaches ET_0, in arid regions, ET_0 exceeds P, and where ET_a is mainly controlled by water, evapotranspiration ratio (ET_a/Pre) tends to 1.

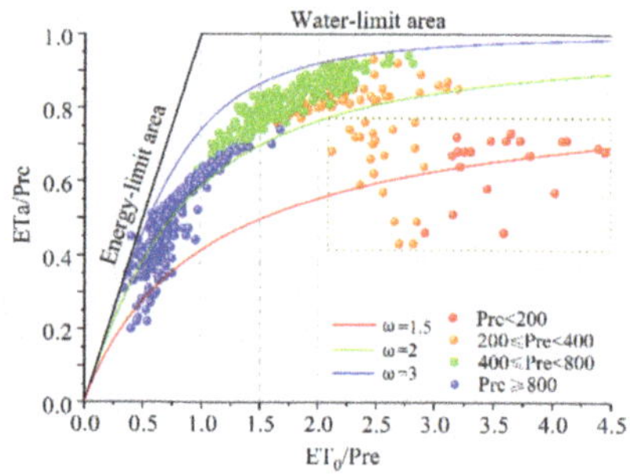

Figure 14. Relationship among ET_a-Pre-ET_0 in different climatic regions of China. The ratio of mean annual ET_a to Pre as a function of the index of dryness (ET_0/Pre) for different values of plant-available water coefficient ω.

Evapotranspiration ratio at some stations in the arid and semi-arid zones (in the lower right dashed box) is less than 1; as such, this may be attributed to the way different kinds of vegetation use soil water and the particularity of precipitation conversion to evaporation in arid areas. In the arid region, small precipitation events are just able to wet the soil surface and are quickly evaporated, whereas large precipitation events increase potential water losses from the ecosystem through runoff or deep soil water percolation [13]. Therefore, evaporation can be lower than precipitation in arid regions, this is consistent with the study of Yang et al. [82] in northern China. In addition, evapotranspiration ratio in arid regions also shows a large fluctuation, which may be related to the large variation in precipitation in arid regions [83,84]. The amount of precipitation in arid and semi-arid regions of China has a large variation [85]; the sparser the precipitation in a region, the greater the variation, and interannual fluctuations in precipitation and regional differences cause large fluctuations in evapotranspiration.

4.2. *Complexity of PUE Change in Transition Zone*

Figure 15 shows a schematic diagram of the controlling factors of PUE in different climatic regions in China, which shows the opposite change of PUE in arid and humid regions and the complexity of PUE change in transition zones.

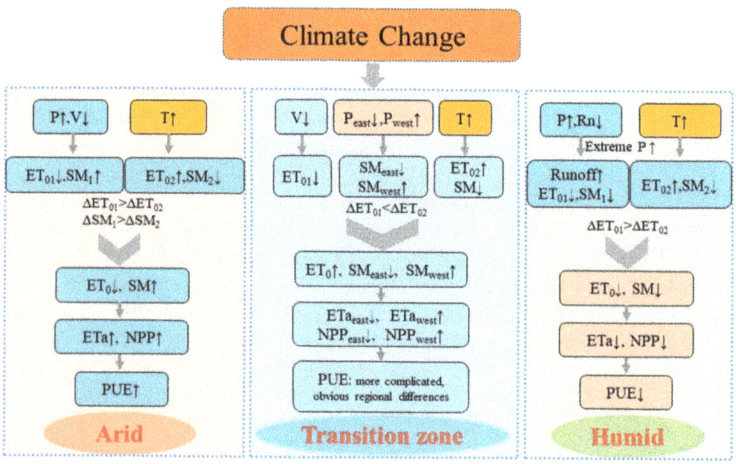

Figure 15. Schematic diagram that illustrates the control factors of the PUE in different climatic regions of China.

Warming and humidification in Northwest China [86] have a great impact on the increased ET_a, NPP, and PUE in arid areas. Warming and humidification increase the amount of water that can evaporate, thereby increasing ET_a. Temperature increase results in earlier onset of the greening period while delaying the yellowing period, enhancing vegetation activities, then increasing NPP. Multiple research based on empirical observations and process-based models have also confirmed that aboveground production in arid ecosystems has exhibited an increasing trend [87]. Decreased ET_a in the humid region is related to the decrease in ET_0, which is consistent with the evaporation paradox [88]. In the southern humid region with sufficient water supply, changes in ET_a and ET_0 are consistent [89]. Thus, when the ET_0 decreases, ET_a also decreases. Rainfall is abundant in humid areas. Vegetation productivity is positively correlated with air temperature and negatively correlated with precipitation. In recent decades, the increase in precipitation in humid regions has reduced photosynthetically active radiation. Enhanced radiation restrictions reduce vegetation NPP.

Factors influencing ET_a, NPP, and PUE in the transition zone are more complex because this region is a competitive zone for water and energy [90]. In the highland region in the western part of the transition zone, NPP increases along with warming and humidification, and a significant decrease in precipitation in the eastern region of the transition zone and the warming and drying caused by temperature increases are the main reasons for the NPP decrease [91]. At the same time, the region is affected by the interannual fluctuation of the intensity of the East Asian summer monsoon, so the interannual and interdecadal precipitation fluctuations in the region are large [92].

4.3. Transformation Characteristics of PUE

The transition zone exhibited the highest PUE. PUE reached its highest value of 2.2 $g\cdot m^{-2}\cdot mm^{-1}$ in the area where the annual precipitation was 414 mm. The unimodal PUE distribution, which first increases and then decreases with increasing precipitation, has been confirmed by other studies. Paruelo et al. [8] indicated that in American temperate grasslands with 200–1200 mm precipitation, PUE first increased and then decreased with increasing precipitation, peaking at 475 mm. PUE in extremely arid and extremely humid regions is low. Hu et al. [5] reported that the PUE initially exhibited a rising trend but subsequently decreased as precipitation increased from 200 to 1200 mm. PUE peaked at 400–500 mm. Huxman et al. [6], Lauenroth and Paruelo [7], and Yu et al. [11] also showed that in regions with annual precipitation less than 600 mm, PUE increased with increasing precipitation. In humid regions with annual precipitation above 650 mm, PUE decreased with increasing precipitation, and when the annual precipitation was above 1500 mm, PUE was approximately constant. Zhang et al. [9] reported that the spatiotemporal PUE pattern in alpine grasslands in northern Tibet initially increased in the arid region and subsequently decreased along the precipitation gradient toward the humid region, reaching a peak at approximately 500 mm precipitation.

The PUE distribution pattern with precipitation in this study is consistent with the above studies, and different precipitation thresholds of the maximum PUE conversion point may be partially attributed to the different research methods and study regions. In addition, the distribution patterns of annual PUE with ΔT, ΔSM, and ΔU changes in different precipitation climate types also indicate that the PUE transition interval coincides with the northern edge of the monsoon. Therefore, the results of this paper are reasonable.

Our research shows that transition regions with limited rainfall have the strongest NPP and PUE processes, and the response of the carbon fluxes in these fragile regions deserves more attention.

5. Conclusions

Based on the improvement of the ET_a model, this study characterized the responses of ET_a, NPP, and PUE to climate change in different climatic regions of China, revealed the

PUE conversion characteristics with the precipitation distribution, and clarified the driving force of PUE changes in different climate regions. The main conclusions are as follows:

The improved ET_a model fully reflects the energy limitation on ET_a in humid regions, and the estimated ET_a is more reasonable and reliable. The distribution of ET_a and NPP in China shows a gradually increasing trend from northwest to southeast, and the trends of ET_a and NPP both change from an increase to a decrease from the arid to the humid region. ET_a and NPP fluctuations in arid regions are mainly controlled by water, and the increase in precipitation and soil moisture is the main reason. ET_a and NPP in humid regions are mainly controlled by energy. ET_a in the transition zone is affected by both water and energy, and regional differences in ET_a and NPP changes in the transition zone are large.

There was a conversion zone of PUE in mainland China. Arid and humid regions had the lowest PUE, and the transition zone with annual precipitation of 200–600 mm had the highest PUE. In the past 58 years, PUE in arid regions has exhibited an increasing trend, whereas PUE in the transition zone generally exhibited a slightly decreasing trend. PUE displayed a decreasing trend in most of the humid regions.

PUE changes in arid regions are dominated by water conditions, whereas changes in energy in humid regions largely determine PUE changes. The transition zone is the conversion zone where the prevailing factor transitions from water to energy. PUE changes are caused by the interaction of energy, water, and dynamic factors. Among them, soil moisture plays the most prominent role, followed by temperature and wind velocity.

Supplementary Materials: The following supporting information can be downloaded at: https://www.mdpi.com/article/10.3390/rs14102467/s1, Table S1: Regression equation between environmental factors and PUE in each sub regions (annual averaged air temperature (Tmean), net radiation (Rn), wind velocity (U), relative humidity (RH), and soil moisture (SM), respectively).

Author Contributions: Conceptualization, P.Y. and Q.Z.; methodology, P.Y., S.W., J.Y. and H.Z.; software, S.W., W.W. and J.W.; formal analysis, S.W. and P.Y.; writing—original draft preparation, S.W.; writing—review and editing, S.W., P.Y. and Q.Z.; visualization, S.W., X.R. and J.W.; funding acquisition, P.Y. and S.W. All authors have read and agreed to the published version of the manuscript.

Funding: This research was funded by the National Natural Science Foundation of China (Grant No. U2142208, 41975016), the Basic Research Innovation Group Project of Gansu Province (Grant No. 20JR5RA121), the Natural Science Foundation of Gansu Province (Grant No. 21JR7RA698), and the Foundation of drought Meteorological Science Research (Grant No. IAM202001).

Institutional Review Board Statement: Not applicable.

Informed Consent Statement: Not applicable.

Data Availability Statement: The GLDAS_Noah025_M.2.0 and 2.1 datasets can be accessed online (https://disc.gsfc.nasa.gov/datasets?page=1&project=GLDAS (accessed on 25 February 2021)); The MOD17A3 surface vegetation NPP data provided by the EOS/MODIS (TERRA satellite) of NASA can be accessed at https://earthdata.nasa.gov/ (accessed on 2 November 2020); and China FLUX ET_a data are available at http://rs.cern.ac.cn/data/initDRsearch?classcode=SYC_A02 (accessed on 19 December 2019).

Conflicts of Interest: The authors declare no conflict of interest.

References

1. Piao, S.L.; Zhang, X.P.; Chen, A.P.; Liu, Q.; Lian, X.; Wang, H.X.; Peng, S.S.; Wu, X.C. The impacts of climate extremes on the terrestrial carbon cycle: A review. *Sci. China Earth Sci.* **2019**, *62*, 1551–1563. [CrossRef]
2. IPCC. Climate Change 2021: The Physical Science Basis. M/OL]. 2021 [2021-08-09]. Available online: https://www.ipcc.ch/report/ar6/wg1/downloads/report/IPCC_AR6_WGI_Full_Report.pdf (accessed on 9 August 2021).
3. Mu, Q.Z.; Zhao, M.S.; Running, S.W.; Liu, M.L.; Tian, H.Q. Contribution of increasing CO_2 and climate change to the carbon cycle in China's ecosystems. *J. Geophys. Res. Biogeosci.* **2015**, *113*, G01018. [CrossRef]
4. Xu, X.T.; Piao, S.L.; Wang, X.H.; Chen, A.P.; Ciais, P.; Myneni, R.B. Spatio-temporal patterns of the area experiencing negative vegetation growth anomalies in China over the last three decades. *Environ. Res. Lett.* **2012**, *7*, 035701. [CrossRef]

5. Hao, Y.B.; Wang, Y.F.; Cui, X.Y. Drought stress reduces the carbon accumulation of the Leymus chinensis steppe in Inner Mongolia, China. *Chin. J. Plant Ecol.* **2010**, *34*, 898–906. (In Chinese)
6. Yuan, W.P.; Cai, W.W.; Chen, Y.; Liu, S.G.; Dong, W.J.; Zhang, H.C.; Yu, G.R.; Chen, Z.Q.; He, H.L.; Guo, W.D.; et al. Severe summer heatwave and drought strongly reduced carbon uptake in Southern China. *Sci. Rep.* **2016**, *6*, 18813. [CrossRef] [PubMed]
7. Xu, X.J.; Zhou, G.M.; Liu, S.G.; Du, H.Q.; Mo, L.F.; Shi, Y.J.; Jiang, H.; Zhou, Y.F.; Liu, E.B. Implications of ice storm damages on the water and carbon cycle of bamboo forests in southeastern China. *Agric. For. Meteorol.* **2013**, *177*, 35–45. [CrossRef]
8. Turner, N.C. Agronomic options for improving rainfall-use efficiency of crops in dryland farming systems. *J. Exp. Bot.* **2004**, *55*, 2413–2425. [CrossRef]
9. Keenan, T.F.; Hollinger, D.Y.; Bohrer, G.; Dragoni, D.; Munger, J.W.; Schmid, H.P.; Richardson, A.D. Increase in forest water-use efficiency as atmospheric carbon dioxide concentrations rise. *Nature* **2013**, *499*, 324–327. [CrossRef]
10. Wang, L.M.; Li, M.Y.; Wang, J.X.; Li, X.G.; Wang, L.C. An analytical reductionist framework to separate the effects of climate change and human activities on variation in water use efficiency. *Sci. Total Environ.* **2020**, *727*, 138306. [CrossRef]
11. Bai, Y.F.; Wu, J.G.; Qi, X.; Pan, Q.M.; Huang, J.H.; Yang, D.L.; Han, X.G. Primary production and rain use efficiency across a precipitation gradient on the Mongolia Plateau. *Ecology* **2008**, *89*, 2140–2153. [CrossRef]
12. Hu, Z.M.; Yu, G.R.; Fan, J.W.; Zhong, H.P.; Wang, S.Q.; Li, S.G. Precipitation-use efficiency along a 4500-km grassland transect: Rain-use efficiency in Chinese grasslands. *Glob. Ecol. Biogeogr.* **2010**, *19*, 842–851.
13. Huxman, T.E.; Smith, M.D.; Fay, P.A.; Knapp, A.K.; Shaw, M.R.; Loik, M.E.; Smith, S.D.; Tissue, D.T.; Zak, J.C.; Weltzin, J.F. Convergence across biomes to a common rain-use efficiency. *Nature* **2004**, *429*, 651–654. [CrossRef] [PubMed]
14. Lauenroth, W.K.; Paruelo, B. Patterns of production and precipitation-use efficiency of winter wheat and native grasslands in the central great plains of the United States. *Ecosystems* **2000**, *3*, 344–351. [CrossRef]
15. Paruelo, J.M.; Lauenroth, W.K.; Burke, I.C.; Sala, O.E. Grassland precipitation-use efficiency varies across a resource gradient. *Ecosystems* **1999**, *2*, 64–68. [CrossRef]
16. Zhang, X.K.; Du, X.D.; Zhu, Z.M. Effects of precipitation and temperature on precipitation use efficiency of alpine grassland in northern Tibet, China. *Sci. Rep.* **2020**, *10*, 20309. [CrossRef]
17. Khan, R.S.; Bhuiyan, M.A.E. Artificial Intelligence-Based Techniques for Rainfall Estimation Integrating Multisource Precipitation Datasets. *Atmosphere* **2021**, *12*, 1239. [CrossRef]
18. Derin, Y.; Yilmaz, K.K. Evaluation of multiple satellite-based precipitation products over complex topography. *J. Hydrometeorol.* **2014**, *15*, 1498–1516. [CrossRef]
19. Mei, Y.W.; Nikolopoulos, E.I.; Anagnostou, E.N.; Borga, M. Evaluating satellite precipitation error propagation in runoff simulations of mountainous basins. *J. Hydrometeorol.* **2016**, *17*, 1407–1423. [CrossRef]
20. Chen, T.; Bao, A.M.; Jiapaer, G.; Guo, H.; Zheng, G.; Jiang, L.; Chang, C.; Tuerhanjiang, T. Disentangling the relative impacts of climate change and human activities on arid and semiarid grasslands in Central Asia during 1982–2015. *Sci. Total Environ.* **2019**, *653*, 1311–1325. [CrossRef]
21. Yu, G.R.; Song, X.; Wang, Q.F.; Liu, Y.F.; Guan, D.X.; Yan, J.H.; Sun, X.M.; Zhang, L.M.; Wen, X.F. Water use efficiency of forest ecosystems in eastern China and its relations to climatic variables. *New Phytol.* **2008**, *177*, 927–937. [CrossRef]
22. Li, H.X.; Liu, G.H.; Fu, B.J. Spatial variations of rain-use efficiency along a climate gradient on the Tibetan Plateau: A satellite-based analysis. *Int. J. Remote Sens.* **2013**, *34*, 7487–7503. [CrossRef]
23. Cramer, W.; Kicklighter, D.W.; Bondeau, A.; Iii, B.M.; Churkina, G.; Nemry, B.; Ruimy, A.; Schloss, A.L. The Participants of the Potsdam NPP Model Intercomparison, Comparing global models of terrestrial net primary productivity (NPP): Overview and key results. *Glob. Chang. Biol.* **1999**, *5* (Suppl. S1), 1–15. [CrossRef]
24. McGuire, A.D.; Melillo, J.M.; Kicklighter, D.W.; Joyce, L.A. Equilibrium Responses of Soil Carbon to Climate Change: Empirical and Process-Based Estimates. *J. Biogeogr.* **1995**, *22*, 785–796. [CrossRef]
25. Piao, S.L.; Fang, J.Y.; Guo, Q.H. Application of casa model to the estimation of Chinese terrestrial net primary productivity. *Acta Phytoecol. Sin.* **2001**, *25*, 603–608. (In Chinese)
26. Turner, D.P.; Ritts, W.D.; Cohen, W.B.; Gower, S.T.; Running, S.W.; Zhao, M.S.; Costa, M.H.; Kirschbaum, A.A.; Ham, J.M. Evaluation of MODIS NPP and GPP products across multiple biomes. *Remote Sens. Environ.* **2006**, *102*, 282–292. [CrossRef]
27. Zhang, Q.; Yang, J.H.; Wang, W.; Ma, P.L.; Lu, G.Y.; Liu, X.Y.; Yu, H.P.; Fang, F. Climatic Warming and Humidification in the Arid Region of Northwest China: Multi-Scale Characteristics and Impacts on Ecological Vegetation. *J. Meteor. Res.* **2021**, *35*, 113–127. [CrossRef]
28. Li, Z.; Yan, Z.W. Homogenized China daily mean/maximum/mini-mum temperature series 1960–2008. *Atmos. Ocean. Sci. Lett.* **2009**, *2*, 237–243.
29. Zhang, Y. *Energy and Water Budget of a Poplar Plantation in Suburban Beijing*; Beijing Forestry University: Beijing, China, 2010. (In Chinese)
30. Zhu, G.F.; Lu, L.; Su, Y.H.; Wang, X.F.; Cui, X.; Ma, J.Z.; Zhang, K.; Li, C.B. Energy flux partitioning and evapotranspiration in a sub-alpine spruce forest ecosystem. *Hydrol. Process.* **2014**, *28*, 5093–5104.
31. Wang, Z.Y. *Energy Balance and Water Vapor Flux of Snail Control and Schistosomiasis Prevention Forests Ecosystem in Yangtze River Beach Land*; Chinese Academy of Forestry: Beijing, China, 2008. (In Chinese)
32. Wilske, B.; Lu, N.; Wei, L.; Chen, S.P.; Zha, T.G.; Liu, C.F.; Xu, W.T.; Noormets, A.; Huang, J.H.; Wei, Y.F.; et al. Poplar plantation has the potential to alter the water balance in semiarid Inner Mongolia. *J. Environ. Manag.* **2009**, *90*, 2762–2770. [CrossRef]

33. Guo, L. *The Variations of Water Use Efficiency and Evapotranspiration over a Plantation in Southern Part of Hilly Areas of North-China Master*; Chinese Academy of Forestry: Beijing, China, 2010. (In Chinese)
34. Lin, Y.; Wang, G.X.; Guo, J.Y.; Sun, X.Y. Quantifying evapotranspiration and its components in a coniferous subalpine forest in Southwest China. *Hydrol. Process.* **2012**, *26*, 3032–3040. [CrossRef]
35. Han, S.; Huang, L.L.; Wang, Z.Y.; Wei, Y.; Zhang, X.D. Ecosystem respiration and its controlling factors in the riparian wetland of Yangtze River. *Acta Ecol. Sin.* **2009**, *29*, 3621–3628. (In Chinese)
36. Wang, W.Z.; Zhao, Z.H.; Kang, W.X.; Tian, D.L.; Xiang, W.H.; Yan, W.D. Characteristics of latent heat flux over Cunninghamia lanceolata plantations in Huitong county. *J. Cent. South Univ. Fores. Tech.* **2011**, *31*, 192–197. (In Chinese)
37. Cui, S. *Study on the CO_2 Flux of a Larch Plantation in NE China by the Micrometeorological Method*; Northeast Forestry University: Harbin, China, 2007. (In Chinese)
38. Liu, S.M.; Xu, Z.W.; Zhu, Z.L.; Jia, Z.Z.; Zhu, M.J. Measurements of evapotranspiration from eddy-covariance systems and large aperture scintillometers in the Hai River Basin, China. *J. Hydrol.* **2013**, *487*, 24–38. [CrossRef]
39. Lin, E.J.; Jiang, H.; Chen, Y.F. Water vapor flux variation and net radiation for a Phyllostachys violascens stand in Taihuyuan. *J. Zhejiang Agric. For. Univ.* **2013**, *30*, 313–318. (In Chinese)
40. Dong, G.; Guo, J.X.; Chen, J.Q.; Sun, G.; Gao, S.; Hu, L.J.; Wang, Y.L. Effects of spring drought on carbon sequestration, evapotranspiration and water use efficiency in the Songnen meadow steppe in Northeast China. *Ecohydrology* **2011**, *4*, 211–224. [CrossRef]
41. Chen, S.P.; Chen, J.Q.; Lin, G.H.; Zhang, W.L.; Miao, H.X.; Wei, L.; Huang, J.H.; Han, X.G. Energy balance and partition in Inner Mongolia steppe ecosystems with different land use types. *Agric. For. Meteorol.* **2009**, *149*, 1800–1809. [CrossRef]
42. Gu, S.; Tang, Y.H.; Cui, X.Y.; Du, M.Y.; Zhao, L.; Li, Y.Q.; Xu, S.X.; Zhou, H.K.; Kato, T.; Qi, P.T.; et al. Characterizing evapotranspiration over a meadow ecosystem on the Qinghai-Tibetan Plateau. *J. Geophys. Res.* **2008**, *113*, D08118. [CrossRef]
43. Shao, C.L.; Chen, J.Q.; Li, L.H. Grazing alters the biophysical regulation of carbon fluxes in a desert steppe. *Environ. Res. Lett.* **2013**, *8*, 025012. [CrossRef]
44. Wang, Y.L.; Zhou, G.S.; Wang, Y.H. Environmental effects on net ecosystem CO_2 exchange at half-hour and month scales over Stipa krylovii steppe in northern China. *Agric. For. Meteorol.* **2008**, *148*, 714–722. [CrossRef]
45. Liu, H.Z.; Feng, J.W. Seasonal and interannual variations of evapotranspiration and energy exchange over different land surfaces in a semi-arid area of China. *J. Appl. Meteorol. Climatol.* **2012**, *51*, 1875–1888.
46. Yue, P.; Zhang, Q.; Zhang, L. Biometeorological effects on carbon dioxide and water-use efficiency within a semiarid grassland in the Chinese Loess Plateau. *J. Hydrol.* **2020**, *590*, 125520. [CrossRef]
47. Wang, W.Z.; Xu, Z.W.; Liu, S.M.; Li, X.M.; Ma, M.G.; Wang, J.M. The characteristics of heat and water vapor fluxes over different surfaces in the Heihe River Basin. *Adv. Earth Sci.* **2009**, *24*, 714–723. (In Chinese)
48. Liu, K. *Multi-Time Scale Characteristics of Evapotranspiration of Artificial Caragana Korshinskii Forests in Desert Steppe*; Ningxia University: Yinchuan, China, 2018. (In Chinese)
49. Liu, R.; Li, Y.; Wang, Q.X. Variations in water and CO_2 fluxes over a saline desert in western China. *Hydrol. Process.* **2012**, *26*, 513–522. [CrossRef]
50. Zhang, F.; Zhou, G.S.; Wang, Y.; Yang, F.L.; Nilsson, C. Evapotranspiration and crop coefficient for a temperate desert steppe ecosystem using eddy covariance in Inner Mongolia, China. *Hydrol. Process.* **2012**, *26*, 379–386. [CrossRef]
51. Li, J.M.; Cai, H.; Cheng, Q.; Qiao, C.L.; Chu, H.; Chen, D.D.; Xu, S.X.; Zhao, X.Q.; Zhao, L. Characterizing the evapotranspiration of a degraded grassland in the Sanjiangyuan Region of Qinghai province. *Acta Pratacult. Sin.* **2012**, *21*, 223–233. (In Chinese)
52. Wu, J.K.; Chen, J.W.; Wu, H.; Zhang, S.Q.; Gao, M.J.; Qin, H. Comparative study of evapotranspiration in an alpine meadow in the upper reach of Shulehe River Basin. *Sci. Geogr. Sin.* **2013**, *33*, 97–103. (In Chinese)
53. Yang, F.L.; Zhang, Q.; Wang, R.Y.; Wang, S.; Yue, P.; Wang, H.L.; Zhao, H. Characteristics of evapotranspiration and crop coefficient of agroecosystems in semi-arid area of Loess Plateau, Northwest China. Chin. *J. Appl. Ecol.* **2013**, *24*, 1209–1214. (In Chinese)
54. Zhou, G.S.; Wang, Y.; Zhou, L. Dynamics of carbon budgets in typical corn and rice ecosystems in Liaohe delta. In Proceedings of the Low Carbon Agriculture Symposium, Beijng, China, 17 June 2010; pp. 265–271. (In Chinese).
55. Chen, H.; Huang, J.H. Study on in-situ evapotranspiration measurement and its influential factors in farmland in Southern China. *Water Resour. Prot.* **2017**, *33*, 79–86. (In Chinese)
56. Wu, C.; Wang, G.Q.; Hao, Z.C.; Gu, Y.; Liu, P.Y.; Yang, Q.L. Study on the variation regulation and influencing factors of farmland evapotranspiration in gully region of Changwu tableland. *J. Water Res. Water Eng.* **2017**, *28*, 37–42. (In Chinese)
57. Guo, C.M.; Ren, J.Q.; Zhang, T.L.; Yu, H. Dynamic Change of Evapotranspiration and Influenced Factors in the Spring Maize Field in Northeast China. *Chin. J. Agrometeorol.* **2016**, *37*, 400–407. (In Chinese)
58. Guo, Y.D.; Song, C.C.; Zhang, J.S.; Wang, L.L.; Sun, L. Influence of wetland reclamation on land-surface energy exchange and evapotranspiration in the Sanjiang plain, Northeast China. *Agric. For. Meteorol.* **2021**, *296*, 108214. [CrossRef]
59. Zhou, L.; Zhou, G.S.; Liu, S.S.; Sui, X.H. Seasonal contribution and interannual variation of evapotranspiration over a reed marsh (Phragmites australis) in Northeast China from 3−year eddy covariance data. *Hydrol. Process.* **2010**, *24*, 1039–1047. [CrossRef]
60. Guo, H.Q.; Noormets, A.; Zhao, B.; Chen, J.Q.; Sun, G.; Gu, Y.J.; Bo, L.; Chen, J.K. Tidal effects on net ecosystem exchange of carbon in an estuarine wet-land. *Agric. For. Meteorol.* **2009**, *149*, 1820–1828. [CrossRef]
61. Yan, G.Y. *Tidal Influence on Energy Balance and Evapotranspiration of Mangrove Ecosystem in Subtropical Area*; Xiamen University: Xiamen, China, 2012. (In Chinese)

62. Chen, H.; Lu, W.Z.; Yan, G.; Yang, S.; Lin, G.H. Typhoons exert significant but differential impacts on net ecosystem carbon exchange of subtropical mangrove forests in China. *Biogeosciences* **2014**, *11*, 5323–5333. [CrossRef]
63. Rodell, M.; Houser, P.R.; Jambor, U.; Gottschalck, J.; Mitchell, K.; Meng, C.J.; Arsenault, K.; Cosgrove, B.; Radkovich, J.; Bosilovich, M.; et al. The Global Land Data Assimilation System. *Bull. Amer. Meteor. Soc.* **2004**, *85*, 381–394. [CrossRef]
64. Lieth, H.; Box, E. Evapotranspiration and primary production. In *Thornthwaite W Memorial Model*; Publications in Climatology; C.W. Thornthwaite Associates: Centerton-Elmer, NJ, USA, 1972; Volume 25, pp. 37–46.
65. Zhou, G.S.; Zhang, X.S. A Natural Vegetation NPP Model. *Chin. J. Plan. Ecolo.* **1995**, *19*, 193–200. (In Chinese)
66. Wang, K.C.; Dickinson, R.E. A review of global terrestrial evapotranspiration: Observation, modeling, climatology, and climatic variability. *Rev. Geophys.* **2012**, *50*, RG2005. [CrossRef]
67. Zhang, K.; Kimball, J.S.; Running, S.W. A review of remote sensing based actual evapotranspiration estimation. *WIREs Water* **2016**, *3*, 834–853. [CrossRef]
68. Allen, R.G.; Pereira, L.S.; Rees, D.; Smith, M. Crop evapotranspiration—Guidelines for computing crop water requirements. In *Irrigation and Drainage*; FAO: Rome, Italy, 1998; pp. 15–28.
69. McCuen, R.H. A sensitivity and error analysis of procedures used for estimating evaporation. *Water. Res. Bull.* **1974**, *10*, 486–498. [CrossRef]
70. Zhang, Q.; Yang, Z.S.; Hao, X.C.; Yue, P. Conversion features of evapotranspiration responding to climate warming in transitional climate regions in northern China. *Clim. Dyn.* **2019**, *52*, 3891–3903. [CrossRef]
71. Wang, W.; Cui, W.; Wang, X.J.; Chen, X. Evaluation of GLDAS−1 and GLDAS−2 Forcing Data and Noah Model Simulations over China at the Monthly Scale. *J. Hydrometeorol.* **2016**, *17*, 2815–2833. [CrossRef]
72. Zhang, M.L.; Lai, R.; Zhao, Y.Y.; Jiang, W.L.; Chen, Q.G. Estimating net primary production of natural grassland and its spatio-temporal distribution in China. *Sci. Total Environ.* **2016**, *553*, 184–195. [CrossRef] [PubMed]
73. Churkina, G.; Running, S.W.; Schloss, A.L. The Participants of the Potsdam NPP Model Intercomparison, Comparing global models of terrestrial net primary productivity (NPP): The importance of water availability. *Glob. Chang. Biol.* **1999**, *5* (Suppl. S1), 46–55. [CrossRef]
74. Fu, C.B.; Dan, L.; Feng, J.M.; Peng, J.; Ying, N. Temporal and spatial variations of total cloud amount and their possible relationships with temperature and water vapor over China during 1960 to 2012. *Chin. J. Atmos. Sci.* **2019**, *43*, 87–98. (In Chinese)
75. Heinsch, F.A.; Zhao, M.S.; Running, S.W.; Kimball, J.S.; Nemani, R.R.; Davis, K.J.; Bolstad, P.V.; Cook, B.D.; Desai, A.R.; Ricciuto, D.M.; et al. Evaluation of remote sensing based terrestrial productivity from MODIS using regional tower eddy flux network observations. *IEEE Trans. Geosci. Remote Sens.* **2006**, *44*, 1908–1925. [CrossRef]
76. Xu, X.C.; Zhang, X.Z.; Dai, E.F.; Song, W. Research of trend variability of precipitation intensity and their contribution to precipitation in China from 1961 to 2010. *Geogr. Res.* **2014**, *33*, 1335–1347. (In Chinese)
77. Du, J.Q.; Shu, J.M.; Zhang, L.B. Analysis of ecosystem degradation and recovery using precipitation use efficiency and NDVI in the headwater catchment of the Yellow River basin. *Acta Ecol. Sin.* **2012**, *32*, 3404–3413. (In Chinese)
78. Liang, F.; Xia, X.A. Long-term trends in solar radiation and the associated climatic factors over China for 1961–2000. *Ann. Geophys.* **2005**, *23*, 2425–2432. [CrossRef]
79. Budyko, M.I. *The Heat Balance of the Earth's Surface*; Stepanova, N.A., Translator, Eds.; United States Department of Commerce: Washington, DC, USA, 1958; 259p. (In Russian)
80. Fu, B.P. On the calculation of the evaporation from land surface. *Sci. Atmos. Sin.* **1981**, *5*, 23–31. (In Chinese)
81. Zhang, L.; Dawes, W.R.; Walker, G.R. Response of mean annual evapotranspiration to vegetation changes at catchment scale. *Water Resour. Res.* **2001**, *37*, 701–708. [CrossRef]
82. Yang, D.F.; Sun, Z.B.; Liu, Z.Y.; Cong, Z.T.; Ni, G.H.; Lei, Z.D. Analyzing spatial and temporal variability of annual water-energy balance in nonhumid regions of China using the Budyko hypothesis. *Water Resour. Res.* **2007**, *43*, W04426. [CrossRef]
83. Gherardi, L.A.; Sala, O.E. Enhanced precipitation variability decreases grass and increases shrub-productivity. *Proc. Natl. Acad. Sci. USA* **2015**, *112*, 12735–12740. [CrossRef] [PubMed]
84. Knapp, A.K.; Fay, P.A.; Blair, J.M.; Collins, S.L.; Smith, M.D.; Carlisle, J.D.; Harper, C.W.; Danner, B.T.; Lett, M.S.; McCarron, J.K. Rainfall variability, carbon cycling, and plant species diversity in a mesic grassland. *Science* **2002**, *298*, 2202–2205. [CrossRef] [PubMed]
85. Huang, J.J.; Zhang, Q.; Zhang, S.; Chen, X.H. Information entropy-based analysis of spatial and temporal variation in precipitation in Xinjiang. *Acta Ecol. Sin.* **2017**, *37*, 4444–4455. (In Chinese)
86. Yang, J.H.; Zhang, Q.; Lu, G.Y.; Liu, X.Y.; Wang, D.W.; Liu, W.P.; Yue, P.; Zhu, B. Climate Transition from Warm-dry to Warm-wet in Eastern Northwest China. *Atmosphere* **2021**, *12*, 548. [CrossRef]
87. Ahlstrom, A.; Raupach, M.R.; Schurgers, G.; Smith, B.; Arneth, A.; Jung, M.; Reichstein, M.; Canadell, J.G.; Friedlingstein, P.; Jain, A.K.; et al. The dominant role of semi-arid ecosystems in the trend and variability of the land CO_2 sink. *Science* **2015**, *348*, 895–899. [CrossRef]
88. Cong, Z.T.; Yang, D.W.; Ni, G.H. Does evaporation paradox exist in China? *Hydrol. Earth Syst. Sci.* **2008**, *13*, 357–366. [CrossRef]
89. Brutsaert, W.; Parlange, M.B. Hydrologic cycle explains the evaporation paradox. *Nature* **1998**, *396*, 30. [CrossRef]
90. Zhang, Q.; Yang, Z.S.; Hao, X.C.; Yue, P. Transition features of surface evapotranspiration responding to climate warming with spatial precipitation–based climate type in Northern China. *Chin. Sci. Bull.* **2018**, *63*, 1035–1049. (In Chinese) [CrossRef]

91. Liu, Y.Y.; Yue, Y.; Wang, Q.; Du, X.L.; Li, J.L.; Gang, C.C.; Zhou, W.; Wang, Z.Q. Evaluating the responses of net primary productivity and carbon use efficiency of global grassland to climate variability along an aridity gradient. *Sci. Total Environ.* **2019**, *652*, 671–682. [CrossRef]
92. Sun, L.; Shen, B.Z.; Sui, B.; Huang, B.H. The influences of East Asian Monsoon on summer precipitation in Northeast China. *Clim. Dyn.* **2017**, *48*, 1647–1659. [CrossRef]

remote sensing

MDPI

Article

Diurnal Variation in Cloud and Precipitation Characteristics in Summer over the Tibetan Plateau and Sichuan Basin

Bangjun Cao [1,2], Xianyu Yang [1,*], Boliang Li [1], Yaqiong Lu [3] and Jun Wen [1]

1 School of Atmospheric Science, Chengdu University of Information Technology, Chengdu 610225, China; caobj@cuit.edu.cn (B.C.); pakleung1997@gmail.com (B.L.); jwen@cuit.edu.cn (J.W.)

2 Institute for Geophysics and Meteorology, University of Cologne, 50969 Cologne, Germany

3 Institute of Mountain Hazards and Environment, Chinese Academy of Sciences, Chengdu 610041, China; yaqiong@imde.ac.cn

* Correspondence: xyang@cuit.edu.cn; Tel.: +86-199-4082-6090

Abstract: The diurnal variation in precipitation and cloud parameters and their influencing factors during summer over the Tibetan Plateau (TP) and Sichuan Basin (SB) were investigated using the Hydro-Estimator satellite rainfall estimates, ground observations, and ERA5 dataset. The precipitation and cloud parameters show diurnal propagation over the SB during the mei-yu period in contrast to such parameters over the TP. The diurnal maximum precipitation from the Hydro-Estimator satellite and cloud ice and liquid water content (cloud LWC and IWC) from the ERA5 dataset are concentrated in the early evening, while their diurnal minimums manifest in the morning. Cloud LWC accounts for more than 60% of the total water during almost the entire diurnal cycle over the inner TP and SB during the mei-yu period. The IWC accounts for more than 60% of the total water in the late afternoon over the edge of the SB and TP. The cloud base height (CBH) above ground level (AGL), the lifting condensation level (LCL) AGL, and the zero degree level AGL are almost equal over the TP during the summer period. The zero degree level AGL over the SB is higher than that over the TP because the air temperature lapse rate over the TP is larger. The thickness of liquid water cloud over the SB is larger than that over the TP. The correlation analysis shows that the CBH AGL and LCL AGL over the TP are related to the dewpoint spread, but less so over the SB because of the stronger turbulence and lower air density over the TP than the SB. Convective available potential energy has a larger impact on precipitation over the TP than the SB. The cloud LWC makes a larger contribution to the precipitation over the SB than over the TP, which is related to the mean zonal wind and diurnal cycle of low-level winds. The precipitation at the edge of the TP and SB (i.e., the steep downstream slope) is largely influenced by the ice water contained within clouds owing to the convergence rising motion over the slopes.

Keywords: Tibetan Plateau; Sichuan Basin; steep slope; precipitation; cloud; diurnal variation

Citation: Cao, B.; Yang, X.; Li, B.; Lu, Y.; Wen, J. Diurnal Variation in Cloud and Precipitation Characteristics in Summer over the Tibetan Plateau and Sichuan Basin. *Remote Sens.* **2022**, *14*, 2711. https://doi.org/10.3390/rs14112711

Academic Editors: Jing Wei and Kai Qin

Received: 15 April 2022
Accepted: 30 May 2022
Published: 5 June 2022

1. Introduction

The Tibetan Plateau (TP) (26°00′–39°47′N, 73°19′–104°47′E) is often referred to as 'the roof of the world' owing to its average elevation exceeding 4000 m, or 'the Asian water tower' [1] because several of Asia's major rivers such as the Yangtze, Yellow, and Lancang originate from the region. The precipitation over the TP is important to these rivers, and the heat supply of the TP serves as an important energy source of the atmosphere [1–3]. The TP has also a profound impact on the precipitation over its surrounding and downstream areas [4–7]. The Sichuan Basin (SB) is a deep basin with an elevation on average 2500 m lower than that of the TP. Remote sensing and ground observations and modeling of the diurnal cycle of cloud and precipitation over the TP are important to understand the weather and climate processes over the TP and surrounding areas. Recently, the second (TIPEX2) and third (TIPEX3) Tibetan Plateau Atmospheric Experiments were carried out [8] and obtained large quantities of observational data from the TP to the SB, especially ground radar observations of cloud and precipitation. These observational data and analysis provide a basis for further studies on the mechanisms of cloud and precipitation and the improvement of parameterization schemes for cloud and precipitation physical processes.

According to ground- and satellite-based observations as well as simulation results in previous research, the total precipitation is <400 mm during the entire summer (June–August) over the TP, which is smaller than that over the surrounding areas [9]. Moreover, the monthly averaged precipitation rate is ~0.3 mm h^{-1}; the precipitation rate in precipitation events is 1~20 mm h^{-1}; the averaged daily precipitation is <15 mm over the TP [10,11]. The cloud top height usually exceeds 12 km above ground level (AGL), and sometimes exceeds 16 km AGL over the TP, mainly including mixed-phase and ice-phase processes, in which super-cold water may be contained [9–11]. The average cloud base height (CBH) during summer over the TP exceeds 1.5 km AGL, which is larger than that over the plains and SB.

The summer precipitation in China can be divided into two stages: the mei-yu period and the midsummer period [12–16]. Precipitation during summer shows an obvious diurnal cycle, peaking in the evening, with the greatest change occurring over the central TP [17–22]. The eastern foothills of the TP are dominated by nocturnal rainfall before midsummer [23]. The diurnal cycle of precipitation from the TP to downstream areas shows diurnal propagation during the pre-mei-yu period. However, this diurnal propagation from the TP to downstream areas disappears during midsummer. The zonal wind weakens from the pre-mei-yu period to midsummer. In addition, the precipitation over the valleys of the Himalaya mainly occurs from midnight to the sunrise [24]. High quantities of cloud cover occur mainly over the ridges and then move to the valleys. Convective activity mainly occurs at night over the valleys [10]. Turbulence and convective cloud over the TP develop more easily than in the surrounding areas because the air density is lower. A convective cloud develops after sunrise, reaching a maximum from late afternoon to early morning the following day, as does convective available potential energy (CAPE) [6,14]. The diurnal cycle in monsoonal flow, the sea–land breeze, boundary-layer flow, low-level jet, aerosols, and inertial oscillation in the mid-level horizontal wind field in the mid-troposphere (~500 hPa) are the key factors that influence the diurnal variation in precipitation over the EASM region [25–27].

Satellite precipitation observation can obtain the precipitation on a global scale, which is better than conventional measurements made by rain (and snow) gauges and surface-based weather radar observations. Many advanced satellite algorithms have been released that make use of infrared and passive microwave data, for example, the Precipitation Estimation from Remotely Sensed Information using Artificial Neural Networks-Climate Data Record (PERSIANN-CDR) [28], the Integrated Multi-Satellite Retrievals for GPM (IMERG) [29], the Tropical Rainfall Measuring Mission (TRMM) Multisatellite Precipitation Analysis (TMPA) [30], the Climate Prediction Center Morphing technique (CMORPH) [31], and the Hydro-Estimator (HE) Satellite Rainfall Estimates [32]. Mountainous regions represent a major challenge for these satellite data products. Over highly complex topography such as the Andes area, HE provides the most stable result, which could be associated with the best performance of HE on the development of precipitation from warmer and relatively lower clouds [33]. The HE with orographic correction to some extent captures the spatial distribution and timing of diurnal convective events over a mountainous region [34]. In addition, spatial distribution in cloud optical thickness and the cloud water path derived from satellite retrievals over the TP were closely associated with the increase in water-vapor transport flux divergence [35].

However, there is little research on the differences in the diurnal cycle of precipitation between HE satellite products and ground observations datasets, as well as the differences between the TP and SB. Moreover, the macroscopic properties of clouds such as cloud cover and cloud liquid or ice water content (LWC and IWC, respectively) and their relationships with surface thermal effects have received minimal attention. HE satellite products do not provide cloud microphysical parameters, while reanalysis data can provide cloud microphysical parameters. Therefore, it is a good choice to combine reanalysis data and high-precision satellite data such as HE satellite products to analyze the distribution characteristics of cloud microphysical parameters over the TP and SB. Focusing on the diurnal cycle of cloud physical parameters, this paper seeks to answer the following questions on the basis of satellite, ground, and cloud radar observations and a reanalysis dataset: (1) What are the phase differences of water within clouds during the diurnal cycles over the TP and SB, and how do these change diurnally? (2) What are the possible mechanisms responsible for the phase differences of precipitation and cloud? Following this introduction and a description of the data (Section 2) and methods (Section 3) employed in this study, the diurnal variation in precipitation and cloud parameters (CBH AGL, cloud cover, cloud IWC and LWC) over the TP and SB using ECMWF Reanalysis v5 (ERA5) and the ground and HE are investigated in Section 4. Additionally, the relationships of the precipitation rate in HE, and cloud cover and cloud IWC and LWC from ERA5 with surface thermal effects are investigated. This research is important for addressing the bias of precipitation in the diurnal cycle during ground and satellite observations.

2. Data

2.1. *Hydro-Estimator Satellite Rainfall Estimates*

The rainfall rate data from Hydro-Estimator (HE) Satellite Rainfall Estimates covering June, July, and August 2014–2020 were used in this study. The HE rainfall rate estimates were produced using the data from NOAA's Geostationary Operational Environmental Satellites and also using available geostationary data over Europe, Africa, and Asia. The datasets were downloaded from ftp://ftp.star.nesdis.noaa.gov/pub/smcd/emb/f_f/hydroest/world/world/archive/ (accessed on 1 January 2022). The horizontal resolution of HE dataset is ~4 km, and the temporal resolution is 1 h. The HE dataset has been widely used in various studies [21–23]. HE provides the most stable result for several kinds of algorithms that combine infrared and passive microwave data over mountainous regions [32–34].

2.2. *Ground Observational Data*

The data from 50 ground meteorological observation sites belonging to the China Meteorological Administration were chosen, covering June, July, and August 2014–2020. Among them, 36 were located in the TP region and 14 in the SB region. The data included precipitation, air temperature at 2 m, humidity, and surface pressure, with a temporal resolution of 1 h, that were employed. Detailed station information is provided in Tables S1 and S2 in the Supporting Information. Additionally, the data from 12 radiosondes, including temperature, wind, pressure, and lifting condensation level (LCL) were chosen. The remaining observations were from TIPEX3 and a research program entitled 'The interaction between the earth and atmosphere of the TP and its influence on the weather and climate in the downstream' [4,7] (Figure 1). Cloud radar data, including CBH, cloud cover, and LWC (temporal resolution: 1 min), at the sites of Naqu (30.46°N, 90.59°E; 4730 m above mean sea level (MSL)), Yushu (33.01°N, 96.56°E; 3689 m MSL), and Linzhi (29.46°N, 94.44°E; 3326 m MSL), from July and August 2014–2020, were also used. The LWC was retrieved by using the cloud radar data with the equation LWC = $3z^{0.5}$, where LWC is a power relationship with reflectivity z. For details about the cloud radar instrument and the cloud radar LWC retrievals, see [36,37]. The locations of the observation sites are shown in Figure 1.

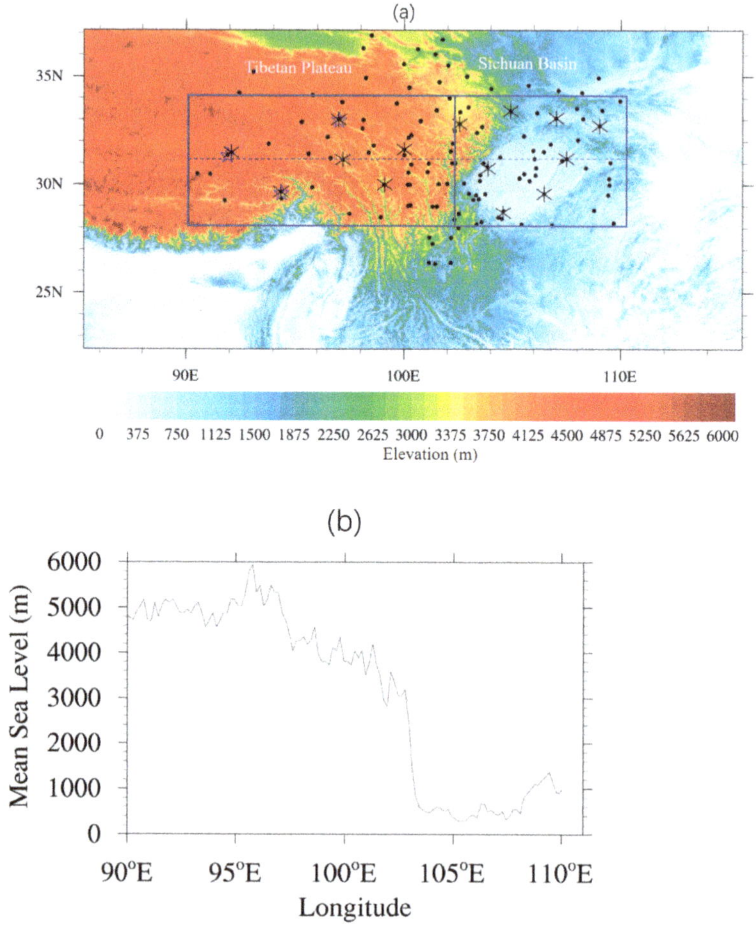

Figure 1. (**a**) Topography of the Tibetan Plateau and Sichuan Basin, in which the black dots indicate the locations of the ground meteorological observation sites of the China Meteorological Administration; '×' represents the radiosonde stations; and pentagrams represent the cloud radar observation sites. (**b**) Elevation goes along the line of 31.2°N marked with the dashed blue line in (**a**).

2.3. ERA5 Reanalysis Dataset

We obtained hourly estimates of u-wind, v-wind, temperature, precipitation, CBH, cloud cover, cloud IWC, cloud LWC, dewpoint spread, zero degree level, and CAPE from the ERA5 reanalysis dataset. ERA5 combines vast amounts of historical observations into global estimates using advanced modeling and data assimilation systems. ERA5 has been widely used in various studies [38]. The ERA5 precipitation rate is greater than that in the observations when the precipitation rate is less than 10 mm day^{-1} over the TP [39]. The data were downloaded from https://www.ecmwf.int/en/forecasts/datasets/reanalysis-datasets/era5 (accessed on 1 January 2022) and spanned the period 2014–2020 for the summer months of June, July, and August. The horizontal resolution of ERA5 is ~31 km.

3. Methods

The summer precipitation in China can be divided into two stages: the mei-yu period and the midsummer period. The 'mei-yu' rain, also called plum rain or the East Asian rainy season, is caused by precipitation along a persistent stationary front known as the

Meiyu front for nearly two months during the late spring and early summer in East Asia. These weather systems can produce heavy rainfall and flooding. The typical mei-yu period is generally at the beginning of mid or late June and at the end of early or middle July [12–16]. Therefore, two time periods were chosen: (1) the mei-yu period, during 1–25 June in the monsoon phase and (2) the midsummer period, during 1 July–10 August. In terms of the study domain, the region of the eastern TP and its downstream area (28°–34° N, 90°–110° E) were chosen (Figure 1) [15], which were then further separated into two subregions with different elevations (framed areas in Figure 1): 90°–100° E and 100°–110° E, which represented the TP and SB. To describe the diurnal cycle during summer, the ERA5 and observational data were grouped into the mei-yu and midsummer seasons. The precipitation feature (PF) number is defined as the number of hours with precipitation in the diurnal cycle during the observational period when the precipitation observed was larger than 0.02 mm h^{-1}. The numbers of the PF in the two subregions during the two different seasons are listed in Table 1.

Table 1. The number of precipitation features (Unit: hour) from the Hydro-Estimator Satellite Rainfall Estimates dataset in the two subregions within 28°–34°N shown in Figure 1 in each of the two chosen seasons.

	Tibetan Plateau	Sichuan Basin
Mei-yu	4531	2687
Midsummer	8145	3556

The ground and cloud radar observations were temporally averaged to 1 h. Local time (LT) was defined as Coordinated Universal Time (UTC) + 7 h. To match the ground observations with the satellite and ERA5 data in spatial terms, four grid values near the ground observations including precipitation and cloud parameters at different levels from the ERA5 dataset and satellite were interpolated using the bilinear interpolation method to produce the value at the ground observation site [40,41]. We use Hovmöller diagrams [23] to show the diurnal cycle of CBH, precipitation, dewpoint spread, IWC, and LWC and their changes with latitude. Typically, longitude is plotted along the x-axis, and time is recorded on the ordinate; then, the contour values of a named physical field are presented through color or shading. In addition, the height of the LCL can be calculated as $Z_{lcl} = 123(T - T_d)$, where T is the air temperature at 2 m, and T_d is the dewpoint temperature, in which the LCL is determined by the dewpoint spread.

4. Results

4.1. Diurnal Cycle of Precipitation and Cloud

During the mei-yu period, the diurnal variation in precipitation from HE satellite estimates shows phase propagation from west to east over the SB (103°–110° E), while no such propagation can be seen over the TP (90°–102° E) (Figure 2a), which is similar to the results of [23]. The diurnal maximum precipitation rate over the TP is concentrated between 1500 LT and 2100 LT, and the diurnal minimum precipitation rate lies within 0000–1200 LT. Over the SB, meanwhile, the diurnal maximum precipitation rate occurs at night, and then decreases to a minimum in the morning, which was also found in previous studies (e.g., [42]). The result for the PF number percentage is similar to that of the precipitation amount, except the maximum PF number percentage occurs at the boundary of the TP and SB (Figure 2c), which is related to the convergence rising motion over the slopes in the afternoon and rising motion in the center of the SB at night. During the midsummer period, the diurnal variation in the precipitation rate shows no propagation from the TP to SB, which is different from the mei-yu period. The diurnal maximum precipitation rate is concentrated between 1500 LT and 2100 LT, while the diurnal minimum precipitation rate lies within 0000–0600 LT, which is similar to that during the mei-yu period (Figure 2b). The average precipitation rate over the SB is larger than that over the TP. The result for the PF number percentage is similar to that of the precipitation amount (Figure 2d). The reasons of the diurnal propagation of rainfall are related to the decreases in the zonal wind profile from the mei-yu to midsummer period and the diurnal cycle of low-level winds over the SB [23].

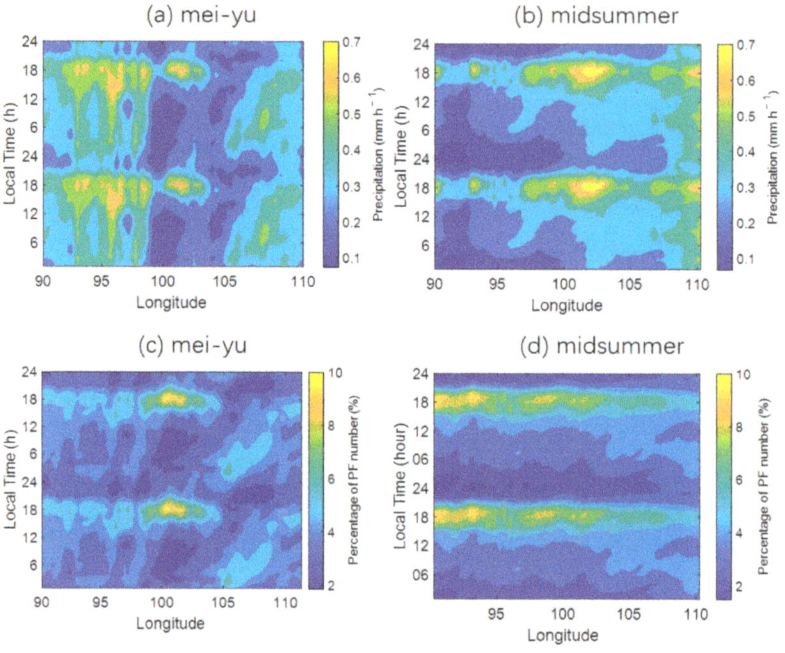

Figure 2. The (**a**,**b**) diurnal variation in precipitation and (**c**,**d**) precipitation feature number percentage during the (**a**,**c**) mei-yu and (**b**,**d**) midsummer period from the Hydro Estimator satellite rainfall estimates.

The precipitation from HE satellite estimates and ground observations are compared in Figure S1. Over the TP, the maximum precipitation rate from the HE satellite is 3 h ahead of that of the ground observations during the mei-yu and midsummer periods, and the precipitation rate from HE satellite is larger than that observed on the ground. Over the SB, the maximum precipitation rate from the HE satellite is also 3 h ahead of that of the ground observations during the mei-yu period, while it is in-phase with ground observations during the midsummer period. The result for the PF number percentage is comparable to that of the precipitation rate.

The diurnal variations in cloud cover, LWC, and IWC from ERA5 are shown in Figure 3. During the mei-yu period, the diurnal variation in cloud cover (Figure 3a), LWC (Figure 3c), and IWC (Figure 3e) shows propagation from west to east over the SB, while no diurnal propagation is apparent over the TP. The diurnal maximum cloud cover and LWC over the TP is concentrated between 1800 LT and 0300 LT, while over the SB they lie within 1500–2400 LT. The diurnal minimum cloud cover and LWC lie within 1000–1500 LT over both the TP and SB. The cloud cover and LWC at the border of the SB and TP are smaller than those over the inner parts of the two regions. The cloud cover and LWC in the SB are larger than those over the TP. The diurnal maximum IWC over the TP is concentrated between 1500 LT and 2100 LT, while the range for the diurnal minimum IWC is 0600–1200 LT. Over the SB, the diurnal maximum IWC is in the mid-afternoon to early evening (1500–2100 LT) and then decreases to a minimum during the morning (0600–1200 LT). The cloud IWC over the edge of TP and SB is larger than that over the inner TP and SB.

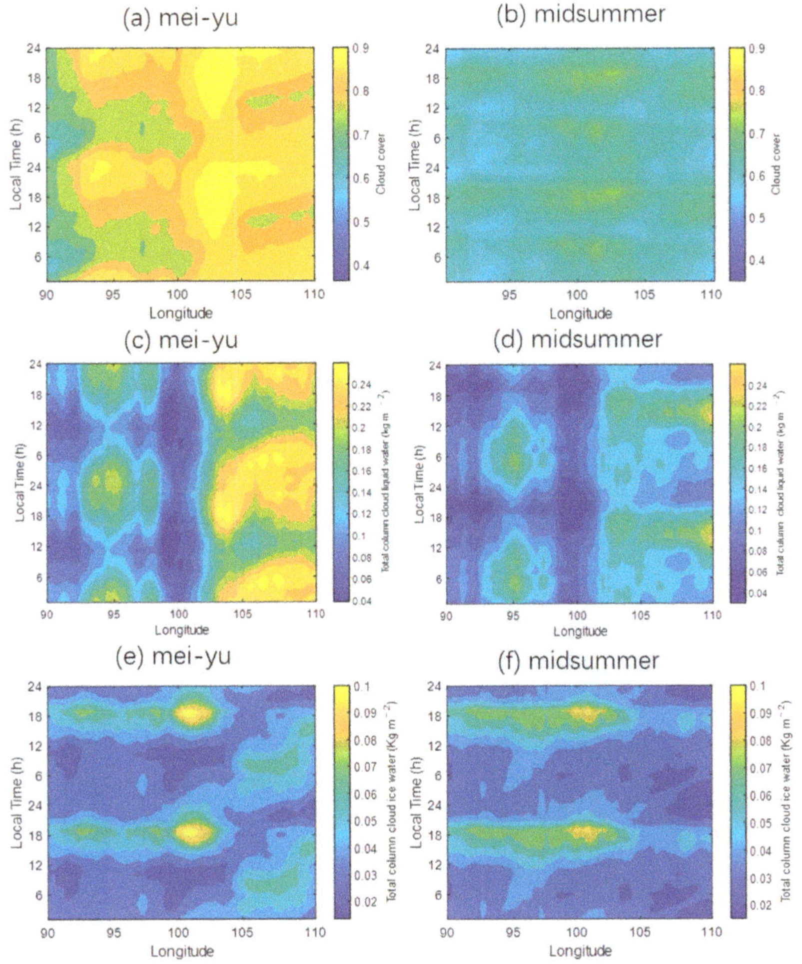

Figure 3. Diurnal variation in (**a**,**b**) cloud cover, (**c**,**d**) total column cloud liquid water content, and (**e**,**f**) cloud ice water content from ERA5 over the Tibetan Plateau and Sichuan Basin during the (**a**,**c**,**e**) mei-yu and (**b**,**d**,**f**) midsummer periods.

During the midsummer period, the diurnal variation in cloud cover (Figure 3b), LWC (Figure 3d), and IWC (Figure 3f) shows no propagation from the TP to SB. Over the TP, the diurnal maximum cloud cover and LWC lie within 1500–2100 LT and 0000–1200 LT, respectively, while the ranges for the diurnal minimum cloud cover and LWC are 1000–1500 LT and 1800–2100 LT. The cloud cover and LWC at the border of the SB and TP are smaller than those over the inner parts of the TP and SB. The cloud cover and LWC over the SB are larger than over the TP. Over both the TP and SB, the diurnal maximum IWC occurs in the mid-afternoon to early evening (1500–2100 LT) and reaches a diurnal minimum in the morning (2400–1200 LT).

The ratios of cloud LWC and IWC to total cloud water were separately calculated, and the results are presented in Figure 4. The results show that LWC accounts for more than 60% of total water during almost the entire diurnal cycle over the inner TP and SB during the mei-yu period, except for the period 1500–1800 LT at the edge of the SB and TP. The IWC accounts for more than 60% at around 1800 LT at the edge of the TP and SB, which is related to the convergence rising motion over the slopes at the edge of the two regions. The proportions of LWC and IWC during midsummer are similar to those during the mei-yu period. However, there are two IWC centers during the midsummer period. One is in the period 1500–1800 LT at the edge of the SB and TP, and the other is located in the region 92°–94°E during 1700–2000 LT, which is related to development of the westerly jet and southern water vapor transportation.

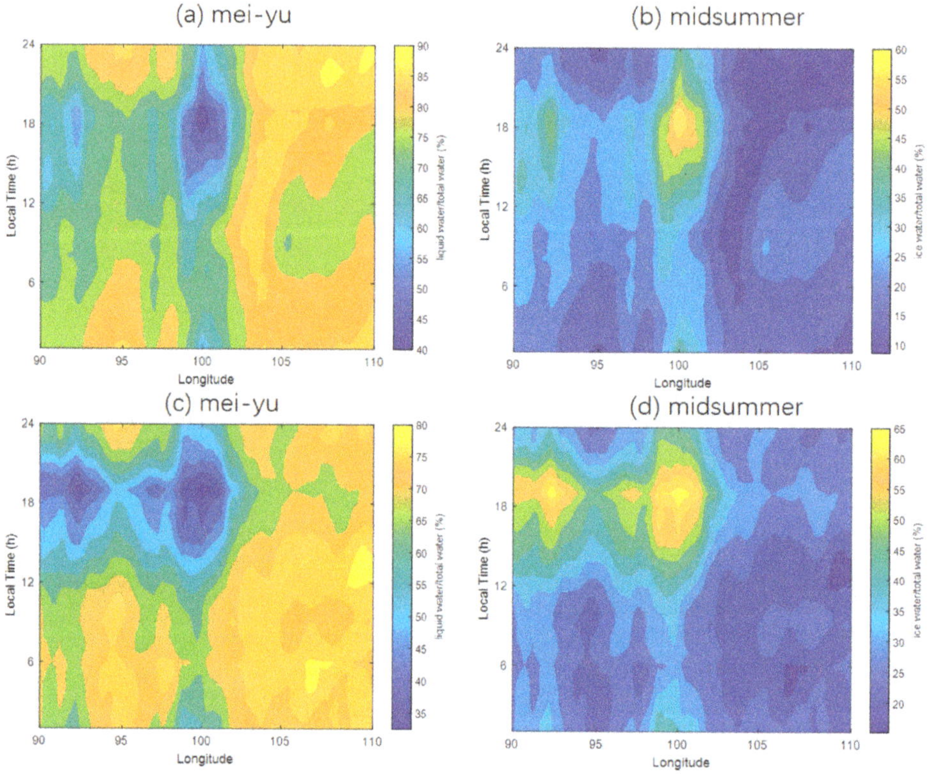

Figure 4. Ratio of cloud (**a**,**c**) liquid water and (**b**,**d**) ice water to total cloud water during the (**a**,**b**) mei-yu and (**c**,**d**) midsummer periods.

During the mei-yu period, the cloud LWC mainly distributes between the 900 hPa and 450 hPa level in the daytime and the 850–450 hPa level in the nighttime over the SB, while it mainly distributes between the 600 hPa and 450 hPa level in the daytime and the 600–400 hPa level in the nighttime over the TP (Figure 5a,b). The cloud LWC in the night is larger than that in the daytime over the SB. During the midsummer period, it is mainly similar to that over the SB and TP (Figure 5c,d) except that the cloud LWC during the midsummer period is smaller than that in the mei-yu period.

During the mei-yu period, the cloud IWC mainly distributes between the 450 hPa and 150 hPa level in the daytime and the 500–100 hPa level in the nighttime over the SB, while it mainly distributes between the 500 hPa and 180 hPa level in the daytime and the 450–100 hPa level in the nighttime over the TP (Figure 6a,b). The cloud IWC in the night is smaller than that in the daytime over the SB. During the midsummer period, it is mainly similar to that over the SB and TP except that the cloud IWC during the midsummer period is larger than that in the mei-yu period (Figure 6c,d). The cloud IWC in the daytime during the midsummer period mainly is concentrated over the edge of the SB and TP.

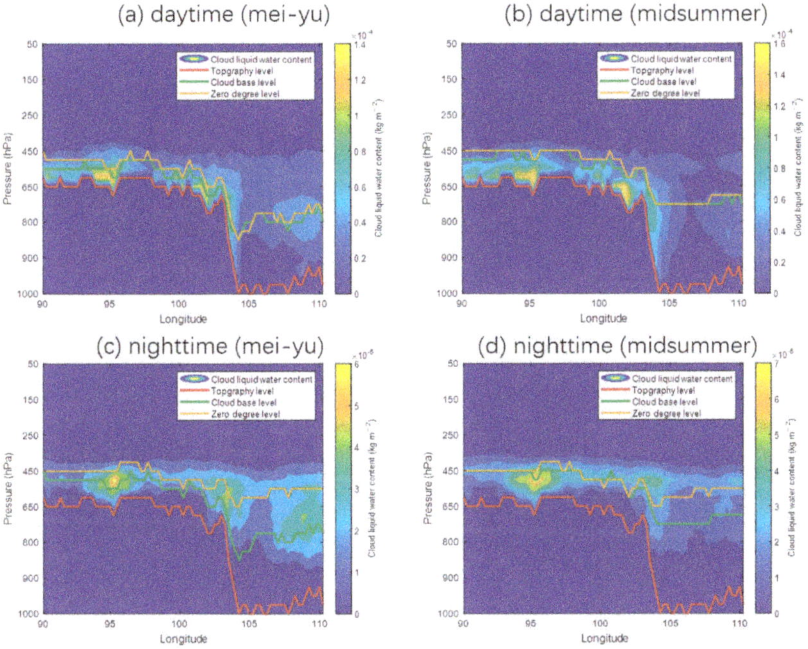

Figure 5. The pressure level of cloud liquid water content (shaded color), cloud base (green line), zero degree level (orange line), and the topography (red line) along with longitude during the (**a**,**b**) mei-yu and (**c**,**d**) midsummer periods during the (**a**,**c**) daytime and (**b**,**d**) nighttime.

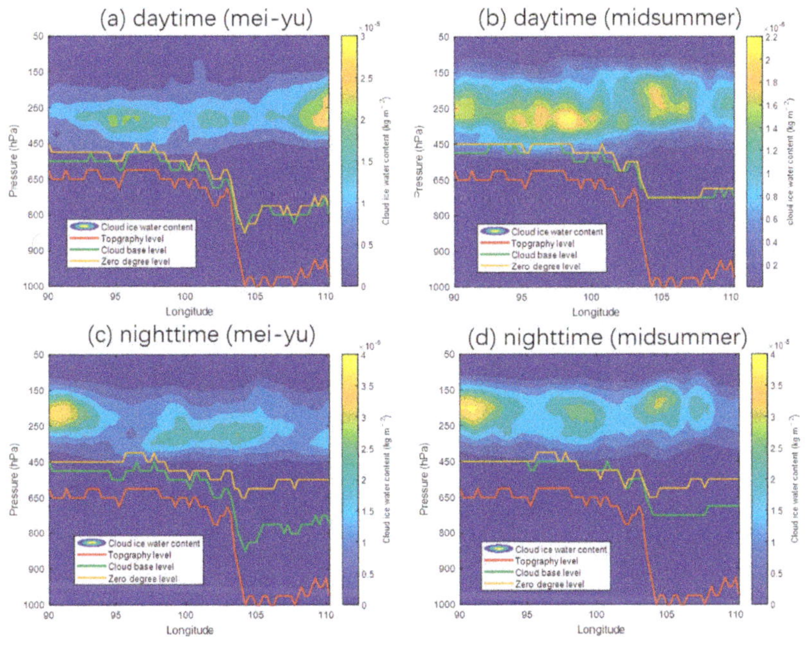

Figure 6. Similar with Figure 5, but for cloud ice water content.

The diurnal cycle of CBH AGL in ERA5 is presented in Figure 7. During the mei-yu period, the diurnal variation in CBH AGL shows propagation from west to east over the SB, while no propagation is apparent over the TP. The diurnal maximum CBH AGL over the TP is concentrated between 1500 LT and 2400 LT, while the diurnal minimum CBH AGL lies within 0900–1200 LT. The CBH AGL in the early evening is higher than that in the daytime over the SB, which is consistent with the nocturnal maximum precipitation. The CBH AGL over the SB is higher than that over the TP (Figure 7a). During the midsummer period, the diurnal variation in CBH AGL shows no propagation from the TP to SB. The diurnal maximum CBH AGL over the TP is concentrated between 2100 LT and 0300 LT, while the diurnal minimum lies within 0700–1500 LT. Over the SB, the diurnal maximum CBH AGL lies within the period 2100–0600 LT, while the diurnal minimum is within 1500–2100 LT (Figure 7b). The average CBH AGL observed by the cloud radar at Yushu, Naqu, and Linzhi is close to that in ERA5 (Figure S2). The maximum CBH AGL from ERA5 is in-phase with the observations averaged from the three sites during the mei-yu (Figure S2a) and midsummer (Figure S2b) periods over the TP.

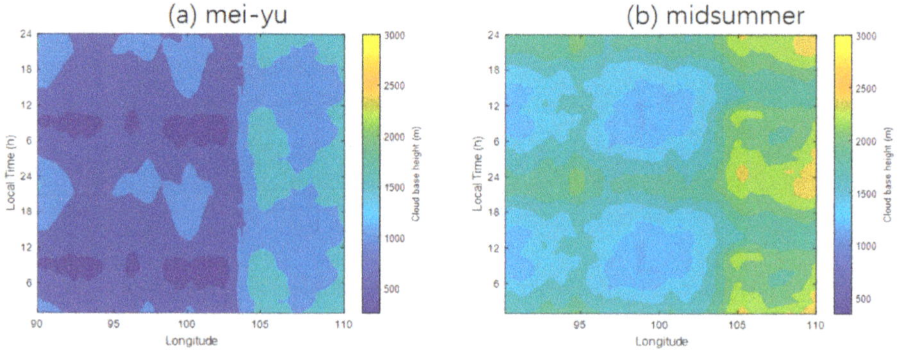

Figure 7. Diurnal variation in cloud base height from ERA5 over the Tibetan Plateau and Sichuan Basin during the (**a**) mei-yu and (**b**) midsummer periods.

The CBH AGL is compared with the zero degree level AGL in Figure 5. The zero degree level AGL is much higher than the CBH AGL over the TP and SB during the mei-yu period, while CBH AGL nears the zero degree level AGL over the TP during midsummer. The latter is related to the air temperature lapse rate, which is larger over the TP than the SB. The zero degree level MSL over the TP is higher than that over the SB. The thickness of cloud LWC over the SB is larger than that near the edge of the TP during the mei-yu and midsummer periods. The zero degree level MSL increases while the CBH remains stable with the increase in distance from the TP. However, super-cooled liquid water can exist down to −40 °C.

4.2. Factors Influencing the Formation of Cloud over the TP and SB

During the mei-yu period, the diurnal variation in dewpoint spread shows no propagation from morning to night over the SB and TP (Figure 8a,b). As is well known, the dewpoint spread is mainly influenced by the solar radiation and surface heating. The diurnal maximum dewpoint spread is concentrated between 1500 LT and 2100 LT, and the diurnal minimum dewpoint spread lies within 0000–0900 LT in the region 90°–100°E during mei-yu period (Figure 8a). During the midsummer period, the diurnal variation in dewpoint spread shows no propagation from the TP to SB. The diurnal maximum dewpoint spread is concentrated between 1500 LT and 2100 LT, while the range for the diurnal minimum is 0600–1000 LT (Figure 8b). The dewpoint spread over the SB is larger than that over the TP.

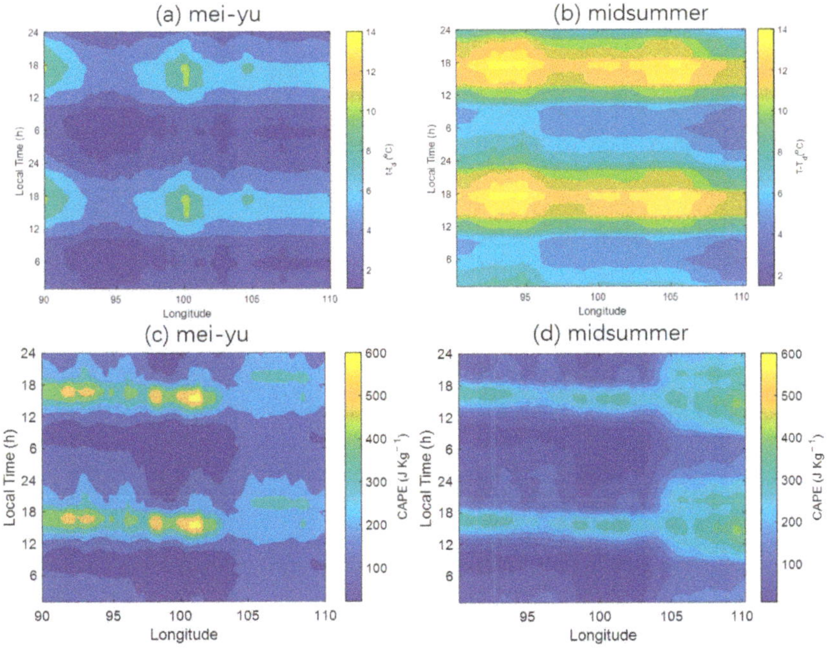

Figure 8. Diurnal variation in (**a**,**b**) dewpoint spread and (**c**,**d**) convective available potential energy (CAPE) from ERA5 over the Tibetan Plateau and Sichuan Basin during the (**a**,**c**) mei-yu and (**b**,**d**) midsummer periods.

A comparison of the dewpoint spread between the observation and ERA5 is given in Figure S3. Over both the TP and SB, the timing of the diurnal maximum dewpoint spread is similar to that in the observations during both the mei-yu and midsummer period. The dewpoint spread from ERA5 is larger than that in the observations, especially during the early evening. Some studies show that the dewpoint spread makes a profound contribution to the LCL (e.g., [43]). In this study, the CBH is close to the LCL during the summer period with large amounts of LWC during that same season (Table 2).

Table 2. Comparison of cloud base height (CBH) from ERA5 dataset and lifting condensation level (LCL) above ground level (AGL) during summer over the Tibetan Plateau (TP) and Sichuan Basin (SB).

	TP		**SB**	
	Mei-Yu	**Midsummer**	**Mei-Yu**	**Midsummer**
CBH AGL (m)	822	901	2556	2341
LCL AGL (m)	753	856	2468	2390

To investigate the effect of dewpoint spread on the CBH AGL, a correlation analysis between the two at each longitudinal grid point in ERA5 was carried out (Figure 9a). Over the TP, the correlation coefficient reaches up to 0.8 at the 5% significance level. However, it decreases at the edge of the TP and SB and becomes gradually negative, reaching a maximum negative value over the inner part of the SB. According to the above-mentioned results, CBH AGL propagates diurnally from the TP to SB, but the dewpoint spread does not. Therefore, the dewpoint spread makes a profound contribution to the CBH over the TP, but little contribution to the CBH AGL over the SB because of the stronger turbulence and lower air density over the TP than the SB [44].

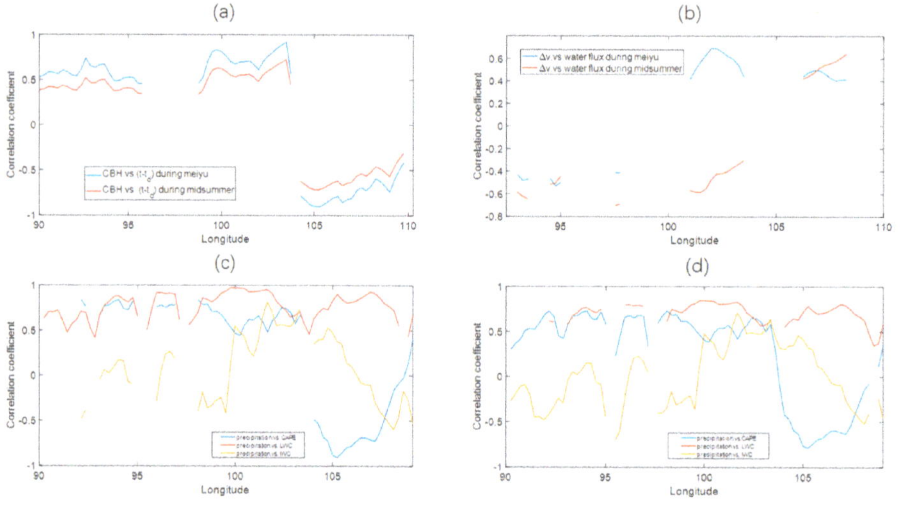

Figure 9. Correlation between (**a**) dewpoint spread and CBH and (**b**) water vapor flux and ΔV during the mei-yu (blue line) and midsummer (red line) periods. (**c**,**d**) Correlation of ΔV, CAPE, cloud LWC and IWC with precipitation rate during the (**c**) mei-yu and (**d**) midsummer periods.

From the spatial distribution of the water vapor flux, the transportation of water vapor can be seen to be mainly from the west (Figure 10), which is similar to the results of [45]. The water vapor flux decreases from the mei-yu period to the midsummer period, especially over the SB. This is related to the decreases in the mean zonal wind profile and low-level winds from the mei-yu to midsummer period.

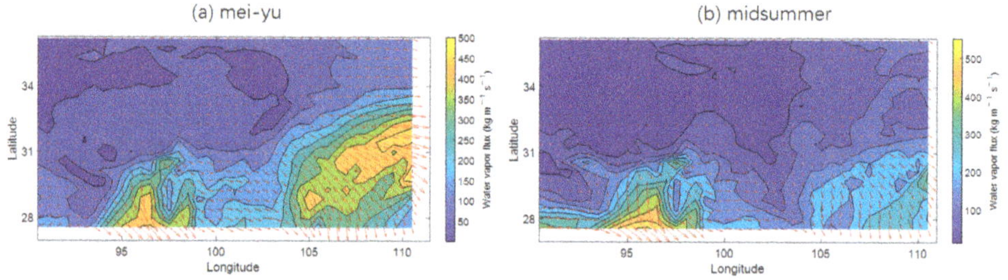

Figure 10. Integral of water vapor flux at 1800 LT from ERA5 over the (**a**) mei-yu and (**b**) midsummer periods.

To investigate the relationship between the total water vapor transportation and westerly wind, the difference in horizontal wind speed between 200 hPa and 500 hPa (ΔV) was calculated (Figure 11); that is, $\Delta V = [u(200) - u(500)]^2 + [v(200) - v(500)]^2$, where u(200) and v(200) refer to u-wind and v-wind at the 200 hPa level, respectively, and u(500) and v(500) refer to the same but at the 500 hPa level. Correlation analysis was carried out between the water vapor flux and ΔV. The ΔV over both the TP and SB decreases from the mei-yu to midsummer period, and the ΔV over the TP is larger than that over the SB during both periods. The ΔV correlates positively with water vapor transportation at the 5% significance level over the TP and SB, and the correlation coefficient over the TP is smaller than that over the SB during the mei-yu and midsummer periods (Figure 9b). The ΔV has a profound impact on the transportation of water vapor over the SB. The correlation coefficient decreases dramatically from the mei-yu to midsummer period in the region of 90°–100° E where the steep slopes and SB are situated.

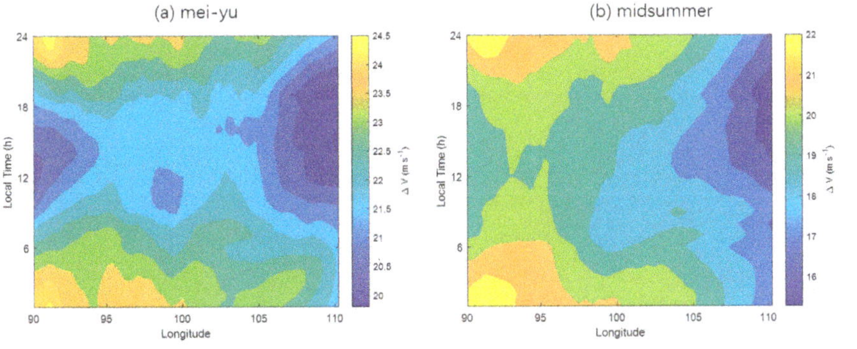

Figure 11. Diurnal variation in ΔV during the (**a**) mei-yu and (**b**) midsummer periods.

4.3. Factors Influencing the Formation of Precipitation over the TP and SB

To investigate the factors influencing the precipitation, the diurnal cycle of CAPE is shown in Figure 8c,d. During the mei-yu period, CAPE shows no propagation over the TP and SB (Figure 8c). The diurnal maximum CAPE occurs in the late afternoon (1500–1900 LT) and early evening (1900–2100 LT) over the TP and SB during the mei-yu period, and the diurnal minimum occurs in the morning (0600–1000 LT). The CAPE over the TP is larger than that over the SB (Figure 8c), which is different from the precipitation and cloud LWC. During the midsummer period, CAPE shows no diurnal propagation over the TP and SB (Figure 8d). The diurnal maximum and minimum CAPE values during the mei-yu period are similar to those during the midsummer period. CAPE is related to the maximum potential vertical velocity of air within an updraft, so higher values of CAPE are an indicator of precipitation.

To investigate the influence of diurnal variation in CAPE, cloud LWC and IWC on precipitation, correlation analysis was performed. During the mei-yu period, the correlation coefficient between CAPE and precipitation over the TP is larger than that over the SB. The correlation coefficient reaches a maximum (0.8) at the edge of the TP and SB at the 5% significance level (Figure 9c). The correlation coefficient between CAPE and precipitation during the midsummer period is smaller than that during the mei-yu period (Figure 9d). These results indicate that CAPE has a larger impact on precipitation over the TP than over the SB.

The correlation coefficient between cloud IWC and precipitation over the TP and SB is larger than 0.7, with a significance level of 5%, during the mei-yu period. The correlation coefficient between cloud LWC and precipitation over the SB during the midsummer period (Figure 9d) is smaller than that during the mei-yu period, which decreases with longitude from west to east due to the diurnal propagation of LWC and precipitation (Figure 9c), while the opposite is the case for cloud IWC. However, the correlation between the cloud IWC and precipitation over the edge of the TP and SB is better than over the inner parts of the TP and SB due to the convergence rising motion over the slopes. The above results indicate that the cloud LWC makes a large contribution to precipitation over the SB. The precipitation over the edge of the TP and SB (i.e., the steep downstream slopes) is influenced by the cloud IWC.

5. Conclusions

The diurnal variations in precipitation in the HE Satellite Rainfall Estimates and cloud parameters (CBH AGL, cloud cover, cloud LWC and IWC, and dewpoint spread) in the ERA5 dataset were analyzed over the TP and SB. Results show that the precipitation and cloud parameters show diurnal propagation from morning to night over the SB during the mei-yu period, while no such diurnal propagation is apparent over the TP during both the mei-yu and midsummer periods. The precipitation, cloud LWC and IWC, and cloud cover

over the TP are smaller than over the SB. The dewpoint spread over the TP is larger than over the SB. The diurnal maximum precipitation and cloud LWC and IWC are concentrated in the early evening, while the diurnal minima occur in the morning.

The precipitation from the HE Satellite is larger than that in the observations. Over the SB, the maximum precipitation from the HE Satellite is 3 h ahead of observations. The result for the PF number percentage is similar to that of the precipitation amount. During both the mei-yu and midsummer period, the cloud LWC from ERA5 at night is larger than that in the daytime. Cloud LWC accounts for more than 60% of total water during almost the entire diurnal cycle over the inner parts of the TP and SB during the mei-yu period except for late afternoon at the edge of the SB and TP. The cloud IWC accounts for more than 60% during early evening at the edge of the TP and SB.

The CBH AGL, LCL AGL, and zero degree level AGL are almost equal over the TP during the summer period. The zero degree level AGL over the SB is higher than that over the TP because the air temperature lapse rate over the TP is larger than that over the SB. The thickness of the LWC over the SB is larger than that over the TP. The dewpoint spread makes a profound contribution to the CBH AGL over the TP but little contribution to the CBH AGL over the SB because of the stronger turbulence and lower air density over the TP. CAPE has a larger impact on precipitation over the TP than over the SB. The cloud LWC makes a large contribution to the precipitation over the SB, which is related to the mean zonal wind profile and diurnal cycle of low-level winds. The precipitation over the edge of the TP and SB (i.e., the steep downstream slopes) is influenced by the cloud IWC owing to the convergence rising motion over the slopes.

Although the distribution of cloud LWC and IWC over the TP and SB was obtained in this study, the distribution of supercooled water and ice level over the edge of TP and SB is still unclear and needs to be investigated. In addition, the results of ERA5 data were uncertainty limited by the resolution of ERA5 data, especially concerning quite an extreme region. The current cloud radar sites are very sparse, making it impossible directly to observe the cloud microphysical parameter data over the entire TP and SB. In the future, the accuracy of the reanalysis dataset and observations over the TP and SB at night should be improved. More cloud radar observations at different locations over the TP and SB are needed in the future. The phase of water within cloud needs to be studied using different kinds of observational and model data.

Supplementary Materials: The following supporting information can be downloaded at https://www.mdpi.com/article/10.3390/rs14112711/s1: Table S1: The information of ground meteorological observation sites of China Meteorological Administration; Table S2: The information of radiosonde station of China Meteorological Administration; Figure S1: The precipitation from Hydro Estimator satellite estimates (blue line) and ground observations (red line) during (a,b) meiyu and (c,d) midsummer period over the (a,c) Tibetan Plateau and (b,d) Sichuan Basin. Blue shaded area is the standard error of precipitation from Hydro Estimator satellite estimates, and red shaded area is the standard error of precipitation from ground observations. The percentage of precipitation feature during (e,f) mei-yu and (g,h) midsummer period over the (e,g) Tibetan Plateau and (f,h) Sichuan Basin and their error bar. Blue shaded area is the standard error of the percentage of precipitation feature from Hydro Estimator satellite estimates, and red shaded area is the standard error of percentage of precipitation feature from ground observations; Figure S2: The comparison of cloud base height from ERA5 dataset (blue line) and cloud radar observations (red line) and their standard error (blue shaded area for ERA5 and red shaded area for observations) during the (a) mei-yu and (b) midsummer period; Figure S3: The comparison of dew point spread from the ERA5 dataset (blue line) and ground observations (red line) and their standard error (blue shaded area for ERA5 dataset, and red shaded area for observations) during the (a,b) mei-yu and (c,d) midsummer period over the (a,c) Tibetan Plateau and (b,d) Sichuan Basin.

Author Contributions: Conceptualization, Y.L.; methods, B.C.; software, B.C.; validation, Y.L.; writing—original draft preparation, B.C.; writing—review and editing, X.Y.; visualization, B.L.; supervision, J.W.; project administration, X.Y.; funding acquisition, X.Y. All authors have read and agreed to the published version of the manuscript.

Funding: This research was funded by the National Science Foundation of China (NSFC) (41975130, 42175174), Strategic Priority Research Program of Chinese Academy of Sciences (XDA20050102), the Second Tibetan Plateau Scientific Expedition and Research (STEP) program (2019QZKK0102), the Research Foundation of Chengdu University of Information Technology (KYTZ201810), and the Project of Science and Technology Plan of Sichuan (2019YJ0408).

Data Availability Statement: The ERA5 data employed in this study are available from https://www.ecmwf.int/en/forecasts/datasets/reanalysis-datasets/era5 (accessed on 1 January 2022). The observational data are available from the third Tibetan Plateau Atmospheric Experiment (TIPEX3; https://data.cma.cn/site/article/id/28986.html (accessed on 1 January 2022)).

Conflicts of Interest: The authors declare no conflict of interest.

Abbreviations

TP	Tibetan Plateau
SB	Sichuan Basin
EASM	East Asian summer monsoon
ECMWF	European Centre for Medium-Range Weather Forecasts
ERA5	ECMWF Reanalysis v5
PERSIANN-CDR	Precipitation Estimation from Remotely Sensed Information using Artificial Neural Networks-Climate Data Record
IMERG	Integrated Multi-satellite Retrievals for GPM
TMPA	Tropical Rainfall Measuring Mission (TRMM) Multisatellite Precipitation Analysis
CMORPH	Climate Prediction Center Morphing technique
HE	Hydro-Estimator Satellite Rainfall Estimates
LT	Local time
UTC	Coordinated Universal Time
PF	Precipitation Feature
AGL	Above ground level
MSL	Mean sea level
CBH	Cloud base height
LCL	Lifting condensation level
CAPE	Convective available potential energy
LWC	Cloud liquid water content
IWC	Cloud ice water content

References

1. Xu, X.; Lu, C.; Shi, X.; Gao, S. World water tower: An atmospheric perspective. *Geophys. Res. Lett.* **2008**, *35*, L20815. [CrossRef]
2. Fu, Y.; Liu, G.; Wu, G.; Yu, R.; Xu, Y.; Wang, Y.; Li, R.; Liu, Q. Tower mast of precipitation over the central Tibetan Plateau summer. *Geophys. Res. Lett.* **2006**, *33*, L05802. [CrossRef]
3. Boos, W.R.; Kuang, Z. Dominant control of the South Asian monsoon by orographic insulation versus plateau heating. *Nature* **2010**, *463*, 218–222. [CrossRef] [PubMed]
4. Fu, Y.; Ma, Y.; Zhong, L.; Yang, Y.; Guo, X.; Wang, C.; Xu, X.; Yang, K.; Xu, X.; Liu, L.; et al. Land surface processes and summer cloud-precipitation characteristics in the Tibetan Plateau and their effects on downstream weather: A review and perspective. *Natl. Sci. Rev.* **2020**, *7*, 500–515. [CrossRef] [PubMed]
5. Kukulies, J.; Chen, D.; Curio, J. The role of mesoscale convective systems in precipitation in the Tibetan Plateau region. *J. Geophys. Res. Atmos.* **2021**, *126*, e2021JD035279. [CrossRef]
6. Barton, E.; Taylor, C.; Klein, C.; Harris, P.; Meng, X. Observed soil moisture impact on strong convection over mountainous Tibetan Plateau. *J. Hydrometeorol.* **2021**, *22*, 561–572. [CrossRef]
7. Houze, R.A.; Wilton, D.C.; Smull, B.F. Monsoon convection in the Himalayan region as seen by the TRMM precipitation radar. *Q.J.R. Meteorol. Soc.* **2007**, *133*, 1389–1411. [CrossRef]
8. Zhao, P.; Xu, X.; Chen, F.; Guo, X.; Zheng, X.; Liu, L.; Hong, Y.; Li, Y.; La, Z.; Peng, H. The Third Atmospheric Scientific Experiment for understanding the earth-atmosphere coupled system over the Tibetan Plateau and its effects. *Bull. Am. Meteorol. Soc.* **2017**, *99*, 757–776. [CrossRef]
9. Bai, A.J.; Liu, C.; Liu, X. Diurnal variation of summer rainfall over the Tibetan Plateau and its neighboring regions revealed by TRMM multi-satellite precipitation analysis. *Chin. J. Atmos. Sci.* **2008**, *51*, 518–529. [CrossRef]
10. Feng, J.; Liu, L.; Wang, Z. The statistic characteristics of radar echo and precipitation and some thermodynamic variables at Naqu in Qinghai-Xizang Plateau. *Plateau Meteorol.* **2002**, *21*, 368–374. (In Chinese)

11. Zhao, P.; Xiao, H.; Liu, J.; Zhou, Y. Precipitation efficiency of cloud and its influencing factors over the Tibetan plateau. *Int. J. Climatol.* **2022**, *42*, 416–434. [CrossRef]
12. Tao, S.; Chen, L. A review of recent research on the East Asian summer monsoon in China. In *Monsoon Meteorology*; Chang, C.P., Krishnamurti, T.N., Eds.; Oxford University Press: Oxford, UK, 1987; pp. 60–92.
13. Ding, Y. Summer monsoon rainfalls in China. *J. Meteor. Soc. Jpn.* **1992**, *70*, 373–396. [CrossRef]
14. Chen, G. Large-scale circulations associated with the East Asian summer monsoon and the mei-yu over South China and Taiwan. *J. Meteor. Soc. Jpn.* **1994**, *72*, 959–983. [CrossRef]
15. Ding, Y.; Chan, J.C.-L. The East Asian summer monsoon: An overview. *Meteor. Atmos. Phys.* **2005**, *89*, 117–142.
16. Chen, G. Research on the phenomena of Meiyu during the past quarter century: An overview. In *The East Asian Monsoon*; Chang, C.-P., Ed.; World Scientific: Singapore, 2004; pp. 357–403.
17. Liu, L.; Zheng, J.; Zheng, R.; Cui, Z.; Hu, Z.; Wu, S.; Dai, G.; Wu, Y. Comprehensive radar observations of clouds and precipitation over the Tibetan Plateau and preliminary analysis of cloud properties. *J. Meteorol. Res.* **2015**, *29*, 546–561. [CrossRef]
18. Mei, Y.; Hu, Z.; Huang, X. A study of convective clouds in the Tibetan Plateau based on dual polarimeteic radar observations. *Acta Meteor Sin.* **2018**, *76*, 1014–1028. [CrossRef]
19. Liu, L.; Feng, J.; Chu, R.; Zhou, Y.; Ueno, K. The diurnal variation of precipitation in monsoon season in the Tibetan Plateau. *Adv. Atmos. Sci.* **2002**, *19*, 365–378. [CrossRef]
20. Fujinami, H.; Yasunari, T. The seasonal and intraseasonal variability of diurnal cloud activity over the Tibetan Plateau. *J. Meteor. Soc. Jpn.* **2001**, *79*, 1207–1227. [CrossRef]
21. Ji, X.L.; Wu, H.M.; Huang, A.N.; Zhao, W.; Yang, W. Characteristics of the precipitation diurnal variation over Qinghai-Tibetan Plateau in summer. *Plateau Meteor.* **2017**, *36*, 1188–1200. (In Chinese)
22. Zhao, C.; Liu, L.; Wang, Q.; Qiu, Y.; Wang, Y.; Wu, X. MMCR-based characteristic properties of non-precipitating cloud liquid droplets at Naqu site over Tibetan Plateau in July 2014. *Atmos. Res.* **2017**, *190*, 68–76. [CrossRef]
23. Xu, W.; Zipser, E.J. Diurnal variations of precipitation, deep convection, and lightning over and east of the eastern Tibetan Plateau. *J. Clim.* **2011**, *24*, 448–465. [CrossRef]
24. Fu, Y.; Pan, X.; Xian, T.; Liu, G.; Zhong, L.; Liu, Q.; Ma, M. Precipitation characteristics over the steep slope of the Himalayas in rainy season observed by TRMM PR and VIRS. *Clim. Dyn.* **2018**, *51*, 1971–1989. [CrossRef]
25. Gong, Y.F.; Ji, L.R.; Duan, T.Y. Precipitation character of rainy season of Qinghai-Xizang Plateau and onset over East Asia monsoon. *Plateau Meteor.* **2004**, *23*, 313–322. (In Chinese)
26. Chen, G.; Sha, W.; Iwasaki, T. Diurnal variation of precipitation over southeastern China: 2. Impact of the diurnal monsoon variability. *J. Geophys. Res.* **2009**, *114*, D21105. [CrossRef]
27. Yu, R.; Li, J.; Chen, H.; Yuan, W. Progress in studies of the precipitation diurnal variation over contiguous China. *Acta Meteorol. Sin.* **2014**, *72*, 948–968. [CrossRef]
28. Sorooshian, S.; Hsu, K.L.; Gao, X.; Gupta, H.V.; Imam, B.; Braithwaite, D. Evaluation of PERSIANN system satellite-based estimates of tropical rainfall. *Bull. Am. Meteorol. Soc.* **2000**, *81*, 2035–2046. [CrossRef]
29. Huffman, G.J.; Bolvin, D.T.; Nelkin, E.J. Integrated Multi-satellite Retrievals for GPM (IMERG) Technical Documentation. NASA/GSFC Code 612; 2015. Available online: http://pmm.nasa.gov/sites/default/files/document_files/IMERG_doc.pdf (accessed on 16 September 2019).
30. Huffman, G.J.; Bolvin, D.T.; Nelkin, E.J.; Wolff, D.B.; Adler, R.F.; Gu, G.; Hong, Y.; Bowman, K.P.; Stocker, E.F. The TRMM multisatellite precipitation analysis (TMPA): Quasi-global, multiyear, combined-sensor precipitation estimates at fine scales. *J. Hydrometeorol.* **2007**, *8*, 38–55. [CrossRef]
31. Joyce, R.J.; Janowiak, J.E.; Arkin, P.A.; Xie, P. CMORPH: A method that produces global precipitation estimates from passive microwave and infrared data at high spatial and temporal resolution. *J. Hydrometeorol.* **2004**, *5*, 487–503. [CrossRef]
32. Scofield, R.A.; Kuligowski, R.J. Status and outlook of operational satellite precipitation algorithms for extreme-precipitation events. *Mon. Wea. Rev.* **2003**, *18*, 1037–1051. [CrossRef]
33. Hobouchian, M.P.; Salio, P.; Skabar, Y.G.; Daniel, V.; Rene, G. Assessment of satellite precipitation estimates over the slopes of the subtropical Andes. *Atmos. Res.* **2017**, *190*, 43–54. [CrossRef]
34. Yucel, I.; Kuligowski, R.J.; Gochis, D.J. Evaluating the hydro-estimator satellite rainfall algorithm over a mountainous region. *Int. J. Remote Sens.* **2011**, *32*, 7315–7342. [CrossRef]
35. Liu, J. Analysis on cloud microphysical property over Qinghai-Xizang Plateau using satellite data. *Plateau Meteor* **2013**, *32*, 38–45. (In Chinese)
36. Cadeddu, M.; Liljegren, J.; Turner, D. The atmospheric radiation measurement (ARM) program network of microwave radiometers: Instrumentation, data, and retrievals. *Atmos. Meas. Tech.* **2013**, *6*, 2359–2372. [CrossRef]
37. Zhao, C.; Xie, S.; Klein, S. Toward understanding of differences in current cloud retrievals of ARM ground-based measurements. *J. Geophys. Res.* **2012**, *117*, D10206. [CrossRef]
38. Guo, J.P.; Zhang, J.; Yang, K.; Liao, H.; Zhang, S.D.; Huang, K.M.; Lv, Y.M.; Shao, J.; Yu, T.; Tong, B.; et al. Investigation of nearglobal daytime boundary layer height using high-resolution radiosondes: First results and comparison with ERA5, MERRA-2, JRA-55, and NCEP-2 reanalyses. *Atmos. Chem. Phys.* **2021**, *21*, 17079–17097. [CrossRef]
39. Cao, B.J.; Lyu, S.H.; Zhang, Y.; Yang, X.; Li, B.; Li, M. Factors influencing diurnal variations of cloud and precipitation in the Yushu area of the Tibetan Plateau. *J. Meteor. Res.* **2022**, *36*, 311–325. [CrossRef]

40. Chen, M.; Chen, M.X.; Fan, S.Y. The real-time radar radial velocity 3DVar assimilation experiments for application to an operational forecast model in North China. *Acta Meteor. Sinica* **2014**, *72*, 658–677. (In Chinese) [CrossRef]
41. Wang, Y.; Yang, Y.; Wang, C.H. Improving forecasting of strong convection by assimilating cloud-to-ground lightning data using the physical initialization method. *Atmos. Res.* **2014**, *150*, 31–41. [CrossRef]
42. Zhang, Y.; Xue, M.; Zhu, K.; Zhou, B. What is the main cause of diurnal variation and nocturnal peak of summer precipitation in Sichuan Basin, China? The key role of boundary layer low- level jet inertial oscillations. *J. Geophys. Res. Atmos.* **2019**, *124*, 2643–2664. [CrossRef]
43. Craven, J.P.; Jewell, R.E.; Brooks, H.E. Comparison between observed convective cloud-base heights and lifting condensation level for two different lifted parcels. *Wea. Forecast* **2002**, *17*, 885–890. [CrossRef]
44. Wang, Y.; Zeng, X.; Xu, X.; Welty, J.; Lenschow, D.H.; Zhou, M.; Zhao, Y. Why are there more summer afternoon low clouds over the Tibetan Plateau compared to eastern China? *Geophys. Res. Lett.* **2020**, *47*, e2020GL089665. [CrossRef]
45. Wu, G.X.; Duan, A.; Liu, Y.; Mao, J.; Ren, R.; Bao, Q.; He, B.; Liu, B.; Hu, W. Tibetan Plateau climate dynamics: Recent research progress and outlook. *Natl. Sci. Rev.* **2015**, *2*, 110–116. [CrossRef]

remote sensing

MDPI

Article

Analysis of Surface Energy Changes over Different Underlying Surfaces Based on MODIS Land-Use Data and Green Vegetation Fraction over the Tibetan Plateau

Jie Ma [1], Xiaohang Wen [1,*], Maoshan Li [1], Siqiong Luo [2], Xian Zhu [3], Xianyu Yang [1] and Mei Chen [1]

1 Plateau Atmosphere and Environment Key Laboratory of Sichuan Province, College of Atmospheric Sciences, Chengdu University of Information Technology, Chengdu 610021, China; 3200101058@stu.cuit.edu.cn (J.M.); lims@cuit.edu.cn (M.L.); xyang@cuit.edu.cn (X.Y.); 2018012049@stu.cuit.edu.cn (M.C.)
2 Key Laboratory of Land Surface Process and Climate Change in Cold and Arid Regions, Northwest Institute of Eco-Environment and Resources, Chinese Academy of Sciences, Lanzhou 730000, China; lsq@lzb.ac.cn
3 School of Atmospheric Sciences, Sun Yat-Sen University, Zhuhai 519000, China; zhux53@mail.sysu.edu.cn
* Correspondence: wxh@cuit.edu.cn

Abstract: To better predict and understand land–atmospheric interactions in the Tibetan Plateau (TP), we used Moderate Resolution Imaging Spectroradiometer (MODIS)-based land-use data and the MODIS-derived green vegetation fraction (GVF) to analyze the variation trend over the TP. The in situ observations from six flux stations ("BJ" (the BJ site of Nagqu Station of Plateau Climate and Environment), "MAWORS" (the Muztagh Ata Westerly Observation and Research Station), "NADORS" (the Ngari Desert Observation and Research Station), "NAMORS" (the Nam Co Monitoring and Research Station for Multisphere Interactions), "QOMS" (the Qomolangma Atmospheric and Environmental Observation and Research Station), and "SETORS" (the Southeast Tibet Observation and Research Station for the Alpine Environment)) at the Chinese TP Scientific Data Center were used to study the surface energy variation characteristics and energy distribution over different underlying surfaces. Finally, we used observation data to verify the applicability of the ERA-5 land reanalysis data to the TP. The results showed that the annual GVF steadily declined from the southeast parts to the northwest parts of the TP, and the vegetation coverage rate was highest from June to September. The sensible heat flux (H), latent heat flux (LE), net surface radiation (*Rn*), and four-component radiation (solar downward shortwave radiation (*Rsd*), surface upward shortwave radiation (*Rsu*), atmospheric downward longwave radiation (*Rld*), and surface upward longwave radiation (*Rlu*)) reached their maxima in summer at each station. *Rld* did not change significantly with time; all other variables increased during the day and decreased at night. The interannual variation in H and LE shows that latent heat exchange was the dominant form of energy transfer in BJ, MAWORS, NAMORS, and SETORS. By contrast, sensible heat exchange was the main form of energy transfer in NADORS and QOMS. The Bowen ratio was generally low in summer, and some sites had a maximum in spring. The surface albedo exhibited a "U" shape, decreasing in spring and summer, and increasing in autumn and winter, and reaching the lowest value at noon. Except for SETORS, ERA-5 Land data and other flux stations had high simulation accuracy and correlation. Regional surface energy changes were mainly observed in the eastern and western parts of the TP, except for the maximum of H in spring; the maximum values of other heat fluxes were concentrated in summer.

Keywords: Tibetan Plateau; ERA-5 reanalysis data; surface energy; land–atmospheric interaction; different underlying surfaces

Citation: Ma, J.; Wen, X.; Li, M.; Luo, S.; Zhu, X.; Yang, X.; Chen, M. Analysis of Surface Energy Changes over Different Underlying Surfaces Based on MODIS Land-Use Data and Green Vegetation Fraction over the Tibetan Plateau. *Remote Sens.* **2022**, *14*, 2751. https://doi.org/10.3390/rs14122751

Academic Editor: Antonino Maltese

Received: 20 April 2022
Accepted: 6 June 2022
Published: 8 June 2022

Publisher's Note: MDPI stays neutral with regard to jurisdictional claims in published maps and institutional affiliations.

1. Introduction

The Tibetan Plateau (TP) is the highest plateau in the world and the largest plateau in western China, known as the "roof of the world" and "the third pole of the earth," with an elevation between 3000 to 5000 m [1]. The plateau's high and towering terrain and complex underlying surface features significantly impact the plateau monsoon, water vapor cycle, and atmospheric vertical circulation, affecting climate change in East Asia [2–6]. In the 1970s, the first Scientific Expedition and Research on the Tibetan Plateau began. The main goal was to elucidate the history of geological development and the causes of plateau uplifting, to study the effects of uplifting on the ecological environment and human activities on local climate, and to look into the characteristics of natural conditions and resources, as well as the directions and routes for their exploitation and modification [7,8]. The Second Tibetan Plateau Scientific Expedition and Research will be based on the First Tibetan Plateau Scientific Expedition and Research, highlighting the change as the theme of investigation and research, to determine the law of change, evaluate and predict the future trend of change, and carry out ten scientific expeditions and research missions [7]. The study of the westerly monsoon synergy's evolution law, variation characteristics, and driving mechanism, as well as greater knowledge of the land–atmospheric interactions, precipitation efficiency, and the impact on the Sichuan Basin and its climate effect are all essential for revealing environmental changes on the TP [9–12]. Land–atmospheric interactions and local climate effects are the primary focus of this study. Specifically, the transfer and exchange of heat, momentum, water vapor, and carbon dioxide fluxes between the land surface and atmosphere are essential components of atmospheric interactions [13,14]. Energy and material transport are essential forcing fields for the development of convection in the atmospheric boundary layer. The thermodynamic and dynamical effects of the TP on the atmosphere are mainly influenced by the free air flow through the near-layer and boundary layer of the TP [15–18]. Land surface parameters such as green vegetation, soil texture, and soil moisture are essential factors that affect changes in surface energy flux over the TP. Moreover, owing to the wide area, complex vegetation types, and high altitude, the underlying surface characteristics significantly affect the water–energy cycle between the land surface and atmosphere. The scarce distribution of meteorological observation stations on the TP could pose a challenge to understanding the effects of the above-mentioned factors [19–21].

Ma et al. first analyzed the radiation characteristics of the period before and after the monsoon in the Nagqu area using radiation observations from the 1998 Intensification Observation Period (IOP). Observations were then compared with parameterized remote sensing results [22–24]. Li et al. found that sensible heat flux (H) is the primary energy source providing heat from the land surface to the atmosphere before the monsoon's outbreak, whereas latent heat flux (LE) is the main source of atmospheric warming during the monsoon season [25]. Studies have found that climate change in the TP exhibits a consistent warming trend at different timescales, and grasslands in semi-arid areas are highly sensitive to temperature and precipitation changes [26]. Studies have found that the H on the interannual variability of the TP shows a trend of weakening and falling at a rate of 2% per decade, with climate change and reduced wind speed over the TP identified as the causes of this phenomenon. However, the plateau's warming rate is higher than at the same latitude in eastern China, which remains unexplained [27–30]. Except for the Yarlung Zangbo River Basin, the LE was found to increase on the TP. This may be due to the increase in the net surface radiation (Rn) from the wetter forest cover underlying surface and the high soil moisture content caused by agricultural irrigation [31,32].

Studies have shown that diurnal variations in surface upward shortwave radiation (Rsu) and soil heat flux in alpine meadows are larger than those in banana plantations [33]. Net longwave radiation can affect soil-water freezing and its duration [34], the near-surface soil freeze–thaw process, heat storage, and melting of snow. Vegetation growth and non-growth periods affect surface energy non-closure [35,36]. The surface energy flux of Qomolangma has clear diurnal and seasonal variation trends that are greatly affected by the southwest monsoon. The response of the surface albedo to changes in rainfall has a lag effect. In winter, the vegetation cover in most areas of the TP is reduced, snow is present on the surface, and the surface albedo is often at the annual highest value [37,38]. Based on the analysis of the surface radiation observation data from the BJ site of Nagqu Station of Plateau Climate and Environment, Muztagh Ata Westerly Observation and Research Station (MAWORS), Ngari Desert Observation and Research Station (NADORS), Nam Co Monitoring and Research Station for Multisphere Interactions (NAMORS), Qomolangma Atmospheric and Environmental Observation and Research Station (QOMS), and Southeast Tibet Observation and Research Station for the Alpine Environment (SETORS), it was found that the *Rsu* and surface albedo of all stations decreased on the whole. The atmospheric downward longwave radiation (*Rld*), surface upward longwave radiation (*Rlu*), net surface radiation (*Rn*), ground surface temperature, and air temperature at most observation stations showed an upward trend at the interannual scale. The amplitude of *Rlu* was more significant than that of the downward long-wave radiation. *Rn* often reaches a maximum in late spring and early summer in the Ngari area [24,39]. The variation in characteristics of the surface energy flux with time at each station has been analyzed in detail; however, the surface energy distribution has yet to be discussed further.

The above studies considered the surface energy variation characteristics of the TP in numerous ways. However, the majority of these studies focused on the TP's eastern part, with only a few addressing the western part. In addition, the majority of these studies used short-term or limited-period data, with only a few studies studying the long-term changes in land surface energy and heat fluxes. In this study, we used Normalized Difference Vegetation Index (NDVI) data from MODIS, ERA-5 Land reanalysis data, and long-term flux observation station data from six sites (BJ, MAWORS, NADORS, NAMORS, QOMS, and SETORS) in the TP of the Second Tibetan Plateau Scientific Expedition and Research to examine the long-time series variation characteristics and energy distribution differences of the surface energy fluxes on different underlying surfaces over the TP.

2. Data and Methods

2.1. Data

2.1.1. Observation Data

In this study, the observed data regarding hourly integrated land–atmospheric interactions on the TP from 2005 to 2016 were obtained from the Chinese Science Data Center of the TP, which integrates the following six stations: MAWORS, NADORS, BJ, NAMORS, QOMS, and SETORS. Specifically, the following were obtained: hourly meteorological, solar radiation, eddy covariance (EC), and soil moisture, and heat data from the six field sites from 2005 to 2016, including multi-layer gradient observation data composed of wind direction, wind speed, air temperature, relative humidity, precipitation, air pressure, multi-layer soil temperature and moisture data, soil heat flux data, four-component radiation, EC turbulent flux data composed of LE and H, and carbon dioxide flux data [40]. In this study, the data used for analysis were the LE, H, and solar radiation components. The H and LE data were collected using an EC system for observation. The EC systems comprise a sonic anemometer (Campbell, CSAT3) and a fast-response gas analyzer (Li-COR, Li-7500) and were installed at 3.02 m, 2.3 m, 2.75 m, 3.06 m, 3.25 m, and 3.04 m above the ground of BJ, MAWORS, NADORS, NAMORS, QOMS, and SETORS, respectively. The CNR1 and NR01 (Kipp&Zonen) four-component radiation observation systems were used to collect radiation measurements at QOMS, SETORS, NADORS, and MAWORS. The NR01 (Vaisala) four-component radiation observation system was used to measure the Namco station, and the error range was within $\pm 10\%$. A solar radiation measurement system (CM21, Kipp&Zonen, and PIR, Eppley) was used to measure the surface radiation component; this system can measure shortwave radiation from the surface and longwave radiation from the atmosphere, with error ranges of $\pm 2\%$ and ± 5 W/m^2, respectively. Local time was used in this study (UTC+8) [40]. The data from 2005 were not included in this study because of discontinuity caused by many missing readings from this year.

2.1.2. ERA-5 Reanalysis Data

The ERA-5 reanalysis data are the fifth generation of global climate atmospheric reanalysis data from the European Center for Medium-Range Weather Forecasts (ECMWF) [41].

ERA-5 combines vast amounts of historical observations into global estimates using advanced modeling and data assimilation systems. The data cover the Earth on a 30 km grid and resolve the atmosphere using 137 levels from the surface up to a height of 80 km. The ERA-5 Land dataset used in this study was a replay of the land component forced by meteorological fields and offers great improvements in precision for land applications [42]. ERA-5 Land dataset coverage was from 1950 to the present time, and was regridded to a spatial resolution of $0.1^{\circ} \times 0.1^{\circ}$. The monthly averaged LE, H, downward and upward radiation, surface albedo, and net radiation from 2006 to 2016 were used in this study.

2.1.3. Land-Use Type Data and GVF Data

Land-use type data obtained by MODIS (Terra and Aqua) were also used in this study to assess the underlying surfaces of the TP. The complete MODIS land-use database contains five different land-use datasets: the IGBP dataset [43], the University of Maryland Data Set (UMD) of 14 classes [44], 10 types of MODIS LAI/FPAR algorithm dataset [45], 8 biological datasets [46] and 12 types of plant functional classifications [47,48]. The MODIS data used in this study were obtained from a 21-category IGBP database with a resolution of 5 km.

The GVF was calculated by using MOD13A3 Level 3 monthly 1 km Vegetation Indices data (https://appeears.earthdatacloud.nasa.gov/task/area, accessed on 30 September 2021) and also upscaling to 5 km resolution. The GVF is obtained using the relationship by Gutman and Ignatov (1998) [49]:

$$\text{NDVI} = (\text{NDVI} - \text{NDVI}_{\text{min}})/(\text{NDVI}_{\text{max}} - \text{NDVI}_{\text{min}}) \tag{1}$$

where NDVI_{min} and NDVI_{max} are bare soil without vegetation (LAI→0) and dense vegetation (LAI→∞), which contain the minimum and maximum NDVI values over the TP, respectively.

2.2. *Analysis Method*

The Formula calculation of *Rn* was as follows:

$$Rn = (Rsd + Rld) - (Rsu + Rlu) \tag{2}$$

In formula (2), the *Rsd* is the downward solar radiation, *Rsu* is the upward radiation, *Rld* is the atmospheric downward longwave radiation, and *Rlu* is the upward longwave radiation (W/m^2).

In the error analysis of the ERA-5 data and observation data, as the temporal and spatial resolution of the radiation flux observation data of each station was different from that of the ERA-5, a monthly average processing method was adopted to unify the temporal resolution of all data. Bilinear interpolation was used to interpolate the ERA-5 to the positions of the observation stations. ERA-5 data for the six field sites were obtained using this method. As the ERA-5 monthly mean data are in J/m^2 and the cumulative period is 24 h, dividing by the cumulative period expressed in seconds converts the units to W/m^2 following the observed data. The formulas for calculating the shortwave and longwave radiation from the land surface upward in the ERA-5 data are as follows:

$$Rsu = Rsd \times fal \tag{3}$$

$$Rlu = Rld - str \tag{4}$$

where *fal* is surface albedo and *str* is surface net thermal radiation.

Three error measures were selected to validate the ERA-5 data: the correlation coefficient (R), bias, and root mean square error (*RMSE*).

R is a statistical indicator that reflects the closeness of the correlation between the variables. The value of R is between −1 and 1. If the coefficient is positive, then the two variables are positively correlated. If the coefficient is negative, then the correlation is negative. The greater the absolute value, the stronger the correlation.

$$R = \frac{\sum_{i=1}^{n}(x_i - \overline{x})(y_i - \overline{y})}{\sqrt{\sum_{i=1}^{n}(x_i - \overline{x})^2 \sum_{i=1}^{n}(y_i - \overline{y})^2}} \tag{5}$$

The bias describes the difference between the simulated and actual values. In this study, a positive deviation means that the reanalysis data overestimate the observed value, and a negative deviation means that the reanalysis data underestimate the observed value. The calculation formula is as follows:

$$Bias = \frac{\sum_{i=1}^{n}(y_i - x_i)}{n} \tag{6}$$

The *RMSE* is extremely sensitive to the maximum or minimum error response in a set of measurements, so it can better reflect the measurement accuracy. A smaller value indicates a higher accuracy. The calculation formula is as follows:

$$RMSE = \sqrt{\frac{\sum_{i=1}^{n}(y_i - x_i)^2}{n}} \tag{7}$$

where y_i is the predicted value of the reanalysis data, x_i is the observed value, and n is the number of measurements.

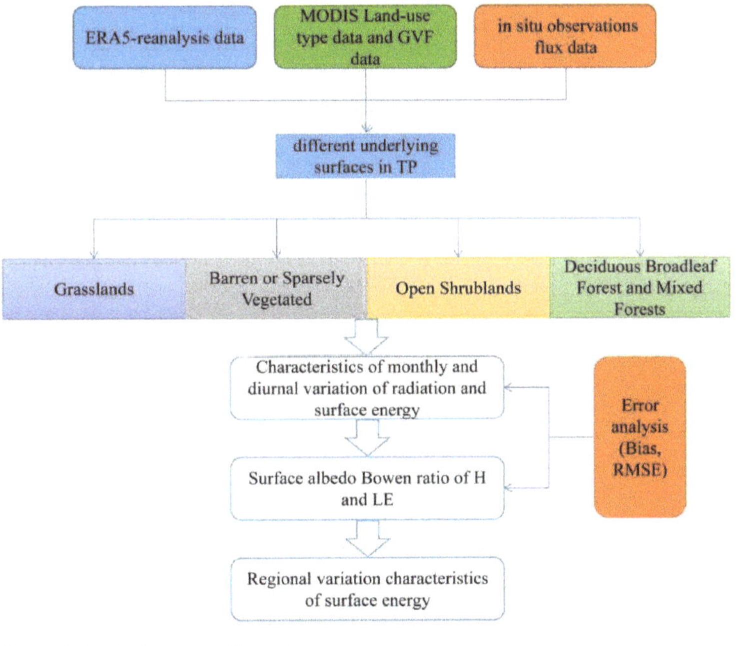

Figure 1. Analysis process flow chart.

The data analysis and processing in this study were conducted as follows: first, ERA-5 reanalysis data, MODIS land-use and NDVI data, and flux site observation data were collected, and the data over different underlying surfaces were pre-processed. The underlying surface of the TP was then divided into four main types: Grasslands; Barren or Sparsely Vegetated Lands; Open Shrublands; Deciduous Broadleaf Forest and Mixed Forests. Based on the feedback effect of energy and water on the atmosphere, we analyzed the monthly variation characteristics of radiation, surface energy flux, the Bowen ratio (β), and surface albedo parameters, and calculated the *RMSE* and bias error. Finally, the distribution characteristics of the ERA-5 data over the TP were obtained, and the applicability of this data was verified. The flow chart of the analysis process is shown in Figure 1, and a schematic diagram of the land-use types, site locations, and elevation on the TP is shown in Figure 2.

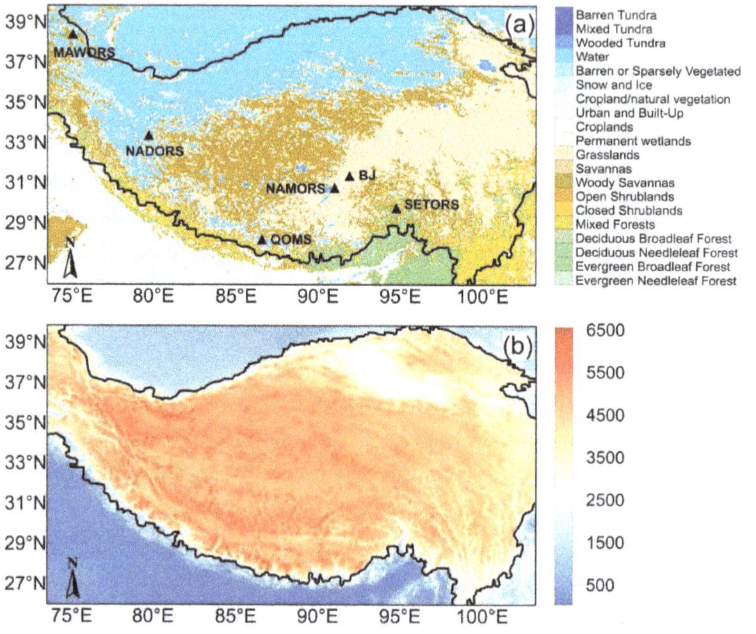

Figure 2. (**a**) Distribution of land-use/vegetation types and site locations (symbol) on the TP; (**b**) Elevation diagram on the TP.

3. Results and Analysis

3.1. Monthly Variation Characteristics of GVF

GVF is an important land surface parameter in land–atmospheric interaction processes and is defined as a part of photosynthetically active green canopy intercepting a midday downward solar grid cell [50]. In the Noah land surface model (LSM), the seasonal variation of GVF also defines the variation of other surface physical characteristics, such as LAI, albedo, roughness length, and surface emissivity [51]. The vegetation distribution is related to precipitation and temperature, and humid and warm areas are conducive to vegetation growth [52]. The annual distribution of GVF gradually decreased from southeast to northwest over the TP. Due to sufficient precipitation and higher temperature in the southeast than in the west TP, vegetation coverage is higher throughout the year. Vegetation is sparse in the northwestern region of the TP. GVF showed an obvious seasonal variation trend, rising in May and gradually decreasing in September (Figure 3e–i). From June to September, the vegetation coverage rate of the TP reached 40−60% (Figure 3f–i).

Of the six sites studied in this paper, MAWORS and NADORS are distributed in the northwest of the TP (Figure 2a), and because of their geographical location, the underlying surface of the two stations is predominantly barren or sparsely vegetated (Table 1), and the GVF is low (Figure 3). The QOMS is located in the south of the TP, and the underlying surface is dominated by barren or sparsely vegetated land (Figure 2a and Table 1). NAMORS and BJ are located in the middle of the TP and the underlying surface is mainly grassland [40]. (Figure 2a and Table 1). As can be seen from Figure 3, their GVF is high from

June to September but low in other months (Figure 3f–i). SETORS is located in the area with the highest annual GVF in the TP (Figures 2a and 3). The underlying surface types are broadleaf forests and mixed forests, which have little influence on seasonal changes.

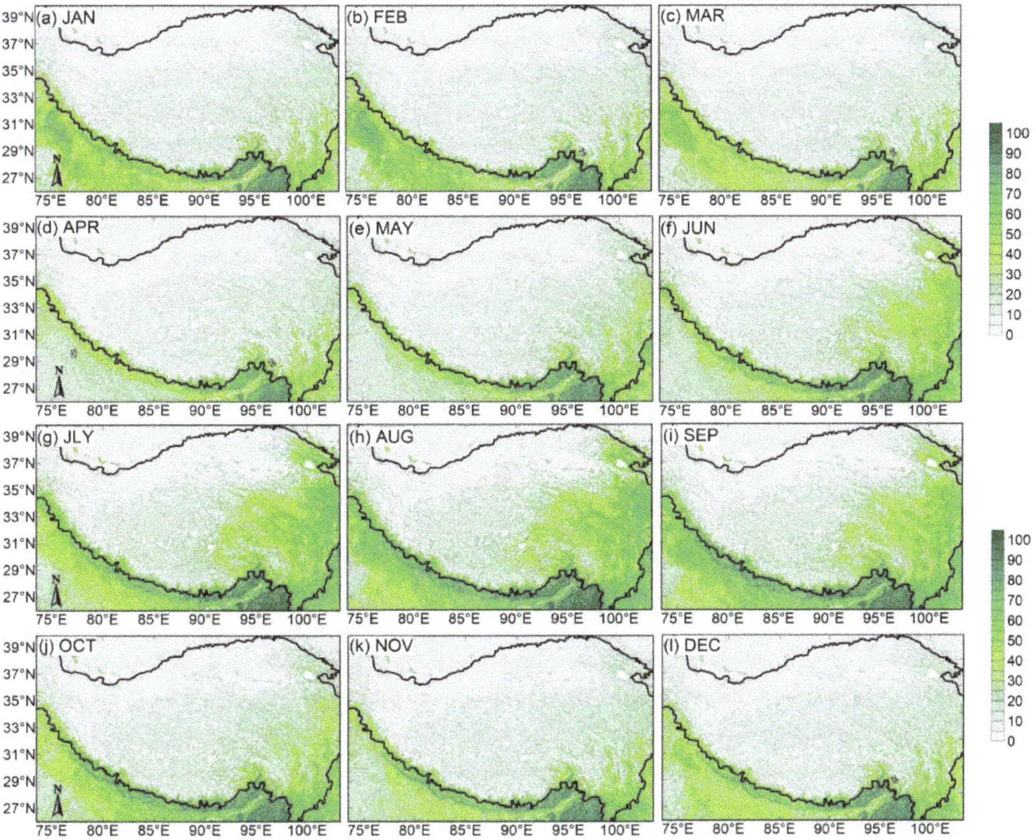

Figure 3. Spatial distribution characteristics of green vegetation fractions in different months on the Tibetan Plateau (Unit: %).

3.2. Variation Characteristics of Surface Energy

3.2.1. Seasonal Variation Characteristics of Surface Energy

The monthly energy variation characteristics of the six stations were different, but also had some similarities. In this study, the four seasons were divided as follows: spring from March to May, summer from June to August, autumn from September to November, and winter from December to the following January. The highest values of H were observed in the spring, decreased in summer, increased to varying degrees in the autumn, and decreased again in winter. After each station's H achieved its maximum value in spring, the time at which it started decreasing varied, with the SETORS station being the earliest. The LE showed a unimodal change. Before the outbreak of the southeast monsoon on the TP, the LE value was minimal. Precipitation rose in summer, soil moisture increased, latent heat exchange was intense, and LE increased rapidly, with the maximum value exceeding 100 W/m^2 (Figure 4a). In autumn, it gradually decreased and reached a minimum value during winter. The difference between H and LE at NADORS and QOMS stations in summer was smaller than for the other four stations (Figure 4c,e). The four-component radiation data showed a single-peak variation, and the solar shortwave downward radiation began to decrease in summer. The *Rsu* values at BJ, QOMS, and NAMORS in spring and winter increased with different amplitudes (Figure 4a,d,e). Longwave radiation increased in spring and summer and decreased in autumn and winter. *Rn* increased in spring, reached a maximum in summer, and decreased in autumn and winter. Possible errors were observed in the longwave radiation values at SETORS, meaning these data were not taken into consideration in analysis.

Table 1. Description of geographic features of six sites.

Site	Latitude, Longitude	Elevation (m)	Land Cover	Initial Observation Time of the Instrument (Radiations/EC)
BJ	31.37° N, 91.90° E	4509	Grasslands	2006
MAWORS	38.41° N, 75.04° E	3668	Barren or Sparsely Vegetated and Open Shrublands	2010
NADORS	33.39° N, 79.70° E	4270	Barren or Sparsely Vegetated	2009/2005
NAMORS	30.77° N, 90.99° E	4730	Grasslands	2005
QOMS	28.21° N, 86.56° E	4298	Barren or Sparsely Vegetated	2005/2007
SETORS	29.77° N, 94.73° E	3327	Deciduous Broadleaf Forest and Mixed Forests	2007

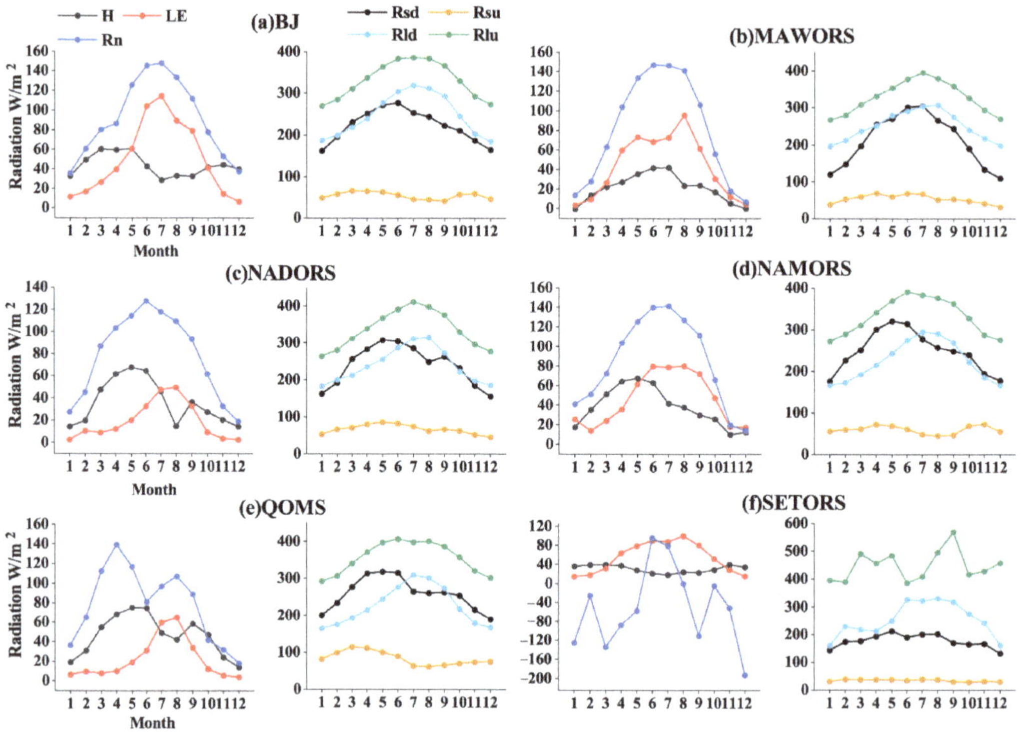

Figure 4. Annual average monthly variation of surface energy at six sites.

3.2.2. Diurnal Variation Characteristics of Surface Energy

To illustrate the diurnal variation of surface energy fluxes in different regions of the TP, Figures 5 and 6 show the diurnal variation of H, LE, *Rn*, and four-component radiation at the six stations in summer and winter. As shown in Figure 5, all three variables reached their peak values at approximately 14:00. Sunshine is highest when the solar altitude is high, and the surface obtains more energy. In summer, *Rn* increased gradually from 7:00 to 9:00, peaked at approximately 14:00, and decreased to its lowest value in a day at 23:00. The peak value of *Rn* in summer was more significant than that in winter, with a difference of approximately 250 W/m^2. The variation in H and LE was the same as that of *Rn*, and the maximum LE could be more than 200 W/m^2. The diurnal variations of H and LE increased at sunrise and decreased at sunset. The LE was generally greater than H in the summer. However, the opposite was true for NADORS and QOMS (Figure 5c,e) because the underlying surface of the two stations comprises barren or sparsely vegetated land, meaning the latent heat exchange is not intense, resulting in an LE that is lower than H. The maximum difference in H and LE between BJ and SETORS could reach more than 100 W/m^2 (Figure 5a,f) because the underlying surface of both stations is covered by dense vegetation, with high precipitation and high soil moisture. The diurnal variation trend of each variable at each station in winter was the same as that in summer, but the peak values at all three stations were lower than those in summer. H was higher than LE in winter because the plateau area was in the non-growing period, and vegetation was reduced.

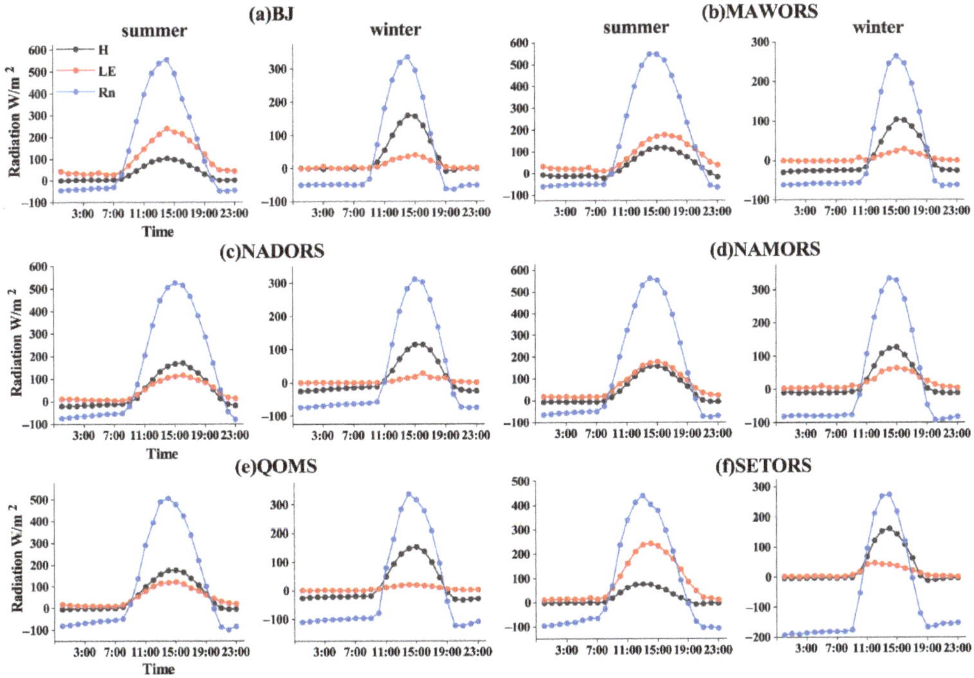

Figure 5. Diurnal variation of monthly mean H, LE, and *Rn* in summer and winter.

Figure 6 shows the diurnal variation of the four-component radiation at the six stations in summer and winter. It can be observed from Figure 6 that the variation trends of the four-component radiation in summer and winter were the same. While *Rld* did not vary significantly with time during summer, the other three variables all increased at sunrise and reached a peak at approximately 14:00, then gradually decreased and reached their lowest levels at 23:00. The surface heat was mainly obtained from *Rsd*, reaching a maximum of 900 W/m^2 or more at noon. There was a significant difference between *Rsd* and *Rsu* in the summer, with a maximum of approximately 700 W/m^2 (Figure 6e). However, the difference decreased in winter, with a maximum of approximately 600 W/m^2, due to the reduced solar radiation in winter. The difference between *Rlu* and *Rld* was smaller than that for shortwave radiation, about 100 W/m^2, while the difference was approximately 200 W/m^2 at SETORS (Figure 6f).

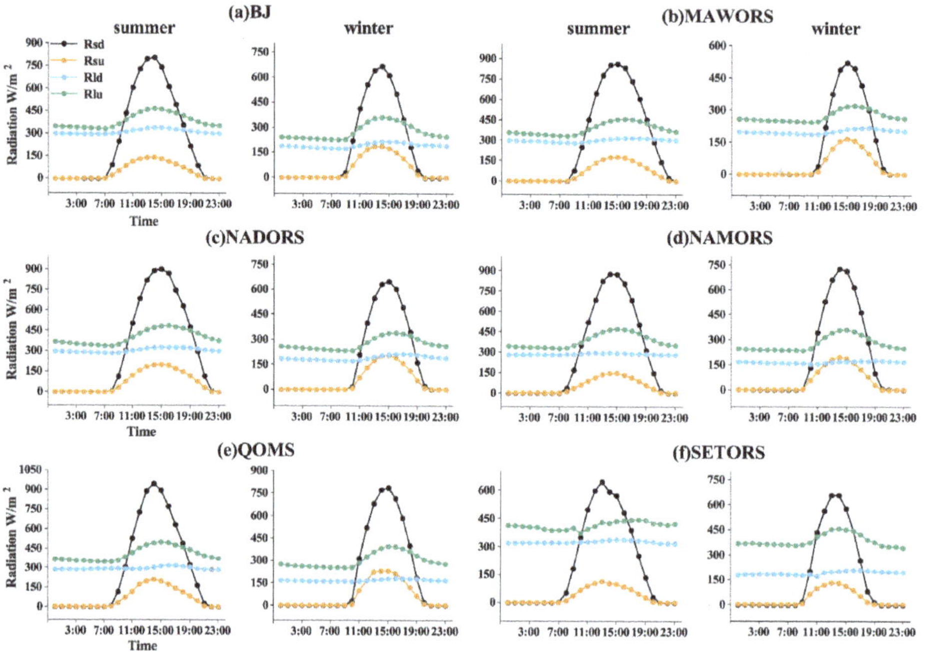

Figure 6. Diurnal variation of monthly mean four-component radiation in summer and winter at six sites.

3.3. *Surface Energy Budget and Distribution*

3.3.1. Land Surface Albedo

Land surface albedo is an essential factor affecting the surface energy budget and distribution, and is mainly determined by two factors: the underlying surface conditions and solar altitude. The albedo was calculated using observational data from 8:00 to 20:00 LT. As shown in Figure 7, the variation of the surface albedo presents as a "U" shaped curve, and was higher in the morning and evening and lower at noon. The solar altitude angle was higher at noon and the surface reflected a minor level of *Rsd*. When the solar altitude angle is low, longwave radiation makes up a major part of the solar radiation that reaches the earth surface. The land surface is highly reflective of longwave radiation. The lowest value varied between 0.2 and 0.4 at each station, and the albedo change at each station differed with the season. Whereas, at NADORS and SETORS, the seasonal variation was relatively insignificant (Figure 7c,f), the surface albedo at the other stations gradually decreased from January to May, reaching the lowest values in July or August, and then gradually increased. The main reason for this is that the albedo rises in winter due to heavy snow cover but falls in spring and summer when the snow melts and vegetation grows, resulting in a decreased albedo. For MAWORS and QOMS (Figure 7b), the underlying surface is barren or sparsely vegetated and the surface albedo should be high. However, there was still sparse vegetation growth in summer, which may be the reason for the decrease in the surface albedo of these two stations in summer. SETORS maintains a low surface albedo throughout the year because of the lush vegetation coverage and little influence of seasonal variation on the station (Figure 7f).

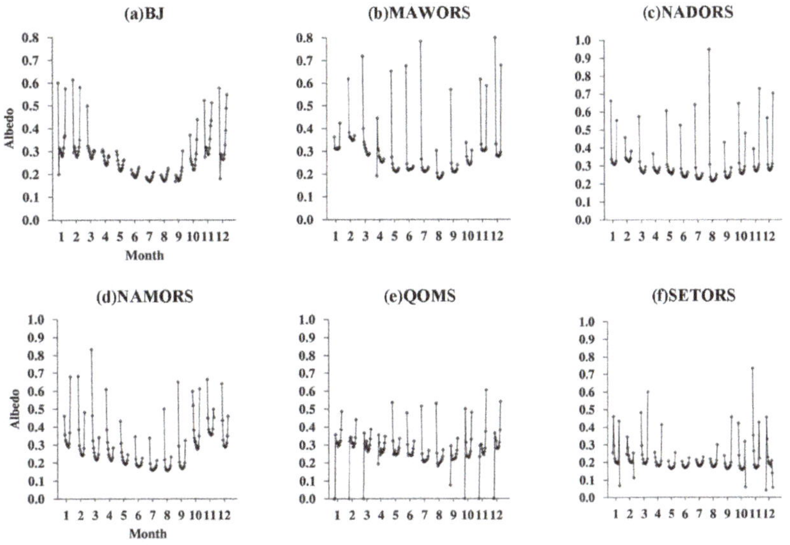

Figure 7. Diurnal variation of monthly mean surface albedo at six sites.

3.3.2. Surface Energy Distribution

To further illustrate the effects of the different routes of energy transfer on the different underlying surfaces, Figures 8 and 9 show the distribution of the surface energy. As can be seen from Figure 8, latent heat played a leading role in energy exchange in BJ, MAWORS, NAMORS, and SETORS (Figure 8a,b,d,f), while sensible heat was the dominant source of surface energy exchange in NADORS and QOMS (Figure 8c,e). Since the underlying surface of BJ and SETORS has dense vegetation, the transpiration of plants was more notable than the soil heat source effect, and latent heat was the main source of energy transfer [53]. For some years, the primary route of energy transfer at NAMORS was sensible heat, and plant transpiration was less than the soil heat source effect, resulting in a weaker latent heat exchange versus sensible heat exchange.

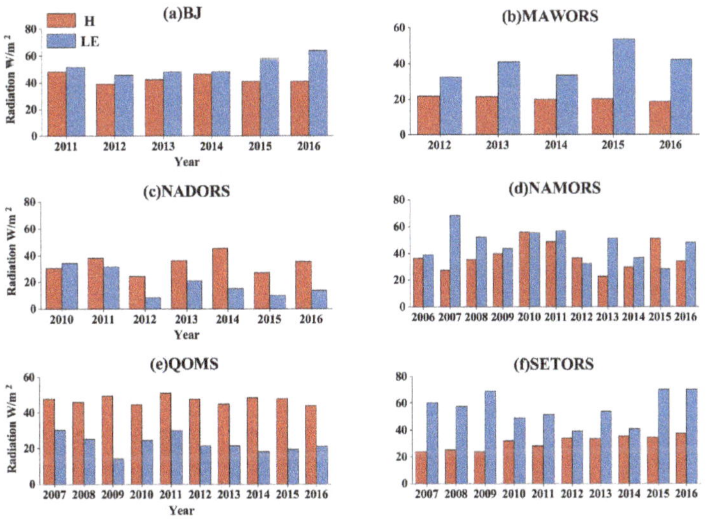

Figure 8. Comparison of annual mean values of H and LE at six sites.

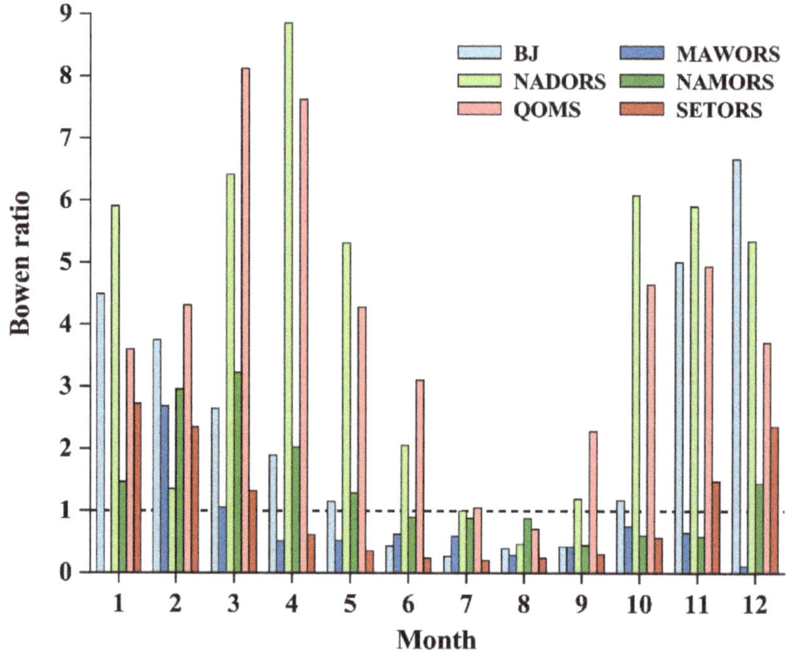

Figure 9. Monthly variation of Bowen ratio (H/LE) at six sites (dashed line indicates that H = LE).

The Bowen ratio (β) is defined as the ratio of H to LE at the surface. A higher value of β indicates greater sensible heat exchange; otherwise, the latent heat exchange is higher. As shown in Figure 9, sensible heat exchange was the main form of energy transfer in winter at all stations except MAWORS. The main reason for this is that plants are in the non-growing phase in winter, resulting in a decline in vegetation and soil hydrothermal conditions [54]. The seasonal variation trends in BJ and SETORS were identical, decreasing in spring and increasing in autumn. Latent heat was the main energy distribution process in BJ in summer, whereas sensible heat was the main process during the other seasons. The variation range at SETORS was smaller than that of BJ, and the energy distribution was mainly latent heat, except in winter. The heat exchange between the land surface and the atmosphere at MAWORS was dominated by latent heat exchange throughout the year, which is consistent with the comparison of the annual mean values of LE and H. β was less than 1 at NADORS during July and August only, and greater than 1 for all other months. The maximum value reached was 8.8, indicating that the energy transfer at NADORS occurred mainly via sensible heat exchange, which accounted for a large proportion. The β levels were relatively seasonally balanced at NAMORS. The monthly variation of β at QOMS was the same as that at NADORS because the underlying surface is mainly desert with sparse vegetation and weak latent heat exchange.

3.4. Error Analysis of ERA-5 Land Data and Observation Data

As the ERA-5 data are reanalysis data, they can be affected by many factors that may cause them to deviate from actual observation values. Error analysis of the ERA-5 data and the observation data from the six stations was carried out to determine whether the ERA-5 data had a high degree of accuracy and could be used to study the surface energy changes of the entire TP region.

Table 2 lists the R between the observed data from the six stations and the ERA-5 data. We can see that, of all the variables, *Rld* has the highest correlations, almost all above 0.9, followed by *Rsd*. H exhibits the lowest correlation, mostly under 0.5, followed by *Rsu*.

Of all the stations, the correlation between the ERA-5 data and the observed values is highest at MAWORS, and lowest at SETORS, and H is actually negatively correlated. The low correlation may be attributed to underlying surface conditions. Actual underlying surface conditions are more complex than those simulated by the ERA-5 data, leading to uncertainty in the predicted values.

Table 2. R between ERA-5 data and observed values (* indicates a failure to pass the significance test).

Variable \ Site	BJ	MAWORS	NADORS	NAMORS	QOMS	SETORS
Rsd	0.94	0.95	0.96	0.70	0.79	0.57
Rsu	0.39	0.52	0.44	0.31	0.60	0.26
Rld	0.94	0.98	0.98	0.97	0.98	0.41
Rlu	0.95	0.96	0.94	0.87	0.79	0.11 *
H	0.49	0.77	0.56	0.37	0.62	−0.50
LE	0.90	0.76	0.81	0.41	0.77	0.84
Rn	0.85	0.91	0.90	0.78	0.30	0.17 *

Surface heat flux is mainly limited by soil temperature and moisture (vegetation cover, atmospheric conditions, and soil physical characteristics). Soil temperature and moisture are greatly affected by precipitation, especially in arid regions [55–57]. Figure 10 shows the monthly variation in the bias between the ERA-5 reanalysis data and observations. It can be seen from the picture that the longwave radiation values of the ERA-5 data underestimated the observed values (except for SETORS) (Figure 10c,d). The bias variation ranges of H and *Rld* are relatively minimal, and the variation range is within 60 W/m^2 (Figure 10c,e). All stations show the same variation in the shortwave radiation bias during spring and summer, which becomes larger in spring and smaller in summer, indicating that the predicted value gradually approaches the observed value in summer, and then reaches a minimum in autumn and winter (Figure 10a,b). The bias of *Rld* did not change significantly with time (except at SETORS). The bias values are approximately 50 W/m^2 lower than the observed values at MAWORS and NADORS, while they are approximately 30 W/m^2 lower at BJ, QOMS, and NAMORS, among which NAMORS shows the smallest bias (Figure 10c). The bias of *Rlu* changes significantly in spring and summer, but little in autumn and winter (Figure 10d). The bias of H was large in the first five months, reached a maximum bias in March at most stations, and then gradually decreased (Figure 10e). Precipitation uncertainty leads to a bias between the surface and soil moisture, leading to a greater uncertainty regarding the LE levels in the ERA-5 data. The predicted LE is closest to the observed value in summer because there is more precipitation and high soil moisture on the TP at this time, and the uncertainty of the ERA-5 data is decreased (Figure 10f). *Rn* showed a high underestimation to different degrees from January to June. There was little change at other stations from July to December, except for QOMS, which had a relatively high estimate in November (Figure 10g). As mentioned above, there are some errors in the longwave radiation data from SETORS; these resulted in an abnormal bias fluctuation, which will not be discussed here.

As *RMSE* can better reflect the accuracy of data, this study also used *RMSE* as an index to evaluate the accuracy of the ERA-5 data. Figure 11 shows the monthly variation in the *RMSE* of the ERA-5 data and the observed data. Generally speaking, the accuracy is highest for BJ and lowest for MAWORS. As the MAWORS is located in a barren or sparsely vegetated area, the precipitation is low, and the soil temperature and soil moisture are greatly affected by precipitation, which adds more uncertainty to the ERA-5 data. The accuracy is higher for summer and lower for spring at most of the stations. The *RMSE* of *Rld* does not change significantly with time (Figure 11c), but the RMSE of LE and H do (Figure 11e,f). The *RMSE* changes in shortwave radiation, *Rlu*, and *Rn* are the same, and their accuracy is higher for summer (Figure 11a,b,d,g).

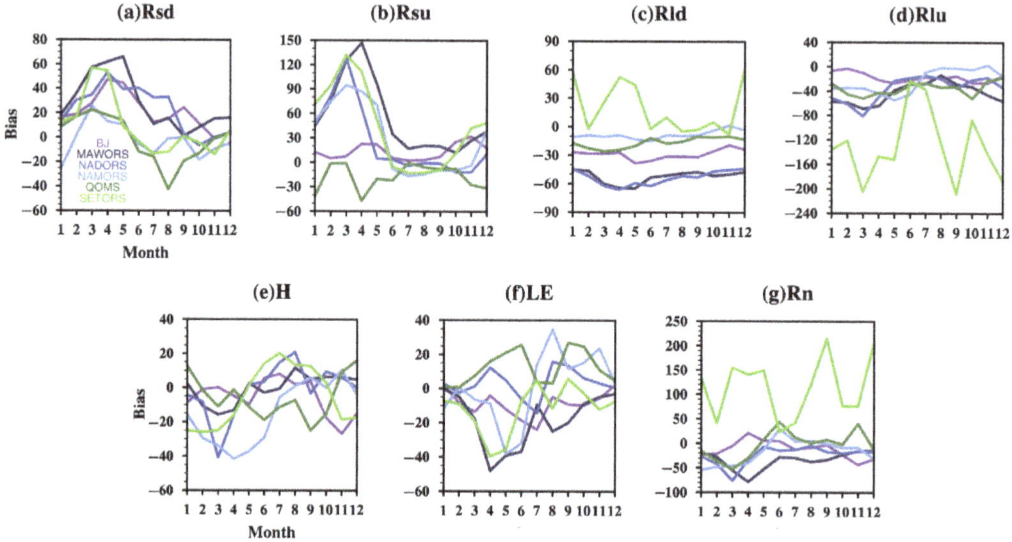

Figure 10. Monthly change of Bias between ERA-5 data and observed values (Unit: W/m^2).

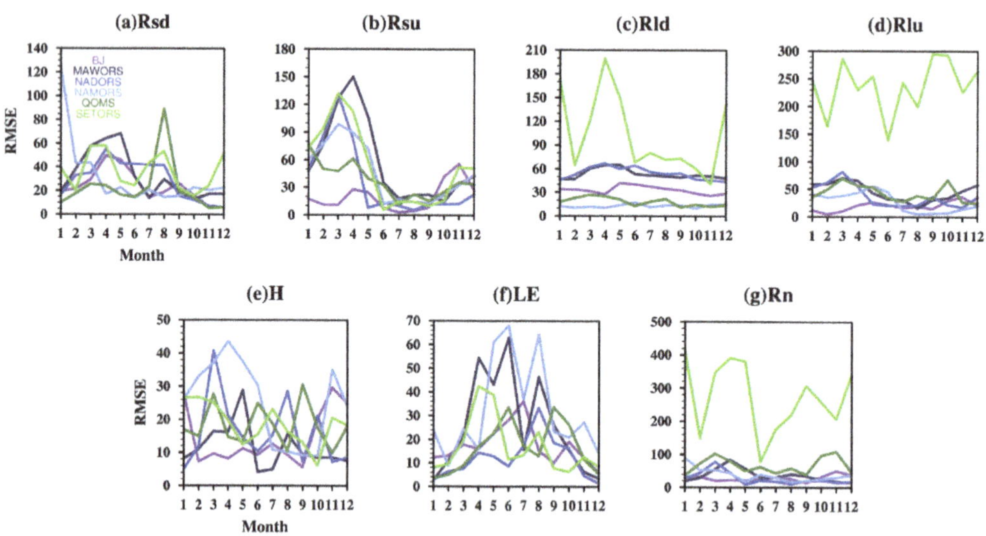

Figure 11. Monthly change of *RMSE* between ERA-5 data and observed values (Unit: W/m^2).

3.5. Energy Variation Characteristics of Surface Area

The changes in surface energy over time at the six stations were analyzed in the previous section. This section considers the changes in regional energy across the TP. Figure 12 shows the seasonal variation in the surface energy of the TP. It can be seen that the LE had an obvious seasonal variation, and the LE in the north and west of the plateau could reach more than 60 W/m^2 in spring and autumn. In most other locations, it was 20–40 W/m^2 (Figure 12a,c). However, in summer, the area with the highest values of LE was primarily located in the east of the plateau, where the maximum could exceed 80 W/m^2 (Figure 12b). This is due to the onset of the summer monsoon, resulting in higher precipitation and lush vegetation in the east of the plateau, and intense latent heat exchange between the surface and the atmosphere, which does not occur in winter. In spring and summer, the area with the highest values of H was mainly concentrated in the west of the plateau, with an average of 40–60 W/m^2 (Figure 12e,f), indicating that the turbulent movement in the west of the plateau was relatively strong at this time. *Rn* reached its peak in summer (Figure 12j), and *Rn* in the north of the plateau was always higher than that in the south, except in summer (Figure 12i,k,l). Generally, the LE was higher in the north and east of the plateau, whereas the maximum H was mainly in the west of the plateau. The energy value of H had a smaller variation range over time compared with that of LE. The *Rn* value in the north of the plateau was higher than that in the south (except in summer).

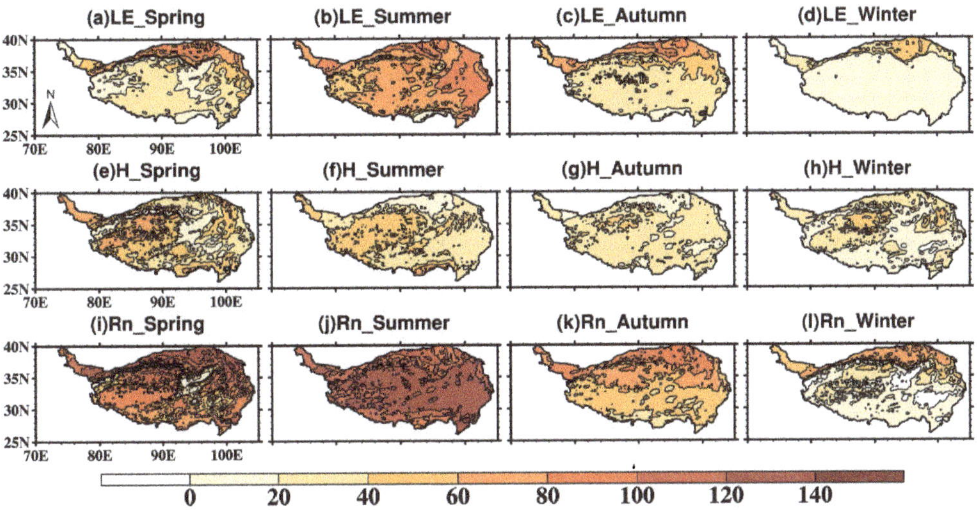

Figure 12. Annual average seasonal variation of ERA-5 data (LE, H, and *Rn*).

Figure 13 shows the seasonal variation in four-component radiation over the TP. As shown in the figure, *Rsd* reached its maximum value in spring, and the high-value area was mainly located in the western and central parts of the TP. At this time, the average *Rsd* on the TP was approximately 300 W/m^2 (Figure 13a). The high-value area was reduced during the summer, during which time the *Rsd* of the plateau was approximately 250 W/m^2 on average (Figure 13b). The radiation value gradually decreased in autumn and winter (Figure 13c,d). However, the decrease in the west of the plateau was smaller than that in the eastern part of the plateau. The *Rsu* value was relatively low throughout the year, between 50 and 100 W/m^2, while that in the north of the plateau was lower than 50 W/m^2 (Figure 13e–h). When combined with Figure 2a, it can be seen that the northern part of the underlying plateau surface is complex, and *Rsd* is relatively low, which may be the reason for the low value of *Rsu* in the north of the plateau. The value of longwave radiation was generally larger than that of shortwave radiation on the plateau. For *Rld*, the seasonal variation was obvious. It increased in spring and reached its maximum in summer, and the radiation value varied between 250 and 300 W/m^2 (Figure 13i,j), decreasing gradually in autumn and winter (Figure 13k,l). The annual value of *Rlu* was greater than that of the first three variables, and the seasonal variation was similar to that of *Rld*. It reached its peak in summer, with levels of more than 350 W/m^2 in most of the plateau and more than 450 W/m^2 in the west of the plateau (Figure 13n). The *Rlu* value in the south of the plateau was always lower than that in the north (Figure 13m–p).

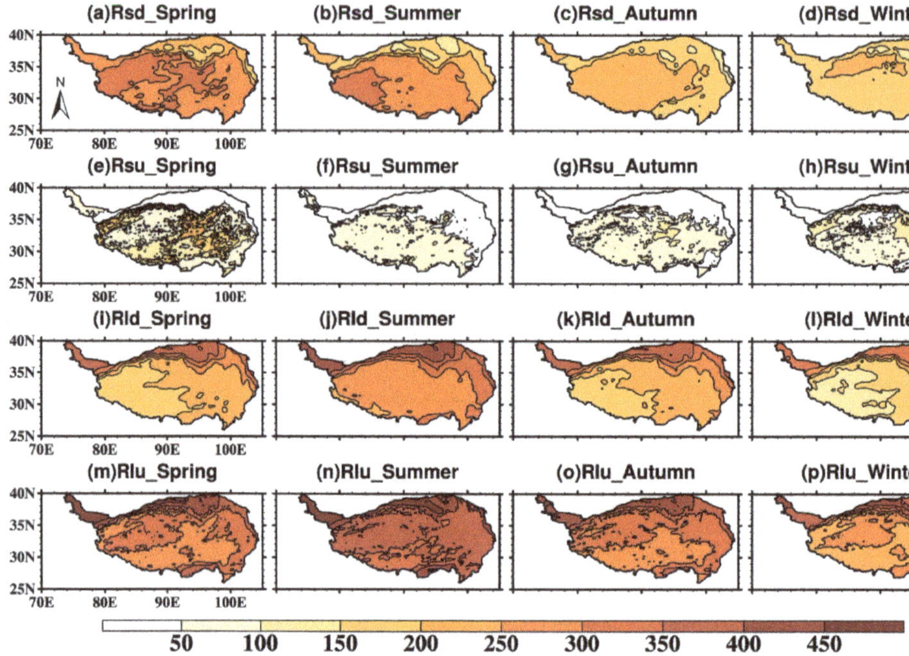

Figure 13. Annual average seasonal variation of ERA-5 four-component radiation (*Rsd*, *Rsu*, *Rld*, and *Rlu*).

4. Discussion

In this paper, MODIS land-use data and NDVI data were used to acquire the underlying surface vegetation types and analyze the distribution of the seasonal variation of GVF over the TP. The geographical location and underlying surface conditions had a great impact on the exchange of surface energy flux. In general, during the vegetation growth period on the TP, the three stations with a higher GVF (BJ, NAMORS, and SETORS), recorded a lower surface albedo, resulting in a decrease in *Rsu* and an increase in *Rn*. The radiation energy was absorbed by the large number of plants and by the soil. Moreover, evaporation from the land surface and vegetation increased, resulting in intensive latent heat exchange. The LE increased rapidly in summer, and played a leading role in surface energy transfer. However, in the low GVF areas (NADORS and QOMS), the surface albedo was always high, causing the surface energy exchange to be dominated by sensible heat. We also found that the relationship between energy distribution and the underlying surface in the MAWORS site area was different from the above mentioned. The MAWORS station is located in the west of the TP, and the underlying vegetation is sparse, but the value of LE was always higher than H throughout the year. The reasons for this phenomenon need further study. After comparative analysis with the observational data, we found that the ERA-5 data have good applicability in the TP. The discrepancy between the ERA-5 radiation data and underlying surface energy flux data was higher in spring and lower in summer over the TP. We preliminarily analyzed the surface radiation and energy variation characteristics of six flux sites in different regions of the TP, and considered the impact of underlying vegetation coverage and land-use types on the energy distribution. However, the contribution of different regions' energy transfer ratios needs to be further examined. The use of ERA-5 reanalysis data to analyze the differences in energy distribution in different regions of the TP also requires further research.

5. Conclusions

Based on the observational data from six stations (BJ, MAWORS, NADORS, NAMORS, QOMS, and SETORS) on the TP, the surface energy variation and energy distribution were studied. The ERA-5 Land data were used to study the regional energy changes of the TP after error analysis, and the following results were obtained:

(1) The annual distribution of GVF gradually decreased from southeast to northwest over the TP. Owing to the influence of precipitation and temperature, vegetation coverage in the southeastern TP is relatively high throughout the year. From June to September, the vegetation coverage rate of the TP reached 40−60%.

(2) Monthly variations in surface energy characteristics included the following. H increased in spring and autumn and decreased in summer and winter. After H reached its maximum value in spring, the decrease began at different times at each station, and was earliest at the SETORS station. The LE increased rapidly in summer, with a maximum value of more than 100 W/m^2, and gradually decreased in autumn and winter. In summer, the difference between H and LE at the NADORS and QOMS stations was lower than that at the other four stations. The four-component surface radiation increased during spring and summer, and decreased in autumn and winter.

(3) The diurnal variation in the surface energy obeyed the following trends. Except for *Rld*, which changed insignificantly over time, these variables began to increase at sunrise, reached their maximum values at noon, and decreased at sunset. LE was generally greater than H in summer, but the opposite was true for NADORS and QOMS. In winter, H was generally greater than LE. Longwave radiation differs from shortwave radiation in that it is more susceptible to solar radiation.

(4) The surface albedo changed in a "U" shape curve, and was high in the morning and evening, and low at noon. Except for NADORS and SETORS, where the surface albedo changed insignificantly with the seasons, all stations showed a gradual decrease in spring, reached their lowest values in summer, and gradually increased in autumn and winter. The interannual variation in H and LE shows that latent heat exchange is the main form of energy transfer in BJ, MAWORS, NAMORS, and SETORS. In contrast, sensible heat played a leading role in surface energy transfer at NADORS and QOMS. The Bowen ratio was generally low in summer, and some sites had a maximum in spring.

(5) The *Rld* value of ERA-5 at each station had the highest correlation with the observed value. The longwave radiation value of ERA-5 was lower than the observed value, and the bias of the shortwave radiation increased in spring and decreased in summer. Among the six stations, the highest precision was observed for BJ.

(6) The LE increased in spring and summer and decreased in autumn and winter, with the highest levels mainly concentrated in the north and east of the plateau (during summer). The high-value area of H was mainly in the west of the plateau. When *Rn* varied with the season, the radiation value in the north of the plateau was always higher than that in the south of the plateau (except in summer). The four components varied significantly with the seasons. *Rld* in the east of the plateau was higher than that in the west, and *Rsd* in the east of the plateau was lower than that in the west. The maximum *Rlu* values were in the northwest and northeast of the plateau.

Author Contributions: Conceptualization, X.W. and J.M.; methodology, X.W.; software, X.W.; validation, J.M., M.L. and S.L.; formal analysis, J.M.; investigation, J.M.; resources, X.W., X.Z., X.Y. and M.C.; data curation, X.W.; writing—original draft preparation, J.M.; writing—review and editing, X.W.; visualization, J.M.; supervision, X.W.; project administration, X.W.; funding acquisition, X.W. All authors have read and agreed to the published version of the manuscript.

Funding: This research was funded by the Second Tibetan Plateau Scientific Expedition and Research Program (STEP) (grant number 2019QZKK010304), the National Natural Science Foundation of China (grant number 41975096), and the Innovation Team Fund of Southwest Regional Meteorological Center, China Meteorological Administration (no grant number).

Data Availability Statement: A long-term (2005-2016) dataset of hourly integrated land–atmosphere interaction observations on the Tibetan Plateau (https://doi.org/10.5194/essd-12-2937-2020, accessed 20 July 2021). ERA5-Land data (10.24381/cds.68d2bb30, accessed on 15 September 2021). MOD13A3 Level 3 monthly 1 km vegetation indices data (https://appeears.earthdatacloud.nasa.gov/task/area, accessed on 30 September 2021).

Acknowledgments: The authors thank the Institute of Tibetan Plateau Research, CAS, for the long-term (2005–2016) radiation observation data, and the ECMWF for the ERA5-Land reanalysis data that provided support for this study.

Conflicts of Interest: The authors declare no conflict of interest.

References

1. Zhang, Y.; Li, B.; Zheng, D. A discussion on the boundary and area of the Tibetan Plateau in China. *Geogr. Res.* **2002**, *21*, 1–8.
2. Ma, Y.; Yao, T.; Hu, Z.; Wang, J. The Cooperative Study on Energy and Water Cycle over the Tibetan Plateau. *Adv. Earth Sci.* **2009**, *24*, 1280.
3. Zhao, T.J.; Zhang, L.X.; Shi, J.C.; Jiang, L.M. A physically based statistical methodology for surface soil moisture retrieval in the Tibet Plateau using microwave vegetation indices. *J. Geophys. Res. Earth Surf.* **2011**, *116*, D08116. [CrossRef]
4. Wu, G.; Mao, J.; Duan, A.; Zhang, Q. Recent progress in the study on the impacts of Tibetan Plateau on asian summer climate. *Acta Meteorol. Sin.* **2004**, *5*, 528–540. [CrossRef]
5. Zhao, P.; Zhou, Z.; Liu, J. Variability of Tibetan Spring Snow and Its Associations with the Hemispheric Extratropical Circulation and East Asian Summer Monsoon Rainfall: An Observational Investigation. *J. Clim.* **2007**, *20*, 3942–3955. [CrossRef]
6. Duan, A.; Xiao, Z.; Wang, Z. Impacts of the Tibetan Plateau winter/spring snow depth and surface heat source on Asian summer monsoon: A review. *Chin. J. Atmos. Sci.* **2018**, *42*, 755–766.
7. Yao, T.; Chen, F.; Cui, P.; Ma, Y.; Xu, B.; Zhu, L.; Zhang, F.; Wang, W.; Ai, L.; Yang, X. From Tibetan Plateau to Third Pole and Pan-Third Pole. *Bull. Chin. Acad. Sci.* **2017**, *32*, 924–931. [CrossRef]
8. Sun, H. Review and prospect of scientific investigation and Research on Qinghai-Tibet Plateau. *Resour. Sci.* **2000**, *22*, 6–8.
9. The Second Tibetan Plateau Scientific Expedition and Research. Available online: http://www.step.ac.cn/info/14568 (accessed on 22 March 2022).
10. Zhao, P.; Xiao, H.; Liu, C.; Zhou, Y. Dependence of Warm Season Cloud-to-Ground Lightning Polarity on Environmental Conditions over Sichuan, Southwest China. *Adv. Meteorol.* **2021**, *2021*, 1500470. [CrossRef]
11. Zhao, P.; Xiao, H.; Liu, C.; Zhou, Y.; Xu, X.; Hao, K. Evaluating a simple proxy for climatic cloud-to-ground lightning in Sichuan Province with complex terrain, Southwest China. *Int. J. Climatol.* **2021**, 1–19. [CrossRef]
12. Zhao, P.; Xiao, H.; Liu, J.; Zhou, Y. Precipitation efficiency of cloud and its influencing factors over the Tibetan plateau. *Int. J. Climatol.* **2021**, *42*, 416–434. [CrossRef]
13. Ma, Y.; Hu, Z.; Wang, B.; Ma, W.; Chen, X.; Han, C.; Li, M.; Zhong, L.; Gu, L.; Sun, F.; et al. The Review of the Observation Experiments on Land-Atmosphere Interaction Progress on the Qinghai-Xizang (Tibetan) Plateau. *Plateau Meteorol.* **2021**, *40*, 1241–1262.
14. Zeng, J.; Zhang, Q. A comparative study of the characteristics of the clear-sky land surface processes over the different un-derlying surfaces in the northern part of China during July-September 2008. *Acta Meteorol. Sin.* **2012**, *70*, 821–836.
15. Ma, Y.; Yao, T.; Wang, J. Experimental Study of Energy and Water Cyclein Tibetan Plateau—The Progress Introduction on the Study of GAME/Tibet and CAMP/Tibet. *Plateau Meteorol.* **2006**, *25*, 344–351.
16. Duan, A.; Li, F.; Wang, M.; Wu, G. Persistent Weakening Trend in the Spring Sensible Heat Source over the Tibetan Plateau and Its Impact on the Asian Summer Monsoon. *J. Clim.* **2011**, *24*, 5671–5682. [CrossRef]
17. Raupach, M. Influences of local feedbacks on land–air exchanges of energy and carbon. *Glob. Change Biol.* **1998**, *4*, 477–494. [CrossRef]
18. Zhao, P.; Li, Y.; Guo, X.; Xu, X.; Liu, Y.; Tang, S.; Xiao, W.; Shi, C.; Ma, Y.; Yu, X.; et al. The Tibetan Plateau surface-atmosphere coupling system and its weather and climate effects: The Third Tibetan Plateau Atmosphere Scientific Experiment. *Acta Mete-Orol. Sin.* **2018**, *76*, 833–860. [CrossRef]
19. Xie, S.; Qu, J.; Xu, X.; Pang, Y. Interactions between freeze–thaw actions, wind erosion desertification, and permafrost in the Qinghai–Tibet Plateau. *Nat. Hazards* **2016**, *85*, 829–850. [CrossRef]
20. Zhang, M.; Wang, J.; Lai, Y. Hydro-thermal boundary conditions at different underlying surfaces in a permafrost region of the Qinghai-Tibet Plateau. *Sci. Total Environ.* **2019**, *670*, 1190–1203. [CrossRef]
21. Yao, J.; Gu, L.; Yang, C.; Chen, H.; Wang, J.; Ding, Y.; Li, R.; Zhao, L.; Xiao, Y.; Qiao, Y.; et al. Estimation of surface energy fluxes in the permafrost region of the Tibetan Plateau based on in situ measurements and the surface energy balance system model. *Int. J. Climatol.* **2020**, *40*, 5783–5800. [CrossRef]
22. Ma, W.; Ma, Y.; Hu, Z.; Li, M.; Wang, J.; Qian, Z. Analyses on surface radiation budget in Northern Tibetan Plateau. *Plateau Meteorol.* **2004**, *23*, 348–352.
23. Ma, W.; Ma, Y.; Hu, Z.; Li, M.; Sun, F.; Gu, L.; Wang, J.; Qian, Z. The contrast between the radiation budget plus seasonal variation and remote sensing over the northern Tibetan Plateau. *J. Arid. Land Resour. Environ.* **2005**, *19*, 109–115.

24. Wang, C.; Ma, Y.; Wang, B.; Ma, W.; Chen, X.; Han, C. Analysis of the Radiation Fluxes over Complex Surfaces on the Tibetan Plateau. *Water* **2021**, *13*, 3084. [CrossRef]
25. Li, M.; Babel, W.; Chen, X.; Zhang, L.; Sun, F.; Wang, B.; Ma, Y.; Hu, Z.; Foken, T. A 3-year dataset of sensible and latent heat fluxes from the Tibetan Plateau, derived using eddy covariance measurements. *Arch. Meteorol. Geophys. Bioclimatol. Ser. B* **2014**, *122*, 457–469. [CrossRef]
26. Zhong, L.; Ma, Y.; Xue, Y.; Piao, S. Climate Change Trends and Impacts on Vegetation Greening Over the Tibetan Plateau. *J. Geophys. Res. Atmos.* **2019**, *124*, 7540–7552. [CrossRef]
27. Wang, S.; Ma, Y. On the simulation of sensible heat flux over the Tibetan Plateau using different thermal roughness length parameterization schemes. *Arch. Meteorol. Geophys. Bioclimatol. Ser. B* **2018**, *137*, 1883–1893. [CrossRef]
28. Han, Y.; Ma, W.; Ma, Y.; Sun, C. Variations of Surface Heat Fluxes over the Tibetan Plateau before and after the Onset of the South Asian Summer Monsoon during 1979–2016. *J. Meteorol. Res.* **2019**, *33*, 491–500. [CrossRef]
29. Yang, K.; Guo, X.; Wu, B. Recent trends in surface sensible heat flux on the Tibetan Plateau. *Sci. China Earth Sci.* **2010**, *54*, 19–28. [CrossRef]
30. Yang, K.; Wu, H.; Qin, J.; Lin, C.; Tang, W.; Chen, Y. Recent climate changes over the Tibetan Plateau and their impacts on energy and water cycle: A review. *Glob. Planet. Chang.* **2014**, *112*, 79–91. [CrossRef]
31. Amatya, P.M.; Ma, Y.; Han, C.; Wang, B.; Devkota, L.P. Recent trends (2003–2013) of land surface heat fluxes on the southern side of the central Himalayas, Nepal. *J. Geophys. Res. Atmos.* **2015**, *120*, 11957–11970. [CrossRef]
32. Han, C.; Ma, Y.; Chen, X.; Su, Z. Trends of land surface heat fluxes on the Tibetan Plateau from 2001 to 2012. *Int. J. Climatol.* **2017**, *37*, 4757–4767. [CrossRef]
33. Ding, Z.; Ma, Y.; Wen, Z.; Ma, W.; Chen, S. A comparison between energy transfer and atmospheric turbulent exchanges over alpine meadow and banana plantation. *Arch. Meteorol. Geophys. Bioclimatol. Ser. B* **2016**, *129*, 59–76. [CrossRef]
34. Li, R.; Zhao, L.; Wu, T.; Wu, X.; Xiao, Y.; Du, Y.; Qin, Y. The impacts of net long-wave radiation on the surface soil thermal regimes over the Qinghai–Tibetan Plateau, China. *Environ. Earth Sci.* **2016**, *75*, 271. [CrossRef]
35. Guo, D.; Yang, M.; Wang, H. Characteristics of land surface heat and water exchange under different soil freeze/thaw condi-tions over the central Tibetan Plateau. *Hydrol. Process.* **2011**, *25*, 2531–2541. [CrossRef]
36. Wang, L.; Li, M.; Lü, Z.; Fu, W.; Shu, L.; Yin, S. Variation Characteristics of Surface Fluxes on Different Underlying Surfaces and Their Relationship with Precipitation in the Canyon Area of Southeast Tibet. *Plateau Meteorol.* **2022**, *41*, 177–189.
37. Zhang, M.; Wang, B.; Wang, D.; Ye, W.; Guo, Z.; Gao, Q.; Yue, G. The effects of rainfall on the surface radiation of permafrost regions in Qinghai-Tibet Plateau: A case study in Beiluhe area. *J. Glaciol. Geocryol.* **2021**, *43*, 1092–1101.
38. Gu, X.; Ma, Y.; Ma, W.; Sun, F. Climatic characteristics of surface radiation flux over the Qinghai-Tibetan Plateau. *Plateau Meteorol.* **2018**, *37*, 1458–1469.
39. Wang, G.; Han, L.; Tang, X. Surface radiation characteristics of the Ali area, Northern Tibetan Plateau. *Environ. Res. Commun.* **2021**, *3*, 041006.
40. Ma, Y.; Hu, Z.; Xie, Z.; Ma, W.; Wang, B.; Chen, X.; Li, M.; Zhong, L.; Sun, F.; Gu, L.; et al. A long-term (2005–2016) dataset of hourly integrated land–atmosphere interaction observations on the Tibetan Plateau. *Earth Syst. Sci. Data* **2020**, *12*, 2937–2957. [CrossRef]
41. Hersbach, H.; Bell, B.; Berrisford, P.; Horányi, A.; Sabater, J.M.; Nicolas, J.; Radu, R.; Schepers, D.; Simmons, A.; Soci, C. Global reanalysis: Goodbye ERA-Interim, hello ERA5. *ECMWF Newsl.* **2019**, *159*, 17–24.
42. Muñoz Sabater, J. ERA5-Land monthly averaged data from 1981 to present. *Copernic. Clim. Change Serv. (C3S) Clim. Data Store (CDS)* **2019**, *146*, 1999–2049.
43. Friedl, M.A.; Sulla-Menashe, D.; Tan, B.; Schneider, A.; Ramankutty, N.; Sibley, A.; Huang, X. MODIS Collection 5 global land cover: Algorithm refinements and characterization of new datasets. *Remote Sens. Environ.* **2010**, *114*, 168–182. [CrossRef]
44. Lotsch, A.; Tian, Y.; Friedl, M.A.; Myneni, R. Land cover mapping in support of LAI and FPAR retrievals from EOS-MODIS and MISR: Classification methods and sensitivities to errors. *Int. J. Remote Sens.* **2003**, *24*, 1997–2016. [CrossRef]
45. Myneni, R.B.; Hoffman, S.; Knyazikhin, Y.; Privette, J.L.; Glassy, J.; Tian, Y.; Wang, Y.; Song, X.; Zhang, Y.; Smith, G.R.; et al. Global products of vegetation leaf area and fraction absorbed PAR from year one of MODIS data. *Remote Sens. Environ.* **2002**, *83*, 214–231. [CrossRef]
46. Running, S.W.; Loveland, T.R.; Pierce, L.L.; Nemani, R.R.; Hunt, E.R., Jr. A remote sensing based vegetation classification logic for global land cover analysis. *Remote Sens. Environ.* **1995**, *51*, 39–48. [CrossRef]
47. Bonan, G.B. The land surface climatology of the NCAR Land Surface Model coupled to the NCAR Community Climate Model. *J. Clim.* **1998**, *11*, 1307–1326. [CrossRef]
48. Wen, X.; Liu, C.; Luo, S. Comparative Analysis of water-energy cycle processes Based on HRADC data over Different Un-derlying Surfaces in Qinghai-Tibet Plateau. *Front. Earth Sci.* **2020**, *8*, 434. [CrossRef]
49. Gutman, G.; Ignatov, A. The derivation of the green vegetation fraction from NOAA/AVHRR data for use in numerical weather prediction models. *Int. J. Remote Sens.* **1998**, *19*, 1533–1543. [CrossRef]
50. Vahmani, P.; Ban-Weiss, G.A. Impact of remotely sensed albedo and vegetation fraction on simulation of urban climate in WRF-urban canopy model: A case study of the urban heat island in Los Angeles. *J. Geophys. Res. Atmos.* **2016**, *121*, 1511–1531. [CrossRef]

51. Yan, D.; Liu, T.; Dong, W.; Liao, X.; Luo, S.; Wu, K.; Zhu, X.; Zheng, Z.; Wen, X. Integrating remote sensing data with WRF model for improved 2-m temperature and humidity simulations in China. *Dyn. Atmos. Oceans* **2020**, *89*, 101127. [CrossRef]
52. Xu, X.; Chen, H.; Levy, J.K. Spatiotemporal vegetation cover variations in the Qinghai-Tibet Plateau under global climate change. *Sci. Bull.* **2008**, *53*, 915–922. [CrossRef]
53. Guo, X.; Wang, K.; Li, L.; Zhang, H.; Ma, L.; Yao, Z.; Zhang, W.; Hu, Z.; Zheng, X. Surface Energy Exchanges and Evapotranspiration of an Alpine Meadow on the Zoige Plateau. *Chin. J. Agrometeorol.* **2021**, *42*, 642–656.
54. Wang, S.; Zhang, Y.; Lü, S.; Shang, L.; Zhang, S. Seasonal variation characteristics of radiation and energy budgets in alpine meadow ecosystem in Maqu grassland. *Plateau Meteorol.* **2012**, *31*, 605–614.
55. Coronato, F.; Bertiller, M. Precipitation and landscape related effects on soil moisture in semi-arid rangelands of Patagonia. *J. Arid Environ.* **1996**, *34*, 1–9. [CrossRef]
56. Ge, F.; Sielmann, F.; Zhu, X.; Fraedrich, K.; Zhi, X.; Peng, T.; Wang, L. The link between Tibetan Plateau monsoon and Indian summer precipitation: A linear diagnostic perspective. *Clim. Dyn.* **2017**, *49*, 4201–4215. [CrossRef]
57. Xie, Z.; Wang, B. Summer Atmospheric Heat Sources over the Western–Central Tibetan Plateau: An Integrated Analysis of Multiple Reanalysis and Satellite Datasets. *J. Clim.* **2019**, *32*, 1181–1202. [CrossRef]

remote sensing

MDPI

Article

Land Surface Snow Phenology Based on an Improved Downscaling Method in the Southern Gansu Plateau, China

Lei Wu, Changbin Li *, Xuhong Xie, Jianan Lv, Songbing Zou, Xuan Zhou and Na Shen

Key Laboratory of Western China's Environmental Systems (Ministry of Education), College of Earth and Environmental Sciences, Lanzhou University, No. 222 South Tianshui Road, Chengguan District, Lanzhou 730000, China; wul17@lzu.edu.cn (L.W.); xiexh20@lzu.edu.cn (X.X.); lvjn19@lzu.edu.cn (J.L.); zousb@lzu.edu.cn (S.Z.); zhouxuan2020@lzu.edu.cn (X.Z.); shenn21@lzu.edu.cn (N.S.)

* Correspondence: licb@lzu.edu.cn; Tel.: +86-931-8912843

Abstract: Snow is involved in and influences water–energy processes at multiple scales. Studies on land surface snow phenology are an important part of cryosphere science and are a hot spot in the hydrological community. In this study, we improved a statistical downscaling method by introducing a spatial probability distribution function to obtain regional snow depth data with higher spatial resolution. Based on this, the southern Gansu Plateau (SGP), an important water source region in the upper reaches of the Yellow River, was taken as a study area to quantify regional land surface snow phenology variation, together with a discussion of their responses to land surface terrain and local climate, during the period from 2003 to 2018. The results revealed that the improved downscaling method was satisfactory for snow depth data reprocessing according to comparisons with gauge-based data. The downscaled snow depth data were used to conduct spatial analysis and it was found that snow depth was on average larger and maintained longer in areas with higher altitudes, varying and decreasing with a shortened persistence time. Snow was also found more on steeper terrain, although it was indistinguishable among various aspects. The former is mostly located at high altitudes in the SGP, where lower temperatures and higher precipitation provide favorable conditions for snow accumulation. Climatically, factors such as precipitation, solar radiation, and air temperature had significantly singular effectiveness on land surface snow phenology. Precipitation was positively correlated with snow accumulation and maintenance, while solar radiation and air temperature functioned negatively. Comparatively, the quantity of snow was more sensitive to solar radiation, while its persistence was more sensitive to air temperature, especially extremely low temperatures. This study presents an example of data and methods to analyze regional land surface snow phenology dynamics, and the results may provide references for better understanding water formation, distribution, and evolution in alpine water source areas.

Keywords: land surface snow phenology; statistical downscaling; terrain; climate; southern Gansu Plateau

Citation: Wu, L.; Li, C.; Xie, X.; Lv, J.; Zou, S.; Zhou, X.; Shen, N. Land Surface Snow Phenology Based on an Improved Downscaling Method in the Southern Gansu Plateau, China. *Remote Sens.* **2022**, *14*, 2848. https://doi.org/10.3390/rs14122848

Academic Editors: Massimo Menenti, Yaoming Ma, Li Jia and Lei Zhong

Received: 17 May 2022
Accepted: 11 June 2022
Published: 14 June 2022

1. Introduction

Snow is one of the main forms of water in the cryosphere and is involved in most land surface energy and moisture transport [1–3]. It influences local and regional land–atmospheric processes and circulation [4,5] and is considered an important indicator of environmental changes at multiple scales [6,7]. Variations in snow and its phenology directly affect the formation of mountain discharge and the evolution of water resources in river source areas [8,9], influencing water utilization and supporting local society–economy–ecological sustainability in the middle and lower reaches [10,11]. Quantitative analysis can help better understand the responses of land surface hydrological systems to environmental changes [12–14].

Snow is very sensitive to climate change. According to the 6th Assessment Report of the Intergovernmental Panel on Climate Change (IPCC), the current global air temperature is approximately 1 °C higher than before industrialization. In terms of the predicted average temperature change in the next 20 years, the global temperature rise is expected to reach or exceed 1.5 °C [15,16]. Future warming may lead to abnormal precipitation and accelerated and earlier glacier and snow melts, which, in turn, will affect the distribution and dynamics of snow in time and space [17,18]. Studies have revealed that the spatiotemporal distribution of snow cover shows strong differentiation in China, and relatively stable snow areas are found mainly in northwestern and northeastern China, the Tibetan Plateau, and Inner Mongolia [19–21]. Land surface snow phenology (LSSP), such as snow cover start date, snow melt end date, and snow depth, is better correlated with temperature than other meteorological factors.

Underlying conditions, such as land surface topography and vegetation types, also affect the distribution and dynamics of snow [22,23]. For example, as the temperature gradually decreases with increasing altitude, snow melt slows, making it easier for snow to accumulate [24,25]. The absorption of solar radiation changes with different terrain conditions (i.e., slope and aspect), leading to diverse environmental temperatures and heating, consequentially influencing snowmelt processes [26,27]. In addition, relatively open areas, such as forest edges and sparse woodlands, are prone to snow accumulation, and the opposite is the case in well-covered woodlands due to canopy interception [28–30].

In recent decades, an increasing number of programs have been initiated internationally to facilitate snow research, such as the Climate and Cryosphere Project of the World Climate Research Program (WCRP), the Cold Region Land Surface Processes Experiment carried out by NASA, and the Western Environmental and Ecological Science Research Project, effectively advancing not only the study of snow and its dynamics as the key objects [31–33], but also techniques for snow monitoring and data derivation. In particular, the idea of using optical remote sensing to obtain snow information has made great progress [34–36], a large number of derivations as data products have been released, such as microwave radiometer-based (i.e., AMSR-E) and MODIS-based (i.e., MOD10A1 and MOD10A2) [37–39]. Among them, MODIS-based snow products have high spatial resolutions, can better reflect the distribution of snow cover, and are widely used in regional snow variation-related studies [40–42]. In contrast, passive microwave monitoring-based snow depth data are useful for equivalent evaluation, but their spatial resolution is generally low [43–45]. To obtain high-resolution snow depth information, downscaling of the data is needed. There are two common methods for this purpose: one is based on statistics, and the other is based on deep learning such as machine training [41,46]. The development of downscaling methods is important for snow studies, especially when conducted at smaller scales [47–49].

The southern Gansu Plateau (SGP), located on the northeast edge of the Tibetan Plateau, is an important water source area in the upper reaches of China's Yellow River and Yangtze River. Snow dynamics effectively influence runoff formation and evolution, and mechanistic exploration is beneficial to the scientific planning and utilization of basin water resources [5,50]. Over the past 20 years, river discharge on the SGP have sharply decreased, although few studies on snow phenology and its hydro effectiveness have been published. In view of the above, the objectives of this study are (1) to improve a downscaling method to obtain high-resolution snow depth (SD) data for the analysis of spatiotemporal variations in LSSP on the SGP during the time period from 2003–2018 and (2) to use a geostatistical method to analyze the effects of topographic and climatic factors on the LSSP. Related results may help improve our knowledge of alpine-cold region snow and can provide basic data and methodological support for comprehensive hydrological simulations and predictions in the water source area of large river basins.

2. Study Area

As an important part of the water source area in the upper reaches of the Yellow River, the SGP administratively includes the whole of the Gannan Tibetan Autonomous Prefecture in Gansu Province of China, geographically located between 33°06′N–35°34′N,

100°45′E–104°45′E (Figure 1a). The elevation ranges from 1159~4866 m and averages approximately 3000 m, topographically featuring higher elevations in the northwest and lower elevations in the southeast. The regionally averaged annual air temperature is 1.7 °C, featuring a short frost-free period and plentiful sunshine throughout the year. The annual total precipitation is 620 mm, concentrated in the rainy season from June to September. The relatively lower air temperature and abundant precipitation, corresponding to a typically continental plateau climate, make the SGP naturally develop many tributary systems of the Yellow River (i.e., the Tao River, the Daxia River, etc.) and Yangtze River (i.e., the Bailong River), becoming remarkable in terms of water conservation on the Tibetan Plateau (Figure 1b,c). Along with the increasing intensity of human activities such as cultivation and overgrazing, ecosystems such as grasslands and wetlands become ecologically fragile, and water yield recharge to rivers are reduced, both seriously affecting the protection of regional water resources and ecological security. Due to the significance of snowmelt to SGP hydro-processes, the analysis of snow distribution and dynamics is important for the formation, evolution, rational development, and utilization of water on the SGP and across all the related basins.

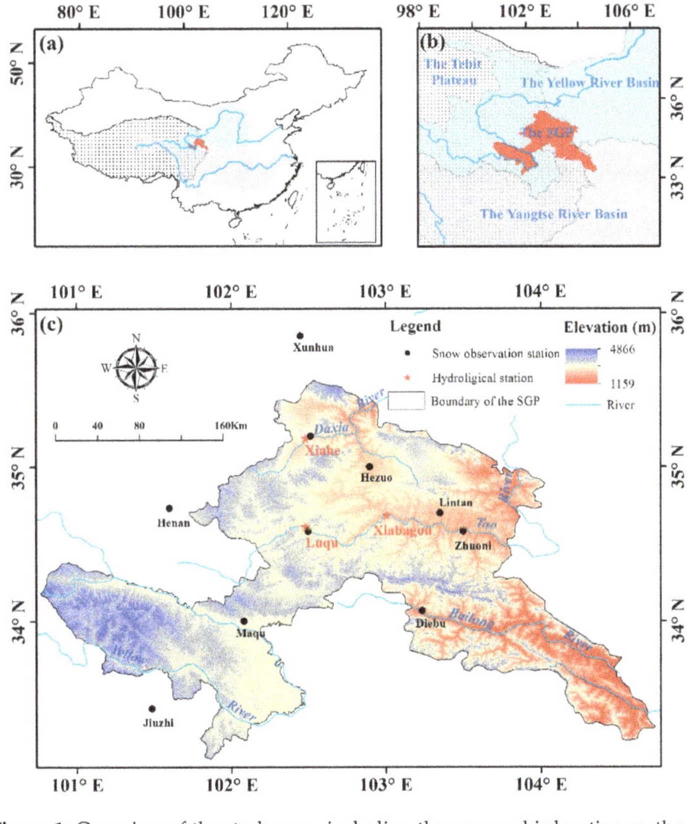

Figure 1. Overview of the study area, including the geographic location on the northeast edge of the Tibetan Plateau (**a**), the water source area in the upper reaches of the Yellow River and Yangtze River (**b**), and the distribution of elevation, stream networks, and snow and hydrological observation stations (**c**).

3. Data and Methods

3.1. Data

The SD data were obtained from the long-term series of the daily snow depth dataset in China (1979–2020), released by the National Tibetan Plateau data center (http://data.tpdc.ac.cn, accessed on 11 October 2021). The data were derived from the inversion of daily passive microwave brightness temperature (SMMR, SSM/I, and SSMI/S), with a spatial resolution of 0.25° [51]. Based on the daily SD data, the monthly and annual maximum SD data were obtained using the maximum value composite (MVC) method, which represented the maximum value in the process of snowmelt and accumulation [52]. The snow cover data were adopted from the MODIS Daily Cloudless 500 m Snow Area Product Dataset over China during the period from 2000 to 2019, released by the National Cryosphere Desert Data Center (http://www.ncdc.ac.cn, accessed on 28 October 2021), used for LSSP calculation [53]. A DEM with a 30 m spatial resolution released by the Chinese Academy of Sciences Geospatial Data Cloud (http://www.gscloud.cn, accessed on 5 July 2018) was used to calculate the slope and aspect across the SGP. The surface net solar radiation data were released by the European Centre for Medium-Range Weather Forecasts (ECMWF) (https://www.ecmwf.int/, accessed on 17 April 2019), with a spatial resolution of 0.25° and a temporal resolution of 3 h. The meteorological data, including precipitation, maximum temperature, and minimum temperature, were from the National Meteorological Information Center of China Meteorological Administration (http://data.cma.cn/, accessed on 21 August 2021). Snow depth observation data came from the National Meteorological Information Center of the China Meteorological Administration and are used to test the accuracy of the downscaling. The time spans of the above data (except for the DEM) were unified to the same period from 2002 to 2018 for simultaneity.

3.2. Methods

3.2.1. SD Data Downscaling

We define the spatial resolution of 0.25° as the lower resolution and that of 500 m as the finer resolution. The purpose of this is to downscale the lower resolution SD data to the finer resolution by considering and integrating multiple influential factors.

(1) Statistical downscaling

Pilot correlation analysis revealed that topographic and geographic factors such as elevation, slope, aspect, longitude, and latitude have significant effects on SD; they affect or regulate climatic process such as snowfall at local or even larger scales [54,55]. All the above factors were calculated and resampled into the lower and finer spatial resolutions. Taking the influential factors at coarse resolution as environmental variables, a multiple linear regression method is used to calibrate the statistical formula based on the original monthly 0.25° SD data (SD_a) as the target variable. Application of the formula results in a simulated series of SD at both the lower (SD_b, Equation (1)) and finer (SD_c, Equation (3)) spatial resolutions. Residuals (Equation (2)) of SD_b to SD_a (ΔSD_{low}) are resampled into the finer 500 m resolution (ΔSD_{high}) using bilinear interpolation. The sums of SD_c and ΔSD_{high} are the statistically downscaled SD data (SD_d, Equation (4)).

$$SD_b = A \times X_{1,low} + B \times X_{2,low} + C \times X_{3,low} + D \times X_{4,low} + E \times X_{5,low} + F \quad (1)$$

$$\Delta SD_{low} = SD_a - SD_b \quad (2)$$

$$SD_c = A \times X_{1,high} + B \times X_{2,high} + C \times X_{3,high} + D \times X_{4,high} + E \times X_{5,high} + F \quad (3)$$

$$SD_d = SD_c + \Delta SD_{high} \quad (4)$$

where X_n (n = 1, 2, 3, 4, 5) represent raster data of elevation, slope, aspect, longitude, and latitude, respectively. Subscripts *low* and *high* denote the lower and finer resolutions in space, respectively.

(2) Improvement based on snow cover data

The original resolution of the SD data is 0.25°, and although it has been geographically and topographically corrected through statistical downscaling, the inaccurate influence is still present. For example, SD occurs in low-altitude regions in warmer months, a situation that seldom occurs, as common sense in the study area dictates. By introducing the spatial distribution probability function of snow, the 500 m MODIS snow cover data are used to modify and improve the precision of the statistical downscaling. Given values of 1 and 0 representing snowy and snowless pixels, respectively, the original MODIS snow cover raster data are converted into binary ones. We define the period from September 1 of one year to August 31 of the next year as a snow hydrological year (SHY), and the accumulation days of snow (*ADS*) in each SHY are calculated from September 2002 to August 2018 at a spatial resolution of 500 m. Then, the cumulative days of snow (*CDS*) in a domain containing 55 × 55 *ADS* grids are counted, approximately corresponding to the spatial resolution of the original SD data (SD_a). The snow distribution probability (*P*) can be determined as follows:

$$P = \frac{ADS * 3025}{CDS} \tag{5}$$

The product of SD_d and P (Equation (6)) is what objective (1) aims at, which is an improvement of the statistically downscaled SD based on snow cover data, comprehensively reflecting a high-resolution SD (SD_{fin}) controlled or influenced by geography, topography, and snow distribution (Figure 2).

$$SD_{fin} = SD_d * P \tag{6}$$

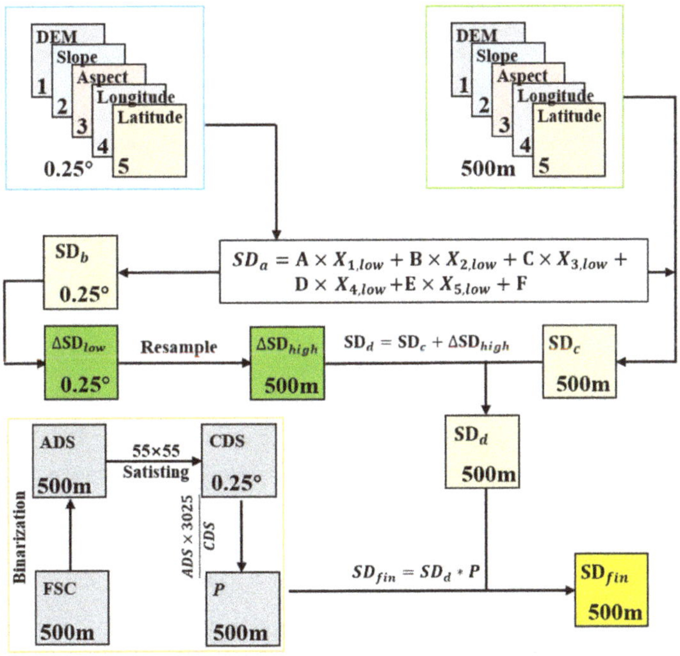

Figure 2. Schematic diagram of the statistical-based method and the improvement from introducing snow distribution probability for downscaling SD data.

3.2.2. LSSP Indicator Extraction

Snow cover maintaining days (SCD), snow cover start date (SCS), and snow cover melt end date (SCM) are calculated based on the binary retreated MODIS snow cover data.

Among them, SCS and SCM are important LSSPs determining SCD [56,57], which represent the dates when a monitored pixel starts accumulating snow and ends melting in a SHY. SCD is the number of days that each pixel is covered by snow in a SHY. The larger the SCD is, the longer and the more the snow reserves.

$$\text{SCS} = F_d - SCD_{bFd} \tag{7}$$

$$\text{SCM} = SCD_{aFd} + F_d \tag{8}$$

$$\text{SCD} = \sum_{i=0}^{N} (S_i) \tag{9}$$

In the above equations, F_d is a fixed number representing the date when the largest snow cover occurs during the period from 2002 to 2018. The statistics resulted in the F_d date being 12 January 2008, the 134th day in the SHY. SCD_{bFd} and SCD_{aFd} represent the numbers of snow cover days before and after F_d, respectively, in each SHY. N is the upper limit for a specified time range, valued as 1 to represent a complete SHY; S_i is the binary retreated pixel value of daily snow cover (snowy or snowless).

3.2.3. Trend Analysis

Sen's slope is selected for series variation amplitude statistics [58,59]. The method can reduce or prevent the impact of data anomalies and omissions when evaluating the trend and range of time series changes. An orderly column is constructed with the change rates of sample sequences of different lengths. Variable testing is then performed statistically according to the given significance level to obtain the value range of the change rates, and the median is used to determine the variation trend and magnitude. The equation is as follows:

$$SS_{ij} = \text{MEDIAN}\frac{(X_j - X_i)}{(j-i)} \tag{10}$$

where SS_{ij} is Sen's slope, X_i and X_j represent the sequential values corresponding to time i and j, respectively, where $1 < i < j < n$ and n is the length of the series.

The Mann–Kendall method [60,61] is a nonparametric test approach used to determine the significance of a trend analysis [62].

$$S = \sum_{i=1}^{n-1} \sum_{j=i+1}^{n} \begin{cases} 1 & y_i - y_j > 0 \\ 0 & y_i - y_j = 0 \\ -1 & y_i - y_j < 0 \end{cases} \tag{11}$$

$$Z = \begin{cases} \frac{S-1}{\sqrt{s(S)}} & S > 0 \\ 0 & S = 0 \\ \frac{S+1}{\sqrt{s(S)}} & S < 0 \end{cases} \tag{12}$$

where y_i and y_j represent the snow phenology indicators in SHY i and j, respectively; n represents the length of the sequence. A positive value of the statistic S indicates an increasing trend of the data series, while a negative one indicates a decreasing trend of the series. The value of Z is in the range of $(-\infty, +\infty)$; for a given confidence interval α, if $|Z| \geq Z_{1-\alpha/2}$, it indicates that there is a significant trend in the data series at confidence level α.

3.2.4. Analysis of Climate-Driven Influences

Correlation-based significance statistics, represented by partial correlation coefficients, are used to examine the strength of climate influences on regional LSSP. To be specific, assuming there are j ($j > 2$) variables (X_{l_1}, X_{l_2}...... X_{l_g}, X_{l_h}, X_{l_i}, X_{l_j}), the formula for the $(j-2)$th order partial correlation coefficient of any two variables X_{l_i} and X_{l_j} is:

$$r_{l_il_j\cdot l_1l_2\cdots l_h} = \frac{r_{l_il_j\cdot l_1l_2\cdots l_{h-1}} - r_{l_il_h\cdot l_1l_2\cdots l_{h-1}} r_{l_jl_h\cdot l_1l_2\cdots l_{h-1}}}{\sqrt{\left(1-r^2_{l_il_h\cdot l_1l_2\cdots l_{h-1}}\right)\left(1-r^2_{l_jl_h\cdot l_1l_2\cdots l_{h-1}}\right)}} \quad (13)$$

where r denotes the correlation coefficient, factors on the right side of the formula are the $(j-3)$th order partial factors.

Four key climatic factors, including precipitation (P), land surface net solar radiation (SSR), maximum air temperature (T_{max}), and minimum air temperature (T_{min}), are selected to investigate the climate-driven strength of the LSSP, the 3rd order partial correlation coefficient is adopted. A t test is used to analyze whether the partial correlation coefficient between LSSP and climatic factors pass the 0.01 significance level (Table 1). The calculation formula is:

$$t = \frac{r^2_{l_4l_5\cdot l_1l_2l_3}}{1-r^2_{l_4l_5\cdot l_1l_2l_3}} \times \frac{n-m-1}{m} \quad (14)$$

where $r_{l_4l_5\cdot l_1l_2l_3}$ is the partial correlation coefficient, m is the number of independent variables, and n is the sequence length. The partitioning criteria are listed in Table 1.

Table 1. Significance-based analysis standards for climate driving forces on LSSP.

Driving Force	*t* Test-P	*t* Test-SSR	*t* Test-T_{max}	*t* Test-T_{min}
[P]	$\lvert t \rvert > t_{0.01}$	$\lvert t \rvert < t_{0.01}$	$\lvert t \rvert < t_{0.01}$	$\lvert t \rvert < t_{0.01}$
[SSR]	$\lvert t \rvert < t_{0.01}$	$\lvert t \rvert > t_{0.01}$	$\lvert t \rvert < t_{0.01}$	$\lvert t \rvert < t_{0.01}$
[P + SSR]	$\lvert t \rvert > t_{0.01}$	$\lvert t \rvert > t_{0.01}$	$\lvert t \rvert < t_{0.01}$	$\lvert t \rvert < t_{0.01}$
......	......	......	......	......
[P + SSR + T_{max} + T_{min}]	$\lvert t \rvert > t_{0.01}$	$\lvert t \rvert > t_{0.01}$	$\lvert t \rvert > t_{0.01}$	$\lvert t \rvert > t_{0.01}$
[NC]	$\lvert t \rvert < t_{0.01}$	$\lvert t \rvert < t_{0.01}$	$\lvert t \rvert < t_{0.01}$	$\lvert t \rvert < t_{0.01}$

Note: *t* test-P, *t* test-SSR, *t* test-T_{max}, and *t* test-T_{min} represent the t significance test of LSSP with P, SSR, T_{max}, and T_{min}, respectively. [P] and [SSR] indicate that LSSP is driven by P or SSR. [P + SSR] indicates that LSSP is driven by both P and SSR. [P + SSR + T_{max} + T_{min}] means that LSSP is conjointly driven by P, SSR, T_{max}, and T_{min}. [NC] means nonclimate-driven. There are four key climatic factors and a total of 16 driving force combinations or single factors. Not all of them are listed due to space limitations.

3.2.5. Sensitivity Analysis

The response of snow variation to climate change is diagnosed using the sensitivity coefficient [63]. The method is widely used in contribution separations of influential factors on hydrological processes. The sensitivity coefficient is calculated as:

$$\varepsilon_x = \frac{\overline{x}}{\overline{y}} \times \frac{\sum_{i=1}^{n}(x-\overline{x})(y-\overline{y})}{\sum_{i=1}^{n}(x-\overline{x})^2} \quad (15)$$

where ε_x is the sensitivity coefficient of y (LSSP) to x (climatic factors), indicating that the ε_x% change in LSSP is caused by 1% variation in a climatic element. $\overline{x}$ and $\overline{y}$ are the multiyear averaged values of x and y. In the following statement, ε_{a-b} represents the sensitivity of LSSP to climatic factors, a is climatic factors such as P, SSR, T_{max}, and T_{min}, b is LSSP indicators such as SD and SCD.

4. Results

4.1. Evaluation of the Improved SD Downscaling Method

Gauge-based SD observations at 10 stations in and near the SGP (Figure 1c) were adopted as references, based on which absolute errors were calculated using SD data before and after downscaling, as shown in Figure 3. The vertical ranges cover all errors, including both the positive and negative ones, while transverse widths indicate occurrent frequencies. It can be seen that the positive downscaled SD (ΔSD_{fin}) errors at all stations and the negative errors at most of the 10 stations are reduced, indicating an effective optimization for the elimination of both the over- and underestimations of the initial SD data. Frequencies of SD error valued at 0 were found to be the highest at all stations, and differences were not clear between the two sets of data, although an overall but slight decrease appeared. As a whole, using the improved downscaling method, the spatial resolution and real representation of SD were verified to be acceptable, and the downscaled data were satisfactory for LSSP analysis in the SGP region.

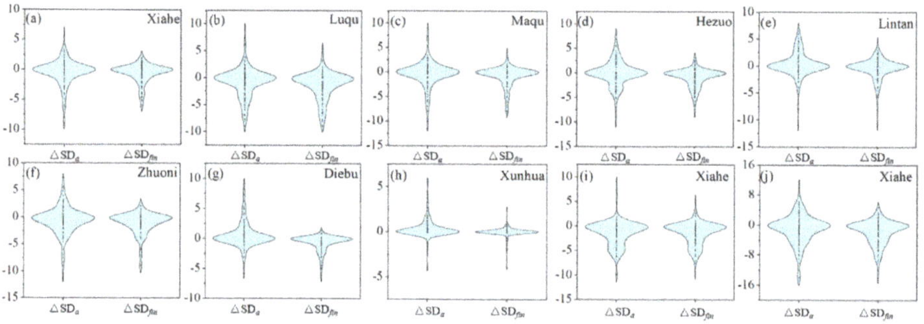

Figure 3. Evaluation of the improved downscaling method based on the differences from the gauge-based SD observations. ΔSD_a and ΔSD_{fin} represent the absolute error between the SD data before and after downscaling, respectively. Subplots (**a**–**j**) correspond to the gauge stations for snow observation. Names of the stations labeled in the right corners of the plots.

4.2. LSSP Characteristics

The binary retreated MODIS snow cover data were used to calculate time-based LSSP values, including the SCS, SCM, and SCD. Together with the downscaled SD data, a spatiotemporal analysis of LSSPs of the SGP from 2002 to 2018 was conducted.

4.2.1. Spatiotemporal Distributions

LSSPs present remarkable heterogeneity in time and space on the SGP. High SD values are mainly found in areas near or at mountain divides, while low SD values are generally distributed in valley areas featuring lower altitudes. For example, in the southwestern and southeastern parts where the main stream of the Yellow River and the Bailong River develop, SD can exceed 18 cm in high altitude zones but less than 10 cm in valleys (Figure 4a). The SCS and SCM show opposite distribution patterns in value space (Figure 4b,c). In or near mountain divides, snow begins to accumulate early (i.e., the earliest is the 34th day in a SHY) but ends melting late (i.e., after the 255th day in a SHY). In most areas, the SCS starts after the 120th day, corresponding to SCM days before the 195th day, and both result in a relatively short SCD time in a SHY, i.e., most of the SCD is quantified to be less than 15 days in the SGP. Overall, the SD, SCS, SCM, and SCD spatially match well in phenology. For example, in areas such as mountain divides featuring deeper snow, accumulation starts earlier, melting ends later, and snow cover lasts longer.

The SD is larger in months from November to the following March than that in the rest of a SHY, the maximum value generally occurs in February, with a multiyear average value of 4.03 cm (Figure 5a). For the presence of snow cover, SCD presents a "multipeak" pattern in a SHY (Figure 5b). Generally, snow cover lasts the longest in November of one year and March of the following year, especially in March, when the existence of snow can reach 8 days. In the context of the regional climate, there is rarely snow cover in the SGP from June to August.

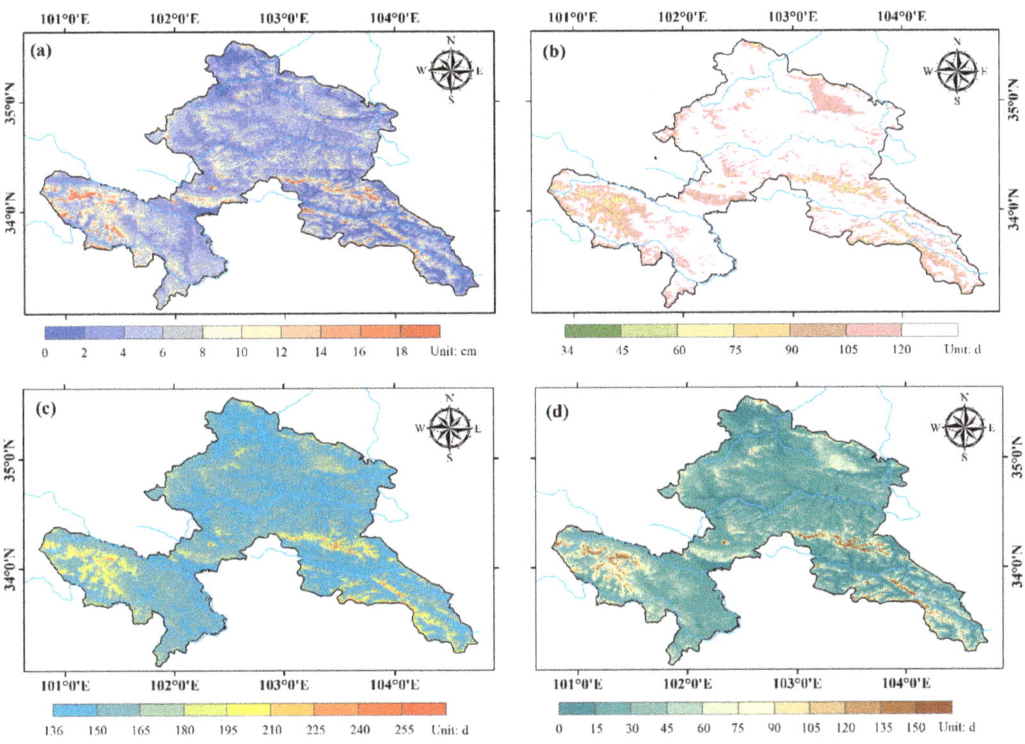

Figure 4. The spatial distribution of LSSPs on the SGP. Subplots (**a**–**d**) are for SD, SCS, SCM, and SCD.

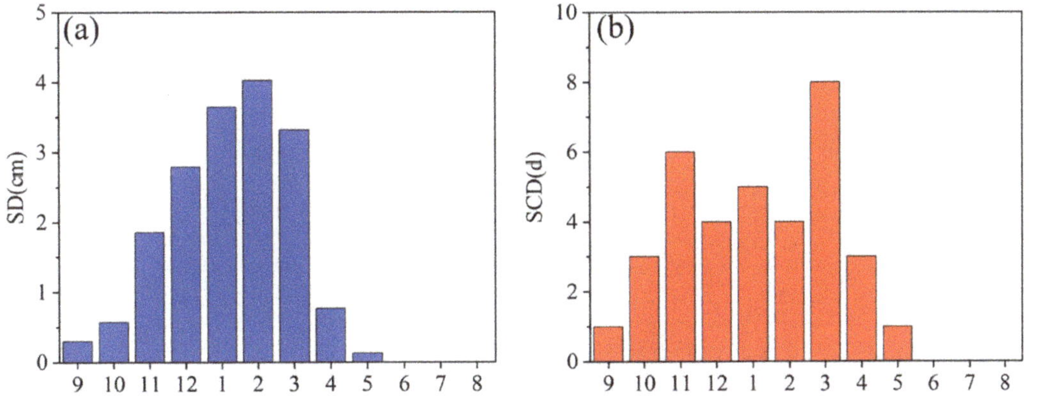

Figure 5. Multiyear averaged monthly distribution of SD (**a**) and SCD (**b**) in SGP (2003–2018).

4.2.2. Spatiotemporal Variations

LSSP variation was found to interannually fluctuate according to quantification of the representative index, including SD, SCS, SCM, and SCD. Generally, higher values of SD and SCD occurred simultaneously, corresponding to smaller SCS values and larger SCM values (Figure 6). During the period from 2003 to 2018, the value ranges of the LSSP index, such as the SD, SCS, SCM, and SCD, were quantified from 2.67 to 10.53 cm, 113 to 134 d, 142 to 188 d, and 16 to 75 d, averaged to 5.15 cm, 123 d, 158 d, and 34 d, respectively. During the statistical period, the maximum SD was found in 2009, when the SCS and SCM were 121 d and 181 d in the SHY, respectively, corresponding to an SCD length of 60 d.

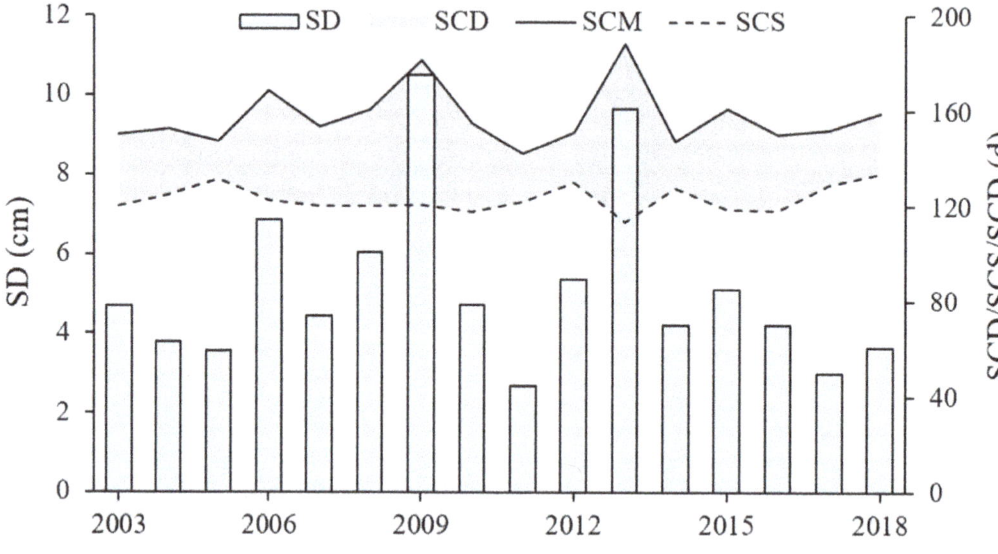

Figure 6. Interannual variations of the four selected LSSP indices on the SGP (2003–2018).

The variation trends and amplitudes of the LSSP during the time period from 2003 to 2018 were measured spatially on the SGP based on Sen's slope method, as shown in Figure 7. A decreasing trend of SD in most of the area occurred, especially in the southwestern part, which is located in the main channel of the Yellow River. A relatively small area had an increase in SD, as in the southeastern part, which dominates the upstream section of the Bailong River. The variation in SD was not significant at $p \geq 0.01$ on most of the SGP, especially in the central and northern parts, similar to the significance test for the spatial statistics of the other three LSSP indices. The areal percentage of the reduced SD occupied 60.36%, and the variation amplitude was overall quantified into a rate of −0.06 cm/a across the whole SGP (Figure 7a). Areas with high altitudes showed significant changes in the SCS and SCM, such as in or near mountain divides. In terms of the areal percentage, areas with increased SCS accounted for 75.10%, indicating that the start date of snow accumulation on most of the SGP presented a delay. Areas with decreased SCM accounted for 65.78%, indicating that most of the snow melt ended earlier. However, the aggregative variation was rated as 0.53 d/a and 0.45 d/a for SCS and SCM, respectively (Figure 7b,c). The amplitude of the former was greater than that of the latter, resulting in an overall reduction in SCD of −0.37 d/a, which was specifically significant in the southwestern and southeastern parts of the SGP. The statistics resulted in a larger areal percentage of 56.90% for the increase in SCD, and the overall reduction was due to the magnitude of the decrease (Figure 7d).

4.3. Terrain Influence on LSSP

Terrain factors, including altitude, slope, and aspect, were calculated and reclassified using value intervals of 500 m, 10°, and 45° for analysis of the topographic influence on the LSSP. Altitude zones between 3000~3500 m and 3500~4000 m are mostly distributed in the region (Figure 8a), while few of the slopes are more than 50° (Figure 8b). The aspect differentiation is not significant, with relatively more differentiation in the northeast and less differentiation in the southeast and south (Figure 8c).

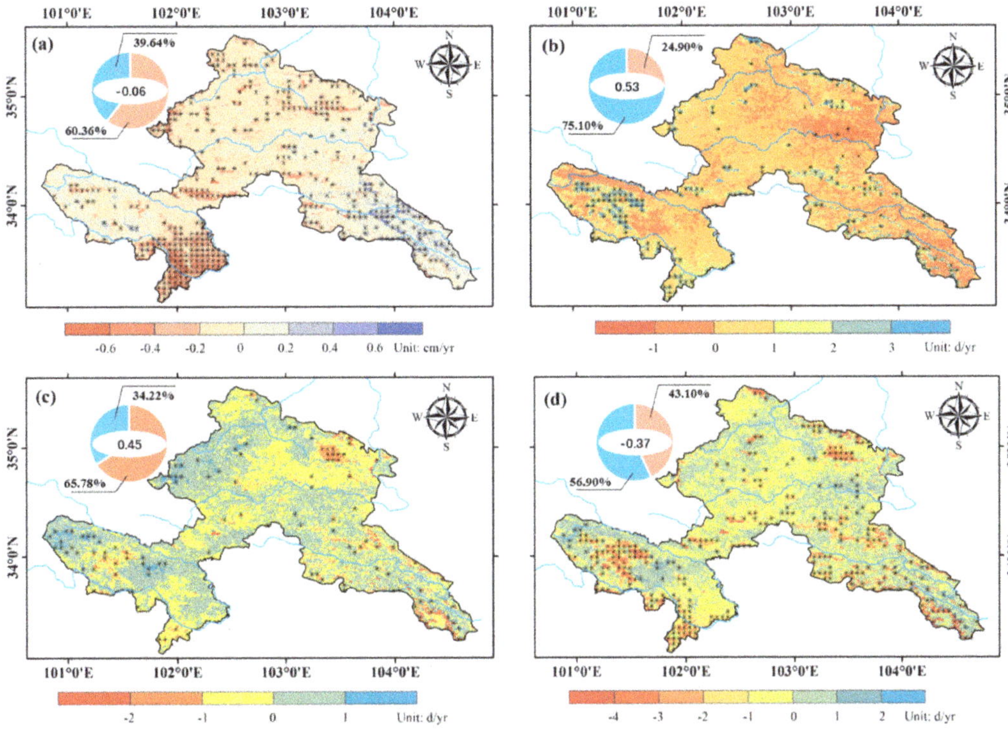

Figure 7. Spatial variation of the four selected LSSP indices across the SGP (2003–2018). Subplots (**a–d**) are for SD, SCS, SCM, and SCD in order). "∗" indicates an area with significantly tested LSSP series ($p \leq 0.01$). Fan diagrams in the upper left corners of the subplots indicate areal proportions of increasing and decreasing, differentiated using blue and orange, while regional averages are indicated by the white part in the middle.

4.3.1. Terrain-Based Distribution

LSSP presented a similar value distribution corresponding to altitude and slope. Statistically, the higher the altitude, or the steeper the slope, the greater the SD (Figure 9a,c), SCM (Figure 9c,f) and SCD (Figure 9d,h), and the smaller the SCS (Figure 9b,g). In particular, the variation amplitudes of the four LSSP indices increased along with the above two terrain factors, indicating that LSSP variability strengthened in regions with higher altitudes and steeper slopes. Generally, snow does not easily accumulate on steep slopes, our analysis showed an inverse pattern. It was found that steep slopes are mostly located in regions with high altitudes on the SGP and are more prone to snowfall, which might be the reason why the higher SD values were more distributed [64]. The influence of aspects on LSSP were statistically close due to the relative equilibrium at the regional scale. South-facing areas, including the southeast, south, and southwest facings, had smaller SD, SCM, and SCD, and larger SCS (Figure 9i–l), reflecting a lower probability of snow accumulation on sunny slopes.

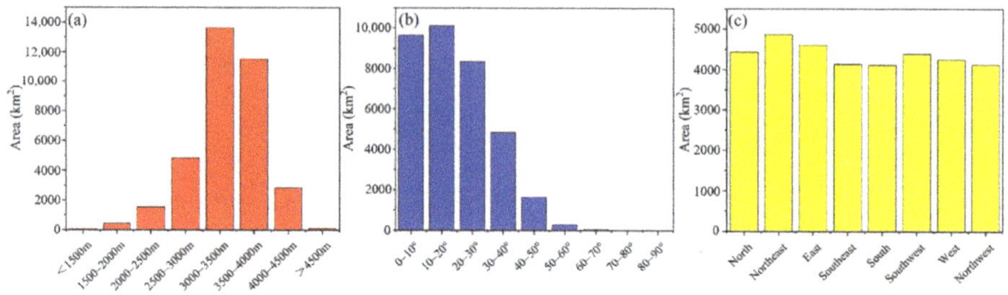

Figure 8. Reclassification of the three selected topographic factors together with their areal statistics across the SGP. Subplots (**a–c**) are for altitude, slope, and aspect, respectively.

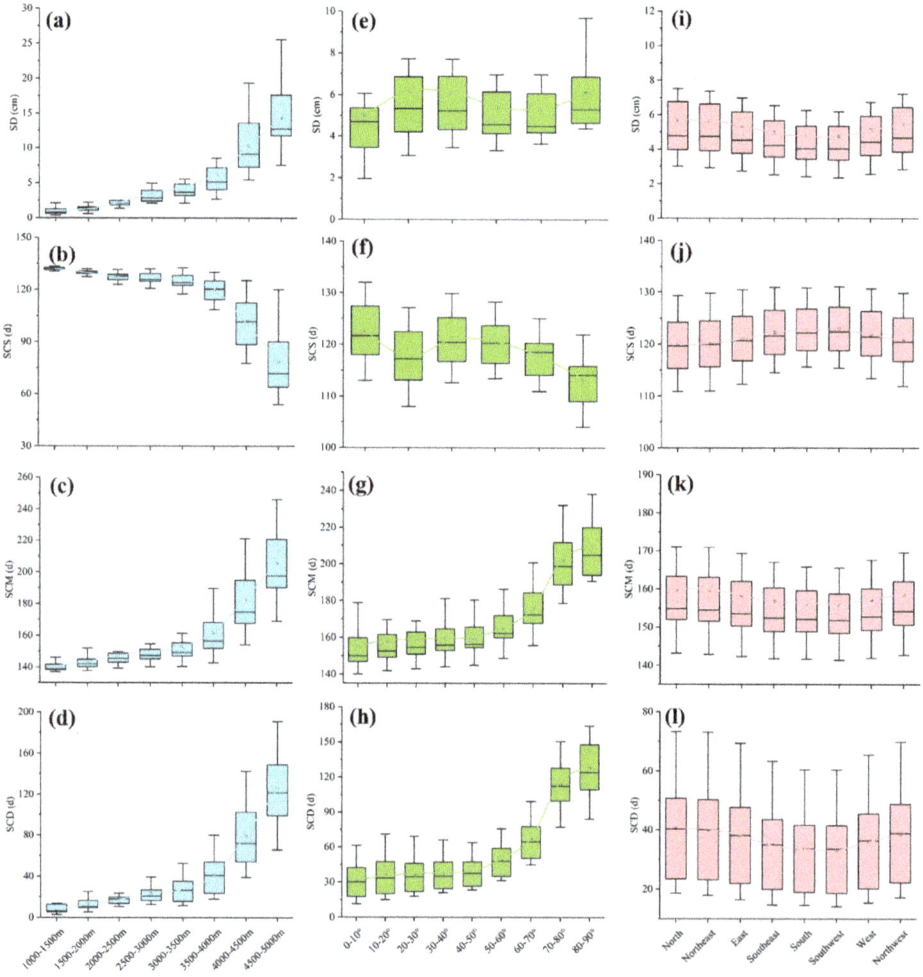

Figure 9. Statistics of LSSP based on terrain. Subplots (**a–d**), (**e–h**) and (**i–l**) correspond to altitude, slope, and aspect, respectively.

4.3.2. Terrain-Based Variation

Statistics were conducted to diagnose the variation difference in the LSSP corresponding to the reclassified terrain. The results revealed that the decrease in SD mainly occurred in regions with altitudes higher than 3000 m, while in other altitude zones, the SD increased. In zones with slopes ranging between 70~80° or less than 40°, the SD was found to decrease, although an increase occurred in other slope zones. All aspects, correspondingly exhibiting almost all altitudes and slopes in zones, were found to decrease with SD. In terms of the time phenology, an increase in SCS, together with a decrease in SCM, was found in most of the altitude zones, resulting in a general decrease in SCD or a shortened snow-maintaining time, especially in regions with altitudes higher than 3000 m, consistent with the variation in SD (Figure 10a). The SCS in all slope zones presented a delay (increase), while the SCM presented a decrease (earlier), except in zones with slopes less than 20° (Figure 10b). Statistics based on aspects resulted in an overall increase in the SCS and SCM, although the variation rates of the latter were less than those of the former (Figure 10c). According to the above, given an altitude of 3000 m and a slope of 20°, LSSPs on the SGP generally presented a pattern of "decrease in higher and steeper areas corresponding to a shortened snow duration, increase in lower and flatter areas corresponding to a relatively lengthened snow duration", which might have a profound relationship with the regional climate change under the warming background.

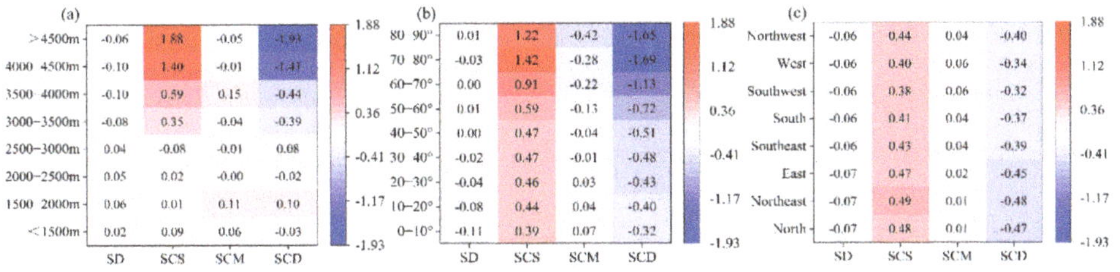

Figure 10. Quantified annual variation in SD (cm/a), SCS (d/a), SCM (d/a), and SCD (d/a) corresponding to the terrain factors of the SGP. Subplots (**a–c**) represent zonal statistics based on the reclassified altitude, slope and aspect, respectively.

4.4. Climate Influence on LSSP

Analysis revealed that the SGP experienced an overall reduction in snow accumulation during the time period from 2003 to 2018, the duration of snow cover moved backward in a SHY. The above, represented by the SD and SCD indices (determined by SCS and SCM), can be considered comprehensive reflections of snow dynamics. Reasons leading to snow variation vary, in which the climate influence plays an important role [25,65]. In particular, regional precipitation (P), surface net solar radiation (SSR), maximum air temperature (T_{max}), and minimum air temperature (T_{min}) are key factors [66] and are thus selected to analyze the responses of the LSSP to regional climate change on the SGP.

4.4.1. Partial Correlation Analysis

The influence of climate change on regional LSSP is mechanically complex. To be more focused, the multicollinearity analysis between factors of climate and terrain are ignored at this stage because it is out of the present study's scope. Given the four selected climatic factors, the singular or synergistic effectiveness of multiple factor combinations on LSSP variation needs to be clarified. Coupling partial correlation (Equation (14)) and significance analysis (Table 1) can help. Among the 16 factor/combinations, the top five categories satisfying the t test standard were sorted by significantly influenced area. The SD order list was [SSR, T_{min}, T_{max}, SSR + T_{min}, P] (Figure 11a), while that of the SCD was [T_{max}, SSR, P, P + SSR + T_{max}, P + T_{max}] (Figure 11b), indicating different impacts of climate on regional LSSP variation, e.g., SD varied remarkably by SSR and T_{min} effectiveness, although SCD was more susceptible to synergistic influences, mechanically showing more complexity. Effectiveness of climatic driving factors on SD and SCD were found to be different, which may be due to local terrain regulations to different LSSP. For example, in the northern part of the SGP featuring lower altitude, the SD is more significantly related to lower air temperature that facilitating more snowfall, while the occurrence of factors with influence that can significantly drive the SCD is less because snow is usually sparse and shortly maintained there. In the southwestern part, the altitude is relatively higher and the air temperature is always lower; the abundant SSR, due to the high altitude, influences SD more, although SCD is dramatically affected by the higher air temperature because the snow event occurs commonly, especially in cold seasons. In summary, the singular effectiveness of the factors was verified to influence the LSSP more over a larger area, which may be related to the zonal differentiation of terrain and climate on the SGP. In other words, key factors affecting regional LSSP might tend to be single, which is in good agreement with the following sensitivity analysis.

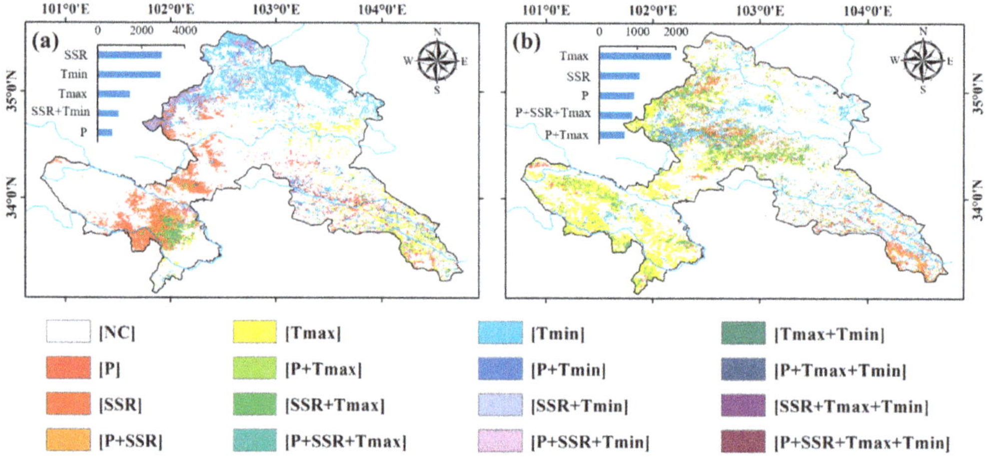

Figure 11. Significance-based diagnosis of the relationship between LSSP and key climatic factors and their combinations on the SGP. Subplots (**a**,**b**) correspond to SD and SCD, respectively. Histograms in the left corners show the top five areas influenced (km^2) and NC indicates an area that failed the significance test.

4.4.2. Sensitivity Analysis

Sensitivity (coefficient) analysis quantifies the percentage variation in the target variable (SD/SCD) caused by a 1% change in the environmental variable (the climatic factors). ε_{P-SD}, ε_{SSR-SD}, $\varepsilon_{Tmax-SD}$ and $\varepsilon_{Tmin-SD}$ represent the sensitivity of SD to P, SSR, T_{max}, and T_{min}, respectively. The results revealed that SD positively or negatively responded to all four climatic factors. Positive facilitation of SD due to increased P was widely characterized on the SGP, accounting for 67.74% of the area (Figure 12a), while the other three factors produced damping effects on SD; along with the increase in SD, the proportions of the area with negative impacts were above 80% (Figure 12b–d). The regional averages of ε_{P-SD}, ε_{SSR-SD}, $\varepsilon_{Tmax-SD}$ and $\varepsilon_{Tmin-SD}$ were 0.71%, −4.12%, −3.22%, and −2.95%, respectively, corresponding to SD variation rates when the climatic factor increased by 1%. Comparatively, SD was more sensitive to SSR on the SGP.

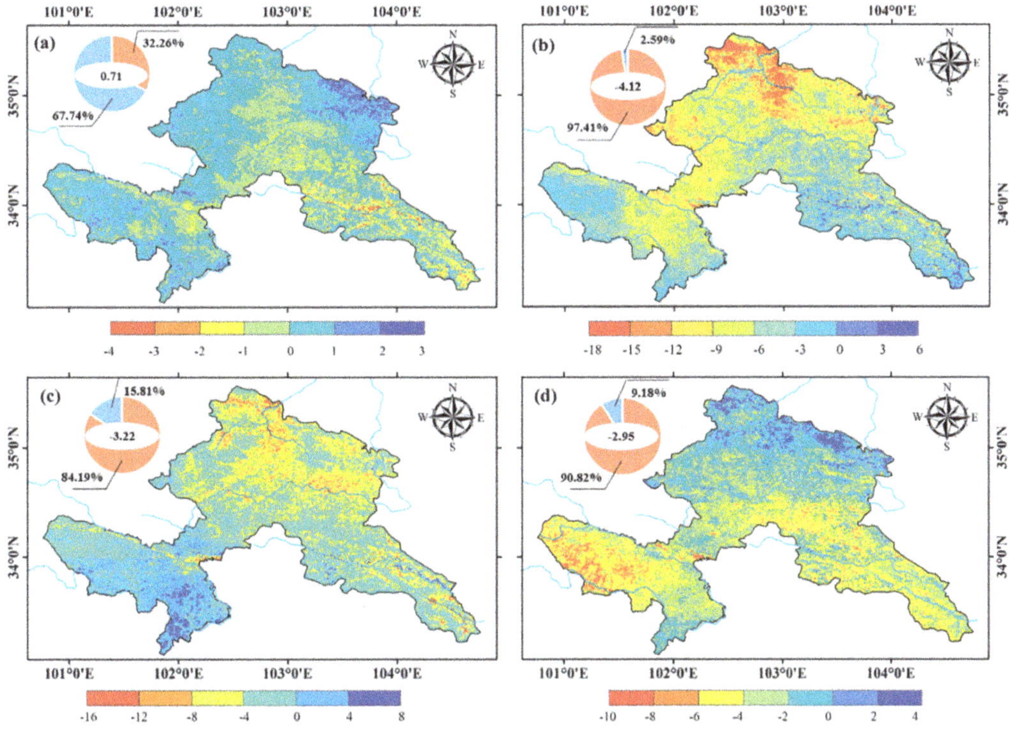

Figure 12. Spatial distribution of the SD sensitivity to the four climatic factors. Subplots (**a**–**d**) correspond to P, SSR, T_{max} and T_{min}, respectively. Fan diagrams in the upper left corners of subplots indicate the areal proportion of positively or negatively influenced and are differentiated using blue and orange, while the regional averages are the white part in the middle.

ε_{P-SCD}, $\varepsilon_{SSR-SCD}$, $\varepsilon_{Tmax-SCD}$, and $\varepsilon_{Tmin-SCD}$ represent the sensitivity of SCD to P, SSR, Tmax, and Tmin, respectively. Similar results were obtained by calculations (Figure 13). In contrast, the regional averages of ε_{P-SCD}, $\varepsilon_{SSR-SCD}$, $\varepsilon_{Tmax-SCD}$, and $\varepsilon_{Tmin-SCD}$ were 1.16%, −3.69%, −3.45%, and −4.01%, respectively, corresponding to SCD variation rates when the climatic factor increased by 1%. Comparatively, SCD was more sensitive to air temperature, especially the lower ones.

Overall, the LSSP on the SGP was influenced by and showed more sensitivity to solar radiation and air temperature. Local precipitation, especially snowfall, provides the matter base for snow formation, distribution, and accumulation. The low sensitivity might be related to the spatiotemporal correspondence and the interactive adaptation between terrain and climate.

4.5. Integrated Effectiveness of Climate and Terrain

To explore the integrated effectiveness of climate and terrain, statistics of LSSP sensitivity (coefficients) to the climate were conducted based on the terrain. The results showed negative values of ε_{P-SD} with respect to the area, mainly at altitudes lower than 3000 m, while positive values were observed in the others, indicating that precipitation tended to positively contribute to SD in higher altitude regions [67]. Values of ε_{P-SCD} were found to be positive in most of the study area except in the altitude range of 1500~2000 m, indicating that it is not easy for snow to form and be maintained in lower altitude regions. The sensitivity coefficients of the two LSSP indices to SSR, T_{max}, and T_{min} were all negative, indicating that an increase in the three climate factors led to a decrease in snow in both amount and persistence at all altitudes (Figure 14a,b). In areas with slopes greater than 30°, the SD sensitivity to precipitation was found to be negative, indicating that snow hardly accumulated on steep slopes. The SCD sensitivity to precipitation was positive but decreased with increasing slope, indicating that the steeper the slope is, the smaller the effect of precipitation on snow persistence. The absolute value of LSSP sensitivity to SSR and T_{max} decreased with increasing slope, while the trend of LSSP sensitivity to T_{min} was opposite, meaning that snow on steeper land is less sensitive to solar radiation and higher air temperature, but more sensitive to lower air temperature (Figure 14c,d). LSSP sensitivities to P and SSR were not found to be significantly different in various aspects, while LSSP sensitivity to T_{max} was found to be stronger on sunny (southward) slopes and LSSP sensitivity to T_{min} was stronger on shaded (northward) slopes (Figure 14e,f).

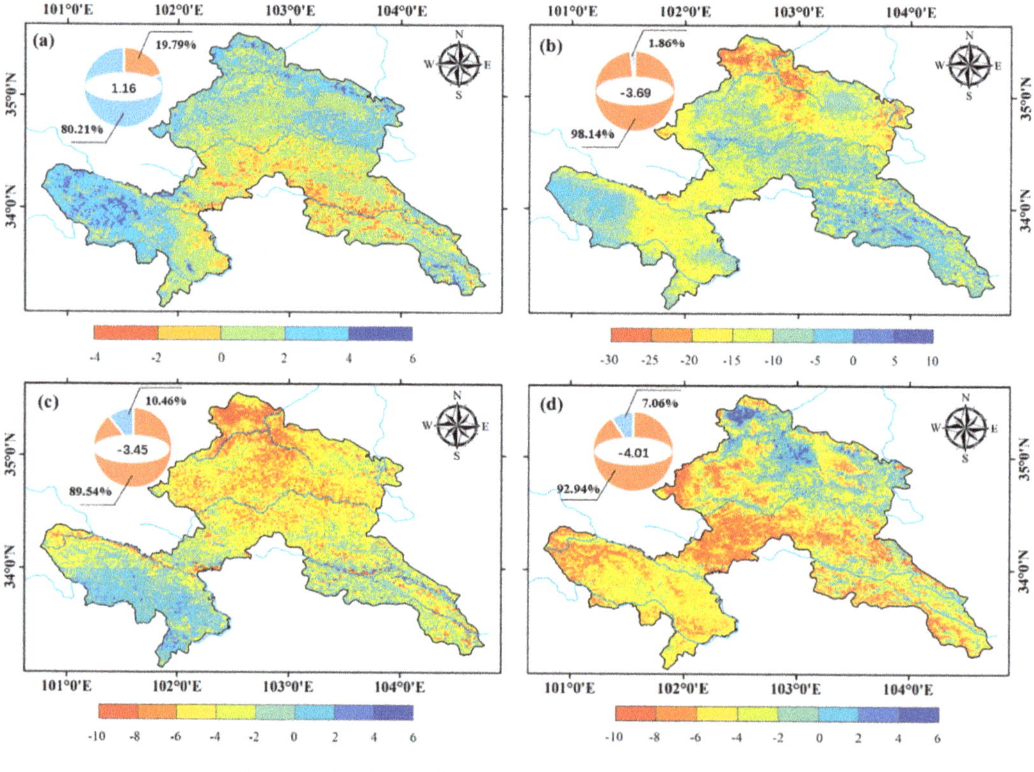

Figure 13. Spatial distribution of SCD sensitivity to the four climatic factors. Subplots (**a–d**) correspond to P, SSR, T_{max} and T_{min}, respectively. Fan diagrams in the upper left corners of subplots indicate the areal proportion of positively or negatively influenced and are differentiated using blue and orange, while the regional averages are indicated by the white part in the middle.

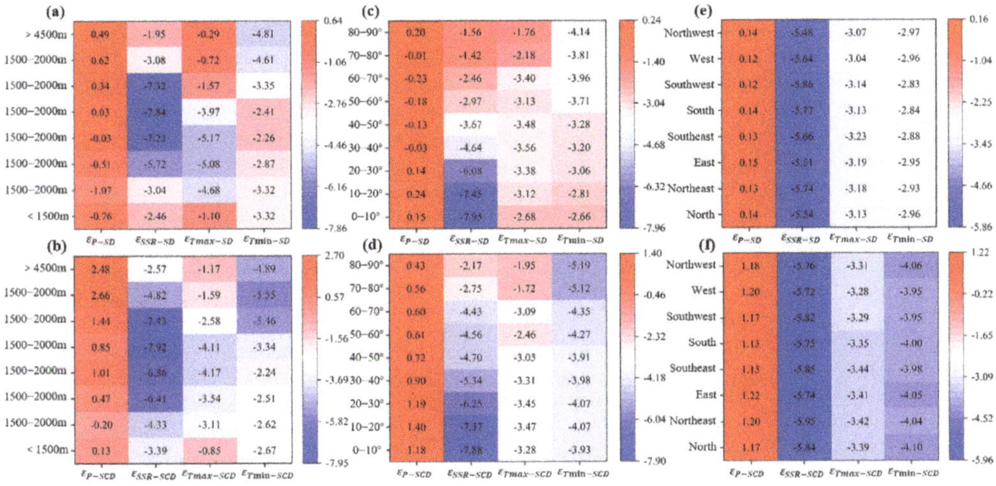

Figure 14. LSSP sensitivity statistics based on combinations of climatic and terrain factors. Subplots (**a**,**b**) are altitude-based, (**c**,**d**) are slope-based, and (**e**,**f**) are aspect-based.

5. Discussion

Located in a midlatitude inland region featuring limitations of circulation transportation, precipitation, especially snowfall in the cold season on the SGP, is relatively scarce [41]. Combined with the influence of land surface topography, snow cover presents remarkable fragmentation [68,69], leading to significant heterogeneity in LSSP variation [70]. For a better understanding, data with a fine spatial resolution are thus essential. To address this issue, we regressively analyzed the relationship between SD and the geographic and topographic factors on the SGP to statistically downscale the SD data product from a coarse spatial resolution of 0.25° to a finer one of 500 m. To eliminate the distortion of the original data, we introduced the MODIS-based spatial probability of snow cover for correction [7,71]. The coupling of the two ideas not only helped advance the downscaling method but also improved the data accuracy of snow depth, which can be considered an example to synthesize methods in data treatment.

Snow formation, accumulation, and redistribution are closely related to land surface terrain [72,73]. We found that on the SGP, the higher the altitude is, the larger the LSSP variability, indicating that snow is more significantly affected by environmental factors at higher altitudes [74]. The slope, together with aspect, is profoundly connected with incidence angle and eolian activities, acting on the amount of solar radiation and energy flux, influencing land surface energy budget, and affecting snow accumulation and melt rates [68,75]. Our study demonstrated the difference in LSSP in terrains with high complexity.

It was found that precipitation and air temperature are the key factors that most influence the regional LSSP [76]. As the basic material source for forming snow, precipitation cannot have an effect on the LSSP unless the air temperature is below the freezing point [77,78], which might be the explanation for the lower sensitivity of the LSSP to precipitation and the more remarkable sensitivity to air temperature [79,80]. Generally, absorption of solar radiation by snow cover relates to land surface albedo [81,82], during which the phase variation of the solar altitude angle in winter and spring plays an important role, dramatically affecting snow melt in a year [83,84]. Our study of the SGP agree with the above findings. The influence of key climatic factors on LSSP was further distinguished, and considerable differences in LSSP in response to the maximum and minimum air temperatures in time and space were quantitatively presented.

In terms of LSSP variation, the chemical composition and physical state of the atmosphere (e.g., aerosol load), structure and density of snow (e.g., crystal size) [85], variation in environmental conditions (e.g., permafrost degradation [82]; land use/cover change [86]), artificial disturbance [87,88], and so on, directly or indirectly have profound impact on snow accumulation and melting and may cause LSSP variability [89]. In particular, a long period with snow is generally influenced by combinations of interactively functioning factors [90]. Our study examined individual effectiveness based on sensitivity analysis, and coupling studies were not conducted. Additionally, LSSP does not vary synchronously with climate, and delays with time often occur [78,91]. Furthermore, land surface vegetation also influences the LSSP [92]. The above processes, together with their effects on regional LSSP, have been neglected and may lead to uncertainties in the present study. It is hoped that shortcomings will be overcome in the future based on the improvement of data and methods.

6. Conclusions

In this study, we developed an improved statistical downscaling method to obtain SD data with higher spatial resolution for a comprehensive study on regional LSSP of the SGP, a specifically selected water source area of China's Yellow River. It was shown that the improved downscaling method can be used to effectively optimize both the spatial resolution and data accuracy of SD. Statistics from 2003 to 2018 revealed an overall reduction in the SGP's LSSP, especially the two main indices of SD and SCD, which varied at negative rates of −0.06 cm/a and −0.37 d/a, respectively. In terms of terrain, the LSSP of the SGP generally presented a pattern of "decrease in higher and steeper areas corresponding to a shortened snow duration, increase in lower and flatter areas corresponding to a relatively lengthened duration". Climatically, given a 1% increase in P, SSR, T_{max} and T_{min}, the regional averages of the SD variation were 0.71%, −4.12%, −3.22%, and −2.95%, respectively, while those of the SCD variation were 1.16%, −3.69%, −3.45%, and −4.01%, respectively. Comparatively, SD was more sensitive to SSR, while SCD was more sensitive to air temperature on the SGP. These findings may be helpful to the awareness of snow hydrology and promote its quantitative analysis in the alpine water source areas of large river basins.

Author Contributions: L.W.: data collection, methodology, writing—original draft. C.L.: supervision, formal analysis, conceptualization, writing—review and editing. X.X.: data analysis. S.Z., J.L., X.Z. and N.S.: writing—review and editing. All authors have read and agreed to the published version of the manuscript.

Funding: This study was supported by the Key Research Programs of Gansu Province in Science and Technology (20ZD7FA005, 21ZD4FA008), the Open Foundation of MOE Key Laboratory of Western China's Environmental System in Lanzhou University as the Fundamental Research Funds for the Central Universities (lzujbky-2020-kb01).

Data Availability Statement: The long-term series of the daily snow depth dataset in China are available in the National Tibetan Plateau data center. These data were derived from the flowing resources available in the public domain: http://data.tpdc.ac.cn. The MODIS Daily Cloudless 500 m Snow Area Product Dataset over China are available in the National Cryosphere Desert Data Center. These data were derived from the flowing resources available in the public domain: http://www.ncdc.ac.cn. The surface net solar radiation data were released by the European Centre for Medium-Range Weather Forecasts (ECMWF) (https://www.ecmwf.int/). The meteorological data were from the National Meteorological Information Center of China Meteorological Administration (http://data.cma.cn/ all accessed on 7 June 2022).

Acknowledgments: We are grateful to the National Meteorological Science Data Center, National Tibetan Plateau Data Center, National Cryosphere Desert Data Center, NASA, and the ECMWF for providing open access to data collections and archives.

Conflicts of Interest: The authors declare that they have no conflict of interest.

Abbreviations

LSSP	Land surface snow phenology
SGP	Southern Gansu Plateau
SD	Snow depth
MVC	Maximum value composite
ECMWF	European Centre for Medium-Range Weather Forecasts
SHY	Snow hydrological year
CDS	Cumulative days of snow
ADS	Accumulation days of snow
SCD	Snow cover maintaining days
SCS	Snow cover start date
SCM	Snow cover melt end date
T_{max}	Maximum air temperature
T_{min}	Minimum air temperature
SSR	Land surface net solar radiation

References

1. Yi, Y.; Liu, S.; Zhu, Y.; Wu, K.; Xie, F.; Saifullah, M. Spatiotemporal heterogeneity of snow cover in the central and western Karakoram Mountains based on a refined MODIS product during 2002–2018. *Atmos. Res.* **2020**, *250*, 105402. [CrossRef]
2. Zhong, X.; Zhang, T.; Kang, S.; Wang, J. Spatiotemporal variability of snow cover timing and duration over the Eurasian continent during 1966–2012. *Sci. Total Environ.* **2021**, *750*, 141670. [CrossRef] [PubMed]
3. Erickson, T.A.; Williams, M.W.; Winstral, A. Persistence of topographic controls on the spatial distribution of snow in rugged mountain terrain, Colorado, United States. *Water. Resour. Res.* **2005**, *41*, W04014. [CrossRef]
4. Bavay, M.; Grunewald, T.; Lehning, M. Response of snow cover and runoff to climate change in high Alpine catchments of Eastern Switzerland. *Adv. Water Resour.* **2013**, *55*, 4–16. [CrossRef]
5. Chen, H.; Chen, Y.; Li, W.; Li, Z. Quantifying the contributions of snow/glacier meltwater to river runoff in the Tianshan Mountains, Central Asia. *Global Planet. Chang.* **2019**, *174*, 47–57. [CrossRef]
6. Tarca, G.; Guglielmin, M.; Convey, P.; Worland, W.R.; Cannone, N. Small-scale spatial–temporal variability in snow cover and relationships with vegetation and climate in maritime Antarctica. *Catena* **2021**, *208*, 105739. [CrossRef]
7. Xie, J.; Jonas, T.; Rixen, C.; Jong, R.; Garonna, I.; Notarnicola, C.; Asam, S.; Schaepman, M.E.; Kneubuhler, M. Land surface phenology and greenness in Alpine grasslands driven by seasonal snow and meteorological factors. *Sci. Total Environ.* **2020**, *725*, 138380. [CrossRef]
8. Thornton, J.; Brauchli, T.; Mariethoz, G.; Brunner, P. Efficient multi-objective calibration and uncertainty analysis of distributed snow simulations in rugged alpine terrain. *J. Hydrol.* **2021**, *598*, 126241. [CrossRef]
9. Lehning, M.; Lowe, H.; Ryser, M.; Raderschall, N. Inhomogeneous precipitation distribution and snow transport in steep terrain. *Water Resour. Res.* **2008**, *44*, W07404.
10. Thapa, S.; Zhang, F.; Zhang, H.; Zeng, C.; Wang, L.; Xu, C.; Thapa, A.; Nepal, S. Assessing the snow cover dynamics and its relationship with different hydro-climatic characteristics in Upper Ganges river basin and its sub-basins. *Sci. Total Environ.* **2021**, *793*, 148648. [CrossRef]
11. Wu, X.; Wang, X.; Liu, S.; Yang, Y.; Xu, G.; Xu, Y.; Jiang, T.; Xiao, C. Snow cover loss compounding the future economic vulnerability of western China. *Sci. Total Environ.* **2021**, *755*, 143025. [CrossRef] [PubMed]
12. Juras, R.; Blocher, J.; Jenicek, M.; Hotovy, O.; Markonis, Y. What affects the hydrological response of rain-on-snow events in low-altitude mountain ranges in Central Europe? *J. Hydrol.* **2021**, *603*, 127002. [CrossRef]
13. Zhang, H.; Immerzeel, W.; Zhang, F.; Kok, R.; Chen, D.; Yan, W. Snow cover persistence reverses the altitudinal patterns of warming above and below 5000 m on the Tibetan Plateau. *Sci. Total Environ.* **2021**, *803*, 149889. [CrossRef] [PubMed]
14. Nepal, S.; Khatiwada, K.; Pradhananga, S.; Kralisch, S.; Samyn, D.; Bromand, M.T.; Jamal, N.; Dildar, M.; Durrani, F.; Rassouly, F.; et al. Future snow projections in a small basin of the Western Himalaya. *Sci. Total Environ.* **2021**, *795*, 148587. [CrossRef]
15. IPCC. *Climate Change 2021: The Physical Science Basis. Contribution of Working Group I to the Sixth Assessment Report of the Intergovernmental Panel on Climate Change*; Cambridge University Press: London, UK, 2021; *in press*.
16. Ishtiaque, A.; Estoque, R.; Eakin, H.; Parajuli, J.; Rabby, Y. IPCC's current conceptualization of 'vulnerability' needs more clarification for climate change vulnerability assessments. *J. Environ. Manag.* **2022**, *303*, 114246. [CrossRef] [PubMed]
17. Zhong, X.; Zhang, T.; Su, H.; Xiao, X.; Wang, S.; Hu, Y.; Wang, H.; Zheng, L.; Zhang, W.; Xu, M.; et al. Impacts of landscape and climatic factors on snow cover in the Altai Mountains, China. *Adv. Clim. Chang. Res.* **2021**, *12*, 95–107. [CrossRef]
18. Collados-Lara, A.; Pardo-Iguzquiza, E.; Pulido- Veelazquez, D. A distributed cellular automata model to simulate potential future impacts of climate change on snow cover area. *Adv. Water Resour.* **2019**, *124*, 106–119. [CrossRef]
19. Yang, W.; Jin, F.; Si, Y.; Li, Z. Runoff change controlled by combined effects of multiple environmental factors in a headwater catchment with cold and arid climate in northwest China. *Sci. Total Environ.* **2021**, *756*, 143995. [CrossRef]

20. Hao, Y.; Sun, F.; Wang, H.; Liu, W.; Shen, Y.; Li, Z.; Hu, S. Understanding climate-induced changes of snow hydrological processes in the Kaidu River Basin through the CemaNeige-GR6J model. *Catena* **2022**, *212*, 106082. [CrossRef]
21. Sun, S.; Kang, S.; Guo, J.; Zhang, Q.; Paudyal, R.; Sun, X.; Qin, D. Insights into mercury in glacier snow and its incorporation into meltwater runoff based on observations in the southern Tibetan Plateau. *J. Environ. Sci.* **2018**, *68*, 130–142. [CrossRef]
22. Smith, T.; Rheinwalt, A.; Bookhagen, B. Topography and climate in the upper Indus Basin: Mapping elevation-snow cover relationships. *Sci. Total Environ.* **2021**, *786*, 147363. [CrossRef] [PubMed]
23. Saydi, M.; Ding, J. Impacts of topographic factors on regional snow cover characteristics. *Water Sci. Eng.* **2020**, *13*, 171–180. [CrossRef]
24. Lehning, M.; Grunewald, T.; Schirmer, M. Mountain snow distribution governed by an altitudinal gradient and terrain roughness. *Geophys. Res. Lett.* **2011**, *38*, L19504. [CrossRef]
25. Tang, G.; Li, S.; Yang, M.; Xu, Z.; Liu, Y.; Gu, H. Streamflow response to snow regime shift associated with climate variability in four mountain watersheds in the US Great Basin. *J. Hydrol.* **2019**, *573*, 255–266. [CrossRef]
26. Zwaaftink, C.; Mott, R.; Lehning, M. Seasonal simulation of drifting snow sublimation in Alpine terrain. *Water Resour. Res.* **2013**, *49*, 1581–1590. [CrossRef]
27. Helbig, N.; Herwijnen, A. Subgrid parameterization for snow depth over mountainous terrain from flat field snow depth. *Water Resour. Res.* **2017**, *53*, 1444–1456. [CrossRef]
28. Zaremehrjardy, M.; Razzavi, S.; Faramarzi, M. Assessment of the cascade of uncertainty in future snow depth projections across watersheds of mountainous, foothill, and plain areas in northern latitudes. *J. Hydrol.* **2021**, *598*, 125735. [CrossRef]
29. Tan, B.; Wu, F.; Yang, W.; He, X. Snow removal alters soil microbial biomass and enzyme activity in a Tibetan alpine forest. *Appl. Soil Ecol.* **2014**, *76*, 34–41. [CrossRef]
30. Qi, J.; Li, S.; Li, Q.; Xing, Z.; Bourque, C.; Meng, F. A new soil-temperature module for SWAT application in regions with seasonal snow cover. *J. Hydrol.* **2016**, *538*, 863–877. [CrossRef]
31. Collados-Lara, A.; Pardo-Iguzquiza, E.; Pulido-Velazquez, D. Assessing the impact of climate change–and its uncertainty–on snow cover areas by using cellular automata models and stochastic weather generators. *Sci. Total Environ.* **2021**, *788*, 147776. [CrossRef]
32. Inatsu, M.; Tanji, S.; Sato, Y. Toward predicting expressway closures due to blowing snow events. *Cold Reg. Sci. Technol.* **2020**, *177*, 103123. [CrossRef]
33. Iseri, Y.; Diaz, A.; Trinh, T.; Kavvas, M.; Ishida, K.; Anderson, M.L.; Ohara, N.; Snider, E.D. Dynamical downscaling of global reanalysis data for high-resolution spatial modeling of snow accumulation/melting at the central/southern Sierra Nevada watersheds. *J. Hydrol.* **2021**, *598*, 126445. [CrossRef]
34. Rittger, K.; Krock, M.; Kleiber, W.; Bair, E.; Brodzik, M.J.; Stephenson, T.R.; Rajagopalan, B.; Bormann, K.J.; Painter, T.H. Multi-sensor fusion using random forests for daily fractional snow cover at 30 m. *Remote Sens. Environ.* **2021**, *264*, 112608. [CrossRef]
35. Han, P.; Long, D.; Li, X.; Huang, Q.; Dai, L.; Sun, Z. A dual state-parameter updating scheme using the particle filter and high-spatial-resolution remotely sensed snow depths to improve snow simulation. *J. Hydrol.* **2021**, *594*, 125979. [CrossRef]
36. Dong, C. Remote sensing, hydrological modeling and in situ observations in snow cover research: A review. *J. Hydrol.* **2018**, *561*, 573–583. [CrossRef]
37. Williamson, S.; Hik, D.S.; Gamon, J.; Jarosch, A.; Anslow, F.; Clarke, G.K.C.; Rupp, T.S. Spring and summer monthly MODIS LST is inherently biased compared to air temperature in snow covered sub-Arctic mountains. *Remote Sens. Environ.* **2017**, *189*, 14–24. [CrossRef]
38. Deng, J.; Huang, X.; Ma, X.; Feng, Q.; Liang, T. Downscaling Algorithm and Verification of AMSR2 Snow Cover Depth Products in North Xinjiang. *Arid Zone Res.* **2016**, *33*, 1181–1188.
39. Dziubanski, D.J.; Franz, K.J. Assimilation of AMSR-E snow water equivalent data in a spatially lumped snow model. *J. Hydrol.* **2016**, *540*, 26–39. [CrossRef]
40. Liang, T.; Huang, X.; Wu, C.; Liu, X.; Li, W.; Guo, Z.; Ren, J. An application of MODIS data to snow cover monitoring in a pastoral area: A case study in Northern Xinjiang, China. *Remote Sens. Environ.* **2008**, *112*, 1514–1526. [CrossRef]
41. Wang, Y.; Huang, X.; Wang, J.; Zhou, M.; Liang, T. AMSR2 snow depth downscaling algorithm based on a multifactor approach over the Tibetan Plateau, China. *Remote Sens. Environ.* **2019**, *231*, 111268. [CrossRef]
42. Tong, R.; Komma, J.; Bloschl, G. Mapping snow cover from daily Collection 6 MODIS products over Austria. *J. Hydrol.* **2020**, *590*, 125548. [CrossRef]
43. Xue, Y.; Janjic, Z.; Dudhia, J.; Vasic, R.; Sales, F. A review on regional dynamical downscaling in intraseasonal to seasonal simulation/prediction and major factors that affect downscaling ability. *Atoms. Res.* **2014**, *147–148*, 68–85. [CrossRef]
44. Stegmann, P.; Tang, G.; Yang, P.; Johnson, B.T. A stochastic model for density-dependent microwave Snow- and Graupel scattering coefficients of the NOAA JCSDA community radiative transfer model. *J. Quant. Spectrosc. Radiat. Transf.* **2018**, *211*, 9–24. [CrossRef]
45. Xiao, X.; Zhang, T.; Zhong, X.; Shao, W.; Li, X. Support vector regression snow-depth retrieval algorithm using passive microwave remote sensing data. *Remote Sens. Environ.* **2018**, *210*, 48–64. [CrossRef]
46. Mhawej, M.; Faour, G.; Fayad, A.; Shaban, A. Towards an enhanced method to map snow cover areas and derive snow-water equivalent in Lebanon. *J. Hydrol.* **2014**, *513*, 274–282. [CrossRef]

47. Hancock, H.; Prokop, A.; Eckerstorfer, M.; Hendrikx, J. Combining high spatial resolution snow mapping and meteorological analyses to improve forecasting of destructive avalanches in Longyearbyen, Svalbard. *Cold Reg. Sci. Technol.* **2018**, *154*, 120–132. [CrossRef]
48. Bland, W.; Helmke, P.; Baker, J. High-resolution snow-water equivalent measurement by gamma-ray spectroscopy. *Agric. For. Meteorol.* **1997**, *83*, 27–36. [CrossRef]
49. Nouri, M.; Homaee, M. Spatiotemporal changes of snow metrics in mountainous data-scarce areas using reanalyses. *J. Hydrol.* **2021**, *603*, 126858. [CrossRef]
50. Li, Z.; Lyu, S.; Chen, H.; Ao, Y.; Zhao, L.; Wang, S.; Zhang, S.; Meng, X. Changes in climate and snow cover and their synergistic influence on spring runoff in the source region of the Yellow River. *Sci. Total Environ.* **2021**, *799*, 149503. [CrossRef]
51. Che, T.; Li, X.; Jin, R.; Armstrong, R.; Zhang, T.J. Snow depth derived from passive microwave remote-sensing data in China. *Ann. Glaciol.* **2008**, *49*, 145–154. [CrossRef]
52. Li, Q.; Yang, T.; Zhou, F.; Li, L. Patterns in snow depth maximum and snow cover days during 1961–2015 period in the Tianshan Mountains, Central Asia. *Atmos. Res.* **2019**, *228*, 14–22. [CrossRef]
53. Dai, L.Y.; Che, T.; Ding, Y.J. Inter-calibrating SMMR, SSM/I and SSMI/S data to improve the consistency of snow-depth products in China. *Remote Sens.* **2015**, *7*, 7212–7230. [CrossRef]
54. Guo, D.; Pepin, N.; Yang, K.; Sun, J.; Li, D. Local changes in snow depth dominate the evolving pattern of elevation-dependent warming on the Tibetan Plateau. *Sci. Bull.* **2021**, *66*, 1146–1150. [CrossRef]
55. Qi, Y.; Wang, H.; Ma, X.; Zhang, J.; Yang, R. Relationship between vegetation phenology and snow cover changes during 2001–2018 in the Qilian Mountains. *Ecol. Indic.* **2021**, *133*, 108351. [CrossRef]
56. Dietz, A.; Kuenzer, C.; Conrad, C. Snow cover variability in Central Asia between 2000 and 2011 derived from improved MODIS daily snow cover products. *Int. J. Remote Sens.* **2013**, *34*, 3879–3902. [CrossRef]
57. Tomaszewska, M.; Nguyen, L.; Henebry, G. Land surface phenology in the highland pastures of montane Central Asia: Interactions with snow cover seasonality and terrain characteristics. *Remote Sens. Environ.* **2020**, *240*, 111675. [CrossRef]
58. Li, Z.; Feng, Q.; Liu, W.; Wang, T.; Cheng, A.; Gao, Y.; Guo, X.; Pan, Y.; Li, J.; Guo, R.; et al. Study on the contribution of cryosphere to runoff in the cold alpine basin: A case study of Hulugou River Basin in the Qilian Mountains. *Global Planet. Change* **2014**, *122*, 345–361.
59. Sen, K.; Kumar, P. Estimates of the regression coefficient based on Kendall's tau. *J. Am. Stat. Assoc.* **1968**, *63*, 1379–1389. [CrossRef]
60. Mann, H.B. Nonparametric tests against trend. *Econometrical* **1945**, *13*, 245–259. [CrossRef]
61. Kendall, M.G. *Rank Correlation Methods*; Griffin: London, UK, 1995.
62. Li, C.; Wang, L.; Wang, W.; Qi, J.; Yang, L.; Zhang, Y.; Wu, L.; Cui, X.; Wang, P. An analytical approach to separate climate and human contributions to basin streamflow variability. *J. Hydrol.* **2018**, *559*, 30–42. [CrossRef]
63. Zheng, H.; Zhang, L.; Zhu, R.; Liu, C.; Sato, Y.; Fukushima, Y. Responses of streamflow to climate and land surface change in the headwaters of the Yellow River Basin. *Water Resour. Res.* **2009**, *45*, 641–648. [CrossRef]
64. Rees, A.; English, M.; Derksen, C.; Toose, P.; Sillis, A. Observations of late winter Canadian tundra snow cover properties. *Hydrol. Process.* **2013**, *28*, 3962–3977. [CrossRef]
65. Kudo, R.; Yoshida, T.; Masumoto, T. Uncertainty analysis of impacts of climate change on snow processes: Case study of interactions of GCM uncertainty and an impact model. *J. Hydrol.* **2017**, *548*, 196–207. [CrossRef]
66. Sade, R.; Rimmer, A.; Litaor, M.; Shamir, E.; Furman, A. Snow surface energy and mass balance in a warm temperate climate mountain. *J. Hydrol.* **2014**, *519*, 848–862. [CrossRef]
67. Li, Y.; Chen, Y.; Li, Z. Climate and topographic controls on snow phenology dynamics in the Tienshan Mountains, Central Asia. *Atmos. Res.* **2020**, *236*, 104813. [CrossRef]
68. Chu, Q.; Yan, G.; Qi, J.; Mu, X.; Li, L.; Tong, Y.; Zhou, Y.; Liu, Y.; Xie, D.; Wild, M. Quantitative analysis of terrain reflected solar radiation in snow-covered mountains: A case study in Southeastern Tibetan Plateau. *J. Geophys. Res. Atmos.* **2021**, *126*, e2020JD034293. [CrossRef]
69. Misra, A.; Kumar, A.; Bhambri, R.; Hartashya, U.; Verma, A.; Dobhal, D.P.; Gupta, A.K.; Gupta, G.; Upadhyay, R. Topographic and climatic influence on seasonal snow cover: Implications for the hydrology of ungauged Himalayan basins, India. *J. Hydrol.* **2020**, *585*, 124716. [CrossRef]
70. Revuelto, J.; Cluzet, B.; Duran, N. Fructus.; Lafaysse, M.; Cosme, E.; Dumont, M. Assimilation of surface reflectance in snow simulations: Impact on bulk snow variables. *J. Hydrol.* **2021**, *603*, 126966. [CrossRef]
71. Wang, S.; Wang, X.; Chen, G.; Yang, Q.; Wang, B.; Ma, Y.; Shen, M. Complex responses of spring alpine vegetation phenology to snow cover dynamics over the Tibetan Plateau, China. *Sci. Total Environ.* **2017**, *593*, 449–461. [CrossRef]
72. Essery, R.; Li, L.; Pnmeroy, J. A distributed model of blowing snow over complex terrain. *Hydrol. Process.* **1999**, *13*, 2424–2438. [CrossRef]
73. Goncharova, O.; Matyshak, G.; Epstein, H.; Sefilian, A.; Bobrik, A. Influence of snow cover on soil temperatures: Meso- and micro-scale topographic effects (a case study from the northern West Siberia discontinuous permafrost zone). *Catena* **2019**, *183*, 104224. [CrossRef]
74. Minder, J.; Letcher, T.; Skiles, S. An evaluation of high-resolution regional climate model simulations of snow cover and albedo over the Rocky Mountains, with implications for the simulated snow-albedo feedback. *J. Geophys. Res. Atmos.* **2016**, *121*, 9069–9088. [CrossRef]

75. Comola, F.; Schaefli, B.; Ronco, P.; Botter, G.; Bavay, M.; Rinaldo, A.; Lehning, M. Scale-dependent effects of solar radiation patterns on the snow-dominated hydrologic response. *Geophys. Res. Lett.* **2015**, *42*, 3895–3902. [CrossRef]
76. You, Q.; Wu, T.; Shen, L.; Pepin, N.; Zhang, L.; Jiang, Z.; Wu, Z.; Kang, S.; AghaKouchak, A. Review of snow cover variation over the Tibetan Plateau and its influence on the broad climate system. *Earth-Sci. Rev.* **2020**, *201*, 103043. [CrossRef]
77. Huang, X.; Deng, J.; Wang, W.; Feng, Q.; Liang, T. Impact of climate and elevation on snow cover using integrated remote sensing snow products in Tibetan Plateau. *Remote Sens. Environ.* **2017**, *190*, 274–288. [CrossRef]
78. Dedieu, J.; Lessard-Fontaine, A.; Ravazzani, G.; Cremonese, E.; Shalpykova, G.; Beniston, M. Shifting mountain snow patterns in a changing climate from remote sensing retrieval. *Sci. Total Environ.* **2014**, *493*, 1267–1279. [CrossRef]
79. Tang, X.; Lv, X.; He, Y. Features of climate change and their effects on glacier snow melting in Xinjiang, China. *Comptes Rendus Geosci.* **2013**, *345*, 93–100. [CrossRef]
80. Dibike, Y.; Eum, H.; Prowse, T. Modelling the Athabasca watershed snow response to a changing climate. *J. Hydrol. Reg. Stud.* **2018**, *15*, 134–148. [CrossRef]
81. Tan, X.; Wu, Z.; Mu, X.; Gao, P.; Zhao, G.; Sun, W.; Gu, C. Spatiotemporal changes in snow cover over China during 1960–2013. *Atmos. Res.* **2019**, *218*, 183–194.
82. Wei, Y.; Wang, S.; Fang, Y.; Nawaz, Z. Integrated assessment on the vulnerability of animal husbandry to snow disasters under climate change in the Qinghai-Tibetan Plateau. *Global Planet. Change* **2017**, *157*, 139–152. [CrossRef]
83. Li, C.; Qi, J.; Wang, S.; Yang, L.; Zou, S.; Zhu, G.; Yang, W. Spatiotemporal characteristics of alpine snow and ice melt under a changing regional climate: A case study in Northwest China. *Quat. Int.* **2015**, *358*, 126–136. [CrossRef]
84. Panda, S.; Dash, S.; Bhaskaran, B.; Pattnayak, K. Investigation of the snow-monsoon relationship in a warming atmosphere using Hadley Centre climate model. *Global Planet. Change* **2016**, *147*, 125–136. [CrossRef]
85. Shenvi, M.; Sandu, C.; Untaroiu, C. Review of compressed snow mechanics: Testing methods. *J. Terramechanics* **2022**, *100*, 25–37. [CrossRef]
86. Mainieri, R.; Favillier, A.; Lopez-Saez, J.; Eckert, N.; Zgheib, T.; Morel, P.; Saulnier, M.; Peiry, J.; Stoffel, M.; Corona, C. Impacts of land-cover changes on snow avalanche activity in the French Alps. *Anthropocene* **2020**, *30*, 100244. [CrossRef]
87. Han, L.; Tsunekawa, A.; Tsubo, M.; He, C.; Shen, M. Spatial variations in snow cover and seasonally frozen ground over northern China and Mongolia, 1988–2010. *Global Planet. Change* **2014**, *116*, 139–148. [CrossRef]
88. Martins, M.S.M.; Valera, C.A.; Zanata, M.; Santos, R.M.B.; Abdala, V.L.; Pacheco, F.A.L.; Fernandes, L.F.S.; Pissarra, T.C.T. Potential Impacts of Land Use Changes on Water Resources in a Tropical Headwater Catchment. *Water* **2021**, *13*, 3249. [CrossRef]
89. Kinnard, C.; Bzeouich, G.; Assani, A. Impacts of summer and winter conditions on summer river low flows in low elevation, snow-affected catchments. *J. Hydrol.* **2022**, *605*, 127393. [CrossRef]
90. Carword, C.; Manson, S.; Bauer, M.; Hall, D. Multitemporal snow cover mapping in mountainous terrain for Landsat climate data record development. *Remote. Sens. Environ.* **2013**, *135*, 224–233.
91. Poulin, A.; Brissette, F.; Leconte, R.; Arsenault, R.; Malo, J. Uncertainty of hydrological modelling in climate change impact studies in a Canadian, snow-dominated river basin. *J. Hydrol.* **2011**, *409*, 626–636. [CrossRef]
92. Frei, A.; Tedesco, M.; Lee, S.; Foster, J.; Hall, D.K.; Kelly, R.; Robinson, D.A. A review of global satellite-derived snow products. *Adv. Space Res.* **2012**, *50*, 1007–1029. [CrossRef]

remote sensing

MDPI

Article

Seasonal Variation in Microphysical Characteristics of Precipitation at the Entrance of Water Vapor Channel in Yarlung Zangbo Grand Canyon

Ran Li, Gaili Wang *, Renran Zhou, Jingyi Zhang and Liping Liu

State Key Laboratory of Severe Weather, Chinese Academy of Meteorological Science, Beijing 100081, China; liran201@mails.ucas.ac.cn (R.L.); zhourenran21@mails.ucas.ac.cn (R.Z.); 3190101053@stu.cuit.edu.cn (J.Z.); liulp@cma.gov.cn (L.L.)

* Correspondence: wanggl@cma.gov.cn

Abstract: Mêdog is located at the entrance of the water vapor channel in the Yarlung Zangbo Grand Canyon (YGC). This area has the largest annual accumulated rainfall totals and precipitation frequency on the Tibetan Plateau (TP). This paper investigates the seasonal variation in raindrop size distribution (DSD) characteristics in Mêdog based on disdrometer observations from 1 July 2019 to 30 June 2020. The DSD characteristics are examined under six rain rate classes and two rainfall types (stratiform and convective) in the winter, premonsoon, monsoon and postmonsoon periods. The highest (lowest) concentration of small raindrops is observed in monsoon (winter) precipitation, whereas large raindrops predominate in premonsoon precipitation. For stratiform rainfall, the mean mass-weighted mean diameter (D_m) exhibits overlooked differences in the four periods, while the mean normalized intercept parameter (N_w) is significantly higher in the monsoon period than in the other three periods. The convective rainfall in the monsoon and postmonsoon periods is characterized by a high concentration of limited-size drops and can be classified as maritime-like. This is probably attributed to abundant warm and humid airflow transported by the Indian Ocean monsoon into Mêdog. The westerly winds prevail over the TP during the premonsoon period, and thereby the premonsoon convective rainfall in Mêdog has a larger mean D_m and a lower mean N_w. In addition, the relationships of radar reflectivity Z and rain rate R for different precipitation types in different periods are also derived. A better understanding of the seasonal variation in the microphysical characteristics of precipitation in Mêdog is important for improving the microphysical parameterization scheme and the precipitation forecast of models on the TP.

Keywords: Tibetan Plateau; raindrop size distribution (DSD); seasonal variation; maritime-like

Citation: Li, R.; Wang, G.; Zhou, R.; Zhang, J.; Liu, L. Seasonal Variation in Microphysical Characteristics of Precipitation at the Entrance of Water Vapor Channel in Yarlung Zangbo Grand Canyon. *Remote Sens.* **2022**, *14*, 3149. https://doi.org/10.3390/rs14133149

Academic Editor: Federico Porcù

Received: 8 April 2022
Accepted: 27 June 2022
Published: 30 June 2022

1. Introduction

The microphysical processes of clouds and precipitation play vital roles in the formation and development of precipitation and the prediction of severe weather. Raindrop size distribution (DSD) is an important feature that characterizes the microphysical process of precipitation [1–3] and is mainly affected by climatic characteristics and precipitation types [3–9]. In recent years, disdrometer DSD measurements have been widely used to study the microphysical characteristics of precipitation [4,10–14]. Many DSD observations and analyses have been carried out in different regions of China. Based on the OTT Particle Size Velocity (PARSIVEL) disdrometer data from Nagqu (4500 m above sea level (ASL)) over the Tibetan Plateau (TP), Chen et al. [15] reported that the discrepancy in DSDs between day and night is nonsignificant in stratiform rainfall but obvious in convective rainfall. The DSDs of different precipitation types (stratiform and convective) between Nagqu over the TP and Yangjiang in southern China were compared and showed that all three gamma parameters for stratiform precipitation over the TP are larger than those in southern China, while the normalized intercept parameter N_w and the shape parameter

μ for convective precipitation are less than those in southern China [16]. DSD statistical analysis was also conducted in Yining, Xinjiang, an arid region of China, and it showed that convective precipitation was neither continental-like nor maritime-like [17]. In addition, the same location will display significant seasonal differences in the microphysical processes of precipitation [18]. The precipitation over the South China Sea (SCS) is dominated by small (midsize) drops during the premonsoon (monsoon) period, while it has the lowest concentration of raindrops in the postmonsoon season [18]. Monsoon precipitation at Thiruvananthapuram, a coastal tropical station in India, has a higher concentration of small drops than in the other three seasons [3]. Krishna et al. [6] found that the mean concentrations of medium and large raindrops in the west monsoon season are higher than those in the east monsoon season in the Palau Islands.

The TP is located in western China, with an average elevation of approximately 4000 m. It is important to the climate and ecosystems of the Asian continent and even the world [19]. The TP is also known as the Water Tower of Asia due to the origination of seven important Asian rivers, including the Yellow River, the Yangtze River, the Yarlung Zangbo River, etc. The westerlies–monsoon synergy zone covers the TP and the surrounding areas. Climate warming has led to anomalies in westerlies–monsoons and an imbalance in the Water Tower of Asia. The Yarlung Zangbo Grand Canyon (YGC), with a total length of 496.3 km and a depth of up to 6009 m [20], is located in the southeast TP and is the largest channel for transporting water vapor to the TP. During the Indian summer monsoon period, warm and wet water vapor is transported northward to the TP along the YGC. The water vapor transport intensity (nearly 2000 g cm^{-1} s^{-1}) is equivalent to that from the south bank of the Yangtze River to the north bank in summer [21]. The YGC plays an important role in climate change in the TP and is a typical unit in the TP climate system.

Mêdog, with a mean altitude of 1200 m, is located at the entrance of the YGC. The humid air from the Indian Ocean flows straight into the gorge, giving Mêdog the most annual accumulated precipitation on the TP [22]. Due to inconvenient transportation and frequent debris flows in the rainy season, in situ observation data are lacking along the YGC, especially in Mêdog. To explore the causes and related mechanisms of water resource changes in the Yarlung Zangbo River basin under the synergistic action of westerlies–monsoons in the southeast TP, a comprehensive cloud precipitation observation test base was established at the Mêdog Climate Observatory (95.32°E, 29.31°N), supported by "the Second Tibetan Plateau Scientific Expedition" and "the Earth-Atmosphere Interaction in the TP and its Influence on the Weather and Climate in the Lower Reaches" projects. A Ka-band cloud radar, a micro rain radar, an OTT PARSIVEL disdrometer and other instruments were deployed at Mêdog National Climate Observatory to obtain the three-dimensional structure of clouds and precipitation characteristics in the YGC. Based on Ka-band cloud radar measurements, the vertical structure characteristics and diurnal variation in clouds over Mêdog in the southeast TP were analyzed [23]. In addition, precipitation in Mêdog was dominated by small and medium drops, and the convective rain in this region could be classified as maritime-like [24]. However, the seasonal variation characteristics of the raindrop spectrum were not analyzed due to the short observation period. In this study, DSD data collected from an OTT PARSIVEL disdrometer during the period of July 2019 to June 2020 were used to study the seasonal variation in microphysical characteristics for different precipitation intensities and precipitation types. In addition, the European Centre for Medium-Range Weather Forecasts (ECMWF) Reanalysis version 5 (ERA5) data, Fengyun-4A (FY-4A) satellite products and automatic weather station (AWS) observations were used to address the possible reasons for the seasonal differences in DSDs in Mêdog. This study aimed to better understand the seasonal variation in the microphysical characteristics of precipitation processes at the entrance of the water vapor channel in the YGC and its relationship with westerlies–monsoon synergy and water vapor transport, which is beneficial for improving the microphysical parameterization scheme and the precipitation forecast of the models in the TP.

The instruments, data and methods adopted in this study are provided in Section 2. The properties of DSDs and microphysics parameters for different rain rate classes and precipitation types in different seasons are reported in Section 3. Section 4 discusses the possible reasons for the seasonal variations in DSDs. The major conclusions are given in the final section.

2. Data and Methods

The proposed research investigation used different datasets to provide an overall evaluation of seasonal variation in raindrop size distribution in Mêdog. The main steps followed along this study are presented in the flow chart depicted in Figure 1.

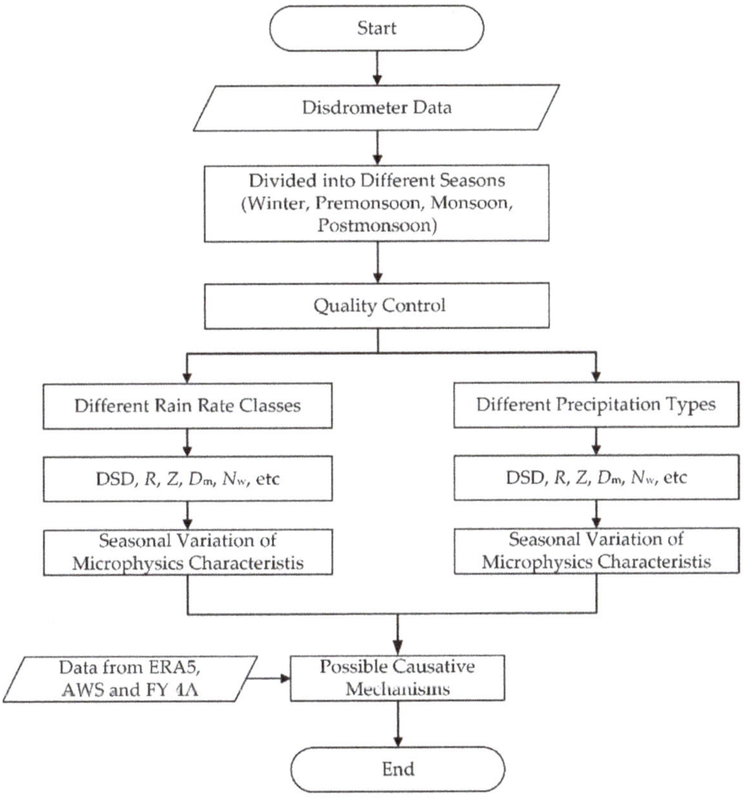

Figure 1. Methodological workflow diagram adopted in this study.

Measurements from a laser-optical PARSIVEL disdrometer [7] in Mêdog with a time resolution of 1 min were used in this study. The position of the Mêdog National Climate Observatory and a picture of the PARSIVEL disdrometer are shown in Figure 2. A disdrometer can simultaneously measure the size and falling speed of hydrometeors. The size and falling speed ranged from 0.06 mm to 24.5 mm and from 0.05 to 20.8 m s^{-1}, respectively, which were divided into 32 nonequidistant bins [25,26].

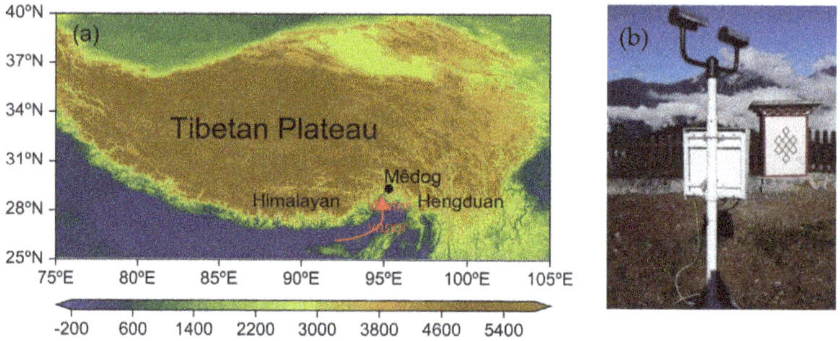

Figure 2. The locations of Mêdog (black solid dot) and Yarlung Zangbo Grand Canyon (YGC), and topography (m) of the Tibetan Plateau (TP) (**a**) and the PARSIVEL disdrometer (**b**). The red arrow indicates water vapor channel in the YGC.

The disdrometer data used in this study were collected from 1 July 2019 to 30 June 2020 and divided into four periods, winter (January–February), premonsoon (March–May), monsoon (June–September) and postmonsoon (October–December) [3], to study the seasonal characteristics of DSDs in Mêdog. During this period, 30 days and 893 min of data were missing due to power outages caused by geological disasters, such as landslides and debris flows.

2.1. Quality Control

In this study, strict quality control was carried out on disdrometer data to eliminate the influence of raindrop classification errors caused by edge landing, strong wind and splashing. Firstly, the falling speed beyond the boundary of ±60% of Beard's empirical speed–diameter relationship was excluded [27]. Given the terrain altitude of Mêdog, the speed–diameter relationship was corrected by multiplying by an air density factor of 1.04 [28]. Figure 3 gives the accumulated raw particle counts by diameter and fall speed observed during the four seasons. Drops beyond a ±60% empirical fall speed–diameter relationship were eliminated. The distribution of fall speed–diameter basically conformed to Beard's empirical fall speed–diameter relationship after quality control.

Secondly, the first and second size bins were removed due to their low signal-to-noise ratio, and the size bins with a number of drops less than 2 or with diameters greater than 6 mm were screened and eliminated [10]. One-minute samples with a total number of raindrops less than 10 or a rainfall rate less than 0.1 mm h^{-1} were regarded as instrument noise and eliminated [3,29]. Good agreements between disdrometer observations and gauge measurements in Mêdog have been reported by Wang et al. [24], although disdrometers tend to underestimate gauged rain. This suggests that disdrometer data could be used to explore the seasonal variation in the microphysical characteristics of precipitation in Mêdog.

A total of 73,707 min of precipitation samples after quality control were collected from the disdrometer at the Mêdog National Climate Observatory, with accumulated rainfall of 1237.57 mm. Table 1 shows the total rain duration and accumulated rainfall amount during the four seasons. As seen from Table 1, Mêdog precipitation mainly occurred in the monsoon period, with the rainfall in this period being 699.92 mm, accounting for approximately 57% of the total, followed by the premonsoon period, accounting for approximately 32%. The rainfall in winter was the lowest, accounting for only approximately 4% of the total rainfall. Rainfall exhibited obvious seasonal variation.

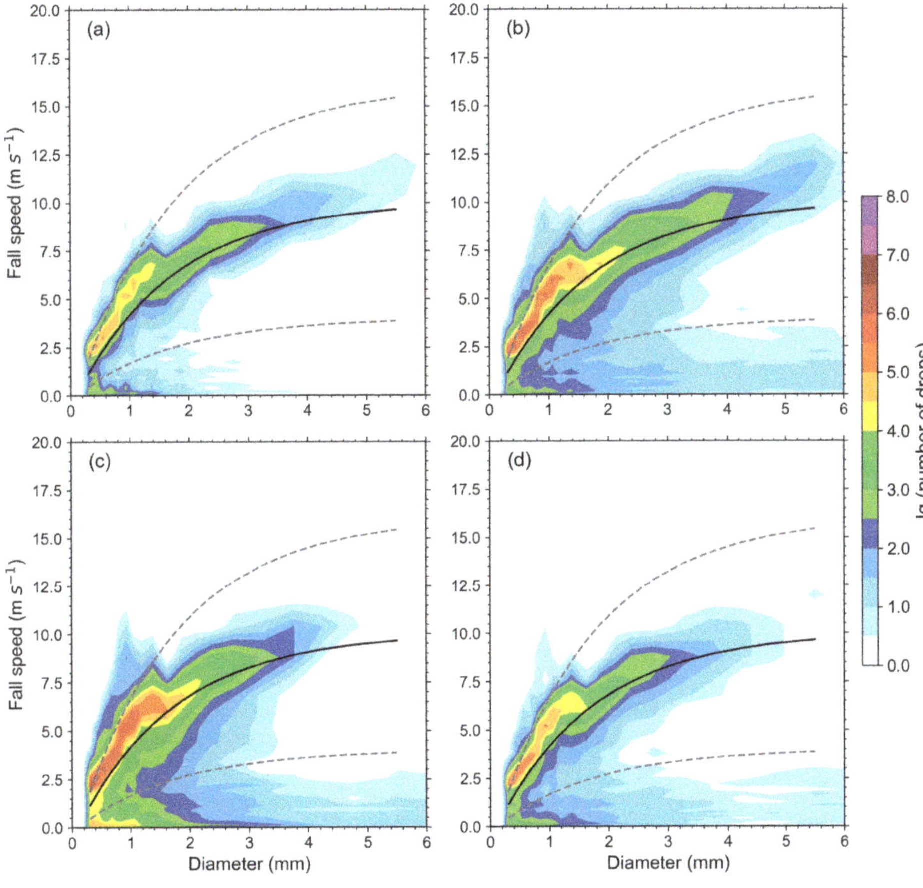

Figure 3. Accumulated raw particle counts by diameter and fall speed were observed during the four seasons: winter (**a**), premonsoon (**b**), monsoon (**c**) and postmonsoon (**d**). Solid black lines indicate the empirical fall speed–diameter relationship. Dashed lines denote the ±60% empirical fall speed–diameter relationship.

Table 1. Total rain duration and accumulated rainfall amount during the four seasons.

Season	Total Rain Duration (min)/Frequency (%)	Accumulated Rainfall (mm)/Percentage (%)
Winter	6153/8.35	48.90/3.95
Pre-mon	24,880/33.75	400.76/32.38
Monsoon	35,538/48.22	699.92/56.56
Post-mon	7136/9.68	87.99/7.11

2.2. Parameter Calculation

The raindrop number ($n_{i,j}$) of the ith size (D_i) and the jth speed (V_j) are measured by a disdrometer. The raindrop number concentration, $N(D_i)$ (m^{-3} mm^{-1}), in the ith size can be calculated as follows:

$$N(D_i) = \sum_{j=1}^{32} \frac{n_{i,j}}{V_j \times S/T \times \Delta D_i} \quad (1)$$

where S (m^2) and T (s) are the sampling area and sampling time and were set to 54 cm^2 and 60 s in this study, respectively. ΔD_i indicates the size interval.

The rainfall rate R $\left(mm\ h^{-1}\right)$, radar reflectivity factor Z $\left(mm^6\ m^{-3}\right)$, total raindrop concentration N_T $\left(m^{-3}\right)$ and liquid water content (LWC, g m^{-3}) can be obtained from the following equations:

$$R = 6\pi \times 10^{-4} \sum_{i=3}^{32} \sum_{j=1}^{32} {D_i}^3 \frac{n_{i,j}}{S \times T} \tag{2}$$

$$Z = \sum_{i=3}^{32} {D_i}^6 N(D_i) \Delta D_i \tag{3}$$

$$N_T = \sum_{i=3}^{32} N(D_i) \Delta D_i, \tag{4}$$

$$LWC = \frac{\pi}{6000} \sum_{i=3}^{32} {D_i}^3 N(D_i) \Delta D_i, \tag{5}$$

In this paper, the gamma distribution model was used to fit the observed DSDs from the disdrometer [30]:

$$N(D) = N_0 D^{\mu} \exp(-\Lambda D) \tag{6}$$

where D (mm) represents the hydrometeor diameter, N_0 $\left(mm^{-\mu-1}\ m^{-3}\right)$ is the intercept parameter, and Λ $\left(mm^{-1}\right)$ and μ indicate the slope parameter and shape parameter, respectively. Λ and μ can be calculated as follows [31]:

$$M_x = \sum_{i=3}^{32} N(D_i) {D_i}^x \Delta D_i, \tag{7}$$

$$G = \frac{M_4^3}{M_3^2 M_6}, \tag{8}$$

$$\mu = \frac{11G - 8 + \sqrt{G(G+8)}}{2(1-G)}, \tag{9}$$

$$\Lambda = (\mu + 4) \frac{M_3}{M_4}. \tag{10}$$

Testud et al. [32] proposed the normalized gamma distribution:

$$N(D) = N_w f(\mu) \left(\frac{D}{D_m}\right)^{\mu} \exp\left[-(4+\mu)\frac{D}{D_m}\right], \tag{11}$$

$$f(\mu) = \frac{\Gamma(4)}{4^4} \frac{(4+\mu)^{4+\mu}}{\Gamma(4+\mu)}, \tag{12}$$

where $\Gamma(x)$ represents a complete gamma function that is defined as follows:

$$\Gamma(x) = \sqrt{2\pi} e^{-x} x^{x-\frac{1}{2}} \tag{13}$$

The mass-weighted mean diameter D_m (mm) and normalized intercept parameter N_w (m^{-3} mm^{-1}) can be used to describe the general DSD characteristics and can be defined as follows [33]:

$$D_m = \frac{M_4}{M_3}, \tag{14}$$

$$N_w = \frac{256}{6} \times \frac{M_3^5}{M_4^4}. \tag{15}$$

2.3. Different Classes in R, D_m and N_T

The DSD data used in this study were divided into the following six categories according to R, D_m and N_T, respectively, as shown in Table 2. Percentages of occurrence and relative contributions to total rainfall were calculated for different categories of R, D_m and N_T. To improve the representativeness of the statistical characteristics, the rainfall rate categories with fewer than 20 samples were excluded.

Table 2. Categories of rain rate (R), mass-weighted mean diameter (D_m) and total raindrop concentration (N_T).

Rain Rate (R)		Mass-Weighted Mean Diameter (D_m)		Total Raindrop Concentration (N_T)	
Variable	**Range, mm h^{-1}**	**Variable**	**Range, mm**	**Variable**	**Range, m^{-3}**
R1	0.1–1	D_m1	<1	N_T1	10–250
R2	1–2	D_m2	1–2	N_T2	250–500
R3	2–5	D_m3	2–3	N_T3	500–750
R4	5–10	D_m4	3–4	N_T4	750–1000
R5	10–20	D_m5	4–5	N_T5	1000–1500
R6	>20	D_m6	>5	N_T6	>1500

2.4. Different Precipitation Types

To further analyze the DSD characteristics in different seasons, the 1 min DSD samples from the disdrometer were also classified into stratiform rainfall and convective rainfall according to a simple method based on the SD σ_R of rainfall rate R [34]. Specifically, for 10 continuous 1 min DSD samples, if $R \geq 5$ mm h^{-1} and $\sigma_R > 1.5$ mm h^{-1}, convective rainfall was distinguished, while when $\sigma_R \leq 1.5$ mm h^{-1}, stratiform rainfall was classified.

In addition to the disdrometer measurements, ECMWF ERA5 reanalysis data, AWS observations and FY-4A products were also used in this study. ERA5 reanalysis data are the fifth-generation ECMWF reanalysis for the global climate and weather for the past 4 to 7 decades recorded by C3S Climate Data Store (CDS, https://cds.climate.copernicus.eu, accessed on 1 September 2021) with a spatial resolution of 0.25° × 0.25°. FY-4A is China's second-generation geostationary meteorological satellite and carries the Advanced Geosynchronous Radiation Imager (AGRI), the Geostationary Interferometric Infrared Sounder (GIIRS) and the Lighting Mapping Imager (LMI). AGRI has 14 channels, and the spatial resolution can reach 0.5–1 km for visible and near-infrared bands and 2–4 km for infrared bands. The AGRI level2 dataset provides the cloud type (CLT), the cloud top height (CTH), the Black Body Temperature (TBB) and other products, which can be obtained at FENGYUN Satellite Data Center (http://satellite.nsmc.org.cn, accessed on 1 September 2021).

3. Result

3.1. Statistical Characteristics

Table 3 shows the maximum, mean and standard deviation (SD) of R, D_m and N_T calculated by using 1 min DSD disdromter samples in different seasons. The maximum rainfall rate of 56.643 mm h^{-1} was in the premonsoon season, indicating that the strongest convective precipitation occurred in the premonsoon season. The mean R was highest in the monsoon season, followed by the premonsoon season, and the weakest in winter. The sequence was in line with that of the accumulated rainfall amount in the four seasons. The low value of SD in all the seasons indicated a small variation in precipitation intensity in Mêdog. The lowest SD in winter may be related to uniform stratiform precipitation and minimal convective precipitation. The higher SD was more or less in the premonsoon and monsoon seasons, showing that convective precipitation mainly occurred in the two seasons.

Table 3. Maximum, mean and standard deviation of R, D_m and N_T during the four seasons.

		Winter	Pre-Mon	Monsoon	Post-Mon
R (mm h^{-1})	Max	7.645	56.650	43.980	24.016
	Mean	0.477	0.966	1.182	0.740
	SD	0.563	1.668	1.733	1.184
D_m (mm)	Max	2.970	3.645	3.168	3.314
	Mean	0.976	1.021	0.900	0.923
	SD	0.258	0.287	0.237	0.236
N_T (m^{-3})	Max	810.352	1914.562	2401.187	1449.321
	Mean	96.262	156.808	279.416	176.428
	SD	57.739	130.039	217.689	167.339

The maximum, mean and SD of D_m were found to be larger in the premonsoon season, which indicates stronger convective actions. During the monsoon season, the mean and SD of D_m were smaller, probably due to precipitation dominated by warm rain processes and the relative consistency of precipitation [3,24]. The maximum, mean and SD of N_T were the highest in the monsoon season, which indicates that the raindrop concentration was the highest with larger dispersion.

In general, the monsoon exhibited the largest values for the mean and SD of R, the smallest values for the mean and SD of D_m, and the highest values for the maximum, mean and SD of N_T. All these features showed that rainfall during the monsoon period is characterized by abundant, smaller drops, which may be attributed to the sufficient warm and humid air flows from the Indian Ocean. In addition, the D_m of premonsoon precipitation registered larger values in the maximum, mean and SD, as well as the highest value of the maximum R. Therefore, stronger convective activities probably occurred in the premonsoon season.

3.2. Seasonal Variation in DSDs

Figure 4 shows the DSDs of different seasons from mean spectra. Drops with $D \leq 1$ mm were regarded as small raindrops, $D > 3$ mm as large raindrops, and $1 < D \leq 3$ mm as medium raindrops [18]. As seen from Figure 4, the DSDs in Mêdog exhibited bimodal distribution with peaks at 0.4 mm and 1.1 mm. This characteristic of the multipeak raindrop spectrum has been discussed [35]. In terms of small raindrops, the highest concentration was in the monsoon season and the lowest was in winter, and the premonsoon season was similar to the postmonsoon season. As raindrop diameter increased, the concentration of medium raindrops was slightly higher in the monsoon and premonsoon seasons than in the postmonsoon and winter seasons. The concentration of large raindrops was the highest (lowest) in the premonsoon (monsoon) season. The results showing that the concentration of large (small) raindrops was the highest in the premonsoon season (monsoon) were consistent with those found in tropical coastal areas, which are also dominated by the Indian Ocean monsoon in summer [3]. Unlike the SCS, Zeng et al. [18] reported that small drops predominate in precipitation during the premonsoon period, while large drops prevail in the postmonsoon season.

To further analyze the DSD characteristics in different seasons, the DSD data used in this study were divided into six categories, as shown in Table 2. The DSDs of different rainfall rate categories from mean spectra for different seasons are shown in Figure 5. As the rainfall rate increased, the spectra width in all seasons became wider, and the difference in DSDs among the four seasons gradually increased. For $R \leq 5$ mm h^{-1} (usually corresponding to stratiform precipitation [13]), the concentration of large raindrops was the highest in winter. Precipitation with $R \geq 10$ mm h^{-1} (usually corresponding to convective rainfall [36,37]) occurred mainly in the premonsoon and monsoon seasons. The concentration of large raindrops in the premonsoon season was significantly greater than that in the monsoon season, indicating stronger convective rainfall in the premonsoon season.

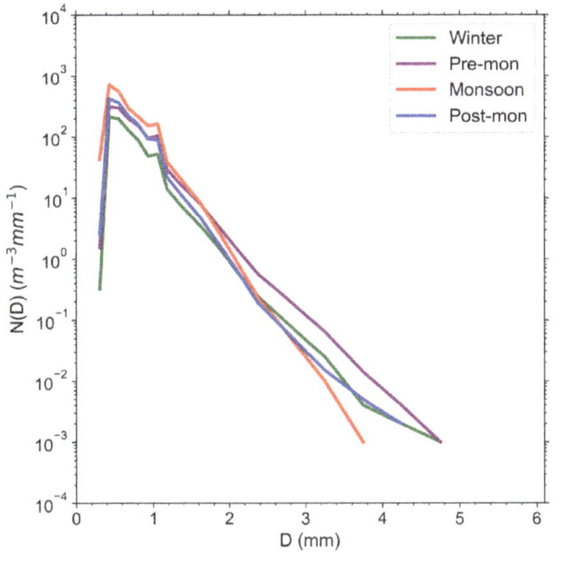

Figure 4. Averaged DSDs during the four seasons.

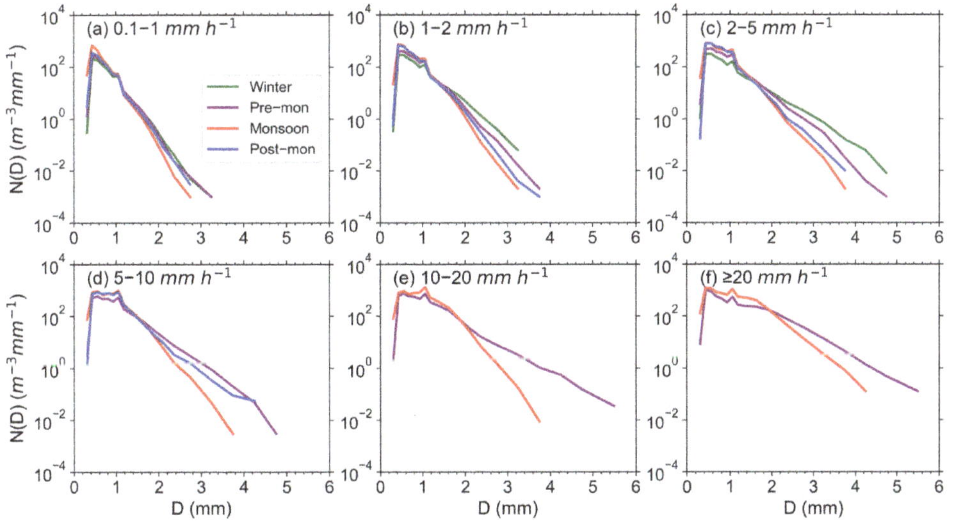

Figure 5. Averaged DSDs for different rain rate categories. (**a**) R1: $0.1 \le R < 1$ mm h^{-1}, (**b**) R2: $1 \le R < 2$ mm h^{-1}, (**c**) R3: $2 \le R < 5$ mm h^{-1}, (**d**) R4: $5 \le R < 10$ mm h^{-1}, (**e**) R5: $10 \le R < 20$ mm h^{-1} and (**f**) R6: $R \ge 20$ mm h^{-1}.

Table 4 gives the average rainfall microphysical parameters for each of the six rainfall rate categories from 1 min DSD samples in different seasons. The mean values of Z, LWC, N_T and D_m increased with increasing rain rate in all seasons. The $\log_{10}(N_w)$ tended to decrease with the increasing rain rate in winter, indicating that the increase in precipitation intensity was mainly attributed to the increase in raindrop size. During other periods, $\log_{10}(N_w)$ tended to increase with increasing rain rate until $R > 20$ mm h^{-1}. For the same rainfall rate categories, monsoon precipitation was characterized by the smallest mean D_m value and the highest mean $\log_{10}(N_w)$ value.

Table 4. Average rainfall microphysical parameters for each of the six rainfall rate classes in the winter, premonsoon, monsoon and postmonsoon seasons.

	Class (mm h^{-1})	Samples	R (mm h^{-1})	Z (dBZ)	LWC (g m^{-3})	N_T (m^{-3})	D_m (mm)	$\log_{10}(N_w)$ (N_w in m^{-3} mm^{-1})	μ
Winter	$0.1 \leq R < 1$	5546	0.332	17.110	0.019	87.880	0.923	3.297	10.037
	$1 \leq R < 2$	444	1.368	27.000	0.066	161.317	1.356	3.222	4.415
	$2 \leq R < 5$	152	2.752	33.175	0.118	192.405	1.729	3.063	2.960
	$5 \leq R < 10$	11	-	-	-	-	-	-	-
	$10 \leq R < 20$	-	-	-	-	-	-	-	-
	$R \geq 20$	-	-	-	-	-	-	-	-
Pre-mon	$0.1 \leq R < 1$	17,796	0.406	17.511	0.024	112.441	0.919	3.395	11.940
	$1 \leq R < 2$	4304	1.401	25.261	0.072	221.116	1.171	3.517	7.066
	$2 \leq R < 5$	2338	2.905	30.741	0.138	311.241	1.394	3.508	5.732
	$5 \leq R < 10$	352	6.420	35.292	0.287	456.934	1.574	3.626	6.271
	$10 \leq R < 20$	63	13.795	40.876	0.574	639.370	1.864	3.667	6.528
	$R \geq 20$	27	32.590	46.225	1.211	736.530	2.336	3.525	5.495
Monsoon	$0.1 \leq R < 1$	22,605	0.423	16.254	0.028	205.271	0.797	3.729	14.108
	$1 \leq R < 2$	7248	1.418	23.496	0.079	324.584	1.021	3.787	8.802
	$2 \leq R < 5$	4710	2.954	27.882	0.153	463.498	1.134	3.889	2.853
	$5 \leq R < 10$	782	6.636	31.735	0.327	732.233	1.205	4.114	10.112
	$10 \leq R < 20$	154	13.132	36.175	0.609	900.956	1.381	4.141	9.568
	$R \geq 20$	39	26.385	42.061	1.130	1095.38	1.743	4.004	6.803
Post-mon	$0.1 \leq R < 1$	5678	0.370	16.568	0.023	127.712	0.874	3.462	13.122
	$1 \leq R < 2$	968	1.378	24.120	0.076	303.602	1.069	3.701	8.052
	$2 \leq R < 5$	412	2.844	28.385	0.147	456.403	1.165	3.848	8.263
	$5 \leq R < 10$	62	6.664	33.567	0.320	670.792	1.330	3.981	8.917
	$10 \leq R < 20$	13	-	-	-	-	-	-	-
	$R \geq 20$	3	-	-	-	-	-	-	-

3.3. Distribution of D_m, R, and N_T

Figure 6 shows the percentage of occurrence (bar) and relative contribution to the total rainfall (line) for the different D_m bins in the four seasons. Mêdog rainfall in all the seasons was dominated by raindrops with $D_m < 2$ mm. The distribution of the occurrence frequency of D_m was similar in all seasons except for a slight difference in the premonsoon season. The occurrence frequency of D_m1 was the highest, followed by D_m2 for all four seasons. During the winter, monsoon and postmonsoon seasons, the occurrence percentage of D_m1 was more than 60%, whereas it was less than 60% during the premonsoon period. On the other hand, the percentage occurrence of D_m2 was more than 40% in the premonsoon season, while it was approximately 30% in the other three seasons. The deceased D_m1 in the premonsoon season was compensated by the increase in D_m2. This result indicated that the occurrence frequency of larger raindrops was higher in the premonsoon season than in the other three seasons.

The distribution of the relative contribution to the rainfall totals was different from that of the occurrence frequency. The D_m2 category produced a greater contribution to rainfall by 50–70%, although it had a lower occurrence frequency than the D_m1 category. The rainfall rate was proportional to the third power of raindrop diameter. The larger raindrops with $2 \leq D_m < 3$ mm only contributed to the total rainfall by approximately 5% in the winter and premonsoon seasons, and there were hardly any larger raindrops with $2 \leq D_m < 3$ mm during the monsoon and postmonsoon periods.

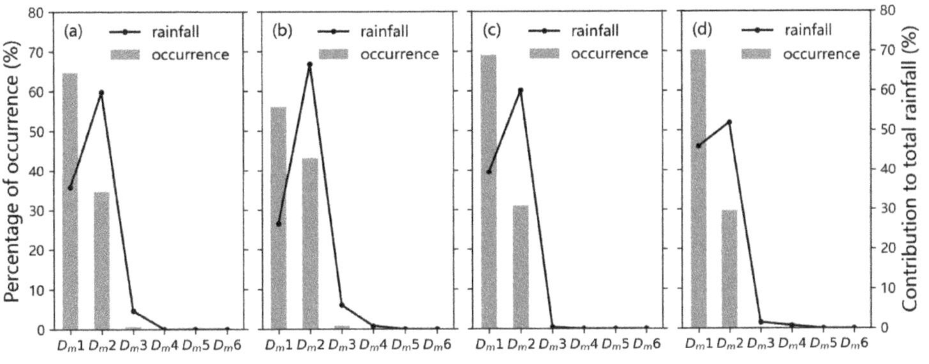

Figure 6. Percentages of occurrence (bar) and relative contributions to the total rainfall (line) from the different D_m bins in the four seasons: winter (**a**), premonsoon (**b**), monsoon (**c**) and postmonsoon (**d**).

The percentage of occurrence (bar) and relative contribution to the total rainfall (line) from the different rain rate categories in the four seasons are shown in Figure 7. Weak rainfall with $R < 1$ mm h^{-1} was dominant in the four seasons, which was evident from the occurrence frequency of the R1 category exceeding 60%, especially more than 80% in the winter season. The occurrence frequencies of R2 and R3 were higher in the monsoon season than in the other three seasons. Considering the relative contribution to total rainfall, the relative contribution to rainfall by R1 was largest and exceeded 60% in the winter season. Similarly, the R1 category also made the largest relative contribution to rainfall in the postmonsoon season. However, the relative contributions to rainfall by the R1, R2, and R3 categories were comparable in the premonsoon season. During the monsoon season, the R3 category made the highest contribution to total rainfall, although its occurrence frequency was lower than that of the R1 and R2 categories.

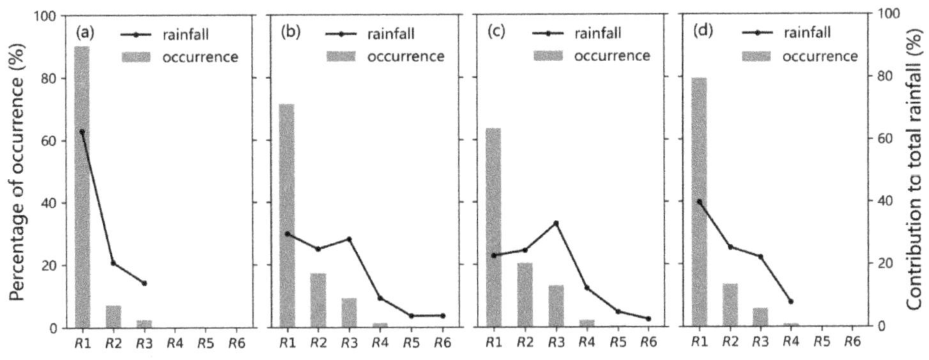

Figure 7. Percentages of occurrence (bar) and relative contributions to the total rainfall (line) from the different R bins in the four seasons: winter (**a**), premonsoon (**b**), monsoon (**c**) and postmonsoon (**d**).

Figure 8 shows the percentage of occurrence (bar) and relative contribution to the total rainfall (line) from the different N_T classes in the four seasons. The occurrence frequencies decreased with the increase in drop number, and the N_T1 class predominated in the four seasons. The drop concentration in Mêdog was mostly below 250 m^{-3}, followed by 250–500 m^{-3}, and a drop concentration of more than 1000 m^{-3} rarely occurred. The occurrence frequency of N_T1 was lower in the monsoon season (approximately 57%) than in the other three seasons (an average of approximately 88%), and the occurrence frequency of N_T2 in the monsoon season (approximately 30%) was higher than in the other three seasons (an average of approximately 9.5%). The relative contribution to the total rainfall monotonically decreased with increasing N_T in all seasons except the monsoon season. During the monsoon season, N_T2 made a larger relative contribution (38%) to the total rainfall than the N_T1 class (28%).

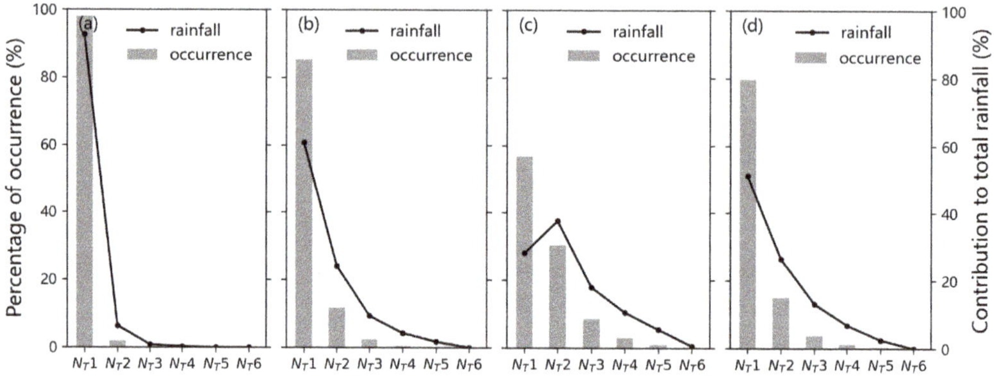

Figure 8. Percentages of occurrence (bar) and relative contributions to the total rainfall (line) from the different N_T bins in the four seasons: winter (**a**), premonsoon (**b**), monsoon (**c**) and postmonsoon (**d**).

3.4. Characteristics of DSDs in Stratiform and Convective Rainfall

Previous studies have shown that the microphysical process of stratiform rainfall is significantly different from that of convective rainfall [34,38]. Therefore, the 1 min DSD samples were classified into stratiform rainfall and convective rainfall. Consequently, the stratiform/convective precipitation samples/percentages were 6130/5 (99.6%/0.1%), 24,155/286 (97.1%/1.1%), 33,468/763 (94.2%/2.1%) and 6930/69 (97.1%1.0%) in the winter, premonsoon, monsoon and postmonsoon seasons, respectively. Considering only five samples, the DSD of convective rainfall in winter was not considered.

Figure 9 shows the DSDs of stratiform rainfall and convective rainfall from mean spectra during different periods. Compared to stratiform rainfall, convective rainfall had a broader spectrum width and a higher concentration of drops. Bimodal distribution could also be seen in both DSDs of stratiform rain and convective rain, and the concentration of the second peak at 1.1 mm was comparable to that of the first peak at 0.4 mm for convective rainfall.

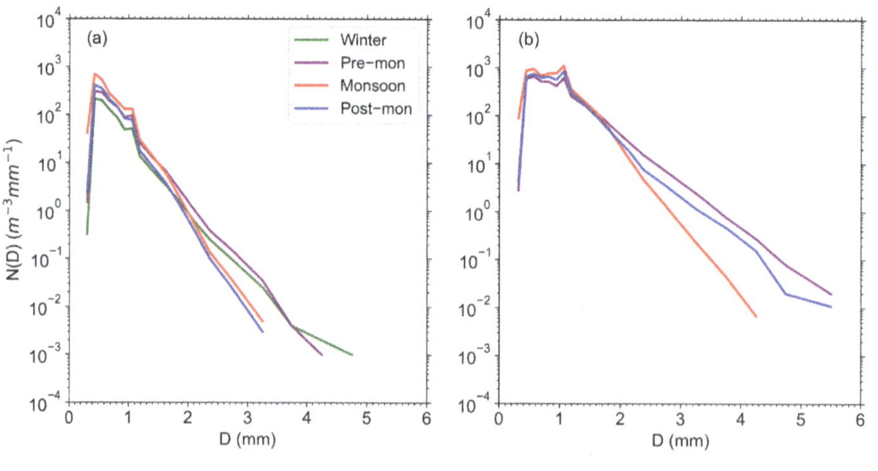

Figure 9. The mean DSDs of stratiform rain (**a**) and convective rain (**b**) for different seasons.

For stratiform rainfall (Figure 9a), the DSD peaked at 0.4 mm in all four seasons and then decreased rapidly. The precipitation in the monsoon season (winter) was characterized by a higher (lower) concentration of drops with sizes less than 1.1 mm. The winter and premonsoon precipitation had higher concentrations of drops with sizes larger than 2.1 mm than the monsoon and postmonsoon precipitation. The precipitation in the four seasons had comparable concentrations of drops with diameters of 1.1–2.1 mm. For convective rainfall (Figure 9b), the highest concentration of raindrop diameters less than 1.1 mm occurred in the monsoon season, and the highest concentration of drops larger than 1.7 mm appeared in the premonsoon season. On the other hand, convective rain in the monsoon season had the lowest concentration of larger drops with $D > 2$ mm, and the lowest concentration of small drops occurred in the premonsoon season. The concentrations of raindrops with sizes of 1.1–1.7 mm were very similar for the three seasons considered.

The average microphysical parameters of stratiform rain and convective rain from 1 min DSD samples during the four periods are given in Table 5. For stratiform rain, the mean LWC, N_T, R, N_w and μ were the highest in the monsoon season. The largest mean D_m value was observed in the premonsoon season, followed by the winter season, and the smallest mean D_m value was observed in the monsoon season. The highest mean Z in the premonsoon season was mainly attributed to the largest D_m because the reflectivity factor is proportional to the sixth power of drop diameter. During the winter period, the lowest mean R and LWC were probably related to the lowest concentration of drops. For convective rain, the mean R, Z, LWC and D_m were the largest in the premonsoon period. The highest mean values of N_T, N_w and μ were found in the monsoon period, followed by the postmonsoon period, and the lowest mean values of N_T and N_w were found in the premonsoon period.

Table 5. The average microphysical parameters of stratiform rain and convective rain in different seasons.

Rain Types	Samples	R (mm h^{-1})	Z (dBZ)	LWC (g m^{-3})	N_T (m^{-3})	D_m (mm)	$\log_{10}(N_w)$ (N_w in m^{-3} mm^{-1})	μ
Winter_Str	6130	0.47	21.50	0.02	95.43	0.97	3.28	9.45
Pre-mon_Str	24,155	0.82	23.50	0.04	149.03	1.01	3.42	10.46
Monsoon_Str	33,468	0.95	21.99	0.05	258.47	0.89	3.75	12.24
Post-mon_Str	6930	0.62	20.08	0.04	166.54	0.91	3.51	12.17
Winter_Con	5	-	-	-	-	-	-	-
Pre-mon_Con	286	10.86	39.50	0.45	554.82	1.67	3.70	7.07
Monsoon_Con	763	9.11	34.44	0.43	805.81	1.26	4.14	10.14
Post-mon_Con	69	9.02	37.24	0.41	646.12	1.49	3.88	8.44

Figure 10 shows the average $\log_{10}(N_w)$ versus average D_m value (along with $\pm\sigma$ SD bars) for stratiform rain and convective rain during different periods. The two outlined squares represent the maritime-like and continental-like convective events reported by Bringi et al. [34]. In general, the average D_m versus average $\log_{10}(N_w)$ showed evident seasonal differences in Mêdog. In terms of convective rain, monsoon precipitation had the smallest (highest) mean D_m ($\log_{10}(N_w)$) value of 1.26 mm (4.14), while premonsoon precipitation was characterized by the largest (lowest) mean D_m ($\log_{10}(N_w)$) value of 1.67 mm (3.70). Convective rain in the monsoon and postmonsoon seasons was similar to maritime-like events, exhibiting smaller D_m and higher $\log_{10}(N_w)$. The convective precipitation during the monsoon and postmonsoon seasons also conformed to the C–S separation line from Thompson et al. [5] for the tropics. Convective events during the premonsoon period were considered to be between maritime- and continental-like events. For stratiform rain, the average D_m versus $\log_{10}(N_w)$ values appeared on the left side (underside) of the C–S separation line, as reported by Bringi et al. [34] (Thompson et al. [5]). The differences in the mean D_m values among the four seasons were relatively slight, whereas the mean $\log_{10}(N_w)$ displayed an evident discrepancy. For example, the mean

$\log_{10}(N_w)$ value of 3.75 in the monsoon season was much higher than that in the winter period, with a value of 3.28.

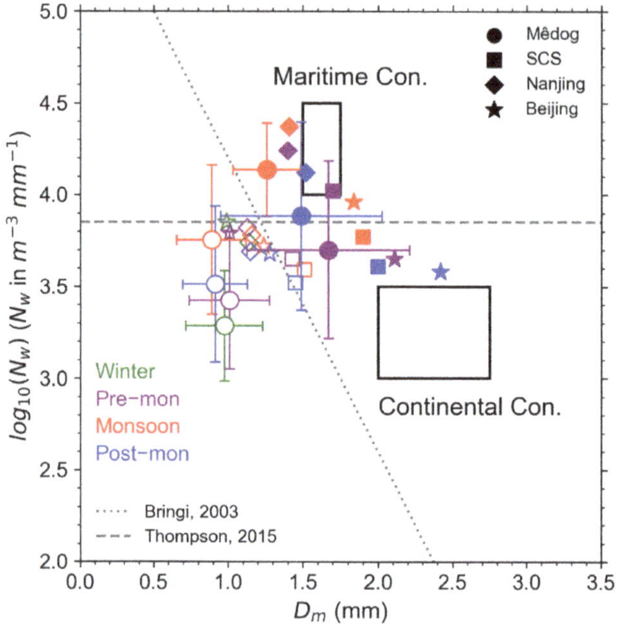

Figure 10. Scatterplots of averaged $\log_{10}(N_w)$ versus D_m (along with $\pm\sigma$ standard deviation bars) for stratiform (blank symbols) and convective (full symbols) precipitation in Mêdog (circles), SCS (squares), Nanjing (diamond) and Beijing (star) during the winter (green), premonsoon (purple), monsoon (red) and postmonsoon (blue) seasons. The two outlined squares represent (left) the maritime and (right) continental types of convective systems reported by Bringi et al. [34]. The dotted line and dashed lines represent the C–S separation line by Bringi et al. [34] and Thompson et al. [5].

For comparison with other regions in China, Figure 10 is also superimposed with the mean D_m and $\log_{10}(N_w)$ values of different seasons from previous studies, including the SCS [18], Nanjing [39] and Beijing [40]. Compared with these regions, the stratiform precipitation in Mêdog showed smaller mean D_m and mean $\log_{10}(N_w)$ values in all seasons, except the monsoon season, which had a similar mean $\log_{10}(N_w)$ value. The mean D_m value of the premonsoon convective precipitation in Mêdog was similar to that in the SCS, and the mean $\log_{10}(N_w)$ value was similar to that in Beijing. Mêdog convective rain in the monsoon season was similar to Nanjing, which may have been due to abundant water vapor in the two regions during this period. The mean D_m ($\log_{10}(N_w)$) of Mêdog convective rain was much smaller (higher) than that in the SCS and Beijing in the monsoon season. This finding was probably related to the predominant warm (cold) rain processes in Mêdog (SCS and Beijing). Similarly, the postmonsoon convective cluster in Mêdog was similar to Nanjing but had a smaller (higher) D_m ($\log_{10}(N_w)$) than Beijing and the SCS.

3.5. The μ–Λ Relationships

The μ–Λ relationship is closely related to the DSD and varies with rain types, climatic characteristics and terrain [38,41]. Zhang et al. [38] proposed the quadratic fitting formula in Florida as follows:

$$\Lambda = 0.0365\,\mu^2 + 0.735\,\mu + 1.935 \tag{16}$$

To minimize the scatter, the samples in Mêdog with rain rates > 5 mm h^{-1} and drop counts > 300 were used to derive μ and Λ [15,38]. Figure 11 shows the scatterplots of μ

and Λ for three seasons due to minimal convective precipitation in winter. The fitted μ–Λ relationships for the premonsoon, monsoon and postmonsoon periods are given as follows:

$$\Lambda = 0.0148\,\mu^2 + 0.786\,\mu + 1.916 \tag{17}$$

$$\Lambda = 0.0056\,\mu^2 + 0.949\,\mu + 1.716, \tag{18}$$

and

$$\Lambda = 0.0250\,\mu^2 + 0.665\,\mu + 2.674. \tag{19}$$

The μ–Λ relationships in Mêdog exhibited little variation among the different periods, especially for $\Lambda < 13$. The shape factor μ in the postmonsoon season gradually became lower than that in other seasons when $\Lambda > 13$, which may be related to the few samples of convective precipitation with increasing Λ during the postmonsoon period. Notably, the μ–Λ relationships in different seasons were similar to the Florida (subtropical environment) relationship reported by Zhang et al. [38]. This finding might indicate that climatic characteristics may play an important role in the determination of the μ–Λ relationship.

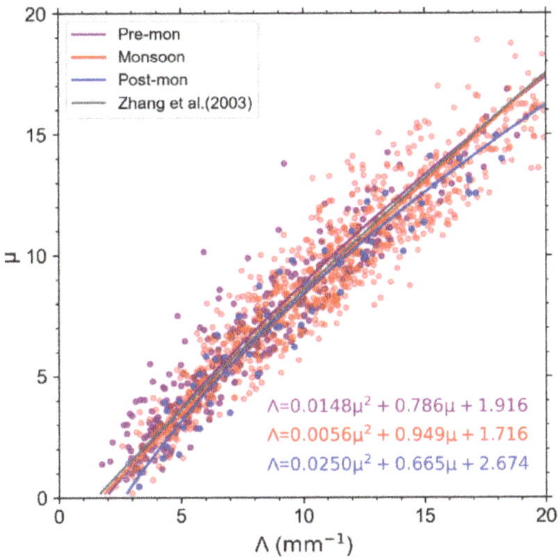

Figure 11. Scatterplots of μ versus Λ and the empirical fitting relationships for samples with rain rates > 5 mm h^{-1} and drop counts > 300 during the premonsoon, monsoon and postmonsoon seasons. The colored solid lines are the fitted empirical μ–Λ relationships in different seasons, and the gray solid line represents the empirical μ–Λ relationship of Zhang et al. [38].

3.6. Quantitative Precipitation Estimation (QPE)

An important application of DSD is quantitative precipitation estimation (QPE). The power-law relationship of $Z = AR^b$ is widely used in radar meteorology and changes with rainfall type, atmospheric conditions and geographic location [42]. The new-generation weather radar system in China uses the empirical relationships of $Z = 300R^{1.4}$ and $Z = 200R^{1.6}$ to describe midlatitude convection [43] and stratiform precipitation [44], respectively. Wu and Liu [16] proposed that coefficient A (exponent b) is 170.7 (1.31) and 69.83 (1.83) for summer convection precipitation and stratiform precipitation in Nagqu, respectively, based on disdrometer measurements. Wang et al. [24] gave the relationships of $Z = 114.79R^{1.34}$ and $Z = 53.69R^{1.71}$ for convection precipitation and stratiform precipitation in rainy seasons in Mêdog, respectively. The equivalent radar reflectivity factor (Z_e, in $mm^6\ m^{-3}$) based on observed DSDs can be expressed according to Zhang et al. [45]:

$$Z_e = \frac{4\lambda^4}{\pi^4 |K_w|^2} \int_{D_{min}}^{D_{max}} |f(D)|^2 N(D) dD \quad (20)$$

where λ indicates the radar wavelength and was set to 5 cm, considering that C-band Doppler weather radars were deployed over the TP. K_w is the water dielectric factor, and $|K_w|^2$ is set to 0.93 by convention. $f(D)$ is the backscattering amplitude for a raindrop of size D, which is calculated by using the extended boundary condition method (EBCM) [46].

Considering the evident seasonal variation in DSD characteristics in Mêdog, the Z–R relationships for the four seasons are discussed in this section. Figure 12 shows the scatter plots of Z and R superimposed with the fitted Z–R relationships using the least squares method for stratiform rain and convective rain, respectively. The fitted coefficient A and exponent b for different rainfall types in the four seasons are given in Table 6. Following Zeng et al. [18], the normalized mean biases (NBs) of the fitted Z–R relations and empirical relations at midlatitudes for different precipitation types were calculated to evaluate the accuracies of different Z–R relationships (Table 7).

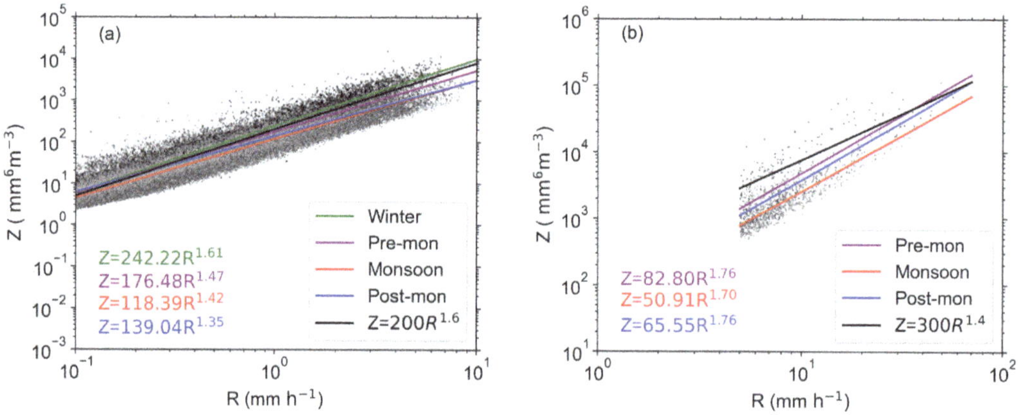

Figure 12. Scatterplots of radar reflectivity factor (Z) and rain rate (R) and fitted Z–R relationships using the least squares method (solid lines) for stratiform rain (**a**) and convective rain (**b**).

Table 6. Fitted radar reflectivity and rain rate (Z–R) relationships for stratiform and convective rain types in the four seasons.

Season	Stratiform Rainfall		Convective Rainfall	
	A	b	A	b
Winter	242.22	1.61	-	-
Pre-mon	176.48	1.47	82.80	1.76
Monsoon	118.39	1.42	50.91	1.70
Post-mon	139.04	1.35	65.55	1.76

Table 7. NB (%) values of the fitted Z–R relationships and empirical relationships for different precipitation types in the four seasons.

Season	Stratiform Rainfall		Convective Rainfall	
	Fitted Z–R	$Z = 200R^{1.6}$	Fitted Z–R	$Z = 300R^{1.4}$
Winter	7.91	21.51	-	-
Pre-mon	9.97	−1.74	7.26	−12.27
Monsoon	6.14	−27.24	2.98	−51.38
Post-mon	7.92	−14.32	11.19	−26.87

For stratiform precipitation, a small discrepancy in fitted *Z–R* relationships among the different seasons could be noted. Winter precipitation had larger A and b values than those of the empirical relationship in midlatitudes, while precipitation in other seasons had smaller A and b values. This result may be related to the fact that winter stratiform precipitation had more (less) large (small) drops than in other seasons. The empirical relationship of $Z = 200R^{1.6}$ underestimated rainfall in the premonsoon, monsoon and postmonsoon seasons by 1.74%, 27.24%, and 14.32%, respectively, while it overestimated winter rainfall by 21.51%. The fitted *Z–R* relationships reduced the NB to less than 10% for all of the considered seasons.

For convective precipitation, the fitted coefficient A (exponent b) in the premonsoon, monsoon and postmonsoon seasons was much less (larger) than that of the empirical relation at midlatitudes. Monsoon precipitation had a minimum coefficient A (50.91) and exponent b (1.70), which might have been attributed to the large number of small raindrops during this period. That is, the same reflectivity factor would derive the highest rain rate in the monsoon season. Given a radar reflectivity factor value of 40 dBZ, the corresponding rainfall rates were 15.24 mm h^{-1}, 22.33 mm h^{-1} and 18.20 mm h^{-1} in the premonsoon, monsoon and postmonsoon seasons, respectively. The empirical relationship of $Z = 300R^{1.4}$ underestimated convective rainfall up to 51.38% in the monsoon season, followed by 26.87% in the postmonsoon season and then 12.27% in the premonsoon season. However, the fitted *Z–R* relationships significantly reduced the NB in all of the considered seasons. In particular, the NB decreased from 51.38% to 2.98% in the monsoon season. The distinct seasonal variation in DSDs in Mêdog convective rain determined the evident discrepancy in *Z–R* relationships among the different seasons. Therefore, the fitted *Z–R* relationships for different seasons could significantly improve the accuracy of radar-based QPEs.

4. Discussion

The significant seasonal variations in DSDs in Mêdog could provide a better understanding of the microphysical process of precipitation at the entrance of the vapor channel in the YGC and improve the parameterization schemes in numerical models over the TP. The possible causative mechanisms of the distinct DSD variability over seasons may be addressed from the standpoint of the meteorological environments of rainfall [13]. To explore the possible causes of seasonal variations in the DSD in Mêdog, meteorological conditions of rainy days from ERA5 reanalysis data, AWS and TBB and CTH products of the FY-4A satellite were collected. The lifting condensation level (LCL), 0 °C isotherm layer height, CTH, TBB probability density function, surface wind speed box diagram and the vertical integral of water vapor flux of rainy days in the four seasons are shown in Figures 13 and 14.

Due to the lack of radiosonde and ceilometer observations in Mêdog, the LCLs calculated from AWS data were approximately considered as cloud base height (CBH). The average LCL was calculated using the surface temperature (T), surface dew point temperature (T_d) and surface pressure (p) according to the empirical formula (Equations (21)–(23)) given by Barnes [47]:

$$T_{LCL} = T_d - (0.001296T_d + 0.1963)(T - T_d), \tag{21}$$

$$p_{LCL} = p[(T_{LCL} + 273.15)/(T + 273.15)]^{\frac{7}{2}}, \tag{22}$$

$$LCL = 18{,}400(1 + at)\log\left(\frac{p}{p_{LCL}}\right), \tag{23}$$

T_{LCL} and p_{LCL} indicate the temperature and pressure at LCL height, respectively. a = 1/273, $t = T_{LCL}\text{-}T$ (unit: °C). The calculated average LCL heights were 0.12 km, 0.13 km, 0.16 km and 0.20 km in the winter, premonsoon, monsoon and postmonsoon periods, respectively, exhibiting a negligible difference in the four seasons. The average heights of the 0 °C isotherm layer from ERA5 in the winter, premonsoon, monsoon and postmonsoon periods were 1.53 km, 2.67 km, 4.01 km and 2.81 km, respectively, and the average CTHs were 5.13 km, 6.64 km, 6.97 km and 5.39 km, respectively.

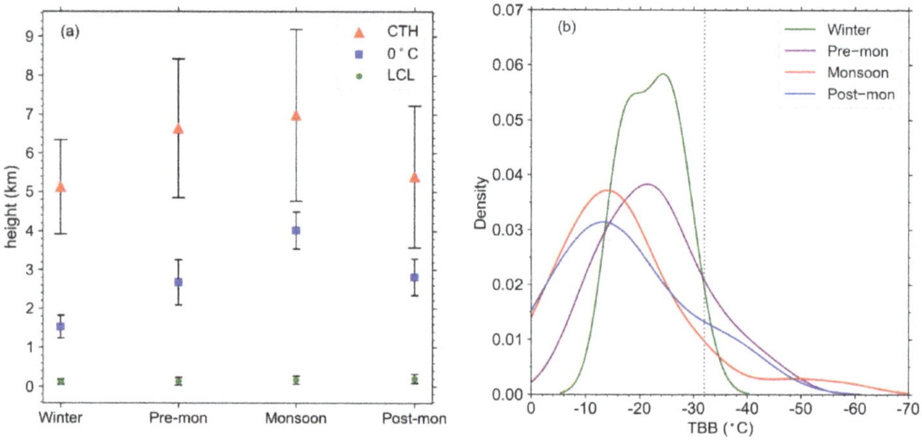

Figure 13. The lifting condensation level (LCL), 0 °C isotherm layer height, cloud top height (CTH) and standard deviation (**a**). TBB probability density function (**b**). The dashed line in Figure 13b denotes the TBB temperature of −32 °C.

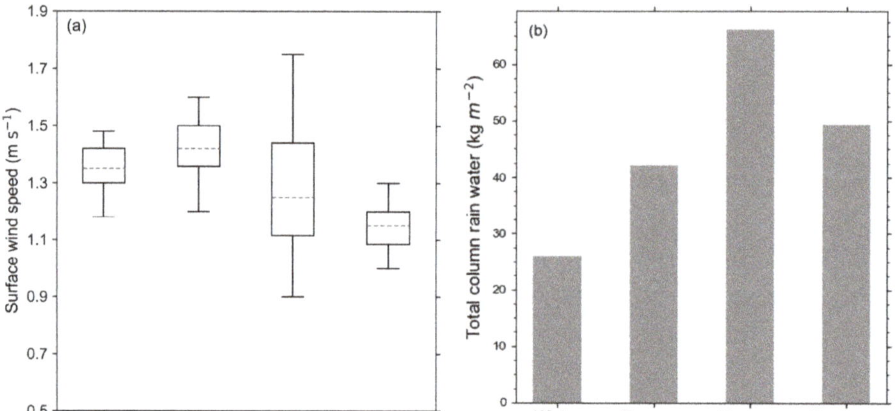

Figure 14. Surface wind speed box diagram (**a**) and the vertical integral of water vapor flux (**b**) in different seasons.

Clouds between LCL and the 0 °C isotherm layer level are defined as warm clouds, and those between the 0 °C isotherm layer level and CTH are considered cold clouds [18]. The cloud rain process is predominant during the winter precipitation period, which is evident from the significant cold cloud depth of 3.60 km compared to the relatively short warm cloud depth of 1.41 km. The microphysical and dynamic mechanisms (e.g., updraft, particle formation and particle growth processes) in the cold rain process are different from those in the warm rain process, leading to significant discrepancies in DSD characteristics [39]. Ice crystals grow quickly above the 0 °C isotherm level in the winter precipitation process. The higher concentration of large drops found in winter precipitation may be attributed to melted ice particles, such as low-density, large snow particles, and/or graupel (e.g., Figures 5 and 9a). In addition, wind and humidity are two important meteorological elements affecting the evaporation process [48]. Stronger evaporation is expected in the winter season due to a larger wind speed and less water vapor (Figure 14), reducing the concentration of small raindrops (e.g., Figures 5 and 9a).

The premonsoon precipitation was characterized by a high concentration of large drops (e.g., Figures 4, 5 and 9). The average cold cloud depth (3.97 km) was much larger than the average warm cloud depth (2.54 km) in the premonsoon season, indicating that the cold rain process is also predominant in this period. The melted ice particles (e.g., graupel and/or snow particles) could result in the formation of larger drops [49]. Convective activity frequently occurs in the premonsoon season, as evidenced by the probability density function (PDF) of TBB (Figure 13b), which is often used to assess the intensity of convective activity [50]. The smaller the TBB value is, the deeper the development of convective clouds. The threshold of TBB ≤ -32 °C is often used to differentiate the development of convection. The probability of TBB ≤ -32 °C was the highest in the premonsoon season, indicating that intense convective activity occurs more frequently during this period. The westerly winds prevail over the TP during this period, and cold air masses can easily invade the middle to upper troposphere. In addition, solar radiation causes an increase in surface heating in the daytime. This destabilization of the troposphere would be beneficial to the formation of dry convection in the premonsoon season [51].

In addition, the largest surface wind speed (e.g., Figure 14a) among the four seasons may lead to relatively strong evaporation in the premonsoon season, which was partly responsible for the relatively low concentration of small drops. Thus, a higher concentration of larger raindrops and a lower concentration of small raindrops were observed for higher rainfall rate categories (e.g., $R > 5$ mm h^{-1}) and convective rainfall types (e.g., Figures 5d–f and 9b). Therefore, the intensity increase in premonsoon precipitation was more attributed to the increase in drop diameter (e.g., Table 4).

During the monsoon season, although the CTH was highest, the average thickness of warm clouds (3.85 km) was significantly larger than that of cold clouds (2.96 km) (Figure 13a). Therefore, monsoon rainfall was dominated by warm rain processes, which tended to produce higher concentrations of small raindrops owing to collisional and coalescence processes (e.g., Figures 4, 5 and 9). A large amount of water vapor is carried to Mêdog by the Indian Ocean monsoon in this season (Figure 14b), which is conducive to the formation of warm clouds and the production of abundant small raindrops. Weak evaporation is expected in the monsoon season due to the smaller wind speed and wet environment, contributing to the production of small raindrops. The increase in precipitation intensity in the monsoon season may be mainly attributed to the significant increase in the concentration of raindrops (Table 4).

The postmonsoon precipitation had less rainfall total and was also characterized by a higher concentration of small drops (i.e., Figures 4 and 5b–d). Although the mean depth of warm clouds (2.61 km) was similar to that of cold clouds (2.58 km) in this season, Figure 14 exhibits humid and weak wind atmospheric conditions, which are favorable to the production of small drops.

5. Conclusions

In this paper, the seasonal variation in DSDs and microphysical parameters among the winter, premonsoon, monsoon, and postmonsoon periods were investigated using PARSIVEL disdrometer data from July 2019 to June 2020 in Mêdog, which is located in the southeast of the TP and at the entrance of the vapor channel in the YGC. In addition, EAR5 reanalysis data, FY-4A satellite products and AWS observations were used to address the possible causative factors for the distinct seasonal variation in DSDs. The conclusions of this study are outlined as follows:

(1) Precipitation mainly occurs during the monsoon period in Mêdog, contributing approximately 57% to the annual rainfall totals, and small drops are dominant in the four seasons. Weak rainfall (i.e., $R < 1$ mm h^{-1}) with small drops (i.e., $D_m < 1$ mm) and low concentrations (i.e., $N_T < 250$ m^{-3}) occurs frequently in the four seasons in Mêdog.

However, taking the contributor to rainfall into account, drops with $1 \leq D_m < 2$ mm are the largest contributor in the four seasons, and the weak rainfall with $R < 1$ mm h^{-1} is the largest contributor in Mêdog except during the monsoon season, during which, rainfall with $2 \leq R < 5$ mm h^{-1} is the largest contributor. For the average spectrum of the four seasons, the monsoon season precipitation has the narrowest spectrum width and is characterized by the highest (lowest) concentration of small (large) drops. The winter (postmonsoon) precipitation has the lowest (highest) concentration of small (large) drops. In terms of rain rate classes, a higher (lower) concentration of small (large) raindrops can be found in the monsoon season for all the considered rainfall rate classes in this study. More large drops and fewer small drops are observed in winter precipitation with $R < 5$ mm h^{-1}. For heavy rainfall (i.e., $R > 5$ mm h^{-1}), the premonsoon precipitation exhibits a higher concentration of large drops.

(2) Mêdog stratiform precipitation in the four seasons has a similar mean D_m value of approximately 1.0 mm but exhibits a distinct difference in the mean value of $\log_{10}(N_w)$. Monsoon stratiform rain has the highest mean $\log_{10}(N_w)$ value of 3.75, followed by post-monsoon rain, and the winter season has the lowest mean $\log_{10}(N_w)$ value of 3.28. The convective rainfall during the monsoon season is characterized by the highest concentration of limited-size drops and is identified as maritime-like. Premonsoon convective rain has predominantly larger drops than other seasons. The largest mean D_m (1.67 mm) and the lowest mean $\log_{10}(N_w)$ (3.70) are observed in the premonsoon convective rainfall, which could be considered a transition between maritime-like and continental-like conditions.

(3) The relationships of μ–Λ and Z–R corresponding to different seasons were also fitted. The μ–Λ relationships of the different periods show little discrepancy. The fitted Z–R relationships for stratiform precipitation exhibit little seasonal variation, and winter stratiform rain has a larger coefficient A and exponent b. The fitted Z–R relationships for convective precipitation show evident discrepancies among the premonsoon, monsoon, and postmonsoon periods. The Z–R relationship in monsoon convective rainfall has a smaller (larger) coefficient A (exponent b) than in other seasons, indicating a higher rain rate in monsoon convective precipitation for a given radar reflectivity. The empirical relationship of $Z = 300R^{1.4}$ at midlatitudes would cause the severe underestimation of convective rain in Mêdog, especially during the monsoon period.

(4) The possible causative meteorological environments responsible for the seasonal variation in DSDs in Mêdog were discussed. Westerlies prevail over the whole TP in the premonsoon season, and rainfall is dominated by cold rain processes, resulting in the formation of large raindrops via the melting of frozen particles. In addition, less water vapor and a larger wind speed contribute to stronger evaporation, which probably leads to a lower concentration of small drops in the premonsoon precipitation. During the monsoon period, abundant warm and humid mass air intrudes from the Indian Ocean into Mêdog, and warm rain processes prevail in this period, producing many small raindrops via active collision and coalescence processes. Atmospheric conditions are characterized by humid and weak winds in the postmonsoon season, which is favorable to the production of small drops.

Notably, this work focused on the seasonal variation in DSD based on disdrometer data in Mêdog. The parameters of gamma distribution model of DSDs are trying to be used to improve the microphysical parameterization scheme of precipitation in the local numerical model. The detailed performance of the model will be evaluated later. Furthermore, disdrometer observations at more locations over the TP will be used to explore the temporal and spatial variation in DSDs. In addition, the vertical structure of DSDs in different seasons will be explored in future research by jointly using K-band Micro Rain Radar and X-band dual-polarization radar observations.

Author Contributions: Conceptualization, G.W. and L.L.; methodology, R.L. and G.W.; software, R.L., R.Z. and J.Z.; writing—original draft preparation, R.L.; writing—review and editing, R.L., G.W.; funding acquisition, G.W. All authors have read and agreed to the published version of the manuscript.

Funding: This project was funded by the Second Tibetan Plateau Scientific Expedition and Research (STEP) program, grant number 2019QZKK0105 and the National Key R & D Projects, grant number 2018YFC1505702.

Acknowledgments: We thank Suolang Zhaxi and Ting Wang for their help in the maintenance of the disdrometer at Mêdog National Climate Observatory.

Conflicts of Interest: The authors declare no conflict of interest.

References

1. Seela, B.K.; Janapati, J.; Lin, P.-L.; Reddy, K.K.; Shirooka, R.; Wang, P.K. A Comparison Study of Summer Season Raindrop Size Distribution Between Palau and Taiwan, Two Islands in Western Pacific. *J. Geophys. Res. Atmos.* **2017**, *122*, 11787–11805. [CrossRef]
2. Han, Y.; Guo, J.; Yun, Y.; Li, J.; Guo, X.; Lv, Y.; Wang, D.; Li, L.; Zhang, Y. Regional variability of summertime raindrop size distribution from a network of disdrometers in Beijing. *Atmos. Res.* **2021**, *257*, 105591. [CrossRef]
3. Sreekanth, T.S.; Varikoden, H.; Sukumar, N.; Mohan Kumar, G. Microphysical characteristics of rainfall during different seasons over a coastal tropical station using disdrometer. *Hydrol. Process.* **2017**, *31*, 2556–2565. [CrossRef]
4. Tang, Q.; Xiao, H.; Guo, C.; Feng, L. Characteristics of the raindrop size distributions and their retrieved polarimetric radar parameters in northern and southern China. *Atmos. Res.* **2014**, *135–136*, 59–75. [CrossRef]
5. Thompson, E.J.; Rutledge, S.A.; Dolan, B.; Thurai, M. Drop Size Distributions and Radar Observations of Convective and Stratiform Rain over the Equatorial Indian and West Pacific Oceans. *J. Atmos. Sci.* **2015**, *72*, 4091–4125. [CrossRef]
6. Krishna, U.V.M.; Reddy, K.K.; Seela, B.K.; Shirooka, R.; Lin, P.-L.; Pan, C.-J. Raindrop size distribution of easterly and westerly monsoon precipitation observed over Palau islands in the Western Pacific Ocean. *Atmos. Res.* **2016**, *174–175*, 41–51. [CrossRef]
7. Löffler-Mang, M.; Joss, J. An Optical Disdrometer for Measuring Size and Velocity of Hydrometeors. *J. Atmos. Ocean. Technol.* **2000**, *17*, 130–139. [CrossRef]
8. Nzeukou, A.; Sauvageot, H.; Ochou, A.D.; Kebe, C.M. Raindrop Size Distribution and Radar Parameters at Cape Verde. *J. Appl. Meteorol. Climatol.* **2004**, *43*, 90–105. [CrossRef]
9. Lavanya, S.; Kirankumar, N.V.P. Classification of tropical coastal precipitating cloud systems using disdrometer observations over Thumba, India. *Atmos. Res.* **2021**, *253*, 105477. [CrossRef]
10. Huo, Z.; Ruan, Z.; Wei, M.; Ge, R.; Li, F.; Ruan, Y. Statistical characteristics of raindrop size distribution in south China summer based on the vertical structure derived from VPR-CFMCW. *Atmos. Res.* **2019**, *222*, 47–61. [CrossRef]
11. Radhakrishna, B.; Rao, T.N.; Rao, D.N.; Rao, N.P.; Nakamura, K.; Sharma, A.K. Spatial and seasonal variability of raindrop size distributions in southeast India. *J. Geophys. Res.* **2009**, *114*, D04203. [CrossRef]
12. Pu, K.; Liu, X.; Wu, Y.; Hu, S.; Liu, L.; Gao, T. A comparison study of raindrop size distribution among five sites at the urban scale during the East Asian rainy season. *J. Hydrol.* **2020**, *590*, 125500. [CrossRef]
13. Wu, Z.; Zhang, Y.; Zhang, L.; Lei, H.; Xie, Y.; Wen, L.; Yang, J. Characteristics of Summer Season Raindrop Size Distribution in Three Typical Regions of Western Pacific. *J. Geophys. Res. Atmos.* **2019**, *124*, 4054–4073. [CrossRef]
14. Hopper, L.J.; Schumacher, C.; Humes, K.; Funk, A. Drop-Size Distribution Variations Associated with Different Storm Types in Southeast Texas. *Atmosphere* **2019**, *11*, 8. [CrossRef]
15. Chen, B.; Hu, Z.; Liu, L.; Zhang, G. Raindrop Size Distribution Measurements at 4,500 m on the Tibetan Plateau During TIPEX-III. *J. Geophys. Res. Atmos.* **2017**, *122*, 11092–11106. [CrossRef]
16. Wu, Y.; Liu, L. Statistical characteristics of raindrop size distribution in the Tibetan Plateau and southern China. *Adv. Atmos. Sci.* **2017**, *34*, 727–736. [CrossRef]
17. Zeng, Y.; Yang, L.; Tong, Z.; Jiang, Y.; Zhang, Z.; Zhang, J.; Zhou, Y.; Li, J.; Liu, F.; Liu, J.; et al. Statistical Characteristics of Raindrop Size Distribution during Rainy Seasons in Northwest China. *Adv. Meteorol.* **2021**, *2021*, 6667786. [CrossRef]
18. Zeng, Q.; Zhang, Y.; Lei, H.; Xie, Y.; Gao, T.; Zhang, L.; Wang, C.; Huang, Y. Microphysical Characteristics of Precipitation during Pre-monsoon, Monsoon, and Post-monsoon Periods over the South China Sea. *Adv. Atmos. Sci.* **2019**, *36*, 1103–1120. [CrossRef]
19. Gao, Y.; Zhou, X.; Wang, Q.; Wang, C.; Zhan, Z.; Chen, L.; Yan, J.; Qu, R. Vegetation net primary productivity and its response to climate change during 2001–2008 in the Tibetan Plateau. *Sci. Total Environ.* **2013**, *444*, 356–362. [CrossRef]
20. Zhang, W.; Zhang, L.; Zhou, T. Interannual Variability and the Underlying Mechanism of Summer Precipitation over the Yarlung Zangbo River Badin. *Chin. J. Atmos. Sci.* **2016**, *40*, 965–980. (In Chinese) [CrossRef]
21. Gao, D. Expeditionary studies on the Moisture Passage of the Yarlung Zangbo River. *Chin. J. Nat.* **2008**, *5*, 301–303. (In Chinese)
22. Wang, J.; De, Q.; Dan, Z.; Wang, Y.; De, J.; Chen, G.; Tian, Y. Characteristic and Causal Analysis of Concentrated and Frequently Occurring Intense Rainfall in Tibet in 2012–2018. *Meteorol. Sci. Technol.* **2021**, *49*, 211–217.
23. Zhou, R.; Wang, G.; Zhaxi, S. Cloud vertical structure measurements from a ground-based cloud radar over the southeastern Tibetan Plateau. *Atmos. Res.* **2021**, *258*, 211–217. (In Chinese) [CrossRef]
24. Wang, G.; Zhou, R.; Zhaxi, S.; Liu, S. Raindrop size distribution measurements on the Southeast Tibetan Plateau during the STEP project. *Atmos. Res.* **2021**, *249*, 105311. [CrossRef]
25. Yuter, S.E.; Kingsmill, D.E.; Nance, L.B.; Löffler-Mang, M. Observations of Precipitation Size and Fall Speed Characteristics within Coexisting Rain and Wet Snow. *J. Appl. Meteorol. Climatol.* **2006**, *45*, 1450–1464. [CrossRef]

26. Thurai, M.; Petersen, W.A.; Tokay, A.; Schultz, C.; Gatlin, P. Drop size distribution comparisons between Parsivel and 2-D video disdrometers. *Adv. Geosci.* **2011**, *30*, 3–9. [CrossRef]
27. Beard, K.V. Terminal Velocity and Shape of Cloud and Precipitation Drops Aloft. *J. Atmos. Sci.* **1976**, *33*, 851–864. [CrossRef]
28. Atlas, D.; Srivastava, R.; Sekhon, R. Doppler Radar Characteristics of Precipitation at Vertical Incidence. *Rev. Geophys. Space Phys.* **1973**, *11*, 1–35. [CrossRef]
29. Wen, L.; Zhao, K.; Zhang, G.; Liu, S.; Chen, G. Impacts of Instrument Limitations on Estimated Raindrop Size Distribution, Radar Parameters, and Model Microphysics during Mei-Yu Season in East China. *J. Atmos. Ocean. Technol.* **2017**, *34*, 1021–1037. [CrossRef]
30. Ulbrich, C.W.; Atlas, D. Rainfall Microphysics and Radar Properties: Analysis Methods for Drop Size Spectra. *J. Appl. Meteorol. Climatol.* **1998**, *37*, 912–923. [CrossRef]
31. Kozu, T.; Nakamura, K.J.J.A.O.T. Rainfall Parameter Estimation from Dual-Radar Measurements Combining Reflectivity Profile and Path-integrated Attenuation. *J. Atmos. Ocean. Technol.* **1991**, *8*, 259–270. [CrossRef]
32. Testud, J.; Oury, S.; Black, R.A.; Amayenc, P.; Dou, X. The Concept of "Normalized" Distribution to Describe Raindrop Spectra: A Tool for Cloud Physics and Cloud Remote Sensing. *J. Appl. Meteorol.* **2001**, *40*, 1118–1140. [CrossRef]
33. Smith, P.L. Raindrop Size Distributions: Exponential or Gamma–Does the Difference Matter? *J. Appl. Meteorol.* **2003**, *42*, 1031–1034. [CrossRef]
34. Bringi, V.N.; Chandrasekar, V.; Hubbert, J.; Gorgucci, E.; Randeu, W.L.; Schoenhuber, M. Raindrop Size Distribution in Different Climatic Regimes from Disdrometer and Dual-Polarized Radar Analysis. *J. Atmos. Sci.* **2003**, *60*, 354–365. [CrossRef]
35. Steiner, M.; Waldvogel, A. Peaks in Raindrop Size Distributions. *J. Atmos. Sci.* **1987**, *44*, 3127–3134. [CrossRef]
36. Tokay, A.; Short, D.A. Evidence from Tropical Raindrop Spectra of the Origin of Rain from Stratiform versus Convective Clouds. *J. Appl. Meteorol. Climatol.* **1996**, *35*, 355–371. [CrossRef]
37. Wang, G.; Li, R.; Sun, J.; Xu, X.; Zhou, R.; Liu, L. Comparative Analysis of the Characteristics of Rainy Season Raindrop Size Distributions in Two Typical Regions of the Tibetan Plateau. *Adv. Atmos. Sci.* **2022**, *39*, 1062–1078. [CrossRef]
38. Zhang, G.; Vivekanandan, J.; Brandes, E.A.; Meneghini, R.; Kozu, T. The Shape–Slope Relation in Observed Gamma Raindrop Size Distributions: Statistical Error or Useful Information? *J. Atmos. Ocean. Technol.* **2003**, *20*, 1106–1119. [CrossRef]
39. Wen, L.; Zhao, K.; Wang, M.; Zhang, G. Seasonal Variations of Observed Raindrop Size Distribution in East China. *Adv. Atmos. Sci.* **2019**, *36*, 346–362. [CrossRef]
40. Luo, L.; Guo, J.; Chen, H.; Yang, M.; Chen, M.; Xiao, H.; Ma, J.; Li, S. Microphysical Characteristics of Rainfall Observed by a 2DVD Disdrometer during Different Seasons in Beijing, China. *Remote Sens.* **2021**, *13*, 2303. [CrossRef]
41. Chen, B.; Yang, J.; Gao, R.; Zhu, K.; Zou, C.; Gong, Y.; Zhang, R. Vertical Variability of the Raindrop Size Distribution in Typhoons Observed at the Shenzhen 356-m Meteorological Tower. *J. Atmos. Sci.* **2020**, *77*, 4171–4187. [CrossRef]
42. Rosenfeld, D.; Ulbrich, C.W. Cloud Microphysical Properties, Processes, and Rainfall Estimation Opportunities. *Meteorol. Monogr.* **2003**, *30*, 237–258. [CrossRef]
43. Fulton, R.A.; Breidenbach, J.P.; Seo, D.-J.; Miller, D.A.; O'Bannon, T. The WSR-88D Rainfall Algorithm. *Weather Forecast.* **1998**, *13*, 377–395. [CrossRef]
44. Marshall, J.; Palmer, W. The Distribution of Raindrops with Size. *J. Atmos. Sci.* **1948**, *5*, 165–166. [CrossRef]
45. Zhang, G.; Vivekanandan, J.; Brandes, E. A method for estimating rain rate and drop size distribution from polarimetric radar measurements. *IEEE Trans. Geosci. Remote Sens.* **2001**, *39*, 830–841. [CrossRef]
46. Barber, P.; Yeh, C. Scattering of electromagnetic waves by arbitrarily shaped dielectric bodies. *Appl. Opt.* **1975**, *14*, 2864–2872. [CrossRef]
47. Barnes, S.L. An Empirical Shortcut to the Calculation of Temperature and Pressure at the Lifted Condensation Level. *J. Appl. Meteorol. Climatol.* **1968**, *7*, 511. [CrossRef]
48. McVicar, T.R.; Roderick, M.L.; Donohue, R.J.; Li, L.T.; Van Niel, T.G.; Thomas, A.; Grieser, J.; Jhajharia, D.; Himri, Y.; Mahowald, N.M.; et al. Global review and synthesis of trends in observed terrestrial near-surface wind speeds: Implications for evaporation. *J. Hydrol.* **2012**, *416–417*, 182–205. [CrossRef]
49. Dolan, B.; Fuchs, B.; Rutledge, S.A.; Barnes, E.A.; Thompson, E.J. Primary Modes of Global Drop Size Distributions. *J. Atmos. Sci.* **2018**, *75*, 1453–1476. [CrossRef]
50. Maddox, R.A. Mesoscale convective complexes. *Bull. Am. Meteorol. Soc.* **1980**, *61*, 1374–1387. [CrossRef]
51. Fujinami, H.; Yasunari, T. The Seasonal and Intraseasonal Variability of Diurnal Cloud Activity over the Tibetan Plateau. *J. Meteorol. Soc. Jpn. Ser. II* **2001**, *79*, 1207–1227. [CrossRef]

remote sensing

Article

The Observed Impact of the South Asian Summer Monsoon on Land-Atmosphere Heat Transfers and Its Inhomogeneity over the Tibetan Plateau

Hongyi Li [1], Libo Zhou [2] and Ge Wang [3,*]

1 China Meteorological Administration Training Centre, Beijing 100081, China; lihongyi@cma.gov.cn
2 State Key Laboratory of Atmospheric Boundary Layer Physics and Atmospheric Chemistry & Department of Lower Atmosphere Observation and Research, Institute of Atmospheric Physics, Chinese Academy of Sciences, Beijing 100029, China; zhoulibo@mail.iap.ac.cn
3 Institute of Plateau Meteorology, China Meteorological Administration, Chengdu 610072, China
* Correspondence: wg800110@aliyun.com

Abstract: To promote Tibetan meteorological research, the third Tibetan Plateau (TP) Experiment for atmospheric sciences (TIPEX III) has been carried out over the plateau region since 2014, with near-surface heat fluxes measured at different sites. Using the observational data of near-surface heat fluxes measured at 8 plateau stations in TIPEX III, as well as the ECMWF ERA Interim reanalysis data, the land-atmosphere heat transfers over different regions of TP and their responses to the South Asian summer monsoon (SASM) during active/break periods were investigated. Inhomogeneity was found in the land-atmosphere heat transfers over the plateau, with large differences among plateau stations. During the observation period, the daily averaged total heat transfer (the sum of sensible and latent heat flux) varied from 70.2 to 101.2 Wm^{-2} among the 8 plateau stations, with the sensible heat flux from 18.8 to 60.1 Wm^{-2} and the latent heat flux from 10.1 to 74.7 Wm^{-2}. These heat transfers were strongly affected by the SASM evolution, but with strong inhomogeneities over the plateau stations. Overall, the more southern station locations exhibited more SASM impacts. The land-atmosphere heat transfers (the total, sensible and latent heat fluxes) were greatly weakened/strengthened during the SASM active/break period at Namco (southeast plateau), Baingoin (central plateau), Lhari (central plateau), and Nagqu (central plateau), which were closely related to the weakened/strengthened radiation conditions. However, the SASM impacts were quite small or even negligible for the other plateau stations, which complicated our conclusions, and further investigations are still needed.

Keywords: Tibetan Plateau; South Asian summer monsoon; land-atmosphere heat transfer; inhomogeneity

Citation: Li, H.; Zhou, L.; Wang, G. The Observed Impact of the South Asian Summer Monsoon on Land-Atmosphere Heat Transfers and Its Inhomogeneity over the Tibetan Plateau. *Remote Sens.* **2022**, *14*, 3236. https://doi.org/10.3390/rs14133236

Academic Editors: Massimo Menenti, Yaoming Ma, Li Jia and Lei Zhong

Received: 9 May 2022
Accepted: 30 June 2022
Published: 5 July 2022

1. Introduction

As the "Third Pole of the Earth" and the "atmospheric water tower", the Tibetan Plateau (TP) has an area of about 2.5 million square kilometers and an average altitude greater than 4000 m. The strong solar radiation on this elevated surface makes the plateau a heating source for the atmosphere in the middle troposphere. This surface heating not only impacts the local weather and synoptic situations over the plateau, but also affects the climate and environment of East Asia and the Northern Hemisphere [1–4]. For example, Yanai et al. revealed that Tibetan surface heating has a great influence on the onset of the South Asian summer monsoon [2], and Zhou et al. noted that the possible impacts of the plateau surface heating could be expanded to the Northern Atlantic Ocean [4]. Zou showed the role of surface heating in the formation of the Tibetan ozone low [5].

To understand the surface heating over the Tibetan Plateau, many scientific experiments have been carried out, in particular, the First and Second Tibetan Plateau Atmospheric Scientific Experiments. Using the observational data, Ye and Gao studied the land-air heat exchange and noted that sensible heat transfer dominates the surface heating over the plateau, especially in summer [1]. Gao et al., Bian et al., Li et al. and Zou et al. observed the land-air heat exchange at Nagqu (Central Tibet), Qamdo (East Tibet), Gerze (West Tibet), and the northern slope of Mt. Everest (South Tibet), respectively, in spring and early summer [6–9]. They showed that the total turbulent heat fluxes (defined as the sum of sensible and latent heat flux) over East, West, and Central Tibet were in the range of 80.0–144.0 Wm^{-2}, with a higher sensible heat transfer in the range of 43.0–86.0 Wm^{-2}, and a lower latent heat transfer with the range of 28.0–59.0 Wm^{-2}. However, Zou et al. found that the near-surface heat transfer in the Southeast Tibet in early summer was significantly different from that in the other Tibetan regions [10], with a total heat flux of 86.3 Wm^{-2}, a sensible heat flux of 22.9 Wm^{-2} and a latent heat flux of 63.4 Wm^{-2}. The latent heat transfer dominates the land-air heat exchange in Southeast Tibet.

The South Asian summer monsoon (SASM) is an important component of the Asian monsoon, which has a great influence on the atmospheric processes over Asia [11–15]. The SASM usually starts in late May or early June, which is characterized by the formation of cyclonic vortices in the Bay of Bengal or in the Southeast Arabian Sea [11,16,17]. After its onset, the SASM develops during the summer and autumn, with several active and break periods observed, which are characterized by the heavy and light rainfalls associated with the different monsoon troughs over South Asia [12,18–20]. The SASM decays in late September or early October. The SASM mainly affects the Indian Peninsula and Indo-China Peninsula, and the affected areas could extend northwards to the Qinghai-Tibet Plateau and Southwest China [13,16]. Gao et al. showed that the precipitation over Southeast Tibet can be affected by the monsoon [21]. Zhou et al. [17,22–25] and Li et al. [26] found that the local atmospheric properties in the Himalayas and Southeast Tibet region are closely related to the SASM evolutions. Most recently, Zou et al. [9,10] and Zhou et al. [27] revealed that the land-air heat transfer in the Himalayas and Yarlung Zangbo River Valley in Nyingchi is strongly affected by the SASM. Due to the lack of observation data, the above studies mainly focused on the analysis of a single site or a typical underlying surface, but there was a lack of studies on the influence of the South Asian summer monsoon on the land-air exchange at other different sites in the Tibetan Plateau.

The network of plateau observation stations is sparse; the representativeness of observation stations is limited by the complex topography and underlying surface characteristics. The study of land-air interaction under the complex terrain of the plateau is much more difficult than that in other areas, the observation time, space, and physical-property variables are very limited. Because of this limitation, the third Tibetan Plateau Experiment for atmospheric science (TIPEX III) has been organized by China Meteorological Administration since June 2014, with nine boundary-layer observation stations established in the central, western, and southeastern parts of the plateau [28]. The observation sites are more widely distributed, and the data are the latest and most comprehensive, which provides an important data basis for the study of land-air energy exchange over the Tibetan Plateau. With this observational data, Wang et al. analyzed the surface parameters and near-surface turbulent fluxes over TP [29]. Most previous studies on the SASM impacts are from one or two in-situ stations in the south or Southeast Tibet, while the SASM impacts may be the largest. In this study, a total of eight stations covering different regions of TP were applied, aiming to study the different phases of SASM (active/break periods) impacts on land-atmosphere heat transfer over different plateau regions. In addition, previous studies suggested the great impacts of SASM on the local TP heating, as well as the regional climate over the Southeast Tibet and South Tibet; however, whether these impacts extend northwards was not clear until now. Thus, one of our purposes was to understand the SASM-affected area and extension. In this paper, data and methods are introduced in Section 2, and the SASM evolution (transition of active and break phases), and its possible influences on the land-air heat transfers over different plateau regions are presented in Section 3. The discussion and conclusions are given in Sections 4 and 5, respectively.

2. Data and Methods

The data used in this paper are from the third Tibetan Plateau (TP) Experiment for atmospheric science (TIPEX III) from late July to early September, 2014. During the experimental period, 9 observation stations were installed over the plateau regions. These stations are Ali, Nagqu, Amdo, Nyainrong, Biru, Baingoin, Lhari, Nyingchi, and Namco (see Figure 1 for the topography and Table 1 for the detailed station locations). At each station, the radiation fluxes (downward shortwave radiation flux and net radiation flux) were measured by a 4-component net radiometer (NR01, Hukseflux Thermal Sensors, Delftechpark, The Netherlands), and the land-atmosphere heat transfers (sensible and latent heat fluxes) were measured by a 3-D ultrasonic anemometer (CSAT3, Campbell Scientific, Inc., Logan, UT, USA). These raw data were calculated as the averaging interval of 30 min for analysis in this paper. In this paper, the total heat flux is defined as the sum of sensible and latent heat flux.

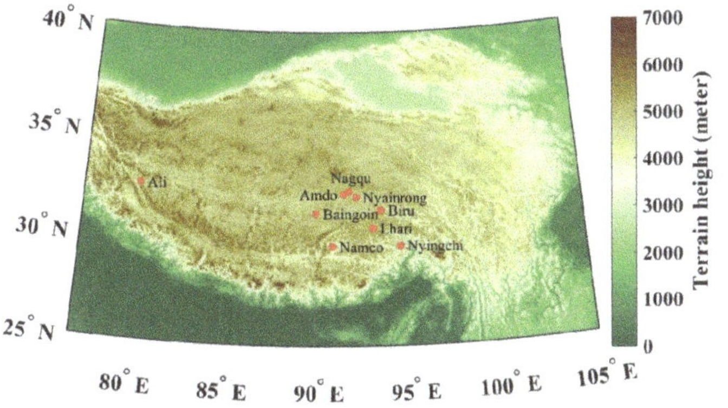

Figure 1. Topography of the Tibetan Plateau, with 9 plateau stations being denoted by red dots. It should be noted that the Nyingchi station was excluded from our study after the data quality control steps were completed.

Table 1. Station locations over the plateau regions.

Stations	Location	Elevation (m)	Plateau Regions
Ali	80.1°E, 32.5°N	4350	northwest plateau
Nagqu	91.9°E, 32.4°N	4509	central plateau
Amdo	91.6°E, 32.2°N	4695	central plateau
Nyainrong	92.3°E, 32.1°N	4730	central plateau
Biru	93.7°E, 31.5°N	4408	central plateau
Baingoin	90.1°E, 31.4°N	4700	central plateau
Lhari	93.2°E, 30.7°N	4500	central plateau
Namco	91.0°E, 29.8°N	4730	southeast plateau
Nyingchi	94.7°E, 29.8°N	3327	southeast plateau

EDDYPRO (version 5.1) software (from Li-COR Corporation) is also used for turbulent flux data quality control [29–31]. After quality control steps were performed, Nyingchi station was excluded from our study due to the missing data of more than 30%. For the following analysis, 29 July–26 August was selected as the observational period when data were available from all 8 stations.

In addition to the observational data, the large-scale reanalysis data from ECMWF (European Centre for Medium-Range Weather Forecasts) Interim were also used, including wind and specific humidity, with a horizontal resolution of $0.75^{\circ} \times 0.75^{\circ}$. The interpolated outgoing long-wave radiation (OLR) data from NOAA (National Oceanic and Atmospheric

Administration) were applied to illustrate the convection conditions, with a horizontal resolution of $1.0^\circ \times 1.0^\circ$.

3. Results

3.1. SASM Evolution and Synoptic Situations

The SASM evolution could have great impacts on the weather and climate in Asia through general circulation changes [1,17,21,23,27,32]. The onset of SASM usually occurs at the end of May or early June in South Asia [33]. To investigate the impacts of the SASM on the land-air heat exchange processes over TP, the SASM evolution during the TIPEX III experiment in 2014 is first analyzed.

To characterize the SASM evolution in the TIPEX III in 2014, a SASM index (SASMI) from Wang et al. [34] is adopted in this study. The SASMI is defined by the standardized difference of averaged zonal wind speeds at 850 hPa from two regions, 5–15°N, 40–80°E, and 20–30°N, 70–90°E. A large positive SASMI corresponds to a strong monsoonal circulation, while a large negative SASMI corresponds to a weak monsoon. Figure 2 presents the daily variations in SASMI during the observation period from 1 May to 30 September 2014. This figure shows that the SASMI turns positive on 6 June, and then begins with a sudden increase to a maximum on 11 June 2014, with the maximum SASMI value being larger than the average value (5.0 m/s). At this time, a strong cyclonic circulation prevailed over the Arabian Sea at 850 hPa (figure not shown), which represents the SASM onset in 2014. Thereafter, SASM experienced several active periods during 11–16 July, 22 July, 29 July–1 August, 5 August, and 27 August–2 September, with positive SASMI exceeding the standard deviation, and break periods during 21–23 June, 28–30 June, 13–15 August, 24–25 August, with negative SASMI values exceeding the standard deviation. Considering the observation period, 29 July–1 August, and 5 August were selected as the SASM active period, and 13–15 August, and 24–25 August were selected as the SASM break period. In order to minimize the influence of the solar altitude angle, the active and break periods were selected close to each other. In the following studies, the land-air heat transfers, as well as radiation fluxes, will be averaged for the SASM active and break periods, to investigate their responses to the SASM evolution.

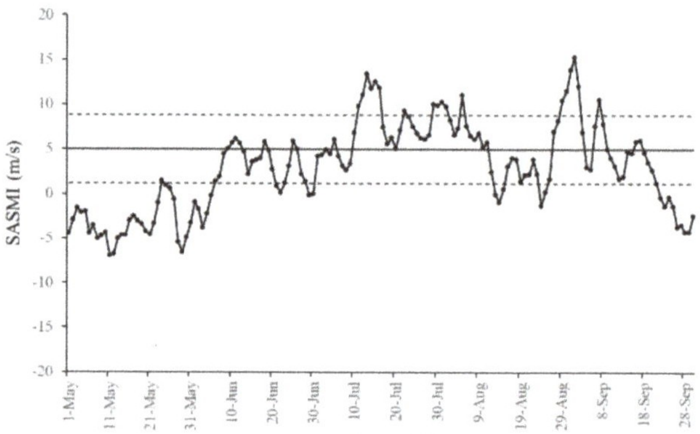

Figure 2. Variations in the SASM index (SASMI), with the averaged SASMI value (5.0 m/s) shown as a solid line, dashed lines represent one standard deviation (3.8 m/s) above and below the mean value.

To illustrate convection related to the SASM evolution, the outgoing long-wave radiation (OLR) was averaged for the entire observational period and for the SASM active and break periods; their distributions are shown in Figure 3. The low OLR values represent strong convection and vice versa. During the observational period (Figure 3a), there were three strong convective activity centers (with OLR values lower than 190 Wm^{-2}) in the eastern part of the Bay of Bengal, the northeastern part of India and the central portion of

TP. During the SASM active period (Figure 3b), the strong convective activities over the Bay of Bengal and North India both intensified and extended northward, with central OLR values less than 160 Wm^{-2}. The convection over the central TP became more severe and enlarged, with central OLR values lower than 160 Wm^{-2} and covering almost the entire TP region. During the break period (Figure 3c), however, all the three convections moved southwards, and the convective activity center over the TP as shown in Figure 3a retreated to the south of TP. Therefore, obvious differences can be found in the OLR distributions between the SASM active and break periods, especially over the TP regions, representing the dominant strong and weak convections there.

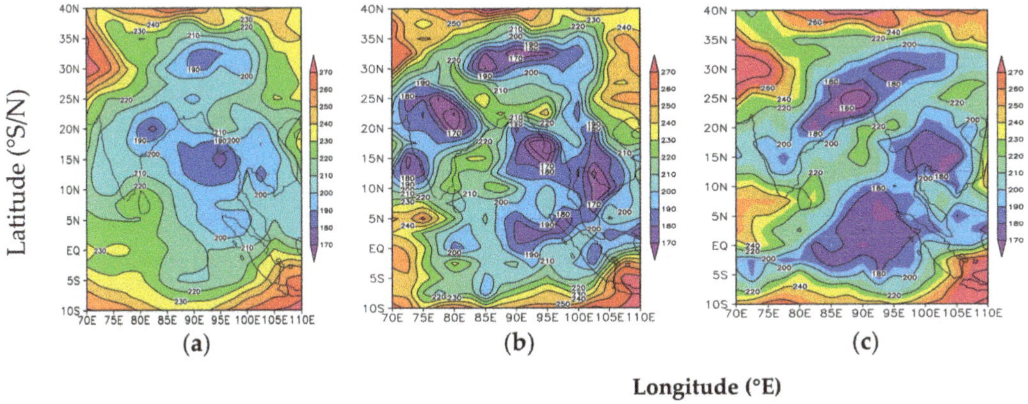

Figure 3. Outgoing long-wave radiation (OLR) distributions over the South Asia and TP regions, averaged for the (**a**) observation period, (**b**) SASM active period, and (**c**) SASM break period, with a contour interval of 10 Wm^{-2}.

Figure 4 presents the averaged wind and specific humidity fields at 500 hPa in the entire observation period, SASM active and break periods. During the observation period (Figure 4a), a cyclone formed over Central India and the Bay of Bengal, with the highest specific humidity greater than 4.5 g/kg. A westerly existed in the west TP, bringing dry air masses there (with the specific humidity less than 3.0 g/kg). A southwesterly prevailed over the south and central plateau regions, with specific humidity values greater than 5.5 g/kg. During the SASM active period (Figure 4b), the cyclone over Central India and the Bay of Bengal intensified, and moved northward and westward, associated with an enhancement of moisture (central specific humidity greater than 5.5 g/kg). An obvious cyclonic circulation appeared over the main body of the plateau. The dry westerly prevailed over the west TP weakened. The southwesterly existed in the south and Central TP became stronger, leading to higher moisture levels over the entire TP, with central values greater than 7.0 g/kg. During the SASM break period (Figure 4c), the cyclone with high water vapor over Central India and the Bay of Bengal as shown in Figure 4b disappeared. The dry westerly over the west and southwest of TP strengthened significantly, with moisture values below 2.0 g/kg. The weakened southwesterly led to the retreat and shrinking of the high moisture center over the south and Central TP.

3.2. The Impacts of SASM Evolution on the Radiation Heat Transfers

From the above results, large differences between the SASM active and break periods were observed from the synoptic situations, including the convection, wind, and moisture fields over the South Asian and TP regions. In the following study, the impacts of SASM evolution on land-air heat exchange processes, as well as the radiation conditions will be covered.

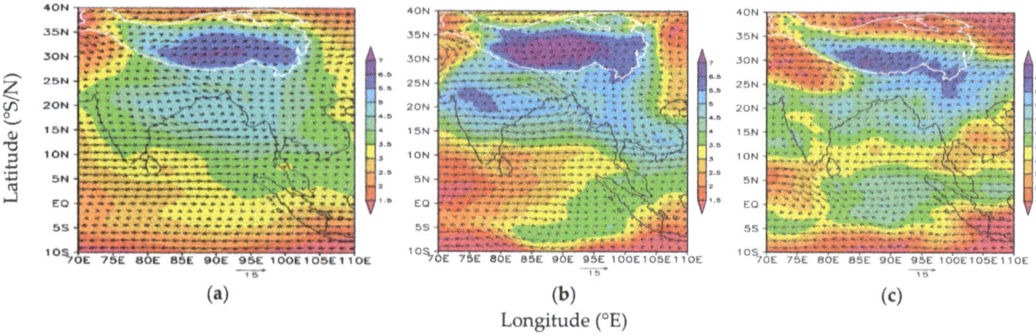

Figure 4. Horizontal distributions of wind (arrows, units: m/s) and specific humidity. (shadings, units: g/kg) at 500 hPa averaged for (**a**) the observation period, (**b**) the SASM active period, and (**c**) the SASM break period.

Figure 5 shows the diurnal variations of the downward shortwave radiation flux (DR), averaged for the observations period, SASM active, and break periods. The diurnal variations of DR are similar for all stations, showing increases at approximately 06:00 LST (local standard time) near sunrise, reaching a maximum at noon, and decreasing to almost zero at approximately 18:00 LST near sunset. However, large DR amplitude differences were found among the 8 plateau stations. The strongest DR occurred over Ali station (northwest plateau), with daily average and maximum values of 319.2 and 1007.8 Wm^{-2}, respectively, which is mainly due to the low precipitation and less moisture in the air. Ali station is located in the northwest of the Qinghai-Tibet Plateau, and Ali is a very dry area with very little rainfall. According to the observation data, Ali station had no precipitation from July to September in 2014, resulting in low water vapor content in the air. The second strong DR is found over Namco station (southeast plateau), with daily averaged and maximum values of 222.9 and 846.3 Wm^{-2}, respectively. For the other stations, the DR differences are noticeably smaller, and the difference of the daily averaged values was less than 20 Wm^{-2}, with a variation between 187.3 Wm^{-2} and 206.1 Wm^{-2}. During the SASM active/break periods, the DR is greatly weakened/strengthened at most plateau stations. For example, the daily averaged value of DR at the Baingoin station was 194.2 Wm^{-2} during the observation period, and the DR varies from 133.2 Wm^{-2} (weakened by 31.4%) during the SASM active period and 234.5 Wm^{-2} (strengthened by 20.8%) during the break period, respectively. The weakened/strengthened DR is closely related to the strong/weak convections over the plateau region during the SASM active/break periods (see Figure 3). The strong/weak convections can result in more/less cloudiness, which further affects the solar radiation by blocking/enhancing effects [17,27,32].

Figure 6 presents the net radiation fluxes (NR) for the 8 stations over the TP during the observations, SASM active, and break periods. The NR patterns exhibit similar diurnal variations as those of the DR, with positive values (net heating) during daytime from approximately 06:00 LST to approximately 18:00 LST and negative values (net cooling) for the other times of the day over most of the plateau stations. NR amplitude differences are also seen among the 8 plateau stations. The strongest net radiation also occurs at Namco station (southeast plateau), with daily averaged and maximum values of 133.2 and 609.2 Wm^{-2}, respectively. The second strongest NR is found over Lhari station (central plateau), with a diurnally averaged value of 120.0 Wm^{-2} and a maximum of 511.6 Wm^{-2}. Over the other stations, the NR differences are quite small; the difference in the daily averaged value was about 10 Wm^{-2}, with a range from 102.8 to 112.2 Wm^{-2}. The NR is also weakened/strengthened during the SASM active/break periods at most of the stations compared with the observational mean. As with DR, the largest effects of SASM on the NR occurred at Baingoin station, the daily averaged value of NR was 102.8 Wm^{-2} during the observation period, while the NR was 66.9 Wm^{-2} during the SASM active period, which weakened by 34.9%, and the NR was 128.8 Wm^{-2} during the SASM break period, which strengthened by 25.3%.

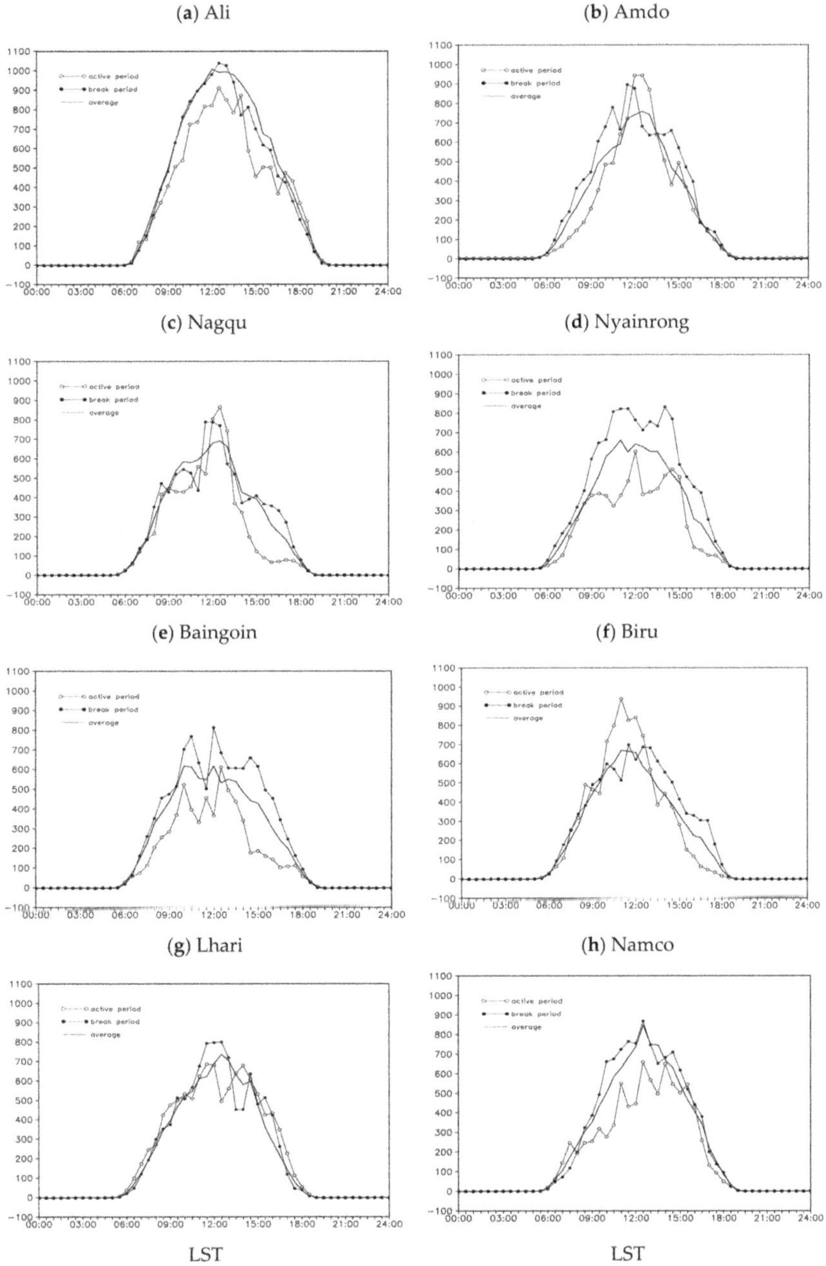

Figure 5. Diurnal variation of the downward short-wave radiation flux (DR) (units: Wm^{-2}) from 8 stations, averaged for the observations, SASM active, and break periods.

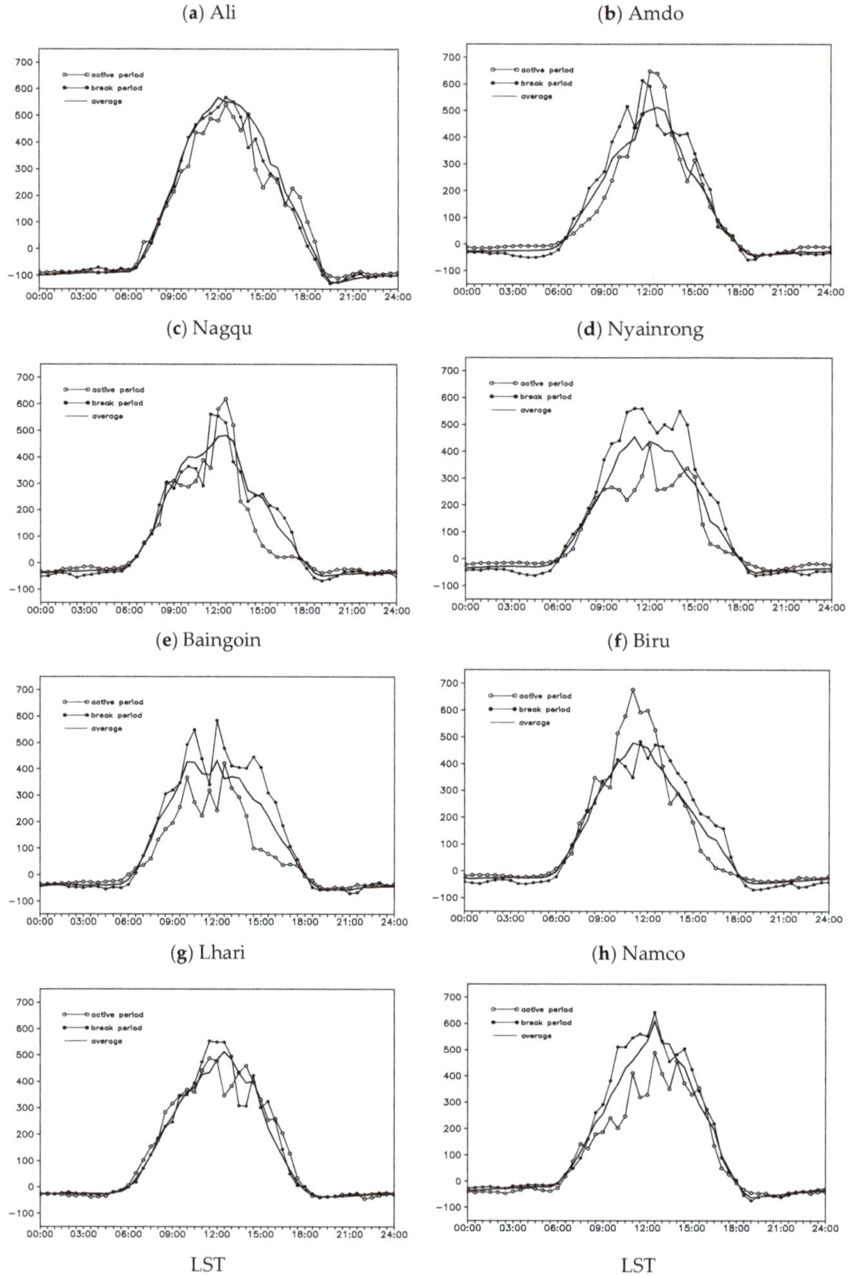

Figure 6. Same as for Figure 5, but for the net radiation flux (NR) (units: Wm^{-2}).

3.3. *The Impacts of SASM Evolution on the Turbulent Heat Transfers*

Figure 7 shows the diurnal variations in sensible heat flux (SH), averaged for the observations, SASM active, and break periods. Driven by the net heating (see Figure 6), the plateau releases heat into the atmosphere during the daytime (positive SH values) and receives heat from the atmosphere at night (negative SH values). Following the diurnal variation in radiation, the sensible heat flux increases from early morning at approximately 6:00 LST, reaches a maximum at noon, and decreases later in the day. Differences are clearly found in the SH averaged and maximum values among the 8 plateau stations, despite the similar diurnal variations. During the observation period, the largest sensible heat transfer occurs over Ali station (northwest plateau), with daily averaged and maximum values of 60.1 (see Table 2) and 197.4 Wm^{-2}, respectively. The second-largest SH is found over Namco station (southeast plateau), with daily averaged and maximum values of 28.2 and 118.9 Wm^{-2}, respectively. Over the other stations, the diurnally averaged SH varies from 20.0 to 26.5 Wm^{-2}, and the daily maximum values vary from 73.1 to 98.2 Wm^{-2}. The smallest SH occurs over the Biru station (central plateau), with averaged and maximum values of 18.8 and 77.5 Wm^{-2}, respectively. Our results are consistent with recent study results. The recent research reveals that SH in the central TP in August is generally between 5 and 40 Wm^{-2}, with an average of 18 Wm^{-2}, while SH in the western TP is between 40 and 70 Wm^{-2}, with an average of 56 Wm^{-2}, these results are also smaller than that in the past [29,35]. Ye and Gao [1] estimated the July-August-mean intensity of SH is 60–80 Wm^{-2} over the central TP and 150–190 Wm^{-2} in the western TP, and Yang and Guo [36] estimated SH in July-August is 50–60 Wm^{-2} in the central TP and 75–90 Wm^{-2} in the western TP, remarkably larger compared to the new findings. This result indicates that SH has been possibly overestimated by the previous studies when calculating SH using the bulk transfer method, which is based on the larger values of the bulk transfer coefficient for sensible heat [28].

During the SASM active period, the response of land-to-atmosphere sensible heat transfer exhibited great differences among the 8 stations. For example, the daily averaged SH weakened at most stations, with the amplitude of weakening varying from −1.7% to −41.4%. The largest weakening occurred at Namco station (southeast plateau), and the SH was 16.5 Wm^{-2} over the SASM active period, which was weakened by 41.4% from its daily averaged value of 28.2 Wm^{-2} during the observation period. The smallest weakening happened at Biru station (central plateau), with a weakening of 1.7% from its daily averaged value. At stations Amdo and Nyainrong, the daily averaged SH values strengthened, with the amplitudes increasing by 12.3% and 13.2%, respectively. During the SASM break period, the SH largely strengthens over all stations, and with the strengths varying from 0.8% (at Ali) to 45.3% (at Nyainrong). Therefore, the sensible heat transfers over the plateau region can be affected by the SASM evolution, with the weakened/strengthened amplitudes over most stations during the SASM active/break periods, which are closely related to the weakened/strengthened radiation conditions [9,10,17,23,27,32]. However, these SASM impacts on the sensible heat transfer exhibit large inhomogeneity over the plateau regions. Overall, the more southerly stations received more SASM impacts. The larger SASM impacts on sensible heat transfers occurred at stations Namco, Baingoin, and Lhari, and the SH differences between the SASM active and break periods were greater than 38% of the daily averaged values. The SASM impacts also extended westward and northward to the Ali and Nagqu stations, with the SH differences between the SASM active and break periods reaching 21.4% and 34.6% of their daily averaged values, respectively. However, the SASM impact appeared to be negligible at Amdo station (south of Nagqu station), which complicated our results.

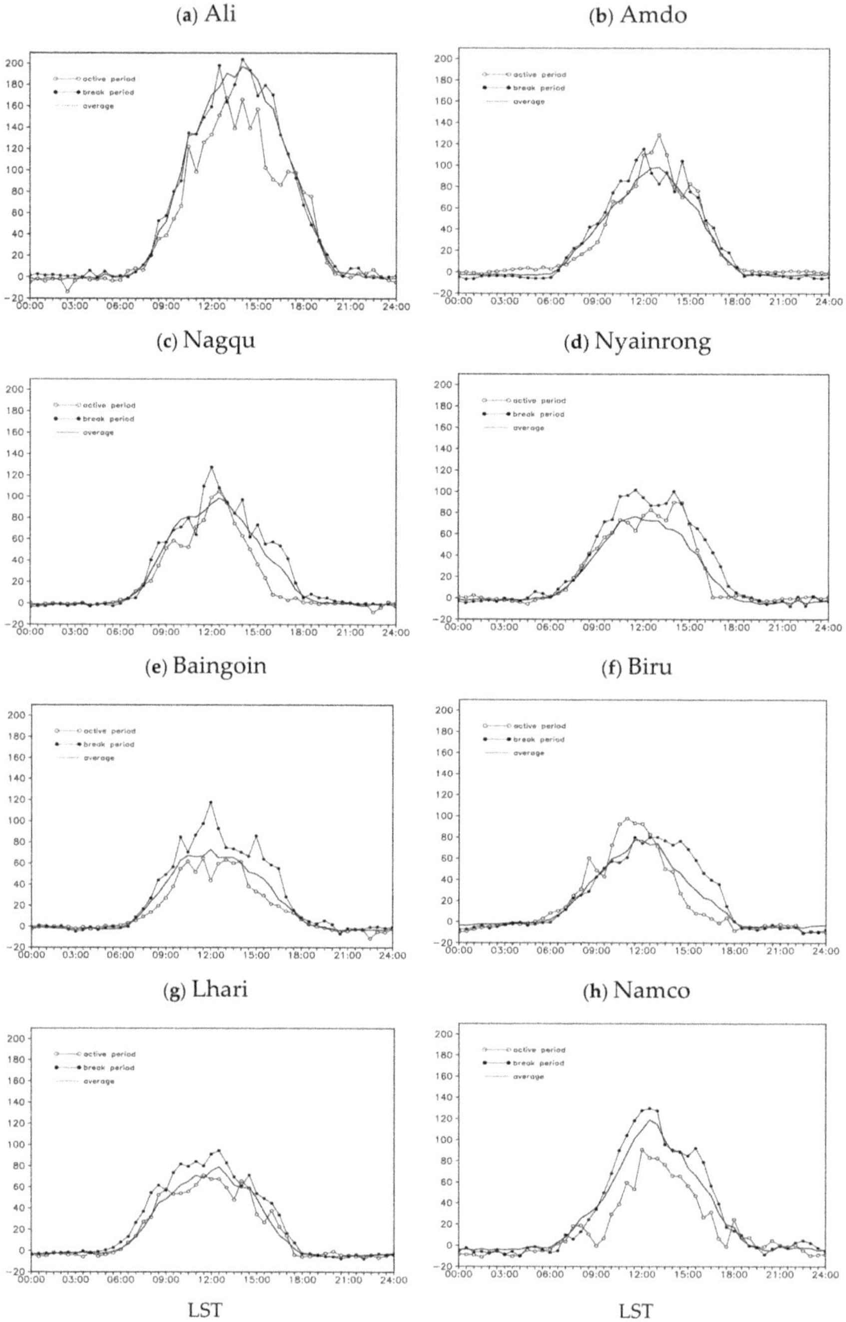

Figure 7. Diurnal variations of the sensible heat flux (SH) (units: Wm^{-2}) from 8 stations, averaged for the observations and SASM active and break periods.

Table 2. Sensible heat flux (SH) (Wm^{-2}) over the 8 plateau stations, averaged for the observations and SASM active and break periods. The bracketed values denote the SH percentage increase (decrease) of the daily averaged value, in which positive (negative) values mean increasing (decreasing).

SH/Stations	Ali	Nagqu	Amdo	Nyainrong	Biru	Baingoin	Lhari	Namco
Observation	60.1	26.5	24.5	20.0	18.8	20.2	20.0	28.2
SASM active	47.7 (−20.5%)	20.9 (−21.2%)	27.5 (12.3%)	22.6 (13.2%)	18.5 (−1.7%)	16.0 (−20.8%)	18.6 (−6.8%)	16.5 (−41.4%)
SASM break	60.6 (0.8%)	30.1 (13.4%)	27.0 (10.1%)	29.0 (45.3%)	21.7 (15.4%)	27.7 (37.6%)	26.3 (31.7%)	30.9 (9.7%)

Figure 8 presents the diurnal variation of latent heat flux (LH), averaged for the observations, SASM active, and break periods. Differing from the sensible heat flux, the latent heat over the TP is always transferred upwards (positive LH values) during the entire day. In addition, the amplitude of LH was much larger than that of SH over most of the plateau stations and that was consistent with previous results [37]. Obvious differences are seen in the LH daily averaged and maximum values among the 8 plateau stations. During the observation period, the largest latent heat transfer occurred at Nagqu station (central plateau), with daily averaged and maximum values of 74.7 (see Table 3) and 238.6 Wm^{-2}, respectively. The smallest LH occurred at Ali station (northwest plateau) due to small amounts of precipitation there, with daily averaged and maximum values of 10.1 and 27.2 Wm^{-2}, respectively. Over the other stations, the diurnally averaged LH varied from 53.0 to 74.4 Wm^{-2}, and the maximum values varied from 152.8 to 245.7 Wm^{-2}.

Table 3. Latent heat flux (LH) (Wm^{-2}) over the 8 plateau stations, averaged for the observations and SASM active and break periods. The bracketed values denote the LH increasing (decreasing) percentage of the daily averaged value, in which the positive (negative) values mean increasing (decreasing).

LH/Stations	Ali	Nagqu	Amdo	Nyainrong	Biru	Baingoin	Lhari	Namco
Observation	10.1	74.7	74.4	66.6	59.3	55.4	53.0	65.0
SASM active	18.9 (87.4%)	56.8 (−24.0%)	78.7 (5.9%)	72.8 (9.4%)	69.5 (17.3%)	47.5 (−14.3%)	48.3 (−8.8%)	59.1 (−9.0%)
SASM break	5.9 (−41.2%)	77.5 (3.8%)	73.5 (−1.1%)	77.6 (16.6%)	63.5 (7.1%)	62.9 (13.6%)	57.8 (9.0%)	63.0 (3.0%)

Compared with the impacts of the SH, the SASM impacts on LH were relatively small and complicated. At stations Namco, Lhari, and Baingoin, while the SASM impacts on the SH were large, but the SASM impacts on LH were smaller, with the LH differences between the SASM active and break periods varying from 12.0% to 27.9% of their daily averaged values. The small impacts over these stations could be closely related to the high moisture conditions there. The SASM impacts could also extend to the north plateau, and Nagqu station with a large LH difference (27.8% of the daily averaged value) between the SASM active and break periods. However, the same as for SH, the SASM impacts on LH seem negligible over station Amdo. It should be noted that the SASM impact over Ali could be ignored due to the small LH value despite having the largest LH response amplitude of the daily averaged values.

The total heat transfer (TH) is defined as the sum of SH and LH. Figure 9 shows the diurnal variation in TH, averaged for the observations, SASM active, and break periods. Clear differences can be seen in the TH magnitudes among the 8 plateau stations. During the observation period, the largest TH occurred at Nagqu station (central plateau), with daily averaged and maximum values of 101.2 (see Table 4) and 336.7 Wm^{-2}, respectively. The smallest TH occurred over Ali station (northwest plateau), with daily averaged and maximum values of 70.2 and 220.3 Wm^{-2}, respectively. Over the other stations, the diurnally averaged TH varied from 73.0 to 98.8 Wm^{-2}, and the daily maximum values varied from 225.9 to 343.9 Wm^{-2}.

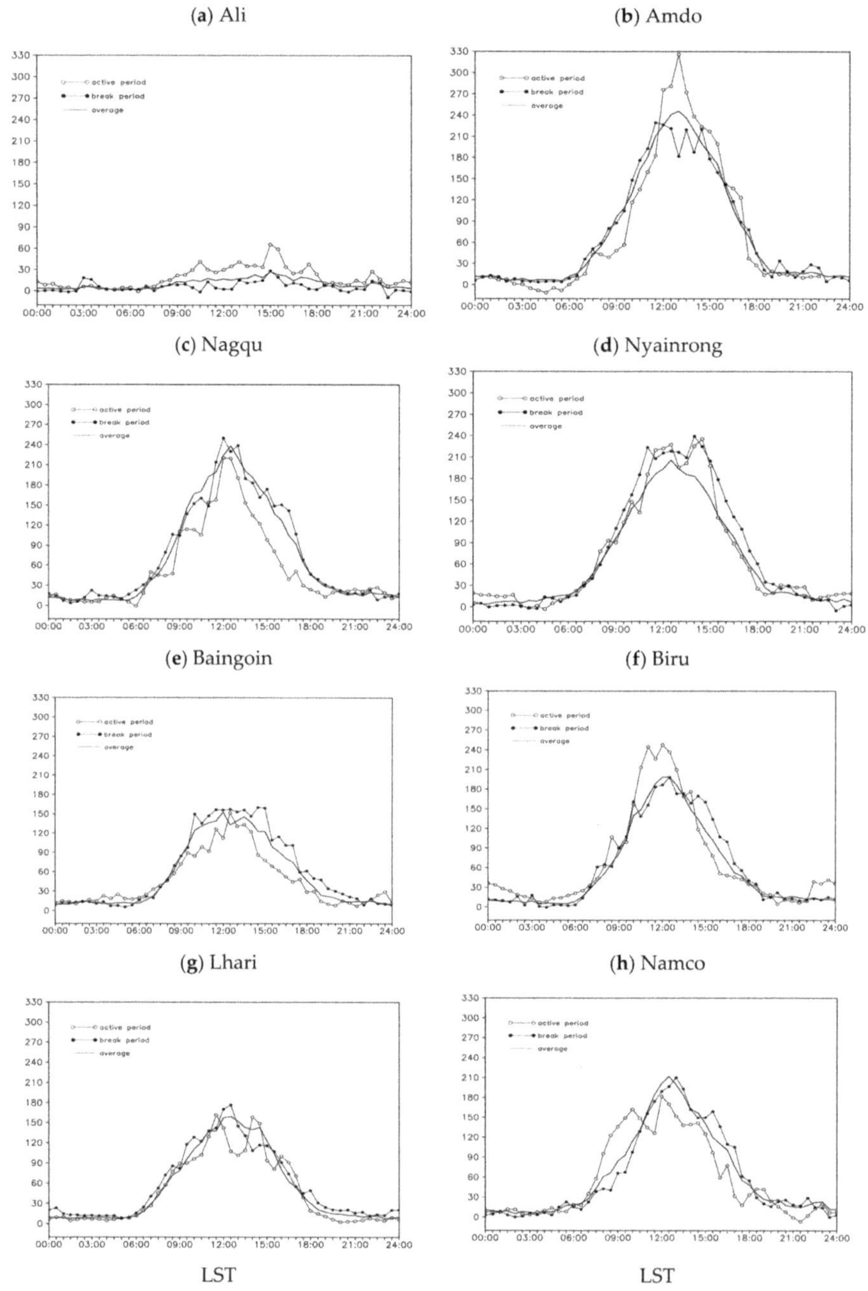

Figure 8. Same as Figure 7, but for latent heat flux (LH) (units: Wm^{-2}).

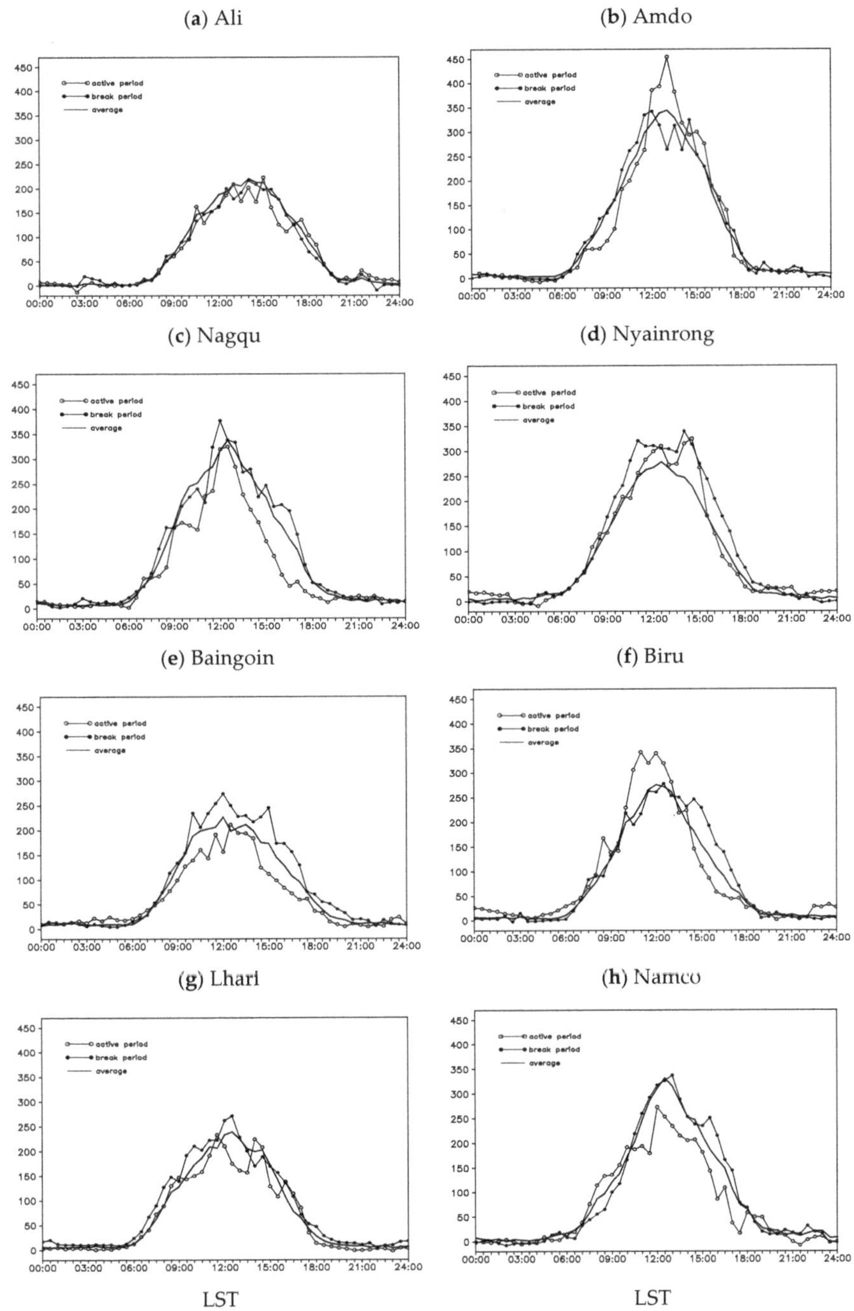

Figure 9. Same as Figure 7, but for total heat flux (TH) (units: Wm^{-2}).

Table 4. Total heat flux (TH) (Wm^{-2}) over the 8 plateau stations, averaged for the observations and SASM active and break periods. The bracketed values denote the TH increasing (decreasing) percentage of the daily averaged value, in which the positive (negative) values mean increasing (decreasing).

TH/Stations	Ali	Nagqu	Amdo	Nyainrong	Biru	Baingoin	Lhari	Namco
Observation	70.2	101.2	98.8	86.5	78.1	75.5	73.0	93.1
SASM active	66.7 (−5.0%)	77.7 (−23.3%)	108.7 (10.0%)	95.4 (10.2%)	88.0 (12.7%)	63.4 (−16.0%)	67.0 (−8.3%)	75.6 (−18.8%)
SASM break	66.5 (−5.2%)	107.6 (6.3%)	100.5 (1.7%)	106.6 (23.2%)	85.2 (9.1%)	90.6 (20.0%)	84.1 (15.2%)	93.9 (0.8%)

From Table 4, the SASM impacts on the TH can be clearly seen at stations Namco, Baingoin, Lhari, and Nagqu, with a weakened/strengthened magnitude during the SASM active/break period, and the TH differences in daily averaged value between the SASM active and break periods are with a range between 19.6% and 36.6%. The largest impacts occurred at Baingoin station, with the TH difference reaching 36.0% of the daily averaged value between the SASM active and break periods. In comparison, the SASM impacts on TH over the other plateau stations were quite small or even negligible.

4. Discussion

In theory, there are multi-scale atmospheric motions, especially over the complex TP regions, which interact with each other. The local land-atmosphere heat transfers over TP could be possibly affected by the large-scale circulations, such as the surrounding South Asia monsoon systems. Theoretically, the monsoon could affect the local heat transfer by adjusting the local radiation conditions through cloud and precipitation variations, which have been proved by previous studies [9,23,27].

From our results, not all stations show consistent impacts from SASM. For example, the Biru station has an opposite variation of heat flux during the different SASM stages, compared with other stations at similar latitudes, such as Baingoin and Lhari. The reason is still unclear; this may have connections with orographic peculiarities and microclimatic features, and which could be partly leading to the different local heating forces there. The problem also needs to be further studied using longer-time observation data in the future.

The spatial differences between various sites could be possibly not only affected by the SASM activities. Our study presented a clue that the local land-air exchange processes could be strongly affected by large-scale circulations, such as the SASM evolution, with inhomogeneous distributions over the whole TP region. The local topographic effects should be carefully considered for future studies.

Due to the hard living and working conditions, the observational data are quite limited over Tibet, especially over the northern and western parts of the plateau. Thus, as shown in our introduction, most previous studies focused mainly on the SASM impacts over the south or the southeast Tibet, one reason is due to the largest impacts there which were shown from our studies, and another reason is no observational data over the other plateau regions. During the most recent year, the Third Tibetan Plateau Atmospheric Scientific Experiment provided us with an opportunity to make a comprehensive study on the possible SASM impacts on different regions of TP. Although our conclusions are based on only one year of data (2014), the physical mechanism is the same for the other year, in spite of year-to-year variations. The SASM evolution could result in cloud and advection variations, adjusting the local radiation conditions, and further affecting the local heat transfers. Therefore, the inhomogeneities of SASM impacts should be confirmed for each year, but varied with different amplitudes and monsoon extended regions. In the future, based on more accumulated data, further investigations and more evidence are still needed for the robustness of conclusions.

5. Conclusions

Using the observation data from TIBEX III, as well as the large-scale reanalysis data from ECMWF Interim and from NOAA, the different impacts of SASM on the land-atmosphere heat exchange processes between active and break periods over different regions of TP were investigated. During the observation period (29 July to 26 August), the land-atmosphere heat transfer exhibited strong inhomogeneous distributions over the plateau regions. The daily average total heat transfer varied from 70.2 to 101.2 Wm^{-2} over the 8 plateau stations, with the sensible heat flux ranging from 18.8 to 60.1 Wm^{-2} and the latent heat flux with a variation between 10.1 and 74.7 Wm^{-2}. The latent heat transfer values larger than the sensible heat transfer values over most of the plateau regions are mainly related to the strong convection that prevailed over the plateau during the observation period (Figure 3a), which caused the high moisture conditions (Figure 4a). The land-atmosphere heat transfer can be largely affected by the SASM evolution, but with strong inhomogeneity over the plateau stations. Overall, the more southerly stations received more SASM impacts. The land-atmosphere heat transfers (the total, sensible, and latent heat fluxes) are greatly weakened/strengthened during the SASM active/break period at the Namco (southeast plateau), Baingoin (central plateau), Lhari (central plateau), and Nagqu (central plateau) stations, with the sensible heat flux differences between the SASM active and break periods varying from 34.6% to 58.4% of the daily averaged values, with a range between 12.0% and 27.9% for the latent heat flux and with a variation between 19.6% and 36.0% for the total heat transfer. These significant SASM influences could be closely related to the weakened/strengthened radiation conditions [9,10,17,23,27,32]. However, the impacts of SASM during active/break periods become complicated over the other plateau stations. For example, the impact of SASM on Ali station is mainly reflected in the influence of sensible heat flux, because sensible heat flux plays a dominant role in the total heat transfer. The different phases of SASM impacts on the Amdo and Nyainrong stations were quite small or even negligible, which complicates our conclusions. Therefore, further investigations are still needed based on more observational data over a long period in the Tibetan Plateau.

Author Contributions: H.L.: Formal analysis, Writing—original draft. L.Z.: Conceptualization, Supervision, Writing—review & editing. G.W.: Formal analysis, Writing—review & editing. All authors have read and agreed to the published version of the manuscript.

Funding: This research was funded by the Strategic Priority Research Program of the Chinese Academy of Sciences (Grant No. XDA19070401), the National Natural Science Foundation of China (Grant 42030611), the Second Tibetan Plateau Scientific Expedition and Research (STEP) program (grant No.2019QZKK0103 and 2019QZKK0105), the National Natural Science Foundation of China (Grant 91937301 and 41830968), the CAS Key Subordinate Project (KGFZD-135-16-023), the Forecaster Special Project of China Meteorological Administration (No.CMAYBY2019-155), the Heavy Rain and Drought-Flood Disasters in Plateau and Basin Key Laboratory of Sichuan Province (SCQXKJYJXMS202116), and the Opening Foundation of Plateau Atmosphere and Environment Key Laboratory of Sichuan Province (grant No.PAEKL-2020-C7).

Data Availability Statement: The data that support the findings of this study are available from the first author upon request (Hongyi Li, lihongyi@cma.gov.cn).

Acknowledgments: We appreciate the access to the ECMWF Interim and NOAA datasets. The authors appreciate all of the hard work done by researchers attending the third Tibetan Plateau (TP) Experiment (TIPEX III).

Conflicts of Interest: The authors declare that they have no known competing financial interests or personal relationships that could have appeared to influence the work reported in this paper.

References

1. Ye, D.; Gao, Y. *Meteorology of the Qinghai-Xizang Plateau*; Science Press: Beijing, China, 1979; p. 279.
2. Yanai, M.; Li, C.; Song, Z. Seasonal heating of the Tibetan Plateau and its effects on the evolution of the Asian summer monsoon. *J. Meteorol. Soc. Jpn.* **1992**, *70*, 319–351. [CrossRef]

3. Wu, G.; Zhang, Y. Tibetan Plateau forcing and the timing of the monsoon onset over South Asia and the South China Sea. *Mon. Weather Rev.* **1998**, *126*, 913–927. [CrossRef]
4. Zhou, X.; Zhao, P.; Chen, J.; Chen, L.; Li, W. Impacts of thermodynamic processes over the Tibetan Plateau on the Northern Hemispheric climate. *Sci. China Ser. D Earth Sci.* **2009**, *52*, 1679–1693. [CrossRef]
5. Zou, H. Seasonal variation and trends of TOMS ozone over Tibet. *Geophys. Res. Lett.* **1996**, *23*, 1029–1032. [CrossRef]
6. Gao, Z.; Wang, J.; Ma, Y.; Kim, J.; Choi, T.; Lee, H.; Asanuma, J.; Su, Z. Calculation of near-surface layer turbulent transport and analysis of surface thermal equilibrium features in Nagqu of Tibet. *Phys. Chem. Earth B* **2000**, *25*, 135–139. [CrossRef]
7. Bian, L.; Gao, Z.; Xu, Q.; Lu, L.; Cheng, Y. Measurements of turbulence transfer in the near-surface layer over the Southeastern Tibetan Plateau. *Bound. Layer Meteorol.* **2002**, *102*, 281–300. [CrossRef]
8. Li, J.; Hong, Z.; Luo, W.; Li, A.; Zhao, Y. A study of surface fluxes in Gerze area, Qinghai-Xizang Plateau. *Chin. J. Atmos. Sci.* **1999**, *23*, 142–151.
9. Zou, H.; Ma, S.; Zhou, L.; Li, P.; Li, A. Measured turbulent heat transfer on the northern slope of Mt. Everest and its relation to the south Asian summer monsoon. *Geophys. Res. Lett.* **2009**, *36*, L09810. [CrossRef]
10. Zou, H.; Li, P.; Ma, S.; Zhou, L.; Zhu, J. The local atmosphere and the turbulent heat transfer in the eastern Himalayas. *Adv. Atmos. Sci.* **2012**, *29*, 435–440. [CrossRef]
11. Li, C.; Yanai, M. The onset and interannual variability of the Asian summer monsoon in relation to land-sea thermal contrast. *J. Clim.* **1996**, *9*, 358–375. [CrossRef]
12. Webster, P.; Magaña, V.; Palmer, T.; Shukla, J.; Tomas, R.; Yanai, M.; Yasunari, T. Monsoons: Processes, predictability, and the prospects for prediction. *J. Geophys. Res.* **1998**, *103*, 14451–14510. [CrossRef]
13. Hsu, H.; Terng, C.; Chen, C. Evolution of large-scale circulation and heating during the first transition of Asian summer monsoon. *J. Clim.* **1999**, *12*, 793–810. [CrossRef]
14. Goswami, B.; Mohan, R.A.; Xavier, P.; Sengupta, D. Clustering of low pressure systems during the Indian summer monsoon by intraseasonal oscillations. *Geophys. Res. Lett.* **2003**, *30*, 1431. [CrossRef]
15. Goswami, B.; Xavier, P. Potential predictability and extended range prediction of Indian summer monsoon breaks. *Geophys. Res. Lett.* **2003**, *30*, 1966. [CrossRef]
16. He, H.; McGinnis, J.; Song, Z.; Yanai, M. Onset of the Asian summer monsoon in 1979 and the effect of the Tibetan Plateau. *Mon. Weather Rev.* **1987**, *115*, 1966–1995. [CrossRef]
17. Zhou, L.; Zou, H.; Ma, S.; Li, P. Observed impact of the South Asian summer monsoon on the local meteorology in the Himalayas. *Acta Meteorol. Sin.* **2012**, *26*, 205–215. [CrossRef]
18. Krishnamurti, T.; Ardanuy, P. The 10 to 20 day westward propagating mode and breaks in the monsoon. *Tellus* **1980**, *32*, 15–26. [CrossRef]
19. Krishnamurthy, V.; Shukla, J. Intraseasonal and interannual variability of rainfall over India. *J. Clim.* **2000**, *13*, 4366–4377. [CrossRef]
20. Ding, Y.; Sikka, D. Synoptic systems and weather. In *The Asian Monsoon*; Wang, B., Ed.; Praxis Publishing Ltd.: Chichester, UK, 2006; pp. 131–202.
21. Gao, D.; Zou, H.; Wang, W. Influences of Brahmaputra river water passage on the precipitation. *Mountain Res.* **1985**, *3*, 239–249.
22. Zhou, L.; Zou, H.; Ma, S.; Li, P.; Wu, R. Impacts of the South Asian Summer Monsoon on Variations of Surface Wind on the Northern Slope of Mt. Qomolangma. *Plateau Meteorol.* **2007**, *26*, 1173–1186.
23. Zhou, L.; Zou, H.; Ma, S.; Li, P. Study on impact of the South Asian summer monsoon on the down-valley wind on the north slope of Mt. Everest. *Geophys. Res. Lett.* **2008**, *35*, L14811. [CrossRef]
24. Zhou, L.; Zhu, J.; Zou, H.; Ma, S.; Li, P.; Zhang, Y.; Huo, C. Atmospheric moisture distribution and transport over the Tibetan Plateau and the impacts of the South Asian summer monsoon. *Acta Meteorol. Sin.* **2013**, *27*, 819–831. [CrossRef]
25. Zhou, L.; Li, F.; Zhu, J.; Zou, H.; Ma, S.; Li, P. Vertical structures of atmospheric properties in Southeast Tibet during the South Asian summer monsoon in 2013. *J. Meteorol. Res.* **2016**, *30*, 258–264. [CrossRef]
26. Li, J.; Li, Y.; Jiang, X.; Gao, D. Characteristics of land–atmosphere energy exchanges over complex terrain area of southeastern Tibetan Plateau under different weather conditions. *Chin. J. Atmos. Sci.* **2016**, *40*, 777–791.
27. Zhou, L.; Zou, H.; Ma, S.; Li, F.; Zhu, J.; Li, P.; Zhang, Y. The observed impacts of South Asian summer monsoon on the local atmosphere and the near surface turbulent heat exchange over the Southeast Tibet. *J. Geophys. Res. Atmos.* **2015**, *120*, 11509–11518. [CrossRef]
28. Zhao, P.; Xu, X.; Chen, F.; Guo, X.; Zheng, X.; Liu, L.; Hong, Y.; Li, Y.; La, Z.; Peng, H.; et al. The third atmospheric scientific experiment for understanding the earth–atmosphere coupled system over the Tibetan plateau and its effects. *Bull. Amer. Meteor. Soc.* **2018**, *99*, 757–776. [CrossRef]
29. Wang, Y.; Xu, X.; Liu, H.; Li, Y.; Li, Y.; Hu, Z.; Gao, X.; Ma, Y.; Sun, J.; Lenschow, D.H.; et al. Analysis of land surface parameters and turbulence characteristics over the Tibetan Plateau and surrounding region. *J. Geophys. Res. Atmos.* **2016**, *121*, 9540–9560. [CrossRef]
30. Tanner, C.; Thurtell, G. *Anemoclinometer Measurements of Reynolds Stress and Heat Transport in the Atmospheric Surface Layer*; Research and Development Technical Report, ECOM 66-G22-F; Department of Soil Science, University of Wisconsin: Madison, WI, USA, 1969; p. 82.

31. Webb, E.K.; Pearman, G.I.; Leuning, R. Correction of flux measurements for density effects due to heat and water vapour transfer. *Q. J. R. Meteor. Soc.* **1980**, *106*, 85–100. [CrossRef]
32. Zhou, L.; Zou, H.; Ma, S.; Li, P.; Zhu, J.; Huo, C. Vertical air mass exchange driven by the local circulation on the northern slope of Mount Everest. *Adv. Atmos. Sci.* **2011**, *28*, 217–222. [CrossRef]
33. Soman, M.; Kumar, K. Space-time evolution of meteorological features associated with the onset of Indian summer monsoon. *Mon. Weather. Rev.* **1993**, *121*, 1177–1194. [CrossRef]
34. Wang, B.; Wu, R.; Lau, K. Interannual variability of the Asian summer monsoon: Contrasts between the Indian and the western North Pacific–East Asian monsoons. *J. Clim.* **2001**, *14*, 4073–4090. [CrossRef]
35. Zhao, P.; Li, Y.; Guo, X.; Xu, X.; Liu, Y.; Tang, S.; Xiao, W.; Shi, C.; Ma, Y.; Yu, X.; et al. The Tibetan Plateau surface-atmosphere coupling system and its weather and climate effect: The Third Tibetan Plateau Atmospheric Scienticfic Experiment. *Acta Meteorol. Sin.* **2018**, *76*, 833–860. [CrossRef]
36. Yang, K.; Guo, X.; He, J.; Qin, J.; Koike, T. On the climatology and trend of the atmospheric heat source over the Tibetan Plateau: An experiments-supported revisit. *J. Clim.* **2011**, *24*, 1525–1541. [CrossRef]
37. Zhao, P.; Chen, L. Climatic features of atmospheric heat source/sink over the Qinghai-Xizang Plateau in 35 years and its relation to rainfall in China. *Sci. China Ser. D Earth Sci.* **2001**, *44*, 858–864. [CrossRef]

remote sensing

Article

Risk Assessment of Snow Disasters for Animal Husbandry on the Qinghai–Tibetan Plateau and Influences of Snow Disasters on the Well-Being of Farmers and Pastoralists

Jinjian Li [1], Yujia Zou [2], Yufang Zhang [2,3,4,*], Shanlei Sun [5] and Xiaobin Dong [3]

1 School of Atmospheric Sciences, Plateau Atmosphere and Environment Key Laboratory of Sichuan Province, Chengdu University of Information Technology, Chengdu 610225, China; ljj@cuit.edu.cn
2 Sichuan Provincial Agricultural Meteorological Center, Chengdu 610072, China; 20141212282@mail.nuist.edu.cn
3 Faculty of Geographical Science, Beijing Normal University, Beijing 100875, China; xbdong@bnu.edu.cn
4 Water-Saving Agriculture in Southern Hill Area Key Laboratory of Sichuan Province, Chengdu 610066, China
5 Key Laboratory of Meteorological Disaster, Ministry of Education/International Joint Research Laboratory on Climate and Environment Change, Nanjing University of Information Science and Technology, Nanjing 210044, China; sun.s@nuist.edu.cn
* Correspondence: 202131051064@mail.bnu.edu.cn

Abstract: In the context of global warming, meteorological disasters occur more frequently in various regions which exert increasing influences on human life. Snow disasters are some of the natural disasters that most seriously affect the development of husbandry on the Qinghai–Tibetan Plateau (QTP), so it is necessary to explore their spatio-temporal variations and perform comprehensive risk assessment. Based on the daily snow depth data set in China, obtained by inversion of satellite remote sensing data, the spatio-temporal variation characteristics of snow disasters on the QTP from 1980 to 2019 were studied. The regional difference in the comprehensive risks of snow disasters for the husbandry on the QTP was evaluated from four perspectives, i.e., the risk of hazard factors, sensitivity of hazard-inducing environments, vulnerability of hazard-affected bodies, and disaster prevention and mitigation capacity. The farmer and pastoralist well-being (FPWB) index in five typical regions was constructed to discuss the possible influences of snow disasters on the FPWB since the 21st century. Results show that, in the last 40 years, the frequency, duration, average snow depth, and grade of snow disasters on the QTP all exhibited significant interannual and interdecadal variabilities, and they also displayed a declining long-term trend. The comprehensive risk of snow disasters for the husbandry on the QTP is low in the north while high in the south. The high-risk zone accounts for 1.54% of the total and is mainly located in Kashgar City in the north-western end of the QTP; the sub-high-risk and medium-risk zones are mainly found in the south of the plateau and are distributed in a tripole pattern, separately covering 15.96% and 16.32% of the total area of the plateau; the north of the plateau mainly belongs to low-risk and sub-low-risk zones, which separately account for 43.06% and 23.12% of the total area of the plateau. Since the beginning of the 21st century, the FPWB in five typical regions, namely, Kashgar (I), Shigatse (II), Nagqu (III), Qamdo (IV), and Yushu (V), has been increasing, while the risk of snow disasters has gradually decreased. Every 1% decrease in the risk of snow disasters corresponded to 0.186%, 0.768%, 0.378%, 0.109%, and 0.03% increases in the FPWB index in the five regions. Snow disasters affect FPWB mainly by directly or indirectly damaging material resources (livestock inventories and meat production) and social and financial resources.

Keywords: Qinghai–Tibetan Plateau; snow disaster; risk assessment; climate change

Citation: Li, J.; Zou, Y.; Zhang, Y.; Sun, S.; Dong, X. Risk Assessment of Snow Disasters for Animal Husbandry on the Qinghai–Tibetan Plateau and Influences of Snow Disasters on the Well-Being of Farmers and Pastoralists. *Remote Sens.* **2022**, *14*, 3358. https://doi.org/10.3390/rs14143358

Academic Editors: Massimo Menenti, Yaoming Ma, Li Jia and Lei Zhong

Received: 26 May 2022
Accepted: 8 July 2022
Published: 12 July 2022

1. Introduction

The sixth assessment report (AR6) released by the Intergovernmental Panel on Climate Change (IPCC) pointed out that extreme climate events have occurred more frequently in the context of global warming compared with pre-industrial-revolution levels [1]. The meteorological disasters and derived disasters triggered by extreme climate events have caused increasingly large losses and higher disaster risks which seriously affect society, the economy, and human life, so the topic has attracted wide attention [2]. The Qinghai–Tibetan Plateau (QTP), located in a mid-to-low-latitude region in the northern hemisphere, is the highest plateau in the world. At an average altitude of above 4000 m, animal husbandry is one of the economic pillars for residents on the QTP [3]. However, husbandry on the QTP is highly susceptible to natural disasters due to the special weather conditions and types of vegetation [4]. Among the natural disasters, snow disasters have become one of the leading meteorological disasters in winter and spring in alpine pastoral regions due to their long duration and wide range of influence. Continuous snowfall is frequent in alpine pastoral regions in winter and spring, and, at the same time, the snow persists for a long time due to the low temperature and readily covers short forage grass [4]. As a result, livestock feeding on forage grass may die of frost and starvation, which greatly threatens the livelihoods and property of local farmers and pastoralists and influences the productivity of husbandry. Meanwhile, previous research showed that the frequency and hazard of snow disasters on the QTP have also risen in the context of global climate change [5,6]. Therefore, assessing the risk of snow disasters based on determining the spatio-temporal distribution characteristics of snow disasters on the QTP is of significance for disaster prevention and protection of the husbandry on the plateau.

In recent years, much research into the spatio-temporal variation characteristics of snow disasters on the QTP has been performed using various data and technological means. Based on occurrence records and observations at meteorological stations, previous studies found that snow disasters on the QTP during winter and spring are mainly caused by abnormal snow accumulation from November to the following March [7]. In addition, snow disasters showed obvious interdecadal variations and a significant variation in the early 1990s. The frequency of snow disasters has shown an increasing trend since the 1990s [8,9], and the Lhoka City in the Tibetan Autonomous Region (Lhoka) in the south-west and the border between southern Qinghai Province and Sichuan Province are two centers with high frequencies of snow disasters [10]. With the climate warming over the plateau, the snow depth and the number of snow cover days for the majority of the QTP show a decreasing trend [11], and the decrease in the snow depth is more significant in the high-altitude areas [12]. Although the above studies achieved certain goals, the research conclusions were quite different due to the sparsely distributed nature of the meteorological stations on the QTP and differences in the selected meteorological stations and research areas [11,13–15]; because remote sensing data can provide snow information with high spatio-temporal resolution, they are widely used for the inversion and monitoring of snow [16–18], assessment of snow disasters [19,20], and early warning of snow disasters [21] in areas with sparse meteorological stations. These works greatly improved popular perception of the variations in, and possible drivers of, snowfall, and some scholars also used remote sensing data to explore the occurrence of snow disasters on the QTP. For example, Yin et al. [4] used AVHRR archival reflectance products to find that the grade of snow disaster on the QTP reduced from 1982 to 2012. No matter which data were used, most studies focused on the spatio-temporal variation characteristics of snow disasters. However, snow disasters, as one of the natural disasters that most greatly affects animal husbandry on the QTP, exert remarkable influences on all aspects of society, the economy, and people's livelihood. Therefore, snow disasters need to be comprehensively studied from the perspective of risk assessment, in addition to the existing studies which discussed the influences of snow disasters on livestock in typical regions of the QTP [22,23]. Meanwhile, researchers have used human well-being to characterize the material and spiritual satisfaction of residents in recent years. Because of the close relations between the people's livelihood and governmental decisions, human well-being has recently been paid much heed by many researchers [24,25]. Numerous studies were conducted on human well-being from multiple perspectives, including studies of variation characteristics and influence factors [26–30]; however, most studies focused on social and ecological topics. Natural disasters may affect material supply, living environment, and even life and property security of residents, so their influences on residents' physical and mental health cannot be underestimated. However, there are few studies on the influences of natural disasters on human well-being.

Therefore, this research mainly aims to determine regional differences in the comprehensive risk degree of snow disasters for husbandry on the QTP by comprehensively considering the risk of hazard factors, sensitivity of hazard-inducing environments, vulnerability of hazard-affected bodies, and disaster prevention and mitigation capacity from the perspective of the risk assessment of snow disasters. This is based on analysis of spatio-temporal variation characteristics of snow disasters on the QTP. Then, the farmer and pastoralist well-being (FPWB) index is constructed to evaluate possible influences of snow disasters on FPWB on the QTP since the beginning of the 21st century. The research results provide a theoretical basis for making policies to prevent snow disasters and selecting policies for FPWB on the QTP.

2. Materials and Methods

2.1. Definition of Snow Disasters

Snow disasters on the QTP mainly occur from October to the following May, so this time period was selected for calculating snow disasters. For time recording, the period from October 1979 to May 1980 was used as a statistical time period, recorded as of the year of 1980, which was divided into last winter (from October 1979 to February 1980) and this spring (from March to May 1980). Other years were recorded in the same way, thus, obtaining snow disasters over 40 years from 1980 to 2019. According to previous research and relevant meteorological standards [4,31,32], the snow disasters of last winter and this spring on the QTP were graded following criteria in Tables 1 and 2. Based on the criteria, the grade, duration, and average snow depth of snow disasters were summarized. Therein, the highest grade of snow disaster was taken as the annual grade of snow disaster. For example, if three snow disasters occurred in a year, including a slight, a moderate, and an extremely heavy event, then the year was recorded as having had an extremely heavy snow disaster. The sum of durations of several snow disasters in a year was recorded as the duration of snow disasters. The average snow depth was the average value during the snow disasters.

Table 1. Division criteria for snow disasters of last winter.

Grade of Snow Disaster	Snow Depth/mm	Snow Duration/d
Slight	[2, 5]	[11, 20]
	(5, 10]	[5, 10]
Moderate	[2, 5]	[21, 40]
	(5, 10]	[11, 20]
	(10, 20]	[5, 10]
Heavy	[2, 5]	(40,)
	(5, 10]	[21, 40]
	(10, 20]	[11, 20]
Extremely heavy	(5, 10]	(40,)
	(10, 20]	(20,)
	(20,)	(15,)

Table 2. Division criteria for snow disasters of this spring.

Grade of Snow Disaster	Snow Depth/cm	Snow Duration/d
Slight	[2, 5]	[6, 10]
	(5, 10]	[3, 5]
Moderate	[2, 5]	[11, 20]
	(5, 10]	[6, 10]
	(10, 20]	[3, 5]
Heavy	[2, 5]	(20,)
	(5, 10]	[11, 20]
	(10, 20]	[6, 10]
Extremely heavy	(5, 10]	(20,)
	(10, 20]	(10,)
	(20,)	(8,)

2.2. Risk Assessment Method of Snow Disasters

Snow disasters are a type of natural disaster. In risk assessment, the comprehensive risk of snow disasters is reflected by the risk of hazard factors, sensitivity of hazard-inducing environments, vulnerability of hazard-affected bodies, and disaster prevention and mitigation capacity according to the risk-forming theory of relevant natural disasters. The disaster risk is expressed as follows:

$$D = f(H, S, V, R) \tag{1}$$

where D, H, S, V, and R separately represent the disaster risk, risk of hazard factors, sensitivity of hazard-inducing environments, vulnerability of hazard-affected bodies, and disaster prevention and mitigation capacity; f is the function relationship.

When assessing the risk of snow disasters on the QTP, the following equation was used:

$$FDVI = \left(E^{WE}\right)V^{WV}\left(S^{WS}\right)(10 - R)^{WR} \tag{2}$$

where $FDVI$ represents the comprehensive risk index of snow disasters, and its value can be used to characterize the risk degree of snow disasters for husbandry on the QTP; the larger its value, the higher the risk of snow disasters. E, V, S, and R separately denote indices of various assessment factors, including the hazard factor, hazard-inducing environment, hazard-affected body, and disaster prevention and mitigation capacity; WE, WV, WS, and WR represent weights of various assessment factors, which are determined using the analytic hierarchy process (AHP). Weights of various factors are listed in Table 3.

In the calculation, various factors contain several different indexes, each of which has a different dimension and order of magnitude. Therefore, Equation (3) is used to normalize the various indices to ensure the comparability of various indices; thereafter, the indices lie within the range 0.5–1.

$$A_{ij} = 0.5 + 0.5 \times \frac{a_{ij} - min_i}{max_i - min_i} \tag{3}$$

where A_{ij} denotes the normalized value of the ith index at the jth station (or grid); a_{ij} is the value of the ith index at the jth station (or grid); max_i and min_i separately represent the maximum and minimum of the ith index.

Finally, the natural breaks method was adopted to grade the comprehensive indices of snow disasters for husbandry as high-risk, sub-high-risk, medium-risk, low-risk, and sub-low-risk zones.

Table 3. Risk assessment indices for snow disasters on the QTP and their weights.

Index	Rule Hierarchy (Weight)	Scheme Layer (Weight)
Risk assessment of snow disasters on the QTP	Hazard factors (0.534)	Duration (0.141)
		Snow depth (0.141)
		Grade of snow disasters (0.455)
		Frequency (0.263)
	Hazard-inducing environments (0.108)	Slope (0.159)
		Slope aspect (0.252)
		Altitude (0.589)
	Hazard-affected bodies (0.282)	Crop-sown area (0.081)
		Livestock inventories at the end of a year (0.378)
	Disaster prevention and mitigation capacity (0.076)	GDP (0.5)
		Net income of rural residents (0.5)

2.3. Establishment of the FPWB Index

Human well-being is used to characterize the living conditions of people, involving health, happiness, and affluence of materials. Early research on human well-being was mainly dedicated to economics and sociology. In recent years, research on human well-being has been gradually heeded by scholars in ecology and geology with the promotion of the idea of sustainable development. Meanwhile, characterization of human well-being has also gradually expanded from a single economic index to the ecological system. According to differences in research foci, human well-being is also divided into objective and subjective dimensions. This research focused on well-being of farmers and pastoralists (shorted as FPWB) according to sources of income, living styles, and the factors influencing the economy of residents on the QTP. To characterize FPWB, the FPWB index on the QTP was established by combining the conceptual framework of objective well-being and the concept of livelihood capital.

The FPWB index is composed of various factors. This research selected key factors that are closely related to the life of farmers and pastoralists from the agricultural part in provincial statistical yearbooks. These factors can be grouped into the following four aspects: natural resources, human resources, material resources, and social and financial resources, and indices contained in each level are listed in Table 4. The indices are quantified using the weighted comprehensive evaluation method, and their weights are determined by the AHP. In this way, the FPWB index can be expressed by Equation (4):

$$FPWB = V_1W_1 + V_2W_2 + V_3W_3 + V_4W_4 \quad (4)$$

where $FPWB$ represents the farmer and pastoralist well-being; W_1, W_2, W_3, and W_4 separately denote the four aspects that constitute the $FPWB$ index, namely, natural resources, human resources, material resources, and social and financial resources; and V_1, V_2, V_3, and V_4 are weights of each level of assessment, which are determined using the AHP. The final weights are listed in Table 4. Likewise, each index is also normalized because each level of assessment involves different indices that are in different units and dimensions and must be normalized to reach the goal of eliminating differences and making the indices comparable.

Table 4. Components of the FPWB index on the QTP.

Index	Rule Hierarchy (Weight)	Scheme Layer (Weight)
FPWB index	Human resources (0.126)	The number of rural households (0.5) The number of employees in farming, forestry, animal husbandry, and fishery (0.5)
	Natural resources (0.222)	Crop-sown area
	Material resources (0.574)	Total power of agricultural machinery (0.081) Total grain output (0.163) Livestock inventories at the end of a year (0.378) Meat production (0.378)
	Social and financial resources (0.077)	Gross output of farming, forestry, animal husbandry, and fishery

2.4. Data Sources

Snow data: the snow depth long time-series data set in China (1979–2019) was provided by the National Tibetan Plateau Data Center (TPDC). The data set was obtained by inversion of SMMR (1979–1987), SSM/I (1987–2007), and SSMI/S (2008–2019) daily EASE-Grid brightness temperature data processed by the National Snow and Ice Data Center of the United States with a spatial resolution of 25 km. The data set has been widely proved to be reliable, and its development is described elsewhere [33–35].

Socio-economic data: socio-economic data, including the number of rural households, the number of employees in farming, forestry, animal husbandry, and fishery, GDP, and net income of rural residents, were extracted from statistical yearbooks of Qinghai Province, the Tibetan Autonomous Region, Sichuan Province, Gansu Province, and the Xinjiang Uygur Autonomous Region.

3. Results

3.1. Spatio-Temporal Variation of Snow Disasters

3.1.1. Temporal Variation Characteristics

According to the above division criteria for the grade of snow disaster, the frequency and areal proportion of snow disasters on the QTP from 1980 to 2019 were summarized (Figure 1). Over the past 40 years, 36,330 snow disasters happened at 724 grid points on the QTP, that is, 908 snow disasters every year on average. The annual average frequency was the lowest, at only 687, in the 2010s (2010–2019), while that in the 1990s (1990–1999) was relatively high at about 1034. Although the climate tendency rate shows that the overall frequency of snow disasters declined significantly (−94.9 times/decade, $p < 0.01$), the variations showed interdecadal characteristics. The frequency of snow disasters increased with volatility (74.2 times/decade, $p < 0.05$) in the 1980s and 1990s, while, after entering the 21st century, it exhibited a monotonic decreasing trend (−285.5 times/decade, $p < 0.01$). Variations of the frequency of different grades of snow disaster also showed remarkable differences: 354 slight snow disasters happened annually on average, with the lowest number in 2018 (168 times), while the largest number occurred in 1999 (622). The frequency of slight snow disasters showed a decreasing trend in the long run (−16.8 times/decade) (not passing the significance test). However, it also demonstrated a tendency to increase first, then decrease during the aforementioned interdecadal variation; the frequency increased significantly in the 1980s and 1990s by 186 times/decade ($p < 0.01$), while it significantly declined since the beginning of the 21st century by −168.4 times/decade ($p < 0.01$). Variations in the frequency of moderate snow disasters also showed similar characteristics: the frequency declined significantly (−19.4 times/decade, $p < 0.05$); it significantly increased before the beginning of the 21st century (39.5 times/decade, $p < 0.05$) and decreased significantly (−67.5 times/decade, $p < 0.01$) thereafter. In terms of the long-term variation trend, the frequencies of heavy and extremely heavy snow disasters were, separately, −28.7 times/decade ($p < 0.01$) and −29.1 times/decade ($p < 0.01$) without significant interdecadal variation.

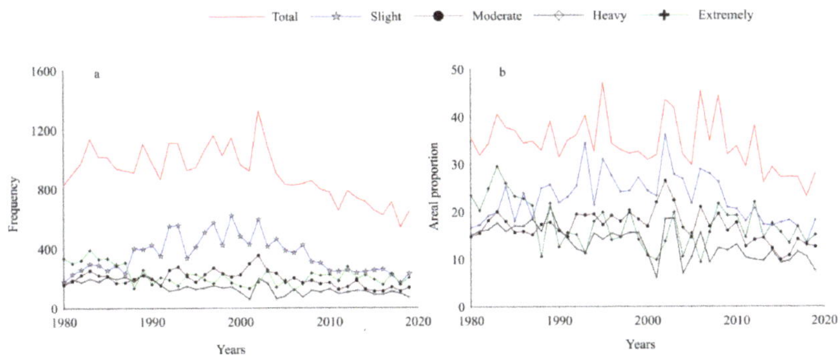

Figure 1. Variation of the frequency (**a**) and areal proportion of snow disasters (**b**) on the QTP over past 40 years.

Regarding the areal proportion of snow disasters, an average of 34.3% of the area of the QTP suffered from snow disasters every year, while the areal proportion of snow disasters shrank at −1.9%/decade ($p < 0.05$). This decline mainly occurred since the 2010s. Although the areal proportions of different grades of snow disaster changed with certain differences over the last 40 years, they all showed a decreasing trend. The areal proportions of slight and moderate snow disasters decreased slightly and did not pass the significance test, while those of heavy and extremely heavy snow disasters declined significantly by −2.1%/decade and −1.8%/decade, separately, passing the significance tests at the 0.01 and 0.05 levels.

As to the long-term variations in the annual frequencies of different grades of snow disasters at various grid points (Figure 2), the long-term variation trends of different grades of snow disasters and that of all snow disasters showed similar spatial distribution over the last 40 years. That is, the frequency of snow disasters showed a decreasing trend for the majority of the QTP in the context of global warming, with areas with the most significant decrease distributed mainly in the center and south of the Tibetan Autonomous Region and the north of Qinghai Province. The areas with an increasing frequency of snow disasters were dispersed and relatively concentrated in the north-western and south-eastern parts of the QTP with a slightly increasing trend.

The duration of snow disasters and snow depth are also important indices used for measuring snow disasters. Figure 3 illustrates the regional average variations of the duration of snow disasters and snow depth on the QTP over the past 40 years and the corresponding spatial distribution of long-term variation trends. The figure shows that the average duration of snow disasters in the past 40 years was 116 d, with the longest being 151 d (1982) and the shortest being 69 d (2006). The average snow depth was 73 mm, with the deepest being 89 mm (1992) and the shallowest being 60 mm (2006). Compared with variations of the snow depth (Figure 3c), the duration of snow disasters showed more obvious interdecadal variation characteristics. It can be seen from Figure 3a that the duration of snow disasters showed a significant decreasing trend before the beginning of the 21st century (−22.7 d/decade, $p < 0.01$); the snow disasters lasted for the shortest time in 2006, while their duration rapidly lengthened in the several years following until 2010 when the duration showed stable volatility. Correspondingly, the snow depth did not fluctuate greatly over the past 40 years and did not have significant interdecadal variation; however, the long-term variation trends of the duration of snow disasters and snow depth at different grid points (Figure 3b,d) exhibited significant spatial differences, and the two showed similar spatial distributions. The majority of the area of the QTP was found to have decreased duration of snow disasters and snow depth. The areas with the most significant decrease in the duration were mainly distributed in the center and south of the Tibetan Autonomous Region. The duration of snow disasters in these high-value centers was found to have a decreasing trend in the range of −40 to −67 d/decade. In addition, a sub-high-value center was found in the north of Qinghai Province, where the duration of snow disasters decreased in a trend from −20 to −40 d/decade. The areas with the most significant decreasing trend (−1 to −2 cm/decade) of snow depth were mainly concentrated in the south of the Tibetan Autonomous Region. Areas where the duration of snow disasters showed an increasing trend were dispersed, with relatively concentrated areas in the north-west and south-east of the QTP, varying within 20 d/decade. Areas with increasing snow depth where the snow depth increase did not exceed 10 mm/decade, were more dispersed.

The annual grade of snow disaster is the highest grade of snow disaster in a year. In this way, the variations in regional average grades of snow disaster and the spatial distribution of long-term variations over the past 40 years were calculated (Figure 4); the multi-year variations in the grade of snow disaster on the QTP showed significant interannual volatilities and a slight decreasing trend over the past 40 years (not passing the significance test). However, the grade tended to decrease then increase in different sections, somewhat akin to the variations in the duration of snow disasters. That is, the grade of snow disasters exhibited a significant decreasing trend before the early 21st century, growing significantly for several years thereafter, and showing slight volatility in the 2010s. The long-term variations at different grid points were found to have an uneven spatial distribution; the grade of snow disaster slightly rose in most areas of the QTP, with the most significant increase at the south-eastern margin of the plateau. Areas where the grade of snow disaster declined were mainly distributed at the southern margin of the plateau and in the region of the Qaidam Basin, particularly the former, where the grade declined most significantly.

3.1.2. Spatial Distribution

From the spatial distribution of the overall frequency of snow disasters on the QTP over the past 40 years (Figure 5a), it can be seen that snow disasters were very unevenly spread across the plateau. In the west of the QTP, there is an obvious, low-value center from Ngari Prefecture in the Tibetan Autonomous Region (hereinafter shorted to Ngari) to Hotan Prefecture in the Xinjiang Uygur Autonomous Region (Hotan) to northern Nagqu County in the Tibetan Autonomous Region (Nagqu) to northern Yushu Tibetan Autonomous Prefecture in Qinghai Province (Yushu). Fewer than 10 snow disasters happened in most areas along the zone in the past 40 years, and most areas did not suffer from any snow disasters. In addition, there is also a low-value center with very few snow disasters in the north-east of the QTP from Xining City in Qinghai Province (Xining) to Haidong Prefecture in Qinghai Province (Haidong) to the Tibetan Autonomous Prefecture of Huangnan in Qinghai Province (Huangnan) to Gannan Tibetan Autonomous Prefecture in Gansu Province (Gannan) to northern Aba Tibetan and Qiang Autonomous Prefecture in Sichuan Province (Aba). There are three relatively concentrated high-value centers and one sub-high-value center with relatively high frequencies of snow disasters. The areas with the highest frequency of snow disasters are mainly found in the west of Shigatse City in the Tibetan Autonomous Region (Shigatse) and south-eastern Ngari in the south of the QTP, extending north-eastward to the center of Nagqu. This zone is where snow disasters happened most frequently on the QTP, with as many as 150 to 252 snow disasters over the past 40 years. Eastern Yushu in the middle of the QTP is also a high-value center and had around 150 to 200 snow disasters. Moreover, there is also a high-value center that had 150 to 200 snow disasters in the north-western end of the QTP from Kashgar Prefecture to Kizilsu Kirgiz Autonomous Prefecture (both in the Xinjiang Uygur Autonomous Region, Kashgar and Kizilsu Kirgiz). Snow disasters also happened frequently in the south-east of the QTP, which is a sub-high-value zone, with cumulative snow disasters amounting to 50 to 150 therein.

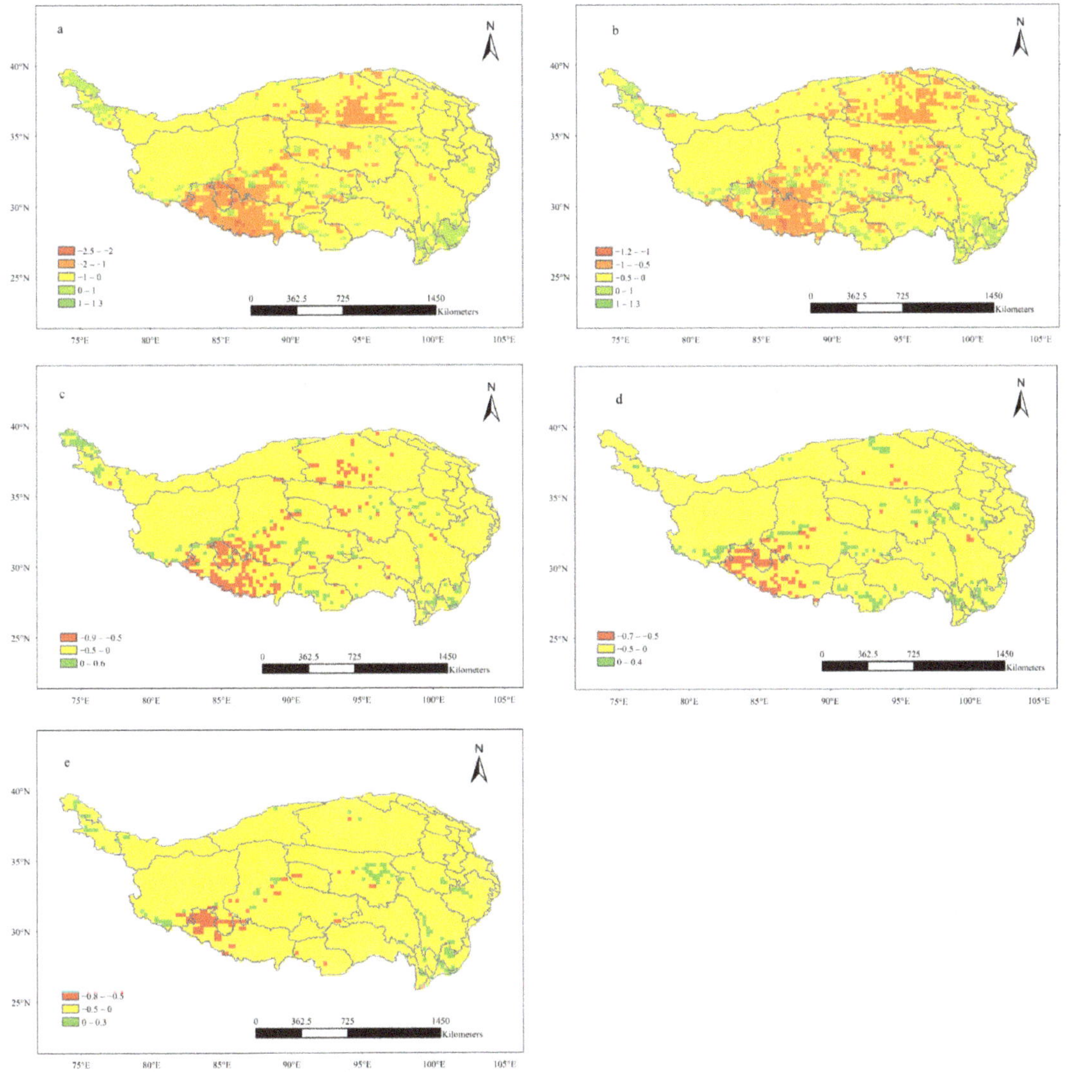

Figure 2. Spatial distribution of long-term variation trends for annual frequencies of different grades of snow disasters ((**a**–**e**) represent total snow disasters, slight, moderate, heavy, and extremely heavy snow disasters, respectively).

According to the spatial distribution of frequencies of slight, moderate, and heavy snow disasters (Figure 5b–d), despite different frequencies of different grades of snow disaster, the spatial distribution of the frequencies of these snow disasters was similar to that of the overall frequency of all snow disasters. That is, there are three relatively concentrated high-value centers, one sub-high-value center, and two low-value zones. In comparison, the frequency of extremely heavy snow disasters showed a very different spatial distribution (Figure 5e), mainly occurring in relatively decentralized high-value centers, and the relatively concentrated high-value centers are mainly located in two regions: the south-eastern end of the QTP and the north-western end of the plateau along Kashgar to Kizilsu Kirgiz. The cumulative numbers of extremely heavy snow disasters in the two high-value centers were mainly between 50 and 95.

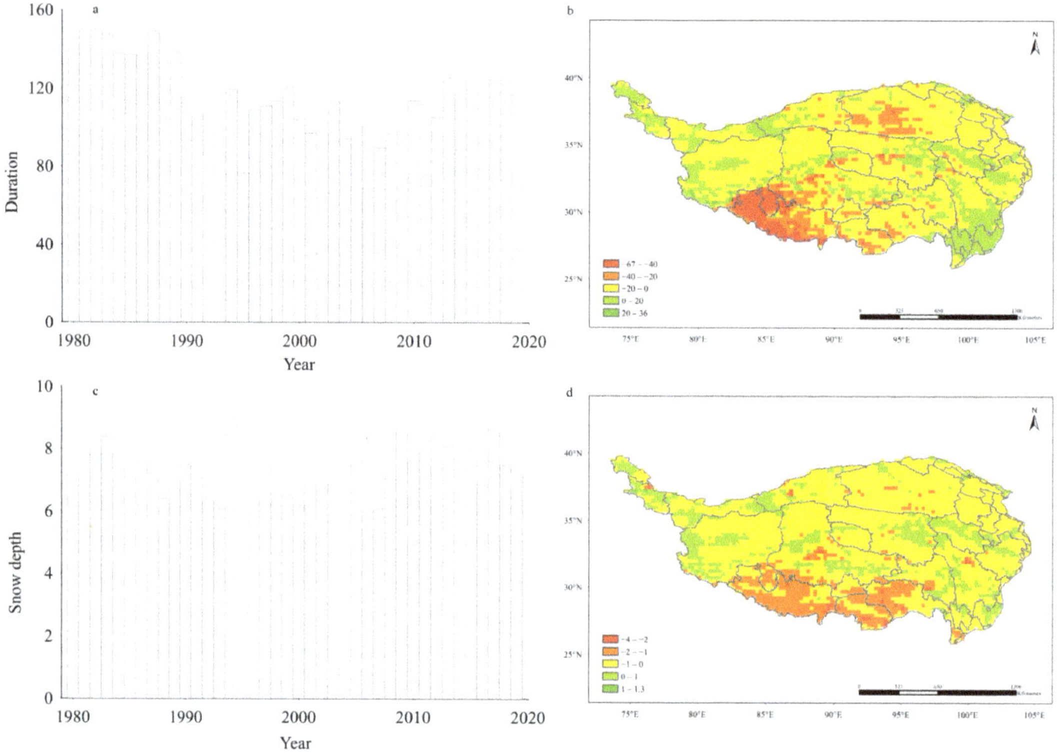

Figure 3. The regional average variation sequences and the spatial distribution of long-term variations of the duration of snow disasters and snow depth ((**a**,**b**) represent the duration of snow disasters; (**c**,**d**) represent the snow depth, respectively).

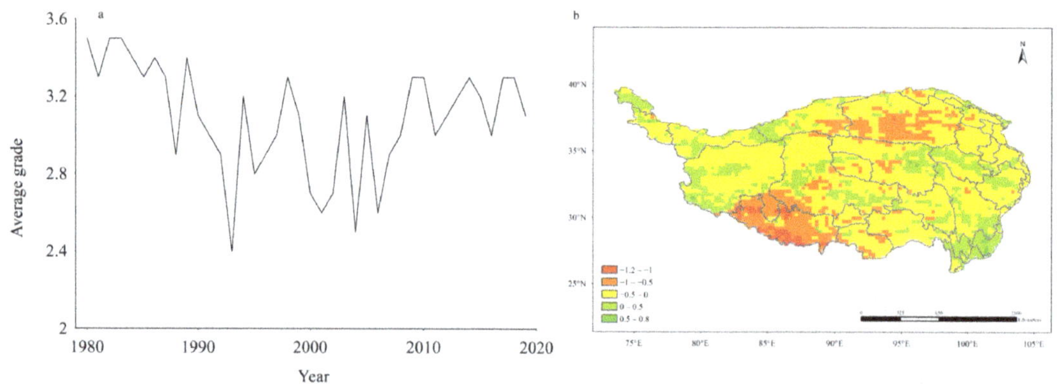

Figure 4. Regional average grades of snow disasters (**a**) and spatial distribution of the long-term variations (**b**).

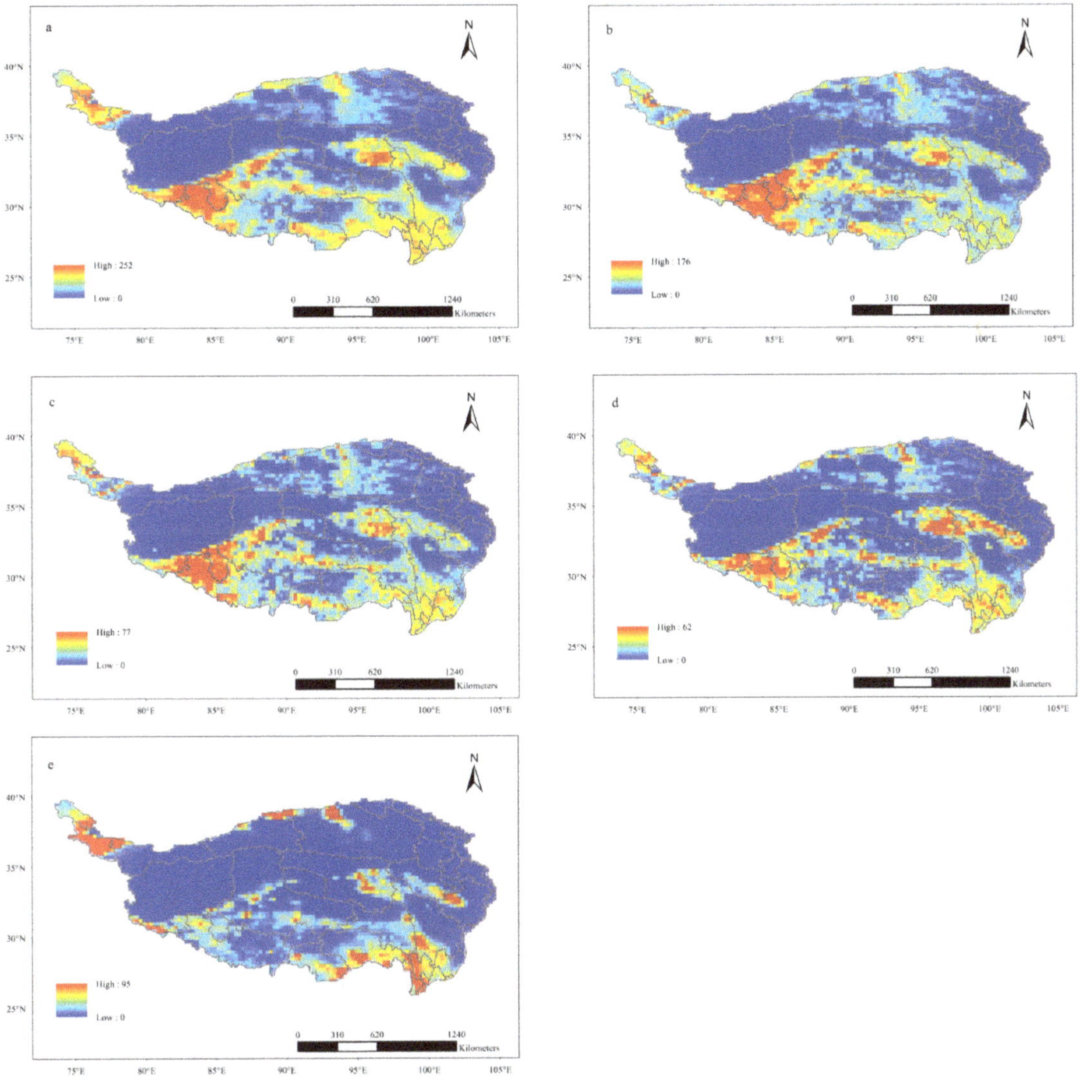

Figure 5. Spatial distribution of frequencies of different grades of snow disasters on the QTP over the past 40 years ((**a–e**) represent total snow disasters, slight, moderate, heavy, and extremely heavy snow disasters, respectively).

Figure 6 shows the spatial distribution of the average duration of snow disasters and the corresponding average snow depth on the QTP over the past 40 years. The low-value zones are distributed in areas consistent with the frequency of snow disasters, while the high-value zones are distributed in different areas. The area with the longest average duration of snow disaster was in the north-western end of the QTP from Kashgar to Kizilsu Kirgiz, and the average duration of snow disaster in the high-value center was between 150 and 229 d. High-value centers with a long duration of snow disaster are also present in the south-east and the middle (eastern Yushu) of the QTP. Although snow disasters occur frequently in the south of the QTP, from the western Shigatse–south-eastern Ngari line, the duration does not tend to be any longer than in the aforementioned areas, so it is a sub-high-value zone. The spatial distribution of the average snow depth was similar to that of the duration of snow disasters; the high-value center is located in the north-western end of the QTP along the Kashgar–Kizilsu Kirgiz line, followed by the south-east of the plateau.

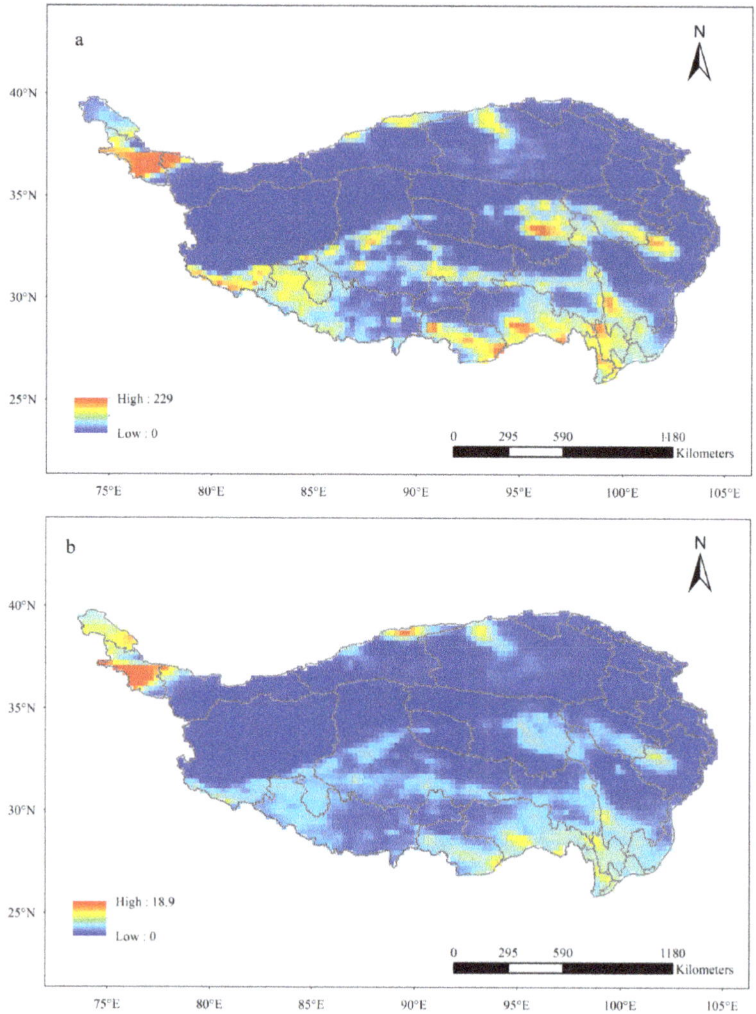

Figure 6. Spatial distribution of the average duration of snow disasters (**a**) and the average snow depth (**b**) on the QTP over the past 40 years.

The spatial distribution of multi-year average grades of snow disaster on the QTP over the past 40 years (Figure 7) indicates that the average grades of snow disaster were distributed with multiple high-value centers. The highest-value center is in the north-western end of the QTP from Kashgar to Kizilsu Kirgiz, where the multi-year average grade of snow disasters always reached a level concomitant with extremely heavy snow disasters every year. There is also an area with a relatively high average grade of snow disaster in the south-east of the QTP, with the core area being from Lhoka City–Nyingchi City in the Tibetan Autonomous Region (Nyingchi) to Diqing Tibetan Autonomous Prefecture in Yunnan Province (Diqing). In that area, heavy snow disasters may take place. The zone from Yushu to the Tibetan Autonomous Prefecture of Golog in Qinghai Province (Golog) to Aba in the center of the QTP is also a continuous area with a high average grade of snow disaster, with snow disasters in the central zone reaching a heavy grade. The zone with a high average grade of snow disasters and covering a large area is found in the south of the QTP along the western Shigatse–south-eastern Ngari line, with a moderate grade of snow disaster on average.

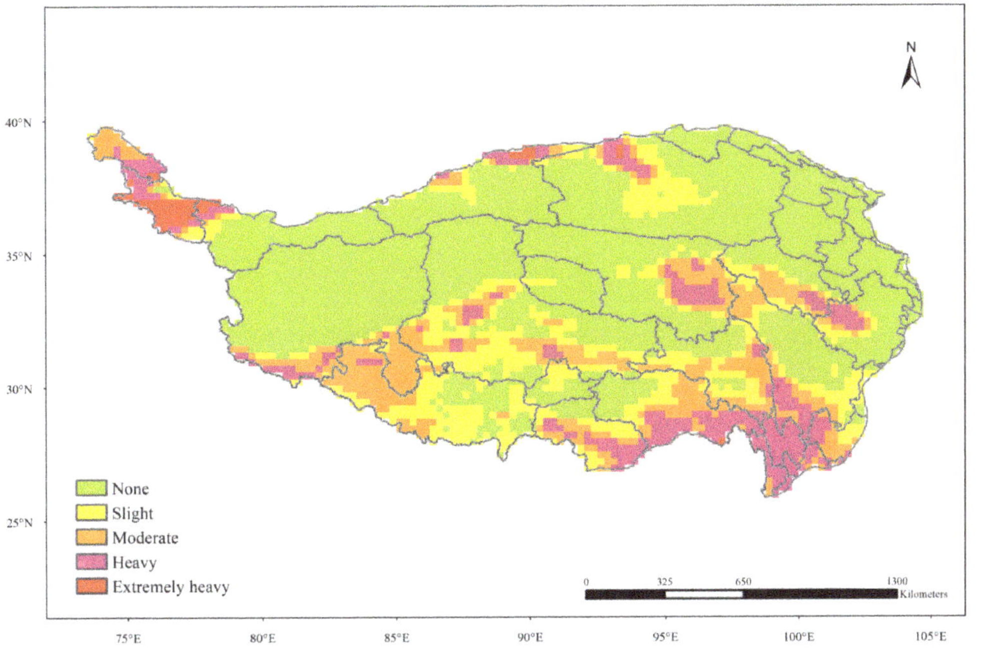

Figure 7. Spatial distribution of the average grade of snow disaster on the QTP.

3.2. Risk Assessment of Snow Disasters for Husbandry

The risk of snow disasters for animal husbandry on the QTP was assessed from four perspectives: hazard factors, hazard-inducing environments, hazard-affected bodies, and disaster prevention and mitigation capacity. Four hazard factors were selected in the research, including the duration, snow depth, frequency, and grade of snow disaster (spatio-temporal variation characteristics of each factor are provided above). By using the weighted comprehensive evaluation method, the risk of hazard factors of snow disasters for animal husbandry on the QTP was zoned. This mainly reflects the intensity and probability of hazard factors that cause snow disasters to affect animal husbandry and is the leading precondition for snow disasters. The larger the risk of hazard factors, the greater the intensity thereof and the higher the probability of ensuing damage. Figure 8a shows the spatial distribution of the risk of hazard factors of snow disasters for animal husbandry on the QTP; the risk of hazard factors is low in the north while high in the south on the whole, having multiple high-value centers. The high-risk zone of snow disasters is mainly concentrated in Kashgar in the north-western end of the QTP, which is characterized by a long duration of snow, large snow depth, and high grade of snow disaster. The sub-high-value zone is mainly concentrated in the south-east of the QTP, including southern Lhoka and Nyingchi, Nujiang Lisu Autonomous Prefecture in Yunnan Province (Nujiang), Diqing, and southern Garzê Tibetan Autonomous Prefecture in Sichuan Province (Garzê). Snow disasters in the zone are mainly characterized by a high grade and a high frequency of extremely heavy snow disasters.

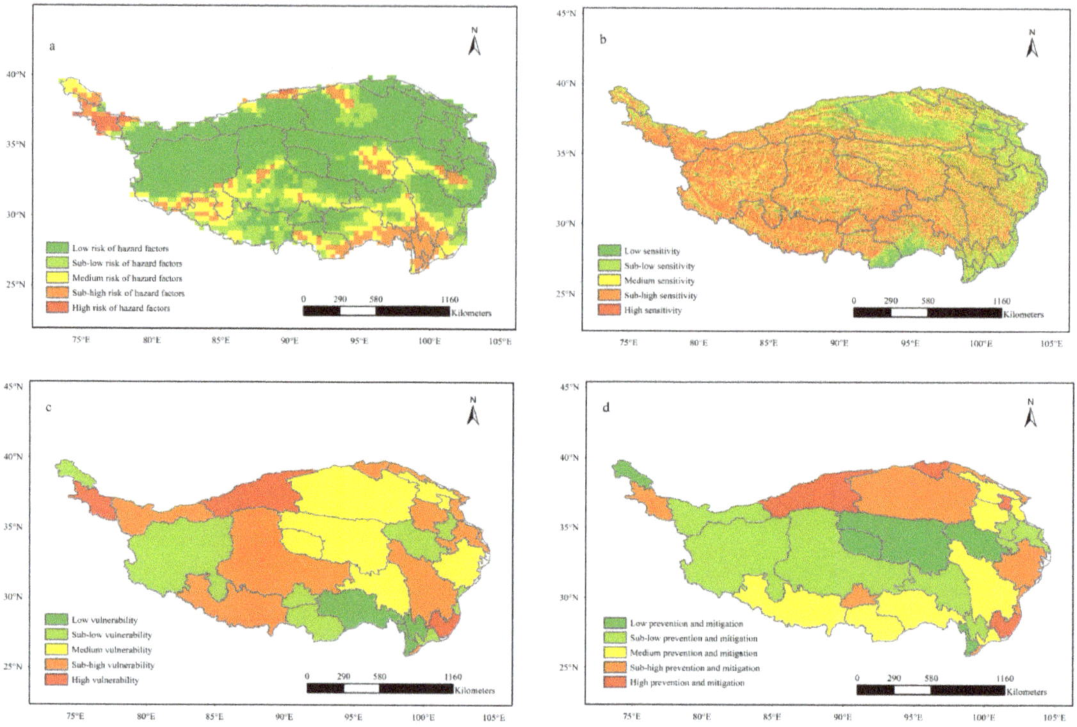

Figure 8. Zoning of the risk of hazard factors (**a**), sensitivity of hazard-inducing environments (**b**), vulnerability of hazard-affected bodies (**c**), and disaster prevention and mitigation capacity (**d**) of snow disasters for husbandry on the QTP.

In terms of the hazard-inducing environment, three factors, including the altitude, slope, and slope aspect, were mainly considered. Research showed that, under same or similar conditions, the combination of snow and landform may further aggravate the influences of snow disasters and cause certain secondary disasters. Therein, small topographic factors, such as the altitude, slope, and slope aspect, exert more significant influences. Therefore, the three factors, i.e., altitude, slope, and slope aspect, were selected to analyze the sensitivity of the hazard-inducing environment of snow disasters for animal husbandry on the QTP (Figure 8b). The QTP lies at a high altitude, on the whole, and features steep mountains, so the hazard-inducing environment of snow disasters for animal husbandry is of high sensitivity, and areas of low sensitivity are mainly located in the north-east and the south-eastern margin.

The degree of damage caused by snow disasters is, in fact, closely related to the body affected by snow disasters. The loss caused by snow disasters is not only dependent on the intensity, duration, and frequency of the disasters, but also is greatly affected by the hazard-affected bodies. Generally, the higher the vulnerability of hazard-affected bodies, the more easily these bodies are affected and the greater the loss (and vice versa). The present research mainly focused on snow disasters and their effects on animal husbandry, so crop-sown area and livestock inventories were selected as indices representing the vulnerability of hazard-affected bodies. Zones with a large crop-sown area are mainly located in the northern and eastern QTP, in which Kashgar has the largest crop-sown area, followed by the Bayingolin Mongol Autonomous Prefecture in the Xinjiang Uygur Autonomous Region (Bayingolin). Two zones have large livestock inventories: one is Kashgar–Hotan–Bayingolin–Nagqu–Shigatse in the west of the QTP, and the other is Garzê–Aba in the east of the plateau. By combining these two indices, the zoning of the vulnerability of hazard-affected bodies in snow disasters for the husbandry on the QTP can be obtained (Figure 8c). Areas of high vulnerability of hazard-affected bodies are mainly distributed in Kashgar–Hotan–Bayingolin–Nagqu–Shigatse in the west of the plateau and Garzê, Hainan Tibetan Autonomous Prefecture in Qinghai Province (Hainan), and Gannan in the east.

The disaster prevention and mitigation capacities refer to both disaster resistance and post-disaster resilience, which are mainly represented by the local level of economic development and the economic capability of farmers and pastoralists. In this research, the (municipal/prefecture) GDP and the per capita net income of farmers and pastoralists were selected as indices to reflect the local disaster prevention and mitigation capacity, thus, finally obtaining a zoning map of the capacity (Figure 8d). As shown in the figure, areas with high disaster prevention and mitigation capacity include Xining, Bayingolin, Liangshan Yi Autonomous Prefecture in Sichuan Province (Liangshan), Aba, Kashgar, and Haixi.

The comprehensive risk index of snow disasters for husbandry on the QTP was calculated using Equation (2) for comprehensive risk assessment according to the weights of the four factors (the hazard factors, hazard-inducing environments, hazard-affected bodies, and disaster prevention and mitigation capacity) (Table 3). The comprehensive risk was graded using the natural breaks method, finally attaining the zoning map for the comprehensive risk of snow disasters for animal husbandry on the QTP (Figure 9). The risk of snow disasters can be divided into five grades: high-risk, sub-high-risk, medium-risk, low-risk, and sub-low-risk zones. The high-risk zone accounts for about 1.54% of the total area of the QTP, mainly located in Kashgar at the north-western end of the plateau; the sub-high-risk and medium-risk zones are mainly found in the south of the plateau and are distributed in a tripole pattern, separately accounting for 15.96% and 16.32% of the plateau. The three "poles" are located along the Lhoka–Nyingchi–Nujiang–Diqing–southern Garzê line in the south-east, western Shigatse in the south, and Yushu in the hinterland of the QTP. The northern QTP is mainly dominated by low-risk and sub-low-risk zones, which separately cover 43.06% and 23.12% of the plateau.

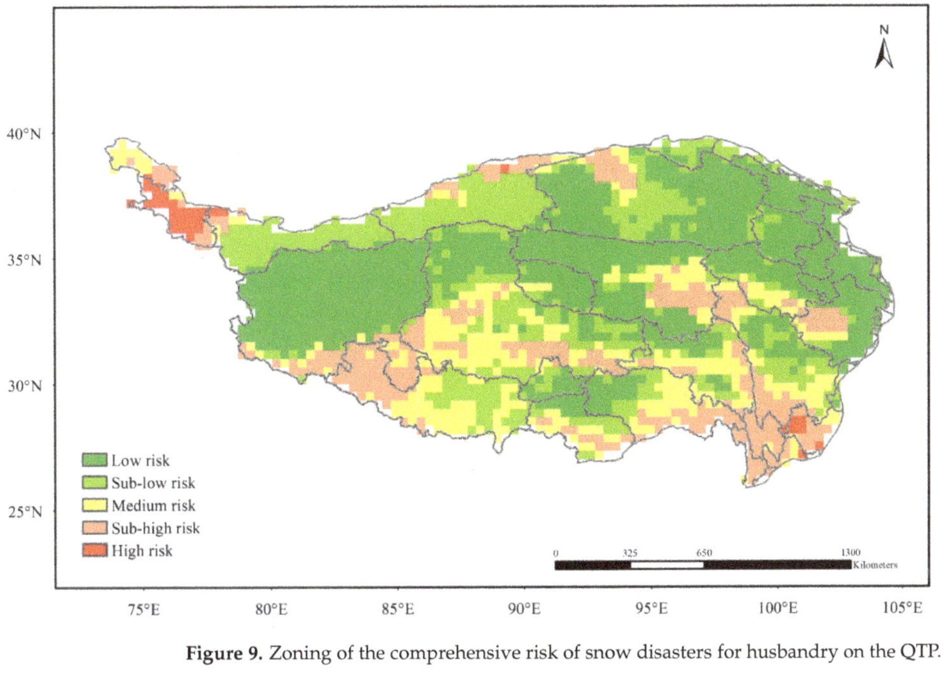

Figure 9. Zoning of the comprehensive risk of snow disasters for husbandry on the QTP.

3.3. Possible Influences of Snow Disasters on FPWB

To further explore the possible influences of major natural disasters, represented by snow disasters on FPWB, several typical regions (Figure 10) were selected for discussion based on the aforementioned analysis of the spatial-temporal variations of snow disasters in the QTP and zoning of the comprehensive risk of snow disasters for animal husbandry. These regions were Kashgar (I), Shigatse (II), Nagqu (III), Qamdo (IV), and Yushu (V). Among these regions, Kashgar represents the high-risk zone and is characterized by the high risk of hazard factors and high vulnerability of hazard-affected bodies while also having favorable disaster prevention and mitigation capacity. Shigatse mainly contains sub-high-risk and medium-risk zones, characterized by the moderate risk of hazard factors and moderate disaster prevention and mitigation capacity, while showing high sensitivity in its hazard-inducing environments and high vulnerability of hazard-affected bodies. Nagqu and Qamdo, sharing the similar comprehensive risk of snow disasters for animal husbandry and the risk of hazard factors, both belong to the medium-risk zone with above moderate vulnerability of hazard-affected bodies and poor disaster prevention and mitigation capacity. Western Yushu is a vast, depopulated zone, while the more heavily populated eastern Yushu region shows a sub-high comprehensive risk of snow disasters and a sub-high risk of hazard factors, moderate vulnerability of hazard-affected bodies, and poor disaster prevention and mitigation capacity.

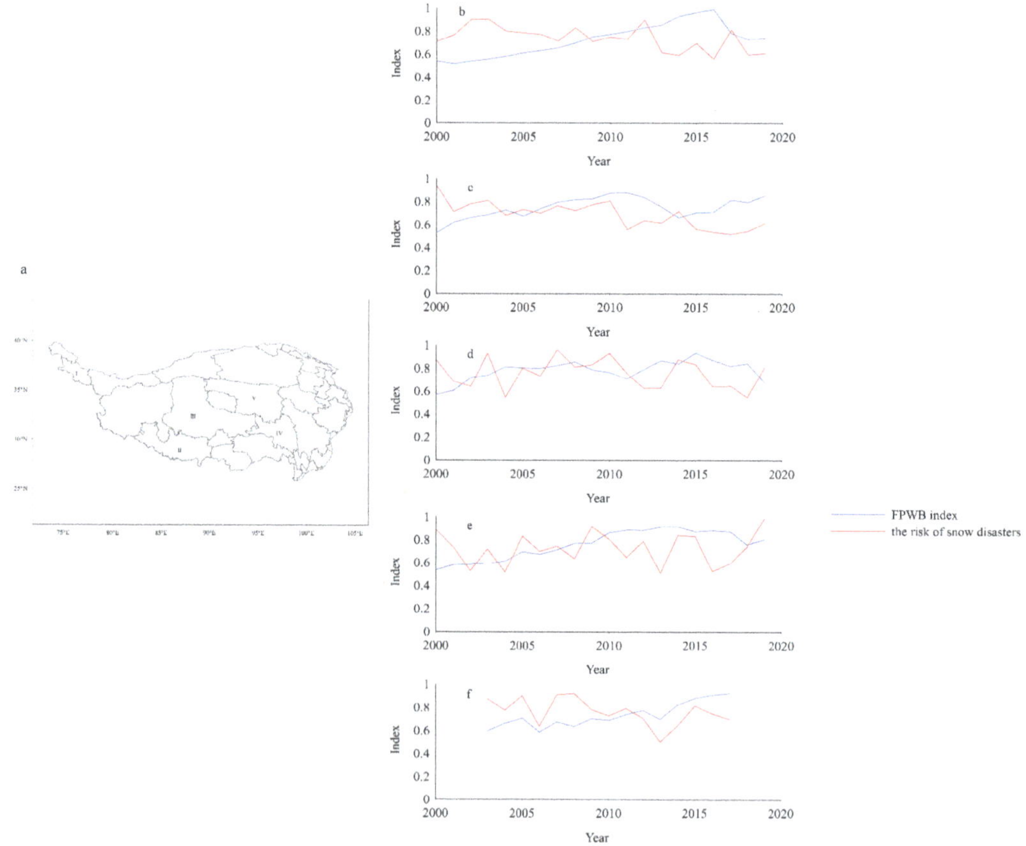

Figure 10. Typical regions (**a**) and variations of the FPWB index and the risk of snow disasters for every typical region ((**b–f**) represent I, II, III, IV, and V, respectively).

According to the method of calculation of the FPWB index, variations in the FPWB index in various typical regions from 2000 to 2019 were calculated (the period from 2003 to 2017 was used for Yushu due to problems encountered in data acquisition). Meanwhile, the risk of hazard factors was used as the comprehensive assessment model for the severity of snow disasters. Variations in the risk of snow disasters of these typical regions in the same period were calculated, and the FPWB indices and variations of the risk of snow disasters are illustrated in Figure 10. From the long-term variation, the FPWB in these typical regions showed a significant ascending trend, which passed the significance test at the 0.01 level. The result indicates that the FPWB on the QTP has improved significantly since the 21st century, and the average growth rate of FPWB in the five typical regions is 3.6%, 1.9%, 2.8%, 1.3%, and 2.3%, respectively. The risk of snow disasters in regions I, II, and V exhibited a significant decreasing trend, while that in regions III and IV did not show significant variation. By using least-squares regression, the contribution of variations in the risk of snow disasters to the FPWB from 2000 to 2019 was quantitatively estimated (Table 5). The results suggest that the risk of snow disasters had an adverse effect on variations of the FPWB. Every 1% increase in the risk of snow disasters in several typical regions corresponded to 0.186%, 0.768%, 0.378%, 0.109%, and 0.03% decreases in the FPWB index. Snow disasters affect the FPWB mainly by directly or indirectly impairing material resources (livestock inventories and meat production) and social and financial resources. Similar results were found in the research by Qiu et al. [22]. Because animal husbandry on the QTP is relatively unsophisticated and mainly depends on individual management of farmers and pastoralists, the area is far from realizing mechanization, intensification, and modernization of its agricultural practices; it is heavily dependent on prevailing meteorological conditions. Natural disasters, represented by snow disasters, directly affect the livelihood of local farmers and pastoralists and exert adverse impacts on the FPWB. Once a snow disaster occurs, it is generally accompanied by low-temperature weather, and the snow cannot be removed for a long time. This, on the one hand, directly affects livestock and poultry (often killing and injuring many animals and birds); on the other hand, a snow disaster also causes insufficient supply of forage grass, which affects later feeding and management and even causes death of livestock and poultry. At the same time, winter and spring are seasons with a high incidence of animal disease epidemics. Once a snow disaster occurs, the insufficient supply of forage grass may also lead to undernutrition and decreased immunity of livestock and poultry, rendering them more susceptible to infection, thus, influencing livestock inventories and meat production. In addition, in the process of snow disasters, the huge and rapid snowfall frequently causes collapse of livestock housing and breeding sheds, inducing deaths and injuries of livestock, bringing an economic loss to farmers and pastoralists and even causing causalities among farmers. Apart from the direct influences affecting development of animal husbandry, snow disasters also directly affect the life and production activities of residents, even causing major economic losses. For example, snow and ice heavily damage electric power facilities and transportation, hindering daily transportation of animal husbandry products, affecting the whole supply chain.

Table 5. Contribution of variations of the risk of snow disasters to the FPWB.

	FPWB	Material Resources	Livestock Inventories	Meat Production	Social and Financial Resources
I	−0.186	−0.552	−0.451	−0.239	−0.759
II	−0.768	−0.601	−0.524	−0.511	−1.121
III	−0.378	−0.947	−0.871	−0.466	−1.054
IV	−0.109	−0.032	−0.172	−0.223	−0.284
V	−0.03	−0.047	−0.12	−0.043	−0.044

4. Discussion

Snow disasters are one of the most important meteorological hazards on the QTP, and their spatial and temporal characteristics and influencing factors have received more and more attention in recent years; however, some findings are constrained by the plateau's poor distribution of meteorological stations. In this paper, we used remote sensing data to analyze the spatial and temporal occurrence patterns of snow disasters on the QTP. Results demonstrated that, during the past 40 years, the frequency, duration, average snow depth, and grade of snow disaster on the QTP have all exhibited significant interannual and interdecadal variabilities, as well as a declining long-term trend, which is consistent with some previous studies [4,11,12,19]. In addition to the direct influence of local climatic conditions [12], many circulation factors also modulate the occurrence of snow disasters on the QTP through westerly winds and atmospheric bridges. These factors all contribute to the variability of snow disasters on interannual to interdecadal time scales. Huang et al. [36] suggested that snow disasters were more likely to occur when the west wind belt and the polar vortex in the eastern hemisphere are stronger and the East Asian trough and subtropical high are more westerly, and vice versa. SST, as an important external forcing factor, also profoundly affects the occurrence of snow disasters, and the equatorial central-east Pacific, the tropical Indian Ocean, and the North Atlantic are some of the more critically affected areas [14,37–39]. In addition, the positive North Atlantic Oscillation (NAO) can excite Rossby waves, which can strengthen the Indo–Myanmar trough on the southern side of the plateau and promote snowfall on the QTP, leading to more snow disasters [40,41]. Furthermore, Arctic sea ice, as an important external forcing factor affecting extreme weather and climate events at medium latitudes [42], also has an important modulating effect on the occurrence of snow disasters on the Tibetan plateau. A positive anomaly of Arctic sea ice can enhance the meridional temperature gradient, which excites upward-propagating and equatorward-propagating anomalous Rossby waves, leading to an anomalous dipole pattern of atmospheric circulation over the polar regions and Eurasia, enhancing the zonal advection and meridional convergence of atmospheric moisture fluxes over the plateau and favoring snow disasters [43,44]. Other circulation factors, such as El Niño (ENSO) and the Arctic Oscillation (AO), can also influence the occurrence of snow disasters through cyclonic circulation propagating along the westerly wind belt [14,37,45,46]. We also discussed the possible influences of snow disasters on the farmer and pastoralist well-being (FPWB) since the 21st century. Since World War II, with the ensuing economic development, academics, decision-makers, and practitioners around the world have paid close attention to human well-being, and gross domestic product (GDP) was once the dominant measure of human well-being [47,48]. However, since GDP is mainly a reflection of economic indicators, it is hoped that other factors, such as social, humanistic, and ecological factors, can be absorbed into the human well-being evaluation system. The Millennium Ecosystem Assessment [49] provides a useful framework for the study of ecosystem services as an influencing factor on human well-being, making explicit the close relationship between ecosystem services and human well-being and ushering in a new era of human well-being research. Within this framework, a large number of studies revealed the characteristics of changes in human well-being in different regions and at different time scales, as well as their main drivers [50–53]. However, current research on human well-being tends to be a holistic concept, and there is no uniform definition of the meaning of well-being for different groups of people. Farmers and pastoralists are the majority of the inhabitants of the QTP, and it is of great practical importance to discuss their well-being to improve the well-being of the plateau people. Generally, farmers' happiness increases with job satisfaction and income, while land as an important means of production for farmers is closely related to farmers' income [54], and the stronger the farmers' willingness to retire from farming, the worse their happiness is [55]; moreover, farmers' self-rated health status has a positive impact on their well-being [56]. Here, the FPWB index on the QTP was established by combining the sources of income, living styles, and economic factors, and exploring the impact of snow disasters on the well-being of farmers and pastoralists. The results (Figure 10 and Table 5) show that snow disasters have a certain negative impact on the FPWB, but we only discussed the impact of a single meteorological disaster, and it is necessary to systematically explore the systematic impact of other meteorological disasters on the FPWB in the future. Meanwhile, it should be noted that the improvement of FPWB on the QTP has been stagnant or even declining in recent years. This may be related to the increase of extreme weather and climate events caused by climate change which affect the living environment, life, and property of farmers and pastoralists. On the other hand, the advancement of urbanization on the QTP may attract a large number of young people to work and live in cities, which indirectly leads to the decline of human resources, production level, and production capacity in agriculture and livestock, thus, affecting the overall situation of farmer and pastoralist well-being.

5. Conclusions

The spatio-temporal variation characteristics of snow disasters on the QTP over the past 40 years were investigated based on the daily snow depth data set in China obtained by the inversion of the satellite remote sensing data provided by the national TPDC. The results show that the frequency, duration, average snow depth, and grade of snow disaster had similar spatial distributions, being low in the north while high in the south. The high-value center in the north is mainly located in Kashgar–Kizilsu Kirgiz at the north-western end of the QTP, which is the zone with the most numerous heavy snow disasters and the highest frequency of snow disaster on the plateau. The high-value center in the south has a tripole distribution pattern, and the three "poles" are located in Lhoka–Nyingchi–Nujiang–Diqing–southern Garzê in the south-east, western Shigatse in the south, and Yushu in the hinterland of the plateau. It can be seen from the interannual and interdecadal variations in the frequency, duration, average snow depth, and grade of snow disaster on the QTP over the past 40 years that various indices all have obvious interannual and interdecadal variabilities. Meanwhile, the long-term variations of various indices also showed a decreasing trend despite certain spatio-temporal differences. On this basis, geographic information system (GIS) technology was used to zone the comprehensive risk of snow disasters on the QTP. The results indicated that the high-risk zone accounts for 1.54% of the plateau, mainly in Kashgar at the north-western end of the plateau. The sub-high-risk and medium-risk zones are located in the south of the plateau and are distributed in a tripole pattern, separately covering 15.96% and 16.32% of the QTP. The north of the QTP is dominated by low-risk and sub-low-risk zones, which separately account for 43.06% and 23.12% of the plateau. Finally, five typical regions, Kashgar (I), Shigatse (II), Nagqu (III), Qamdo (IV), and Yushu (V), were selected to discuss the possible influences of snow disasters on FPWB since the 21st century. The results implied that every 1% increase in the risk of snow disasters corresponded to 0.186%, 0.768%, 0.378%, 0.109%, and 0.03% decreases in the FPWB. Snow disasters affect the FPWB mainly by directly and indirectly damaging material resources (livestock inventories and meat production) and social and financial resources.

Author Contributions: Conceptualization, J.L., Y.Z. (Yufang Zhang) and X.D.; Structure and methodology, J.L., Y.Z. (Yujia Zou) and Y.Z. (Yufang Zhang); Data and software, J.L. and S.S.; Formal analysis and original writing, J.L. and S.S.; Funding acquisition, Y.Z. (Yufang Zhang). All authors have read and agreed to the published version of the manuscript.

Funding: This research was funded by the Second Tibetan Plateau Scientific Expedition and Research (STEP) program (no. 2019QZKK0608 and 2019QZKK0103).

Data Availability Statement: The snow depth long time-series data set in China (1979–2019) was provided by National Tibetan Plateau Data Center (http://data.tpdc.ac.cn, accessed on 21 June 2020). The other data generated for this study are available on request to the corresponding authors.

Acknowledgments: We also want to appreciate the support from China Scholarship Council.

Conflicts of Interest: The authors declare that the research was conducted in the absence of any commercial or financial relationships that could be construed as a potential conflict of interest.

References

1. IPCC. *Climate Change 2021: The Physical Science Basis. Contribution of Working Group I to the Fifth Assessment Report of the Intergovernmental Panel on Climate Change*; Cambridge University Press: Cambridge, UK, 2021.
2. Qin, P. More than six billion people encountering more exposure to extremes with 1.5 °C and 2.0 °C warming. *Atmospheric Res.* **2022**, *273*, 106165. [CrossRef]
3. Dong, S.; Sherman, R. Enhancing the resilience of coupled human and natural systems of alpine rangelands on the Qinghai-Tibetan Plateau. *Rangel. J.* **2015**, *37*, i–iii. [CrossRef]
4. Yin, H.; Cao, C.; Xu, M.; Chen, W.; Ni, X.; Chen, X. Long-term snow disasters during 1982–2012 in the Tibetan Plateau using satellite data. *Geomat. Nat. Hazards Risk* **2017**, *8*, 466–477. [CrossRef]
5. Wang, Y.; Wang, J.; Li, S.; Qin, D. Vulnerability of the Tibetan Pastoral Systems to Climate and Global Change. *Ecol. Soc.* **2014**, *19*, 8–18. [CrossRef]
6. Qin, D.H. *China National Assessment Report on Risk Management and Adaptation of Climate Extremes and Disasters*; Science Press: Beijing, China, 2015.
7. Wei, Z. Persistence of the Snow Anomalies on the Qinghai-Tibetan Plateau. *J. Glaciol. Geocryol.* **2001**, *23*, 225–230. (In Chinese)
8. Zhou, L.; Li, H.; Wang, Q. The basic characteristics of heavy snowstorm process and snow disaster distribution in eastern pastoral sreas of Qinghai-Xizang Plateau. *Plateau Meteorol.* **2000**, *19*, 450. (In Chinese)
9. Dong, A.; Ju, Z.; Xin, X.; Zhou, L. The singular spectrum analysis of snow damage in Eastern Qinghai-Xizang Plateau. *Plateau Meteorol.* **2001**, *20*, 214. (In Chinese)
10. Gao, Y.; Qiu, J. Study on the characteristics and distribution pattern of major natural disasters on the Qinghai-Tibet Plateau. *J. Arid. Land Resour. Environ.* **2011**, *25*, 101–106. (In Chinese)
11. Xu, W.; Ma, L.; Ma, M.; Zhang, H.; Yuan, W. Spatial–temporal variability of snow cover and depth in the Qinghai–Tibetan Plateau. *J. Clim.* **2017**, *30*, 1521–1533. [CrossRef]
12. Guo, D.; Pepin, N.; Yang, K.; Sun, J.; Li, D. Local changes in snow depth dominate the evolving pattern of elevation-dependent warming on the Tibetan Plateau. *Sci. Bull.* **2021**, *66*, 1146–1150. [CrossRef]
13. Chen, L.; Wu, R. Interannual and decadal variations of snow cover over Qinghai-Xizang Plateau and their relationships to summer monsoon rainfall in China. *Adv. Atmos. Sci.* **2000**, *17*, 18–30.
14. You, Q.; Kang, S.; Ren, G.; Fraedrich, K.; Pepin, N.; Yan, Y.; Ma, L. Observed changes in snow depth and number of snow days in the eastern and central Tibetan Plateau. *Clim. Res.* **2011**, *46*, 171–183. [CrossRef]
15. Tan, X.; Wu, Z.; Mu, X.; Gao, P.; Zhao, G.; Sun, W.; Gu, C. Spatiotemporal changes in snow cover over China during 1960–2013. *Atmos. Res.* **2019**, *218*, 183–194.
16. Xu, W.; Ma, H.; Wu, D.; Yuan, W. Assessment of the Daily Cloud-Free MODIS Snow-Cover Product for Monitoring the Snow-Cover Phenology over the Qinghai-Tibetan Plateau. *Remote Sens.* **2017**, *9*, 585. [CrossRef]
17. Malmros, J.K.; Mernild, S.H.; Wilson, R.; Tagesson, T.; Fensholt, R. Snow cover and snow albedo changes in the central Andes of Chile and Argentina from daily MODIS observations (2000–2016). *Remote Sens. Environ.* **2018**, *209*, 240–252. [CrossRef]
18. Zhang, C.; Mou, N.; Niu, J.; Zhang, L.; Liu, F. Spatio-Temporal Variation Characteristics of Snow Depth and Snow Cover Days over the Tibetan Plateau. *Water* **2021**, *13*, 307. [CrossRef]
19. Shijin, W.; Lanyue, Z.; Yanqiang, W. Integrated risk assessment of snow disaster over the Qinghai-Tibet Plateau. *Geomat. Nat. Hazards Risk* **2019**, *10*, 740–757. [CrossRef]
20. Che, T.; Hao, X.; Dai, L.; Li, H.; Huang, X.; Xiao, L. Snow cover variation and its impacts over the Qinghai-Tibet Plateau. *Bull. Chin. Acad. Sci.* **2019**, *34*, 1247–1253.
21. Gao, J.; Huang, X.; Ma, X.; Feng, Q.; Liang, T.; Xie, H. Snow Disaster Early Warning in Pastoral Areas of Qinghai Province, China. *Remote Sens.* **2017**, *9*, 475. [CrossRef]
22. Qiu, X.; Yang, X.; Fang, Y.-P.; Xu, Y.; Zhu, F. Impacts of snow disaster on rural livelihoods in southern Tibet-Qinghai Plateau. *Int. J. Disaster Risk Reduct.* **2018**, *31*, 143–152. [CrossRef]
23. Ye, T.; Liu, W.; Chen, S.; Chen, D.; Shi, P.; Wang, A.; Li, Y. Reducing livestock snow disaster risk in the Qinghai–Tibetan Plateau due to warming and socioeconomic development. *Sci. Total Environ.* **2022**, *813*, 151869. [CrossRef]
24. Summers, J.K.; Smith, L.M.; Case, J.L.; Linthurst, R.A. A Review of the Elements of Human Well-Being with an Emphasis on the Contribution of Ecosystem Services. *Ambio* **2012**, *41*, 327–340. [CrossRef] [PubMed]
25. King, M.F.; Reno, V.F.; Novo, E.M.L.M. The Concept, Dimensions and Methods of Assessment of Human Well-Being within a Socioecological Context: A Literature Review. *Soc. Indic. Res.* **2014**, *116*, 681–698. [CrossRef]
26. Bieling, C.; Plieninger, T.; Pirker, H.; Vogl, C.R. Linkages between landscapes and human well-being: An empirical exploration with short interviews. *Ecol. Econ.* **2014**, *105*, 19–30. [CrossRef]
27. Naeem, S.; Chazdon, R.L.; Duffy, J.E.; Prager, C.; Worm, B. Biodiversity and human well-being: An essential link for sustainable development. *Proc. R. Soc. B Boil. Sci.* **2016**, *283*, 20162091. [CrossRef]
28. Balmford, A.; Bond, W. Trends in the state of nature and their implications for human well-being. *Ecol. Lett.* **2005**, *8*, 1218–1234. [CrossRef]

29. Shackleton, R.T.; Shackleton, C.M.; Kull, C.A. The role of invasive alien species in shaping local livelihoods and human well-being: A review. *J. Environ. Manag.* **2019**, *229*, 145–157. [CrossRef]
30. Reyes-Riveros, R.; Altamirano, A.; De La Barrera, F.; Rozas-Vásquez, D.; Vieli, L.; Meli, P. Linking public urban green spaces and human well-being: A systematic review. *Urban For. Urban Green.* **2021**, *61*, 127105. [CrossRef]
31. Guo, X.; Li, L.; Liu, C.; Li, B. Spatio-temporal Distribution of Snow Disaster in Qinghai Plateau During 1961-2008. *Clim. Change Res.* **2010**, *6*, 332–337. (In Chinese)
32. *Qinghai Province Local Standards DB63/T372—2001*; Meteorological Disaster Standards. Qinghai Quality and Technical Supervision Bureau: Xining, China, 2001; pp. 6–8. (In Chinese)
33. Che, T.; Li, X.; Jin, R.; Armstrong, R.; Zhang, T. Snow depth derived from passive microwave remote-sensing data in China. *Ann. Glaciol.* **2008**, *49*, 145–154. [CrossRef]
34. Dai, L.Y.; Che, T.; Ding, Y.J. Inter-calibrating SMMR, SSM/I and SSMI/S data to improve the consistency of snow-depth products in China. *Remote Sens.* **2015**, *7*, 7212–7230. [CrossRef]
35. Che, T. *Research on Passive Microwave Remote Sensing Inversion of Snow Accumulation and Snow Data Assimilation Methods*; Institute of Cold and Arid Regions Environment and Engineering, Chinese Academy of Sciences: Lanzhou, China, 2006. (In Chinese)
36. Huang, X.; Tang, S.; Ciwang, D. Variation of the snow disasters under global warming and its relationship with general circulation over Tibetan Plateau. *Plateau Meteorol.* **2018**, *37*, 325–332. (In Chinese)
37. Shaman, J.; Tziperman, E. The Effect of ENSO on Tibetan Plateau Snow Depth: A Stationary Wave Teleconnection Mechanism and Implications for the South Asian Monsoons. *J. Clim.* **2005**, *18*, 2067–2079. [CrossRef]
38. Yu, W.; Liu, Y.; Yang, X.-Q.; Wu, G.; He, B.; Li, J.; Bao, Q. Impact of North Atlantic SST and Tibetan Plateau forcing on seasonal transition of springtime South Asian monsoon circulation. *Clim. Dyn.* **2021**, *56*, 559–579. [CrossRef]
39. Yuan, C.; Tozuka, T.; Yamagata, T. IOD influence on the early winter tibetan plateau snow cover: Diagnostic analyses and an AGCM simulation. *Clim. Dyn.* **2012**, *39*, 1643–1660. [CrossRef]
40. Liu, Y.; Chen, H.; Wang, H.; Qiu, Y. The Impact of the NAO on the Delayed Break-Up Date of Lake Ice over the Southern Tibetan Plateau. *J. Clim.* **2018**, *31*, 9073–9086. [CrossRef]
41. Xiaoge, X.; Zhou, T.; Yu, R. Increased Tibetan Plateau snow depth: An indicator of the connection between enhanced winter NAO and late-spring tropospheric cooling over East Asia. *Adv. Atmos. Sci.* **2010**, *27*, 788–794. [CrossRef]
42. Platov, G.; Krupchatnikov, V.; Gradov, V.; Borovko, I.; Volodin, E. Analysis of the Northern Hemisphere Atmospheric Circulation Response to Arctic Ice Reduction Based on Simulation Results. *Geosciences* **2021**, *11*, 373. [CrossRef]
43. Chen, Y.; Duan, A.; Li, D. Connection between winter Arctic sea ice and west Tibetan Plateau snow depth through the NAO. *Int. J. Clim.* **2021**, *41*, 846–861. [CrossRef]
44. Chen, Y.; Duan, A.; Li, D. Atmospheric Bridge Connecting the Barents Sea Ice and Snow Depth in the Mid-West Tibetan Plateau. *Front. Earth Sci.* **2020**, *8*, 265. [CrossRef]
45. Cannon, F.; Carvalho, L.M.V.; Jones, C.; Bookhagen, B. Multi-annual variations in winter westerly disturbance activity affecting the Himalaya. *Clim. Dyn.* **2015**, *44*, 441–455. [CrossRef]
46. Dou, J.; Wu, Z. Southern Hemisphere Origins for Interannual Variations of Snow Cover over the Western Tibetan Plateau in Boreal Summer. *J. Clim.* **2018**, *31*, 7701–7718. [CrossRef]
47. Kummu, M.; Taka, M.; Guillaume, J.H.A. Gridded global datasets for Gross Domestic Product and Human Development Index over 1990–2015. *Sci. Data* **2018**, *5*, 180004. [CrossRef] [PubMed]
48. Costanza, R.; Kubiszewski, I.; Giovannini, E.; Lovins, H.; McGlade, J.; Pickett, K.E.; Ragnarsdóttir, K.V.; Roberts, D.; De Vogli, R.; Wilkinson, R. Development: Time to leave GDP behind. *Nature* **2014**, *505*, 283–285. [CrossRef] [PubMed]
49. Millennium Ecosystem Assessment. *Ecosystems and Human Well-Being*; Island Press: Washington, DC, USA, 2005; Volume 5, p. 563.
50. Hickel, J. The sustainable development index: Measuring the ecological efficiency of human development in the anthropocene. *Ecol. Econ.* **2020**, *167*, 106331. [CrossRef]
51. Clark, W.A.V.; Yi, D.; Huang, Y. Subjective well-being in China's changing society. *Proc. Natl. Acad. Sci. USA* **2019**, *116*, 16799–16804. [CrossRef]
52. Liu, L.; Wu, J. Ecosystem services-human wellbeing relationships vary with spatial scales and indicators: The case of China. *Resour. Conserv. Recycl.* **2021**, *172*, 105662. [CrossRef]
53. Oswald, A.J.; Wu, S. Objective Confirmation of Subjective Measures of Human Well-Being: Evidence from the U.S.A. *Science* **2010**, *327*, 576–579. [CrossRef]
54. Duc, N.M. Farmers' satisfaction with aquaculture—A logistic model in Vietnam. *Ecol. Econ.* **2008**, *68*, 525–531. [CrossRef]
55. Peel, D.; Berry, H.L.; Schirmer, J. Farm exit intention and wellbeing: A study of Australian farmers. *J. Rural Stud.* **2016**, *47*, 41–51. [CrossRef]
56. Mzoughi, N. Do organic farmers feel happier than conventional ones? An exploratory analysis. *Ecol. Econ.* **2014**, *103*, 38–43. [CrossRef]

remote sensing

Article

Evaluation of Albedo Schemes in WRF Coupled with Noah-MP on the Parlung No. 4 Glacier

Lian Liu [1], Massimo Menenti [2,3] and Yaoming Ma [1,4,5,6,7,8,*]

1 Land-Atmosphere Interaction and Its Climatic Effects Group, State Key Laboratory of Tibetan Plateau Earth System, Resources and Environment (TPESRE), Institute of Tibetan Plateau Research, Chinese Academy of Sciences, Beijing 100101, China
2 State Key Laboratory of Remote Sensing Science, Aerospace Information Research Institute, Chinese Academy of Sciences, Beijing 100101, China
3 Department of Geoscience and Remote Sensing, Faculty of Civil Engineering and Geosciences, Delft University of Technology, 2628 Delft, The Netherlands
4 College of Earth and Planetary Sciences, University of Chinese Academy of Sciences, Beijing 100049, China
5 College of Atmospheric Science, Lanzhou University, Lanzhou 730000, China
6 National Observation and Research Station for Qomolongma Special Atmospheric Processes and Environmental Changes, Dingri 858200, China
7 Kathmandu Center of Research and Education, Chinese Academy of Sciences, Beijing 100101, China
8 China-Pakistan Joint Research Center on Earth Sciences, Chinese Academy of Sciences, Islamabad 45320, Pakistan
* Correspondence: ymma@itpcas.ac.cn

Abstract: Meteorological variables (e.g., air temperature (T2), radiation flux, and precipitation) determine the evolution of glacier mass and characteristics. Observations of these variables are not available with adequate spatial coverage and spatiotemporal resolution over the Tibetan Plateau. Albedo is the key factor of net radiation and is determined by the land cover and snow-related variables. This study focuses on evaluating the performance of the albedo parameterization scheme in WRF coupled with Noah-MP in terms of glacio-meteorological variables, by conducting experiments applying the standard surface albedo scheme with the default vegetation and corrected to ice cover and the modified glacial albedo scheme to the Parlung No. 4 Glacier in the 2016 ablation season. In situ glacio-meteorological element observations and MODIS-retrieved albedo are selected to assess the performance of the model. The key results are as follows. First, compared to the air temperature bias of 1.56 °C in WRF applying the standard surface albedo scheme and the default vegetation cover, realistic land-use categories considerably reduce the model warm bias on the glacier. The model using realistic land-use categories yields similar T2 diurnal patterns to the observations, with a mean bias of only −0.5 °C, no matter which glacial albedo scheme is implemented. Second, the default glacial albedo scheme gives a rather high albedo value of 0.68, causing an apparent underestimation of the net shortwave radiation and net radiation; the modified glacial albedo scheme gives a mean albedo value of 0.35, close to the in situ observations, helping to relieve underestimations of net shortwave radiation and net radiation. Compared with the MODIS albedo of the glacier, WRF applying the default glacial albedo scheme apparently overestimates the albedo with a mean error of 0.18, while WRF applying the modified glacial albedo scheme slightly underestimates the albedo with a mean error of only −0.08. Third, the mean net radiation flux (142 W m^{-2}) and high ground heat flux (182 W m^{-2}) values that were estimated by WRF applying the corrected land cover and the modified glacial albedo scheme result in the heating of the glacier surface and subsurface, causing ice melt and the liquid water content to increase more quickly and preferentially, equating to an estimated ice thickness decrease of 1 m by mid-June in the ablation region. Our study confirms the ability of the WRF model to reproduce glacio-meteorological variables as long as a reasonable glacial albedo scheme and the corrected land cover is applied and provides a theoretical reference for researchers that are committed to further improvement of the glacial albedo scheme.

Citation: Liu, L.; Menenti, M.; Ma, Y. Evaluation of Albedo Schemes in WRF Coupled with Noah-MP on the Parlung No. 4 Glacier. *Remote Sens.* **2022**, *14*, 3934. https://doi.org/10.3390/rs14163934

Academic Editor: Yi Luo

Received: 20 June 2022
Accepted: 9 August 2022
Published: 13 August 2022

Keywords: WRF; glacier; temperature; albedo; radiation; Tibetan Plateau

1. Introduction

The Tibetan Plateau (TP) has a complex topography and unique geographical environment, with a mean elevation of approximately 4000 m above sea level (a.s.l.). It is known as the Earth's 'third pole' [1], containing many of the world's middle- and low-latitude glaciers. According to the Second Chinese Glacier Inventory, there are 48,571 glaciers in China, with a total area of 51,480 km^2 and estimated water reserves of 5600 km^3 (http://news.sciencenet.cn/htmlnews/2015/1/310736.shtm, accessed on 1 June 2020), of which approximately 80% are found on the TP. Glaciers represent an important land surface type, and their glacier–atmosphere interaction affects the exchange of water and energy in the land–atmosphere system. Very complex physical feedback mechanisms link glaciers and the climate system [2,3]. Therefore, glaciers are considered to be natural indicators and sensitive recorders of climatic and environmental changes [4,5].

As a significant component of the cryosphere, mountain glaciers have attracted unprecedented attention, in particular in regard to their mass balance [6–10]. Under global warming, glaciers on the TP have been retreating and shrinking for decades, a trend that has accelerated in recent years [11–14]. It has been noted that glaciers that are located in the southeastern TP and central Himalayas have retreated rapidly, while those that are located in the Karakoram and Eastern Pamirs have retreated slowly, revealing the great spatial variability in glacier mass balance across the whole TP [15]. Glacier mass balance change has an important impact on the availability of glacial meltwater to recharge the surrounding rivers and lakes of the Yangtze River basin. The retreat of glaciers has contributed to rising lake levels in regions with extensive glacier coverage, such as the Nam Co Lake and Selin Co Lake [16], and has contributed to global mean sea level rise [17,18].

Glacier mass balance has been observed sparsely and far from comprehensively over the topographically-complex TP. Previous research has been limited to a small number of glaciers, including the Qiyi, Xiaodongkemadi, and Parlung glaciers [15,16]. Most investigations of glacier mass balance have depended on energy-based models [6,10] and remote sensing retrievals [9,19]. Glacio-meteorological variables (i.e., near-surface air temperature, precipitation, wind speed, relative humidity, and radiation fluxes) greatly affect the glacier mass balance and are essential factors in mass balance models. The glacial meteorology, point energy, and mass balance of Parlung No. 4 Glacier has previously been investigated [20–23], revealing that net radiation fluxes (especially net shortwave radiation) govern the surface melt of the glacier, with net shortwave radiation contributing 98% of the surface melt. The temperature index model has been proven to be applicable for mass balance and ablation modeling when incorporating solar irradiance [20]. Modeling the spatial distribution of glacier mass balance requires distributed glacio-meteorological forcing, but this is difficult to implement owing to the sparse and uneven distribution of in situ observation stations across the TP. Also, collecting valid in situ measurements of glacio-meteorological variables and energy balance is difficult owing to logistical problems that are associated with the harsh, high-elevation environment of the TP.

Land–atmosphere interactions are evident at the interface of glaciers and the lower atmosphere and drive the rapid response of glaciers to surrounding environmental changes. Temperature, precipitation, and general atmospheric circulation are essential factors influencing changes in glacier mass [15,24]. As one of several coupled atmosphere–land surface models (LSMs), the advanced Weather Research and Forecasting (WRF) model [25] is a good candidate for estimating the glacio-meteorological variables that are required to force glacier mass balance models (e.g., a distributed energy and mass balance model). Numerous studies have evaluated the ability of WRF to produce forcing data for glaciological studies with a correct representation of the glacierized area [26–30]. Great efforts have been made to estimate glacio-meteorological variables using WRF coupled with an LSM (e.g.,

Noah, multi-physics Noah (Noah-MP), and Rapid Update Cycle (RUC)). These schemes have been used to drive the physically-based, distributed glacier energy and mass balance models that were developed to estimate mountain glacier mass variability in dynamically-downscaled, offline, or interactive coupling simulations [26,27,31]. Provisional results have indicated that WRF-modelled meteorological variables at high spatial resolution (i.e., 1 km) can be used to force distributed simulations of Kersten Glacier mass balance with acceptable accuracy [26]. This method has also been successful in simulating the Zhadang Glacier, a small alpine glacier, although feedbacks from the glacier surface mass change into the regional atmospheric forcing were neglected [31]. Models of the interactive coupling between WRF and glacier mass balance have shown promise in studying glacier mass variability [27,32]. However, previous studies of glacier mass balance are seldom based only on the WRF model although the mass accumulation and ice melt as well as energy budget had been involved in this model, leaving more possibility in future research.

The land surface type is a significant factor affecting land surface properties (e.g., emissivity, albedo, and roughness length), and in WRF has an important influence on the modeled surface and near-surface meteorological variables (e.g., temperature, radiation, albedo, wind speed, and snowmelt). However, the statistical land-use product in WRF is wrong to match the Parlung No. 4 Glacier land cover. In addition, snow albedo determining the surface energy budget and influences the glacier mass balance, undergoes large variations during the snow melting and accumulation periods, which is essential for the ice surface energy and mass balance because of its strong controls on the length of the accumulation and ablation seasons. It is significantly affected by many parameters, i.e., snow depth and age, snow cover, surface temperature, cloud cover fraction, wind speed, positive accumulated degrees days, solar zenith angle, and impurities [33–37]. Generally, the snow albedo schemes depend on the observation data and involve the empiric parameters with the most important to be the maximum and minimum albedo. From the review of the currently existing snow albedo schemes [33], many glacial albedo schemes use the minimum albedo about 0.5, which is mostly suitable for the thick ice but not suitable for relatively thin ice. The maximum prescribed snow albedo is usually 0.8–0.85, but the fresh snow albedo is observed up to 0.95. What's more, the simplest snow albedo schemes apply constant values of albedo for different land cover. Other schemes depend on temperature to account for snow metamorphism and snow thinning. More sophisticated schemes consider the snow-related variables and solar zenith angle (for example Biosphere-Atmosphere Transfer Scheme (BATS) [38] and LSM [39]) and impurities [40,41]. It is revealed that the deposition of absorbing aerosols decreases the snow albedo of the Himalayan region by 0.15 ± 0.13, causing a positive radiative effect of 14 ± 13 W m^{-2} and an increase of the surface temperature by 1.33 ± 1.2 °C as well as the reduction of the snow cover fraction by $7 \pm 11\%$ [42]. Therefore, the choice of the snow albedo scheme has a considerable impact on the simulations of both weather and climate and the glacier energy and mass balance.

In addition, glaciers in the Himalayas are mostly sensitive to monsoon-related precipitation perturbations and summer air temperature, which are highly linked to albedo owing to the crucial snow–albedo feedback in summer [43]. In this study, we conducted three numerical experiments using WRF, with one applying the default land surface type (open shrub-land) and the other two adjusted to the snow and ice type and modified albedo scheme on Parlung No. 4 Glacier in the ablation season. Glacio-meteorological variables (near-surface air temperature, wind speed, precipitation, albedo, and radiation fluxes) were modeled and evaluated based on in situ observations and satellite-retrieved albedo at the glacier. This preliminary work is helpful in assessing the two-way coupling of WRF and glacier mass balance models when estimating the mass change of maritime mountain glaciers.

2. Data and Methodology

2.1. Study Area and In-Situ Measurements

Parlung No. 4 Glacier is located at the southeast margin of the TP (Figure 1). This geographic region is strongly dominated by the South Asian monsoon and receives frequent precipitation after the onset of the Indian summer monsoon; thus, glaciers here are of the maritime type. It is a debris-free glacier with an area of 11.7 km^2 and a length of 8 km [22]. Many high-quality glacio-meteorological and mass balance observations are available for Parlung No. 4 Glacier, and these have allowed detailed study of the glacier [20–23]. The glacial mass change in the ablation zone is a crucial component of the whole glacier mass balance. Therefore, this study focuses on the assessment of the glacio-meteorological variables simulations in the ablation zone of the Parlung No. 4 Glacier.

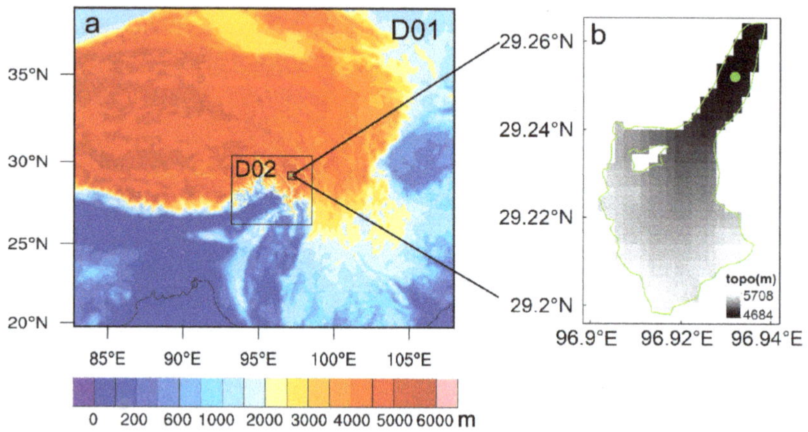

Figure 1. (**a**) WRF domains and model topography. (**b**) Terrain elevation from the WRF model, shaded in units of meters with the green line denoting the glacier boundary and the green solid circle denoting the observation site in the ablation zone of Parlung No. 4 Glacier.

An automatic weather station (AWS) is installed at 29.25°N, 96.93°E, at an elevation of 4800 m a.s.l. in the ablation zone of Parlung No. 4 Glacier (Figure 1). Specific meteorological variables including air temperature at 2 m height (T2), and upward and downward shortwave and longwave radiation, are collected hourly by a CR1000 Campbell Scientific data logger. The hourly precipitation is measured by a Geonor T-200B weighing bucket gauge. The observed T2, components of the radiation fluxes, and precipitation in summer were used to evaluate the numerical estimates in our three experiments. These observational data were obtained from the TP scientific data center website and are freely available from the National Tibetan Plateau Data Center (https://data.tpdc.ac.cn/en/, accessed on 1 July 2020). The observed surface albedo was calculated as the ratio of upward shortwave radiation flux to solar irradiance. Detailed instrumental information and descriptions of the effects of the observed meteorology and surface energy fluxes on the glacier in the ablation season have already been provided [20,22].

2.2. Model Configuration and Experimental Design

The WRF model was developed through a partnership of the National Center for Atmospheric Research (NCAR), the National Oceanic and Atmospheric Administration (NOAA), the United States Air Force, the Naval Research Laboratory, the University of Oklahoma, and the Federal Aviation Administration (The model can be downloaded from https://www2.mmm.ucar.edu/wrf/users/, accessed on 1 July 2020). It is a state-of-the-art atmospheric modeling system, comprising of a fully compressible and non-hydrostatic model with a terrain-following pressure vertical coordinate and Arakawa C-grid horizontal coordinates [25]. It uses Runge–Kutta second- and third-order integration in the time schemes, and second- to sixth-order integration in the advection schemes. The options for atmospheric and land surface processes can be chosen with a broad range of grid sizes, from tens of meters to thousands of kilometers.

WRF has the capacity to estimate glacio-meteorological variables in the low boundary layer and can successfully force distributed energy and mass balance models of mountain glaciers [27,31]. In order to investigate the performance of WRF with respect to glacio-meteorological variables above maritime mountain glaciers in the ablation season, the more recent WRF version 4.3.1 was applied to Parlung No. 4 Glacier in summer 2016. We configured 3 nested domains, with the inner-most domain covering the glacier and its surroundings (Figure 1). The model was centered on 29.23°N, 96.92°E, with spatial resolutions of 12.5, 2.5, and 0.5 km. All the domains were set to 50 terrain-following vertical levels, stretching from the surface to 50 hPa. A dataset from the Interim European Centre for Medium-Range Weather Forecasts (ECMWF) Re-Analysis (ERA-Interim) [44], which is found to have the closest agreement with in situ measurements of air temperature in the Tibetan Plateau [45], was chosen to produce the initial and boundary meteorological conditions with a 0.25° × 0.25° horizontal resolution and a six-hourly intervals. The Noah-MP LSM includes a separate glacier treatment and improved snow physics, with up to three layers in the snowpack, representing improvements over the original Noah scheme [46,47]; it also features a modified two-stream radiation transfer scheme, which considers the three-dimensional canopy structure to calculate radiation fluxes that are reflected, absorbed, and transmitted by vegetation. This LSM uses a 'tile' approach to calculate albedo, a key factor in the energy budget, considering bare ground, vegetation canopy, and snow cover [48]. The Noah-MP coupled with WRF has been shown to provide suitable robust precipitation estimates across the TP [49]; hence, this scheme was chosen for our current study. The model was run from 1 May to 1 October 2016, producing three-hourly output meteorological variables. The first month of simulation was regarded as the model spin-up. The physics schemes that were selected when using WRF and the multi-physical parameterization schemes from Noah-MP are detailed in Table 1.

Table 1. Detailed options that were selected in WRF coupled with the Noah-MP LSM.

Simulations Time Period	From 1 May to 1 October 2016
Nest	Three nested domains (two-way)
Projection	Mercator
Center of domain	29.23°N, 96.92°E
Resolution	12.5 km, 2.5 km, 0.5 km
Microphysics	Thompson scheme
Longwave radiation	RRTM scheme
Shortwave radiation	Dudhia scheme
Surface layer	Revised MM5 Monin-Obukhov scheme
Planetary boundary layer	YSU scheme
Cumulus parameterization	Grell-Devenyi ensemble scheme in domain 1
Land surface	Noah-MP
Canopy stomatal resistance	Ball-Berry scheme
Soil moisture factor for stomatal resistance	Noah scheme
Runoff and groundwater	TOPMODEL with groundwater
Surface layer drag coefficient	Monin–Obukhov
Soil permeability option	Linear effect, more permeable
Radiation transfer	Two-stream applied to vegetated fraction
Ground surface albedo option	CLASS (Canadian Land Surface Scheme)
Precipitation partitioning between snow and rain	Jordan scheme
Snow/soil temperature time scheme	Semi-implicit
Noah-MP glacier treatment	Includes phase change

Land-use type strongly influences radiation fluxes and near-surface air temperatures [50]. The default static land-use in the current WRF version 4.3.1 is the Moderate Resolution Imaging Spectroradiometer (MODIS) land-use product with 30 arc-seconds spatial resolution. This land-use product incorrectly classifies land cover within the ablation zone of Parlung No. 4 Glacier as open shrub-land. To evaluate the importance of land surface type and albedo-related parameters in the accurate estimation of meteorological elements, numerical experiments were conducted: the first experiment used the default land-use product and default albedo scheme of CLASS [51] as the control experiment (CTL); the second experiment (Sens1) used the true land-use type of snow and ice for the extent of the glacier, default bare ice albedo (visible = 0.8, near infrared = 0.55, background albedo = 0.55), and snow albedo scheme of CLASS; the third experiment (Sens2) was similar to Sens1, but corrected bare ice albedo according to previous results (visible = 0.5, near infrared = 0.2 [52], background albedo = 0.23 [22]), and additionally included snow age and solar zenith angle in the CLASS scheme according to the principle of the BATS snow albedo scheme [53]. The CLASS (Equations (1)–(4)) snow albedo scheme and snow cover fraction that were used in the CLASS scheme are described in the Sections 2 and 3.4 of the technical description of Noah-MP [53] in the following equations:

$$\alpha_1 = 0.55 + (\alpha_{old} - 0.55)e^{\frac{-0.01dt}{3600}} \quad (1)$$

$$f_{sn} = \tanh\left(\frac{h_{sn}}{2.5z_0\left(\frac{\rho_{sn}}{\rho_{new}}\right)^{f_m}}\right) \quad (2)$$

$$\alpha_s = \alpha_1 + f_{sn}(0.84 - \alpha_1) \quad (3)$$

$$\alpha_{sd1} = \alpha_{sd2} = \alpha_{si1} = \alpha_{si2} = \alpha_s \quad (4)$$

where α_{old} is the albedo of the last time step (dt). f_{sn} is fractional snow cover. h_{sn} is snow depth in unit of m. ρ_{sn} is the bulk density of snow in unit of kg m^{-3}. ρ_{new} is the fresh snow density with the value of 100 kg m^{-3}. z_0 is the snow surface roughness length with the value of 0.002 m. f_m is melting factor determining the curves in melting season which is adjustable and sets to 1.0 in Noah-MP. α_s is the albedo of snow. α_{sd1} and α_{sd2} denote the direct albedo of snow for visible and near infrared bands, respectively, and α_{si1} and α_{si2} denote the diffuse albedo of snow for visible and near infrared bands, respectively.

The BATS (Equations (5)–(9)) snow albedo scheme is described in the Section 3.3 of the technical description of Noah-MP [53] in the following equations:

$$Z_c = \frac{1.5}{1 + 4\cos Z} - 0.5 \quad (5)$$

$$\alpha_{si1} = 0.95(1 - 0.2A_c) \quad (6)$$

$$\alpha_{si2} = 0.65(1 - 0.5A_c) \quad (7)$$

$$\alpha_{sd1} = \alpha_{si1} + 0.4Z_c(1 - \alpha_{si1}) \quad (8)$$

$$\alpha_{sd2} = \alpha_{si2} + 0.4Z_c(1 - \alpha_{si2}) \quad (9)$$

where Z is the solar zenith angle and A_c is the snow age.

The default snow albedo scheme in the model is developed based on the deep snow with slow melting, which shows a large snow-related simulation deviation on the TP where the snow is shallow and melts rapidly. The insufficient consideration of snow age leading to the lag of melting is the potential reason. In order to more accurately account for the impact of snow age on snow melting on the TP, we attempted to simultaneously consider the snow age in both CLASS and BATS schemes in the Sens2 experiment. However, the albedo of snow for visible and near infrared bands is parameterized to the same value in the CLASS scheme. In reality, the spectral albedo is the different values according to the spectral albedo measurements with a higher albedo for visible band and a lower for near infrared band

and with different spectral albedo curves for fresh snow and old snow [54]. In terms of spectral albedo that is related to snow age in the BATS scheme, multiplicative factors of 0.95 in Equation (6) and 0.65 in Equation (7) represent the diffuse fresh snow albedo for visible and near infrared bands, respectively, which corresponds to about 1.2 and 0.8 times the prescribed fresh snow albedo value of 0.8 for broadband in the model. Therefore, we boldly modified multiplicative factors of 0.95 to $\alpha_s \times 1.2$ in Equation (6) and 0.65 to $\alpha_s \times 0.8$ in Equation (7) when integrating the CLASS and BATS snow albedo schemes in Sens2 experiment. Eventually, the Equations (1)–(3) and (5), the modified Equations (6)–(9) were used ordinally to calculate the spectral snow albedo for the direct and diffuse irradiance in the Sens2 experiment.

The different initial surface conditions and the applied snow albedo schemes in the ablation zone of the glacier in three experiments are outlined in Table 2.

Table 2. Initial surface conditions in the ablation zone of the glacier and snow albedo schemes that were used in our experiments (vis = visible, nir = near infrared).

Experiment	Land-Use Categories	Vegetation Fraction/%	Soil Categories	Background Albedo	Bare Ice Albedo	Snow Albedo Scheme
CTL	open shrub-land	30.5	loam	0.16	-	CLASS
Sens1	snow and ice	0	land-ice	0.55	vis = 0.8, nir = 0.55	CLASS
Sens2	snow and ice	0	land-ice	0.23	vis = 0.5, nir = 0.2	combined CLASS and BATS

2.3. Evaluation of Model Performance

In order to assess the performance of the model in the ablation region of the glacier, the root-mean-square error (RMSE), mean absolute deviation, and the correlation coefficient (CC) between the in situ observations and model estimates were used to evaluate the model performance in terms of the glacio-meteorological variables (T2 and shortwave/longwave radiation). Also, we applied linear regression to the observed and modeled net radiation. Pearson linear cross-correlation was chosen to calculate the CC, and the t-test was chosen to test the significance of the correlation. A significance level of 0.01 was specified in this study. In addition, space remote sensing instrument developed by National Aeronautics and Space Administration, MODIS is used to monitoring global climate change. The product of MOD09GA Version 6 (available from: https://lpdaac.usgs.gov/products/mod09gav006/, accessed on 1 July 2020) provides an estimate of the daily surface spectral reflectance of MODIS/Terra in bands 1 to 7 with a spatial resolution of 500 m. This product was used to assess the performance of the model across the whole glacier, including the main body of the glacier. Quality assurance information regarding the quality control code and atmospheric condition flag of the MOD09GA product were considered to achieve the ideal quality (ideal quality of bands, no cloud, low aerosol quantity) broadband albedo product, which can be estimated using Liang's algorithm [55]:

$$\alpha_{short} = 0.160\alpha_1 + 0.291\alpha_2 + 0.243\alpha_3 + 0.116\alpha_4 + 0.112\alpha_5 + 0.081\alpha_7 - 0.0015 \quad (10)$$

where α_{short} is the surface broadband albedo, and α_1 to α_7 represent the surface reflectance in MODIS bands 1 to 7, respectively. The spectral coverage for MODIS bands 1 to 7 is 0.62–0.67, 0.84–0.87, 0.46–0.48, 0.54–0.56, 1.23–1.25, 1.63–1.65, and 2.11–2.15 μm, respectively.

3. Results

3.1. Air Temperature and Precipitation

The near-surface air temperature and precipitation are essential parameters in forcing glacier mass balance models. These parameters greatly influence the ablation and expansion of mountain glaciers in the summer, with high air temperatures and liquid precipitation accelerating the ablation process and high solid precipitation accelerating the expansion process. The observed and modeled T2 and daily rainfall, and their diurnal variation over the Parlung No. 4 Glacier are shown in Figure 2. Compared with the T2 observations, both Sens1 and Sens2 experiments applying the realistic land cover yielded similar T2 values and diurnal patterns in the 2016 ablation season, while the CTL experiment applying the default land cover greatly overestimated T2. The mean T2 in the ablation season was observed to be 3.68 °C and was estimated to be 5.24 °C in the CTL experiment, 3.17 °C in the Sens1 experiment, and 3.23 °C in the Sens2 experiment. The mean T2 deviation (absolute deviation) reached 1.56 (1.88) °C, −0.51 (1.19) °C, and −0.45 (1.21) °C in the CTL, Sens1, and Sens2 experiments, respectively. The RMSE of T2 between the observations and CTL (Sens1) (Sens2) estimates was 2.42 (1.51) (1.54) °C and the CC was up to 0.7. On the whole, the model accurately recreated the diurnal cycle of T2 on the glacier. The T2 estimates were significantly correlated with the ground observations at the specified significance level of 0.01. WRF applying the real updated land surface type (Sens1 and Sens2 experiments) successfully reproduced T2 in the ablation zone of the glacier in summer 2016, confirming the significance of using realistic land surface types in model simulations of near-surface meteorological elements (Figure 2a).

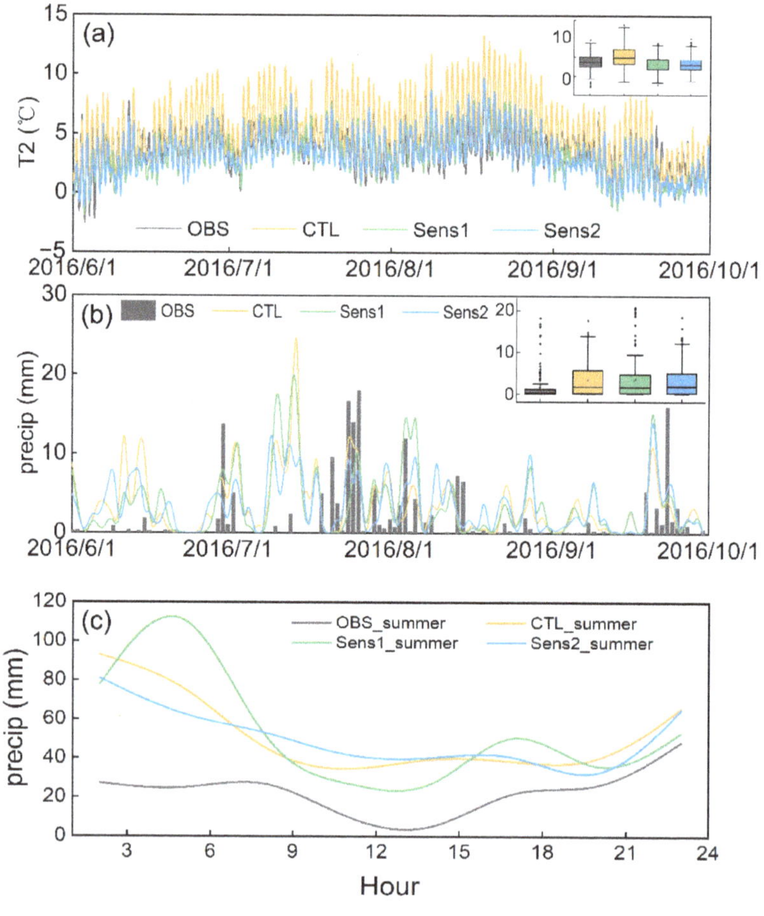

Figure 2. Observed and modeled (**a**) 2 m air temperature (T2), (**b**) daily rainfall, and (**c**) accumulated diurnal variation of rainfall in the ablation zone over the Parlung No. 4 Glacier, with a box whisker diagram in the upper right corner.

Precipitation processes of rainfall and snowfall present different mechanisms for glaciers' mass balance. The heat of rainfall will be transferred to the glacier contributing to the melting of glaciers, while snowfall contributes to the expansion of glaciers by providing solid precipitation particles. Therefore, the precipitation differences for the contributions from liquid precipitation (rain) were investigated first, and then solid precipitation (snow). The observed rainfall mainly occurred in July (89.67 mm) and August (47.21 mm), accounting for 47% and 25%, respectively, of the total rainfall in the ablation season over the Parlung No. 4 Glacier. The modeled rainfall also mainly occurred in July and August. However, the model greatly overestimated light rainfall and underestimated moderate rainfall. For the accumulated rainfall in the ablation season, the observed total rainfall amount was 190.6 mm, the Sens2 estimate was 417.2 mm, and the Sens1 and CTL estimates were approximately 433 mm. The three experiments estimated the total rainfall to be more than double the observed total amount (Figure 2b). In terms of diurnal variation of accumulated rainfall, the observed maximum accumulated rainfall in summer 2016 occurred at nighttime (21:00–23:00 LST; LST = UTC + 8 h), reaching 48.33 mm, and rainfall from the afternoon to sunrise (15:00–08:00 LST) accounted for 92% of the daily rainfall. The model demonstrated its capacity to simulate the diurnal variation curves of rainfall but notably overestimated the rainfall amount, with the maximum accumulated rainfall in summer reaching 92.83 mm (00:00–02:00 LST), 111.06 mm (03:00–05:00 LST), and 80.95 mm (00:00–02:00 LST) for CTL, Sens1, and Sens2 modeling, respectively. For the main rainfall duration period (15:00–08:00 LST), the three experiments modeled rainfall that accounted for more than 80% of the total daily rainfall (Figure 2c).

Old snow melted and small amounts of sleet occurred before 20 September, then snowfall ensued thereafter. The Sens1 and Sens2 experiments successfully estimated snowfall in late September, although Sens1 falsely simulated a large amount of snow before 20 September. The observed and modeled three-hourly snow water equivalent (SWE) in late September and the corresponding accumulated diurnal variation in snowfall and snow melt is shown in Figure 3. Compared with the snowfall observations, the Sens2 experiment reproduced snowfall on 20 September followed by rapid snow melt, but failed to reproduce snowfall on 24 September. The two snowfall events were reproduced by Sens1, but the first snowfall amount was remarkably overestimated and this was followed by slow snow melt. The total snowfall amounts from the ground observations, Sens1, and Sens2 estimates were 11.77, 28.58, and 18.47 mm, respectively, with the model greatly overestimating the snowfall amount (Figure 3a). In addition, the observed snowfall mainly occurred in the morning before 11:00 LST, with the largest amount at sunrise, presenting a single peak pattern. However, both Sens1 and Sens2 simulated the snowfall occurrence as double peaks, with the first high peak occurring at 6:00 LST and the second low peak at 17:00 LST for Sens1, and the first peak occurring at 5:00 LST and the second slightly lower peak at 11:00 LST for Sens2. Remarkably, snow melted at local noon in the estimates of both experiments, and the peak of snow melt from Sens2 lagged behind Sens1 by approximately 2 h (Figure 3b).

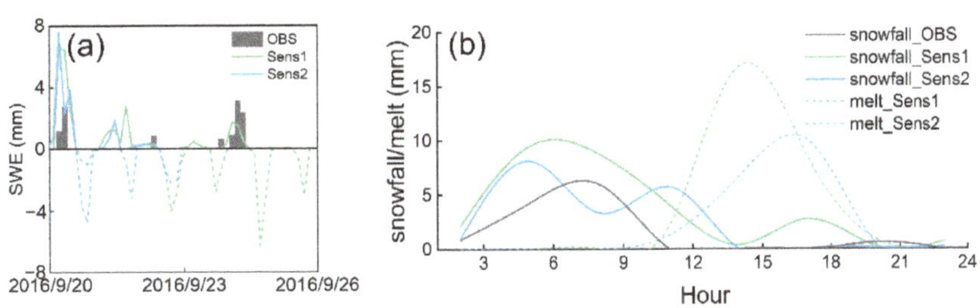

Figure 3. (**a**) Observed and modeled three-hourly SWE in late September over the Parlung No. 4 Glacier (solid line indicates snow accumulation and the dashed line indicates snow melt). (**b**) Corresponding accumulated diurnal variation of snowfall and snow melt estimates.

3.2. Net Radiation Flux and Albedo

Net radiation is a direct and immediate forcing of glacier mass balance, which consists of net shortwave radiation and net longwave radiation; excess energy will melt the snow and ice and affect the near-surface air temperature. A comparison of the net radiation and its components, and the diurnal variation of net radiation between observations and model estimates, is displayed in Figure 4. It is clear to see that similar performance of WRF simulating net longwave radiation was achieved for Sens1 and Sens2 experiments with the similarly temporal pattern and the same range of values (Figure 4a,b). The mean observation of the net longwave radiation was only 13.77 W m^{-2} and the mean estimates reached 30.56 W m^{-2} for both experiments (Sens1, Sens2). Compared with the observed net longwave radiation, the net longwave radiation was rather consistently overestimated by the two experiments (Sens1, Sens2), with an average overestimation value of 17 W m^{-2}. A similar performance of WRF in relation to the net shortwave radiation and the net radiation was achieved, revealing the variation in the net shortwave radiation to be a fundamental factor in the variation in net radiation. Owing to the distinct glacial albedo scheme that was implemented in WRF, the instantaneous net shortwave radiation and net radiation was distributed discretely on both sides of the perfect fitting line (y = x) for Sens2, and concentrated below the perfect fitting line (y = x) for Sens1, with rather large underestimations (Figure 4c–e). The average net shortwave radiation for observation, Sens1, and Sens2 estimates was 137.21, 84.2, and 172.41 W m^{-2}, respectively. The average net radiation was calculated to be 123.44 W m^{-2} using in situ observations, 53.64 W m^{-2} for Sens1 estimates, and 141.85 W m^{-2} for Sens2 estimates. The linear regression lines for Sens1 and Sens2 estimates were y = 0.47x + 19.61 (CC = 0.81) and y = 0.92x + 46.16 (CC = 0.77), respectively, for the net shortwave radiation; and y = 0.45x − 1.95 (CC = 0.83) and y = 0.88x + 30.92 (CC = 0.78), respectively, for the net radiation. These illustrate that Sens2 apparently outperforms Sens1 in estimating the instantaneous net radiation and its dominant component (net shortwave radiation), and shows a consistent conclusion in terms of the daily mean net radiation estimates that Sens2 apparently outperforms Sens1. Owing to the notable discreteness of net shortwave radiation and the net radiation from Sens2 estimates, Sens2 yielded large RMSEs of ~140 W m^{-2} for both the net radiation and net shortwave radiation. Such large RMSEs were also simulated by the Sens1 experiment because of remarkable underestimations of the net shortwave radiation and net radiation. Compared with the observed diurnal variations in the net radiation, the two experiments achieved identical diurnal cycles, with maximum values occurring at local noon. Yet, different peaks were calculated in Sens1 (228.92 W m^{-2}) and Sens2 (503.99 W m^{-2}), the latter being closer to the observed value of 454.62 W m^{-2} (Figure 4f). Coincident high CCs of ~0.8 were calculated between the observations and experimental (Sens1, Sens2) estimates of net radiation and net shortwave radiation, which passed the significance test at the specified 0.01 level.

Surface albedo is a key factor in net radiation and is the main determinant of the energy and mass balance of a glacier. Rapid changes in the albedo arise because of snowfall and the subsequent evolution of the snowpack, with large values associated with fresh snow, and decreasing values as snow melt progresses. Our experiments applied the snow albedo scheme and a bare ground albedo scheme for Sens1 and Sens2, and additionally involved a modified two-stream radiation transfer scheme considering the three-dimensional structure of the canopy in the control experiment. The albedo was calculated from a combination of the reflected radiation of snow, bare ground, and vegetation canopy in the model, and was compared with ground observations and MODIS product (Figure 5). The observed daily albedo declined sharply from 0.75 to 0.24 before 29 June, slightly increased to 0.47 on 3 July, then remained at a rather low, variable value of approximately 0.23 until 19 September. Subsequently, the observed albedo rapidly increased to 0.75 on 20 September, then to 0.78 on 24 September (Figure 5a). The variation in the observed daily albedo indicates that the seasonal snowpack in the ablation zone was melting and did not disappear until late June, when the ice became exposed on the surface. Subsequently, the observed variable low albedo value before 19 September not only results from small amounts of sleet and subsequent snow melt increasing the albedo, but also from rainfall increasing the liquid water content above the glacier's surface and in the meantime decreasing the albedo. In the meanwhile, the observed bare ice albedo fluctuates around 0.23 which was affected by the liquid water content. Therefore, the low ice albedo of 0.23 instead of the minimum albedo is set as the background albedo of the glacier in the Sens2 experiment. The large albedo value in late September resulted from two apparent snowfall events. Owing to the very different treatments of snow, canopy, and bare-ground albedo in the WRF model, the three experiments yielded substantially different surface albedo estimates. The albedo maintained a value of 0.12 in the ablation season in the CTL experiment because the surface type was wrongly classified as open shrub-land. The Sens1 experiment applied the default albedo scheme on the glacier, presenting a very high albedo value of ~0.8 on 3 July and 9, 20, and 21 September, while retaining a constantly high albedo value of 0.68 for the rest of the ablation season. The Sens2 experiment, using the modified albedo scheme on the glacier, yielded a constantly low albedo value of 0.35, which was close to the in situ observation, before increasing to a maximum of 0.78 during snowfall in late September (Figure 5a). Apparently, the albedo error of Sens2 in the ablation season is significantly smaller than that of Sens1, while the error of Sens1 before the middle of June and after the middle of September is significantly lower than that of Sens1. This reveals that the modified glacial albedo scheme for Sens2 is only applicable to the ablation season.

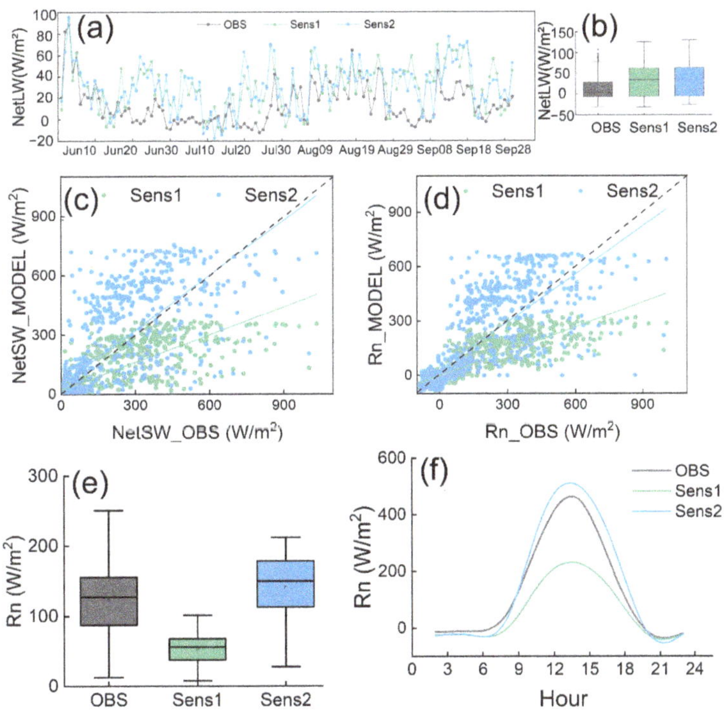

Figure 4. Comparison of the net longwave radiation (**a**) daily mean NetLW; (**b**) instantaneous NetLW, net shortwave radiation; (**c**) instantaneous NetSW; and net radiation (**d**) instantaneous Rn; (**e**) daily mean Rn), and the diurnal variation in instantaneous net radiation (**f**) between the observations (OBS: grey) and model estimates (Sens1: green; Sens2: blue). The grey dashed line denotes the perfect linear fit line of y = x, the green solid line denotes the linear fit between the observation and Sens1 estimates, and the blue solid line denotes the linear fit between the observation and Sens2 estimates in (**c**) and (**d**).

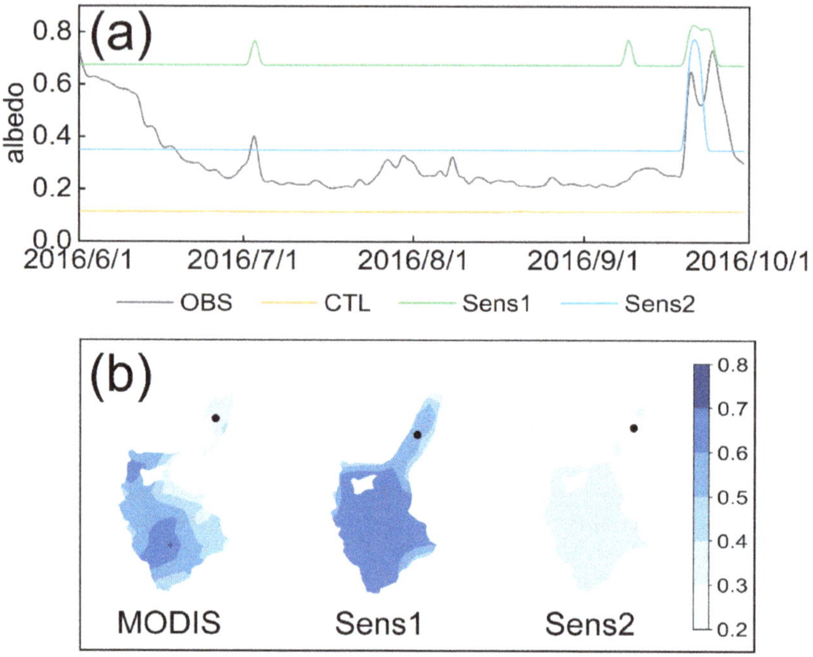

Figure 5. Comparison of model-estimated albedo at 14:00 LST against observations (**a**) and MODIS retrieved product on 21 August 2016 (**b**). The solid black circle indicates the location of the AWS.

In addition, the performance of WRF in relation to the albedo estimates when applying different glacial albedo schemes was also evaluated against the MODIS-retrieved high pixel resolution and the daily albedo product over the whole glacier. MODIS retrieval of albedo on 21 August 2016 was selected to assess the model performance on simulating the spatial distribution of albedo (Figure 5b), because the albedo in the pixels of the glacier was measured in the highest quality on 21 August 2016. It revealed that the glacial albedo was different across the pixels of the glacier in the summer with a low value of 0.2–0.4 in the ablation zone and a high value of 0.5–0.7 in the accumulated zone. The model applying the default glacial albedo scheme greatly overestimated the albedo of the whole glacier particularly in the accumulated zone where the albedo kept to 0.7–0.8. The model applying the modified glacial albedo scheme showed the remarkable advantages in relation to the albedo estimates in the ablation zone, but underestimated the albedo in the accumulated zone. Therefore, in the ablation zone, using the observed background albedo had a large effect on the simulated albedo (Figure 5b). Besides, we used the MODIS-retrieved albedo in pixels of the glacier, where the ideal quality of the spectral bands, no cloud effect, and low aerosol quantity were recognized in the ablation season. The albedo RMSE and the mean error between Sens1 and the MODIS product were 0.22 and 0.18, respectively, indicating an overestimation of the albedo of the glacier when the default glacial albedo scheme was applied in WRF. In contrast, the albedo RMSE and the mean error between Sens2 and the MODIS product were 0.16 and −0.08, respectively, illustrating a slight underestimation of the albedo of the glacier when the modified glacial albedo scheme was applied in WRF. The albedo estimates from WRF were significantly correlated with the MODIS retrievals at the specified significance level of 0.01, with similar CC values of ~0.5.

On the whole, the albedo was largely underestimated with a rather low value of 0.12 kept in the CTL experiment due to wrong land cover that was used in the model. The Sens1 and Sens2 experiments using the updated land cover successfully simulated snowfall events in late September. However, Sens1 gave a rather high albedo, with the average value of 0.68 leading to an underestimation of the net shortwave radiation and net radiation. Sens2 gave a low albedo, with the average value of 0.35 being closer to the in situ observations when little snow appeared, causing an overestimation of the net shortwave radiation and net radiation in early June, followed by underestimation in July and August (Figures 4 and 5a). On average, surface shortwave radiative forcing reached to −141 (−60) W m^{-2} due to albedo changes that were induced by land cover in CTL updated to the one in Sens1 (Sens2), which shows the cooling effect. Surface shortwave radiative forcing reached to 85 W m^{-2} due to glacial albedo changes that were induced by albedo the scheme in Sens1 that was updated to the scheme in Sens2, which shows an apparent heating effect and may cause a considerable ablation of the glacier.

3.3. Turbulent Heat Flux

The surface net radiation determines the energy redistribution between turbulent water vapor and heat fluxes. Albedo is a key factor of net radiation. In our previous analysis, albedo schemes present large impacts on shortwave radiation estimates, but similar T2 is simulated by both Sens1 and Sens2 experiments. Maybe compensating the effects of energy fluxes to/away from the ice surface results in the similar T2. In order to explain this finding reasonably, the modeled daily mean sensible and latent heat fluxes above the Parlung No. 4 Glacier, the ground heat flux beneath the glacier, as well as the daily mean surface heat budget in the ablation season are shown in Figure 6. It shows that the Sens1 and Sens2 experiments simulated similarly low values and consistent daily variations of sensible (−60 to 0 W m^{-2}) and latent (−51 to 36 W m^{-2}) heat fluxes over the entire time. The excess surface energy enters the glacier in the form of ground heat flux. High ground heat flux from Sens1 (15–184 W m^{-2}) and Sens2 (30–295 W m^{-2}) heats the glacier, resulting in surface and subsurface ice melt (Figure 6a). From the boxplots analysis of the daily mean surface heat budget (Figure 6b), the large difference of the net radiation between Sens1 and Sens2 estimates attributes to the large difference of net shortwave radiation that is caused by different glacial albedo schemes. The simulated net radiation, sensible and a proportion of the latent heat flux by both Sens1 and Sens2 experiments are used to heat the surface glacier first and then transmit downwards to heat the subsurface glacier. The main energy contributor to surface and subsurface ice melt from both experiments is the net radiation, supplemented by sensible heating. Similar sensible heat flux was reproduced by both Sens1 and Sens2 experiments, which is highly linked to the temperature difference between the ground and air, and the ground temperature of the glacier kept to 0 °C. These account for the similar T2 that was estimated by WRF applying different glacial albedo schemes. The average ground heat flux from Sens1 was 88 W m^{-2}, which was much lower than that from Sens2 (182 W m^{-2}). This is because the albedo from Sens1 was much higher that from Sens2, causing apparently lower net radiation and reduced energy entering the glacier.

In addition, ice melt leads to an increase in the liquid water content in the glacier. The liquid water content estimates in the different layers in the ablation region of the glacier are displayed in Figure 7. This shows that excess energy is first used to melt the surface ice (the upper layer, 0–0.1 m), then the second layer (0.1–0.4 m), followed by the third layer (0.4–1.0 m), and finally the bottom layer (1.0–2.0 m). Apparently, the ice in the upper layer melts more quickly than that in the bottom layer of the glacier. Compared with Sens1 estimates, more energy from Sens2 estimates enters the glacier, causing ice to melt more quickly and preferentially, which in turn causes the liquid water content to increase more rapidly and preferentially. Both Sens1 and Sens2 experiments illustrated considerable glacier melting in the ablation region, with an ice thickness reduction of 0.4 m by 22 June for the Sens1 estimates, and an ice thickness reduction of 1.0 m by mid-June for the Sens2 estimates (Figure 7).

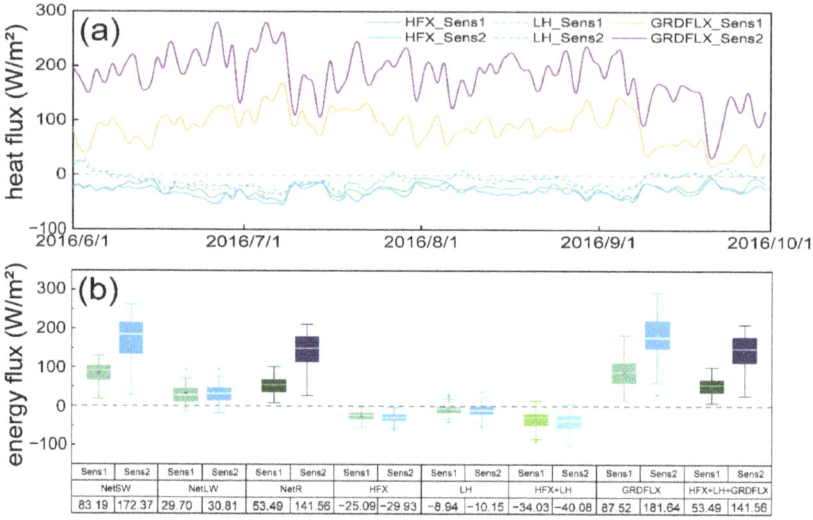

Figure 6. Comparison between the Sens1 and Sens2 estimates of the daily mean near-surface sensible (HFX, + to atmosphere), latent (LH, + to atmosphere), ground (GRDFLX, + to soil) heat fluxes (**a**) and each component of the surface heat budget (**b**), NetSW: net shortwave radiation; NetLW: net longwave radiation; NetR: net radiation).

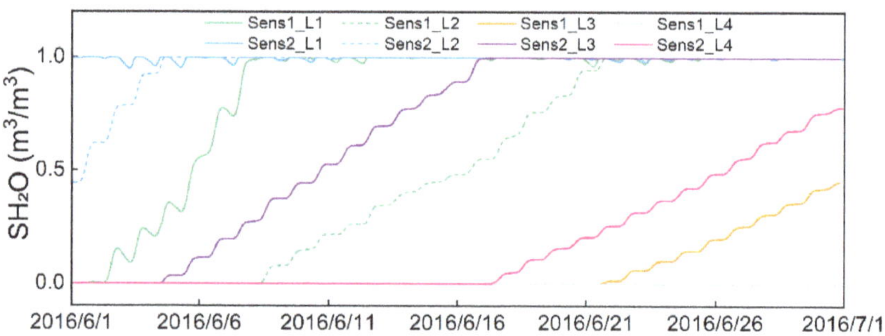

Figure 7. Comparison between the Sens1 and Sens2 estimates of the liquid water content (SH_2O) in the first (L1), second (L2), third (L3), and fourth layer (L4) in June 2016 in the ablation region of the Parlung No. 4 Glacier.

4. Discussion and Conclusions

The glacier mass balance has a great effect on the physical processes in the cryosphere and hydrosphere, and exhibits strong feedback with current climate change [2,3]. However, high quality observations of glacier mass balance and glacio-meteorological variables are sparse across the TP, while the WRF model can provide useful estimates of meteorological conditions (e.g., T2, precipitation, and radiation fluxes). Previous numerical studies of glacier mass balance estimates usually combine the standard WRF model and the distributed glacier energy and mass balance models through dynamically-downscaled, offline, or interactive coupling simulations [26,27,31,32]. The land-atmosphere coupling WRF involves the mass supplement from precipitation processes and ice melt that is related to the energy budget, which provides a possibility to use WRF alone to study glacier mass changes. However, rarely do studies attempt the related research. In addition, as the key factor of surface net radiation, albedo highly links to the energy balance of glaciers. Previous studies often implement the default glacial albedo scheme when using the standard WRF to force the distributed glacier energy and mass balance models, seldom improving the glacial albedo scheme in WRF. In our study, control and sensitivity experiments focusing on the land surface type and glacial albedo scheme were conducted using the WRF model coupled with Noah-MP to simulate glacio-meteorological variables in the 2016 ablation season above the Parlung No. 4 Glacier on the TP. The in situ meteorological element observations and the MODIS-retrieved albedo product were selected to evaluate the model performance. Our study initially tries to modify the glacial albedo scheme and evaluates the capacity of WRF alone to the ablation of a marine glacier over the TP. It is one step forward in understanding the glacier mass balance over the TP.

Surfaces with a high albedo reflect a large fraction of solar irradiance, resulting in only a small amount of shortwave radiation being absorbed by the land surface, thereby reducing the surface net radiation heating, and leading to a lower T2 through turbulent heat exchange between the glacier surface and the near-surface atmosphere. In contrast, low albedo has the opposite effect and promotes a higher T2. In this study, the three experiments were able to obtain not only daily variations, but also the mean diurnal pattern of T2. However, the model overestimated the daily mean T2 throughout the ablation season when using the unrealistic open shrub-land surface type, owing to the very low estimated surface albedo of ~0.12. Vegetation decreases the surface albedo mainly because of the low background albedo of the underlying vegetation cover in the CTL experiment. Such a low albedo contributes to a high surface net radiation through reducing the reflected irradiance, and the high net radiation heats the land surface with open shrub-land covered enhancing the turbulent sensible heat exchange among the land surface, vegetation, and near-surface atmosphere. This leads to the CTL estimated mean air temperature of 1.56 °C higher than the in situ observations. In addition, the vegetation increases not only the evapotranspiration, changing the surface water balance but also the roughness length and friction velocity enhancing the momentum exchange between the land surface and near-surface atmosphere. The apparent warm bias of 1.56 °C was significantly alleviated by sensitivity experiments using a larger glacier albedo, with the mean bias decreasing to −0.5 °C and a 37% improvement in RMSE. Consequently, the realistic geostatic land surface parameters such as land surface type are strongly recommended in the model simulations in the following research.

Albedo is a key factor in surface energy balance and affects the redistribution of surface net radiation between turbulent water vapor and heat fluxes. Noah-MP imposes a minimum ice surface albedo of 0.55 which does not adequately represent the surface conditions in the ablation zone of the glacier. This was the main cause of the large albedo value of 0.68 that was simulated in the model that applied the default glacial albedo scheme. A rather high albedo was retained in the default glacial albedo scheme resulting in significant underestimations of the net shortwave radiation and the net radiation. However, an albedo value of 0.35 was held in the modified glacial albedo scheme; this contributed to relieving the underestimation of not only the net shortwave radiation and net radiation, but also the peak net radiation value at local noon. In addition, our simulation results confirm that the net shortwave radiation is the dominant energy source during the ablation season on the southeast TP [20]. Notably, the albedo error of Sens2 in the ablation season is significantly smaller than that of Sens1, but the error of Sens1 before the middle of June and after the middle of September is significantly lower than that of Sens1, revealing that the modified glacial albedo scheme for Sens2 is only applicable to the ablation season. This is because the observed bare ice albedo in the ablation season is set as the background albedo and used in the modified glacial albedo scheme. Besides, the WRF model cannot precisely grasp the variation of albedo during snowfall processes in early June and late September, even though the model applies the observed background albedo to the modified glacial albedo schemes. This is highly linked to the improper snow albedo scheme and the inappropriate description of the relationship between snow albedo and the underlying ice surface.

In addition, the net longwave radiation was apparently overestimated by 17 W m^{-2} in Sens1 and Sens2 experiments. The overestimation of net longwave radiation was highly related to the cloud effect on the downward longwave radiation. The glacier emitted the same intensity of longwave radiation due to the unified observed and simulated surface temperature of 0 °C. In realistic cloudy conditions, the cloud increased the downward longwave radiation causing a decrease of the net longwave radiation. However, there is more rainfall in both sensitive experiments (Sens1 and Sens2) on days when no rainfall was observed, and clouds may be present in both sensitive experiments but less matching the timing of the observed clouds. More rainfall in WRF sensitive experiments was caused by cloud particles gradually growing into raindrops through complex microphysical processes and falling to the ground. Afterwards, the cloud dissipated accompanied by a decrease of downward longwave radiation. This would result in an increase of the net longwave radiation for both sensitivity experiments. What is more, the complex topography is a potential factor causing large RMSEs of net radiation and its components for both sensitive experiments; our sensitive experiments that were reported herein did not consider the effects of complex local topography on the direct solar irradiance, and also neglected the reflected and diffuse radiation from surrounding mountainous topography.

Both sensitivity experiments (Sens1 and Sens2) generated relatively small and consistent turbulent water vapor and heat exchange values, showing the similarity of model performance in relation to surface sensible and latent heat, despite two different glacial albedo schemes being applied. The sensible and latent heat fluxes were estimated to be a little higher than Yang's results using the energy balance model [22]. For example, the mean downward sensible heat flux was calculated to be 28 W m^{-2} from our experiments and 17 W m^{-2} from Yang's estimate [22]. Compared with WRF using the default glacial albedo scheme, the excess surface energy that was obtained from WRF using the modified glacial albedo scheme heats surface ice and enters the glacier, heating subsurface ice in the form of ground heat flux, which is estimated to be 182 W m^{-2}. This causes surface and subsurface ice melt and the liquid water content to increase more quickly and preferentially in WRF with the modified glacial albedo scheme. Our study confirms net radiation to be the main contributor to surface ice melt, supplemented by sensible heating, and illustrates considerable glacier melting in the ablation region during the ablation season. The ice thickness decreased by 1.0 m by mid-June when applying the modified glacial albedo scheme in WRF, which closely matches the ablation that was measured by stakes and calculated using the surface energy-balance model from Yang's results [22]. However, the ice thickness of glaciers is defined to be 2 m in the current Noah-MP, and the maximum ice thickness reduction is limited to 2 m. Therefore, it is reasonable to model the ablation of glaciers under the condition of ice thickness more than 2 m but the cumulative reduction less than 2 m. In our study, the thickness of the Parlung No. 4 Glacier and the observed cumulative reduction are more than 2 m in the ablation zone in summer 2016 [22]. In order to accurately simulate the large ablation, the ice thickness should be adjusted in the Noah-MP, which was not considered in our study but should be considered further through conducting more numerical tests to ensure the stable calculation during the model's integration. Understanding glacier mass balance anomalies requires quantification and insight into subtle shifts in the energy balance at high altitude glaciers [30]. In order to understand the accumulated and melt regimes of glaciers, further work will focus on the investigation of the roles of each component of glacial surface and subsurface energy balance.

More realistic land-use parameters in the model can reduce the modelled T2 and precipitation biases, and more accurate surface albedo will provide further model bias reductions [56]. This study used the satellite-retrieved spectral albedo and the in situ observed albedo in the ablation zone of the glacier to modify the glacial albedo scheme, which achieved better performance in relation to the glacio-meteorological elements in the focused ablation region. However, the modified albedo scheme is not reasonable in the main area of the glacier because the parameters in the ablation zone cannot represent the parameters in the main area of the glacier. Further improvements in glacial albedo schemes in land–atmosphere coupling models are urgently required in order to obtain better model performance in the estimation of glacio-meteorological variables across the entire glaciers.

The ablation of glaciers has contributed to rising not only the lake levels in the glacier concentration areas [16], but also the global mean sea level [17,18]. The lake and sea level rise will threaten the lives of people who reside by lakes and in the coastal cities. Therefore, the accurate prediction of glacier ablation contributions to the lake and sea level rise is essential in regions with extensive glacier coverage. Our study assesses the ability of WRF itself to estimate the ablation of a mountainous glacier, which is the first step before WRF itself can be widely used in the accurate simulation and prediction of glacier ablation. Our study confirms the potential of the WRF model to generate reasonable glacio-meteorological variables, such as T2, radiation, and ground heat flux except precipitation, as long as a reasonable glacial albedo scheme is applied. These variables are essential to estimate glacier energy and mass changes. Precipitation is another decisive factor in glacier mass changes and different phases of precipitation (i.e., rain and snow) show different mechanisms of glacier mass balance; however, the simulated precipitation was twice that of the in situ observations during the ablation season over the glacier. This is not directly influenced by the glacial albedo scheme but was affected not only by the microphysical and cumulus parameterization schemes in the WRF model, but also by the complex mountainous topography. At present, we have not investigated complex parameters in the microphysics and cumulus schemes, nor the role of terrain and the terrain-induced water vapor transport, which relate to the very high precipitation duration, intensity, and amount in WRF estimates. Therefore, it is important to explore how to improve the parameterizations of precipitation in regional models over complex topography. In the future, we will focus on evaluating and improving the performance of the model with regard to the total amount of summer precipitation over the southeastern TP, through investigation of the model microphysics and cumulus schemes. Also, we will further investigate the roles of mountainous topography and the terrain-induced water vapor from the Indian summer monsoon in increasing precipitation over the southeastern TP.

Author Contributions: Conceptualization, L.L. and M.M.; methodology, L.L.; writing original draft preparation, L.L., M.M. and Y.M.; funding acquisition, Y.M. All authors have read and agreed to the published version of the manuscript.

Funding: This research was funded by the Strategic Priority Research Program of Chinese Academy of Sciences (XDA20060101), the Second Tibetan Plateau Scientific Expedition and Research program (STEP) (2019QZKK0103), and the National Natural Science Foundation of China (91837208).

Data Availability Statement: The ERA-Interim dataset is available from http://apps.ecmwf.int/datasets/data/interim-full-daily/, accessed on 1 July 2020. Meteorological observations on Parlung No. 4 Glacier are available from National Tibetan Plateau Data Center (https://data.tpdc.ac.cn/en/, accessed on 1 July 2020).

Acknowledgments: The first author would like to acknowledge all group members for their help in completing this paper.

Conflicts of Interest: The authors declare no conflict of interest.

References

1. Qiu, J. The third pole. *Nature* **2008**, *454*, 393–396. [CrossRef] [PubMed]
2. Robinson, D.A. Hemispheric snow cover and surface albedo for model validation. *Ann. Glaciol.* **1997**, *25*, 241–245. [CrossRef]
3. Bamzai, A.S.; Shukla, J. Relation between Eurasian snow cover, snow depth, and the Indian summer monsoon: An observational study. *J. Clim.* **1999**, *12*, 3117–3132. [CrossRef]
4. Zemp, M.; Roer, I.; Kääb, A.; Hoelzle, M.; Paul, F.; Haeberli, W. *Global Glacier Changes: Facts and Figures*; World Glacier Monitoring Service: Zürich, Switzerland, 2008.
5. Davaze, L.; Rabatel, A.; Dufour, A.; Hugonnet, R.; Arnaud, Y. Region-Wide Annual Glacier Surface Mass Balance for the European Alps from 2000 to 2016. *Front. Earth Sci.* **2020**, *8*, 149. [CrossRef]
6. Fujita, K.; Ageta, Y. Effect of summer accumulation on glacier mass balance on the Tibetan Plateau revealed by mass-balance model. *J. Glaciol.* **2000**, *46*, 244–252. [CrossRef]

7. Fujita, K.; Ageta, Y.; Pu, J.; Yao, T. Mass balance of Xiao Dongkemadi glacier on the central Tibetan Plateau from 1989 to 1995. *Ann. Glaciol.* **2000**, *31*, 159–163. [CrossRef]
8. Cogley, J.G. Climate science: Himalayan glaciers in the balance. *Nature* **2012**, *488*, 468–469. [CrossRef]
9. Jacob, T.; Wahr, J.; Pfeffer, W.T.; Swenson, S. Recent contributions of glaciers and ice caps to sea level rise. *Nature* **2012**, *482*, 514–518. [CrossRef]
10. Ding, B.; Yang, K.; Yang, W.; He, X.; Chen, Y.; La, Z.; Guo, X.; Wang, L.; Wu, H.; Yao, T. Development of a Water and Enthalpy Budget-based Glacier mass balance Model (WEB-GM) and its preliminary validation. *Water Resour. Res.* **2017**, *53*, 3146–3178. [CrossRef]
11. Ding, Y.; Liu, S.; Li, J.; Shangguan, D. The retreat of glaciers in response to recent climate warming in western China. *Ann. Glaciol.* **2006**, *43*, 97–105. [CrossRef]
12. Kang, S.; Qin, D.; Ren, J.; Zhang, Y.; Kaspari, S.; Mayewski, P.A.; Hou, S. Annual accumulation in the Mt. nyainqentanglha ice core, southern Tibetan plateau, China: Relationships to atmospheric circulation over Asia. *Arct. Antarct. Alp. Res.* **2007**, *39*, 663–670. [CrossRef]
13. Li, X.; Cheng, G.; Jin, H.; Kang, E.; Che, T.; Jin, R.; Wu, L.; Nan, Z.; Wang, J.; Shen, Y. Cryospheric change in China. *Glob. Planet. Change* **2008**, *62*, 210–218. [CrossRef]
14. Yao, T.; Wang, Y.; Liu, S.; Pu, J.; Shen, Y.; Lu, A. Recent glacial retreat in High Asia in China and its impact on water resource in Northwest China. *Sci. China Ser. D-Earth Sci.* **2004**, *47*, 1065–1075. [CrossRef]
15. Yao, T.; Thompson, L.; Yang, W.; Yu, W.; Gao, Y.; Guo, X.; Yang, X.; Duan, K.; Zhao, H.; Xu, B.; et al. Different glacier status with atmospheric circulations in Tibetan Plateau and surroundings. *Nat. Clim. Change* **2012**, *2*, 663–667. [CrossRef]
16. Yao, T.; Pu, J.; Lu, A.; Wang, Y.; Yu, W. Recent glacial retreat and its impact on hydrological processes on the tibetan plateau, China, and sorrounding regions. *Arct. Antarct. Alp. Res.* **2007**, *39*, 642–650. [CrossRef]
17. Meier, M.F.; Dyurgerov, M.B.; Rick, U.K.; O'neel, S.; Pfeffer, W.T.; Anderson, R.S.; Anderson, S.P.; Glazovsky, A.F. Glaciers dominate Eustatic sea-level rise in the 21st century. *Science* **2007**, *317*, 1064–1067. [CrossRef] [PubMed]
18. Hock, R.; De Woul, M.; Radic, V.; Dyurgerov, M. Mountain glaciers and ice caps around Antarctica make a large sea-level rise contribution. *Geophys. Res. Lett.* **2009**, *36*, L07501. [CrossRef]
19. Kääb, A.; Berthier, E.; Nuth, C.; Gardelle, J.; Arnaud, Y. Contrasting patterns of early twenty-first-century glacier mass change in the Himalayas. *Nature* **2012**, *488*, 495–498. [CrossRef]
20. Yang, W.; Guo, X.; Yao, T.; Yang, K.; Zhao, L.; Li, S.; Zhu, M. Summertime surface energy budget and ablation modeling in the ablation zone of a maritime Tibetan glacier. *J. Geophys. Res. Atmos.* **2011**, *116*, D14116. [CrossRef]
21. Yang, W.; Yao, T.; Guo, X.; Zhu, M.; Li, S.; Kattel, D.B. Mass balance of a maritime glacier on the southeast Tibetan Plateau and its climatic sensitivity. *J. Geophys. Res. Atmos.* **2013**, *118*, 9579–9594. [CrossRef]
22. Yang, W.; Yao, T.; Zhu, M.; Wang, Y. Comparison of the meteorology and surface energy fluxes of debris-free and debris-covered glaciers in the southeastern Tibetan Plateau. *J. Glaciol.* **2017**, *63*, 1090–1104. [CrossRef]
23. Zhu, M.; Yao, T.; Yang, W.; Maussion, F.; Huintjes, E.; Li, S. Energy- and mass-balance comparison between Zhadang and Parlung No. 4 glaciers on the Tibetan Plateau. *J. Glaciol.* **2015**, *61*, 595–607.
24. Mölg, T.; Maussion, F.; Scherer, D. Mid-latitude westerlies as a driver of glacier variability in monsoonal High Asia. *Nat. Clim. Change* **2014**, *4*, 68. [CrossRef]
25. Skamarock, W.C.; Klemp, J.B.; Dudhia, J.; Gill, D.O.; Liu, Z.; Berner, J.; Wang, W.; Powers, J.G.; Duda, M.G.; Barker, D.; et al. *A Description of the Advanced Research WRF Model Version 4.3*; (No. NCAR/TN-556+STR); National Center for Atmospheric Research: Boulder, CO, USA, 2021. [CrossRef]
26. Mölg, T.; Kaser, G. A new approach to resolving climate-cryosphere relations: Downscaling climate dynamics to glacier-scale mass and energy balance without statistical scale linking. *J. Geophys. Res. Atmos.* **2011**, *116*, D16101. [CrossRef]
27. Collier, E.; Mölg, T.; Maussion, F.; Scherer, D.; Mayer, C.; Bush, A.B.G. High-resolution interactive modelling of the mountain glacier-atmosphere interface: An application over the Karakoram. *Cryosphere* **2013**, *7*, 779–795. [CrossRef]
28. Collier, E.; Maussion, F.; Nicholson, L.I.; Molg, T.; Immerzeel, W.W.; Bush, A.B.G. Impact of debris cover on glacier ablation and atmosphere-glacier feedbacks in the Karakoram. *Cryosphere* **2015**, *9*, 1617–1632. [CrossRef]
29. Bonekamp, P.N.J.; Collier, E.; Immerzeel, W.W. The Impact of Spatial Resolution, Land Use, and Spinup Time on Resolving Spatial Precipitation Patterns in the Himalayas. *J. Hydrometeorol.* **2018**, *19*, 1565–1581. [CrossRef]
30. Bonekamp, P.N.J.; De Kok, R.J.; Collier, E.; Immerzeel, W.W. Contrasting Meteorological Drivers of the Glacier Mass Balance Between the Karakoram and Central Himalaya. *Front. Earth Sci.* **2019**, *7*, 107. [CrossRef]
31. Mölg, T.; Maussion, F.; Yang, W.; Scherer, D. The footprint of Asian monsoon dynamics in the mass and energy balance of a Tibetan glacier. *Cryosphere* **2012**, *6*, 1445–1461. [CrossRef]
32. Aas, K.S.; Dunse, T.; Collier, E.; Schuler, T.V.; Berntsen, T.K.; Kohler, J.; Luks, B. The climatic mass balance of Svalbard glaciers: A 10-year simulation with a coupled atmosphere-glacier mass balance model. *Cryosphere* **2016**, *10*, 1089–1104. [CrossRef]
33. Pirazzini, R. Challenges in Snow and Ice Albedo Parameterizations. *Geophysica* **2009**, *45*, 41–62.
34. Mölders, N.; Luijting, H.; Sassen, K. Use of atmospheric radiation measurement program data from Barrow, Alaska, for evaluation and development of snow-albedo parameterizations. *Meteorol. Atmos. Phys.* **2008**, *99*, 199–219. [CrossRef]
35. Oerlemans, J.; Knap, W.H. A 1 year record of global radiation and albedo in the ablation zone of Morteratschgletscher, Switzerland. *J. Glaciol.* **1998**, *44*, 231–238. [CrossRef]

36. Pedersen, C.A.; Winther, J.G. Intercomparison and validation of snow albedo parameterization schemes in climate models. *Clim. Dyn.* **2005**, *25*, 351–362. [CrossRef]
37. Schmidt, L.S.; Aðalgeirsdóttir, G.; Guðmundsson, S.; Langen, P.L.; Pálsson, F.; Mottram, R.; Gascoin, S.; Björnsson, H. The importance of accurate glacier albedo for estimates of surface mass balance on Vatnajökull: Evaluating the surface energy budget in a regional climate model with automatic weather station observations. *Cryosphere* **2017**, *11*, 1665–1684. [CrossRef]
38. Dickinson, R.E.; Henderson-Sellers, A.; Kennedy, P.J. *Biosphere-Atmosphere Transfer Scheme (BATS) Version le as Coupled to the NCAR Community Climate Model*; Tech. Rep. NCAR/TN-387+STR; National Center for Atmospheric Research: Boulder, CO, USA, 1993; p. 174
39. Bonan, G.B. *A Land Surface Model (LSM Version 1.0) for Ecological, Hydrological, and Atmospheric Studies: Technical Description and User's Guide*; Tech. Rep. NCAR/TN-417+STR; National Center for Atmospheric Research: Boulder, CO, USA, 1996; 150p.
40. Warren, S.G.; Wiscombe, W.J. A Model for the Spectral Albedo of Snow. II: Snow Containing Atmospheric Aerosols. *J. Atmos. Sci.* **1980**, *37*, 2734–2745. [CrossRef]
41. Willeit, M.; Ganopolski, A. The importance of snow albedo for ice sheet evolution over the last glacial cycle. *Clim. Past* **2018**, *14*, 697–707. [CrossRef]
42. Usha, K.H.; Nair, V.S.; Babu, S.S. Modeling of aerosol induced snow albedo feedbacks over the Himalayas and its implications on regional climate. *Clim. Dyn.* **2020**, *54*, 4191–4210. [CrossRef]
43. Arndt, A.; Scherer, D.; Schneider, C. Atmosphere Driven Mass-Balance Sensitivity of Halji Glacier, Himalayas. *Atmosphere* **2021**, *12*, 426. [CrossRef]
44. Berrisford, P.; Dee, D.P.; Poli, P.; Brugge, R.; Fielding, K.; Fuentes, M.; Kallberg, P.W.; Kobayashi, S.; Uppala, S.; Simmons, A. *The ERA-interim Archive Version 2.0*; European Centre for Medium Range Weather Forecasts: Reading, UK, 2011.
45. Liu, L.; Gu, H.; Xie, J.; Xu, Y. How well do the ERA-Interim, ERA-5, GLDAS-2.1 and NCEP-R2 reanalysis datasets represent daily air temperature over the Tibetan Plateau? *Int. J. Climatol.* **2021**, *41*, 1484–1505. [CrossRef]
46. Niu, G.; Yang, Z.; Mitchell, K.E.; Chen, F.; Ek, M.B.; Barlage, M.; Kumar, A.; Manning, K.; Niyogi, D.; Rosero, E.; et al. The community Noah land surface model with multiparameterization options (Noah-MP): 1. Model description and evaluation with local-scale measurements. *J. Geophys. Res. Atmos.* **2011**, *116*, D12109. [CrossRef]
47. Yang, Z.; Niu, G.; Mitchell, K.E.; Chen, F.; Ek, M.B.; Barlage, M.; Longuevergne, L.; Manning, K.; Niyogi, D.; Tewari, M.; et al. The community Noah land surface model with multiparameterization options (Noah-MP): 2. Evaluation over global river basins. *J. Geophys. Res. Atmos.* **2011**, *116*, D12110. [CrossRef]
48. Niu, G.; Yang, Z. The effects of canopy processes on snow surface energy and mass balance. *J. Geophys. Res. Atmos.* **2004**, *109*, D23111. [CrossRef]
49. Liu, L.; Ma, Y.; Menenti, M.; Zhang, X.; Ma, W. Evaluation of WRF Modeling in Relation to Different Land Surface Schemes and Initial and Boundary Conditions: A Snow Event Simulation Over the Tibetan Plateau. *J. Geophys. Res. Atmos.* **2019**, *124*, 209–226. [CrossRef]
50. Vose, R.S.; Karl, T.R.; Easterling, D.R.; Williams, C.N.; Menne, M.J. Climate—Impact of land-use change on climate. *Nature* **2004**, *427*, 213–214. [CrossRef]
51. Verseghy, D.L. Class-A Canadian land surface scheme for GCMS. I. Soil model. *Int. J. Climatol.* **1991**, *11*, 111–133. [CrossRef]
52. Naegeli, K.; Damm, A.; Huss, M.; Wulf, H.; Schaepman, M.; Hoelzle, M. Cross-Comparison of Albedo Products for Glacier Surfaces Derived from Airborne and Satellite (Sentinel-2 and Landsat 8) Optical Data. *Remote Sens.* **2017**, *9*, 110. [CrossRef]
53. Yang, Z.; Cai, X.; Zhang, G.; Tavakoly, A.A.; Jin, Q.; Meyer, L.H.; Guan, X. The Community Noah Land Surface Model with Multi-Parameterization Options (Noah-MP). Technical Description. 2011. Available online: https://www.jsg.utexas.edu/noah-mp/downloads/ (accessed on 1 January 2022).
54. Dadic, R.; Mullen, P.C.; Schneebeli, M.; Brandt, R.E.; Warren, S.G. Effects of bubbles, cracks, and volcanic tephra on the spectral albedo of bare ice near the Transantarctic Mountains: Implications for sea glaciers on Snowball Earth. *J. Geophys. Res. Earth Surf.* **2013**, *118*, 1658–1676. [CrossRef]
55. Liang, S. Narrowband to Broadband conversions of land surface albedo: I. Algorithms. *Remote Sens. Environ.* **2000**, *76*, 213–238. [CrossRef]
56. Gou, J.; Wang, F.; Jin, K.; Mu, X.; Chen, D. More realistic land-use and vegetation parameters in a regional climate model reduce model biases over China. *Int. J. Climatol.* **2019**, *39*, 4825–4837. [CrossRef]

remote sensing

MDPI

Article

Optical Turbulence Characteristics in the Upper Troposphere–Lower Stratosphere over the Lhasa within the Asian Summer Monsoon Anticyclone

Kun Zhang [1,2], Feifei Wang [1,2], Ningquan Weng [1,2], Xiaoqing Wu [1,2], Xuebin Li [1,2] and Tao Luo [1,2,*]

1 Key Laboratory of Atmospheric Optics, Anhui Institute of Optics and Fine Mechanics, Hefei Institutes of Physical Science, Chinese Academy of Sciences, Hefei 230031, China
2 Advanced Laser Technology Laboratory of Anhui Province, Hefei 230037, China
* Correspondence: luotao@aiofm.ac.cn

Abstract: The high elevation, complex topography, and unique atmospheric circulations of the Tibetan Plateau (TP) make its optical turbulence characteristics different from those in low-elevation regions. In this study, the characteristics of the atmospheric refractive index structure constant (C_n^2) profiles in the Lhasa area at different strength states of the Asian summer monsoon anticyclone (ASMA) are analyzed based on precious in situ sounding data measured over the Lhasa in August 2018. C_n^2 in the upper troposphere–lower stratosphere fluctuates significantly within a few days during the ASMA, particularly in the upper troposphere. The effect of the ASMA on C_n^2 varies among the upper troposphere, tropopause, and lower stratosphere. The stronger and closer the ASMA is to Lhasa, the more pronounced is the "upper highs and lower lows" pressure field structure, which is beneficial for decreasing the potential temperature lapse rate. The decrease in static stability is an important condition for developing optical turbulence, elevating the tropopause height, and reducing the tropopause temperature. However, if strong high-pressure activity occurs at the lower pressure layer, such as at 500 hPa, an "upper highs and lower highs" pressure field structure forms over the Lhasa, increasing the potential temperature lapse rate and suppressing the convective intensity. Being almost unaffected by low-level atmospheric high-pressure activities, the ASMA, as the main influencing factor, mainly inhibits C_n^2 in the tropopause and lower stratosphere. The variations of turbulence intensity in UTLS caused by ASMA activities also have a great influence on astronomical parameters, which will have certain guiding significance for astronomical site testing and observations.

Keywords: Tibetan Plateau; optical turbulence; Asian summer monsoon anticyclone; upper troposphere-lower stratosphere

Citation: Zhang, K.; Wang, F.; Weng, N.; Wu, X.; Li, X.; Luo, T. Optical Turbulence Characteristics in the Upper Troposphere–Lower Stratosphere over the Lhasa within the Asian Summer Monsoon Anticyclone. *Remote Sens.* **2022**, *14*, 4104. https://doi.org/10.3390/rs14164104

Academic Editor: Michael E. Gorbunov

Received: 16 July 2022
Accepted: 18 August 2022
Published: 21 August 2022

1. Introduction

The unique features of the Tibetan Plateau (TP), such as its complex terrain formed by high mountains and valleys, dramatic changes in the atmospheric environment, differences in atmospheric composition, and unique geographical climate and circulation characteristics, form different atmospheric optical turbulence characteristics over the TP from those of low-elevation plain regions [1]. The TP is the major energy source providing sensible and latent heat fluxes to the atmosphere depending on the turbulence processes that occur during land–atmosphere interactions for mass and energy exchanges [2]. The combined effect of the complex terrain of the TP and the heat source enables the development of turbulence in the middle and upper atmosphere over the region.

Strong Asian summer monsoon circulations exist above the TP, including deep convective activities and planetary-scale anticyclones, such as the South Asian high, SAH (hereinafter referred to as the Asian summer monsoon anticyclone (ASMA)) [3,4]. The ASMA is stable and strong in the vertical direction at 70–300 hPa and occupies almost the entire upper troposphere–lower stratosphere (UTLS) area [5], which is closely related to plateau land-atmosphere heat transfer [6,7]. The ASMA is closely related to frequent convective activities, particularly during the period from June to September. The geographic location of the ASMA center varies over periods of a few days or even over longer periods [8]. The coupling of atmospheric circulation and convection that prevails over the TP during the summer results in the frequent occurrence of convective activities in the lower atmosphere. Less details are known about the influence of the ASMA on the thermodynamic structure of the atmosphere in the stratosphere.

The strong convective activities and ASMA on the TP affect the atmospheric components and their distribution in the UTLS of the Asian monsoon region by uplifting the lower atmosphere [9–11]. A turbulent atmosphere is an important transport medium in stratosphere–troposphere exchange (STE). First, convective injections can impact air and aerosol transport from the atmospheric boundary layer (ABL) to the UTLS [12,13]. In contrast, deep convection activities carry low concentrations of ozone and high concentrations of water vapor into the ASMA, which remain inside the ASMA for a period of time, and they are relatively isolated from the outside air and subsequently uploaded to the UTLS [14,15]. Moreover, the vertical distribution of atmospheric turbulence is one of the factors that must be inevitably considered in the astronomical site testing of ground-based astronomical optical telescopes [16]. Atmospheric turbulence is the main reason for the serious degradation of optical imaging quality, and it is also an important indicator for comparing the quality of astronomical sites [17,18].

TP has garnered significant attention as the third pole of the Earth. In the past three decades, researchers have conducted several atmospheric scientific experiments on the TP combined with numerical simulations. Some studies have been conducted mainly on the structure of the ABL over the TP and its surrounding areas, including studies on the occurrence and development of weather systems and the impact of the TP on atmospheric circulation [19,20].

However, the relationship between the optical turbulent structure and meteorological parameters in the UTLS has rarely been studied [21]. The results of observations and numerical simulations obtained in recent years indicate that the ASMA region surrounding Lhasa as the core area is an extremely important area, through which surface pollutants enter the global stratosphere. The transport of these materials into the stratosphere through atmospheric turbulence has important effects on the global climate and environment through microphysical, chemical, and radiative processes [22–24].

At present, most research on atmospheric turbulence structure in UTLS over the TP are studied based on model simulations or reanalysis, because in situ observed turbulent data are scarce. In consequence, details regarding the structure of atmospheric turbulence in the UTLS, STE process, and characteristics of atmospheric composition budgets are still lacking. The widely used measurement techniques of high-altitude atmospheric turbulence characteristics are roughly divided into two categories. One is the path-averaged turbulence intensity measurement technique, such as the optical triangulation method [16,25] and extension technology [26]; the other is the localized turbulence intensity measurement technique, such as the micro-thermal pulsation method. In situ turbulent measurement based on radiosonde is a simple but very reliable, effective, high-precision, and commonly used atmospheric detection method, particularly the micro-thermal sensors mounted on the balloon, which realizes the measurement of atmospheric turbulence in UTLS [27–30]. As such, not only basic atmospheric parameters such as temperature and humidity can be measured, but also the value of atmospheric refractive index structure constant C_n^2 in the middle and upper atmosphere is obtained simultaneously, which is an important parameter to measure turbulence intensity. This study analyzes the reasons for the large short-term fluctuations of C_n^2 in the Lhasa region from the aspects of atmospheric turbulence parameters, the ASMA, high-pressure activities at 500 hPa, and atmospheric circulation.

2. Methods

2.1. Sounding Data

From 3 to 18 August 2018, a thermal turbulent sounding experiment was conducted at the Lhasa Meteorological Bureau (91°06′E, 29°36′N) ("pentagram" in Figure 1) by the Hefei Institutes of Physical Science, Chinese Academy of Sciences. This experiment collected very precious high vertical resolution thermal turbulence sounding data, which provided a reliable basis for the study of the fine atmospheric structure in UTLS and the verification of model simulation over the TP [30]. The sounding balloons were equipped with conventional meteorological sensors to measure the atmospheric temperature (T), humidity, pressure (P), and wind speed, along with two-channels turbulent meteorological radiosondes developed by the Anhui Institute of Optics and Fine Mechanics.

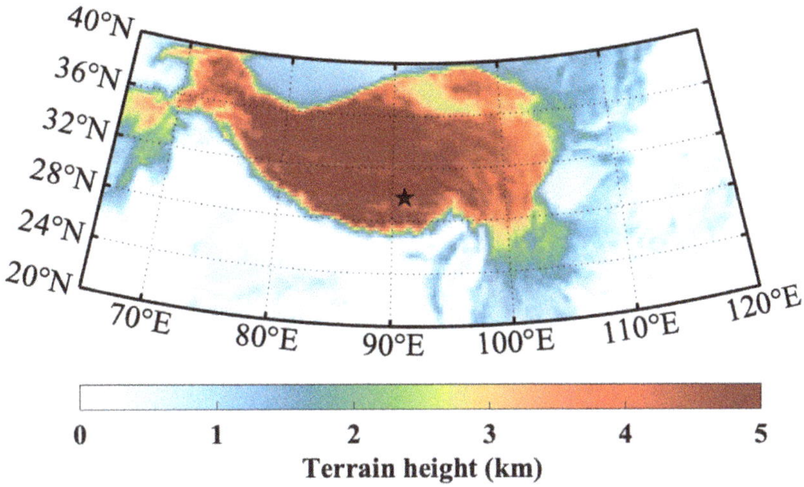

Figure 1. The elevation height of TP (color shaded) and the geographical location of Lhasa (pentagram). This figure was plotted using the Lambert conformal conic projection.

Each thermal turbulence radiosonde comprises two platinum wire probes (15 μm in diameter) separated by a distance of r (=1 m). The platinum wire probes have linear resistance–temperature coefficients. The thermal turbulence radiosondes measure the temperature difference between the distance r and voltage change between the two micro-thermal probes [28–30]. Then, the temperature structure constant (C_T^2) in the inertial subrange can be calculated using the following equation [31]:

$$\left\langle \left[T\left(\vec{x}\right) - T\left(\vec{x} + \vec{r}\right)\right]^2 \right\rangle = C_T^2 r^{\frac{2}{3}} \; (l_0 \ll r \ll L_0) \tag{1}$$

where $T\left(\vec{x}\right)$ and $T\left(\vec{x} + \vec{r}\right)$ denote the temperatures at two points, $\langle \cdots \rangle$ represents the ensemble average, and l_0 and L_0 represent the inner and outer scales of the turbulence, respectively (units of m).

C_n^2, the degree of refractive index fluctuation due to variations in atmospheric temperature and density [18,32], can be calculated by inputting temperature (T) and pressure (P) profiles, according to the relationship between C_n^2 and C_T^2 (Equation (2)):

$$C_n^2 = \left(79 \times 10^{-6} \frac{P}{T^2}\right)^2 C_T^2 \tag{2}$$

The range of the response frequency of the thermal turbulent radiosonde is 0.1–30 Hz, and the minimum measurable standard deviation of the temperature fluctuation does not exceed 0.002 °C. In addition, the vertical resolution of the radiosondes was 30 m.

Five thermal turbulence radiosondes were launched during the experiments at about 19:30 local time (LT). The detailed experimental records are summarized in Table 1. Owing to weather and transmission interference problems, four valid data sets were obtained over 20 km above sea level (ASL, the height below refers specifically to ASL except for special explanations) in height.

Table 1. Detailed records of radiosonde experiments over the Lhasa.

Radiosonde Number	Release Time (LT)	Release Altitude (m ASL)	Termination Altitude (m ASL)	Remark
1	12 August 2018 19:24:36	3653.6	21,810.0	Cloudy
2	13 August 2018 19:28:03	3654.1	23,221.1	Cloudy
3	14 August 2018 19:40:46	3652.8	30,658.8	Cloudy
4	15 August 2018 19:04:16	3658.2	29,956.9	Storm; micro-thermal sensors was destroyed
5	16 August 2018 19:20:15	3652.2	31,206.0	Cloudy

2.2. C_n^2 Integrated Parameters

C_n^2 integrated parameters (the Fried parameter r_0, seeing ε, isoplanatic angle θ_0) are of importance evaluation criteria for the astronomical site testing and the design of adaptive optics systems, defining as:

$$r_0 = [0.423\left(\frac{2\pi}{\lambda}\right)^2 sec\beta \int_0^\infty C_n^2(h)dh]^{-\frac{3}{5}} \quad (3)$$

$$\varepsilon = 5.25\lambda^{-\frac{1}{5}}\left[\int_0^\infty C_n^2(h)dh\right]^{-\frac{3}{5}} = 0.98\frac{\lambda}{r_0} \quad (4)$$

$$\theta_0 = 0.057\lambda^{\frac{6}{5}}\left[\int_0^\infty C_n^2(h)h^{\frac{5}{3}}dh\right]^{-\frac{3}{5}} \quad (5)$$

where, λ (=550 nm for this study) is a given wavelength for visible light, β is the zenith angle.

3. Results

3.1. ASMA Activities during the Experimental Period

SAH has significant multicenter characteristics [33], particularly bimodality [5,34], which is attributed to the warm preference of the SAH. Only the Tibetan mode is considered in this study. Figure 2 shows the distributions of the ECMWF geopotential height at 200-hPa pressure level on 12 August (Figure 2a), 13 (Figure 2b), 14 (Figure 2c), and 16 (Figure 2d), overlaid with the wind fields. The black square in Figure 2 denotes the location of Lhasa, whereas the black circle denotes the center of the SAH that is associated with the largest geopotential height. As shown in Figure 3a, the center of the SAH gradually moved northeastward during the experimental period. After 14 August, two SAH centers formed, with one located over the TP and the other over the Iranian Plateau.

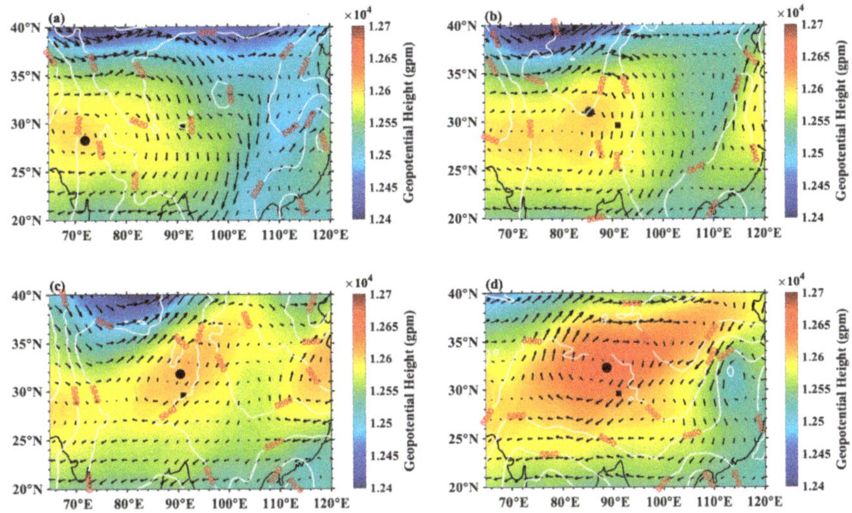

Figure 2. Distribution of the 200 hPa geopotential height (shaded), wind (vector), the 500 hPa geopotential height (contour), and the central region of the ASMA (the black square represents the geographic location of Lhasa, and the black dots represent the strongest negative vortex region. The same symbols are used below). The weather conditions during experimental periods were analyzed using the ERA-interim reanalysis data (http://www.ecmwf.int/, accessed on 17 January 2019) of the European Centre for Medium-Range Weather Forecasts (ECMWF). The horizontal resolutions of the meridional wind and relative vorticity were 2.5° × 2.5° and 0.25° × 0.25°, respectively. (**a**) 12 August 2018; (**b**) 13 August 2018; (**c**) 14 August 2018; and (**d**) 16 August 2018.

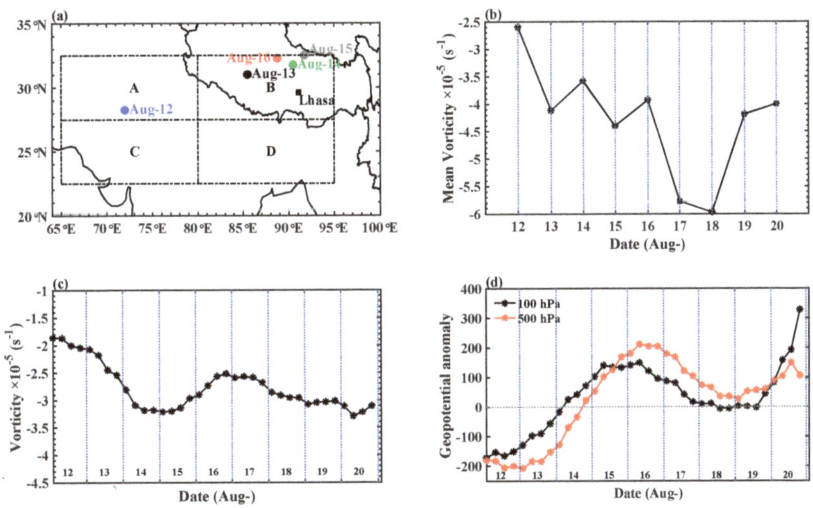

Figure 3. (**a**) The geographic locations of the ASMA centers from 12–16 August 2018. A-area (27.5–32.5°N, 65–80°E), B-area (27.5–32.5°N, 80–95°E), C-area (22.5–27.5°N, 65–80°E), and D-area (22.5–27.5°N, 80–95°E) were divided to identify the ASMA center (the black square represents the geographic location of Lhasa; the same symbols are used below); (**b**) the mean vorticity of 20 × 20 grids with the ASMA center considered as the geometric center, representing the strength of the ASMA; (**c**) the vorticity at 200 hPa over Lhasa from 12–20 August 2018; and (**d**) the geopotential height anomalies at 500 hPa and 100 hPa over Lhasa from 12–20 August 2018.

Figure 3 shows the development of the ASMA and its influence on Lhasa during the experimental period. The geolocation of the strongest ASMA center (dots in Figure 3a) is defined as the position of the greatest potential height in the strongest anticyclone area among the four areas (denoted as the A-area (27.5–32.5° N, 65–80° E), B-area (27.5–32.5° N, 80–95° E), C-area (22.5–27.5° N, 65–80° E), and D-area (22.5–27.5° N, 80–95° E)), according to the literatures [35,36]. In the physical sense, the criterion used to measure the strength of the ASMA is that the potential vorticity (PV) at the center of the anticyclone is smaller than that in the surrounding region [15,37,38]. The intensity of ASMA is shown in Figure 3b. The strength of the high-pressure system over the TP gradually increased with increasing geopotential height. Along with the development of a high-pressure system, the ASMA center moved to the TP after 12 August (onset), and its intensity was generally enhanced (13–15 August, early stage formation). After 16 August, the ASMA was fully established over the TP, with its intensity reaching the highest value.

The variations in relative vorticity at 200 hPa (Figure 3c) and geopotential height anomalies at 500 and 100 hPa over Lhasa (Figure 3d) are also presented. Along with the approach of the ASMA center toward Lhasa, the strength of the PV at 200 hPa over Lhasa increased, reached a maximum on 14 August, and thereafter gradually decreased. The anomalies at 100 hPa over Lhasa (Figure 3d) gradually increased from 12–15 August 2018, indicating that the transition from low to high pressure activity gradually occurred. Although the high-pressure activity at 500 hPa started to form with a one-day lag, it showed a more rapid growth trend and lasted longer. It can also be seen from the 500 hPa geopotential height field (Figure 2) that the high-pressure activity at 500 hPa gradually increased from the 13th, enveloping the entire TP on 16th. Notably, the geopotential height anomaly changed from negative to positive on 14 August, and the subsequent growing tendency was fiercer, which indicated that the high-pressure activity at 500 hPa became abnormally strong from 14 August.

3.2. Characteristics of C_n^2 under Different ASMA Strength States

We selected four representative profiles obtained at night on 12–14 and 16 August 2018, as shown in Figure 4, to analyze the variation characteristics of C_n^2 under different ASMA strength states over the Lhasa. There were strong maximum peaks of C_n^2 at 17–18 km (about 100 hPa). Certainly, a weak and thin maximum peak of Cn2 appeared at around 12 km (about 200 hPa), such as the 16 August 2018.

C_n^2 was largest on 13 August and smallest on 16 August. On 12 August, ASMA had a subtle impact on the C_n^2 profile over Lhasa. As the ASMA center approached Lhasa and its intensity increased, C_n^2 increased correspondingly on 13 and 14 August. Although these two days are in the middle stage of the ASMA, C_n^2 decreased on 14 August to a lower value than that recorded on 13 August in the range of ~15–20 km. With the departure and attenuated intensity of the ASMA, C_n^2 decreased rapidly on 16 August 2018.

In general, when the ASMA intensity was higher or the ASMA center was closer to Lhasa, a more pronounced "upper highs and lower lows" pressure field structure appeared over Lhasa. The stronger the convective activity, the greater the value of C_n^2. However, changes in the low-level pressure field, such as at 500 hPa, may have had a crucial impact on the vertical profile of C_n^2. A turning point occurred on 14 August, when the geopotential anomaly value changed from negative to positive, that is, the pressure field constructed in the UTLS changed from "upper highs and lower lows" to "upper highs and lower highs".

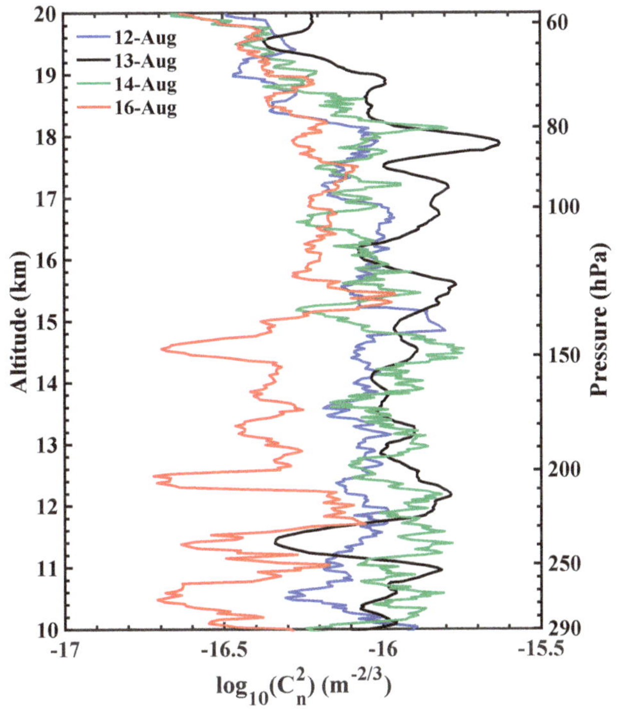

Figure 4. Vertical profiles of the atmospheric refractive index structure constant C_n^2 in the UTLS.

In comparison with that recorded on 16 August 2018, the high-pressure activity at the 500 hPa layer was weaker on 14 August, and the upward movement of the atmosphere was only slightly suppressed [37]. Therefore, C_n^2 on 14 August was higher than that on 16 August, but lower than that on 13 August.

3.3. Contribution of Atmospheric Turbulence in UTLS to the Total Integrated Parameters

The turbulent energy ratio (TER) in the range of 10–20 km, describing the contribution of atmospheric turbulence in this layer to ε and θ_0 of the total layer, can be calculated using the following equation [29]:

$$\mathrm{TER}_\varepsilon = \frac{\varepsilon(i)^{5/3}}{\varepsilon(\mathrm{total})^{5/3}} \times 100\% \quad (6)$$

$$\mathrm{TER}_{\theta_0} = \frac{\theta_0(i)^{5/3}}{\theta_0(\mathrm{total})^{5/3}} \times 100\% \quad (7)$$

where, *i* stands for 10–20 km turbulent layer.

Table 2 summarizes the integrated contribution of seeing (ε) and isoplanatic angle (θ_0) from the range of 10–20 km and the total integrated parameters. The atmospheric turbulence in the range of 10–20 km has a more significant proportion of $\varepsilon(\mathrm{total})$ (more than 60%) and $\theta_0(\mathrm{total})$ (more than 70%) over the Lhasa, which is consistent with the results of Gaomeigu site, Yunnan observatories, Chinese Academy of Science [29]. The $\varepsilon(i)$ ($\theta_0(i)$) differs 0.5″ (0.21″) between 13 and 16 August 2018, and TER_ε, as well as TER_{θ_0}, varies so widely (more than 10%), which indicates that the variations of C_n^2 under different ASMA strength states are related with the astronomical observations.

Table 2. The contribution of seeing (ε) and isoplanatic angle (θ_0) from the range of 10–20 km to total height layer.

		12 August 2018	13 August 2018	14 August 2018	16 August 2018
ε (″)	$\varepsilon(i)$	1.08	1.32	1.19	0.82
	$\varepsilon(\text{total})$	1.30	1.63	1.60	1.11
	TER_ε	73.73%	70.21%	61.09%	59.76%
θ_0 (″)	$\theta_0(i)$	0.44	0.35	0.41	0.56
	$\theta_0(\text{total})$	0.40	0.30	0.35	0.45
	TER_{θ_0}	85.58%	78.82%	77.84%	69.65%

4. Discussion

4.1. Potential Temperature Gradient

Equation (2) indicates that the most critical step in the parameterization scheme of atmospheric turbulence is how to parameterize C_T^2, and C_n^2 can be calculated logically. According to the dimensional analysis, Tatarski [31] defined the atmospheric temperature structure constant as follows:

$$C_T^2 = 1.6\varepsilon_\theta \varepsilon^{-\frac{1}{3}} \tag{8}$$

where ε_θ denotes the molecular diffusivity caused by temperature difference, and ε is the turbulent energy dissipation rate. The energy of atmospheric turbulence mainly originates from the dynamic and thermal effects. The former implies that when there is wind direction and wind speed shear, the turbulent shear stress works on the air micro-clusters, whereas the latter implies that in an unstable atmosphere, the buoyant force works on the air micro-clusters that move vertically to increase the turbulence.

Bougeault and Lacarrere [39] parameterized C_T^2 as follows:

$$C_T^2 = 0.59L^{4/3}\left(\frac{\delta\bar{\theta}}{\delta z}\right)^2 \varnothing_3 \tag{9}$$

where L denotes the Bougeault–Lacarrere mixing length, $\varnothing_3$ is the reversed turbulent Prandtl number, and θ is the potential temperature. Refer to the detailed derivation process in the literature [40].

$$L = \sqrt{\frac{2e}{\frac{g}{\theta}\frac{\delta\bar{\theta}}{\delta z}}} \tag{10}$$

where e denotes the turbulence energy. Parametric Equations (8)–(10) clearly indicate that, the potential temperature gradient is directly related to the buoyancy frequency, turbulent energy dissipation rate, and temperature structure constant. This is an indispensable and important parameter in the parameterization of atmospheric turbulence [41]. Therefore, under different ASMA intensities, the numerical changes in the potential temperature gradient can also reflect the thermal convection intensity.

4.2. Discussion on the Temperature Structure in UTLS

Radiosonde data were used to compare and analyze the atmospheric temperature structure from 12 to 16 August 2018. The temperatures above ~16 km fluctuated significantly during these days, where was stably controlled by ASMA. As shown in Figure 5b, the cold-point tropopause (CPT), which corresponds to the coldest temperature, was higher in the middle of the ASMA than in other stages, and this timing may have coincided with a decrease in static stability [42] and may be related to strong convective activities [43,44].

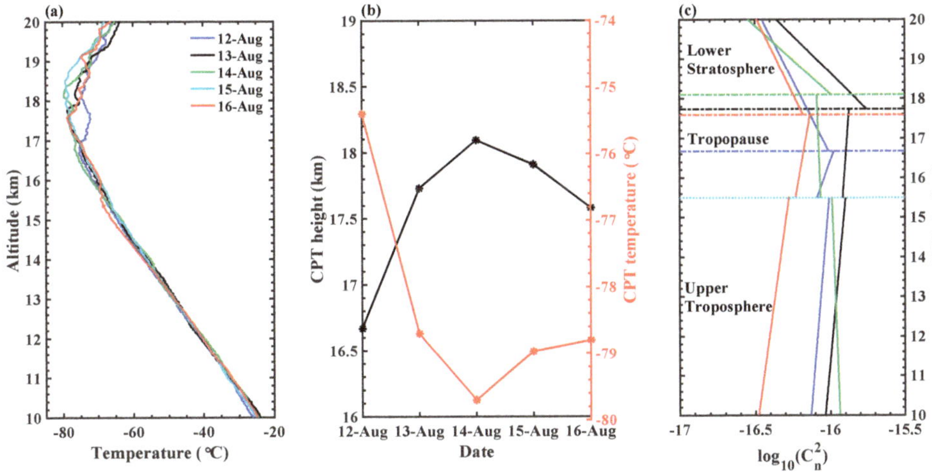

Figure 5. (**a**) Vertical temperature profiles from 12 to 16 August 2018. (**b**) Cold-point tropopause height (the black line) and temperature (the red line) from 12 to 16 August 2018. The lowest layer of the multi-tropopause serves as the CPT height. (**c**) First-order linear piecewise fitting to $\log_{10}(C_n^2)$ within the upper troposphere, tropopause, and lower stratosphere.

Table 3 summarizes the fitting results of $\log_{10}(C_n^2)$ using first-order linear fitting within the upper troposphere, tropopause, and lower stratosphere (Figure 5c), and the coefficients represent the increase rate of $\log_{10}(C_n^2)$, reaching respective maximum value at tropopause layer. However, when the ASMA intensity is large, the $\log_{10}(C_n^2)$ increase rate in the tropopause is weakened, and the $\log_{10}(C_n^2)$ in the lower stratosphere decreases.

Table 3. Increase rate of $\log_{10} C_n^2$ ($m^{7/3}$) within upper troposphere, tropopause, and lower stratosphere.

	12 August 2018	13 August 2018	14 August 2018	16 August 2018
Upper troposphere (10–15.5 km)	0.02	0.02	−0.01	0.04
Tropopause (15.5 km—CPT)	0.09	0.02	−0.01	0.05
Lower stratosphere (CPT—20 km)	−0.13	−0.27	−0.29	−0.13

During the experimental period, the potential temperature profiles varied significantly in the UTLS (Figure 6a), particularly in the upper troposphere. The potential temperature lapse rate (unit: K/km) within three heights in the range 10–16 km was fitted using the first-order linear piecewise method (Table 4). The potential temperature lapse rates in the ranges of 10–11 km (Figure 6b) and 11–12.5 km (Figure 6c) corresponded to the minimum and maximum recorded values on 13 and 16 August 2018, respectively. Under the control of the high-intensity ASMA on 14 August 2018, the potential temperature lapse rate was approximately equivalent to the values recorded on 12 and 16 August. In particular, in the 10–11 km range, the potential temperature lapse rate on 14 August was approximately 2.2 times as high as that on 13 August owing to high-pressure activity at 500 hPa. This shift suppressed the vertical flow of the atmosphere and enhanced the static stability of the UTLS.

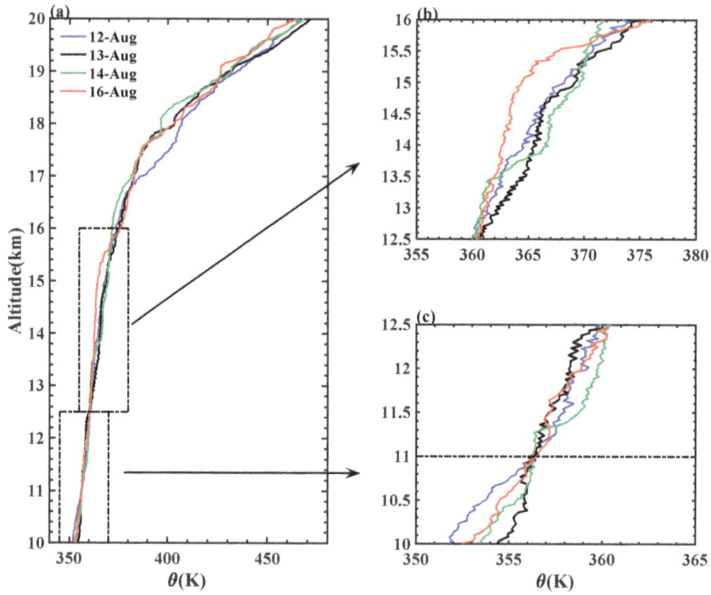

Figure 6. (**a**) Vertical profiles of potential temperature on 12–14 and 16 August 2018. (**b**,**c**) show magnified views of the 12.5–16 km and 10–12.5 km ranges, respectively.

Table 4. Potential temperature lapse rate (K/km) within the range of 10–16 km.

	12 August 2018	13 August 2018	14 August 2018	16 August 2018
10–11 km	4.84	1.48	3.29	3.48
11–12.5 km	2.19	2.03	2.96	2.6
12.5–16 km	3.79	3.65	3.73	3.24

In general, the potential temperature lapse rates for the four days were at the same amplitude within the range of 12.5–16 km. However, the trend of potential temperature on 16 August 2018 was significantly different from that on the other three days within 12.5–16 km. There was a weak thermal inversion layer in the range of 15–16 km on 16 August 2018. The inversion layer was able to block the upward movement of air [45,46], corresponding to a small C_n^2 value in the range of 10–15 km.

Piecewise linear fitting was performed on the potential temperature profiles within the tropopause and lower stratosphere regions at intervals of 1000 m. The CPT height of the TP was approximately 100 hPa, and the potential temperature lapse rate within 2 km below the CPT varied significantly among the four profiles. The potential temperature change rate on 14 August during the ASMA was twice as high as that on 12 and 16 August. This indicates that the stronger the ASMA, the greater the potential temperature lapse rate in tropopause. Above the CPT, C_n^2 cannot escape the fate of being affected by the ASMA, but thermal convection is inhibited in the lower stratosphere. The dynamic and thermal structures of the troposphere and stratosphere are completely different, and this difference is mainly characterized by high stability and weak turbulence in the stratosphere [14,47,48].

The tropopause is the mixed layer between the troposphere and stratosphere and it has dual characteristics of both the troposphere and stratosphere [49–51]. The CPT height corresponds to the minimum saturated water vapor mixing ratio, which is considered to be the upper boundary of the tropical tropopause [52]. Figure 7 shows that the potential temperature lapse rate is completely different between the tropopause and lower stratosphere, mainly showing that the potential temperature lapse rate increases sharply in the lower stratosphere. The turbulence characteristics of the lower stratosphere are hardly affected by the high-pressure activity over the 500-hPa layer, and the ASMA is the strongest influencing factor. Figure 2c shows that the Lhasa area was affected by the decreased ASMA on 12–14 and 16 August 2018. The potential temperature lapse rate in the tropopause area and above the CPT reached a maximum on 14 August and a minimum on 12 August 2018. The area 2 km lower than the CPT belongs to the tropopause mixed layer, and the potential temperature lapse rate in this area was largest on 14 August; the other three days did not differ extensively. It is inferred from the current data that the presence of the ASMA inhibits the vertical movement of the atmosphere in the lower stratosphere and tropopause. However, this conclusion requires additional data for verification.

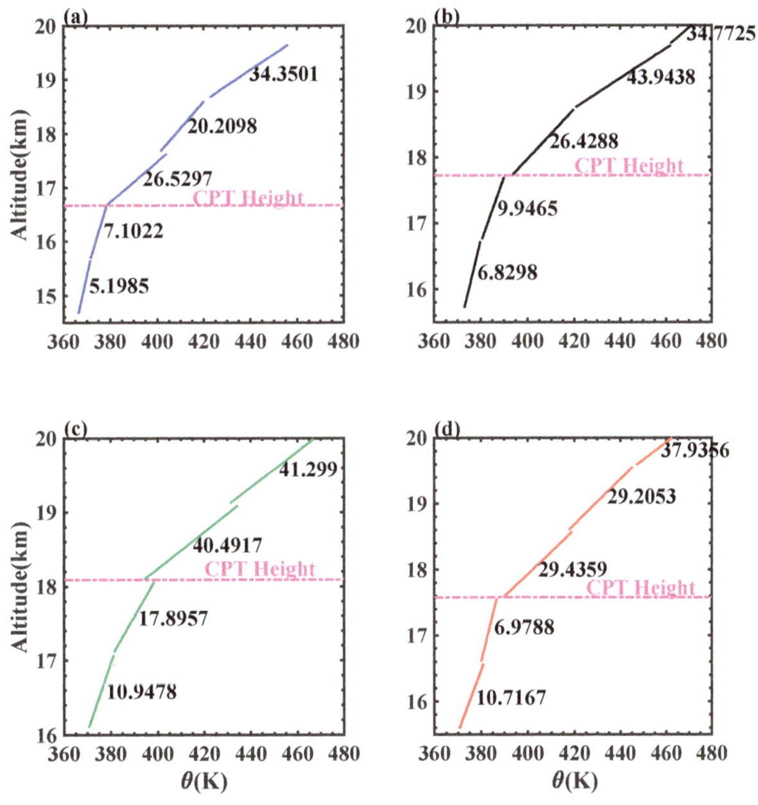

Figure 7. The piecewise fitting lines of potential temperature in tropopause and lower stratosphere. The potential temperature profiles are linearly fitted with the least-square method at intervals of 1000 m. The numbers indicate the coefficients (unit: K/km) of the piecewise fitting, indicating the potential temperature lapse rates. The pink dotted lines represent the corresponding CPT heights. (**a**) 12 August 2018, (**b**) 13 August 2018, (**c**) 14 August 2018, and (**d**) 16 August 2018.

5. Conclusions

In this study, the impacts of the ASMA and high-pressure activities in the 500 hPa layer on C_n^2 were analyzed under different ASMA strength states over the Lhasa during the summer based on precious in situ sounding data.

The atmospheric refractive index structure constant C_n^2 characterizes the optical turbulence intensity, which is directly affected by the atmospheric temperature. The ASMA is a warm, high-pressure system in the upper troposphere that causes the TP to be a strong heat source during the summer, heating the air over the TP. The upper atmosphere diverged to form a high-pressure circulation system, and the lower atmosphere converged to form a low-pressure circulation system. The "upper highs and lower highs" pressure structure enhances the potential temperature lapse rate, which is conducive for the reduction in static stability and development of optical turbulence in the UTLS. However, once strong high-pressure activity exists in the lower pressure layer, such as at 500 hPa, the high-pressure system is dominant from the 500- to 100-hPa layer, constituting an "upper highs and lower highs" pressure field structure. In comparison with the "upper highs and lower lows" pressure field structure observed in most cases, this particular pressure field structure suppresses the vertical potential temperature lapse rate and vertical upward movement, and weakens the atmospheric convective activity. Under the combined action of the ASMA and low-pressure activity over 500 hPa, the potential temperature lapse rate decreased rapidly, and C_n^2 increased by an order of magnitude in the upper-troposphere.

The situations in both the tropopause and lower stratosphere are different from those in the upper troposphere, where atmosphere is almost unaffected by high-pressure activities at 500 hPa. The difference in the potential temperature lapse rate caused by the ASMA is particularly manifested in the region adjacent to the tropopause. The best evidence is that the potential temperature lapse rate in high-intensity ASMA situations is twice as high as that in low-intensity states. In other words, the potential temperature gradient can not only reflect the static structure of the atmosphere represented by buoyancy frequency, but also qualitatively analyze the variation tendency of C_n^2. Under different ASMA intensities, the potential temperature lapse rate is consistent with the variation tendency of C_n^2, and the profile on 13 August 2018, was the most evident.

The tropopause height over the TP is close to the 100-hPa layer, corresponding to the scope of activity of the ASMA. The tropopause height is closely related to the turbulence intensity. Strong turbulence elevates the CPT to a higher position, and the CPT temperature is lowered [42,53]. When the impact of the ASMA is greater, the CPT height rises by approximately 1.5 km.

In general, the ASMA has different mechanisms of influence on the atmospheric refractive index structure constant C_n^2 in the upper troposphere, tropopause, and lower stratosphere. It was found that during high-intensity ASMA, turbulent activity in the tropopause and lower stratosphere (in the upper-troposphere) is suppressed (promoted), which may be not conductive to the STE process (astronomical observations).

The extent of the promotion of convection by ASMA is not only related to the position of its center and strength but is also inseparable from the high-pressure activities of the lower atmosphere. Clearly, because of the limited radiosonde data considered in this study, determining whether our discussion is regional and limited and the reasons for the short-term C_n^2 fluctuations necessitate further exploration and analyses with various and abundant data, such as Stereo-SCIDAR measurements [26].

Author Contributions: K.Z.: methodology, software, data analysis, writing—original draft. F.W.: investigation. X.W.: data curation, project administration. N.W., X.L.: writing—review and editing. T.L.: project administration, funding acquisition, formal analysis, conceptualization, writing—review and editing. All authors have read and agreed to the published version of the manuscript.

Funding: This research was funded by the Strategic Priority Research Program of Chinese Academy of Sciences (Grant No. XDA17010104), and the National Natural Science Foundation of China (Grant Nos. 4157685 and 91752103).

Data Availability Statement: In this section, the data underlying this article cannot be shared publicly due to the confidentiality requirements of the project in the study.

Acknowledgments: Thanks to the reanalysis data provided by the European Centre for Medium-Range Weather Forecasts (ECMWF) for the purposes of this study.

Conflicts of Interest: The authors declare no conflict of interest.

References

1. Lv, Y.; Guo, J.; Li, J.; Cao, L.; Chen, T.; Wang, D.; Chen, D.; Han, Y.; Guo, X.; Xu, H.; et al. Spatiotemporal characteristics of atmospheric turbulence over China estimated using operational high-resolution soundings. *Environ. Res. Lett.* **2021**, *16*, 054050. [CrossRef]
2. Li, M.; Ma, Y.; Ma, W.; Hu, Z.; Ishikawa, H.; Zhongbo, S.; Sun, F. Analysis of turbulence characteristics over the northern Tibetan Plateau area. *Adv. Atm. Sci.* **2006**, *23*, 579–585. [CrossRef]
3. Highwood, E.J.; Hoskins, B.J. The tropical tropopause. *Q. J. R. Meteorol. Soc.* **1998**, *124*, 1579–1604. [CrossRef]
4. Hoskins, B.J.; Rodwell, M.J. A model of the Asian Summer Monsoon: 1. The global-scale. *J. Atmos. Sci.* **1995**, *52*, 1329–1340. [CrossRef]
5. Randel, W.J.; Park, M. Deep convective influence on the Asian summer monsoon anticyclone and associated tracer variability observed with Atmospheric Infrared Sounder (AIRS). *J. Geophys. Res. Atmos.* **2006**, *111*, D12314. [CrossRef]
6. Li, H.; Zhou, L.; Wang, G. The Observed Impact of the South Asian Summer Monsoon on Land-Atmosphere Heat Transfers and Its Inhomogeneity over the Tibetan Plateau. *Remote Sens.* **2022**, *14*, 3236. [CrossRef]
7. Zhao, X.; Liu, C.; Yang, N.; Li, Y. Diurnal and seasonal variations of surface energy and CO2 fluxes over a site in Western Tibetan Plateau. *Atmos* **2020**, *11*, 260. [CrossRef]
8. Wu, G.; He, B.; Liu, Y.; Bao, Q.; Ren, R. Location and variation of the summertime upper-troposphere temperature maximum over South Asia. *Clim. Dyn.* **2015**, *45*, 2757–2774. [CrossRef]
9. Bian, J.; Li, D.; Bai, Z.; Li, Q.; Lyu, D.; Zhou, X. Transport of Asian surface pollutants to the global stratosphere from the Tibetan Plateau region during the Asian summer monsoon. *Natl. Sci. Rev.* **2020**, *7*, 516–533. [CrossRef]
10. Sun, Y.; Chen, Q. Variation of Atmospheric Composition in UTLS during a Strong Convection Process in Tibetan Plateau. *Meteorol. Sci. Technol.* **2017**, *45*, 1083–1089.
11. Tomsche, L.; Pozzer, A.; Ojha, N.; Parchatka, U.; Lelieveld, J.; Fischer, H. Upper tropospheric CH4 and CO affected by the South Asian summer monsoon during the Oxidation Mechanism Observations mission. *Atmos. Chem. Phys.* **2019**, *19*, 1915–1939. [CrossRef]
12. Bergman, J.W.; Jensen, E.J.; Pfister, L.; Yang, Q. Seasonal differences of vertical-transport efficiency in the tropical tropopause layer: On the interplay between tropical deep convection, large-scale vertical ascent, and horizontal circulationes. *J. Geophys. Res. Atmos.* **2012**, *117*, D05302. [CrossRef]
13. Fadnavis, S.; Semeniuk, K.; Pozzoli, L.; Schultz, M.G.; Ghude, S.D.; Das, S.; Kakatkar, R. Transport of aerosols into the UTLS and their impact on the Asian monsoon region as seen in a global model simulation. *Atmos. Chem. Phys.* **2013**, *13*, 8771–8786. [CrossRef]
14. Chen, H.; Bian, J.; Lv, D. Advances and Prospects in the Study of Stratosphere-Troposphere Exchange. *Chin. J. Atmos. Sci.* **2006**, *30*, 813–820.
15. Ploeger, F.; Gottschling, C.; Griessbach, S.; Grooss, J.U.; Guenther, G.; Konopka, P.; Mueller, R.; Riese, M.; Stroh, F.; Tao, M.; et al. A potential vorticity-based determination of the transport barrier in the Asian summer monsoon anticyclone. *Atmos. Chem. Phys.* **2015**, *15*, 13145–13159. [CrossRef]
16. Butterley, T.; Wilson, R.W.; Sarazin, M. Determination of the profile of atmospheric optical turbulence strength from SLODAR data. *Mon. Not. R. Astron. Soc.* **2006**, *369*, 835–845. [CrossRef]
17. Hach, Y.; Jabiri, A.; Ziad, A.; Bounhir, A.; Sabil, M.; Abahamid, A.; Benkhaldoun, Z. Meteorological profiles and optical turbulence in the free atmosphere with NCEP/NCAR data at Oukaimeden—I. Meteorological parameters analysis and tropospheric wind regimes. *Mon. Not. R. Astron. Soc.* **2012**, *420*, 637–650. [CrossRef]
18. Masciadri, E.; Jabouille, P. Improvements in the optical turbulence parameterization for 3D simulations in a region around a telescope. *Astron. Astrophys.* **2001**, *376*, 727–734. [CrossRef]
19. Wang, Y.; Xu, X.; Zhao, T.; Sun, J.; Yao, W.; Zhou, M. Structures of convection and turbulent kinetic energy in boundary layer over the southeastern edge of the Tibetan Plateau. *Sci. China Earth Sci.* **2015**, *58*, 1198–1209. [CrossRef]
20. Yang, K.; Koike, T.; Fujii, H.; Tamura, T.; Xu, X.D.; Bian, L.G.; Zhou, M.Y. The daytime evolution of the atmospheric boundary layer and convection over the Tibetan Plateau: Observations and simulations. *J. Meteorol. Soc. Jpn* **2004**, *82*, 1777–1792. [CrossRef]
21. Chen, X.L.; Ma, Y.M.; Kelder, H.; Su, Z.; Yang, K. On the behaviour of the tropopause folding events over the Tibetan Plateau. *Atmos. Chem. Phys.* **2011**, *11*, 5113–5122. [CrossRef]
22. Fadnavis, S.; Kalita, G.; Kumar, K.R.; Gasparini, B.; Li, J.-L.F. Potential impact of carbonaceous aerosol on the upper troposphere and lower stratosphere (UTLS) and precipitation during Asian summer monsoon in a global model simulation. *Atmos. Chem. Phys.* **2017**, *17*, 11637–11654. [CrossRef]
23. Gu, Y.; Liao, H.; Xu, J.; Zhou, G. The chemical effects on the summertime ozone in the upper troposphere and lower stratosphere over the Tibetan Plateau and the South Asian monsoon region. *Meteorol. Atmos. Phys.* **2019**, *131*, 431–441. [CrossRef]
24. Randel, W.J.; Park, M.; Emmons, L.; Kinnison, D.; Bernath, P.; Walker, K.A.; Boone, C.; Pumphrey, H. Asian monsoon transport of pollution to the stratosphere. *Science* **2010**, *328*, 611–613. [CrossRef]
25. Kovadlo, P.G.; Shikhovtsev, A.Y.; Kopylov, E.A.; Kiselev, A.V.; Russkikh, I.V. Study of the Optical Atmospheric Distortions using Wavefront Sensor Data. *RuPhJ* **2021**, *63*, 1952–1958. [CrossRef]
26. Shepherd, H.W.; Osborn, J.; Wilson, R.W.; Butterley, T.; Avila, R.; Dhillon, V.S.; Morris, T.J. Stereo-SCIDAR: Optical turbulence profiling with high sensitivity using a modified SCIDAR instrument. *Mon. Not. R. Astron. Soc.* **2014**, *437*, 3568–3577. [CrossRef]

27. Hummel, J.R.; Hazen, D.A.; Beland, R.R. Modeling optical turbulence parameters with high resolution radiosonde data. In Proceedings of the Conference on Lasers and Electro-Optics, Baltimore, Maryland, 24 April 1989; p. THK46.
28. Wu, X.; Huang, H.; Qian, X.; Wang, P.; Cui, C. Thermosonde meaurement of temperature structure parameter and temperature spectral power-law exponent profile in the lower stratosphere. *Acta Opt. Sin.* **2014**, *34*, 0501001.
29. Xiaoqing, W.; Xianmei, Q.; Honghua, H.; Ping, W.; Chaolong, C.; Chun, Q. Measurment of seeing, isoplanatic angle, and coherence time by using balloon-borne microthermal probes at Gaomeigu. *Acta Astron. Sin.* **2014**, *55*, 144–153. [CrossRef]
30. Qing, C.; Wu, X.; Li, X.; Luo, T.; Su, C.; Zhu, W. Mesoscale optical turbulence simulations above Tibetan Plateau: First attempt. *Opt. Express* **2020**, *28*, 4571–4586. [CrossRef]
31. Tatarski, V.I. *Wave Propagation in a Turbulent Medium*; McGraw-Hill: New York, NY, USA, 1961.
32. Chen, J.; Chen, X.; Jing, X.; Jia, W.; Yu, Y.; Zhao, S. Ergodicity of turbulence measurements upon complex terrain in Loess Plateau. *Sci. China Earth Sci.* **2021**, *64*, 37–51. [CrossRef]
33. Peng, L.; Sun, Z.; Chen, H.; Zhu, W.; Zeng, G.; Ni, D. Analysis on the Multi-center Structure of Summer South Asia High and Its Thermal Influence Factors. *Chin. J. Atmos. Sci.* **2016**, *40*, 1089–1106.
34. Zhang, Q.; Wu, G.X.; Qian, Y.F. The bimodality of the 100 hPa South Asia High and its relationship to the climate anomaly over East Asia in summer. *J. Meteorol. Soc. Jpn.* **2002**, *80*, 733–744. [CrossRef]
35. Choi, K.-S.; Kim, B.-J.; Zhang, R.; Nam, J.-C.; Park, K.-J.; Kim, J.-Y.; Kim, D.-W. Possible influence of South Asian high on summer rainfall variability in Korea. *Clim. Dyn.* **2016**, *46*, 833–846. [CrossRef]
36. Wei, W.; Zhang, R.; Wen, M.; Kim, B.-J.; Nam, J.-C. Interannual Variation of the South Asian High and Its Relation with Indian and East Asian Summer Monsoon Rainfall. *J. Clim.* **2015**, *28*, 2623–2634. [CrossRef]
37. Duan, T. The characteristics of the atmospheric heating and the atmospheric circulation in southeast Asia during the active periods of 500hPa high pressure over the Qinghai-Xizang plateau in summer. *Acta Meteorol. Sin.* **1994**, *52*, 194–200.
38. Garny, H.; Randel, W.J. Dynamic variability of the Asian monsoon anticyclone observed in potential vorticity and correlations with tracer distributions. *J. Geophys. Res. Atmos.* **2013**, *118*, 13421–13433. [CrossRef]
39. Bougeault, P.; Lacarrere, P. Parameterization of orography-induced turbulence in a mesobeta-scale model. *Mon. Weather Rev.* **1989**, *117*, 1872–1890. [CrossRef]
40. Wang, H.; Yao, Y.; Liu, L. A review of atmoshperic optical turbulence modeling research. *Prog. Astron.* **2012**, *30*, 16. [CrossRef]
41. Wu, S.; Wu, X.; Su, C.; Yang, Q.; Xu, J.; Luo, T.; Huang, C.; Qing, C. Reliable model to estimate the profile of the refractive index structure parameter (Cn2) and integrated astroclimatic parameters in the atmosphere. *Opt. Express* **2021**, *29*, 12454–12470. [CrossRef]
42. Alappattu, D.P.; Kunhikrishnan, P.K. First observations of turbulence parameters in the troposphere over the Bay of Bengal and the Arabian Sea using radiosonde. *J. Geophys. Res. Atmos.* **2010**, *115*, D6. [CrossRef]
43. Feng, S.; Fu, Y.; Xiao, Q. Is the tropopause higher over the Tibetan Plateau? Observational evidence from Constellation Observing System for Meteorology, Ionosphere, and Climate (COSMIC) data. *J. Geophys. Res. Atmos.* **2011**, *116*, D21121. [CrossRef]
44. Yang, S.; Zhou, S. Review of researches on tropopause in recent 30 years. *Meteorol. Sci. Technol.* **2010**, *38*, 145–151. [CrossRef]
45. Kandil, H.A.; Abdelkader, M.M.; Moaty, A.A.; Elhadidi, B.; Sherif, A.O. Simulation of atmospheric temperature inversion over greater CAIRO using the MMS MESO-SCALE atmospheric model. *Egypt J. Remote Sens. Space Sci.* **2006**, *9*, 15–30.
46. Liou, Y.A.; Yan, S.K. Two-year Microwave Radiometric Observations of Low-level Boundary-layer Temperature Inversion Signatures. In Proceedings of the IEEE International Conference on Geoscience & Remote Sensing Symposium, Denver, CO, USA, 31 July 2006–4 August 2006.
47. Gettelman, A.; Forster, P.M.D. A climatology of the tropical tropopause layer. *J. Meteorol. Soc. Jpn.* **2002**, *80*, 911–924. [CrossRef]
48. Schmidt, T.; Wickert, J.; Beyerle, G.; Reigber, C. Tropical tropopause parameters derived from GPS radio occultation measurements with CHAMP. *J. Geophys. Res. Atmos.* **2004**, *109*, D13105. [CrossRef]
49. Fischer, H.; Wienhold, F.G.; Hoor, P.; Bujok, O.; Schiller, C.; Siegmund, P.; Ambaum, M.; Scheeren, H.A.; Lelieveld, J. Tracer correlations in the northern high latitude lowermost stratosphere: Influence of cross-tropopause mass exchange. *Geophys. Res. Lett.* **2000**, *27*, 97–100. [CrossRef]
50. Hoor, P.; Fischer, H.; Lange, L.; Lelieveld, J.; Brunner, D. Seasonal variations of a mixing layer in the lowermost stratosphere as identified by the CO-O3 correlation from in situ measurements. *J. Geophys. Res.* **2002**, *107*, ACL1-1–ACL1-11.
51. Zahn, A.; Brenninkmeijer, C.A.M.; Maiss, M.; Scharffe, D.H.; Crutzen, P.J.; Hermann, M.; Heintzenberg, J.; Wiedensohler, A.; Gusten, H.; Heinrich, G.; et al. Identification of extratropical two-way troposphere-stratosphere mixing based on CARIBIC measurements of O-3, CO, and ultrafine particles. *J. Geophys. Res. Atmos.* **2000**, *105*, 1527–1535. [CrossRef]
52. Dessler, A.E.; Palm, S.P.; Spinhirne, J.D. Tropical cloud-top height distributions revealed by the Ice, Cloud, and Land Elevation Satellite (ICESat)/Geoscience Laser Altimeter System (GLAS). *J. Geophys. Res. Atmos.* **2006**, *111*, D12215. [CrossRef]
53. Khan, A.; Jin, S. Effect of gravity waves on the tropopause temperature, height and water vapor in Tibet from COSMIC GPS Radio Occultation observations. *J. Atmos. Solar-Terr. Phys.* **2016**, *138*, 23–31. [CrossRef]

remote sensing

MDPI

Article

Vertical Structures Associated with Orographic Precipitation during Warm Season in the Sichuan Basin and Its Surrounding Areas at Different Altitudes from 8-Year GPM DPR Observations

Chengfeng Shen [1,2], Guoping Li [1,3,*] and Yuanchang Dong [4]

1 School of Atmospheric Sciences, Chengdu University of Information Technology, Chengdu 610225, China
2 Fujian Meteorological Observatory, Fujian Meteorological Bureau, Fuzhou 350028, China
3 Collaborative Innovation Center on Forecast and Evaluation of Meteorological Disasters, Nanjing University of Information Science & Technology, Nanjing 210044, China
4 Institute of Plateau Meteorology, China Meteorological Administration (CMA)/Heavy Rain and Drought-Flood Disaster in Plateau and Basin Key Laboratory of Sichuan Province, Chengdu 610072, China
* Correspondence: liguoping@cuit.edu.cn; Tel.: +86-28-8596-6386

Abstract: Global precipitation measurement (GPM) is one of the effective means employed to observe orographic precipitation, and its inversed GPM DPR data can be used to study the microphysical structure of precipitation particles. This study considers statistics on convective precipitation (CP) and stratiform precipitation (SP) events over three types of terrain (plains, mountains, and high mountains) using the DPR onboard the GPM Core Observatory from May to September of 2014–2021 to analyze the vertical structure of heavy CP and SP. In mountain areas and high mountain areas, the updraft rendered by topography or seeder-feeder mechanism is not only conducive to the collision and merger of raindrops into large raindrops, but also increases the concentration of small drops, which is the main reason why the occurrence probability of not only large but also small raindrops increases and the horizontal distribution domain of mass weighted average raindrop diameter (D_m) widens. For heavy SP, the occurrence probability of medium-diameter precipitation particles below the freezing height (FzH) over high mountains is greater than those over plains. The precipitation particles above 10 km altitude of high mountains have characteristics, such as lower droplet number concentration and larger diameter, compared with those over plains. Furthermore, the study also analyzed the correlation between storm top altitude (STA) and D_m, water vapor and STA respectively. This study is helpful to further understand the effect of topography on heavy precipitation through cloud microphysical processes and the vertical structure of precipitation.

Keywords: heavy precipitation; GPM; dual frequency spaceborne radar; vertical structure; microphysical characteristic

Citation: Shen, C.; Li, G.; Dong, Y. Vertical Structures Associated with Orographic Precipitation during Warm Season in the Sichuan Basin and Its Surrounding Areas at Different Altitudes from 8-Year GPM DPR Observations. *Remote Sens.* **2022**, *14*, 4222. https://doi.org/10.3390/rs14174222

Academic Editors: Massimo Menenti, Yaoming Ma, Li Jia and Lei Zhong

Received: 26 July 2022
Accepted: 24 August 2022
Published: 27 August 2022

1. Introduction

Precipitation types can be mainly classified as stratiform precipitation (SP) and convective precipitation (CP) [1]. There are significant differences between SP and CP regarding the growth of precipitation particles by aggregation, riming, and coalescence, which is due to the different thermal dynamics and microphysics processes of the two types of precipitation [2,3]. The vertical structure of precipitation can reflect the characteristics of dynamics and microphysical of hydrometeor growth attenuation in the precipitation clouds. These microphysical and thermodynamic processes can affect the precipitation efficiency, and the intensity of surface precipitation, likewise, plays a role in determining precipitation type to some extent [4–7]. The topography has a very important influence on the vertical structure of precipitation and clouds [8–10], manifested in that topographical thermodynamic and dynamic processes affects atmospheric circulation, thereby the initiation and development of rainfall systems was significantly affected.

Sichuan province is located in the interior of southwest China, with the Qinling Mountains to the north, the Yun-Gui Plateau to the south, and the eastern edge of the Qinghai-Tibet Plateau. The terrain of Sichuan is high west and low east that the topography is diverse, and the weather processes are also complex and changeful. Heavy rainfall is one of the meteorological disasters with the highest frequency and the severest personal casualty and property loss in the Sichuan Basin and its surrounding areas. This region not only has high annual average precipitation, but is also prone to short-time heavy precipitation because of the complex terrain. In the past, research on mountain precipitation was difficult due to scarce surface observational data. At the same time, the detection of weather radar in mountainous areas is limited by terrain [11]. The Global Precipitation Measurement (GPM) mission was initiated by the National Aeronautics and Space Administration (NASA) and the Japan Aerospace and Exploration Agency (JAXA), which is the successor to Tropical Rainfall Measuring Mission (TRMM). The GPM observatory carries the first spaceborne dual-frequency phased array precipitation radar (DPR), developed by JAXA and National Institute of Communication Technology (NICT), and a conical-scanning multichannel microwave imager, developed and built by the Ball Aerospace and Technology Corporation under contract with NASA's Goddard Space Flight Center (GSFC). The GPM sensor has higher sensitivity and wider measurement range compared with TRMM instruments, and can provide more accurate precipitation microstructure information [12–14]. The detection of precipitation radar (PR) from satellite is not restricted to the geographical environment. Thus, it is feasible for PR to monitor cloud clusters on the distant seas and oceans, the plateau, or mountains where ground-based PR is hard to deploy, which can effectively make up for the deficiency of ground-based PR. In addition, the large precipitation particles are usually located in the middle and lower layers of cloud cluster, in the upper layer of cloud cluster, radar beam form satellite attenuated more lightly than that from ground-based because of the work type of satellite PR detection is top-down, which is available for obtaining the structural information of the upper layer of cloud cluster [15].

The reliability of GPM DPR data has been evaluated and verified by many scholars. Liao et al. [16] conducted a physical evaluation of the rain profiling retrieval algorithms for DPR on board the GPM Core Observatory satellite and proved that the DPR dual-wavelength algorithm can generally provide accurate rainfall rate. Jin et al. [17] evaluated the applicability of the GPM data in Mount Taishan area, and the results showed that the GPM had the highest accuracy in the mountainous area that could estimate the precipitation system with more accurate accuracy and lower error. Zhang et al. [18] revealed that the GPM DPR inversion product was more capable of revealing structures of both strong and light precipitation through individual cases and statistical analysis. Lasser et al. [19] compared the precipitation observation data measured by GPM DPR with that measured by ground weather stations, and the result showed that the precipitation observation data measured by GPM DPR was basically consistent with the data measured by ground meteorological station. Sun et al. [20] compared the GPM DPR data with the measured results of one-dimensional (1-D) laser-optical particle size velocity (PARSIVEL) disdrometers over the Yangtze-Huai River Valley in central China, and found that the measured results were similar, the mean deviation was relatively small, and the skewness was close to zero. 16 laser-optical PARSIVEL disdrometers deployed in Sichuan were used to validate the GPM-DPR results by Li et al. [21] verified that the DPR data in this area are basically credible. However, it is also mentioned that due to the complex topography of Sichuan Province and the difference of measurement principles between DPR and disdrometers, it is difficult to have identical and ideal conditions to evaluate DPR data by PARSIVEL disdrometers. All of the above comparative evaluation works show that the GPM DPR data are very reliable.

Drop size distribution (DSD) and its spatial variability is not only essential in understanding the microphysical processes that occur at the different stages of a precipitating system, but also useful in microwave communication, soil erosion and landslide triggering studies [22–24]. Seela et al. [25] studied the DSD variability in summer season rainfall between two islands in the western Pacific region and the results indicated that both orography and aerosol loading are responsible for the spatial variability of DSD. Harikumar et al. [26] made a comparative study on the data of four tropical stations in the peninsular India and found that orographic rain has larger drops when compared with non-orographic rain when the rain rate is high. Zwiebel et al. [27] studied the impact of complex terrain located in France on the structure of rainfall and mentioned that the evolution of the DSD over the transition and plain areas is dominated by coalescence and evaporation.

In this study, DPR observations are used to analyze the radar reflectivity factors and the vertical structural characteristic of DSD of SP and CP over different terrain regions (plains, mountains, high mountains) in the Sichuan Basin and its surrounding areas, expecting to obtain the characteristics of precipitation structure, which will help in further understanding the influence of mountain topography on precipitation structure and internal microphysical processes, and it also plays a very important role in deepening scientific recognition of mountain heavy rainfall mechanism. The remainder of this study is arranged as follows: The dataset used and the methodology are given in Section 2. Section 3 provides the results, including the vertical structure characteristics, horizontal distribution characteristics, and the relationship between each physical quantities of stratiform and CP over different types of terrain. The discussion is provided in Section 4 and followed by a summary in Section 5.

2. Data and Methods

GPM Core Observatory takes about 93 min to fly around the earth. The global coverage is from 68°S to 68°N, and the flight heights is at the altitude of 407 km. The GPM radar is able to provide observations of the 3D structure of precipitation from the surface to 22 km altitude. The DPR instrument is made of a Ka-band precipitation radar (KaPR) operating at 35.5 GHz and a Ku-band precipitation radar (KuPR) at 13.6 GHz, and the KuPR and KaPR will provide coaligned 5-km-resolution footprints on Earth's surface. 2A.GPM.DPR is the dataset of DPR Ku and Ka-band radar reflectivity profile and radar-based precipitation retrievals. The dataset carries three radar profiles, including the Ku-band normal scan (NS), the Ka-band matched scan (MS), and the Ka-band high-sensitivity scan (HS). The NS has a nominal vertical range resolution of 250 m with cross-track swath widths of 245 km, the MS has a nominal vertical range resolution of 250 m with cross-track swath widths of 120 km, and the HS has a nominal vertical range resolution of 500 m with cross-track swath widths of 120 km [12].The 2A.GPM.DPR database V06 (covering the period from May to September of 2014–2021) has been used in the study, which can provide detailed precipitation information, including reflectivity factor with attenuation correction (Z_e), DSD, storm top altitude (STA), freezing height (FzH), rain rate (RR), precipitation type, and so on. In single frequency classification (CSF) modules, i.e., in Ku-only and Ka-only modules, precipitation type classification is made by a Vertical profiling method (V-method) and by an Horizontal pattern method (H-method) [28,29]. In the dual frequency module, in place of the V-method, the measured dual frequency ratio (DFR_m) method is used for precipitation type classification and for BB detection in the inner swath [30,31], classifying rain into stratiform, transition, and convective. More details can be found online at https://gpm.nasa.gov/sites/default/files/2019-05/ATBD_DPR_201811_with_Appendix3b.pdf (accessed on 23 August 2022).

The research regions are the Sichuan Basin and its surrounding areas of China (99°E–109°E, 27°N–33°N). ETOPO1 is a 1 arc-minute global relief model that was developed by the National Geophysical Data Center (NGDC), an office of the National Oceanic and Atmospheric Administration (NOAA), which was used to divide the research regions in this study. Fan et al. [32] found the best statistical window of topographic relief in Sichuan to be 9.92 km^2, while the statistical window of topographic relief is defined as 13.69 km^2 in this paper because of the limit of resolution of ETOPO1. Topographic relief is the difference between maximum and minimum of altitude. Referring to related research results [33–35], dividing the research regions into three types of topography by ETOPO1 (Figure 1): 1. Plains with altitude from 0 to 1500 m, topographic relief < 100 m; 2. Mountains with altitude from 500 to 1500 m, topographic relief ≥ 200 m; 3. High mountains with altitude from 1500 to 4000 m, topographic relief ≥ 200 m. In Figure 1b, the white areas represent the place where topographic relief does not meet the requirements of three categories mentioned above, e.g., the white areas of 1500–4000 m above mean sea level represent high-altitude areas with flat terrain as well as with few rainy pixels, therefore not going to study such areas. For statistical analysis, the number of rainy pixel samples occurred in study regions during the period May to September of 2014–2021 was counted based on GPM DPR data (Table 1), and Figure 2 shows the spatial distribution of the number of stratiform and convective rainy pixels.

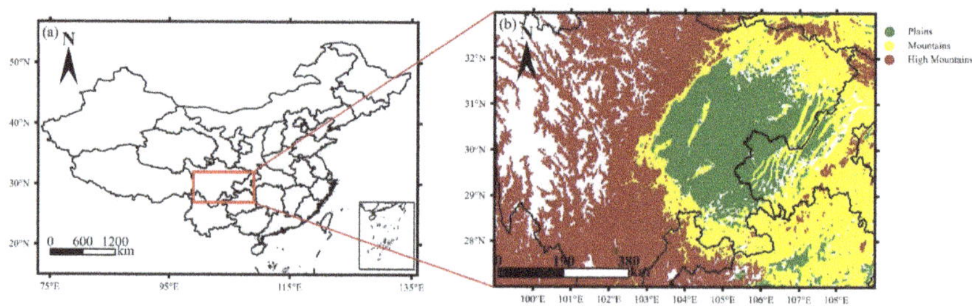

Figure 1. Administrative regional division of the People's Republic of China (**a**), and geographical division of Sichuan Basin and its surrounding areas; (**b**) The green, yellow and brown areas represent the plains, mountains and high mountains respectively.

Table 1. Number of convective and stratiform rainy pixels of different types of terrain in Sichuan Basin and its surrounding detected by GPM DPR from May to September of 2014–2021.

RR (mm/h)	Plain	Mountain	High Mountain
0.5 ≤ RR < 2	2186 (32,530)	4616 (72,504)	6646 (65,973)
2 ≤ RR < 4	2031 (10,851)	4146 (25,204)	3885 (14,688)
4 ≤ RR < 8	1985 (5301)	4263 (12,772)	3481 (5545)
8 ≤ RR < 20	1685 (1674)	3530 (3813)	2126 (1121)
RR ≥ 20	981 (247)	1579 (483)	495 (85)

Numbers in brackets denote SP, and numbers outside brackets denote CP.

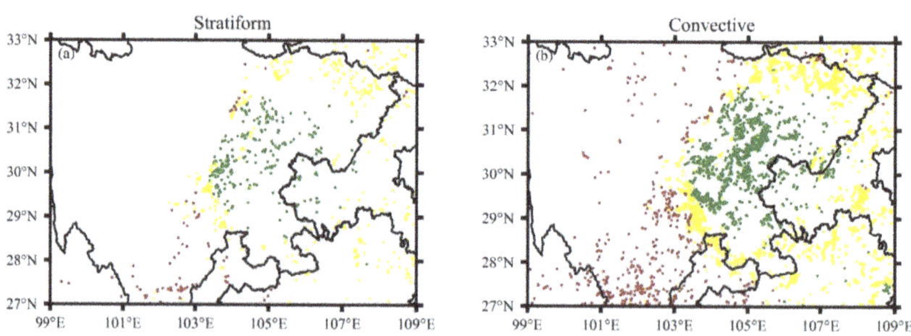

Figure 2. The spatial distribution of heavy stratiform (**a**) and convective (**b**) rainy pixels. The heavy precipitation occurs over plains (green dots), mountains (yellow dots) and high mountains (brown dots).

The contoured frequency by altitude diagram (CFAD) is an effective method to represent the vertical structure of precipitation that has been applied in many studies [36,37]. The entire frequency distribution normalized by dividing by the maximum absolute frequency of the samples within the region of analysis is NCFAD, which permits the comparison of CFADs between regions despite the different absolute frequencies [38]. In this paper, NCFAD is mainly used to statistically analyze the vertical structure characteristics of SP and CP that occurred over plains, mountains and high mountains regions of the Sichuan Basin and its surrounding area. RR $\geq$ 20 mm/h is defined as heavy precipitation [39], and 2A.GPM.DPR retrievals with RR < 0.5 mm/h are discarded from this study because of the limitations of dual-frequency retrieval. For the detailed discussion, the ground precipitation grades were divided into 5 levels, including 0.5 < RR $\leq$ 2 mm/h, 2 < RR $\leq$ 4 mm/h, 4 < RR $\leq$ 8 mm/h, 8 < RR $\leq$ 20 mm/h and RR $\geq$ 20 mm/h. Since shallow rain is archived as CP in DPR retrievals, excluded it from the CP, and only precipitation events in which the phase behavior of near-surface particles is liquid are considered. The STA, FzH, and bright band height in this study are absolute heights altitude relative to sea level, not relative to the surface. In addition, the horizontal distribution of STA and other physical quantity of DPR orbit dataset were gridded at 0.25° $\times$ 0.25°. For example, the average STA denotes the ratio of the sum of STA within the 0.25° $\times$ 0.25° to the total number of samples within that area.

3. Results

3.1. Vertical Distribution Characteristics of Radar Reflectivity Factor

Figure 3 shows the vertical structure of reflectivity factor. In addition to revealing the vertical structure of precipitation, the maximum frequency profile of the Z_e is also a good indicator of its microphysical processes [40]. SP is usually characterized by bright band near FzH, and the bright band is a good indicator of phase change of hydrometeor [41]. The hydrometeors above bright band are mainly ice or snow. In bright band, there are mixed-phase hydrometeors including partially melted ice, snow and raindrops, while it occurs at the liquid phase below bright band. As shown in the maximum frequency profile of Z_e from 6 to 10 km, Z_e of two types of heavy precipitation increases with altitude reducing, indicating that precipitation particles are growing as they are falling. For stratiform heavy precipitation (Figure 3a–c), different to CP, the growth rate of Z_e is very large near the freezing layer that the profile tends to be horizontal, showing particles undergoing a rapid transition from ice to liquid phase near the freezing layer. Figure 4 shows the correlation between the elevation and FzH for heavy rainfall. Only Figure 4e,g,h passed the 0.01 significance level, and the r-value is relatively large in Figure 4h, which suggests that the FzH of the two types of heavy precipitation is hardly affected by topography, except for heavy CP over high mountains.

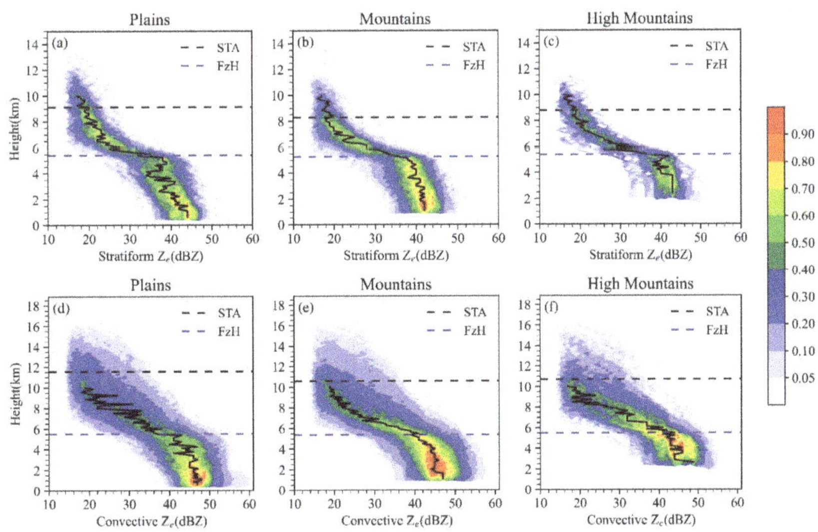

Figure 3. Normalized contoured frequency by altitude diagrams (NCFAD) of reflectivity factor of two types of heavy precipitation over plains (**a**,**d**), mountains (**b**,**e**), and high mountains (**c**,**f**). The color areas and black solid lines indicate occurrence frequency and maximum frequency profile of reflectivity factor, respectively. The black and blue dash lines represent STA and FzH, respectively.

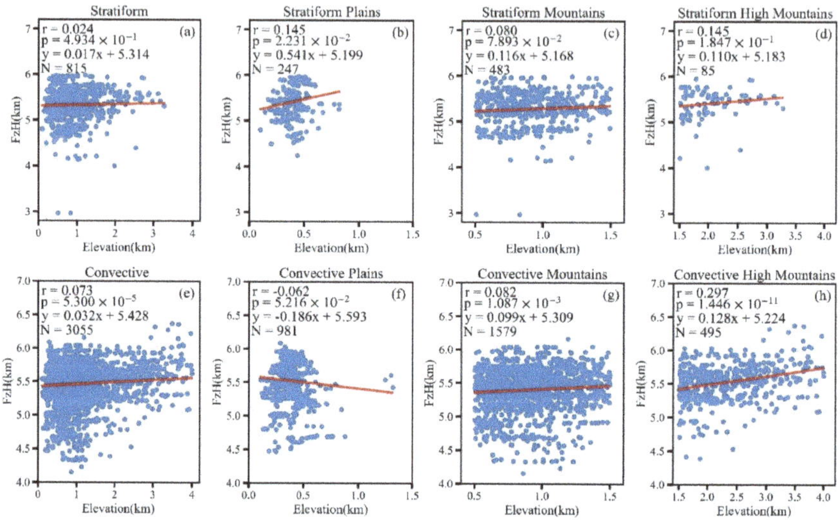

Figure 4. Scatter plots of elevation and FzH of two types of heavy precipitation over (**a**,**e**) three areas, (**b**,**f**) plains, (**c**,**g**) mountains, and (**d**,**h**) high mountains. r represents Pearson correlation coefficient, p represents significance level, y and red solid lines are the regression equation and N represents sample size.

Figure 5 exhibits the mean Z_e profiles with various RR. Z_e increases from STA to the surface with altitude decreasing, and the near-surface Z_e increases with near-surface RR increasing. As shown in Figure 5a–e, the bright band characteristics of SP are remarkable at an altitude of about 6 km, and there is a clear turning point near FzH, which is the microphysical property of the SP particles during the falling process, while Z_e profiles of CP do not have such characteristic. For heavy precipitation, Z_e from FzH to the surface overall increases, indicating that the coalescence of falling particles is more efficient than breakup and evaporation.

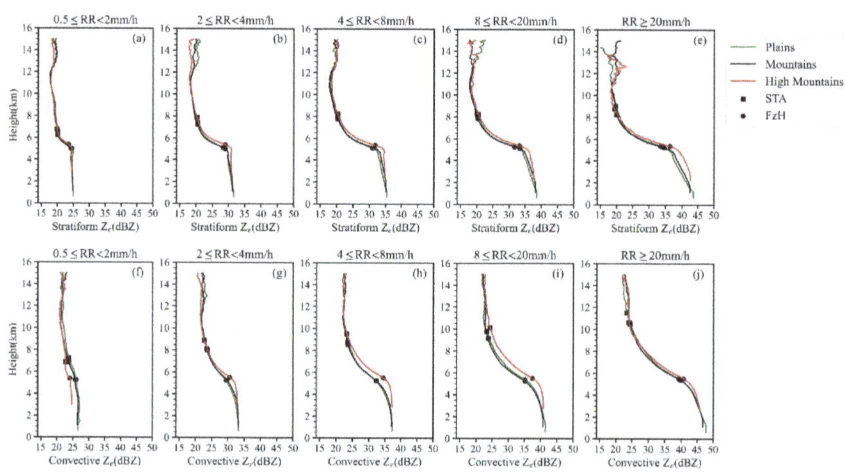

Figure 5. The vertical profile of Z_e with various near-surface RR. (**a**–**e**) represent SP (top panels), (**f**–**j**) represent CP (bottom panels). The green, black and red solid lines represent plains, mountains and high mountains, respectively. The square and round dots represent STA and FzH, respectively.

3.2. Vertical Distribution Characteristics of Mass-Weighted Mean Raindrop Diameter

DSD is a fundamental property of precipitation, which is very important for understanding the microphysical processes occurring in precipitation systems. DSD contains two parameters: mass-weighted mean raindrop diameter (D_m in mm) and normalized DSD intercepts (N_w in mm^{-1}/m^3), where N_w denotes the raindrop number concentration as rainwater content and raindrop size are constant. The relationship between dBN_w and N_w is $dBN_w = 10\ lg(N_w)$. For simplicity, hereinafter, dBN_w is used rather than N_w. Figure 6 shows the NCFAD of D_m of two types of heavy precipitation over different types of terrain. Small precipitation particles ($D_m \leq 1.2$ mm) are mainly concentrated between STA and FzH, and the precipitation particles in this layer mainly exist in the form of ice crystals. With the elevation of terrain, the horizontal distribution domain of D_m increases while the vertical distribution domain decreases. For the same terrain, the horizontal and vertical distribution domain of D_m of heavy SP is smaller than that of heavy CP. In addition, the maximum D_m of the horizontal distribution domain is also smaller than that of CP. For heavy CP, with the elevation increasing, the occurrence probability of large raindrops ($D_m \geq 2.6$ mm) increasing, as shown in Figure 6d–f. It can be clearly seen from Figure 6a–c that in heavy SP, precipitation particles with D_m of 1.3–1.6 mm below FzH of mountains and high mountains have a higher probability of occurrence than that of plains. This is mainly caused by the difference of water vapor content near FzH, which will be discussed in Section 4.

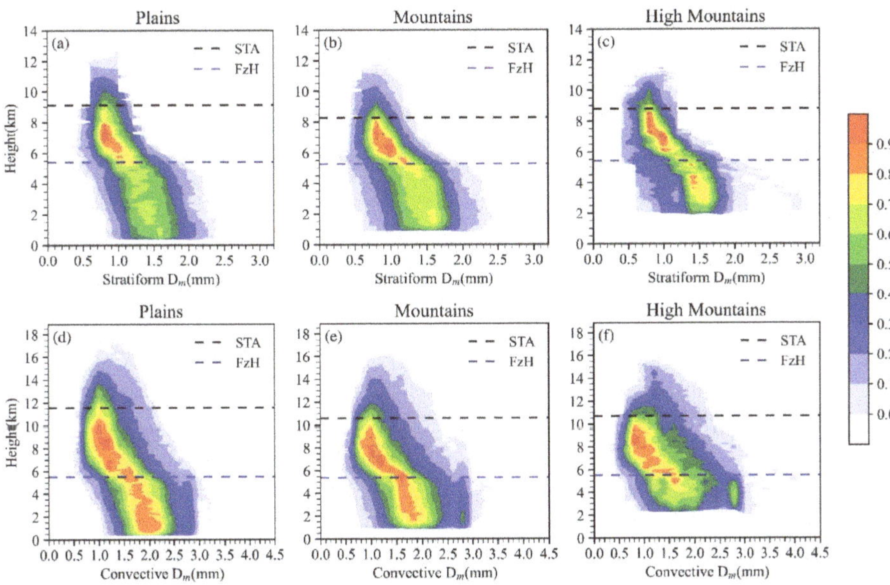

Figure 6. NCFAD of D_m of two types of heavy precipitation over plains (**a**,**d**), mountains (**b**,**e**), and high mountains(**c**,**f**). The color areas indicate occurrence frequency of D_m. The black and blue dash lines represent STA and FzH, respectively.

From D_m profiles of different rain intensity levels (Figure 7), it can be seen that the value of CP near the surface is always greater than SP under the same terrain and rain intensity levels. The D_m profiles of light CP ($0.5 \leq RR < 2$ mm/h) that are shown in Figure 7f are obviously different from that of other rainfall intensity. The D_m profile of light CP over high mountains decreases as altitude decreases from STA to FzH, while that of other rainfall intensity increases as altitude decreases. Different from light SP (Figure 7a), the D_m profile of plains of CP (Figure 7f) decreases as altitude decreases from FzH to surface because of the less water vapor in low-level atmosphere over plains, which tends to reduce the diameter of small raindrops by evaporation. Above an altitude of 10 km, the D_m of heavy precipitation over high mountains are larger than those over plains and mountain (Figure 7e,j), particularly in heavy SP, which is significantly different from that of other rainfall intensity.

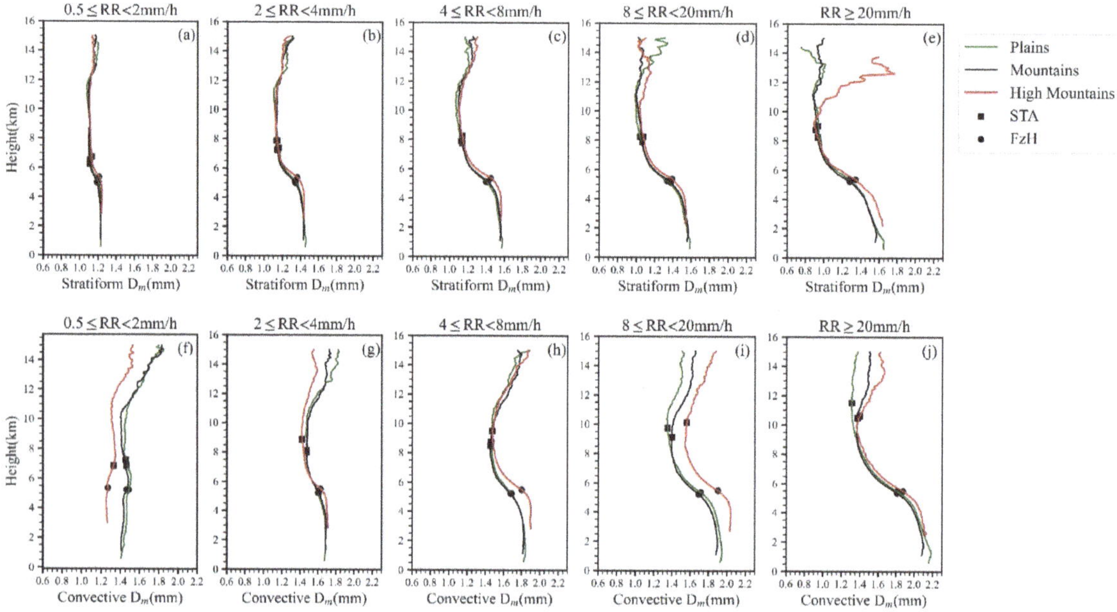

Figure 7. The vertical profile of the mean D_m with various near-surface RR. (**a–e**) represent CP, (**f–j**) represent SP. The green, black and red solid lines represent plains, mountains and high mountains, respectively. The square and round dots represent STA and FzH, respectively.

There is a close relationship between D_m and Z_e. In general, large D_m of precipitation particles correspond to large Z_e. As shown in Figure 7e,j, the near-surface D_m of CP is significantly larger than that of heavy SP, and therefore the near-surface Z_e is also larger than that of SP (Figure 5e,j).

3.3. Vertical Distribution Characteristics of dBN_w

For the same terrain, compared with dBN_w NCFAD of two types of heavy precipitation (Figure 8), it shows that the horizontal distribution domain of dBN_w of heavy SP over plains and mountains is wider than that of CP while the high frequency areas of dBN_w are not as concentrated as those of heavy CP. In contrast, the horizontal distribution domain of dBN_w of heavy SP over high mountains is smaller than that of CP while the high frequency areas are more concentrated. For different types of terrain, the vertical structure of dBN_w of heavy CP is also different, as shown in Figure 8d–f: The higher elevation is, the wider horizontal distribution domain of dBN_w is. Below FzH, the high frequency areas of dBN_w over plains are the most concentrated with 34–38 mm^{-1}/m^3, while those over mountains and high mountains are more scattered, especially over high mountains. The opposite is true for heavy SP. The high frequency areas of dBN_w over mountains and high mountains are more concentrated than those over plains (Figure 8a–c).

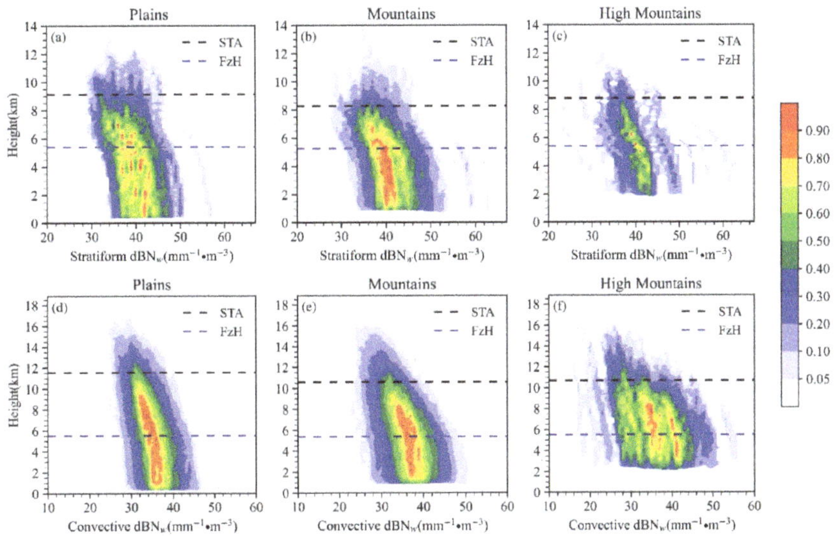

Figure 8. Normalized contoured frequency by altitude diagrams (NCFAD) of dBN_W of two types of heavy precipitation over plains (**a**,**d**), mountains (**b**,**e**), and high mountains (**c**,**f**). The color areas indicate occurrence frequency of dBN_W, the black and blue dash lines represent STA and FzH, respectively.

Figure 9 shows the mean dBN_W profiles with different surface rainfall intensity. In each panel, dBN_W overall increases as altitude decreases. For heavy precipitation (Figure 9e,j), the dBN_W above 10 km over high mountains are obviously smaller than those over plains. Combined with the mean D_m profiles of heavy precipitation (Figure 7e,j), compared with plain areas, precipitation particles over high mountains have the characteristics of lower number concentration and larger scale above 10 km.

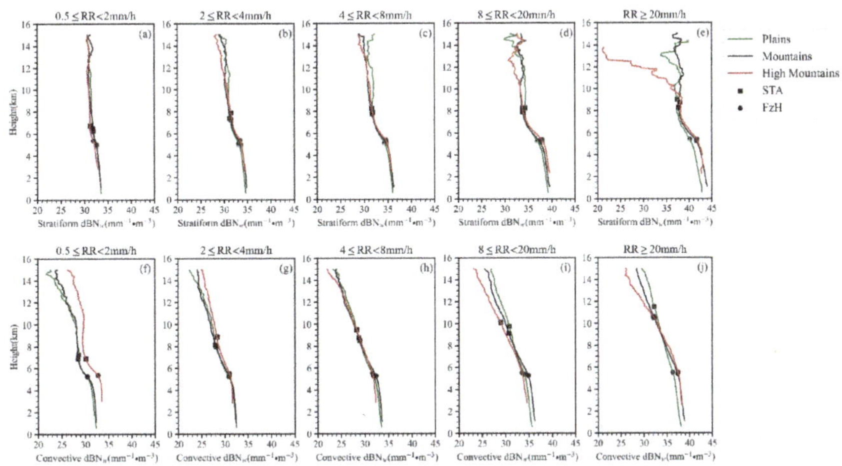

Figure 9. The vertical profile of dBN_W with various near-surface RR. (**a**–**e**) represent SP, (**f**–**j**) represent CP. The green, black and red solid lines represent plains, mountains and high mountains, respectively. The square and round dots represent STA and FzH, respectively.

3.4. Distribution Characteristics of Storm Top Altitude and Water Vapor

In general, for the same type of precipitation, the higher STA is, the heavier precipitation is. STA is usually lower than the cloud top height, and when STA is low, the cloud top height varies widely. The higher STA is, the more similar the cloud top height is to STA [42].

Figure 10 shows the horizontal distribution of STA when heavy precipitation occurred. For the same types of terrain, STA of heavy CP is overall higher than that of heavy SP. For heavy CP, the high value areas of STA over plains (Figure 10d) are mainly located in the central and western of the Sichuan Basin where roughly correspond to the high frequency areas of heavy precipitation. The high value areas of STA over mountains are mainly located in the region where the plain and the mountain meet in the west of the Sichuan Basin (Figure 10e). In general, STA over plains is higher than those over mountains and high mountains, regardless of heavy CP or SP.

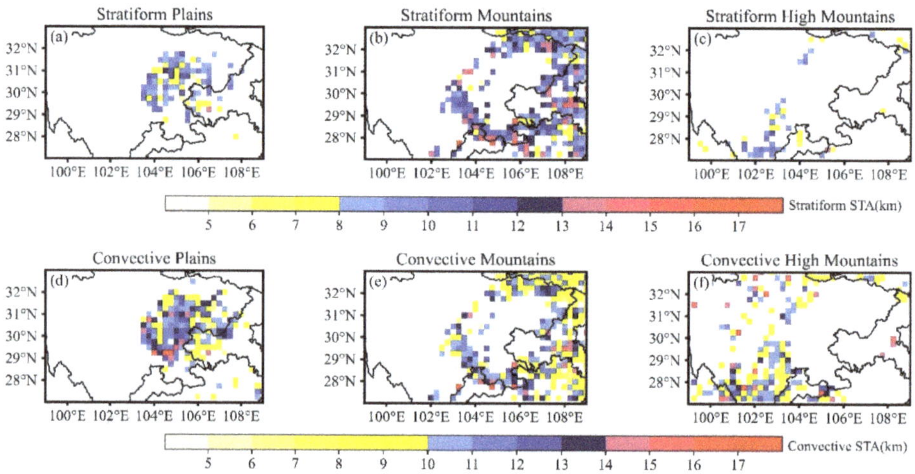

Figure 10. The horizontal Distribution of storm top altitude of two types of heavy precipitation over plains (**a**,**d**), mountains (**b**,**e**), and high mountains (**c**,**f**).

In order to have a more intuitive understanding of the distribution of STA, Figure 11 shows the probability distribution functions (PDFs) of STA over plains, mountains and high mountains with different rain intensity levels. It can be seen that when heavy precipitation occurs, STA is mostly above 6 km altitude (cloud system is deep). In the same terrains, STA of heavy CP is generally higher than heavy SP (Figure 11e,j), which also indicates that convective cloud develops more vigorously than stratus in the vertical direction. When the RR is less than 8 mm/h, there is no significant difference in STA of CP and SP over the same terrain. When the surface RR increased to 8–20 mm/h, STA of precipitation over mountains and high mountains was still roughly the same, but STA of precipitation over plains began to show significant differences. STA of CP over plains was significantly higher than that of SP, and its highest frequency was about 4 km higher than that of SP. When CP occurs, with the increase of RR, the height corresponding to the maximum occurrence frequency of STA also increases. For the two types of precipitation with RR < 20 mm/h, STA of precipitation over high mountains is generally higher than those over plains and mountain. When the RR $\geq$ 20 mm/h for heavy CP, STA over plains is obviously higher than those over mountains and high mountains. STA of heavy CP over plains is mainly concentrated at 10–14 km altitude, while that in mountains and high mountains are mainly concentrated at 9–11 km altitude.

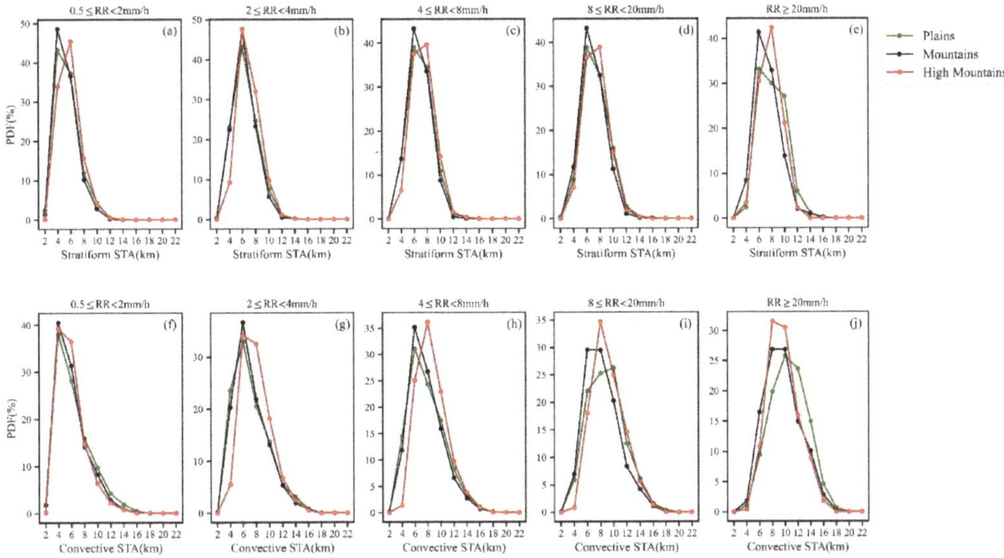

Figure 11. Probability distribution functions (PDFs) of storm top altitude with various near-surface RR of SP (**a**–**e**) and CP (**f**–**j**) over different topographic. The green, black and red solid lines represent plains, mountains and high mountains, respectively.

It is usually considered that convective overshooting is with STA > 10 km, and deep SP is with STA > 8 km. Fu et al. [43] divided STA into three categories, including STA < 5 km, 5–10 km and > 10 km, which represent shallow convective, moderate convective and deep convective (also called convective overshooting) respectively. Likewise, divided STA into below 5 km, 5–8 km and above 8 km, representing shallow stratiform, moderate stratiform and deep stratiform respectively. Statistics on the categories of STA over the Sichuan Basin and its surrounding areas are shown in Table 2. It can be found that with the increase of RR, the proportion of shallow precipitation gradually decreases, while the deep precipitation increases with the increase of RR. For heavy precipitation, the proportion of deep convective over plains is 69.6%, higher than mountains (54.8%) and high mountains (57.2%). For heavy SP, there is little difference in the proportion of deep stratiform between plains and high mountains (64.4% and 65.9%, respectively), both higher than mountains (50.1%).

STA is related to the strength of the updraft and affected by properties of the underlying surface [44]. The updraft not only influences STA, but also plays a significant role in the size of the precipitation particles. Therefore, it is speculated that there may be a correlation between STA and the precipitation particles size, then the correlation between STA and the near-surface D_m was tested for significance (Figure 12). Except for the heavy SP over plains and high mountains that failed with 0.01 significance level, all others passed the significance test, indicating that there is a positive linear correlation between STA and near-surface D_m of heavy precipitation except for the heavy SP over plains and high mountains. For the same types of terrain, the correlation of heavy CP is higher than that of heavy SP, and the correlation of heavy CP over mountains is the most significant, which may be related to different mechanisms of heavy CP and SP. For heavy CP, as shown in Figure 12d–f, the slope of the regression equation of high mountains is the largest, followed by mountains and plains in sequence, which shows that when near-surface D_m grows to the same diameter, STA over high mountains is usually the lowest while that over plain is usually the highest.

Table 2. Statistics on the proportion of shallow convective (stratiform), moderate convective (stratiform) and deep convective (stratiform) over different types of terrain in the Sichuan Basin and its surrounding detected by GPM DPR from May to September of 2014–2021.

RR (mm/h)	0.5 ≤ RR < 2	2 ≤ RR < 4	4 ≤ RR < 8	8 ≤ RR < 20	RR ≥ 20
Shallow convective (stratiform) of plains	15.2% (15.0%)	5.8% (5.4%)	3.7% (2.7%)	1.2% (1.5%)	0.1% (0.0%)
Shallow convective (stratiform) of mountains	17.4% (15.3%)	5.1% (4.0%)	2.4% (1.9%)	0.8% (1.5%)	0.2% (0.6%)
Shallow convective (stratiform) of high mountains	7.9% (4.6%)	0.5% (0.5%)	0.1% (0.2%)	0.0% (0.7%)	0.0% (0.0%)
Moderate convective (stratiform) of plains	68.6% (68.0%)	71.3% (61.2%)	66.3% (50.2%)	52.0% (46.6%)	30.3% (35.6%)
Moderate convective (stratiform) of mountains	70.3% (71.3%)	73.9% (66.3%)	71.5% (55.1%)	65.2% (53.4%)	45.0% (49.3%)
Moderate convective (stratiform) of high mountains	82.7% (74.9%)	71.8% (56.4%)	62.5% (44.0%)	53.4% (43.0%)	42.8% (34.1%)
Deep convective (stratiform) of plains	16.2% (17.0%)	22.9% (33.4%)	30.0% (47.1%)	46.8% (51.9%)	69.6% (64.4%)
Deep convective (stratiform) of mountains	12.3% (13.4%)	21.0% (29.7%)	26.1% (43.0%)	34.0% (45.1%)	54.8% (50.1%)
Deep convective (stratiform) of high mountains	9.4% (20.5%)	27.7% (43.1%)	37.4% (55.8%)	46.5% (56.3%)	57.2% (65.9%)

Numbers in brackets denote SP, and numbers outside brackets denote CP.

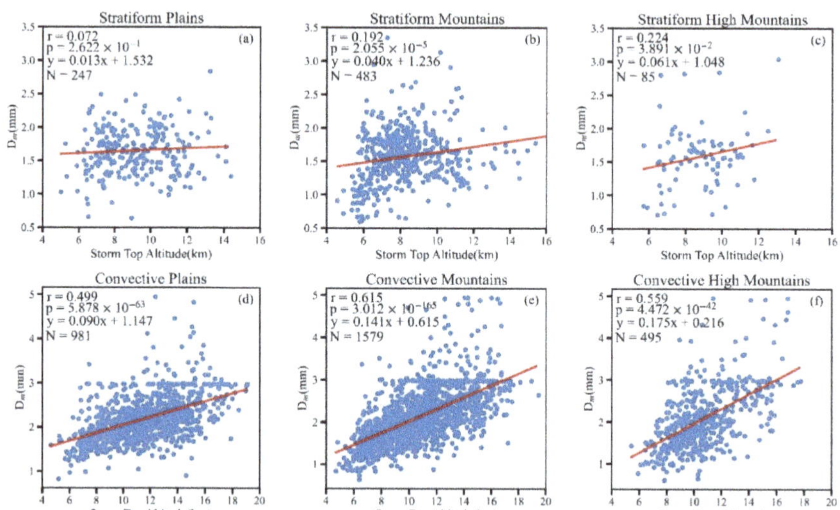

Figure 12. Scatter plots of storm top altitude and D_m of two types of heavy precipitation over plains (**a**,**d**), mountains (**b**,**e**), and high mountains (**c**,**f**). r represents Pearson correlation coefficient, p represents significance level, y and red solid lines are the regression equation and N represents sample size.

Water vapor is one of the important physical quantities that affects precipitation. The momentum, water vapor and heat convergence in meteorological boundary layer all contribute to the rainstorm during the heavy rainfall processes. Figure 13 exhibits the horizontal distribution characteristics of total water vapor from the surface to FzH when heavy precipitation occurs. The maximum value of water vapor is mainly located over the plains, and the water vapor gradually decreases as the elevation increases, showing a stair-like decrease from west to east. The horizontal distribution characteristics of water vapor are roughly similar to STA.

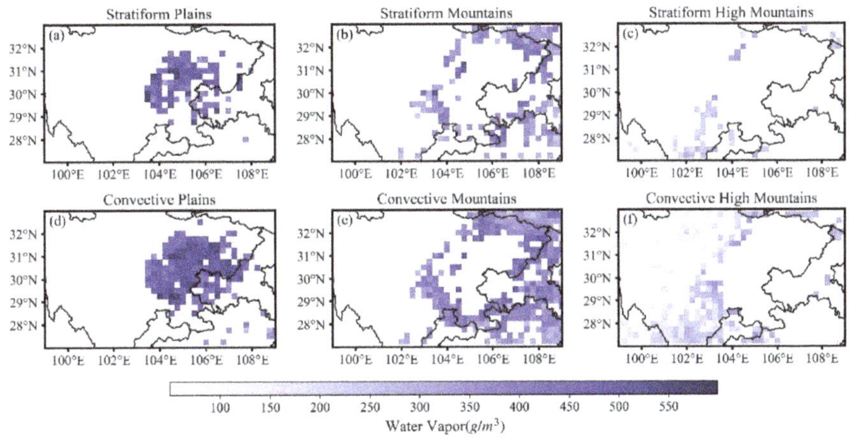

Figure 13. The horizontal Distribution of water vapor of two types of heavy precipitation over plains (**a**,**d**), mountains (**b**,**e**), and high mountains (**c**,**f**).

Figure 14 shows the correlation between the sum of water vapor from the surface to FzH and STA. For the heavy CP and SP over plains and mountains except high mountains passed with 0.01 significance level, which suggests there is a positive linear correlation between the sum of water vapor from surface to FzH and STA. For the same terrain, the correlation of heavy SP is more relevant than that of heavy CP, and it is most significant over plains.

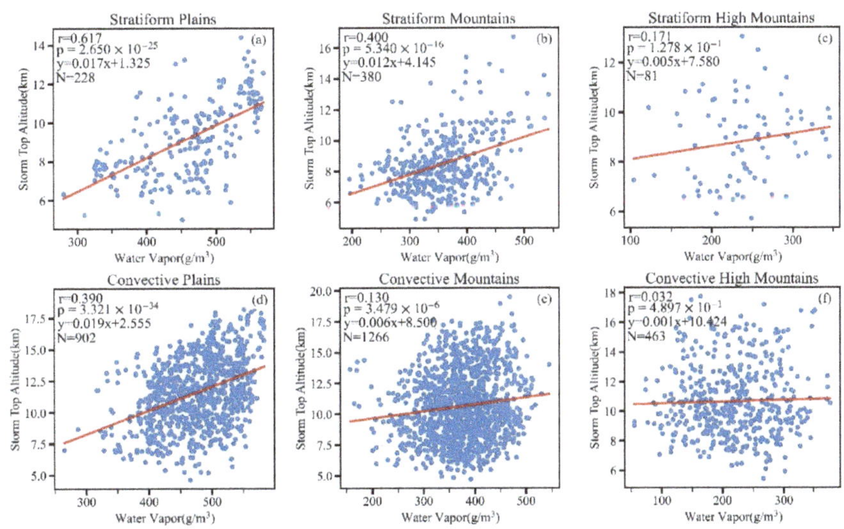

Figure 14. Scatter plots of water vapor and storm top altitude of two types of heavy precipitation over plains (**a**,**d**), mountains (**b**,**e**), and high mountains (**c**,**f**). r represents Pearson correlation coefficient, p represents significance level, y and red solid lines are the regression equation and N represents sample size.

4. Discussion

In this section, we discuss and analyses the reasons for the difference of precipitation particles of heavy precipitation over different terrain, including plains, mountains, and high mountains. For heavy CP, the occurrence probability of larger precipitation particles ($D_m \geq 2.6$ mm) increases with elevation (Figure 6d–f). We suggest that strong updraft is more likely to form due to mountainous topographic lifting, and the strong updrafts tend to hold up the falling raindrops, slowing them down, and can carry some small raindrops to high altitudes to collide and merge with the falling raindrops, which leads to higher collision efficiency. In addition, the strong updraft can also carry sufficient water vapor into the cloud, which is also conducive to the formation of large raindrops. Yan et al. [45] proposed that the strengthening of ascending movement will also cause the snow and graupel particles above FzH to grow rapidly, making it easier to generate large rain droplets. Strong turbulence facilitates raindrops colliding and merging into large raindrops, meanwhile, it is facilitating raindrops breakup into little raindrops. This phenomenon is well evidenced by the wider D_m horizontal distribution over mountains and high mountains than those over plains (Figure 6d–f). Another explanation is due to the seeder–feeder mechanism. In Figure 15, the relationship between near-surface D_m and near-surface dBN_w is shown for the heavy CP over the three areas. For heavy CP, when dBN_w is > 40 mm^{-1}/m^3, it generally corresponds to D_m < 2 mm, and this relationship is best in high mountains. In Figure 8f, it can be clearly observed that the occurrence probability of region with dBN_w > 40 mm^{-1}/m^3 below FzH of high mountains is much higher than those of plains. Wilson and Barros [46] showed that the seeder-feeder mechanism leads to an accelerated growth of small and moderate size raindrops (D_m < 2 mm). This process could explain the enhancement of coalescence and the increase of the concentration of small drops [27]. Yan et al. [45] mentioned that for heavy rain, the cloud ice particles with large number concentrations (>600 L^{-1}) seldom occur, and they are more inclined to gather at moderate concentrations (100–250 L^{-1}) above 9 km over the Tibetan Plateau (roughly corresponds to the high mountains in this study) compared with the northern India and south of the Tibetan Plateau (NIST) and tropical ocean (TO). The cloud ice particles with smaller number concentrations usually correspond to the larger sizes of cloud ice particles. In this study, for heavy precipitation, precipitation particles over high mountains have the characteristics of lower number concentration and larger scale above 10 km. We speculate that there is some connection between the cloud ice particles and precipitation particles. However, this assumption needs further exploration and validation in future studies.

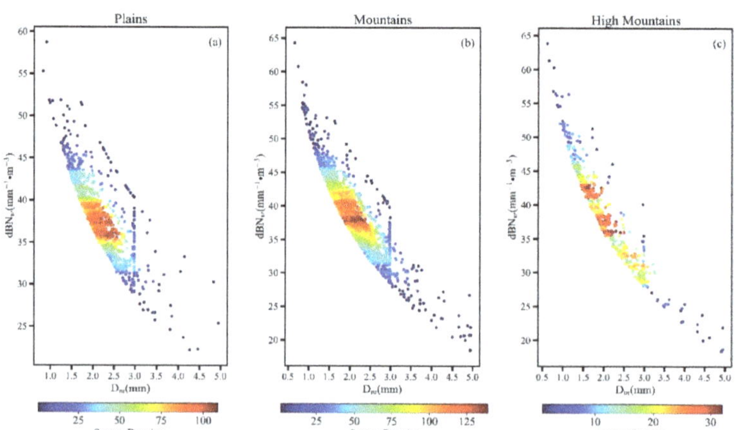

Figure 15. Scatter distribution of near-surface D_m-dBN_w of heavy CP over plains (**a**), mountains (**b**), and high mountains (**c**). The color indicates the scatter density.

For heavy SP, the occurrence probability of precipitation particles D_m below FzH within 1.3–1.6 mm over high mountains are significantly higher than those over plains (Figure 6a,c). The total water vapor near FzH (altitude at 4.5–6.5 km) over high mountains is more sufficient than those over plains (Figure 16). In addition, the underlying surface of the high mountains is closer to FzH, and the updraft is more likely to transport water vapor to FzH, which makes snow and graupel particles above FzH grow rapidly and form larger raindrops more easily. Owing to sufficient water vapor, the occurrence probability of D_m in the range of 1.1–1.3 mm over high mountains is much lower than those over plains, and the D_m over high mountains tends to concentrate in the range of 1.3–1.6 mm (Figure 6a,c).

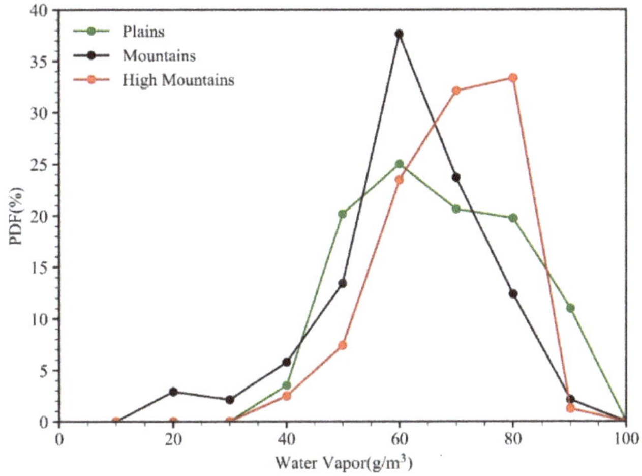

Figure 16. Probability distribution functions (PDFs) of total water vapor content at 4.5–6.5 km of heavy SP of different topographic. The green, black and red solid lines represent plains, mountains and high mountains, respectively.

5. Conclusions

The Sichuan Basin and its surrounding areas are divided into three types of terrain based on ETOPO1, including plains, mountains, and high mountains. The Z_e and the vertical structural of SP and CP particles over the three types of terrain were analyzed by measurements and retrievals from GPM DPR from May to September of 2014–2021. The major conclusions of this study can be summarized as follows.

1. The FzH of the two types of heavy precipitation is hardly affected by topography, except for heavy CP over mountains. The mean Z_e profiles of SP are significantly bent near FzH, which is different from that of CP.
2. For the same types of terrain, the D_m horizontal distribution domain, vertical distribution domain, and the maximum D_m of the horizontal distribution domain of heavy CP are all larger than those of heavy SP. With the increase of elevation, the horizontal distribution domain of D_m of the two types of heavy precipitation increases, while the vertical distribution domain of D_m decreases. For heavy CP, as elevation increases, the occurrence probability of larger precipitation particles ($D_m \geq 2.6$ mm) increases as well as D_m horizontal distribution domain, which is probably related to the strong updraft rendered by topography, while the seeder-feeder mechanism may also enhance particle growth mechanism. For heavy SP, the occurrence probability of medium raindrops (D_m in 1.3–1.6 mm) below FzH over high mountains is higher than those over plains. This is because water vapor over high mountains is more sufficient, the underlying surface is closer to FzH, and the updraft is more likely to transport water vapor to FzH, which makes snow and graupel particles above FzH

grow rapidly, and thus it is easier to form larger raindrops. For heavy precipitation, the higher the elevation, the larger the Dm above 10 km altitude.

3. The mean dBN_w profiles of the two types of precipitation overall increase with the decreases of altitude. For heavy CP, the dBN_w high frequency area of plains is more concentrated than that of mountains and high mountains, while the situation of heavy SP is opposite. Above 10 km altitude, compared with plains, the precipitation particles of heavy precipitation over high mountains have the characteristics of lower number concentration and larger scale.
4. In the same terrain, overall, STA of heavy CP is higher than heavy SP. The distribution characteristics of STA of the two types of heavy precipitation are consistent, and STA of plain is generally higher than that of mountains and high mountains. There is a positive linear correlation between STA and the near-surface D_m for heavy precipitation (excluding heavy SP over plains and high mountains), and the correlation is most significant for heavy CP over mountainous. The slope of the linear regression equation of heavy CP over high mountainous terrain is the largest, followed by mountains and plains, indicating that when precipitation particles grow to the same size over different terrains, the STA of high mountainous is the smallest, while that of plains is the largest.
5. For heavy precipitation, the total distribution of water vapor from ground to FzH is the most sufficient over plains, followed by over the mountains, and the smallest over high mountains, which is caused by the difficulty of transporting water vapor to mountains and high mountains areas due to the effect of altitude and terrain barrier. For heavy precipitation (excluding heavy precipitation over high mountains), there is also a positive linear correlation between STA and the total water vapor in the altitude layer from surface to FzH, and the correlation of heavy SP over plains is the most significant.

The present study mainly focuses on heavy rain DSD, by comparing the DSD characteristics at low-altitude flat terrain and high-altitude complex terrain, leads to evidence and understanding about the effect of Sichuan's unique orography on rain physics. It is very clear from the present study that for heavy precipitation, high-altitude complex terrain has a higher occurrence probability of larger raindrops than plains in Sichuan. This situation is critical because larger rain drops could cause more soil erosion, which leads to the triggering of landslides [26]. Therefore, in the present study on the orographic effect of rain DSD, it would also be useful and throw more light on landslide triggering mechanisms. Likewise, preliminary connections between some physical quantities would help improve the current understanding of the effects of Sichuan's unique topography on microphysical precipitation processes. The current work is a preliminary study focusing on some facts about the characteristics and differences due to different terrains, as well as the internal relationships and mechanisms that remain to be studied further in the future.

Author Contributions: Conceptualization, C.S. and G.L.; methodology, C.S. and G.L.; software, C.S.; validation, C.S. and G.L.; formal analysis, C.S.; investigation, C.S.; resources, G.L.; data curation, C.S.; writing—original draft preparation, Y.D. and G.L.; writing—review and editing, G.L.; visualization, C.S.; supervision, G.L.; project administration, G.L.; funding acquisition, G.L. All authors have read and agreed to the published version of the manuscript.

Funding: This research was funded by the National Natural Science Foundation of China, grant No. 42175002, No. 91937301, No. 42075013.

Data Availability Statement: 2A.DPR V06 dataset is publicly available. The dataset can be found here: https://storm.pps.eosdis.nasa.gov/storm/ (accessed on 5 February 2022).

Acknowledgments: The authors sincerely thank the reviewers for their constructive comments and editorial suggestions, which have greatly contributed to the quality of the paper.

Conflicts of Interest: The authors declare no conflict of interest.

References

1. Robert, A., Jr. Houze Cloud Clusters and Large-Scale Vertical Motions in the Tropics. *J. Meteorol. Soc. Jpn. Ser. II* **1982**, *60*, 396–410. [CrossRef]
2. Houze, R.A., Jr. Mesoscale Convective Systems. *Rev. Geophys.* **2004**, *42*, RG4003. [CrossRef]
3. Cifelli, R.; Rutledge, S.A. Vertical Motion, Diabatic Heating, and Rainfall Characteristics in North Australia Convective Systems. *Q. J. R. Meteorol. Soc.* **1998**, *124*, 1133–1162. [CrossRef]
4. Hobbs, P.V. Research on Clouds and Precipitation: Past, Present, and Future, Part I. *Bull. Am. Meteorol. Soc.* **1989**, *70*, 282–285. [CrossRef]
5. Zipser, E.J.; Lutz, K.R. The Vertical Profile of Radar Reflectivity of Convective Cells: A Strong Indicator of Storm Intensity and Lightning Probability? *Mon. Weather. Rev.* **1994**, *122*, 1751–1759. [CrossRef]
6. Pruppacher, H.R.; Klett, J.D. Microstructure of Atmospheric Clouds and Precipitation. In *Microphysics of Clouds and Precipitation*; Springer: Dordrecht, The Netherlands, 2010; pp. 10–73. ISBN 978-0-306-48100-0.
7. Chang, W.-Y.; Lee, W.-C.; Liou, Y.-C. The Kinematic and Microphysical Characteristics and Associated Precipitation Efficiency of Subtropical Convection during SoWMEX/TiMREX. *Mon. Weather. Rev.* **2015**, *143*, 317–340. [CrossRef]
8. Wu, G.; Liu, Y.; Zhang, Q.; Duan, A.; Wang, T.; Wan, R.; Liu, X.; Li, W.; Wang, Z.; Liang, X. The Influence of Mechanical and Thermal Forcing by the Tibetan Plateau on Asian Climate. *J. Hydrometeorol.* **2007**, *8*, 770–789. [CrossRef]
9. Boos, W.R.; Kuang, Z. Dominant Control of the South Asian Monsoon by Orographic Insulation versus Plateau Heating. *Nature* **2010**, *463*, 218–222. [CrossRef]
10. Zhang, A.; Fu, Y.; Chen, Y.; Liu, G.; Zhang, X. Impact of the Surface Wind Flow on Precipitation Characteristics over the Southern Himalayas: GPM Observations. *Atmos. Res.* **2018**, *202*, 10–22. [CrossRef]
11. Wen, Y.; Kirstetter, P.; Hong, Y.; Gourley, J.J.; Cao, Q.; Zhang, J.; Flamig, Z.; Xue, X. Evaluation of a Method to Enhance Real-Time, Ground Radar–Based Rainfall Estimates Using Climatological Profiles of Reflectivity from Space. *J. Hydrometeorol.* **2016**, *17*, 761–775. [CrossRef]
12. Hou, A.Y.; Kakar, R.K.; Neeck, S.; Azarbarzin, A.A.; Kummerow, C.D.; Kojima, M.; Oki, R.; Nakamura, K.; Iguchi, T. The Global Precipitation Measurement Mission. *Bull. Am. Meteorol. Soc.* **2014**, *95*, 701–722. [CrossRef]
13. Mugnai, A.; Di Michele, S.; Smith, E.; Baordo, F.; Bauer, P.; Bizzarri, B.; Joe, P.; Kidd, C.; Marzano, F.; Tassa, A.; et al. Snowfall Measurements by Proposed European GPM Mission. In *Advances in Global Change Research*; Springer: Dordrecht, The Netherlands, 2007; Volume 28, pp. 655–674, ISBN 978-1-4020-5834-9.
14. Kulie, M.; Bennartz, R. Utilizing Spaceborne Radars to Retrieve Dry Snowfall. *J. Appl. Meteorol. Climatol.* **2009**, *48*, 2564–2580. [CrossRef]
15. Fu, Y. Satellite-borne active and passive instruments for remote sensing of heavy rain in China:A review. *Torrential Rain Disasters* **2019**, *38*, 554–563. (In Chinese) [CrossRef]
16. Liao, L.; Meneghini, R. Physical Evaluation of GPM DPR Single- and Dual-Wavelength Algorithms. *J. Atmos. Ocean. Technol.* **2019**, *36*, 883–902. [CrossRef] [PubMed]
17. Jin, X.; Shao, H.; Zhang, C.; Yan, Y. The Applicability Evaluation of Three Satellite Products in Tianshan Mountains. *J. Nat. Resour.* **2016**, *31*, 2074–2085. (In Chinese) [CrossRef]
18. Zhang, A. Structural Features of Precipitating Clouds observed by Geostationary Satellite and Dual-frequency Radar. Ph.D. Thesis, University of Science and Technology of China, Anhui, China, 2019.
19. Lasser, M.O.S.; Foelsche, U. Evaluation of GPM-DPR Precipitation Estimates with WegenerNet Gauge Data. *Atmos. Meas. Tech.* **2019**, *12*, 5055–5070. [CrossRef]
20. Sun, Y.; Dong, X.; Cui, W.; Zhou, Z.; Fu, Z.; Zhou, L.; Deng, Y.; Cui, C. Vertical Structures of Typical Meiyu Precipitation Events Retrieved From GPM-DPR. *J. Geophys. Res. Atmos.* **2020**, *125*, e2019JD031466. [CrossRef]
21. Li, J.; Zheng, J.; Liu, Y.; Cheng, Z.; He, J.; Ren, T.; Chen, S. A Study on Vertical Structure and Macro- to Micro-Characteristics and Differences of Precipitation in Sichuan Basin and the Surrounding Areas. *Acta Meteorol. Sin.* **2022**, *80*, 205–223. (In Chinese) [CrossRef]
22. Radhakrishna, B.; Saikranthi, K.; Rao, T.N. Regional Differences in Raindrop Size Distribution within Indian Subcontinent and Adjoining Seas as Inferred from Global Precipitation Measurement Dual-Frequency Precipitation Radar. *J. Meteorol. Soc. Jpn. Ser. II* **2020**, *98*, 573–584. [CrossRef]
23. Verma, A.K.; Jha, K.K. Rain Drop Size Distribution Model for Indian Climate. *Indian J. Radio Space Phys.* **1996**, *25*, 15–21.
24. Sasi Kumar, V.; Sampath, S.; Vinayak, P.V.S.S.K.; Harikumar, R. Rainfall Intensity Characteristics at Coastal and High Altitude Stations in Kerala. *J. Earth Syst. Sci.* **2007**, *116*, 451–463. [CrossRef]
25. Seela, B.K.; Janapati, J.; Lin, P.-L.; Reddy, K.K.; Shirooka, R.; Wang, P.K. A Comparison Study of Summer Season Raindrop Size Distribution Between Palau and Taiwan, Two Islands in Western Pacific. *J. Geophys. Res. Atmos.* **2017**, *122*, 11787–11805. [CrossRef]
26. Harikumar, R. Orographic Effect on Tropical Rain Physics in the Asian Monsoon Region. *Atmos. Sci. Lett.* **2016**, *17*, 556–563. [CrossRef]
27. Zwiebel, J.; Van Baelen, J.; Anquetin, S.; Pointin, Y.; Boudevillain, B. Impacts of Orography and Rain Intensity on Rainfall Structure. The Case of the HyMeX IOP7a Event. *Q. J. R. Meteorol. Soc.* **2016**, *142*, 310–319. [CrossRef]

28. Awaka, J.; Iguchi, T.; Okamoto, K. Early Results on Rain Type Classification by the Tropical Rainfall Measuring Mission (TRMM) Precipitation Radar. In Proceedings of the 8th URSI Commission F Triennial Open Symposium on Wave Propagation and Remote Sensing, Aveiro, Portugal, 22–25 September 1998; pp. 143–146.
29. Awaka, J.; Iguchi, T.; Okamoto, K.; Trmm, P.R. Standard Algorithm 2A23 and Its Performance on Bright Band Detection. *J. Meteorol. Soc. Jpn. Ser. II* **2009**, *87A*, 31–52. [CrossRef]
30. Le, M.; Chandrasekar, V. Precipitation Type Classification Method for Dual-Frequency Precipitation Radar (DPR) Onboard the GPM. *IEEE Trans. Geosci. Remote Sens.* **2013**, *51*, 1784–1790. [CrossRef]
31. Le, M.; Chandrasekar, V. Hydrometeor Profile Characterization Method for Dual-Frequency Precipitation Radar Onboard the GPM. *IEEE Trans. Geosci. Remote Sens.* **2013**, *51*, 3648–3658. [CrossRef]
32. Fan, J.; Zhang, Z.; Li, L. Mountain Demarcation and Mountainous Area Divisions of Sichuan Province. *Geogr. Res.* **2015**, *34*, 65–73. (In Chinese) [CrossRef]
33. Gao, X. The Subjective and Objective Classification of Geomorphologic Forms. *J. Mt. Sci.* **2004**, *22*, 261–266. (In Chinese) [CrossRef]
34. Zhong, J.; Lu, T. Optimal Statistical Unit for Relief Amplitude in Southwestern China. *Bull. Soil Water Conserv.* **2018**, *38*, 175–181,186. (In Chinese) [CrossRef]
35. Yang, B. The Research of Mountains Ontology and Digital Classification in the framework of "Digital Mountains"—A Case Study in Sichuan Province. Ph.D. Thesis, Chengdu University of Technology, Sichuan, China, 2009.
36. Yuter, S.E.; Houze, R.A. Three-Dimensional Kinematic and Microphysical Evolution of Florida Cumulonimbus. Part I: Spatial Distribution of Updrafts, Downdrafts, and Precipitation. *Mon. Weather. Rev.* **1995**, *123*, 1921–1940. [CrossRef]
37. Zhang, A.; Chen, Y.; Zhang, X.; Zhang, Q.; Fu, Y. Structure of Cyclonic Precipitation in the Northern Pacific Storm Track Measured by GPM DPR. *J. Hydrometeorol.* **2020**, *21*, 227–240. [CrossRef]
38. Houze, R.A., Jr.; Wilton, D.C.; Smull, B.F. Monsoon Convection in the Himalayan Region as Seen by the TRMM Precipitation Radar. *Q. J. R. Meteorol. Soc.* **2007**, *133*, 1389–1411. [CrossRef]
39. Chen, B.; Gao, W.; Zhou, X. Synoptic Characteristic Analysis of Short-time Hard Rain in Southwest Sichuan Basin. *Plateau Mt. Meteorol. Res.* **2016**, *36*, 14–20. (In Chinese)
40. Cao, Q.; Hong, Y.; Gourley, J.J.; Qi, Y.; Zhang, J.; Wen, Y.; Kirstetter, P.-E. Statistical and Physical Analysis of the Vertical Structure of Precipitation in the Mountainous West Region of the United States Using 11+ Years of Spaceborne Observations from TRMM Precipitation Radar. *J. Appl. Meteorol. Climatol.* **2013**, *52*, 408–424. [CrossRef]
41. Sánchez-Diezma, R.; Zawadzki, I.; Sempere-Torres, D. Identification of the Bright Band through the Analysis of Volumetric Radar Data. *J. Geophys. Res. Atmos.* **2000**, *105*, 2225–2236. [CrossRef]
42. Fu, Y.; Feng, J.; Zhu, H.; Li, R.; Liu, D. Structures of a thermal convective precipitation system happened in controlling of the western subtropical pacific high. *Acta Meteorol. Sin.* **2005**, *63*, 750–761. (In Chinese) [CrossRef]
43. Fu, Y.; Cao, A.; Li, T.; Feng, S.; Zheng, Y.; Liu, Y.; Zhang, A. Climatic Characteristics of the Storm Top Altitude for the Convective and Stratiform Precipitation in Summer Asia Based on Measurements of the TRMM Precipitation Radar. *Acta Meteorol. Sin.* **2012**, *70*, 436–451. (In Chinese) [CrossRef]
44. Fu, Y.; Feng, J.; Zhu, H.; Li, R.; Liu, D. Precipitation Structures of a Thermal Convective System Happened in the Central Western Subtropical Pacific Anticyclone. *J. Meteorol. Res.* **2006**, *20*, 232–243.
45. Yan, Y.-F.; Wang, X.-C.; Liu, Y.-M. Cloud Vertical Structures Associated with Precipitation Magnitudes over the Tibetan Plateau and Its Neighboring Regions. *Atmos. Ocean. Sci. Lett.* **2018**, *11*, 44–53. [CrossRef]
46. Wilson, A.M.; Barros, A.P. An Investigation of Warm Rainfall Microphysics in the Southern Appalachians: Orographic Enhancement via Low-Level Seeder–Feeder Interactions. *J. Atmos. Sci.* **2014**, *71*, 1783–1805. [CrossRef]

remote sensing

MDPI

Article

Determination of Long-Term Soil Apparent Thermal Diffusivity Using Near-Surface Soil Temperature on the Tibetan Plateau

Bing Tong [1], Hui Xu [1], Robert Horton [2], Lingen Bian [3] and Jianping Guo [1,*]

1 State Key Laboratory of Severe Weather, Chinese Academy of Meteorological Sciences, Beijing 100081, China
2 Agronomy Department, Iowa State University, Ames, IA 50011, USA
3 Chinese Academy of Meteorological Sciences, Beijing 100081, China
* Correspondence: jpguo@cma.gov.cn

Abstract: The knowledge of soil apparent thermal diffusivity (k) is important for investigating soil surface heat transfer and temperature. Long-term k determined using the near-surface soil temperature is limited on the Tibetan Plateau (TP). The main objective of this study is to determine k with a conduction–convection method using the near-surface soil temperature measured at three sites during 2014–2016 on the TP. The hourly, daily, and monthly k values of the 0.0 m to 0.20 m layer were obtained. The hourly and daily k values ranged from 0.3×10^{-6} m^2 s^{-1} to 1.9×10^{-6} m^2 s^{-1} at the wet site, and from 1.0×10^{-7} m^2 s^{-1} to 4.0×10^{-7} m^2 s^{-1} at the two dry sites. For the monthly timescale, k ranged from 0.4 ($\pm$0.0) $\times 10^{-6}$ m^2 s^{-1} to 1.1 ($\pm$0.2) $\times 10^{-6}$ m^2 s^{-1} at the wet site, and varied between 1.7 ($\pm$0.0) $\times 10^{-7}$ m^2 s^{-1} and 3.3 ($\pm$0.2) $\times 10^{-7}$ m^2 s^{-1} at the two dry sites. The k was not constant over a day, and it varied seasonally to different degrees at different sites and years. The variation of k with soil moisture (θ) appeared to be roughly similar for unfrozen soil at these sites and years, namely, k increased sharply before reaching the peak as θ increased, and then it tended to be stable or varied slightly with further increases in θ. This variation trend was consistent with previous studies. However, the relationship between k and θ changed when soil temperature was below 0 °C, because ice had higher k than water. The correlation coefficients (r) between k and θ ranged from 0.37 to 0.80, and 0.80 to 0.92 on hourly and monthly timescales, respectively. The monthly and annual k values were significantly correlated (r: 0.73~0.93) to the Normalized Difference Vegetation Index (NDVI). The results broaden our understanding of the relationship between in situ k and θ. The presented values of k at various timescales can be used as soil parameters when modeling land–atmosphere interactions at these TP regions.

Keywords: soil thermal diffusivity; conduction–convection method; soil temperature; soil water content; Normalized Difference Vegetation Index (NDVI)

Citation: Tong, B.; Xu, H.; Horton, R.; Bian, L.; Guo, J. Determination of Long-Term Soil Apparent Thermal Diffusivity Using Near-Surface Soil Temperature on the Tibetan Plateau. *Remote Sens.* **2022**, *14*, 4238. https://doi.org/10.3390/rs14174238

Academic Editors: Massimo Menenti, Yaoming Ma, Li Jia and Lei Zhong

Received: 19 July 2022
Accepted: 25 August 2022
Published: 28 August 2022

1. Introduction

Soil apparent thermal diffusivity (k), defined as the ratio of soil thermal conductivity (λ) to volumetric heat capacity (Cv), is the parameter that describes the rate of transmission of temperature change within the soil [1]. It is associated with transient processes of heat conduction and intra-porous convection [2,3]. Knowledge of k is essential for estimating soil temperature, which plays an important role in regulating land surface processes [4–8], estimating soil heat flux [3,9,10], and simulating permafrost extent [11].

Soil apparent thermal diffusivity can be measured in a laboratory with a commercial (e.g., KD2, METER Group, Inc., Pullman, WA, USA) or homemade instrument (e.g., heat pulse probe, [12,13]), or estimated by k empirical models using soil texture, moisture (θ), bulk density, soil organic matter, temperature, vegetation index, etc. (e.g., [14–17]). Obtaining k by sampling and analyzing soil from multiple depths and locations is highly valuable, but it is also time consuming, invasive, difficult to repeat over time, and costly [18]. Heat pulse probe, commonly used in laboratory, has been recognized as reliable tool for measuring soil thermal properties in the field (e.g., [19,20]). The small sampling volume and relatively sophisticated equipment setup, however, limit its extensive application in field conditions [17]. The k empirical models require various soil specific parameters and some of the parameters are difficult to determine in situ, which limits their use.

Besides, there are numerous studies focused on determining k for time periods (e.g., diurnal, or annual) using near surface soil temperature measured in situ at multiple depths (e.g., [3,21–28]). Conduction is the primary heat transfer mechanism in soil [29]. Some methods are based on solutions of one-dimensional conduction heat transfer equations with constant diffusivity and soil upper boundary described by a sinusoidal function [30], by two harmonics [31,32], or by a Fourier series [21]. Horton et al. [21] examined these k methods and reported that the harmonic method performed best. Since soil is an intra-porous medium, soil heat transfer occurs via a complex combination of conductive and convective heat transfer processes [3]. Considering this, some researchers developed methods of estimating k based on the solution of a one-dimensional soil conduction–convection heat transfer equation with the soil upper boundary described by a sinusoidal function [23,24,33], or by a Fourier series [34]. Wang et al. [35] compared the six k methods (the amplitude, phase, arctangent, logarithmic, conduction–convection, and harmonic) at a site in the Loess Plateau of China, and suggested that the harmonic method performed best, and the conduction–convection method followed. Compared with the harmonic method, the conduction–convection method had a less accurate description of the upper boundary temperature, but it included more physics in the soil heat transfer process. By comparisons of the soil temperature estimation at the 0.10 m depth in the permafrost regions of Qinghai-Xizang (Tibet) Plateau, Hu et al. [36] found the conduction–convection method [23] performed better than the conduction methods (amplitude and phase method). Compared with the k method proposed by Hu et al. [34], the k method proposed by Gao et al. [23] is easier to calculate and has an explicit mathematical expression for k. Some researchers also developed analytical [37] and numerical [38–40] solutions for estimating the k of nonuniform soils. Due to some limitations and complexities, these methods have not been widely used in various fields. The k methods mentioned above with soil temperature can only obtain daily or longer timescale k values. The shorter timescales, e.g., hourly k values, remain unknown.

The Tibet Plateau (TP) is known as the "third pole" of the earth, with an average altitude of over 4000 m [41]. Land–atmosphere interactions over the TP play a crucial role in controlling the regional and global climate [42–45]. The soil thermal properties exert import roles in soil heat and water transfer, which have significant effects on land–atmosphere interactions [46,47]. Wang et al. [48] derived the soil thermal properties of the 0.025–0.075 m layer with soil data measured over two years at a cold semi-desert site on the western TP, and examined their relationships with soil moisture. They calculated k as the ratio of λ/Cv, which were determined by soil temperature and heat fluxes measured at two depths [49]. Gao [26] determined the k of the 0.015–0.05 m layer with the conduction–convection method at BJ on the TP during day of year (DOY) 195 to 258, 1998. Using the same k method, recently Zhou et al. [28] determined the daily k with soil temperatures measured in the 0.8 m and 3.2 m soil layer at 39 weather sites in the TP during 1980 to 2001 and examined the spatio-temporal variation of k.

Due to the harsh natural conditions and limited observations in the TP, few studies have focused on soil apparent thermal diffusivity determined with long-term in situ near-surface soil data. To fill the gap, the objective of this study is to determine long-term soil apparent thermal diffusivity using the conduction–convection method [23], with the near-surface soil temperature measured at three sites during 2014–2016 over the TP. The specific objectives are twofold, (1) to determine k at different timescales (e.g., hourly, daily, and monthly) over 2014–2016 at the three sites; (2) to examine the relationship between k and soil moisture at different timescales. The conduction–convection method is chosen to determine k because it considers more physics in the soil heat transfer process compared to conduction methods, and it provides an explicit mathematical expression for k as a function of soil temperature amplitude and phase. To obtain sub-daily k, the Dynamic Harmonic Regression (DHR, [50,51]) is used to extract sub-daily soil temperature parameters (amplitude and phase) from the soil temperature series data. In order to indirectly investigate the effect of the soil moisture on k from the vegetation aspect, the monthly relationship between k and the Normalized Difference Vegetation Index (NDVI) is also investigated.

2. Materials and Methods

2.1. Site Description

The data used in this study are from three TP sites: the Bijie site of Nagqu Station of Plateau Climate and Environment, Chinese Academy of Sciences (NPCE-BJ, hereinafter abbreviated as BJ) in the central part of the TP, the Qomolangma Atmospheric and Environmental Observation and Research Station, CAS (QOMS) in the south part of the TP, and the Ngari Desert Observation and Research Station, CAS (NADORS) in the northwestern part of the TP. The geographic characteristics of the three sites are listed in Table 1, and the locations of the three sites are shown in Figure 1. The BJ site is in a flat, open prairie except for the north, where a low hill stands. This site is well covered with grass and a canopy height up to 5 cm. The other two sites are barren and the ground is relatively flat and open, with sparse and short vegetation. Additional information about the sites can be found in Ma et al. [52].

Table 1. List of the geographic characteristics of the three sites.

Site	Latitude	Longitude	Elevation (m)	Land Cover	Soil Type
BJ	31.37°N	91.90°E	4509	Alpine meadow	Sandy silt loam
QOMS	28.36°N	86.95°E	4298	Alpine desert	Sand and gravel
NADORS	33.39°N	79.70°E	4270	Alpine desert	Sand and gravel

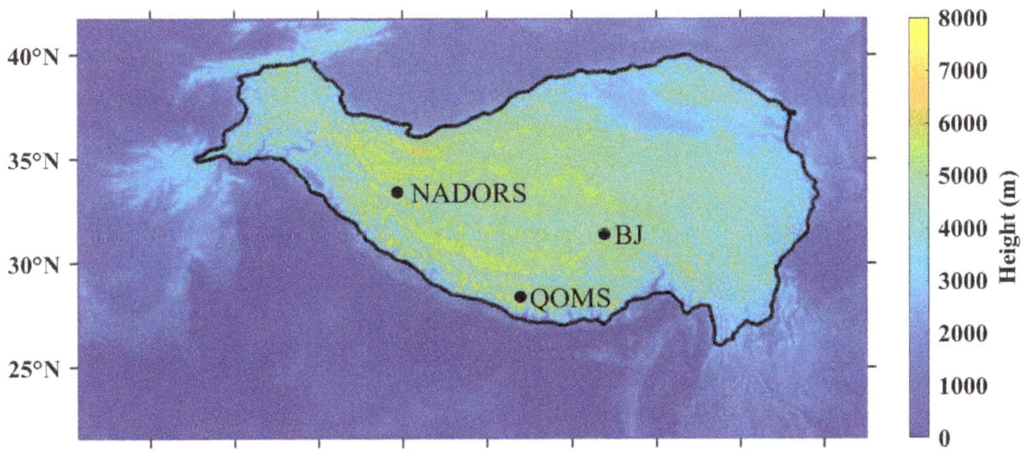

Figure 1. Geographical distribution of the observational sites (BJ, QOMS, and NADORS) over the TP (the colored shading denotes topography above sea level).

2.2. Observations

The soil data used in this study, measured at the period of 2014–2016, are shared by Ma et al. [52] at the Science Data Bank (https://doi.org/10.11922/sciencedb.00103 (accessed on 26 August 2021), Ma et al. [52]), and the National Tibetan Plateau Data Center (https://doi.org/10.11888/Meteoro.tpdc.270910 (accessed on 26 August 2021)). The data were hourly values. The information of soil sensors used at each site is presented in Table 2. Note that at 2014 BJ the soil temperature and moisture were measured at the depths of 0.04 m, 0.10 m and 0.20 m, and 0.04 m and 0.20 m, respectively. Both the soil temperature and moisture were measured at the depths of 0.05 m, 0.10 m, and 0.20 m at 2015–2016 BJ.

Table 2. The sensors and observation information at each site.

Site	Variables	Senser Models	Manufactures	Depths	Accuracy	Units
BJ	Tsoil	TS-301/TR-219L	Okazak/ Tri-Tronics	0.04/0.10/0.20 m; 0.05/0.10/0.20 m	Unknown	°C
	θ	CS616-L	Campbell	0.04/0.20 m; 0.05/0.10/0.20 m	±2.5% θ	$m^3\ m^{-3}$
QOMS	Tsoil	Model 107	Campbell	0.0/0.10/0.20 m	≤±0.01 °C	°C
	θ	CS616	Campbell	0.0/0.10/0.20 m	±2.5% θ	$m^3\ m^{-3}$
NADORS	Tsoil	CSI 109	Campbell	0.0/0.20 m	≤±0.03 °C	°C
	θ	CS616	Campbell	0.0/0.20 m	±2.5% θ	$m^3\ m^{-3}$

The NDVI data were obtained from MYD13C2-v006, provided on a monthly basis, with a spatial resolution of 0.05° × 0.05° (https://giovanni.gsfc.nasa.gov/giovanni/, accessed on 15 April 2022).

2.3. The Method Used to Determine Soil Apparent Thermal Diffusivity

Expanding the heat conduction equation presented by Van Wijk and De Vries [30], Gao et al. [23] presented the conduction and convection heat transfer equation with an assumption that the soil apparent thermal diffusivity was vertically homogenous, as follows:

$$\frac{\partial T}{\partial t} = k\frac{\partial^2 T}{\partial z^2} + W\frac{\partial T}{\partial z} \tag{1}$$

where T (°C) is soil temperature, t (s) is the time, and z (m) is the vertical coordinate positive downward; k ($m^2\ s^{-1}$) is soil apparent thermal diffusivity, W ($m\ s^{-1}$) is the apparent convection parameter. With a boundary condition described with the sine function of soil temperature, Gao et al. [23] obtained an analytical solution of this heat transfer equation, and derived the equations of k, as follows:

$$k = -\frac{(z_1 - z_2)^2 \omega \ln(A_1/A_2)}{(\Phi_1 - \Phi_2)\left[(\Phi_1 - \Phi_2)^2 + \ln^2(A_1/A_2)\right]} \tag{2}$$

where z_1 (m) and z_2 (m) are the measurement depths of soil temperature; A_1 (°C) and A_2 (°C) are soil temperature amplitude at the depths of z_1 and z_2, respectively, Φ_1 (rad) and Φ_2 (rad) are soil temperature phase at the depths of z_1 and z_2, respectively; ω ($=2\pi/P = 7.292 \times 10^{-5}\ rad^{-1}$) is the angular velocity of the Earth's rotation; and P ($=24 \times 3600$ s) is the harmonic period of the soil temperature.

2.4. Data Processing

Equation (2) is the conduction–convection method for determining k, which is the same equation derived by McCallum et al. [53] and Luce et al. [54] for saturated soil. In order to determine k at various timescales, we first derived hourly A and Φ of daily soil temperature at two depths using the DHR (See Appendix A) from the Captain toolbox [50,51], and then put them into Equation (2) to obtain hourly k. The daily and monthly values of k were obtained by averaging the hourly values.

To ensure the quality of k, the first 3 days of soil k data from the beginning and end of the data collection periods were discarded due to the edge effects of digital filtering with DHR [55,56]. Gordon et al. [57] suggested that data from any time series that have amplitudes below the sensor resolution should be treated with suspicion. To minimize the amount of suspicion, we deleted the data when the soil temperature amplitude at 0.20 m depth was below the values of 0.5 °C, 0.1 °C and 0.1 °C at BJ, QOMS and NADORS, respectively. After data deletions, approximately 95–100%, 96–100%, and 78–82% of the original data remained for analyses at BJ, QOMS and NADORS, respectively.

The soil moisture measured at the 0.10 m depth was used to represent the soil water status in the 0.0 m to 0.20 m layer. The θ measured at the 0.04 m depth was used as the soil water condition in the interest layer for 2014 BJ. For NADORS, the θ at the 0.10 m depth was calculated by the arithmetic mean of θ measured at the surface and the 0.20 m depth. Note that here θ indicates liquid water content in the soil, and the ice content is not measured. Ice content is qualitatively discussed based on soil temperature and initial soil liquid water content as described in the Section 4.2.

3. Results

3.1. The Variations of Soil Moisture

Figure 2 shows the variation of soil moisture (θ) over 2014–2016 at the BJ, QOMS, and NADORS. At BJ, θ varied greatly for most months except autumn and winter (October–December) 2015 and winter and spring (January–May) 2016. During non-winter periods, θ varied with rainfall; while during other periods, θ fluctuations were mainly attributed to soil thawing and freezing processes with temperature (e.g., spring (March–April), 2014; autumn (November), 2016). Soil temperatures fluctuated around the freezing temperature (Figure S1), resulting in a water phase transition between liquid and ice. The minimum θ values were similar for 2014–2016, around 0.05 $m^3\ m^{-3}$ at BJ, while the maximum θ values differed for the three years and occurred in summer, with about 0.33 $m^3\ m^{-3}$ in July–August 2014, 0.28 $m^3\ m^{-3}$ in August 2015, and 0.25 $m^3\ m^{-3}$ in July 2016, respectively.

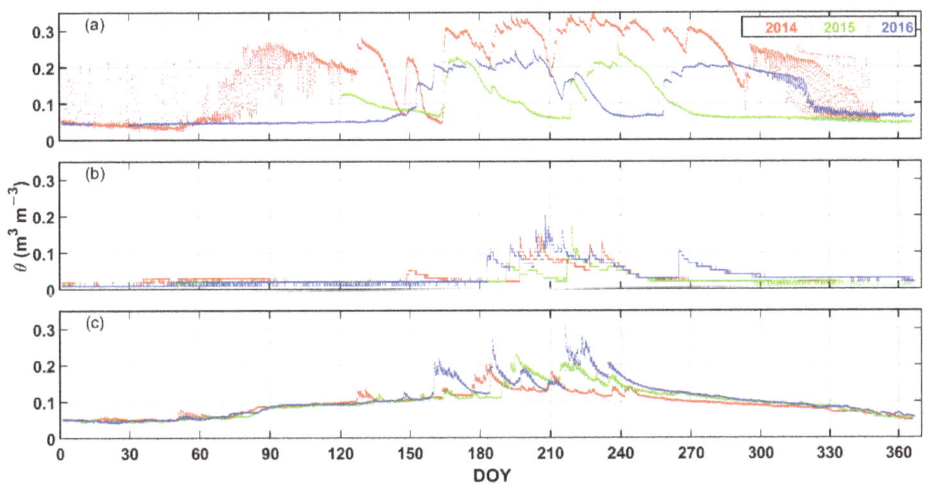

Figure 2. The variations of hourly soil moisture (θ, $m^3\ m^{-3}$) at the depth of 0.10 m at (**a**) BJ, (**b**) QOMS, and (**c**) NADORS for 2014–2016. Note that θ was measured at 0.04 m depth at BJ 2014, θ at NADORS were the arithmetic mean of soil moisture measured at the depths of 0.00 m and 0.20 m, and θ were directly measured at the 0.10 m depth for other cases.

Compared to BJ, the soil at QOMS and NADORS had distinct wet and dry cycles. The θ varied greatly in the summer and remained relatively constant during other periods. Besides, no large fluctuations in the θ were measured in winter at the two sites. The reason was that the θ decreased to a low value before winter, therefore, no large fluctuations in the θ occurred as soil temperature varied around the freezing temperature. The θ ranged from 0.01 m^3 m^{-3} to 0.20 m^3 m^{-3} at QOMS, and 0.05 m^3 m^{-3} to 0.30 m^3 m^{-3} at NADORS.

Overall, the soil at QOMS was driest, followed by NADORS and BJ. Increases in the θ were sharp after rain, and decreases in the θ were relatively slow as soil water evaporated.

3.2. The Variations of Soil Apparent Thermal Diffusivity

The hourly soil temperature amplitude (A) and phase (Φ) at two depths were extracted from soil temperature time series using DHR. For simplicity, only the 2014 results at the three sites are presented as illustrated in Figure 3, and the other results are shown in Figures S2–S4. The data were removed as A at 0.20 m depth was less than the corresponding threshold value at each site.

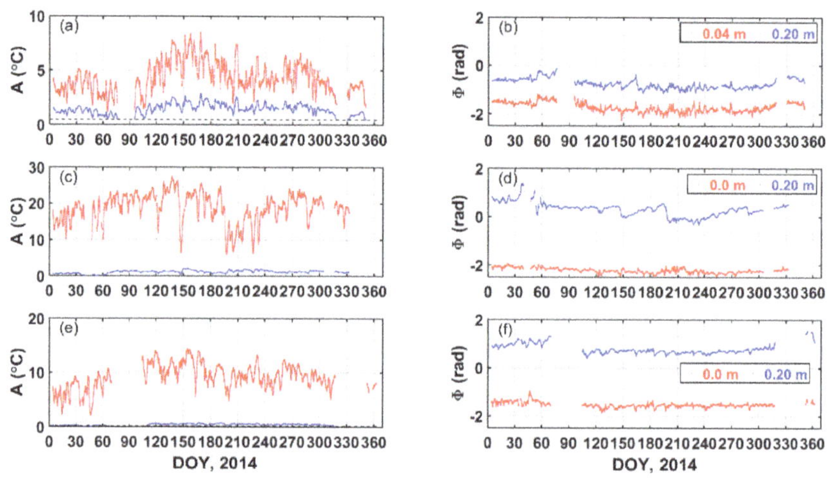

Figure 3. The variations of hourly soil temperature amplitude (A, °C) and phase (Φ, rad) at two depths of 0.04 m (or 0.0 m) and 0.2 m at (**a**,**b**) BJ, (**c**,**d**) QOMS, and (**e**,**f**) NADORS for 2014.

It is obvious that A varied with time, and its fluctuation at the shallow depth was much larger than that at the deeper depth (Figure 3a,c,e). As soil is heated by solar radiation at daytime and soil temperature generally decreases exponentially with depth [21], as does A. Compared to other periods, A was relatively small in winter and wet periods. Compared with A, the fluctuation of Φ was relatively small, and the deeper the depth, the larger the Φ. Since soil temperature phase shifts vary linearly with depth [21].

Among the sites, the differences in A and Φ at the two depths were larger at QOMS and NADORS than those at BJ. The soils at QOMS and NADORS were dry (corresponding to relatively small k, see the following results). Thus, soil temperature changes were transmitted slowly through the soil, resulting in large differences in A and Φ between the two depths at the two sites. Tong et al. [58] derived the relationship between conduction–convection k and $\ln(A_1/A_2)$ and $(\Phi_2 - \Phi_1)$ by taking partial derivatives, finding that when $\ln(A_1/A_2)$ is constant, k increases as $(\Phi_2 - \Phi_1)$ decreases; when $(\Phi_2 - \Phi_1)$ is constant, k increases (decreases) with increasing $\ln(A_1/A_2)$ when $(\Phi_2 - \Phi_1) > \ln(A_1/A_2)$ $[(\Phi_2 - \Phi_1) < \ln(A_1/A_2)]$.

After obtaining hourly A and Φ values for soil temperature at the first depth and the 0.20 m depth, hourly k was determined with Equation (2). The daily k was also obtained by averaging the hourly values over a day. The variations of the hourly and daily k for 2014–2016 at the three sites are shown in Figure 4.

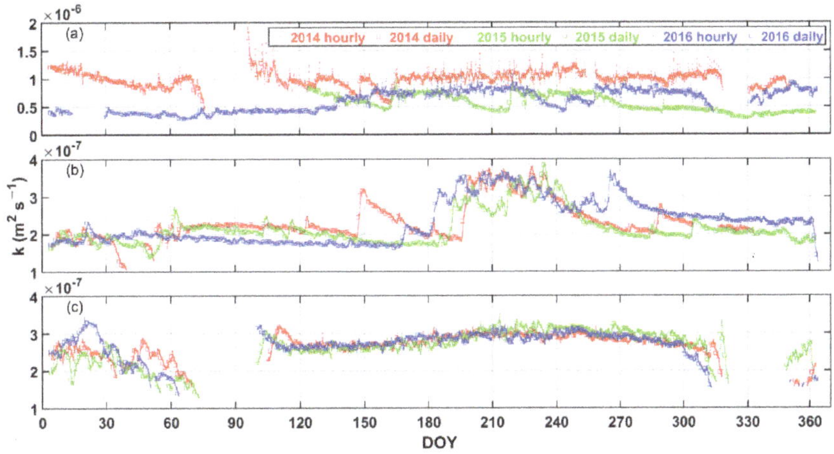

Figure 4. The variations of hourly (marked with dot) and daily (marked with square) soil apparent thermal diffusivity (k, $m^2\ s^{-1}$) of the 0.0 m to 0.20 m layer at (**a**) BJ, (**b**) QOMS, and (**c**) NADORS for 2014–2016.

Generally, k varied with time to varying degrees at different sites and years. At BJ and QOMS, k had an obvious seasonal variation (Figure 4a,b), and it roughly varied with θ during the wetting periods (compare Figure 2a,b to Figure 4a,b).

At BJ, k values in 2014 were larger than those in 2015 and 2016 except for some time in June. The 2014 k varied around $1.0 \times 10^{-6}\ m^2\ s^{-1}$, and the minimum and maximum values were about $0.5 \times 10^{-6}\ m^2\ s^{-1}$ and $1.9 \times 10^{-6}\ m^2\ s^{-1}$, respectively. For 2015–2016, the k ranged from $0.3 \times 10^{-6}\ m^2\ s^{-1}$ to $0.9 \times 10^{-6}\ m^2\ s^{-1}$. For QOMS, k fluctuated between $1.0 \times 10^{-7}\ m^2\ s^{-1}$ to $4.0 \times 10^{-7}\ m^2\ s^{-1}$ over 2014–2016, and except for summer time, k varied slightly and was relatively small at most time. The k varied almost exclusively with θ, namely, large k corresponded to large θ (compare Figure 2b to Figure 4b).

Compared to BJ and QOMS, the differences in k at NADORS among the three years were relatively small, and k varied slightly during the non-winter periods, although the θ obviously varied. From January to mid-March, k tended to decrease from $3.3 \times 10^{-7}\ m^2\ s^{-1}$ to $1.2 \times 10^{-7}\ m^2\ s^{-1}$; while it was relatively stable in spring–mid autumn (May-October), ranging from $2.5 \times 10^{-7}\ m^2\ s^{-1}$ to $3.2 \times 10^{-7}\ m^2\ s^{-1}$.

Figure 4 shows that k is not always constant throughout a day, and it can change drastically when the soil is wetted (e.g., DOY 218 in 2015 at BJ; DOY 150 in 2014 at QOMS).

The monthly variations of k and θ for 2014–2016 at the three sites are further examined in Figure 5. The values of the monthly k (mean ± one standard deviation) are listed in Table 3.

At BJ, the monthly median k fluctuated around $1.0 \times 10^{-6}\ m^2\ s^{-1}$ in 2014. While the 2015 monthly k was around $7.0 \times 10^{-7}\ m^2\ s^{-1}$ in May–September and decreased to about $4.0 \times 10^{-7}\ m^2\ s^{-1}$ in October–December, the monthly k was around $4.0 \times 10^{-7}\ m^2\ s^{-1}$ in February–May and increased to $7.5 \times 10^{-7}\ m^2\ s^{-1}$ after May in 2016. For QOMS, the monthly k peaked in August 2014-2015, with a median value of $3.2 \times 10^{-7}\ m^2\ s^{-1}$, and varied around $2.0 \times 10^{-7}\ m^2\ s^{-1}$ for most other months. In July and August 2016, the monthly k value was the largest, at $3.2 \times 10^{-7}\ m^2\ s^{-1}$, after which it decreased almost linearly until December ($2.3 \times 10^{-7}\ m^2\ s^{-1}$). The monthly k was relatively stable at around $1.8 \times 10^{-7}\ m^2\ s^{-1}$ before July.

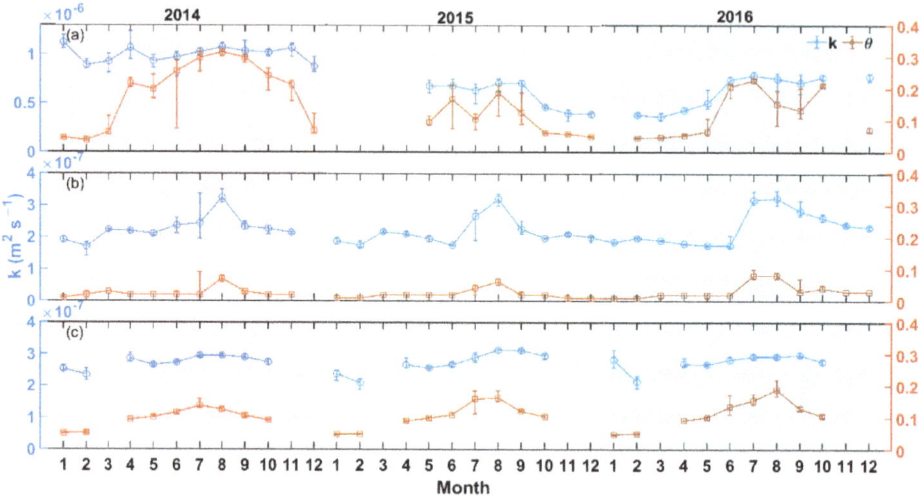

Figure 5. The monthly variations (25th, median, and 75th percentiles) of soil apparent thermal diffusivity (k, $m^2\ s^{-1}$) and moisture (θ, $m^3\ m^{-3}$) at (**a**) BJ, (**b**) QOMS, and (**c**) NADORS for 2014–2016. Data are deleted when there are less than 15 days available in a month.

Table 3. The monthly (mean ± one standard deviation) $k \times 10^7$ ($m^2\ s^{-1}$) at each site.

Month	BJ			QOMS			NADORS		
	2014	2015	2016	2014	2015	2016	2014	2015	2016
1	11.2 ± 0.8	- [1]	-	1.9 ± 0.1	1.9 ± 0.1	1.9 ± 0.1	2.6 ± 0.1	2.3 ± 0.2	2.9 ± 0.3
2	9.0 ± 0.6	-	3.8 ± 0.3	1.7 ± 0.3	1.8 ± 0.2	1.9 ± 0.1	2.4 ± 0.2	2.1 ± 0.3	2.2 ± 0.3
3	9.0 ± 1.4	-	3.7 ± 0.5	2.2 ± 0.1	2.2 ± 0.1	1.9 ± 0.0	-	-	-
4	11.2 ± 2.3	-	4.3 ± 0.2	2.2 ± 0.0	2.1 ± 0.1	1.8 ± 0.0	2.9 ± 0.2	2.7 ± 0.2	2.8 ± 0.2
5	9.1 ± 1.1	6.9 ± 0.8	5.2 ± 1.0	2.2 ± 0.3	2.0 ± 0.0	1.7 ± 0.0	2.7 ± 0.1	2.6 ± 0.1	2.7 ± 0.1
6	9.3 ± 1.6	6.5 ± 1.2	7.4 ± 0.7	2.4 ± 0.3	1.8 ± 0.0	1.9 ± 0.0	2.8 ± 0.1	2.7 ± 0.1	2.8 ± 0.1
7	10.3 ± 0.9	6.1 ± 1.2	7.9 ± 0.6	2.7 ± 0.7	2.5 ± 0.5	3.2 ± 0.3	3.0 ± 0.1	2.9 ± 0.2	2.9 ± 0.1
8	10.8 ± 0.9	6.7 ± 1.3	6.9 ± 1.5	3.3 ± 0.2	3.2 ± 0.3	3.2 ± 0.3	3.0 ± 0.1	3.1 ± 0.1	2.9 ± 0.1
9	10.6 ± 1.0	6.7 ± 0.9	7.1 ± 1.2	2.4 ± 0.2	2.3 ± 0.3	2.9 ± 0.3	2.9 ± 0.1	3.1 ± 0.1	3.0 ± 0.1
10	10.1 ± 0.7	4.6 ± 0.2	7.7 ± 0.5	2.4 ± 0.2	2.0 ± 0.0	2.6 ± 0.1	2.8 ± 0.1	2.9 ± 0.1	2.7 ± 0.2
11	10.3 ± 1.2	3.9 ± 0.6	-	2.2 ± 0.0	2.1 ± 0.0	2.4 ± 0.0	-	-	-
12	8.8 ± 0.8	3.8 ± 0.2	7.6 ± 0.8	-	1.9 ± 0.1	2.3 ± 0.0	-	-	-

[1] indicates there is no data.

Differing from BJ and QOMS, the monthly k at NADORS varied little, about $3.0 \times 10^{-7}\ m^2\ s^{-1}$ for most of the non-winter period (April-October), although θ did vary during this period. It decreased from January to February over the 3-year period, ranging from $3.0 \times 10^{-7}\ m^2\ s^{-1}$ to $1.0 \times 10^{-7}\ m^2\ s^{-1}$.

Overall, the variation trends of monthly k were roughly similar to those of θ, except for BJ 2014 and NADORS 2016. The monthly k at BJ ranged from $0.4(\pm 0.0) \times 10^{-6}\ m^2\ s^{-1}$ to $1.1(\pm 0.2) \times 10^{-6}\ m^2\ s^{-1}$, from $1.7(\pm 0.0) \times 10^{-7}\ m^2\ s^{-1}$ to $3.3(\pm 0.2) \times 10^{-7}\ m^2\ s^{-1}$ at QOMS, and from $2.1(\pm 0.3) \times 10^{-7}\ m^2\ s^{-1}$ to $3.1(\pm 0.1) \times 10^{-7}\ m^2\ s^{-1}$ at NADORS (Table 3).

3.3. The Relationship between Soil Apparent Thermal Diffusivity and Moisture

Figure 6 shows how k varies with θ on an hourly timescale in 2014–2016 at the sites. Overall, the trends of unfrozen soil k versus θ at the three sites were roughly similar, i.e., k increased rapidly to a maximum value with increasing θ and then tended to be constant or decrease slightly as θ increased further. The values of θ corresponding to peak k values were different, and the θ values at QOMS were less than those at BJ.

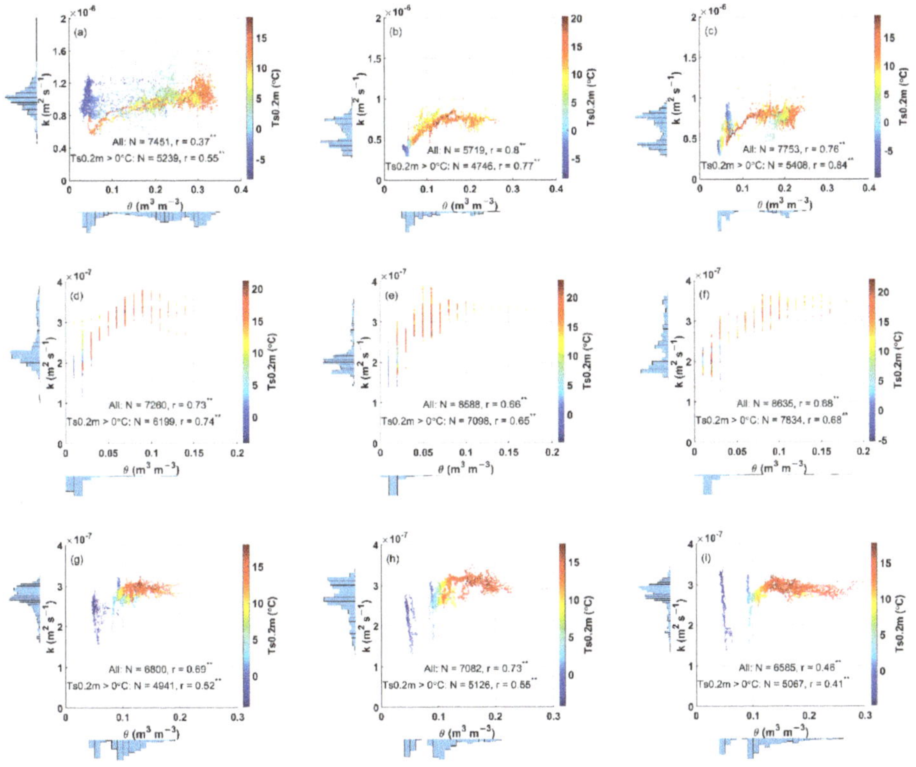

Figure 6. The variation of soil apparent thermal diffusivity (k, $m^2\ s^{-1}$) with soil moisture (θ, $m^3\ m^{-3}$) at 0.10 m depth on an hourly timescale for 2014 (in the 1st column), 2015 (in the 2nd column) and 2016 (in the 3rd column) at (**a–c**) BJ, (**d–f**) QOMS, and (**g–i**) NADORS, respectively. The color bar indicates the value of soil temperature at the depth of 0.20 m (Ts0.2m), and the larger the temperature, the redder the color. The marker size suggests the amplitude of Ts0.2m (A0.2m), and the larger the value of A0.2m, the larger the marker. The correlation coefficients (r) between k and θ are provided for all of the data and for the data when Ts0.2m > 0 °C, respectively. "**" indicates *p*-value < 0.01. The probability distributions of k and θ are shown on the y-axis and x-axis sides, respectively.

Interestingly, the relationship between k and θ did not appear to be significant when Ts0.2m < 0 °C and θ < 0.1 $m^3\ m^{-3}$, i.e., k fluctuated greatly within a narrow range of θ (e.g., see the blue points in Figure 6a,c,g–i). The correlation coefficients (r) between k and θ for all of the data ranged from 0.66 to 0.80 except for BJ 2014 (r = 0.37) and NADORS 2016 (r = 0.46). Under the condition of Ts0.2m > 0 °C, the r coefficients changed to different degrees at each site and year, and the changes depended on the location of the data of Ts0.2m ≤ 0 °C on the curve. Overall, without including data for Ts0.2m ≤ 0 °C, the r coefficients decreased at NADORS, increased at BJ, and changed slightly at QOMS, respectively.

Note that the probability distributions of k were different, especially at different sites. The same is true for θ. Compared to k, the probability distribution of θ for a given site was more consistent during 2014–2016. There was an obvious gap between 0.05 $m^3\ m^{-3}$ to 0.10 $m^3\ m^{-3}$ at NADORS (Figure 6g–i), since the data in the θ range were deleted according to the standard of A0.2m < 0.1 °C, as mentioned in Section 2.4.

The relationship between k and θ on a daily timescale (Figure S5) was similar to that on an hourly timescale. Few studies have investigated the relationship between k and θ for soil below and above the freezing temperature simultaneously.

We furthermore investigated the relationship between median k and θ on a monthly timescale, as shown in Figure 7. Similar to the hourly results, the monthly k tended to increase with θ when θ was relatively small, reached a peak value and then became relatively stable as θ increased further.

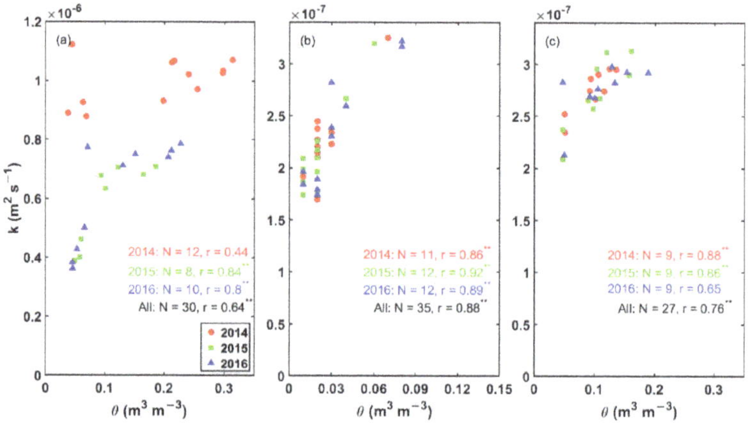

Figure 7. Same as Figure 6, but for the relationship between median k and θ on a monthly timescale. The number of months per year (N) and corresponding correlation coefficients (r) are given. "**" indicates p-value < 0.01, and there is no label after r if r is not significant (p-value > 0.05). The data for a month are deleted when the number of days is less than 15.

Overall, monthly k was significantly correlated to θ regardless of site and year, except for 2014 BJ and 2016 NADORS. The significant (p-value < 0.01) r ranged from 0.80 (2016 BJ) to 0.92 (2015 QOMS) on a monthly timescale, and from 0.64 (NADORS) to 0.88 (QOMS) on an annual timescale. The r coefficients of k vs. θ on a monthly timescale were larger than those on an hourly timescale. This could be explained because the effect of frozen soil on the relationship between k and θ was greatly reduced on a monthly timescale.

In order to indirectly investigate the effect of soil moisture on k from the vegetation aspect, the relationship between monthly k and NDVI is also examined as shown in Figure 8. The ranges of NDVI were different at the three sites, and the maximum NDVI at BJ (0.51) was approximately 2.5 times that of QOMS (0.22) and NADORS (0.18).

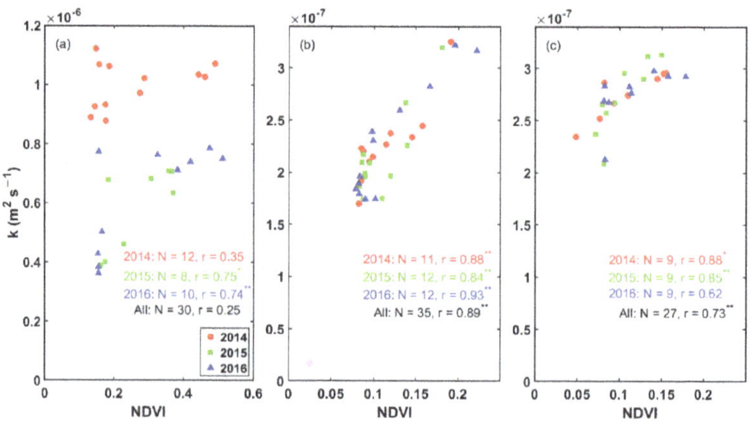

Figure 8. Same as Figure 7, but for the relationship between k and NDVI. "*" indicates p-value < 0.05, "**" indicates p-value < 0.01, and there is no label after r if r is not significant (p-value > 0.05).

Interestingly, the monthly k had a similar correlation with NDVI as it did with θ. At QOMS and NADORS, the r coefficients of k vs. NDVI were close to those of k vs. θ ($\Delta r < 0.08$), while the r coefficients of k vs. NDVI were smaller than those of k vs. θ at BJ.

4. Discussion

4.1. The Variations of Soil Apparent Thermal Diffusivity

With k methods based on the solution of the soil heat transfer equation, normally only daily or longer timescale k values were obtained in previous studies. In this study, we used a conduction–convection method combined with DHR to obtain hourly k. The daily and monthly k were also provided for 2014–2016 at the three sites (Figures 4 and 5 and Table 3). Figure 4 visually shows that in a day, k is not necessarily constant over a day, and it can change drastically when the soil become wetter suddenly. The knowledge of k with higher temporal resolution may have great implications in improving soil heat and water models. The issue of underestimating soil temperature at night with the conduction-convection method [24] may be largely resolved by using hourly k values instead of daily k values as inputs. Besides, higher temporal resolution k may help to improve the modeling of permafrost distributions [11].

Figure 4 also indicates that k had obvious seasonal variations at BJ and QOMS, while it did not vary greatly during wetting at NADORS. The soil types for both QOMS and NADORS were sandy and gravel, but θ at NADORS during wetting (> 0.10 $m^3\ m^{-3}$) was greater than that at QOMS (Figure 2b,c). Previous studies suggested that k variations during wetting depended on the magnitude of θ, and k was insensitive to changes in θ when θ reached certain thresholds (e.g., 0.1~0.2 $m^3\ m^{-3}$ for sand soils, [16]), which could explain why k did not vary much during wetting at NADORS.

The k values at QOMS and NADORS were much less than those at BJ (Figures 4 and 5). In addition to the soil moisture (Figure 2), the soil bulk density at QOMS and NADORS was expected to differ from that at BJ, since their soil types were sand and gravel, whereas BJ's soil type was sandy silt loam. Previous studies indicated that k varied with bulk density as well as soil moisture [17,18,59–61]. Therefore, the combined effects of the soil texture, moisture and bulk density resulted in the relative magnitude of k at the three sites. At BJ, the k values in 2014 were much larger than those at 2015 and 2016 (Figure 6a–c). The main reason may be due to differences in soil bulk density and soil moisture content.

With the conduction–convection method, Gao [26] determined the daily k ranging from 0.1×10^{-6} to 2.0×10^{-6} $m^2\ s^{-1}$ at BJ during DOYs 195 to 258, 1998, which was within the range of k at BJ for 2014–2016 in this study. Additionally, with the conduction–convection method, Zhou et al. [28] determined daily k based on soil temperatures measured at 0.8 m and 3.2 m depths in 39 weather sites in the TP during 1980 to 2001. They reported that the magnitude of k in most areas of the TP was 10^{-7} $m^2\ s^{-1}$, and relatively high k values were obtained at the central and eastern parts of the plateau with an order of magnitude of 10^{-6} $m^2\ s^{-1}$. Our results that k at BJ are larger than those at QOMS and NADORS are broadly consistent with the findings of Zhou et al. [28]. The soil thermal properties were determined at a cold semi-desert site on the western TP for about two years by Wang et al. [48]. They calculated k as the ratio of λ/C_V, which were determined by soil temperature and heat fluxes measured at two layers according to the method proposed by Zhang and Huang [49]. Their daily k values ranged from 3.0×10^{-7} $m^2\ s^{-1}$ to 9.0×10^{-7} $m^2\ s^{-1}$, which was larger than our k values at QOMS and NADORS, and less than those at BJ. The differences in k were attributed to the differences in soil texture, moisture, and the method of determining k. Even with a well-calibrated soil heat flux sensor, it is difficult to measure the soil heat flux accurately [62–65], because the soil heat and moisture fluxes are disturbed [66]. Heat flux plates measure only sensible heat as it moves past the plate by means of the temperature gradient, which exists across the plate. Latent heat, which is hidden in the evaporative process, is not detected [67]. Therefore, cautions should be exercised when determining soil thermal properties using soil heat flux plate data.

Using monthly soil temperature data measured in the 0.2 m to 3.2 m soil layer at the northern permafrost regions, Zhu et al. [11] obtained k with a conduction method, ranging from 0.2×10^{-7} m^2 s^{-1} to 2.0×10^{-6} m^2 s^{-1}. The minimum k values occurred at sites with large soil organic carbon (SOC, 60~70 kg C m^{-3}), and they were less than our minimum results. A possible reason is that there is more SOC at the northern permafrost regions than at our sites, and the k of SOC is an order of magnitude smaller than that of typical soil minerals [68].

Few studies provided long-term k using a conduction–convection method on different timescales over the TP. The long-term k values obtained in this study on different timescales can be used as input for land surface models over this region.

4.2. The Relationship between Soil Apparent Thermal Diffusivity and Soil Moisture

Normally, k increases rapidly with increasing θ to a maximum and then decreases with further increases in θ. This is explained by the fact that the heat capacity increases linearly with θ, whereas the thermal conductivity experiences its most rapid rise at low θ, leading to the ratio of thermal conductivity to heat capacity to have an internal maximum as a function of θ [29,69]. The increase rate of k with θ differs for different soil textures. A laboratory experiment showed that for sandy soils, k increased rapidly with θ when θ was less than 0.1~0.2 m^3 m^{-3}, and then it remained stable or slightly decreased; while the variation of k with θ was smaller for silty and clay-textured soils [16].

Figure 6 shows that although the probability distributions of k and θ differed, especially at different sites, the variation trends of unfrozen soil k versus θ were roughly similar, which was consistent with previous studies reported based on laboratory measured data under controlled conditions (e.g., [16,17,29,59]) and in situ data (e.g., [22,25,70]). We should be aware that this trend of k versus θ reported with laboratory measurements generally only applies when the soil temperature is room temperature, since measurements in the laboratory are usually conducted at room temperature. However, if some ice is present in a soil layer, the k versus θ relationship is expected to deviate from this trend, since ice has a higher k value than water (1.1 vs. 0.14×10^{-6} m^2 s^{-1}, [68]). That is why there are some "outliers" in the trend of k versus θ as mentioned above, i.e., k varied greatly within a narrow range of θ, as the blue point when θ < 0.1 m^3 m^{-3} shown in Figure 6a,c,g–i.

About 2/3 blue points in Figure 6c (when θ ranged from 0.06 m^3 m^{-3} to 0.075 m^3 m^{-3}) appeared on DOYs 332–364. To elucidate this phenomenon, the variations of soil temperature, θ and k over DOYs 300–365, 2016, at BJ are shown in Figure 9. One can see that on DOYs 332–364 soil liquid water content, θ in the 0.05–0.20 m layer showed a downward trend, and as the soil temperature dropped to the freezing point, more and more ice was expected to form. At the same time, θ at the depth of 0.10 m varied slightly, while k tended to increase greatly. Therefore, one can see that k had a wide range from 0.6×10^{-6} m^2 s^{-1} to 0.95×10^{-6} m^2 s^{-1} with a narrow range of θ in Figure 6c. Note that almost no water loss is assumed since DOY 300 here, as soil water evaporation is generally small when soil temperature is below freezing. Therefore, the decrease in liquid water content in the soil layer is expected to be a result of increasing ice content.

For Figure 6i, the "outliers" (when θ < 0.054 m^3 m^{-3} and k ranged from 1.4×10^{-7} m^2 s^{-1} to 3.4×10^{-7} m^2 s^{-1}) appeared on DOYs 1–63, 2016 at NADORS. The variations of soil temperature, θ and k over DOYs 1–70, 2016, at NADORS are shown in Figure 10. In contrast to Figure 9, the soil temperature tended to increase from DOYs 1 to 63, resulting in an increasing amount of time that soil temperature was above freezing (Figure 10a). We expect that ice content in the 0–0.20 m soil layer tended to decrease over time, although θ appeared to vary slightly. Figure 10 also shows that as the ice content decreased with increasing soil temperature, k tended to decrease, eventually varying from approximately 1.4×10^{-7} m^2 s^{-1} to 3.4×10^{-7} m^2 s^{-1}. Therefore, one can see that k varied greatly with a narrow range of θ in Figure 6i. Figures 9 and 10 explains why large k changes occurred over a small range of θ in Figure 6a,c,g–i.

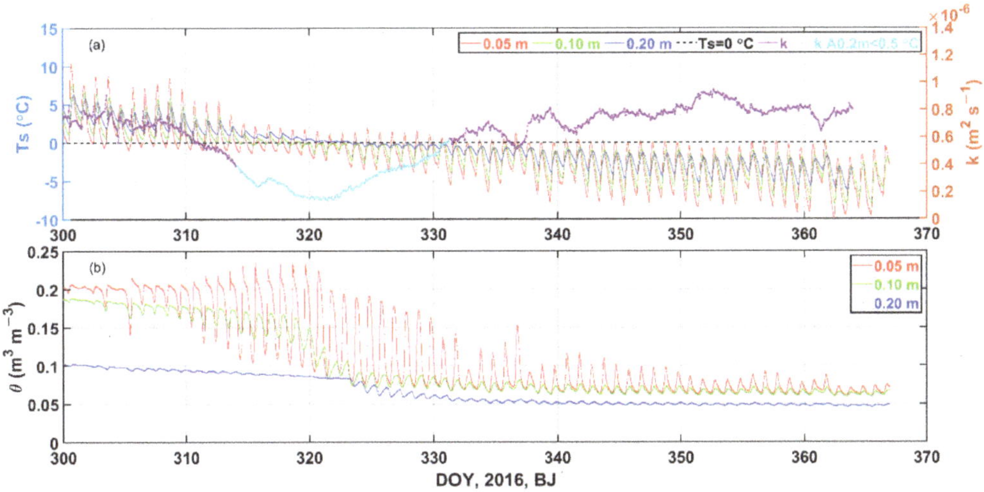

Figure 9. The (**a**) soil temperature (Ts, °C) at the depths of 0.05 m, 0.10 m and 0.20 m, and soil apparent thermal diffusivity (k, $m^2\ s^{-1}$), and (**b**) soil moisture (θ, $m^3\ m^{-3}$) at the depths of 0.05 m, 0.10 m and 0.20 m over DOYs 300–365, 2016, at BJ. The labels of k are marked in cyan when the soil amplitude at the 0.2 m depth is less than 0.5 °C.

Compared to BJ and NADORS, there were no apparent "outliers" at QOMS (Figure 6d–f). The soil at QOMS was driest among the three sites, and θ decreased to the minimum value (<0.03 $m^3\ m^{-3}$) in September from the summer (Figure 2b), and small θ lasted until winter. Some laboratory experiments indicated that the unfrozen water content of freezing soil was largely controlled by the initial volumetric water content [71], which may prove why there was almost no ice after September at QOMS. Ochsner and Baker [20] presented some in situ measurements of soil thermal properties across a full range of temperatures encountered in freezing and thawing soil, and the measurement and model both showed that for temperatures between −5 °C and 0 °C, soil thermal properties were strongly temperature dependent. They explained that temperature dependence was primarily the result of latent heat transfer processes when water underwent a phase change. Therefore, although soil temperature changed from positive to negative at QOMS in autumn, the "frozen" soil had a lesser effect on k since little ice was produced. Thus, few "outliers" existed during the cold months at QOMS (Figure 6d–f).

To indirectly investigate the effect of soil moisture on k from the vegetation aspect, the relationship between monthly k and NDVI was also examined (Figure 8), which is similar to the relationship between k to θ (Figure 7). The reason may be that NDVI is closely related to θ at these sites. The r coefficients between NDVI and θ ranged from 0.73 to 0.92 on a monthly timescale, and from 0.69 to 0.82 on an annual timescale (Table 4). Compared to the other two sites, more vegetation was present in BJ, so the NDVI of BJ represented more vegetation information than bare soil. Therefore, the correlation between k and NDVI was weaker than that between k and θ at BJ, while closer to that between k and θ at the other two sites, as shown in Figures 7 and 8.

With in situ soil temperature, we determined k by the conduction–convection method for 2014–2016 at three TP sites and examined the relationship between k and θ. Our results indicated that k had a clear relationship with θ for unfrozen soil, but the relationship changed when the soil temperature was less than 0 °C and the initial θ was not too small. These findings broaden our understanding of the relationship between in situ k and θ.

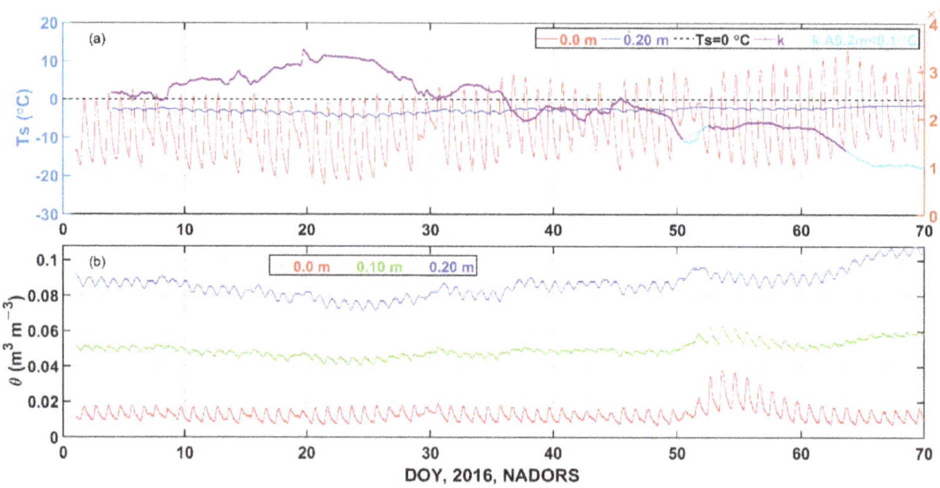

Figure 10. The (**a**) soil temperature (Ts, °C) at the depths of 0.0 m and 0.20 m, and soil apparent thermal diffusivity (k, m^2 s^{-1}), and (**b**) soil moisture (θ, m^3 m^{-3}) at the depths of 0.0 m, 0.10 m and 0.20 m over DOYs 0–70, 2016, at NADORS. The labels of k are marked in cyan when the soil amplitude at the 0.2 m depth is less than 0.1 °C.

Table 4. Correlation coefficient (r) of the monthly soil moisture and NDVI for 2014-2016 at the three sites.

Site	2014	2015	2016	All
BJ	0.82 ** [1]	0.75 * [2]	0.86 **	0.69 **
QOMS	0.73 *	0.87 **	0.92 **	0.86 **
NADORS	0.87 **	0.88 **	0.91 **	0.86 **

[1] ** *p*-value < 0.01, [2] * *p*-value < 0.05.

4.3. Limitations

In this study, long-term k was determined for frozen and unfrozen conditions. Due to the uncertainty of the k method, several k values were removed during the transition periods with soil thawing/freezing when soil temperature variations were low, as mentioned in Section 2.4. Another method or sensor is needed to determine k during soil thawing/freezing periods.

Our results indicated a potential effect of ice content on the relationship between k and θ, while the amount of ice content was inferred by combining the variations of soil temperature and soil moisture, rather than direct measurements. The ice content measured in situ is vital to quantify the relationship between k and soil moisture (including liquid and ice phases). The thermo-TDR sensor is a candidate for in situ measurement of both liquid and ice contents, since its performance was satisfactory in laboratory experiments reported in previous studies (e.g., [72–74]). In addition, thermo-TDR was also used in the field to measure soil thermal properties during thawing and freezing (e.g., [20]). By using actively heated fiber Bragg grating (AH-FBG) sensors, Wu et al. [75] measured the ice content of frozen soil in laboratory. The AH-FBG sensor integrates the functions of active heating and temperature measurement, which can accurately detect the thermal response of frozen soil [75]. We recommend using soil thermocouples and thermo-TDR sensors or only AH-FBG sensors for soil temperature, water content (liquid and ice phase) and thermal property measurements over multiple thawing and freezing cycles to more deeply explore the time variations of k and its relationship with water content.

In this study, we did not examine the effects of other soil factors (e.g., soil texture, SOC) on k due to a lack of data. Zhu et al. [11] suggested that SOC is the dominate factor (among soil texture, bulk density, moisture, and SOC) controlling the variability of diffusivity at 200 sites in high latitude regions, and k is a strong predictor for simulated permafrost extent. In future investigations, additional soil factors should be included in the study of long-term variations of soil thermal properties.

5. Conclusions

Based on in situ soil temperature data measured at three TP sites (BJ, QOMS, and NADORS), we determined the hourly, daily, and monthly soil apparent thermal diffusivity values of the 0.0 m to 0.20 m layer for 2014–2016 by using a conduction–convection method combined with DHR. The hourly, daily, and monthly k values of the 0.0 m to 0.20 m layer were obtained. The hourly and daily k values ranged from 0.3×10^{-6} m^2 s^{-1} to 1.9×10^{-6} m^2 s^{-1} at BJ, and from 1.0×10^{-7} m^2 s^{-1} to 4.0×10^{-7} m^2 s^{-1} at QOMS and NADORS. The monthly k ranged from $0.4(\pm0.0) \times 10^{-6}$ m^2 s^{-1} to $1.1(\pm0.2) \times 10^{-6}$ m^2 s^{-1} at BJ, from $1.7(\pm0.0) \times 10^{-7}$ m^2 s^{-1} to $3.3(\pm0.2) \times 10^{-7}$ m^2 s^{-1} at QOMS, and from $2.1(\pm0.3) \times 10^{-7}$ m^2 s^{-1} to $3.1(\pm0.1) \times 10^{-7}$ m^2 s^{-1} at NADORS. The results suggested that k was not constant over a day, and k showed seasonal variations. The variations of k with θ appeared to be roughly similar for unfrozen soil at these sites and years, namely, k increased sharply before it reached a peak value as θ increased, and then it tended to be stable or varied slightly with further increases in θ. The correlation coefficients (r) between k and θ ranged from 0.37 to 0.80, and 0.80 to 0.92 on hourly and monthly timescales, respectively. However, the relationship between k and θ changed when soil temperature was below 0 °C. Our results also suggested that the k and NDVI values were significantly related on monthly and annual timescales, with r ranging from 0.73 to 0.93. These results broaden our understanding of the relationship between in situ k and θ. The presented values of k at various timescales can be used as soil parameters when modeling land–atmosphere interactions at these TP regions.

Supplementary Materials: The following supporting information can be downloaded at: https://www.mdpi.com/article/10.3390/rs14174238/s1, Figure S1. The variations of soil temperature at (a) BJ, (b) QOMS, (c) NADORS, respectively. Figure S2. The amplitude (A, °C) and phase (Φ, rad) of soil temperature 2015–2016 at BJ. Figure S3. The amplitude (A, °C) and phase (Φ, rad) of soil temperature 2015–2016 at QOMS. Figure S4. The amplitude (A, °C) and phase (Φ, rad) of soil temperature 2015–2016 at NADORS. Figure S5. The variation of soil apparent thermal diffusivity (k, m^2 s^{-1}) with soil moisture (θ, m^3 m^{-3}) on a daily timescale in 2014 (in the 1st column), 2015 (in the 2nd column) and 2016 (in the 3rd column) at (a–c) BJ, (d–f) QOMS, and (g–i) NADORS, respectively.

Author Contributions: B.T.: formal analysis, writing—original draft, writing—review and editing, funding acquisition. H.X.: formal analysis, writing—review and editing. R.H.: conceptualization, writing—review and editing. L.B.: conceptualization, supervision. J.G.: supervision, funding acquisition, writing—review and editing. All authors have read and agreed to the published version of the manuscript.

Funding: This research was funded by the Second Tibetan Plateau Scientific Expedition and Research program (STEP) (2019QZKK0102), the Natural Science Foundation of China (U2142209), and the China Postdoctoral Science Foundation (2021M703558).

Data Availability Statement: The data that support the findings of this study are available from the Science Data Bank (https://doi.org/10.11922/sciencedb.00103; Ma et al. 2020 [52]) and additionally at the National Tibetan Plateau Data Center (https://doi.org/10.11888/Meteoro.tpdc.270910).

Acknowledgments: We appreciate the access to the NASA datasets. The authors appreciate all of the hard work done by researchers attending the Tibetan Plateau (TP) Experiment.

Conflicts of Interest: The authors declare that they have no known competing financial interests or personal relationships that could have appeared to influence the work reported in this paper.

Appendix A. Determination of Soil Temperature Amplitude and Phase with DHR

Gordon et al. [57] described how to extract soil amplitude and phase information from a soil temperature time series with the Dynamic Harmonic Regression (DHR). DHR, a simplification of the unobserved component model, has the form as follows:

$$y_t = T_t + C_t + e_t \tag{A1}$$

where y_t is the observed soil temperature time series, T_t is a trend or zero-frequency component, C_t is a cyclical component, and e_t is an irregular, white-noise component [50]. The C_t is modeled as a sum of the fundamental signal and its associated harmonics, as follows:

$$C_t = \sum_{i=1}^{N} [a_{i,t} cos(\omega_i t) + b_{i,t} sin(\omega_i t)] \tag{A2}$$

where $a_{i,t}$ and $b_{i,t}$ are stochastic time-varying parameters and ω_i (I = 1:N) are the fundamental frequency and its harmonics (ω_1) up to the Nyquist frequency (ω_N). This DHR model is a non-stationary extension of the discrete Fourier transform, where the amplitude (A) and phase (Φ) of the soil temperature for each time series vary with time. Identification of the time-varying parameters is achieved in a stochastic state formulation using two-step Kalman filtering and fixed-interval smoothing [50].

After obtaining the time-varying parameters, the A and Φ of any harmonic component at discrete time can be calculated by the following equations:

$$A_{i,t} = \sqrt{a_{i,t}^2 + b_{i,t}^2} \tag{A3}$$

$$\Phi_{i,t} = tan^{-1}(a_{i,t}/b_{i,t}) \tag{A4}$$

where $A_{i,t}$ and $\Phi_{i,t}$ are the amplitude and phase for the component with frequency ω_i at time t, respectively.

References

1. Horton, R. Soil thermal diffusivity. In *Methods of Soil Analysis: Part 4—Physical Methods;* Soil Science Society of America Book Series; Dane, J.H., Topp, G.C., Eds.; Soil Society of America: Madison, WI, USA, 2002; Volume 5, pp. 349–360.
2. Zhang, T.; Osterkamp, T.E. Considerations in determining thermal diffusivity from temperature time series using finite difference methods. *Cold Reg. Sci. Technol.* **1995**, *23*, 333–341. [CrossRef]
3. Passerat de Silans, A.M.B.; Monteny, B.A.; Lhomme, J.P. Apparent soil thermal diffusivity, a case study: HAPEX-Sahel experiment. *Agric. For. Meteor.* **1996**, *81*, 201–216. [CrossRef]
4. Dickinson, R.E.; Henderson-Sellers, A.; Kennedy, P.J. *Biosphere–Atmosphere Transfer Scheme (BATS) Version 1e as Coupled to the NCAR Community Climate Model;* NCAR Technical Note NCAR/TN-3871STR; NCAR: Boulder, CO, USA, 1993; 88p.
5. Sellers, P.J.; Randall, D.A.; Collatz, G.J.; Berry, J.A.; Field, C.B.; Dazlich, D.A.; Zhang, C.; Collelo, G.D.; Bounoua, L. A revised land surface parameterization (SiB2) for atmospheric GCMs. Part I: Model formulation. *J. Climate* **1996**, *9*, 676–705. [CrossRef]
6. Chen, F.; Dudhia, J. Coupling an advanced land surface–hydrology model with the Penn State–NCAR MM5 modeling system: Part I. Model implementation and sensitivity. *Mon. Wea. Rev.* **2001**, *129*, 569–582. [CrossRef]
7. Dai, Y.; Zeng, X.; Dickinson, R.E.; Baker, I.; Bonan, G.B.; Bosilovich, M.G.; Denning, A.S.; Dirmeyer, P.A.; Houser, P.R.; Niu, G.; et al. The Common Land Model. *Bull. Amer. Meteor. Soc.* **2003**, *84*, 1013–1023. [CrossRef]
8. Xia, Y.; Ek, M.; Sheffield, J.; Livneh, B.; Huang, M.; Wei, H.; Wood, E. Validation of Noah-simulated soil temperature in the North American Land Data Assimilation System phase 2. *J. Appl. Meteor. Climatol.* **2013**, *52*, 455–471. [CrossRef]
9. Gao, Z.; Horton, R.; Liu, H.P. Impact of wave phase difference between soil surface heat flux and soil surface temperature on soil surface energy balance closure. *J. Geophys. Res.* **2010**, *115*, D16112. [CrossRef]
10. Roxy, M.S.; Sumithranand, V.B.; Renuka, G. Estimation of soil moisture and its effect on soil thermal characteristics at Astronomical Observatory, Thiruvananthapuram, south Kerala. *J. Earth Syst. Sci.* **2014**, *123*, 1793–1807. [CrossRef]
11. Zhu, D.; Ciais, P.; Krinner, G.; Maignan, F.; Jornet Puig, A.; Hugelius, G. Controls of soil organic matter on soil thermal dynamics in the northern high latitudes. *Nat. Commun.* **2019**, *10*, 3172. [CrossRef]
12. Bristow, K.L.; Kluitenberg, G.J.; Horton, R. Measurement of soil thermal properties with a dual-probe heat-pulse method. *Soil. Sci. Soc. Amer. J.* **1994**, *58*, 1288–1294. [CrossRef]
13. Ren, T.S.; Noborio, K.; Horton, R. Measuring soil water content, electrical conductivity and thermal properties with a thermo-time domain reflectometry probe. *Soil Sci. Soc. Amer. J.* **1999**, *63*, 450–457. [CrossRef]

14. Campbell, G.S. *Soil Physics with BASIC: Transport Models for Soil–Plant Systems*; Elsevier Science: Amsterdam, The Netherlands, 1985; pp. 26–39.
15. Arkhangel'skaya, T.A. Parameterization and mathematical modeling of the dependence of soil thermal diffusivity on the water content. *Eurasian Soil Sci.* **2009**, *42*, 162–172. [CrossRef]
16. Arkhangelskaya, T.; Lukyashchenko, K. Estimating soil thermal diffusivity at different water contents from easily available data on soil texture, bulk density, and organic carbon content. *Biosyst. Eng.* **2018**, *168*, 83–95. [CrossRef]
17. Xie, X.; Lu, Y.; Ren, T.; Horton, R. An Empirical Model for Estimating Soil Thermal Diffusivity from Texture, Bulk Density, and Degree of Saturation. *J. Hydrometeorol.* **2018**, *19*, 445–457. [CrossRef]
18. Brunetti, C.; Lamb, J.; Wielandt, S.; Uhlemann, S.; Shirley, I.; McClure, P.; Dafflon, B. Probabilistic estimation of depth-resolved profiles of soil thermal diffusivity from temperature time series. *Earth Surf. Dynam.* **2022**, *10*, 687–704. [CrossRef]
19. Heitman, J.L.; Horton, R.; Sauer, T.J.; DeSutter, T.M. Sensible heat observations reveal soil-water evaporation dynamics. *J. Hydrometeorol.* **2008**, *9*, 165–171. [CrossRef]
20. Ochsner, T.E.; Baker, J.M. In situ monitoring of soil thermal properties and heat flux during freezing and thawing. *Soil Sci. Soc. Am. J.* **2008**, *72*, 1025–1032. [CrossRef]
21. Horton, R.; Wierenga, P.J.; Nielsen, D.R. Evaluation of methods for determining the apparent thermal diffusivity of soil near the surface. *Soil. Sci. Soc. Amer. J.* **1983**, *47*, 25–32. [CrossRef]
22. Verhoef, A.; van den Hurk, B.J.J.M.; Jacobs, A.F.G.; Heusinkveld, B.G. Thermal soil properties for vineyard (EFEDA-I) and savanna (HAPEX-Sahel) sites. *Agric. Forest Meteorol.* **1996**, *78*, 1–18. [CrossRef]
23. Gao, Z.; Fan, X.; Bian, L. An analytical solution to one-dimensional thermal conduction-convection in soil. *Soil Sci.* **2003**, *168*, 99–107. [CrossRef]
24. Gao, Z.; Lenschow, D.H.; Horton, R.; Zhou, M.; Wang, L.; Wen, J. Comparison of two soil temperature algorithms for a bare ground site on the Loess Plateau in China. *J. Geophys. Res.* **2008**, *113*, D18105. [CrossRef]
25. Gao, Z.; Tong, B.; Horton, R.; Mamtimin, A.; Li, Y.; Wang, L. Determination of desert soil apparent thermal diffusivity using a conduction-convection algorithm. *J. Geophys. Res-Atmos.* **2017**, *122*, 9569–9578. [CrossRef]
26. Gao, Z. Determination of soil heat flux in a Tibetan short-grass prairie. *Bound Layer. Meteorol.* **2005**, *114*, 165–178. [CrossRef]
27. An, K.; Wang, W.; Zhao, Y.; Huang, W.; Chen, L.; Zhang, Z.; Wang, Q.; Li, W. Estimation from soil temperature of soil thermal diffusivity and heat flux in sub-surface layers. *Bound Layer. Meteorol.* **2016**, *158*, 473–488. [CrossRef]
28. Zhou, Y.; Gao, X.; Li, Z.; Yang, L.; Hui, X. Spatio-temporal distribution of thermal diffusivity in deep soil in Qinghai-Tibetan Plateau. *Acta Pedol. Sin.* **2018**, *5*, 351–359, (In Chinese with English abstract).
29. Jury, W.A.; Horton, R. *Soil Physics*; John Wiley & Sons: Hoboken, NJ, USA, 2004; pp. 161–201.
30. Van Wijk, W.R.; De Vries, D.A. Periodic temperature variations in a homogeneous soil. In *Physics of Plant Environment*; Van Wijk, W.R., Ed.; North Holland Publishing Company: Amsterdam, The Netherlands, 1963; pp. 103–143.
31. Nerpin, S.V.; Chudnovskii, A.F. Physics of the soil. In *Israel Program for Scientific Translations*; Keter Press: Jerusalem, Israel, 1967.
32. Seemann, J. Measuring technology. In *Agrometeorology*; Springer: Berlin/Heidelberg, Germany, 1979; pp. 40–45.
33. Fan, X.; Tang, M. A preliminary study on conductive and convective soil heat flux. *Plateau Meteorol.* **1994**, *13*, 14–19. (In Chinese with English abstract)
34. Hu, G.; Zhao, L.; Wu, X.; Li, R.; Wu, T.; Xie, C.; Qiao, Y.; Shi, J.; Li, W.; Cheng, G. New Fourier-series-based analytical solution to the conduction–convection equation to calculate soil temperature, determine soil thermal properties, or estimate water flux. *Int. J. Heat Mass Tran.* **2016**, *95*, 815–823. [CrossRef]
35. Wang, L.; Gao, Z.; Horton, R. Comparison of six algorithms to determine the soil apparent thermal diffusivity at a site in the Loess Plateau of China. *Soil Sci.* **2010**, *175*, 51–60. [CrossRef]
36. Hu, G.; Zhao, L.; Wu, X.; Li, R.; Wu, T.; Xie, C.; Qiao, Y.; Cheng, G. Comparison of different soil temperature algorithms in permafrost regions of Qinghai-Xizang (Tibet) Plateau of China. *Cold Reg. Sci. Technol.* **2016**, *130*, 1–7. [CrossRef]
37. Lettau, H. Improved models of thermal diffusion in the soil. *Trans. Amer. Geophys. Union* **1954**, *35*, 121–132. [CrossRef]
38. Nassar, I.N.; Horton, R. Determination of the apparent thermal diffusivity of a nonuniform soil. *Soil Sci.* **1989**, *147*, 238–244. [CrossRef]
39. Nassar, I.N.; Horton, R. Determination of soil apparent thermal diffusivity from multiharmonic temperature analysis for nonuniform soils. *Soil Sci.* **1990**, *149*, 125–130. [CrossRef]
40. Ross, P.J. Estimation of nonuniform soil thermal properties by harmonic analysis. *Vadose Zone J.* **2013**, *12*, 1–9. [CrossRef]
41. Qiu, J. The third pole. *Nature* **2008**, *454*, 393–396. [CrossRef]
42. Yang, K.; Koike, T.; Fujii, H.; Tamura, T.; Xu, X.; Bian, L.; Zhou, M. The daytime evolution of the atmospheric boundary layer and convection over the Tibetan Plateau: Observations and simulations. *J. Meteorol. Soc. Jpn. Ser. II* **2004**, *82*, 1777–1792. [CrossRef]
43. Duan, A.M.; Wu, G.X. Role of the Tibetan Plateau thermal forcing in the summer climate patterns over subtropical Asia. *Clim. Dynam.* **2005**, *24*, 793–807. [CrossRef]
44. Xiao, Z.; Duan, A. Impacts of Tibetan Plateau Snow Cover on the Interannual Variability of the East Asian Summer Monsoon. *J. Clim.* **2016**, *29*, 8495–8514. [CrossRef]
45. Li, W.; Guo, W.; Qiu, B.; Xue, Y.; Hsu, P.C.; Wei, J. Influence of Tibetan Plateau snow cover on East Asian atmospheric circulation at medium-range time scales. *Nat. Commun.* **2018**, *9*, 4243. [CrossRef]

46. Peters-Lidard, C.D.; Blackburn, E.; Liang, X.; Wood, E.F. The effect of soil thermal conductivity parameterization on surface energy fluxes and temperature. *J. Atmos. Sci.* **1998**, *55*, 1209–1224. [CrossRef]
47. Santanello, J.A.; Dirmeyer, P.A.; Ferguson, C.R.; Findell, K.L.; Tawfik, A.B.; Berg, A.; Ek, M.; Gentine, P.; Guillod, B.P.; van Heerwaarden, C.; et al. Land–atmosphere interactions: The LoCo perspective. *Bull. Amer. Meteor. Soc.* **2018**, *99*, 1253–1272. [CrossRef]
48. Wang, K.; Wang, P.; Liu, J.; Sparrow, M.; Haginoya, S.; Zhou, X. Variation of surface albedo and soil thermal parameters with soil moisture content at a semi-desert site on the western Tibetan Plateau. *Bound Layer. Meteorol.* **2005**, *116*, 117–129. [CrossRef]
49. Zhang, Q.; Huang, R.H. Parameters of land-surface processes for gobi in north-west China. *Bound. Layer Meteor.* **2004**, *110*, 471–478.
50. Young, P.C.; Pedregal, D.J.; Tych, W. Dynamic harmonic regression. *J. Forecast.* **1999**, *18*, 369–394. [CrossRef]
51. Young, P.C.; Taylor, C.J.; Tych, W.; Pegregal, D.J.; McKenna, P.G. *The Captain Toolbox*; Centre for Research on Environmental Systems and Statistics; Lancaster University, UL: Lancaster, UK, 2010. Available online: http://www.es.lancs.ac.uk/cres/captain (accessed on 19 October 2021).
52. Ma, Y.; Hu, Z.; Xie, Z.; Ma, W.; Wang, B.; Chen, X.; Li, M.; Zhong, L.; Sun, F.; Gu, L.; et al. A long-term (2005–2016) dataset of hourly integrated land–atmosphere interaction observations on the Tibetan Plateau. *Earth Syst. Sci. Data* **2020**, *12*, 2937–2957. [CrossRef]
53. McCallum, A.M.; Andersen, M.S.; Rau, G.C.; Acworth, R.I. A 1-D analytical method for estimating surface water–groundwater interactions and effective thermal diffusivity using temperature time series. *Water Resour. Res.* **2012**, *48*. [CrossRef]
54. Luce, C.H.; Tonina, D.; Gariglio, F.; Applebee, R. Solutions for the diurnally forced advection-diffusion equation to estimate bulk fluid velocity and diffusivity in streambeds from temperature time series. *Water Resour. Res.* **2013**, *49*, 488–506. [CrossRef]
55. Hatch, C.E.; Fisher, A.T.; Revenaugh, J.S.; Constantz, J.; Ruehl, C. Quantifying surface water–groundwater interactions using time series analysis of streambed thermal records: Method development. *Water Resour. Res.* **2006**, *42*, W10410. [CrossRef]
56. Keery, J.; Binley, A.; Crook, N.; Smith, J.W.N. Temporal and spatial variability of groundwater–surface water fluxes: Development and application of an analytical method using temperature time series. *J. Hydrol.* **2007**, *336*, 1–16. [CrossRef]
57. Gordon, R.P.; Lautz, L.K.; Briggs, M.A.; McKenzie, J.M. Automated calculation of vertical pore-water flux from field temperature time series using the VFLUX method and computer program. *J. Hydrol.* **2012**, *420*, 142–158. [CrossRef]
58. Tong, B.; Gao, Z.; Horton, R.; Wang, L. Soil apparent thermal diffusivity estimated by conduction and by conduction-convection heat transfer models. *J. Hydrometeorol.* **2017**, *18*, 109–118. [CrossRef]
59. Van Duin, R.H.A. *The influence of Soil Management on the Temperature Wave near the Surface*; Technical Bulletin 29; Institute for Land and Water Management Research: Wageningen, The Netherlands, 1963.
60. Lu, Y.; Lu, S.; Horton, R.; Ren, T. An empirical model for estimating soil thermal conductivity from texture, water content, and bulk density. *Soil. Sci. Soc. Amer. J.* **2014**, *78*, 1859–1868. [CrossRef]
61. Tong, B.; Kool, D.; Heitman, J.L.; Sauer, T.J.; Gao, Z.; Horton, R. Thermal property values of a central Iowa soil as functions of soil water content and bulk density or of soil air content. *Eur. J. Soil Sci.* **2020**, *71*, 169–178. [CrossRef]
62. Philip, J.R. The theory of heat flux meters. *J. Geophys. Res.* **1961**, *66*, 571–579. [CrossRef]
63. Fuchs, M.; Hadas, A. Analysis and performance of an improved soil heat flux transducer. *Soil Sci. Soc. Amer. J.* **1973**, *37*, 173–175. [CrossRef]
64. Sauer, T.J.; Meek, D.W.; Ochsner, T.E.; Harris, A.R.; Horton, R. Errors in heat flux measurement by flux plates of contrasting design and thermal conductivity. *Vadose Zone J.* **2003**, *2*, 580–588. [CrossRef]
65. Tong, B.; Sauer, T.J.; Gao, Z.; Xiao, X.; Horton, R. Improving soil heat flux accuracy with the Philip correction technique. *J. Hydrometeorol.* **2019**, *20*, 1435–1448. [CrossRef]
66. Van Loon, W.K.P.; Bastings, H.M.H.; Moors, E.J. Calibration of soil heat flux sensors. *Agr. Forest Meteorol.* **1998**, *92*, 1–8. [CrossRef]
67. Gardner, H.R.; Hanks, R.J. Evaluation of the evaporation zone in soil by measurement of heat flux. *Soil Sci. Soc. Amer. Proc.* **1996**, *30*, 425–428. [CrossRef]
68. Farouki, O.T. *Thermal Properties of Soils*; Cold Regions Research and Engineering Lab: Hanover, Germany, 1981.
69. Patten, H.E. *Heat Transference in Soils*; Bulletin 59; U.S. Department of Agriculture Bureau of Soils: Washington, DC, USA, 1909.
70. Guan, X.; Huang, J.; Guo, N.; Bi, J.; Wang, G. Variability of soil moisture and its relationship with surface albedo and soil thermal parameters over the Loess Plateau. *Adv. Atmos. Sci.* **2009**, *26*, 692–700. [CrossRef]
71. Zhang, M.; Zhang, X.; Lu, J.; Pei, W.; Wang, C. Analysis of volumetric unfrozen water contents in freezing soils. *Exp. Heat Transfer.* **2018**, *32*, 426–438. [CrossRef]
72. Tian, Z.; Heitman, J.; Horton, R.; Ren, T. Determining soil ice contents during freezing and thawing with thermo-time domain reflectometry. *Vadose Zone J.* **2015**, *14*, 1–9. [CrossRef]
73. Tian, Z.; Lu, Y.; Horton, R.; Ren, T. A simplified de Vries-based model to estimate thermal conductivity of unfrozen and frozen soil. *Eur. J. Soil Sci.* **2017**, *67*, 564–572. [CrossRef]
74. Zhao, Y.; Si, B. Thermal properties of sandy and peat soils under unfrozen and frozen conditions. *Soil Till. Res.* **2019**, *189*, 64–72. [CrossRef]
75. Wu, B.; Zhu, H.; Cao, D.; Xu, L.; Shi, B. Feasibility study on ice content measurement of frozen soil using actively heated FBG sensors. *Cold Reg. Sci. Technol.* **2021**, *189*, 103332. [CrossRef]

remote sensing

MDPI

Article

Evaluation of Three Air Temperature Reanalysis Datasets in the Alpine Region of the Qinghai–Tibet Plateau

Xiaolong Huang [1,2], Shuai Han [3,*] and Chunxiang Shi [3]

1 Heavy Rain and Drought-Flood Disasters in Plateau and Basin Key Laboratory of Sichuan Province, Chengdu 610072, China
2 Sichuan Meteorological Observation and Data Centre, Chengdu 610072, China
3 National Meteorological Information Center, China Meteorological Administration, Beijing 100081, China
* Correspondence: hans@cma.gov.cn; Tel.: +86-177-1026-2113

Abstract: Surface air temperature is a critical element in the surface–atmosphere interaction, energy exchange, and water cycle. Multi-source fusion reanalysis products (hereafter referred to as reanalysis) have spatiotemporal continuity and broad applicability that can provide key data support for various studies such as glacier melting, soil freeze-thaw and desertification, ecosystem, and climate change in the alpine region of the Qinghai–Tibet Plateau (QTP). Surface air temperature observations collected at 17 weather stations in the High-cold region Observation and Research Network for Land Surface Process and Environment of China (HORN) over the period of 2017–2018 are implemented to evaluate the advanced and widely used surface air temperature reanalysis datasets, which include the European Centre for Medium-Range Weather Forecasts (ECMWF) Fifth Generation Land Surface Reanalysis (ERA5L), the U.S. Global Land Data Assimilation System (GLDAS), and China Meteorological Administration Land Data Assimilation System (CLDAS). Results are as follows: (1) Evaluation results of temporal changes and spatial distribution characteristics indicate that the three reanalysis datasets are consistent with in-situ observations in the alpine region of the QTP. CLDAS is more consistent with observations and can better describe details of temperature distribution and variation than ERA5L and GLDAS. (2) For the evaluation period, CLDAS is 0.53 °C higher than the in-situ observation, while ERA5L and GLDAS are lower than the in-situ observation by −3.45 °C and −1.40 °C, respectively. (3) The accuracy of CLDAS is better than ERA5L and GLDAS under different elevations and land covers. We resampled three reanalysis datasets with a spatial resolution of 0.25° and used the two most common interpolation methods to analyze the impact of spatial resolution and different interpolation methods on the evaluation results. We found that the impact is small. In summary, the three reanalysis datasets all have certain applicability in the alpine region of the QTP, and the accuracy of CLDAS is significantly higher than ERA5L and GLDAS. The results of the present paper have important implications for the selection of reanalysis data in the studies of climate, ecosystem, and sustainable development in the QTP.

Keywords: air temperature; Qinghai–Tibet Plateau; reanalysis dataset; alpine region; applicability

Citation: Huang, X.; Han, S.; Shi, C. Evaluation of Three Air Temperature Reanalysis Datasets in the Alpine Region of the Qinghai–Tibet Plateau. *Remote Sens.* **2022**, *14*, 4447. https://doi.org/10.3390/rs14184447

Academic Editors: Massimo Menenti, Yaoming Ma, Li Jia and Lei Zhong

Received: 11 July 2022
Accepted: 1 September 2022
Published: 6 September 2022

1. Introduction

The Qinghai–Tibet Plateau (QTP) is regarded as the Earth's "Third Pole" and "Asian Water Tower" [1,2]. It is the highest plateau in the world with an average elevation of over 4000 m. The strong dynamic and thermodynamic effects [3] of the QTP significantly affect atmospheric circulations in the northern hemisphere, as well as the Asian monsoon process and the climate patterns in East Asia [4,5], and have extremely important impacts on global climate change [1–14]. Glaciers, frozen soil, meadows, snow, and wetlands are widely distributed in the alpine region of the plateau, where the headwaters of China's major rivers are located. The alpine region of the plateau is an important area of ecological barrier, but it is also an area of harsh climatic conditions [6] and fragile ecological environments [9] with low levels of economic development [5]. Climate change in the QTP and the various impacts it brings have become a frontier and hotspot in earth system science research, which has attracted extensive attention within the scientific community [1,2,14]. Surface air temperature is a key variable in the land-surface–atmosphere interaction and energy exchange, as well as in water cycle processes. It is also an important basis [7,8] for the studies of glacial melting, soil freeze-thaw and desertification, and ecosystems and climate change in the plateau. Due to the vast area of the QTP, the restrictions of transport, and the terrain environment, weather stations are only sparsely distributed in the QTP and mainly concentrated in the eastern and southern parts of the QTP [15]; few stations are located in the western and northern parts of the plateau [16]. To complicate matters furthers, many stations in the QTP are established late and with short sequences of observations, which makes the temperature observations unable to fully reflect the state of the surface air temperature over the entire plateau. Therefore, reanalysis products of temperatures with spatiotemporal continuity and broad applicability are required to provide critical data support [15–19] for climate change and impact studies over the QTP [20,21].

In recent years, several research institutions in the United States, the European Union, China, Japan, and other countries have successfully developed a series of land surface reanalysis systems and multi-source data fusion analysis systems [22–24]. Great progress has been made in the land surface reanalysis dataset. Compared with atmospheric reanalysis datasets, land surface reanalysis products have higher spatiotemporal resolutions and wider application. At present, the most advanced land surface reanalysis datasets include the ECMWF Fifth Generation Land Surface Reanalysis (ERA5L) [25–30], the NASA Global Land Data Assimilation System (GLDAS) [31–33], and the China Meteorological Administration Land Surface Data Assimilation System (CLDAS) [34–38]. These datasets include surface meteorological elements and soil information. A series of research results have been achieved based on the application of these datasets in studies of weather and climate prediction, water resources management and water cycle, etc.

Due to differences in input data, numerical assimilation models, parameterization schemes, and the spatiotemporal resolutions of final products, these reanalysis datasets demonstrate quite different performances in different regions. Therefore, accuracy evaluation and applicability analysis of various reanalysis datasets are a prerequisite for their application. Several studies have evaluated the applicability of CLDAS, ERA5L, and GLDAS in the QTP [39–42]. For example, Han et al. [40] compared surface air temperature from CLDAS and GLDAS with observations collected at 2380 weather stations in China over the period 2010–2015. Their results indicate that surface air temperatures in the two reanalysis datasets are lower than observations in the QTP, while the accuracy and correlation of CLDAS with station observations are better than GLDAS. On different temporal and spatial scales, Huang et al. [41] verified CLDAS, ERA5L, and GLDAS against observations collected at 2265 weather stations in China during 2017–2019. They found that the three aforementioned reanalysis datasets can represent the characteristic temperature changes in the QTP well, although they are lower than observations. CLDAS is highly consistent with station observations, and its accuracy is significantly better than the other two reanalysis datasets. GLDAS is better than ERA5L. Liu et al. [42] selected 32,552 assessment stations that have been fused into the CLDAS system and 12,403 non-assessment stations that are non-fused into the CLDAS system as the data sources for evaluation and conducted dependent and independent verifications of CLDAS hourly temperature data in different regions of China. Results of both dependent and independent verification confirm that CLDAS has a relatively high accuracy and applicability in the QTP. Wang et al. [43] compared GLDAS with China's gridded surface air temperature dataset in the QTP and surrounding areas. They found that GLDAS performs better in arid regions than in sub-humid areas, and that the data are more accurate during 1979–1994 than during 2000–2007.

In summary, the three aforementioned reanalysis datasets all demonstrate a relatively high applicability in the QTP and thus have potential values for weather and climate studies. We also found that the previous applicability studies of temperature reanalysis datasets often use observations collected at operational weather stations of the China Meteorological Administration as reference data, and, while these observations have high accuracy and reliability, the following issues need to be addressed: (1) Many national-level meteorological station observations have been included in the international exchange list, and many of the data have been used as input for assimilation and/or data fusion to produce various reanalysis datasets. Therefore, it is hard to achieve independent results using these data to verify reanalysis products. (2) Most of these weather stations are located in suburbs of cities or areas along highways that are easily accessible. Their coverage of QTP topography and landform types is limited, which makes the evaluation results have limited reference value for assessing the reliability of the reanalysis datasets in the QTP. Based on the aforementioned discussion, the present study uses in-situ observations provided by "China Alpine Region Surface Process and Environmental Monitoring Research Network" [44,45] to evaluate surface air temperature from CLDAS, ERA5L, and GLDAS. The present study reveals some important similarities and differences in comparison to previous studies. Results of the study will be helpful in studies of the special atmospheric, hydrological, and ecological processes in the alpine region of the QTP [46].

2. Data and Methods

2.1. Data

2.1.1. Reanalysis of Surface Air Temperature

1. CLDAS dataset

CLDAS is a land-surface data assimilation system developed in the National Meteorological Information Center of the China Meteorological Administration (CMA) [22,35]. Advanced fusion technology is combined with independent innovations proposed in CMA during the development of CLDAS. Multi-grid variational analysis, spatial grid stitching, discrete ordinate shortwave radiation remote sensing retrievals, terrain correction, ensemble simulations of multiple land surface models (CLM, Noah-MP, CoLM), etc., are combined to produce surface pressure, ground precipitation, temperature, humidity, UV winds, shortwave radiation, surface air temperature and humidity, soil moisture and temperature, etc. The China Land-surface Data Assimilation System Version 2 (CLDAS-V2.0) [22] was released in 2015 and upgraded in 2018. This system can efficiently fuse observations collected at nearly 60,000 weather stations in China with numerical prediction data and satellite remote sensing data, and can release a real-time fused land surface data analysis product on $0.05° \times 0.05°$ grids at 1 h intervals. This product has been widely applied in meteorological and agricultural studies [36,37,41].

2. ERA5L dataset

ERA5L is a high spatiotemporal resolution global land surface reanalysis dataset produced by ECMWF for global land areas. It is a component of the fifth-generation European Reanalysis Product (ERA5) [28] that was developed within the framework of the European Commission Copernicus Climate Change Service (C3S). Based on outputs of numerical simulations of the ECMWF land-surface model, ERA5L is a downscaled dataset from the ERA5 climate reanalysis, and elevation correction for near-surface thermal states is conducted to ensure consistent evolvement of water and energy cycles over the land [27]. ERA5L can be applied for trend and anomaly analysis. ECMWF released the ERA5L product in 2019, which contains data from 1981 until present, with ongoing updates. The historical dataset over 1950–1980 was released in September 2021. With a high spatial resolution of $0.1° \times 0.1°$ and temporal resolution of 1h, as well as long data sequences and data consistency, ERA5L provides a strong support in hydrological study and numerical weather/climate model initialization. It is also widely applied in studies of water resources and land and environment management [29,30,47], etc.

3. GLDAS dataset

GLDAS is produced by the NASA Goddard Space Flight Center (GSFC) and the National Centers for Environmental Prediction (NCEP) of National Oceanic and Atmospheric

Administration (NOAA). Surface observations and satellite remote sensing retrievals are assimilated into the land surface models of Noah, Mosaic, CLM, and VOC to simulate global surface variables [32] (such as soil moisture, land surface temperature, etc.) and fluxes (such as evaporation, sensible heat flux, etc.). GLDAS has been widely applied to global climate change studies and comparative studies with other remote sensing products. GLDAS provides two versions of the dataset (GLDAS-1 [48] and GLDAS-2 [49]). The present study uses GLDAS-2, and the spatial and temporal resolutions of the dataset are 1h and 0.25°, respectively.

Table 1 lists the attributes of the datasets evaluated in the present study, including their spatial and temporal resolutions, coverages, and data download sources.

Table 1. Characteristics of the reanalysis temperature datasets.

Dataset	Areal Coverage	Spatial Resolution	Temporal Resolution	Unit	Website for Download
GLDAS	180°W–180°E; 60°S–90°N	0.25° × 0.25°	3 hourly	K	http://disc.sci.gsfc.nasa.gov/hydrology/data-holdings (accessed on 30 June 2022)
ERA5L	180°W–180°E; 60°S–90°N	0.1° × 0.1°	Hourly	K	https://cds.climate.copernicus.eu/cdsapp#!/dataset/reanalysis-ERA5L?tab=form (accessed on 30 June 2022)
CLDAS	70°E–140°E; 0°–60°N	0.05° × 0.05°	Hourly	K	http://data.cma.cn/ (accessed on 30 June 2022)

2.1.2. In-Situ Temperature Observations in the Alpine Region of the QTP

The in-situ temperature observations used to evaluate the reanalysis datasets are provided by the Tibetan Plateau Data Center of China. The data were downloaded from http://data.tpdc.ac.cn/ (accessed on 3 January 2022). These observations are collected at 17 field observation sites (Figure 1), which are evenly distributed in the alpine region of the QTP. Temperature, precipitation, wind speed and wind direction, relative humidity, radiation, evaporation, etc., are measured. Long-term surface processes and environmental changes are continuously monitored to understand patterns of climate and water resource changes in the headwater areas of the Yangzi River and the Yellow River. This information will be helpful to reveal the changes in ecosystem structure and function, build ecological protection barriers, and grasp the mechanism for the occurrence of natural disasters such as ice and snow freezing and thawing [44]. All the in-situ temperature observations used in the present study are daily mean temperature. Table 2 lists the properties of the in-situ observation sites and related information [45].

2.2. Methods

To quantitatively compare the consistency of the three reanalysis datasets with in-situ observations and evaluate their errors of daily average temperatures, the reanalysis data from 2017–2018 are divided into eight times per day (00, 03, 06, 09, 11, 14, 17, 20 UTC), which is the division used in GLDAS. The arithmetic mean is taken as daily mean temperature for individual reanalysis datasets. Based on the latitude and longitude information of the observation sites, daily mean temperatures from reanalysis are interpolated to the observation sites using the nearest neighbor interpolation method. Two sequences of daily temperatures from the reanalysis and from the in-situ observations, with 11,635 samples in each sequence, are then compared. Pearson correlation coefficient (CC), mean bias error (MBE), root-mean-square error (RMSE), Nash–Sutcliffe efficiency coefficient (NSE) [50,51], Kling–Gupta efficiency (KGE) [52,53], and Willmott's Index of Agreement (WIA) [54] are then calculated to evaluate the accuracy and applicability of CLDAS, ERA5L, and GLDAS temperature data in the alpine region of the QTP. The calculations of the aforementioned indices are as follows:

$$CC = \frac{\sum_{i=1}^{n} (R_i - \overline{R})(S_i - \overline{S})}{\sqrt{\sum_{i=1}^{n} (R_i - \overline{R})^2}\sqrt{\sum_{i=1}^{n} (S_i - \overline{S})^2}} \quad (1)$$

$$MBE = \frac{1}{n}\sum_{i=1}^{n} R_i - S_i \quad (2)$$

$$RMSE = \sqrt{\frac{1}{n}\sum_{i=1}^{n} (R_i - S_i)^2} \quad (3)$$

$$NSE = 1 - \frac{\sum_{i=1}^{n} (R_i - S_i)^2}{\sum_{i=1}^{n} (S_i - \overline{S})^2} \quad (4)$$

$$KGE = 1 - \sqrt{(CC-1)^2 + (\alpha - 1)^2 + (\beta - 1)^2} \text{ with } \alpha = \frac{\overline{R}}{\overline{S}}, \text{ and } \beta = \frac{\sigma_R}{\sigma_S} \quad (5)$$

$$WIA = 1 - \frac{\sum_{i=1}^{n} (R_i - S_i)^2}{\sum_{i=1}^{n} \left(\left|R_i - \overline{S}\right| + \left|S_i - \overline{S}\right|\right)^2} \quad (6)$$

where R_i is the reanalysis temperature interpolated to the observation site, S_i is the in-situ observation at the site, n is the total number of records that participate in the evaluation, and $\overline{R}$ and $\overline{S}$ denote the averages of reanalysis data and observations during the study period, respectively. CC (Equation (1)) represents the correlation between reanalysis and observations with values that range within [−1, 1]. $|CC| = 1$ indicates that the two sequences are completely linearly correlated; CC = 0 means there is no correlation between the two sequences, and $0 < |CC| < 1$ indicates that there is a certain degree of linear correlation between the two. The closer $|CC|$ is to 1, the higher the linear relationship is; the closer $|CC|$ is to 0, the weaker the linear correlation is between the two. CC>0 indicates that the reanalysis and the in-situ observations have the same trends of change, and CC<0 means that they have opposite trends. MBE (Equation (2)) reflects the deviations of reanalysis data from observations. Negative MBE values indicates that the reanalysis data are lower than the observation, and vice versa. RMSE (Equation (3)) shows the overall difference between the reanalysis and the observations, including systematic and non-systematic biases. A closer-to-0 RMSE corresponds to a more accurate reanalysis dataset. NSE (Equation (4)) is widely applied to quantify the prediction ability of hydrological models. It reflects the consistency of two datasets: NSE = 1 indicates that the reanalysis data are completely consistent with the observations; NSE ≤ 0 indicates that the two datasets are inconsistent with each other. The KGE (Equation (5)) is based on a decomposition of the NSE into its constitutive components (correlation, mean bias, and variability bias) and is increasingly used for model calibration and evaluation. σ_R and σ_S are the standard deviations in reanalysis and in-situ observations, respectively. KGE can vary from negative infinity to 1, and KGE = 1 indicates perfect agreement between simulations and observations. WIA (Equation (6)) is similar to NSE, but the denominator of the main term in the equation is the potential maximum difference [54]. The value of WIA ranges between 0 (not consistent) and 1 (perfectly consistent). On the scatterplot of reanalysis versus in-situ observations, both WIA and NSE indicate how close the data points are to the fitted 1:1 line. During the evaluation period, all samples used for evaluation are calculated based on the cumulative results of daily observations.

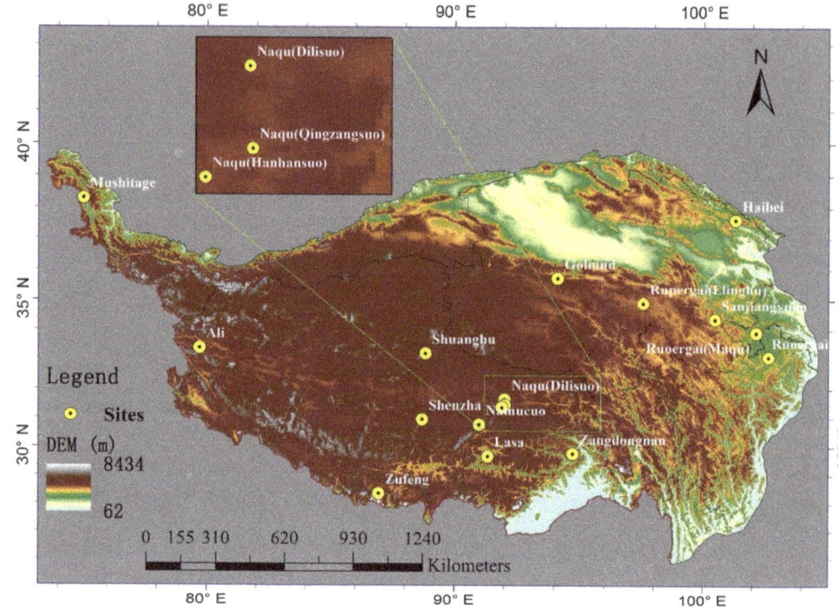

Figure 1. Elevation of the study area and distribution of in-situ observation sites.

Table 2. Properties of in-situ observation sites and related information.

NO.	Name	Longitude (°)	Latitude (°)	Elevation (m)	Height of the Sensor from the Ground (m)	Land Cover Type
1	Zangdongnan	94.7363	29.7593	3326	1.3	Grassland in forests
2	Namucuo	90.9885	30.7740	4730	1.5	Alpine meadow
3	Zufeng	86.9422	28.3590	4276	1.5	Sand and gravel
4	Golmud	94.1333	35.7167	4538	2.0	Alpine meadow
5	Lasa	91.3333	29.6667	3688	1.5	Artificial grassland
6	Mushitage	75.0183	38.2868	4400	1.5	Gravel
7	Ali	79.7013	33.3917	4264	1.5	Desert
8	Rupergai (Elinghu)	97.5588	34.9021	4278	2.0	Alpine meadow
9	Sanjiangyuan	100.4833	34.3667	3958	1.5	Alpine meadow
10	Shenzha	88.7000	30.9500	4675	2.0	Alpine meadow
11	Ruoergai	102.6509	33.1026	3483	2.7	Peatland
12	Ruoergai (Maqu)	102.1515	33.9205	3430	2.0	Alpine meadow
13	Naqu (Dilisuo)	92.0097	31.6437	4602	1.8	Alpine meadow
14	Naqu (Qingzangsuo)	92.0170	31.4410	4500	1.5	Alpine meadow
15	Shuanghu	88.8322	33.2167	4939	2.0	Alpine meadow
16	Haibei	101.3167	37.6167	3220	1.5	Alpine meadow
17	Naqu (Hanhansuo)	91.9000	31.3700	4509	1.5	Alpine meadow

Note: The properties of these sites are derived from the metadata provided in http://data.tpdc.ac.cn (accessed on 18 May 2022). The unit of latitude and longitude in degree, minute, and second is transferred to decimal unit in degree and keeps four decimal places. For a few sites where the metadata is missing, the properties are derived from other information. For example, the elevation at Shuanghu is derived from 90 m-resolution DEM, and the land cover type at Ali is derived from http://www.horn.ac.cn/index.jsp (accessed on 18 May 2022).

3. Results Analysis

3.1. Comparative Analysis of Spatial Distribution Characteristics

Spatial distributions of average temperature for reanalysis datasets and in-situ observations during the study period of 2017–2018 are displayed in Figure 2, which clearly shows that the three reanalysis datasets and the in-situ observations roughly follow the variation of latitude and elevation. Despite the slight differences at local or regional scales, the magnitude and spatial distribution of temperature are basically the same for these datasets (Figure 2a). Temperature gradually increases from north to south, and high temperature centers are located in southeastern Tibet, southwestern Sichuan, and northwestern Yunnan. The Qaidam Basin in the northwest of Qinghai is surrounded by mountains. The average elevation of the basin is about 2600 m, which is lower than the surrounding areas. The annual average temperatures in the three reanalysis datasets in the Qaidam Basin are all significantly higher than those of the surrounding areas. The average elevation of the Kunlun Mountains and Karakoram Mountains located in western Tibet is more than 5500 m, and the temperature is significantly lower than other areas in the same latitude. Compared to ERA5L and GLDAS, CLDAS describes more details of temperature changes with altitude. For example, CLDAS aptly describes the dramatic temperature changes caused by large altitude differences in the Hengduan Mountains region located at the junction of Tibet, Sichuan, and Yunnan, where mountains, valleys, and rivers are intertwined. In contrast, the other two reanalysis datasets can barely reflect this characteristic distribution of temperature in the Hengduan Mountains.

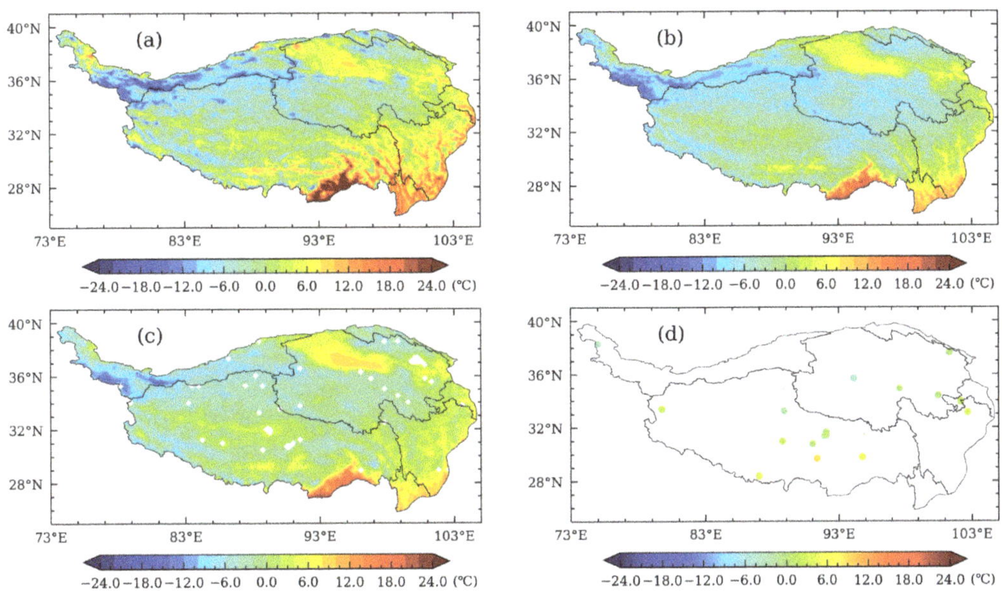

Figure 2. Spatial distributions of annual mean temperature over 2017–2018 ((**a**) CLDAS; (**b**) ERA5L; (**c**) GLDAS; (**d**) in-situ observations).

Spatial distributions of seasonal temperature are displayed in Figures 3–6. In the spring (Figure 3), CLDAS temperature is higher than ERA5L and GLDAS in the entire study area except the Qaidam Basin and the low-elevation region of southern Tibet, where CLDAS is lower than ERA5L and GLDAS. ERA5L and GLDAS show large differences in the spatial distribution of air temperature in the plateau area, though CLDAS has a small difference. ERA5L is also significantly lower than CLDAS and GLDAS in the central QTP. In the summer (Figure 4), the spatial distributions of CLDAS and GLDAS are similar to each other, while ERA5L is obviously lower than the other two reanalysis datasets. In the autumn (Figure 5), CLDAS and ERA5L are closer to each other, while GLDAS is lower than CLDAS and ERA5L in the high-elevation region of western QTP, but higher in the low-elevation region of the southeastern QTP. In the winter (Figure 6), spatial distributions of the three reanalysis datasets are basically consistent, although ERA5L is lower than CLDAS and GLDAS in southeastern Qinghai and northeastern Tibet. Overall, compared to GLDAS and ERA5L, CLDAS is closer to observations and demonstrates higher spatial consistency.

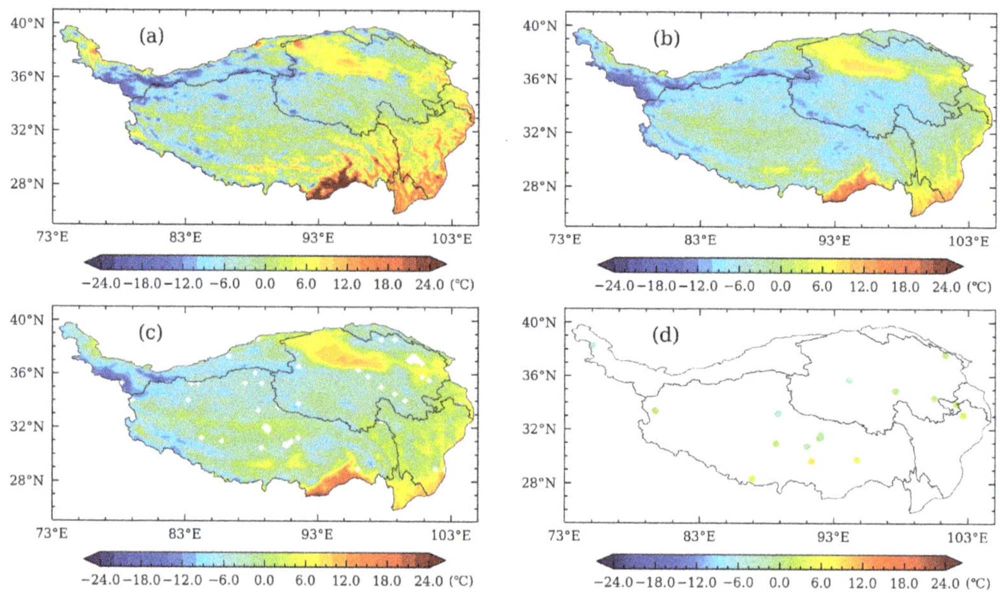

Figure 3. Spatial distributions of spring mean temperature over 2017–2018 ((**a**) CLDAS; (**b**) ERA5L; (**c**) GLDAS; (**d**) in-situ observations).

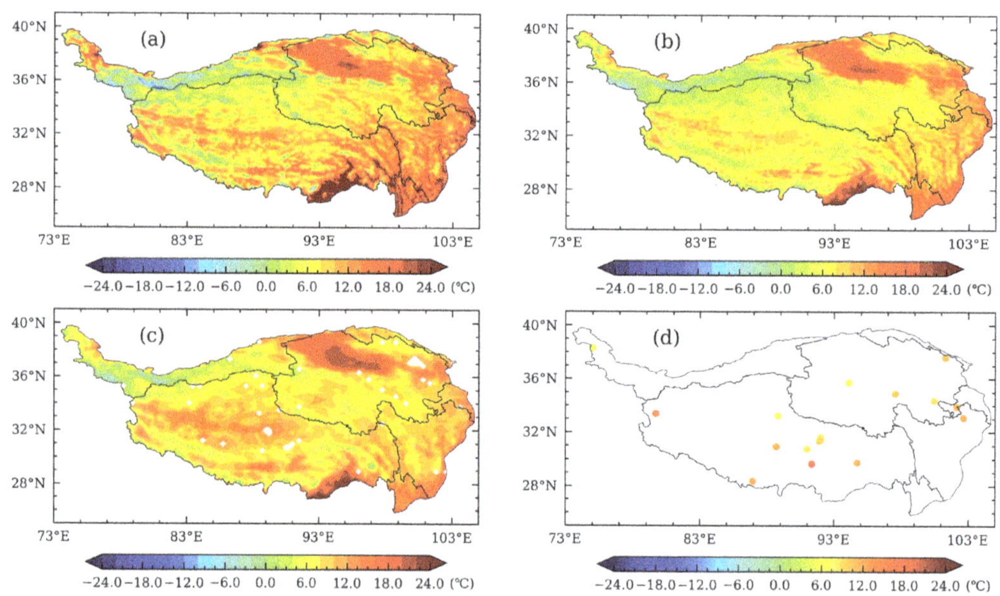

Figure 4. Spatial distributions of summer mean temperature over 2017–2018 ((**a**) CLDAS; (**b**) ERA5L; (**c**) GLDAS; (**d**) in-situ observations).

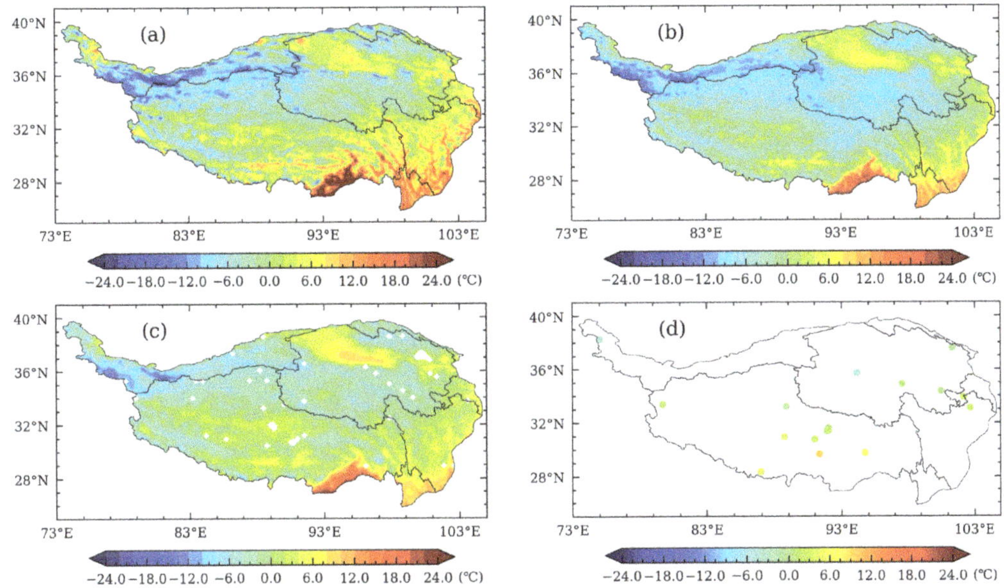

Figure 5. Spatial distributions of autumn mean temperature over 2017–2018 ((**a**) CLDAS; (**b**) ERA5L; (**c**) GLDAS; (**d**) in-situ observations).

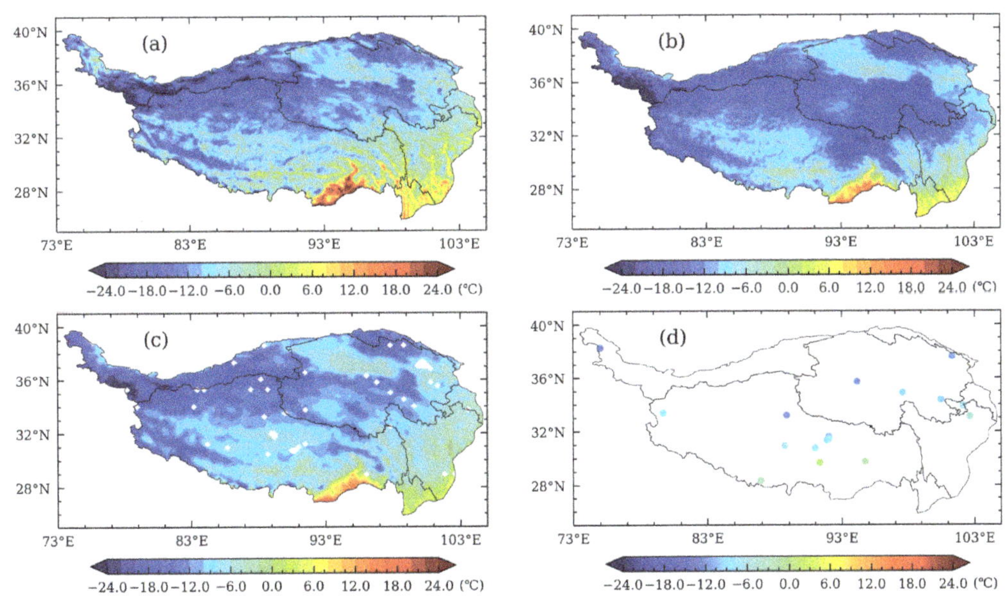

Figure 6. Spatial distributions of winter mean temperature over 2017–2018 ((**a**) CLDAS; (**b**) ERA5L; (**c**) GLDAS; (**d**) in-situ observations).

3.2. Accuracy of the Reanalysis Datasets for the Evaluation Period

Table 3 lists the evaluation results over the period 2017–2018. The mean temperatures of CLDAS, ERA5L, and GLDAS are 1.49 °C, −2.491 °C, and −0.44 °C, respectively. The mean value of CLDAS is the closest to the average of the in-situ observations (0.956 °C).

The correlation coefficient (CC) between CLDAS and the observations is the highest (0.969), followed by the correlation between ERA5L and the observations (0.934); the correlation coefficient between GLDAS and the observations is the lowest (0.92). The MBEs of ERA5L and GLDAS are −3.45 °C and −1.40 °C, respectively, which suggests that temperature is underestimated in the two reanalysis datasets, to a certain degree. Conversely, the MBE of CLDAS is 0.53, which indicates that CLDAS overestimates temperature in those in-situ observation sites. The RMSEs of CLDAS, ERA5L, and GLDAS are 2.18 °C, 4. 83 °C, and 3.64 °C, respectively, which indicates that the errors of CLDAS are smaller than the other two reanalyses. The values of NSE and WAI are close to 1 (the premium value) for all the three reanalysis datasets, suggesting that they are highly consistent with the in-situ observations, especially CLDAS. From the value of KGE, CLDAS is closer to 1, which indicates that it is better than ERA5L and GLDAS. This result agrees with NSE and WIA. In general, CLDAS is noticeably better than GDAS and ERA5L during the evaluation period based on evaluation indices of correlation, bias, and consistency. GLDAS is better than ERA5L, although the differences between them are relatively small.

Table 3. Accuracy evaluation results of CLDAS, ERA5L and GLDAS for the period 2017–2018.

Dataset	Mean Temperature (°C)	CC	MBE (°C)	RMSE (°C)	NSE	KGE	WIA
CLDAS	1.49	0.969	0.534	2.175	0.933	0.44	0.983
ERA5L	−2.491	0.934	−3.447	4.827	0.67	−2.609	0.927
GLDAS	−0.44	0.92	−1.396	3.638	0.813	−0.463	0.952

To better display the consistency of the three reanalysis datasets with the observations during the evaluation period, Figure 7 shows the scatter plots of reanalysis data versus in-situ observations and the results of univariate linear regression. The goodness of fit (R^2) for CLDAS, ERA5L, and GLDAS are 0.939, 0.872, and 0.847, respectively, which indicates that CLDAS is more consistent with in-situ observations. This result agrees with the results shown in Table 3.

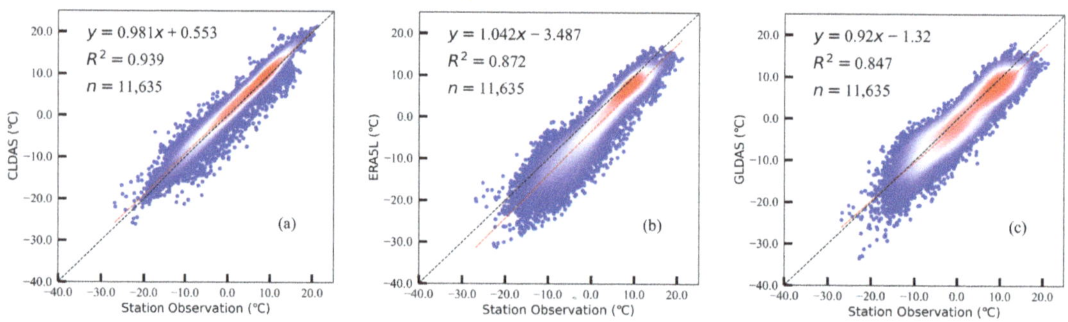

Figure 7. Scatter plots of reanalysis datasets versus in-situ observations ((**a**) CLDAS; (**b**) ERA5L; (**c**) GLDAS; n: total number of samples).

3.3. *Evaluation of Temporal Variation*

3.3.1. Daily Variation

To analyze differences in daily temperature of the reanalysis datasets during the evaluation period, daily average temperatures of CLDAS, GLDAS, ERA5L, and in-situ observations over the evaluation period are displayed in Figure 8a, which shows that the daily variations and temporal changes of surface air temperature are basically consistent between the three reanalysis datasets and observations, and that CLDAS is closer to the observations than GLDAS and ERA5L are. Looking at the time series of daily CC (Figure 8b), we found that in 85. 6% of the days, the CCs of CLDAS with observations are above 0.8.

However, the CCs of GLDAS and ERA5L with observations are below 0.8 in 60.7% and 90.5% of the days, respectively. Furthermore, the magnitude of daily variation of CLDAS is relatively small, which implies a more stable correlation with in-situ observations. The ranges of daily RMSE variation for CLDAS, ERA5L, and GLDAS (Figure 8c) are within 0.61–2.35 °C, 1.97–3.80 °C, and 2.43–3.76 °C, respectively. Note that the daily variation of RMSE for CLDAS is obviously lower than—the other two reanalysis datasets. In 76% of the total days, the RMSE values of GLDAS were lower than those of ERA5L, which indicates that the quality of GLDAS is higher than ERA5L in most days. The time series of daily MBE are displayed in Figure 8d, which shows that the MBEs of CLDAS are closer to the zero line than GLDAS and ERA5L are, which suggests that CLDAS is more consistent with observations than GLDAS and ERA5L. The MBE of CLDAS is positive in 78.5% of the days, whereas the MBE of GLDAS is negative in 90% of the days, and ERA5L is negative throughout the study period. This result indicates that daily temperature is overestimated by CLDAS and underestimated by GLDAS in most of the days, and it is always underestimated by ERA5L. The consistency indices of NSE (Figure 8e), KGE (Figure 8f), and WIA (Figure 8g) of CLDAS are closer to 1 with a smaller range of variation compared to that of ERA5L and GLDAS, which shows that CLDAS is more consistent with observations and demonstrates a higher stability.

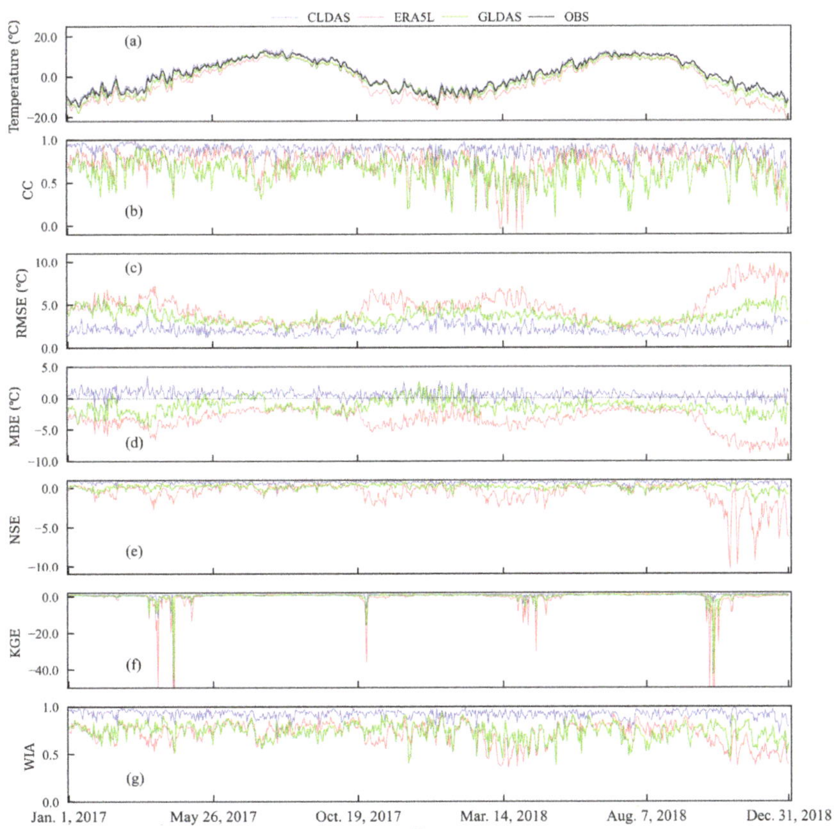

Figure 8. Daily evaluation of during 2017–2018. (**a**) Time series of daily mean temperature; (**b**) CC; (**c**) RMSE; (**d**) MBE; (**e**) NSE; (**f**) KGE; (**g**) WIA.

3.3.2. Monthly Variation

Figure 9 presents characteristic changes in monthly mean errors of the reanalysis datasets. The time series of monthly mean temperature (Figure 9a) indicates that the variation trends of the three reanalysis datasets are similar to that of the observations, i.e., temperature is the lowest in January and gradually increases from then onwards, reaches the highest in July, and then gradually decreases. Monthly CCs for CLDAS are all higher than those for GLDAS and ERA5L (Figure 9b). CCs for GLDAS are higher than for ERA5L in all months except March 2017 and February 2018, when the CCs for GLDAS are slightly lower than those of ERA5L. The RMSEs of CLDAS, ERA5L, and GLDAS (Figure 9c) range between 1.637–3.046 °C, 2.535–8.353 °C, and 2.682–5.054 °C, respectively. Note that the RMSEs of CLDAS are smaller than those of the DLDAS and ERA5L in all months, while the RMSEs of GLDAS are lower than those of ERA5L in all months except August and September of 2017 and July and August of 2018, when the RMSEs of GLDAS are slightly higher than those of ERA5L. Monthly MBE variations (Figure 9d) indicate that CLDAS overestimates monthly mean temperature in all months except December 2018, when it slightly underestimates the monthly mean temperature by −0.025 °C. The largest overestimation of 1.172 °C occurs in March 2017. Monthly MBEs of ERA5L are negative in all months, with the largest negative bias of −7.395 °C occurring in November 2018. Monthly MBEs of GLDAS are negative in all months except December 2017, when the monthly mean temperature of GLDAS is higher than the observation by 0.445 °C. The largest negative bias of GLDAS occurs in March 2017 with the value of −2.993 °C. Monthly consistency indices of NSE (Figure 9e) for CLDAS, ERA5L, and GLDAS are within the ranges of 0.581–0.847, −2.253–0.363, and −0.191–0.541, respectively, and the ranges of KGE (Figure 9f) are 0.027–0.892, −26.714–0.736, and −8.948–0.709, respectively. The indices of WIA (Figure 9g) are within the ranges of 0.903–0.961, 0.545–0.855, and 0.685–0.865, respectively. The lowest value of NSE occurs in either July or August for all the three reanalysis datasets, whereas the lowest value of WIA occurs in either October or August. Compared to the other two reanalysis datasets, monthly values of NSE, KGE, and WIA for CLDAS are closer to one, suggesting that CLDAS is more consistent with observations. GLDAS overall is better than ERA5L, with the exception of a few months.

3.3.3. Seasonal Analysis

Figure 10 displays seasonal error characteristics during the evaluation period. The histograms of seasonal mean air temperature changes from reanalysis datasets and in-situ observations are displayed in Figure 10a, which shows that seasonal temperatures of CLDAS, ERA5L, and GLDAS as well as in-situ observations all present a unimodal feature of being low in winter and high in summer. This result indicates that the three reanalysis datasets can well describe the seasonal variation of temperature in the QTP. Seasonal CCs (Figure 10b) of the three reanalysis datasets with observations are all the highest in autumn, while the CCs of CLDAS and ERA5L with observations are the lowest in winter and higher in spring than in summer. Although the CC of GLDAS with observations is the lowest in summer, the difference between CCs in winter and summer is quite small. Seasonal RMSEs (Figure 10c) of the CLDAS, ERA5L, and GLDAS all gradually increase from the minimum values in summer (1.819 °C, 2.863 °C and 2.828 °C) to the maximum values in winter (2.62 °C, 5.693 °C and 4.451 °C), and then decrease in the spring. Seasonal MBEs are displayed in Figure 10d, which indicates that CLDAS overestimates seasonal mean temperature in all seasons, though the overestimation is relatively small in autumn. Opposite to CLDAS, ERA5L and GLDAS both underestimate seasonal mean temperature, and the underestimation is more severe in ERA5L. The largest negative bias occurs in autumn and the smallest negative bias occurs in summer for both ERA5L and GLDAS. The histograms of seasonal NSE (Figure 10e) and WIA (Figure 10g) show that the consistency of the three reanalysis data with in-situ observations is relatively poor in winter, and is optimal in autumn. However, from the perspective of KGE (Figure 10f), the three reanalysis datasets are worst in spring and better in summer.

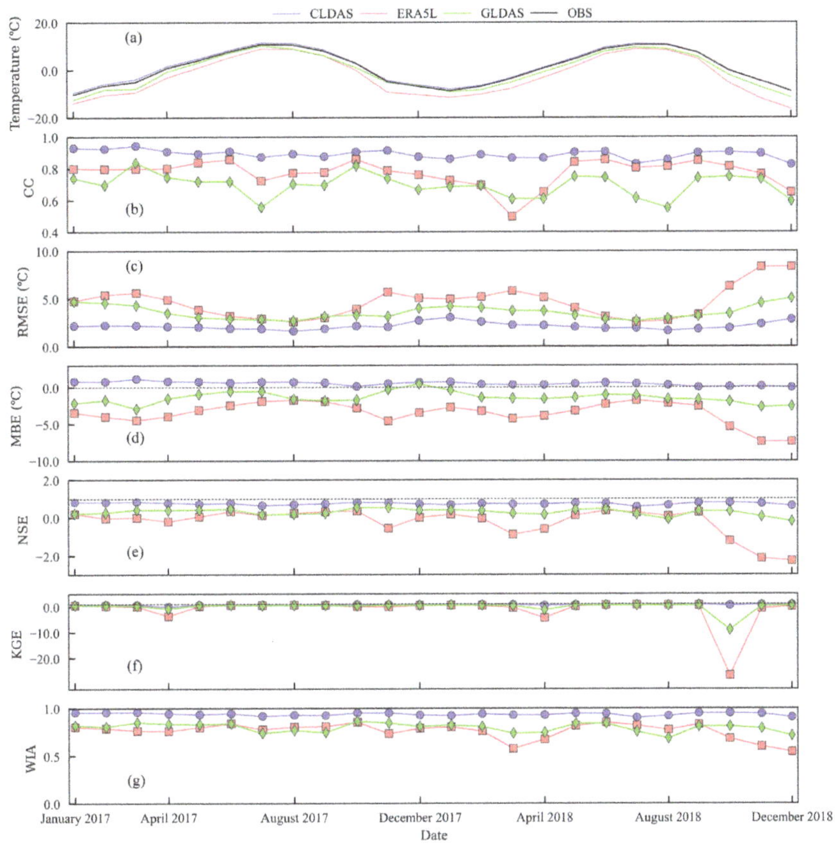

Figure 9. Monthly evaluation of during 2017–2018. (**a**): Series of monthly mean temperature; (**b**) CC; (**c**) RMSE; (**d**) MBE; (**e**) NSE; (**f**) NSE; (**g**) WIA.

3.4. Comparative Reanalysis at Individual Sites

Figure 11 shows box plots of temperature errors of CLDAS, GLDAS, and ERA5L during 2017–2018. The numbers of stations with CC (Figure 11a) higher than 0.95 account for 82.4%, 52.9%, and 70.6% of the total number of stations for CLDAS, ERA5L, and GLDAS, respectively. The lowest CCs, with respective values of 0.902, 0.915, and 0.913 for CLDAS, ERA5L, and GLDAS, all occur at Ruoergai (Elinghu), while the highest CCs occur at Ali and Golmud (0.992 for CLDAS), Haibei (0.974 for ERA5L), and Mushitage (0.979 for GLDAS). RMSEs (Figure 11b) are within the ranges of 1.222–4.289 °C, 2.345–6.076 °C, and 2.366–5.736 °C for CLDAS, ERA5L, and GLDAS, respectively. The largest RMSEs of CLDAS and ERA5L occur at Ruoergai (Elinghu), where the correlation is the lowest. The largest RMSE of GLDAS is found at Lasa. The smallest RMSEs of the three datasets occur at different sites. The box plot of MBE (Figure 11c) shows that CLDAS is lower than observations at only 4 sites, i.e., Ruoerai (Elinghu), Sanjinagyuan, Golmud, and Shenzha, which account for 23.5% of the total observation sites. The largest negative bias (−1.995) is found at Ruoergai (Elinghu), and the largest positive bias (1.99 °C) occurs at Naqu (Hanhansuo). ERA5L data are lower than observations at all sites, and the largest bias is found at Mushitage (−4.968 °C). Positive biases of GLDAS only occur at Ruoergai (Maqu), Naqu (Qingzangsuo), Naqu (Hanhansuo), and Shuanghu, which account for 29.4% of the total stations. The largest positive bias occurs at Shuanghu (2.37 °C); the biases of GLDAS are negative at all other sites with the largest negative bias (−5.385 °C) at Lasa.

The consistency indices of NSE for CLDAS, ERA5L, and GLDAS (Figure 11d) are within the ranges of 0.748–0.974, 0.102–0.919, and 0.219–0.934, respectively, and the ranges of KGE (Figure 11e) are −157.108–0.918, −389.193–0.537, and −138.881–0.882, respectively. The ranges of WIA (Figure 11f) are 0.936–0.994, 0.85–0.978, and 0.839–0.982, respectively. Based on NSE, KGE, and WIA, the consistency of CLDAS is the worst at Ruoergai (Elinghu) and the consistency of GLDAS is the worst at Lasa. The consistency of ERA5L is worst at Zhufeng, Zangdongnan, and Lasa. In summary, the various reanalysis datasets show different qualities and applicability.

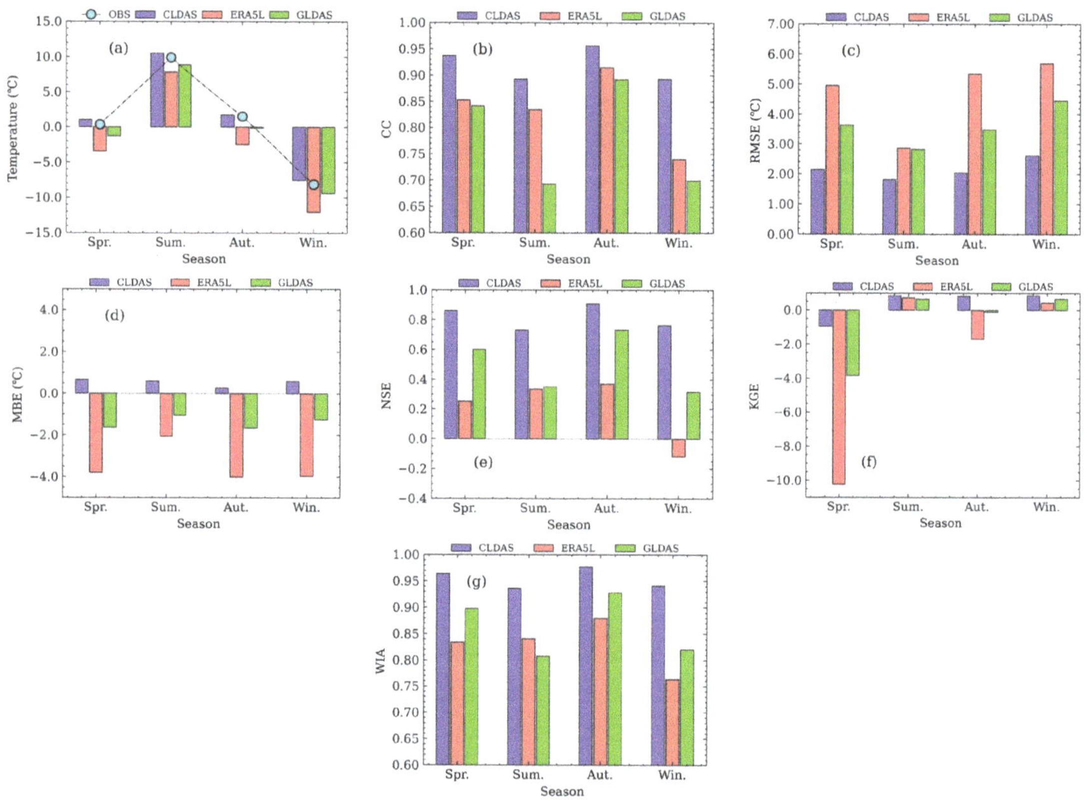

Figure 10. Seasonal evaluation during 2017–2018. (**a**): Seasonal changes of the average temperature; (**b**): CC; (**c**): RMSE; (**d**): MBE; (**e**): NSE; (**f**): KGE; (**g**): WIA.

To intuitively and easily understand the relationship between the consistency and errors of CLDAS, ERA5L, and GLDAS at the 17 observation stations, Taylor diagrams between the three reanalysis datasets and in-situ observations at each individual observation site are displayed in Figure 12. Figure 12a–q show that the standard deviations of CLDAS and GLDAS are relatively large at 11 and 7 sites, respectively, while ERA5L shows greater variability at 15 sites. The correlation coefficient between CLDAS and in-situ observations is larger than those between the other two reanalysis datasets and observations at all observation sites except Ruoergai (Elinghu), where the CC of CLDAS is slightly lower than the CC of GLDAS and ERA5L. The Taylor diagram between the three reanalysis datasets and all the in-situ observations (Figure 12r) indicates that CLDAS is closer to, and more consistent with, observations with smaller deviation.

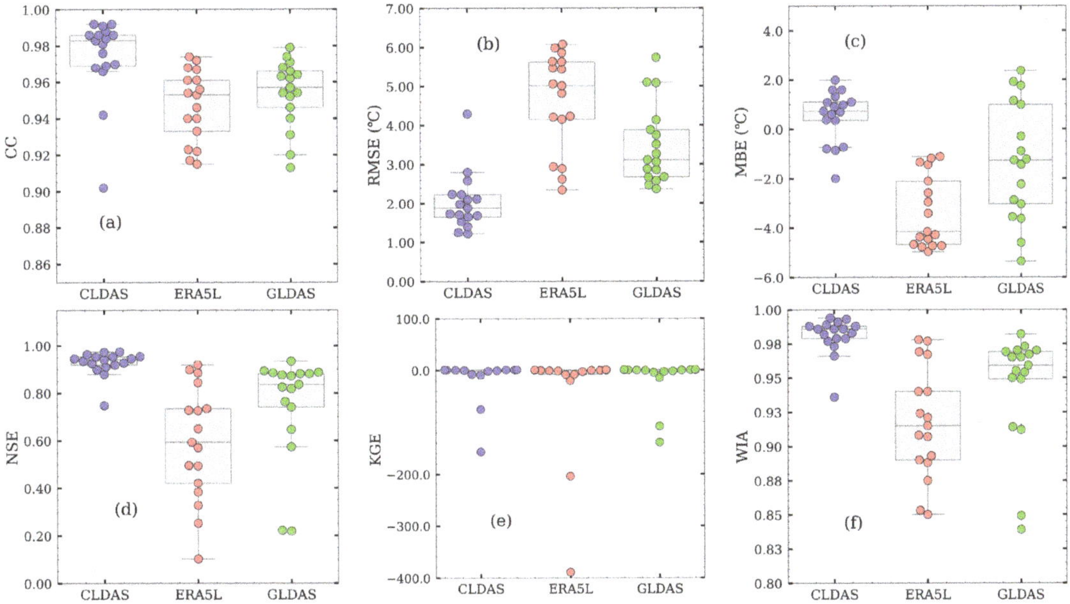

Figure 11. Box plots of temperature errors in CLDAS, ERA5Land GLDAS: (**a**) CC, (**b**) RMSE, (**c**) MBE, (**d**) NSE, (**e**) KGE, (**f**) WIA.

3.5. Comparative Reanalysis at Different Terrain Elevations

To explore the temperature variation characteristics of the three reanalysis datasets in different elevations, the observation sites are divided into four categories of elevation: <3500 m, ≥3500–4000 m, ≥4000–4500 m, and ≥4500 m. Figure 13 show the bias characteristics of the three gridded datasets at different elevations. In terms of evaluation indices (CC, NSE, KGE, and WIA), the consistency between CLDAS and the in-station observations is higher than the other analysis products at any altitude. The MBEs of CLDAS show a positive deviation relative to the observation station, while ERA5L and GLDAS are opposite. The RMSEs of CLDAS are lower than those of the other two reanalysis datasets, and GLDAS is better than ERA5L. Although the CCs of ERA5L is slightly higher than GLDAS, other indices (NSE, KGE and WIA) are relatively lower than GLDAS. Compared to ERA5L and GLDAS, the CLDAS temperature data is less affected by elevation.

3.6. Comparative Reanalysis at Different Land Covers

According to the land cover type, the observation sites are divided into seven categories: alpine meadow (AE), desert (DT), grassland in forests (GF), gravel (GL), peatland (PD), sand and gravel (SG), and artificial grassland (AG). Figure 14 show the bias characteristics of CLDAS, ERA5L, and GLDAS at different land covers. The MBEs of ERA5L and GLDAS showed negative deviation in in-station observations at different land covers, while CLDAS is opposite. The deviation of CLDAS at land cover of artificial grassland is the smallest, as are ERA5L and GLDAS with alpine meadow and peatland, respectively. In essence, the consistency indices (CC, NSE, KGE, and WIA) and deviation (MBE and RMSE) between CLDAS and in-station observations have a small range of variation, and are better than ERA5L and GLDAS in each land cover.

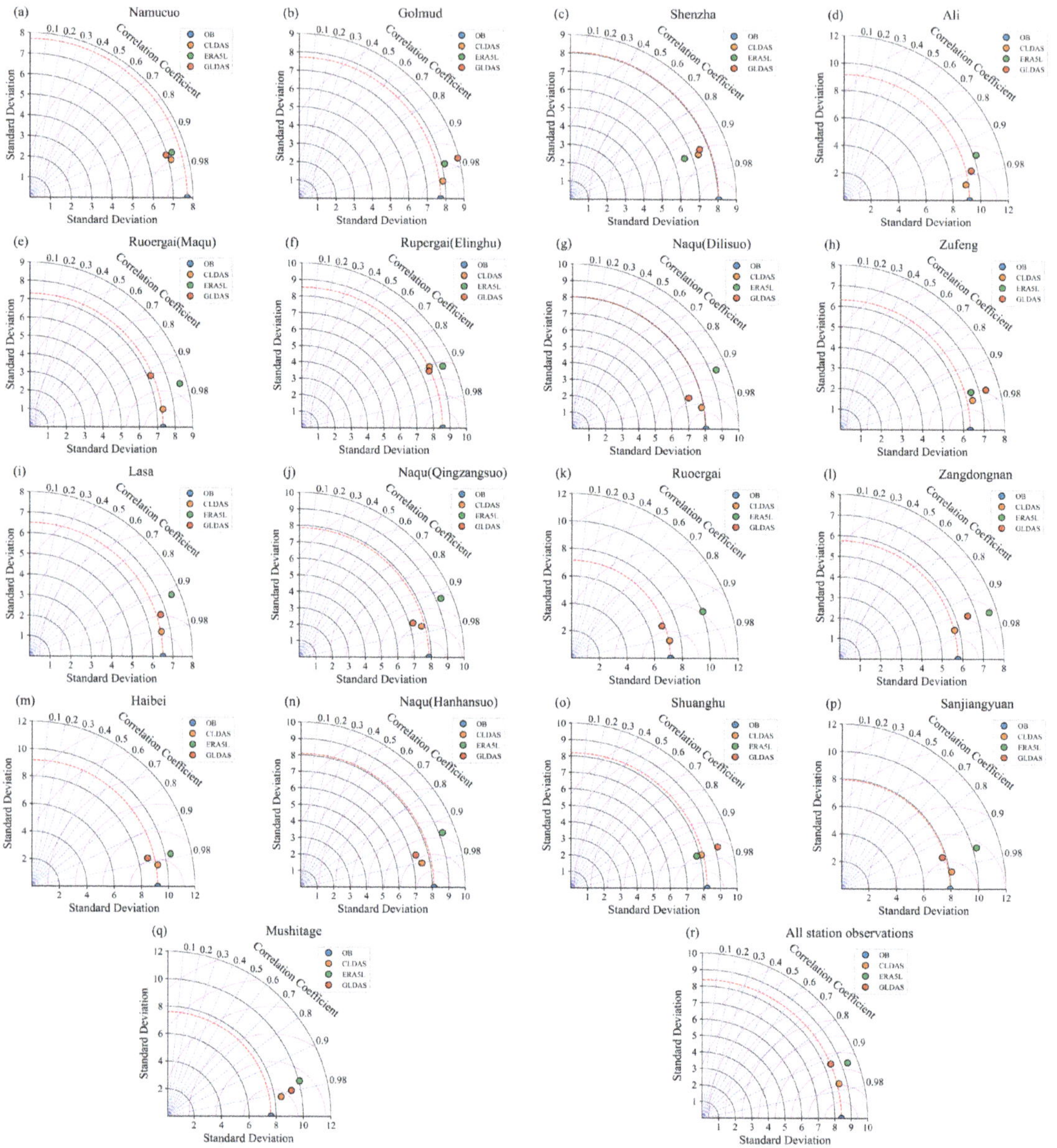

Figure 12. Taylor diagrams of (**a–q**) CLDAS, ERA5L, GLDAS against in-situ observations at 17 stations and (**r**) all observation stations.

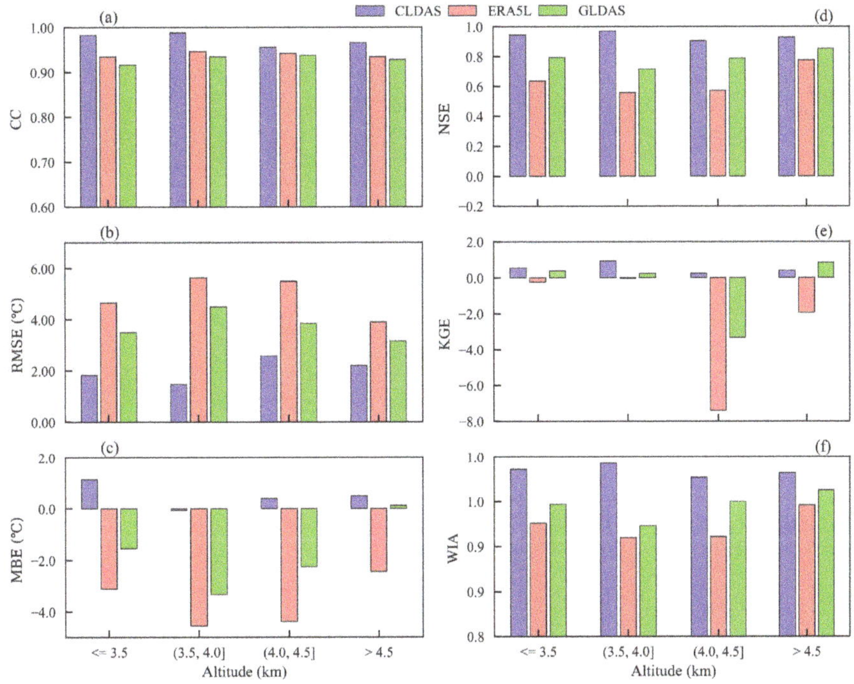

Figure 13. The errors at different altitudes: (**a**) CC; (**b**) RMSE; (**c**) MBE; (**d**) NSE; (**e**) KGE; (**f**) WIA.

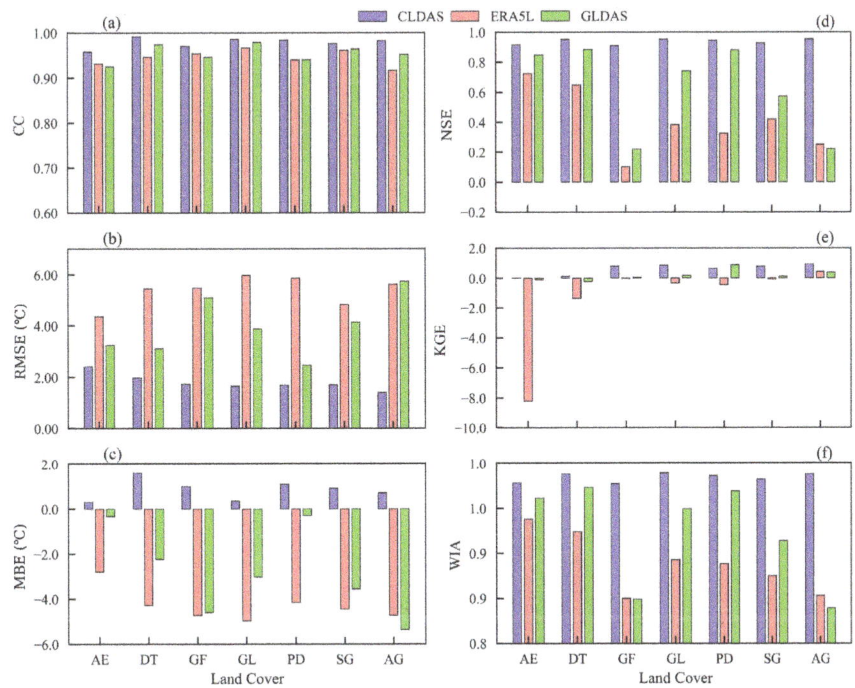

Figure 14. The errors at different land covers: (**a**) CC; (**b**) RMSE; (**c**) MBE; (**d**) NSE; (**e**) KGE; (**f**) WIA.

4. Discussion

Based on the validation results, we found that a certain degree of errors are inevitable in the three reanalysis datasets in comparison to in-situ observations. Surface air temperature in the QTP is not only controlled by regional factors (longitude, latitude), but also affected by other factors such as geographical conditions (altitude, aspect, slope) and the underlying surface (such as vegetation, snow cover), and are further complicated by temperature changes [55–63].

Several previous studies have also found differences of biases between stations and air temperature reanalysis datasets. For example, Huang et al. [41] found that elevation is not the only factor that causes biases in reanalysis datasets. Through the assessment of the slopes of in-situ observation sites, they found that the errors and applicability of CLDAS, ERA5L, and GLDAS increase with the increasing slope of the observation site. Ding et al. [64] found that temperature changes are significantly correlated with the elevation and slope of the observation site, and the complexity of the terrain is the main factor leading to large errors in reanalysis data. Meanwhile, these results also indicate that when using temperature reanalysis data, topographic correction should be performed on the data in order to effectively reduce errors and improve the accuracy and applicability of the reanalysis data. Liu and Long et al. [42,65] evaluated CLDAS; Huang et al. [41] evaluated CLDAS, ERA5L, and GLDAS in China. They found that that the errors of the three reanalysis datasets all increase with increasing monthly average temperatures, and the correlation between the reanalysis data and the in-situ observations gradually decreased, reaching the minimum value in July or August. The correlation and bias then gradually increased with decreasing temperature. However, the present study indicates that the monthly variation of errors in the QTP is not significant, though the seasonal variation is essentially the same with the previous studies of Huang et al. [41], which may be attributed to the differences in the time series of the data used and the division of regions.

Errors that resulted from the approach of evaluation at individual sites can be influenced by a few factors: (1) The spatial scales do not match. The in-situ observations at a specific site only reflect temperature changes in a certain area around it, and, due to the influence of topography, the representativeness of the observations is still limited. In contrast, the reanalysis data at a specific grid represents the average value of the grid. Thus, it is difficult to solve the problem of spatial mismatch [56,59] between in-situ observations and gridded reanalysis data. (2) The difference between the terrain height of the re-analysis grid and the elevation of the station [41,42,66]. If the observation site is located in a valley and its altitude is lower than the altitude of the surrounding grid points of the reanalysis dataset, the evaluation result at this site will generally show a colder deviation; if the site is located at the top of mountains, which is higher than the elevation of the surrounding grids in the reanalysis dataset, the evaluation result will show a warm bias at this site. (3) Systematic errors [67,68] caused by the numerical model or assimilation method. For example, cold errors occur at 70.6% and 100% of the total number of stations for GLDAS and ERA5L, respectively, which may be caused by systematic errors. In addition, errors in input data and errors introduced during the interpolation of reanalysis data (e.g., from Gaussian grids to coordinate grids) are also sources of errors that need to be further verified. Therefore, error characteristics and the applicability of reanalysis data should be fully considered in the application of temperature reanalysis data. Next, we will resample the three reanalysis datasets at the same resolution and use different interpolation methods for evaluation to discuss the influences on the accuracy of the gridded datasets.

4.1. Inpact of Grid Resolutions on the Accuracy of the Reanalysis Datasets

To explore the temperature variation characteristics and accuracy of the three reanalysis datasets at the same resolution, GLDAS and ERA5L are resampled to GLDAS grids with the spatial resolution of 0.25° using the mean value algorithm [55], and thus the three reanalysis datasets have uniform temporal and spatial resolutions. Figure 15 show the spatial distributions of annual mean temperature over 2017–2018. Although remapping

CLDAS and ERA5L reduces the resolution, it still shows more advantages in detail than GLDAS. Figure 16 shows the bias of annual mean temperature between in-situ observations and reanalysis datasets at 17 weather stations. It is easy to find that, at each observation station, CLDAS has a small deviation to in-situ observations compared with ERA5L and GLDAS.

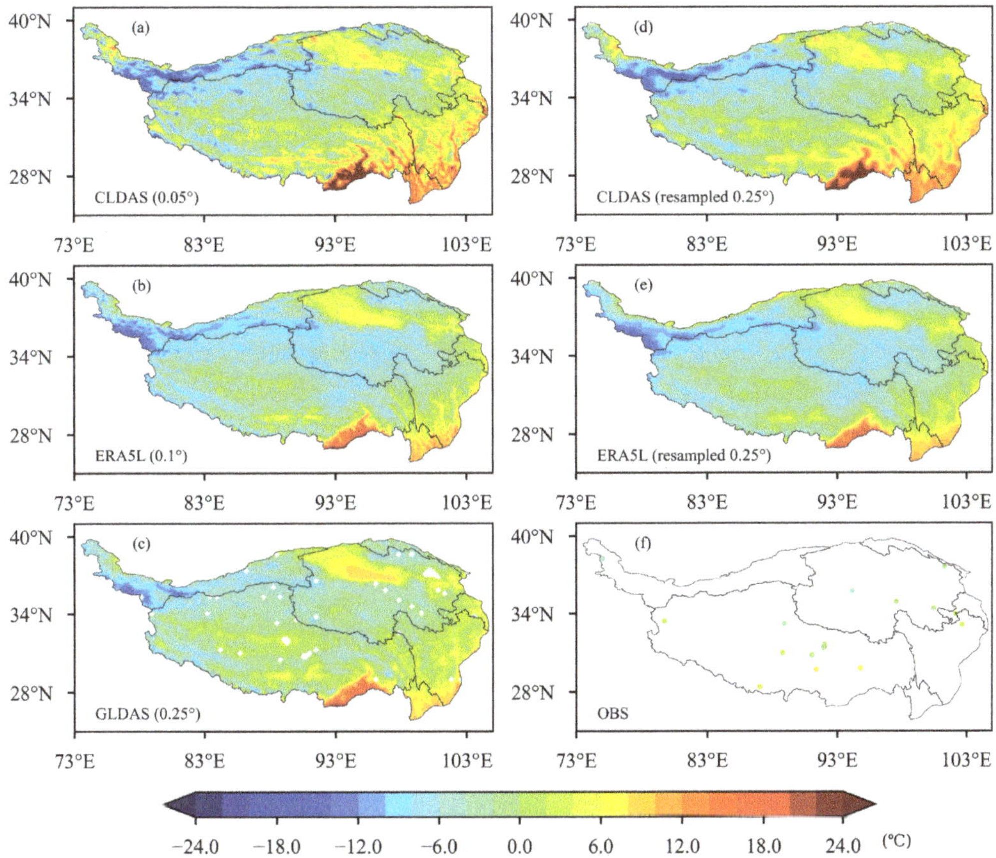

Figure 15. Spatial distributions of annual mean temperature over 2017–2018. (**a**) CLDAS (0.05°); (**b**) ERA5L (0.1°); (**c**) CLDAS (0.25°); (**d**) CLDAS (resampled to 0.25°); (**e**) ERA5L (resampled to 0.25°); (**f**) OBS (in-situ observations).

4.2. *Impact of Interpolation Methods on the Accuracy of the Reanalysis Datasets*

In order to analyze the impact of different interpolation methods on the evaluation results, the two most common interpolation methods, nearest neighbor method and bilinear interpolation method, are used in the present study [40,42]. The results are shown in Table 4. We found that different interpolation methods can have a certain impact on the evaluation results, but the impact is very small. It can also be seen that the bilinear interpolation method also shows that the deviation of CLDAS from the in-situ observations is lower than the other two reanalysis datasets, and that GLDAS is better than ERA5L.

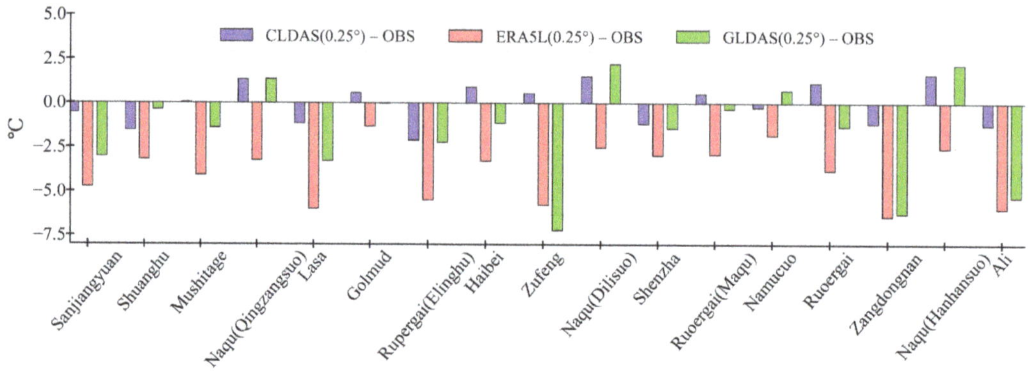

Figure 16. The bias of annual mean temperature between in-situ observations (OBS) and three reanalysis datasets over 2017–2018.

Table 4. Accuracy evaluation results used two interpolation methods for the evaluation period.

Dataset	CC		MBE (°C)		RMSE (°C)		NSE		KGE		WIA	
	Nea	Bil	Nea	Bil	Nea	Bil	Nea	Bil	Nea	Bil	Nea	Bil
CLDAS	0.969	0.968	0.534	0.404	2.175	2.179	0.933	0.933	0.44	0.576	0.983	0.983
ERA5L	0.934	0.933	−3.447	−3.61	4.827	4.942	0.67	0.654	−2.609	−2.779	0.927	0.924
GLDAS	0.92	0.927	−1.396	−1.103	3.638	3.37	0.813	0.839	−0.463	−0.157	0.952	0.958

Note: Nearest neighbor interpolation method (Nea), bilinear interpolation method (Bil).

Although all three reanalysis datasets can accurately reflect the distribution characteristics of air temperature in the alpine region of the QTP, CLDAS performs better overall. It is also better at individual stations and on daily, monthly, and seasonal time scales. One important reason is that in the QTP, CLDAS integrates the observations collected at thousands of surface automatic weather stations [22,35], which is of great benefit to the quality of CLDAS. ERA5L and GLDAS are global reanalysis products, both of which show large deviations from observations, probably due to the lack of observations for assimilation over the QTP. Furthermore, when compared with the other two reanalysis datasets, CLDAS also has higher spatial resolution, which can improve its ability for temperature description [40], especially in complex terrain areas. Using data assimilation and fusion techniques with in-situ observation data, satellite remote sensing data, and numerical model data, the reanalysis system produces regular gridded data with a certain temporal and spatial resolution. This process will introduce some uncertainties [22] into the final data products, which is why deviation of reanalysis from in-situ observations is important.

Based on the aforementioned discussion, we believe that, although there are certain deviations in the temperature reanalysis datasets in the QTP, they still have certain applicability and credibility in the alpine region of the QPT, where observation sites are unevenly distributed with a low density of observations. Thus, these reanalysis datasets have certain reference values. It should be noted that, even though the quality of CLDAS is better than the other two datasets based on careful evaluation of the three reanalysis datasets in the present study, the reanalysis datasets of ERA5L and GLDAS have longer time sequences, larger spatial coverage, and better continuity compared to CLDAS. Due to the extremely complex terrain in the QTP and the short sequences of in-situ observations at the 17 sites, which cannot cover all of the QTP area, the three datasets have their respective advantages and disadvantages in different areas and further studies are necessary for local scale.

5. Conclusions

In this study, the consistency between in-situ observations collected at 17 field observation stations in the alpine region of the QTP and the three reanalysis datasets (CLDAS, GLDAS, and ERA5L) and their deviations from observations from 2017–2018 are evaluated. Major conclusions are as follows:

(1) The spatial distributions of the three reanalysis datasets and the in-situ observations follow the change patterns of latitude and elevation. Temporal variations of average temperature and spatial distributions of temperature in the reanalysis datasets, as well as their correlations with and deviation from in-situ observations, all indicate that the three reanalysis datasets are consistent with observations and demonstrate reasonability. Despite some slight differences in local or regional scales, the magnitudes of the data and their spatial distributions remain consistent.

(2) The spatial distributions of the three reanalysis datasets are consistent, while CLDAS is closer to, and more consistent with, observations than GLDAS and ERA5L are. In the spring, CLDAS temperature is higher than ERA5L and GLDAS over the entire study area except the Qaidam Basin and the low elevation area of southern Tibet. Compared to ERA5L and GLDAS, CLDAS shows smaller differences in spatial distribution. In the summer, spatial distributions of CLDAS and GLDAS are closer, while ERA5L is obviously lower. In the autumn, CLDAS and ERA5L become closer, while GLDAS is relatively low in the high-elevation area of the western QTP but relatively high in the low-elevation area of the southeastern QTP. In the winter, ERA5L is lower than CLDAS and GLDAS in southeastern Qinghai and northeastern Tibet.

(3) Evaluation results on multi-time scales (daily, monthly, and seasonal) and multi-space scales (individual stations, elevations, and land covers) indicate that the accuracy and applicability of CLDAS are discernibly better than the other two datasets. GLDAS is better than ERA5L, but the difference between the two is small. However, the quality of the reanalysis datasets is different at observation sites.

In summary, CLDAS is more consistent with observations than GLDAS and ERA5L are and demonstrates better capability for the description of temperature in the alpine region of the QTP. Despite certain defects and limitations, ERA5L and GLDAS are still reliable and applicable in the alpine area of the QTP where observations are sparse and unevenly distributed. Results of the present paper have great implications for ecosystems and sustainable development studies in the QTP.

Author Contributions: Conceptualization, X.H., S.H. and C.S.; methodology, X.H., S.H. and C.S.; validation, S.H.; data curation, S.H. and C.S.; writing—original draft preparation, X.H.; writing—review and editing, X.H., S.H. and C.S.; visualization, X.H.; supervision, S.H.; funding acquisition, S.H. and C.S. All authors have read and agreed to the published version of the manuscript.

Funding: This research was funded by the National Key Research and Development Program of China (No. 2018YFC1506601), the Key Research and Development Program of Sichuan (No. 2022YFS0541), the Key Techniques and Data Sets of Land Surface Reanalysis in Qinghai Xizang Plateau (No. NMICJY202106), Study on the Fusion of Precipitation and Soil Moisture with Multi-Source Data (No. 2011DFG23150), and the Innovative Development Project of the China Meteorological Administration (No. CXFZ2021Z007).

Acknowledgments: The authors appreciate the National Tibetan Plateau Data Center for providing the meteorological observation data set of HORN in 2017–2018, and thank all the participants in the field observations. We are very grateful for the helpful input and suggestions from the anonymous reviewers and editors.

Conflicts of Interest: The authors declare no conflict of interest.

References

1. Qiu, J. China: The third pole. *Nature* **2008**, *454*, 393–397. [CrossRef] [PubMed]
2. Wu, G.; Liu, Y.; Zhang, Q.; Duan, A.; Wang, T.; Wan, R.; Liu, X.; Li, W.; Wang, Z.; Liang, X. The influence of mechanical and thermal forcing by the Tibetan Plateau on Asian climate. *J. Hydrometeorol.* **2007**, *8*, 770–789. (In Chinese) [CrossRef]
3. Xu, X.; Dong, L.; Zhao, Y.; Wang, Y. Effect of the Asian Water Tower over the Qinghai-Tibet Plateau and the characteristics of atmospheric water circulation. *Chin. Sci. Bull.* **2019**, *64*, 2830–2841. [CrossRef]
4. Deliang, C.; Baiqing, X.; Tandong, Y.; ZhengTang, G.; Peng, C.; FaHu, C.; Zhang, R.; Zhang, X.; Zhang, Y.; Jie, F. Assessment of past, present and future environmental changes on the Tibetan Plateau. *Chin. Sci. Bull.* **2015**, *60*, 3025–3035. (In Chinese) [CrossRef]
5. Ma, Y.; Hu, Z.; Tian, L.; Zhang, F.; Duan, A.; Yang, K.; Zhang, Y.; Yang, Y. Study progresses of the Tibet Plateau climate system change and mechanism of its impact on East Asia. *Adv. Earth Sci.* **2014**, *29*, 207–215. (In Chinese) [CrossRef]
6. Zhang, R.; Fengge, S.; Jiang, Z.; Xuejie, G.; Donglin, G.; Jian, N.; Qinglong, Y.; Cuo, L.; BoTao, Z. An overview of projected climate and environmental changes across the Tibetan Plateau in the 21st century. *Chin. Sci. Bull.* **2015**, *60*, 3036–3047. (In Chinese) [CrossRef]
7. Yao, T.; Yu, W.; Wu, G.; Xu, B.; Yang, W.; Zhao, H.; Wang, W.; Li, S.; Wang, N.; Li, Z. Glacier anomalies and relevant disaster risks on the Tibetan Plateau and surroundings. *Chin. Sci. Bull.* **2019**, *64*, 2770–2782. (In Chinese) [CrossRef]
8. Sheridan, S.C.; Lee, C.C.; Smith, E.T. A comparison between station observations and reanalysis data in the identification of extreme temperature events. *Geophys. Res. Lett.* **2020**, *47*, e2020GL088120. [CrossRef]
9. Tandong, Y.; Shilong, P.; Miaogen, S.; Jing, G.; Wei, Y.; Guoqing, Z.; Yanbin, L.; Yang, G.; Liping, Z.; Baiqing, X. Chained impacts on modern environment of interaction between Westerlies and Indian Monsoon on Tibetan Plateau. *Bull. Chin. Acad. Sci.* **2017**, *32*, 976–984. (In Chinese) [CrossRef]
10. Fu, Y.; Ma, Y.; Zhong, L.; Yang, Y.; Guo, X.; Wang, C.; Xu, X.; Yang, K.; Xu, X.; Liu, L. Land-surface processes and summer-cloud-precipitation characteristics in the Tibetan Plateau and their effects on downstream weather: A review and perspective. *Natl. Sci. Rev.* **2020**, *7*, 500–515. [CrossRef]
11. DingYihui, Z. Intercomparison of the time for climate abrupt change between the Tibetan Plateau and other regions in China. *Chin. J. Atmos. Sci.* **2008**, *32*, 794–805. (In Chinese)
12. Liu, X.; Chen, B. Climatic warming in the Tibetan Plateau during recent decades. *Int. J. Climatol.* **2000**, *20*, 1729–1742. [CrossRef]
13. Hansen, J.; Ruedy, R.; Sato, M.; Lo, K. Global surface temperature change. *Rev. Geophys.* **2010**, *48*, RG4004. [CrossRef]
14. You, Q.; Kang, S.; Chen, D.; Li, J.; Ji, Z. Several research frontiers of climate change over the Tibetan Plateau. *J. Glaciol. Geocryol.* **2021**, *43*, 885–901. (In Chinese) [CrossRef]
15. Duo, C.; Yong, Y.; Jiancan, L.; Ciren, B. Applicability analysis of MERRA surface air temperature over the Qinghai-Xizang Plateau. *Plateau Meteorol.* **2016**, *35*, 337–350. (In Chinese) [CrossRef]
16. Wang, X.; Pang, G.; Yang, M. Precipitation over the Tibetan Plateau during recent decades: A review based on observations and simulations. *Int. J. Climatol.* **2018**, *38*, 1116–1131. [CrossRef]
17. Zhao, T.; Fu, C. Applicability evaluation of surface air temperature from several reanalysis datasets in China. *Plateau Meteorol.* **2009**, *28*, 594–606. (In Chinese)
18. Zheng, R.; Li, D.; Jiang, Y. New characteristics of temperature change over Qinghai-Xizang Plateau on the background of global warming. *Plateau Meteorol.* **2015**, *34*, 1531–1539. (In Chinese) [CrossRef]
19. Han, S.; Shi, C.; Sun, S.; Gu, J.; Xu, B.; Liao, Z.; Zhang, Y.; Xu, Y. Development and Evaluation of a Real-Time Hourly One-Kilometre Gridded Multisource Fusion Air Temperature Dataset in China Based on Remote Sensing DEM. *Remote Sens.* **2022**, *14*, 2480. [CrossRef]
20. Gao, K.; Duan, A.; Chen, D.; Wu, G. Surface energy budget diagnosis reveals possible mechanism for the different warming rate among Earth's three poles in recent decades. *Sci. Bull.* **2019**, *64*, 1140–1143. [CrossRef]
21. Lai, H.-W.; Chen, H.W.; Kukulies, J.; Ou, T.; Chen, D. Regionalization of seasonal precipitation over the Tibetan Plateau and associated large-scale atmospheric systems. *J. Clim.* **2021**, *34*, 2635–2651. [CrossRef]
22. Shi, C.; Pan, Y.; Gu, J.; Xu, B.; Han, S.; Zhu, Z.; Zhang, L.; Sun, S.; Jiang, Z. A review of multi-source meteorological data fusion products. *Acta Meteorol. Sin.* **2019**, *77*, 774–783. (In Chinese) [CrossRef]
23. Xia, Y.; Hao, Z.; Shi, C.; Li, Y.; Meng, J.; Xu, T.; Wu, X.; Zhang, B. Regional and global land data assimilation systems: Innovations, challenges, and prospects. *J. Meteorol. Res.* **2019**, *33*, 159–189. [CrossRef]
24. Jiang, Y.; Han, S.; Shi, C.; Gao, T.; Zhen, H.; Liu, X. Evaluation of HRCLDAS and ERA5 Datasets for Near-Surface Wind over Hainan Island and South China Sea. *Atmosphere* **2021**, *12*, 766. [CrossRef]
25. Hersbach, H.; Bell, B.; Berrisford, P.; Horányi, A.; Sabater, J.M.; Nicolas, J.; Radu, R.; Schepers, D.; Simmons, A.; Soci, C. Global reanalysis: Goodbye ERA-Interim, hello ERA5. *ECMWF Newsl.* **2019**, *159*, 17–24.
26. Hersbach, H.; Bell, B.; Berrisford, P.; Hirahara, S.; Horányi, A.; Muñoz-Sabater, J.; Nicolas, J.; Peubey, C.; Radu, R.; Schepers, D. The ERA5 global reanalysis. *Q. J. R. Meteorol. Soc.* **2020**, *146*, 1999–2049. [CrossRef]
27. Hoffmann, L.; Günther, G.; Li, D.; Stein, O.; Wu, X.; Griessbach, S.; Heng, Y.; Konopka, P.; Müller, R.; Vogel, B. From ERA-Interim to ERA5: The considerable impact of ECMWF's next-generation reanalysis on Lagrangian transport simulations. *Atmos. Chem. Phys.* **2019**, *19*, 3097–3124. [CrossRef]

28. Muñoz-Sabater, J.; Dutra, E.; Agustí-Panareda, A.; Albergel, C.; Arduini, G.; Balsamo, G.; Boussetta, S.; Choulga, M.; Harrigan, S.; Hersbach, H. ERA5-Land: A state-of-the-art global reanalysis dataset for land applications. *Earth Syst. Sci. Data Discuss.* **2021**, *13*, 4349–4383. [CrossRef]
29. Xu, W.; Lei, X.; Chen, S.; Zhang, M.; Jiang, L.; Bao, R.; Guan, X.; Ma, M.; Wei, J.; Feng, A. How well does the ERA5 reanalysis capture the extreme climate events over China? Part II: Extreme temperature. *Front. Environ. Sci.* **2022**, *10*, 756. [CrossRef]
30. Pelosi, A.; Terribile, F.; D'Urso, G.; Chirico, G.B. Comparison of ERA5-Land and UERRA MESCAN-SURFEX reanalysis data with spatially interpolated weather observations for the regional assessment of reference evapotranspiration. *Water* **2020**, *12*, 1669. [CrossRef]
31. Ghazanfari, S.; Pande, S.; Hashemy, M.; Sonneveld, B. Diagnosis of GLDAS LSM based aridity index and dryland identification. *J. Environ. Manag.* **2013**, *119*, 162–172. [CrossRef] [PubMed]
32. Rodell, M.; Houser, P.; Jambor, U.; Gottschalck, J.; Mitchell, K.; Meng, C.-J.; Arsenault, K.; Cosgrove, B.; Radakovich, J.; Bosilovich, M. The global land data assimilation system. *Bull. Am. Meteorol. Soc.* **2004**, *85*, 381–394. [CrossRef]
33. Wu, Z.; Feng, H.; He, H.; Zhou, J.; Zhang, Y. Evaluation of Soil Moisture Climatology and Anomaly Components Derived from ERA5-Land and GLDAS-2.1 in China. *Water Resour. Manag.* **2021**, *35*, 629–643. [CrossRef]
34. Shi, C.; Xie, Z.; Qian, H.; Liang, M.; Yang, X. China land soil moisture EnKF data assimilation based on satellite remote sensing data. *Sci. China Earth Sci.* **2011**, *54*, 1430–1440. [CrossRef]
35. Han, S.; Shi, C.; Jiang, Z.; Xu, B.; Li, X.; Zhang, T.; Jiang, L.; Liang, X.; Zhu, Z.; Liu, J. Development and progress of high resolution CMA land surface data assimilation system. *Adv. Meteorol. Sci. Technol.* **2018**, *8*, 102–108 + 116. (In Chinese) [CrossRef]
36. Sun, S.; Shi, C.; Pan, Y.; Bai, L.; Xu, B.; Zhang, T.; Han, S.; Jiang, L. Applicability assessment of the 1998–2018 CLDAS multi-source precipitation fusion dataset over China. *J. Meteorol. Res.* **2020**, *34*, 879–892. [CrossRef]
37. Wang, Q.; Li, W.; Xiao, C.; Ai, W. Evaluation of High-Resolution Crop Model Meteorological Forcing Datasets at Regional Scale: Air Temperature and Precipitation over Major Land Areas of China. *Atmosphere* **2020**, *11*, 1011. [CrossRef]
38. Han, S.; Shi, C.; Xu, B.; Sun, S.; Zhang, T.; Jiang, L.; Liang, X. Development and Evaluation of Hourly and Kilometer Resolution Retrospective and Real-Time Surface Meteorological Blended Forcing Dataset(SMBFD) in China. *J. Meteorol. Res.* **2019**, *33*, 1168–1181. [CrossRef]
39. Liu, L.; Gu, H.; Xie, J.; Xu, Y.P. How well do the ERA-Interim, ERA-5, GLDAS-2.1 and NCEP-R2 reanalysis datasets represent daily air temperature over the Tibetan Plateau? *Int. J. Climatol.* **2021**, *41*, 1484–1505. [CrossRef]
40. Han, S.; Liu, B.; Shi, C.; Liu, Y.; Qiu, M.; Sun, S. Evaluation of CLDAS and GLDAS datasets for Near-surface Air Temperature over major land areas of China. *Sustainability* **2020**, *12*, 4311. [CrossRef]
41. Huang, X.; Han, S.; Shi, C. Multiscale Assessments of Three Reanalysis Temperature Data Systems over China. *Agriculture* **2021**, *11*, 1292. [CrossRef]
42. Liu, Y.; Shi, C.; Wang, H.; Han, S. Applicability assessment of CLDAS temperature data in China. *Trans. Atmos. Sci.* **2021**, *44*, 540–548. (In Chinese) [CrossRef]
43. Wang, W.; Gao, Y.; Xu, J. Applicability of GLDAS and climate change in the Qinghai-Xizang Plateau and its surrounding arid area. *Plateau Meteorol.* **2013**, *32*, 635–645. (In Chinese) [CrossRef]
44. Peng, P.; Zhu, L. Observations of land surface processes of the Tibetan Plateau based on the field stations network. *Sci. Technol. Rev.* **2017**, *35*, 97–102. (In Chinese) [CrossRef]
45. Peng, P.; Zhu, L. *Meteorological Data of Surface Environment and Observation Network in China's Cold Region (2018)*; National Tibetan Plateau Data Center: Beijing, China, 2020. [CrossRef]
46. Rolland, C. Spatial and seasonal variations of air temperature lapse rates in Alpine regions. *J. Clim.* **2003**, *16*, 1032–1046. [CrossRef]
47. Camargo, L.R.; Schmidt, J. Simulation of multi-annual time series of solar photovoltaic power: Is the ERA5-land reanalysis the next big step? *Sustain. Energy Technol. Assess.* **2020**, *42*, 100829. [CrossRef]
48. Wang, W.; Cui, W.; Wang, X.; Chen, X. Evaluation of GLDAS-1 and GLDAS-2 forcing data and Noah model simulations over China at the monthly scale. *J. Hydrometeorol.* **2016**, *17*, 2815–2833. [CrossRef]
49. Rui, H.; Beaudoing, H. README Document for NASA GLDAS Version 2 Data Products. Available online: https://hydro1.gesdisc.eosdis.nasa.gov/data/GLDAS/README_GLDAS2.pdf (accessed on 10 July 2022).
50. Nash, J.E.; Sutcliffe, J.V. River flow forecasting through conceptual models part I—A discussion of principles. *J. Hydrol.* **1970**, *10*, 282–290. [CrossRef]
51. Moriasi, D.N.; Arnold, J.G.; Van Liew, M.W.; Bingner, R.L.; Harmel, R.D.; Veith, T.L. Model evaluation guidelines for systematic quantification of accuracy in watershed simulations. *Trans. ASABE* **2007**, *50*, 885–900. [CrossRef]
52. Gupta, H.V.; Kling, H.; Yilmaz, K.K.; Martinez, G.F. Decomposition of the mean squared error and NSE performance criteria: Implications for improving hydrological modelling. *J. Hydrol.* **2009**, *377*, 80–91. [CrossRef]
53. Pool, S.; Vis, M.; Seibert, J. Evaluating model performance: Towards a non-parametric variant of the Kling-Gupta efficiency. *Hydrol. Sci. J.* **2018**, *63*, 1941–1953. [CrossRef]
54. Willmott, C.J. On the validation of models. *Phys. Geogr.* **1981**, *2*, 184–194. [CrossRef]
55. Ji, L.; Senay, G.B.; Verdin, J.P.J.J.o.H. Evaluation of the Global Land Data Assimilation System (GLDAS) air temperature data products. *J. Hydrometeorol.* **2015**, *16*, 2463–2480. [CrossRef]
56. You, Q.; Fraedrich, K.; Ren, G.; Pepin, N.; Kang, S. Variability of temperature in the Tibetan Plateau based on homogenized surface stations and reanalysis data. *Int. J. Climatol.* **2013**, *33*, 1337–1347. [CrossRef]

57. You, Q.; Kang, S.; Pepin, N.; Flügel, W.-A.; Yan, Y.; Behrawan, H.; Huang, J. Relationship between temperature trend magnitude, elevation and mean temperature in the Tibetan Plateau from homogenized surface stations and reanalysis data. *Glob. Planet. Chang.* **2010**, *71*, 124–133. [CrossRef]
58. Li, X.; Cheng, G.; Lu, L. Spatial analysis of air temperature in the Qinghai-Tibet Plateau. *Arct. Antarct. Alp. Res.* **2005**, *37*, 246–252. [CrossRef]
59. Yang, J.; Huang, M.; Zhai, P. Performance of the CRA-40/Land, CMFD, and ERA-Interim datasets in reflecting changes in surface air temperature over the Tibetan Plateau. *J. Meteorol. Res.* **2021**, *35*, 663–672. [CrossRef]
60. Yin, H.; Sun, Y.; Donat, M.G. Changes in temperature extremes on the Tibetan Plateau and their attribution. *Environ. Res. Lett.* **2019**, *14*, 124015. [CrossRef]
61. Ding, L.; Zhou, J.; Zhang, X.; Liu, S.; Cao, R. Downscaling of surface air temperature over the Tibetan Plateau based on DEM. *Int. J. Appl. Earth Obs. Geoinf.* **2018**, *73*, 136–147. [CrossRef]
62. Hu, G.; Zhao, L.; Wu, X.; Li, R.; Wu, T.; Su, Y.; Hao, J.J.T.; Climatology, A. Evaluation of reanalysis air temperature products in permafrost regions on the Qinghai-Tibetan Plateau. *Theor. Appl. Climatol.* **2019**, *138*, 1457–1470. [CrossRef]
63. Peng, X.; Frauenfeld, O.W.; Jin, H.; Du, R.; Qiao, L.; Zhao, Y.; Mu, C.; Zhang, T. Assessment of temperature changes on the Tibetan Plateau during 1980–2018. *Earth Space Sci.* **2021**, *8*, e2020EA001609. [CrossRef]
64. Ding, M.; Du, J.; Xiao, T. Applicability of the four reanalysis temperature data in Chang Tang Nature Reserve. *J. Glaciol. Geocryol.* **2020**, *42*, 1046–1056. (In Chinese) [CrossRef]
65. Long, K.; Shi, C.; Han, S.; Huang, X.; Xu, X.; Sun, S.; Gu, J. Quality assessment of High Resolution Temperature Merged Grid Analysis Product in China. *Plateau Mt. Meteorol. Res.* **2019**, *39*, 67–74. (In Chinese) [CrossRef]
66. Shi, G.; Sun, Z.; Qiu, X.; Zeng, Y.; Chen, P.; Liu, C. Comparison of two air temperature gridding methods over complex terrain in China. *Theor. Appl. Climatol.* **2018**, *133*, 1009–1019. [CrossRef]
67. Gao, L.; Bernhardt, M.; Schulz, K. Elevation correction of ERA-Interim temperature data in complex terrain. *Hydrol. Earth Syst. Sci.* **2012**, *16*, 4661–4673. [CrossRef]
68. Wang, A.; Zeng, X. Evaluation of multireanalysis products with in situ observations over the Tibetan Plateau. *J. Geophys. Res. Atmos.* **2012**, *117*, d05102. [CrossRef]

remote sensing

Article

How Well Do CMIP6 Models Simulate the Greening of the Tibetan Plateau?

Jiafeng Liu [1,2] and Yaqiong Lu [1,*]

1 Institute of Mountain Hazards and Environment, Chinese Academy of Sciences, Chengdu 610299, China
2 University of Chinese Academy of Sciences, Beijing 100049, China
* Correspondence: yaqiong@imde.ac.cn; Tel.: +86-02861158015

Abstract: The "warm-humid" climate change across the Tibetan Plateau (TP) has promoted grassland growth and an overall greening trend has been observed by remote sensing products. Many of the current generations of Earth System Models (ESMs) incorporate advanced process-based vegetation growth in the land surface module that can simulate vegetation growth, but the evaluation of their performance has not received much attention, especially over hot spots where projections of the future climate and vegetation growth are greatly needed. In this study, we compare the leaf area index (LAI) simulations of 35 ESMs that participated in CMIP6 to a remote-sensing-derived LAI product (GLASS LAI). The results show that about 40% of the models overestimated the Tibetan Plateau's greening, 48% of the models underestimated the greening, and 11% of the models showed a declining LAI trend. The CMIP6 models generally produced poor simulations of the spatial distribution of LAI trend, and overestimated the LAI trend of alpine vegetation, grassland, and forest, but underestimated meadow and shrub. Compared with other vegetation types, simulations of the forest LAI trend were the worst, the declining trend in forest pixels on the TP was generally underestimated, and the greening of the meadow was underestimated as well. However, the greening of the grassland, was greatly overestimated. For the Tibetan Plateau's averaged LAI, more than 70% of the models overestimated this during the growing seasons of 1981–2014. Similar to the forest LAI trend, the performance of the forest LAI simulation was the worst among the different vegetation types, and the forest LAI was underestimated as well.

Keywords: Coupled Model Intercomparison Project Phase 6 (CMIP6); LAI; LAI trend; Tibetan Plateau

Citation: Liu, J.; Lu, Y. How Well Do CMIP6 Models Simulate the Greening of the Tibetan Plateau? *Remote Sens.* **2022**, *14*, 4633. https://doi.org/10.3390/rs14184633

Academic Editor: Lunche Wang

Received: 1 July 2022
Accepted: 13 September 2022
Published: 16 September 2022

1. Introduction

Vegetation is a critical component of terrestrial ecosystems and is very sensitive to climate change [1–3]. The global average surface temperature increased by 0.85 °C from 1880 to 2012 [4], which triggered phenological changes in different vegetation types in different regions. The increase in temperature, as one of the causes of variation in vegetation, has led to a significant overall change in vegetation, manifested by an increase in the Normalized Difference Vegetation Index (NDVI) during the vegetation growth season in the Northern Hemisphere [5], and the growth rate of NDVI in forests is greater than that of other vegetation types [6–8]. The community structure of snow-meadow vegetation has changed significantly as a result of climate change in Northern Japan over the last 40 years [9]. In the Siberian Mountains, the birch area has increased by 10%, and birch stands and the treeline boundary have moved upslope at a rate of 1.4 m yr^{-1} and 4.0 m yr^{-1}, respectively, since the 1970s with the onset of warming [10]. In China, the zone of tundra vegetation of the Changbai Mountains has been invaded by herbaceous plants with the rising temperature over the last 30 years [11].

As the third pole of the earth, the Tibetan Plateau (TP) is highly sensitive to climate change and has been experiencing a rapid warming of 0.4° 10 yr^{-1} over the last 30 years [12,13] and with precipitation increasing by 1.96 mm 10 yr^{-1} in 1994–2015 [14].

This "warm–humid" trend has led to tremendous changes on the land surface, such as glaciers collapsing [15], permafrost thawing [16], and lakes expanding [17], as well as surface vegetation growth. Liu et al. [18] found that the vegetation coverage on the TP showed a trend of "overall increase and partial degradation" from 1981 to 2005, with the area of improvement much larger than the area of degradation. Wei et al. [19] found that "warm-humid" has a significant promoting effect on the improvement of vegetation on the TP, and Zhang et al. [20] found that the overall NDVI of grassland in the growing season of the TP also shows an increasing trend. Xu et al. [21] used the leaf area index inversion by NOAA–AVHRR to study the temporal and spatial changes in vegetation cover characteristics in the TP, and also found an overall increase in vegetation cover. Zhang et al. [13] found that the green-up dates with the alpine vegetation in the Plateau had a continuous advancing trend with a rate of ~1.04 $d \cdot y^{-1}$ from 1982 to 2011.

Remote sensing, as one of the major tools for studying vegetation's response to climate change [22], was used to study the vegetation on the TP, with various long-term vegetation leaf area index (LAI) datasets derived through satellite remote sensing, such as GLASS LAI [23], GLOBMAP LAI [24], GIMMS LAI [25], and MODIS LAI [5]. Hua et al. (2018) [26] used the GIMMS NDVI dataset (NDVI-3g) to study the temporal and spatial variations in vegetation dynamics controlled by climate on the Tibetan Plateau during 1982–2011 and found that the potential cause of the change in vegetation dynamics might be controlled by the climate, particularly the increasing precipitation and the significant temperature rise in the Central and Southeastern Tibetan Plateau. Although remote sensing products are very useful for understanding historical vegetation variations, satellite remote sensing could not directly measure future vegetation dynamics. Another powerful tool, the state-of-the-art Earth System Models that incorporate a process-based vegetation growth module, can simulate not only historical variations in vegetation but also those in future climate. Zhu et al. [27] built the first pedotransfer function to simulate temporal variations in vegetation coverage (VC) and found that the pedotransfer function more accurately simulated temporal variation in VC than a multiple linear regression in an alpine meadow on the Tibetan Plateau. Lu et al. [28] found that net primary productivity (NPP) and LAI decreased from the southeast to the northwest of the Tibetan Plateau by using the atmosphere–vegetation interaction model (AVIM) to simulate the distribution of LAI and NPP over the Tibetan Plateau. The accuracy of the simulation results varies greatly due to the design and use of the model itself, so it is very important to evaluate the accuracy of the simulation data before using the simulations.

The International Coupled Model Comparison Program (CMIP), proposed by the World Climate Research Program Group, currently in the sixth generation (CMIP6), has been widely used for studying various environmental changes. Tian et al. [29] analyzed changes in the annual mean surface air temperature (SAT) and precipitation, and also the related uncertainties using historical simulations and future projections under the Representative Concentration Pathway scenarios (RCPs) from the CMIP5 models across China and in its seven sub-regions. Zhang et al. [30] demonstrated that there may be a basic spatial scale limit below which it may not be useful to further refine climate model predictions based on an integrated analysis of coupled model simulations and projections from CMIP3 and CMIP5. Using the established linear relationship and monthly temperature simulations from CMIP5 models over the Northern Hemisphere during the 21st century, Xia et al. [31] found the start of the vegetation growing season (SOS) will have advanced by 4.7 days under RCP2.6 (Representative Concentration Pathway) by 2040–2059. After CMIP5, more and more models have incorporated a dynamic vegetation growth module, and therefore evaluating CMIP vegetation simulations has drawn much attention. Anav et al. [32] assessed the ability of 18 Earth system models (ESMs) in CMIP5 and found that most models overestimated the global average LAI and half of the models also overestimated the LAI trend for 1986–2005. Zhao et al. [33] analyzed the changes in projected global LAI from 16 CMIP5 ESMs and 17 CMIP6 ESMs, and found that the CMIP6 models had a better ability to describe the global area-averaged LAI time series.

Lawrence et al. [34] did not evaluate the performance of the simulated global tree height of the CMIPs' ESMs but gave the biases of tree height for the offline simulations of CLM5BGC. Brovkin et al. [35] evaluated the performance of MPI-ESM, and Seller [36] evaluated UKESM1-0-LI in terms of vegetation distribution; both found that the two models overestimated the fraction of tree coverage. Most evaluations have focused on the global scale; few have focused on regional scales such as the Tibetan Plateau. Bao et al. [37] evaluated 12 CMIP5 ESMs for reproducing vegetation cover and LAI over the Tibetan Plateau in 1986–2005, and found that INMCM4, BCC-CSM-1.1M, MPI-ESM-LR, IPSL-CM5A-LR, HadGEM2-ES, and CCSM4 were the best six models for capturing vegetation among the 12 models. CMIP6 has had the largest participation since its implementation [38]. However, how well the CMIP6 models simulate vegetation growth, especially the recent greening of the Tibetan Plateau, is unknown.

LAI is usually defined as half of the total leaf surface area per unit of surface area [39], and NDVI is defined as the ratio of the difference between the near-infrared band (NIR) and the visible red band (R), and the sum of the two bands, $NDVI = (NIR - R)/(NIR + R)$. NDVI is directly obtained from the satellites' reflection information and the real-time variation of vegetation after a simple calculation, which can quantitatively reflect the actual variation of vegetation, including the vegetation structure, the vegetation growth, and the vegetation coverage during the observation period, and is widely used in the field of vegetation remote sensing [40–42]. LAI and NDVI are both important indices for quantifying the vegetation variations, but only LAI could be validated because NDVI is not an output of the dynamic vegetation growth models in CMIP6. LAI, as a key indicator of vegetation growth [43], has been widely used in global climate models, ecological models, hydrological models, and ecosystem productivity models [44]. Therefore, we focused on LAI validations in our work rather than NDVI.

In recent decades, although greening is one of the most important changes in the Tibetan Plateau, few works have particularly focused on the performance of the model simulations on the greening of the Tibetan Plateau. We developed our own ranking method that considered the temporal and spatial simulations' abilities to give an overall assessment of CMIP6 models. We also quantified the growth of different vegetation types. Our goals with this work are to evaluate the performance to simulate the LAI trend and LAI of the CMIP6 model during the growing season and to provide a reference for the selection of simulation data of vegetation changes, aid the research into vegetation in the Tibetan Plateau, and analyze the sources of temporal and spatial error in each model, laying a foundation for model optimization.

2. Data and Methods

2.1. Study Area

The TP [45,46] is located at 26–39°N latitude in Southwest China. Surrounded by high mountains on the edge of the area, the internal topography is complex, including plateaus, basins, glaciers, lakes, and swamps [47]. Its geographical features, such as the high altitude, and the complex and changeable topography, have created special climatic conditions and water and heat distribution in this area, and have also created its distinctive vegetation distribution. As the largest alpine grassland ecosystem in the world, the TP is dominated by meadows and grasslands (Figure 1), concentrated across a wide range of Central Tibet. The vegetation types in Tibet have spatial distribution characteristics that gradually change from southeast to northwest. From southeast to northwest in Tibet, the vegetation types are distributed in the order of forests, shrubs, meadows, grassland, and desert (Figure 1). The dataset is derived from the 1:1 million vegetation data set collected in China in 2001, and it is provided by the National Cryosphere Desert Data Center (http://www.ncdc.ac.cn) (accessed on 9 December 2021).

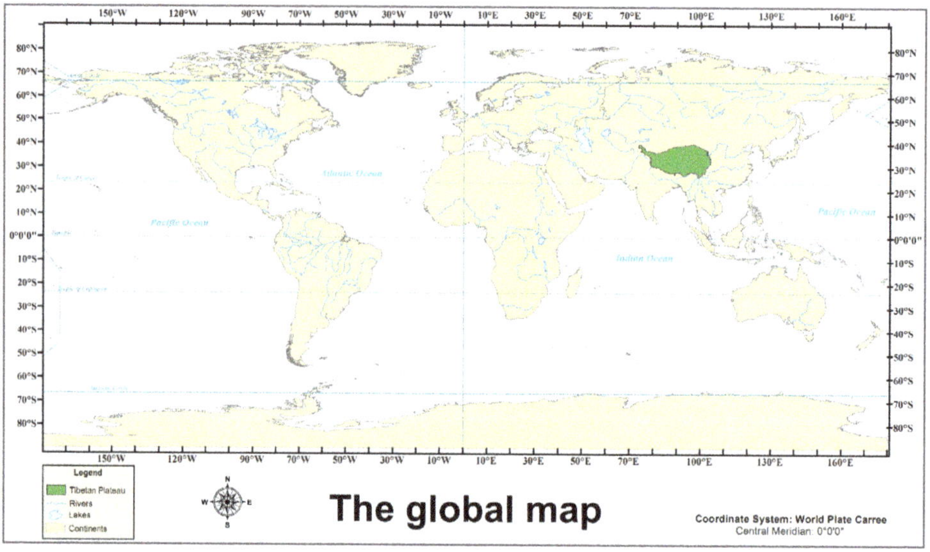

(**a**)

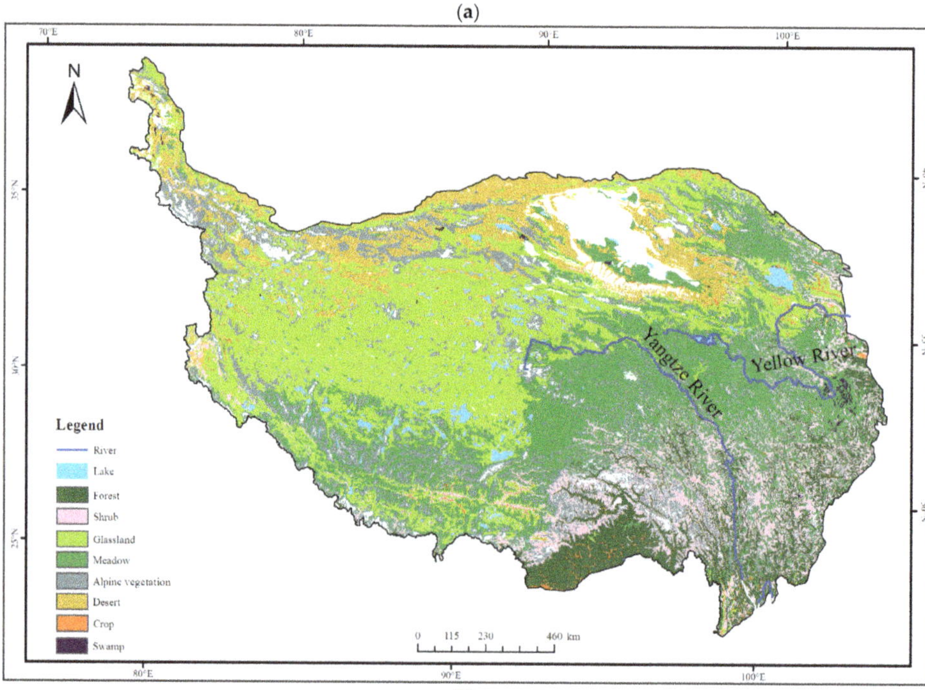

(**b**)

Figure 1. (**a**) The location of the Tibetan Plateau [48] in the world map, and the world map from ArcGIS. (**b**) Distribution of vegetation types on the Tibetan Plateau [49].

2.2. Satellite Data

To evaluate the ability of the 35 models from the CMIP6 to reproduce the LAI over the Tibetan Plateau, the 1981–2018 LAI data from the Global Land Surface Satellite (GLASS) dataset with an eight-day temporal frequency and a $0.5° \times 0.5°$ spatial resolution were used as a benchmark in our study. GLASS LAI uses generalized regression neural networks (GRNNs) to invert LAI from Time-Series AVHRR Surface Reflectance data; the algorithm trains GRNNs using preprocessed AVHRR Time-Series AVHRR Surface Reflectance, and then uses rolling processing to produce time-continuous long-term GLASS LAI products from the preprocessed AVHRR Surface Reflectance [23]. Compared with other LAI datasets, GLASS LAI data have a long observation period, high quality, and good accuracy [50]. They have more complete trajectories than the MODIS LAI product and also show lower uncertainty than the MODIS and CYCLOPES LAI products compared with 20 ground-measured LAI reference maps. Many studies use GLASS LAI as a reference database for research or validation [51–55]. All these factors make it an ideal long-term dynamic LAI observation dataset in this study. The GLASS LAI product (V50) used in this study is available from the University of Maryland and the Center for Global Change Data Processing and Analysis of Beijing Normal University (http://www.glass.umd.edu/Download.html, accessed on 9 March 2021).

2.3. CMIP6 Model Simulations

Thirty-five CMIP6 models with no missing data were selected in this study, and the LAI from outputs of historical simulations for 1850–2014 was used (https://esgf-node.llnl.gov/search/cmip6/, accessed on 16 August 2021).

In order to facilitate the comparison of the simulation and observational data, all simulations were downloaded and converted to a 0.5° × 0.5° spatial resolution by bilinear interpolation from low to high resolution. The overlaps of the GLASS datasets and CMIP6 were 1981–2014, so our analysis focused on 1981–2014. The model's information is shown in Table 1.

2.4. Evaluation Approach

A series of evaluation indicators was applied to quantify the agreement between the observed and simulated LAI and the trend of the CMIP6 models. In this study, we calculated the average LAI during the growing season (May–September) for each year as the average LAI, a linear regression trend of the average LAI from 1981 to 2014 as the trend, and an increasing trend indicated TP greening. We also calculated the monthly average LAI for each month of the growing season, and the TP averaged monthly average LAI during 1981–2014 as the monthly LAI. Then, we calculated the linear regression trend of the monthly average LAI for each month during the growing season from 1981 to 2014, and the TP averaged trend of the monthly average LAI as the monthly LAI trend. We obtained monthly variations from the monthly LAI and the monthly LAI trend during the growing season. In the following, we further describe the metrics used for model evaluation and the method used for ranking the models.

2.4.1. Evaluation Metrics

The spatial correlation (pattern correlation) was used to quantify the correlation between the grid cell trend (or the grid cell average LAI from 1981 to 2014) distribution in the models and observations. Through a combination of the definitions of Bao et al. [37] and Chang et al. [80], the spatial correlation formula for the simulated and observed trends in this study was defined as follows:

$$Pattern\ correlation = \frac{\frac{1}{N}\sum_i^N W_i(M_i-\overline{M})(O_i-\overline{O})}{\sqrt{\frac{1}{N}\sum_i^N W_i(M_i-\overline{M})^2}\sqrt{\frac{1}{N}\sum_i^N W_i(O_i-\overline{O})^2}}. \quad (1)$$

where N is the total number of grid cells under evaluation, M_i and O_i are the simulated and observed trend (or the average LAI from 1981 to 2014) from the CMIP6 models and the GLASS of the grid cell i , and W_i is the area weight of the grid cell i (all grid weights add up to 1) [37]. We calculated W_i in the Pearson correlation coefficient equation as the area of each grid cell associated with the central geographic latitude of each grid cell [37]. In the TP, the variation in W_i is not obvious and the value of W_i can almost be neglected.

Table 1. Model description.

Model	Institute	Land Surface Model	Resolution	Reference
AWI-ESM-1-1-LR	AWI (Germany)	CABLE2.4	250 km	[56]
ACCESS-ESM1-5	CSIRO (Australia)	CABLE2.4	250 km	[57]
BCC-CSM2-MR	BCC (China)	BCC-AVIM2.0	100 km	[58]
BCC-ESM1	BCC (China)	BCC-AVIM2.0	250 km	[58]
CAMS-CSM1-0	China	CoLM	100 km	[59]
CanESM5	CCCMA (Canada)	CLASS3.6-CTEM1.2	500 km	[60]
CanESM5-CanOE	CCCMA (Canada)	CLASS3.6-CTEM1.2	500 km	[60]
CESM2	NCAR (USA)	CLM5	100 km	[61]
CESM2-FV2	NCAR (USA)	CLM5	100 km	[61]
CMCC-CM2-SR5	CMCC (Italy)	CLM4.5	100 km	[62]
CMCC-ESM2	CMCC (Italy)	CLM4.5	100 km	[62]
E3SM-1-0	E3SM-Project (USA)	ELM	100 km	[63]
E3SM-1-1	E3SM-Project (USA)	ELM	100 km	[63]
E3SM-1-1-ECA	E3SM-Project (USA)	ELM	100 km	[63]
EC-Earth3-Veg	EC-Earth-Consortium (Europe)	HTESSEL	100 km	[64]
EC-Earth3-Veg-LR	EC-Earth-Consortium (Europe)	HTESSEL	100 km	[64]
FGOALS-g3	China	CAS-LSM	$2 \times 2°$	[65]
FIO-ESM-2-0	FIO (China)	CLM4.0	100 km	[66]
GFDL-CM4	GFDL (USA)	LM4.0	100 km	[67]
GFDL-ESM4	GFDL (USA)	LM4.1	100 km	[68]
GISS-E2-1-G	GISS (USA)	GISS LSM	250 km	[69]
HadGEM3-GC31-LL	HadGEM (United Kingdom)	JULES	250 km	[70]
HadGEM3-GC31-MM	HadGEM (United Kingdom)	JULES	100 km	[70]
INM-CM4-8	INM (Russia)	INM-LND1	100 km	[71]
INM-CM5-0	INM (Russia)	INM-LND1	100 km	[72]
IPSL-CM6A-LR	IPSL (France)	ORCHIDEE v2	250 km	[73]
KIOST-ESM	KIOST (Korea)	LM3.0	250 km	[74]
MIROC-ES2L	MIROC (Japan)	MATSIRO6.0 +VISIT-e v1	500 km	[75]
MPI-ESM-1-2-HAM	HAMMOZ Consortium (Switzerland, Germany, Finland, UK)	CABLE2.4	250 km	[76]
MPI-ESM1-2-HR	MPI (Germany)	CABLE2.4	100 km	[76]
MRI-ESM2-0	MRI (Japan)	HAL 1.0	100 km	[77]
NorESM2-LM	NCC (Norway)	CLM5	250 km	[78]
NorESM2-MM	NCC (Norway)	CLM5	100 km	[78]
TaiESM1	AS-RCEC (Taiwan, China)	CLM4.0	100 km	[79]
UKESM1-0-LI	MOHC (UK)	JULES-ES-1.0	250 km	[36]

The bias between the simulated and observed grid cell trend (or the grid cell average LAI from 1981 to 2014) was calculated to quantify the main bias between the model simulations and GLASS observations. In our study, we subtracted the observed trend (or the average LAI from 1981 to 2014) from the simulated trend (or the average LAI from 1981 to 2014) to get trend (or the average LAI from 1981 to 2014) bias at the single grid cell i by Equation (2). We thus obtained a value of the bias at every grid cell and the distribution of the bias across the whole study region. The relative bias of grid cell trend (or the grid cell average LAI from 1981 to 2014) was calculated as the ratio of the trend (or the average LAI from 1981 to 2014) bias to the observed trend (or the average LAI from 1981 to 2014) at the grid cell i in Equation (3). We also calculated the TP averaged bias using Equation (4).

$$Bias = M_i - O_i \tag{2}$$

$$Relative_{Bias} = \frac{Bias_i}{O_i} \tag{3}$$

$$Bias_{avg} = \frac{\sum_1^N |M_i - O_i|}{N} \tag{4}$$

The root-mean-square error (RMSE) was used to measure the difference between the simulations and observations. Similar to bias, we calculated the trend (the average LAI from 1981 to 2014) of the two datasets at grid cell i , and then aggregated the results over the entire TP. Next, we converted spatial two-dimensional data of trend (or the average LAI from 1981 to 2014) in simulations and observations into one dimension and calculated the RMSE of the two columns (the simulations and observations) of the one-dimensional data by Equation (5). This RMSE was used in the ranking. Moreover, we used RMSE to quantify the difference in the average LAI from 1981 to 2014 sequence between the simulation and observation during the growing season in 1981–2014 at single grid cell i , and then obtained the distribution of RMSE across the region.

$$RMSE = \sqrt{\frac{\sum_1^N (M_i - O_i)^2}{N}} \tag{5}$$

The ratio of the standard deviation ($Ratio_\delta$) was used to quantify the magnitude of the difference in variation between the simulation and the observation. Similar to RMSE, we first converted the spatial two-dimensional data of the grid cell trend (or the grid cell average LAI from 1981 to 2014) in simulations and observations into one dimension, and then calculated the standard deviation of the simulations and observations by Equation (6), and finally calculated the ratio of the two standard deviations. Furthermore, δ_M and δ_O were the standard deviations of the model simulations and the GLASS observations, respectively. The ratio of trend ($Ratio_{trend}$) was used to quantify the variation of the simulated trend and the observed trend as either overestimation or underestimation. We calculated the $Ratio_{trend}$ by Equation (7); $trend_M$ was the simulated trend, and the $trend_O$ was the GLASS trend. A ratio less than 0 indicated that the trend was not captured, contrary to the trend in GLASS. A ratio greater than 0 but less than 1 indicated that the greening or declined trend was captured, but was underestimated. A ratio greater than 1 indicated an overestimation of the greening or declined trend.

$$Ratio_\delta = \frac{\delta_M}{\delta_O} \tag{6}$$

$$Ratio_{trend} = \frac{trend_M}{trend_O} \tag{7}$$

2.4.2. Significant Test Method

We used two methods for significance testing, the Student's *t*-test and the Mann–Kendall trend test. The Student's *t*-test was used for the significant difference test between simulations and observations. The Mann–Kendall trend test was used to detect whether a time series was steadily increasing/decreasing or unchanging.

2.4.3. Ranking Method

A ranking scheme was developed by Brunke et al. to score the multi-bulk aerodynamic algorithm for calculating the turbulence fluxes on the ocean surface [81]. Decker et al. [82] ranked the bias and standard deviation of error between reanalysis products and flux tower measurements using the same method as Brunke et al. On the basis of Decker et al., Wang et al. [83] extended this ranking approach and increased the statistical parameters to four, including the correlation coefficient (ρ), the standard deviation ratio (σ_r/σ_{obs}), the standard deviation error (σ_d), and the difference (bias) to rank the ability of six kinds of reanalysis data to reproduce climate characteristics over the Tibetan Plateau. Since then, this ranking approach, as a good example of model performance evaluation, has been used in many studies [32,37]. In this study, we adjusted the ranking method used by Wang and Zeng, and the ranking metrics were changed into the spatial correlation (pattern correlation), the bias (Bias), the root mean square error (RMSE), and the ratio of standard deviation ($Ratio_\delta$).

In the context of this study, the simulation with the highest pattern correlation, the lowest bias and RMSE, and the closest ratio was considered to have the best performance for reproducing the trend (or the LAI) over the Tibetan Plateau. The models were ranked from 1 to 35, with 1 being the model with the lowest value in magnitude of bias, RMSE, or | ratio-1 | (or the highest pattern correlation) and 35 being the model with the highest value in magnitude of bias, RMSE, or | ratio-1 | (or the lowest pattern correlation) [82]. We then calculated the total score of the four metrics for a single model and defined the total score as the "error ranking". The higher the model's error ranking, the closer the relationship between the simulations and observations.

3. Result

3.1. The Average Growing Season LAI and Trend

More than 70% of models overestimated and about 28% of models underestimated the area-averaged growing season LAI over the Tibetan Plateau (Figure 2). EC-Earth3-Veg, C-Earth3-Veg-LR, and HadGEM3-GC31-LL showed the smallest average LAI bias with slight underestimations of 0.0066–0.018 $m^2\ m^{-2}$ in comparison with GLASS LAI. CMIP6 models (except FI0-ESM-2-0) incorporating the community land model (hereafter referred to as the CLM family) showed a much larger LAI bias of 2–5.5 $m^2\ m^{-2}$, especially CESM2, CESM2-FV2, NorESM2-LM, and NorESM2-MM (4–5.5 $m^2\ m^{-2}$). CanESM5, CanESM5-CanOE, E3SM-1-0, GISS-E2-1-G, IPSL-CM6A-LR, and KIOST-ESM underestimated the average LAI (0.1–0.40 $m^2\ m^{-2}$), but these underestimations were much smaller than the overestimations of other CMIP6 models.

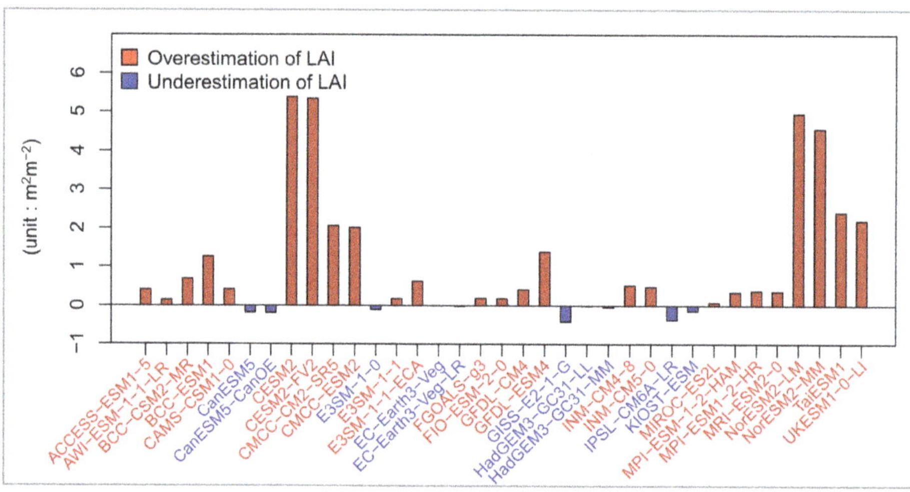

Figure 2. The bias of the area-averaged LAI during the growing season in Tibetan Plateau from 1981 to 2014 between each CMIP6 model and GLASS data.

In Figure 3, we show the ratio of the area-averaged trend between simulations and observations from 1981–2014 in TP. For the Tibetan Plateau LAI trend in 1981–2014, about 40% of the models overestimated the Tibetan Plateau's greening, more than 48% of the models underestimated the greening, and 11% models showed a declining LAI trend (Figure 3). E3SM-1-1 and MPI-ESM-1-2-HAM showed the closest trend estimations among the 35 CMIP6 models. For some CMIP6 models, the overestimation or the underestimation of greening and the area-averaged LAI (in Figure 2) occurred at the same time. For example, CMIP6 models (except for FI0-ESM-2-0) that incorporated CLM also greatly overestimated the greening of the Tibetan Plateau above the GLASS data (2.5–6.5 times higher), while CanESM5 underestimated not only the average LAI but also the greening. However, models such as AWI-ESM-1-1-LR and UKESM1-0-LI overestimated the average LAI but underestimated the greening.

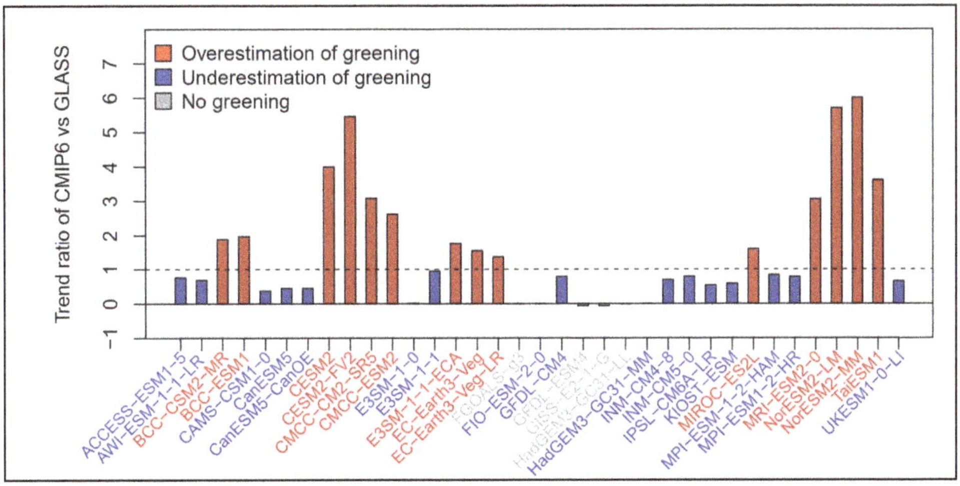

Figure 3. The ratio of the area-averaged LAI trend of the growing season (1981–2014) between each CMIP6 models and the GLASS data.

3.2. *LAI and Trend Monthly Variations*

3.2.1. Monthly Leaf Area Index

The maximum underestimation of LAI mainly occurred in July and August, while the maximum overestimation of LAI varied greatly across different CMIP6 models, and this variation depended greatly on the land surface models incorporated in the different CMIP6 models (Figure 4a). The monthly variation in the bias of the LM family (UKESM1-0-LI, GFDL-CM4, and GFDL-ESM4) was similar for each month of the growing season. Unlike the LM family, the overestimation bias of the CLM family (except for FIO-ESM-2-0) first increased and then remained stable, with the bias in May being the smallest, and the largest being in June or September. The bias of the BCC family showed more complex monthly variation characteristics, with the overestimation bias increasing and then decreasing, and the bias in August being the largest.

Moreover, the good simulations of area-averaged LAI of EC-Earth3-Veg, EC-Earth3-Veg-LR, and HadGEM3-GC31-LL were due to the positive and negative biases in different months cancelling each other out. EC-Earth3-Veg and EC-Earth3-Veg-LR underestimated LAI in May (−0.04 to −0.01 m^2 m^{-2}), July (−0.06 to -0.03 m^2 m^{-2}), and August (−0.1 to −0.05 $m^2 m^{-2}$), while LAI was overestimated in June (0.003–0.022 m^2 m^{-2}) and September (0.02–0.05 m^2 m^{-2}), HadGEM3-GC31-LL overestimated LAI in May (0.014 m^2 m^{-2}) and June (0.024 m^2 m^{-2}), but underestimated LAI in July (−0.016 m^2 m^{-2}), August (−0.09 m^2 m^{-2}), and September (−0.021 m^2 m^{-2}), and these biases partially canceled each other out, making the overall average bias smaller.

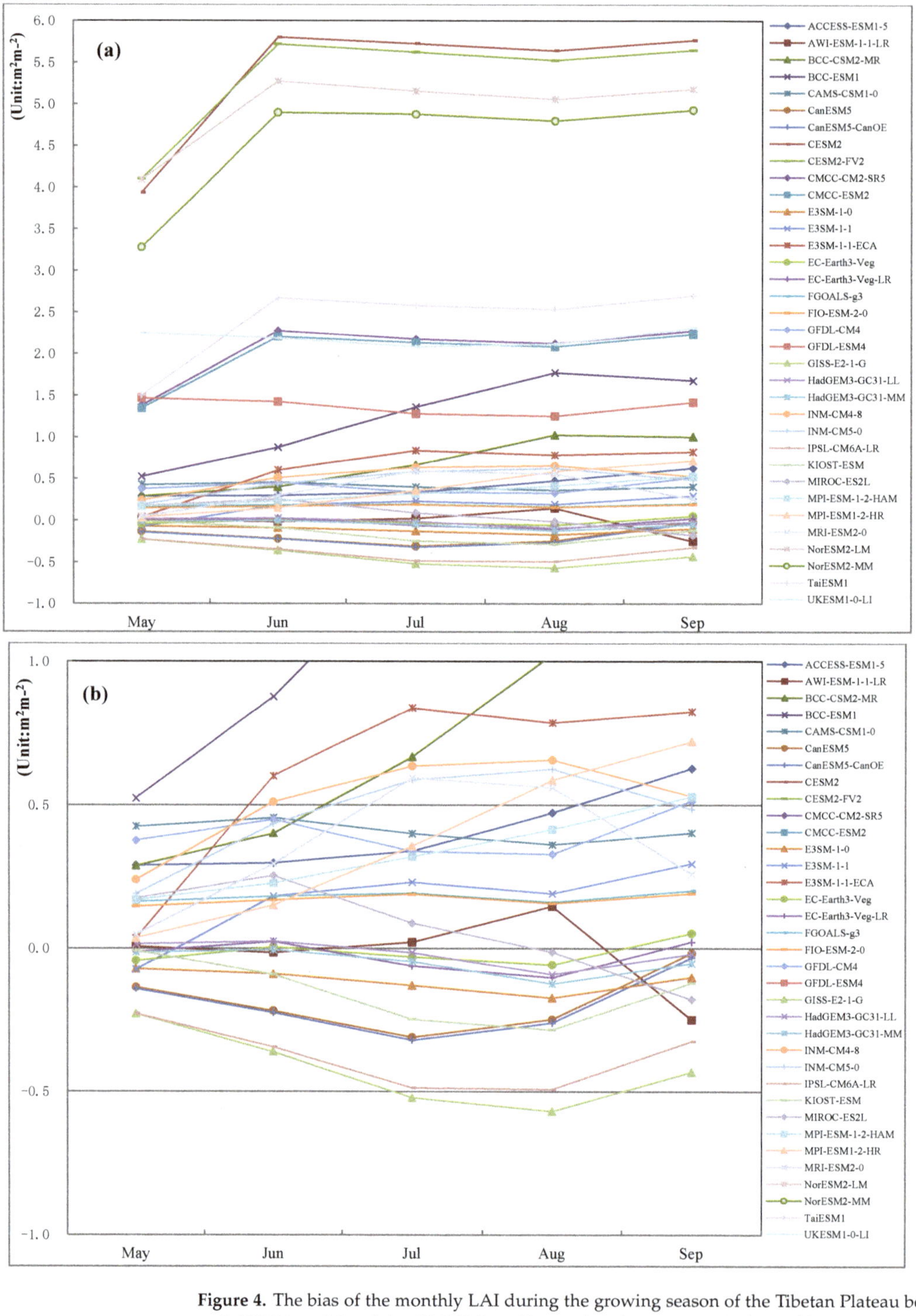

Figure 4. The bias of the monthly LAI during the growing season of the Tibetan Plateau between each CMIP6 model and the GLASS data. The *y*-axis is from −1 to 6 in (**a**), and from −1 to 1 in (**b**).

Although the bias of LAI in May was small, the relative LAI bias was quite large in May (Figure S1). For example, the relative LAI bias of the CLM family (except for FIO-ESM-2-0) was highest in May and June (364–1105%) and then decreased from May or June to August (265–725%), which suggested that improvements at the beginning of growth are key to these models.

3.2.2. Monthly LAI Trend

None of the CMIP6 models captured the monthly LAI trend well, even those models that showed good agreement for the annual LAI trend (Figure 5). The good overall greening simulations of E3SM-1-1, INM-CM5-0, INM-CM4-8, and MPI-ESM1-2-HR were due to the overestimations and underestimations in different months cancelling each other out.

Models that underestimated the greening of the Tibetan Plateau generally had the greatest underestimation in July and August (Figure 5). For example, except for IPSL-CM6A-LR, the monthly error of other models that underestimated the greening of the Tibetan Plateau showed the changes first increasing and then decreasing, and the underestimation error was usually the largest in July and August. However, the models that overestimated the greening of the Tibetan Plateau showed inconsistent monthly variations. For example, the CLM family (except for FIO-ESM-2-0) showed the largest overestimation in May (2.93–18.37) and the smallest overestimations in July (1.22–3.24) and August (1.15–3.04). BCC-CSM2-MR showed the greatest overestimation in September (3.33), while E3SM-1-1-ECA showed the greatest overestimation in June (2.81). The models that did not simulate greening also did not simulate the greening trend for each month of the growing season.

Unlike the large difference between the LAI bias and relative LAI bias, the ratio of the monthly LAI trend and the bias of the monthly LAI trend had consistent variations (Figure S2). The CLM family (except for FIO-ESM-2-0) showed the largest overestimation in May, and the greatest underestimation of LAI trend in July and August.

3.3. LAI Spatial Comparison

3.3.1. Averaged Leaf Area Index for 1981–2014

GLASS LAI gradually decreased from southeast to northwest (Figure 6). The LAI of forests in Southeast TP was larger (2.8–4.8 m^2 m^{-2}), and the LAI dominated by grasslands and shrubs in the central and northwest areas was smaller (0–0.8 m^2 m^{-2}).

Before evaluating the spatial distribution simulation capability, we ranked the performance of the CMIP6 models to capture the LAI spatial distribution based on the evaluation metrics (Table S1), then we presented the LAI spatial distribution results in Figures S3 and 7 by ranking their scores from the best to the worst.

Almost all the CMIP6 models could reproduce a spatially declining pattern from southeast to northwest, but there was still large spatial bias. The pattern correlation of 88% of the models was greater than 0.60 and the highest was 0.934 for HadGEM3-GC31-MM (Figure S3). We also found that the top five models among the 35 CMIP6 models mainly underestimated the LAI, and the underestimation bias mainly came from the alpine forest area and alpine meadow areas in southeast Tibet. The main feature of the model ranked in the middle (ranked 6–20) among the CMIP6 models is that there were both overestimations and underestimations in the region, while the models with lower (after 20) rankings mainly overestimated the LAI, and the overestimation bias was more obvious in the southeast.

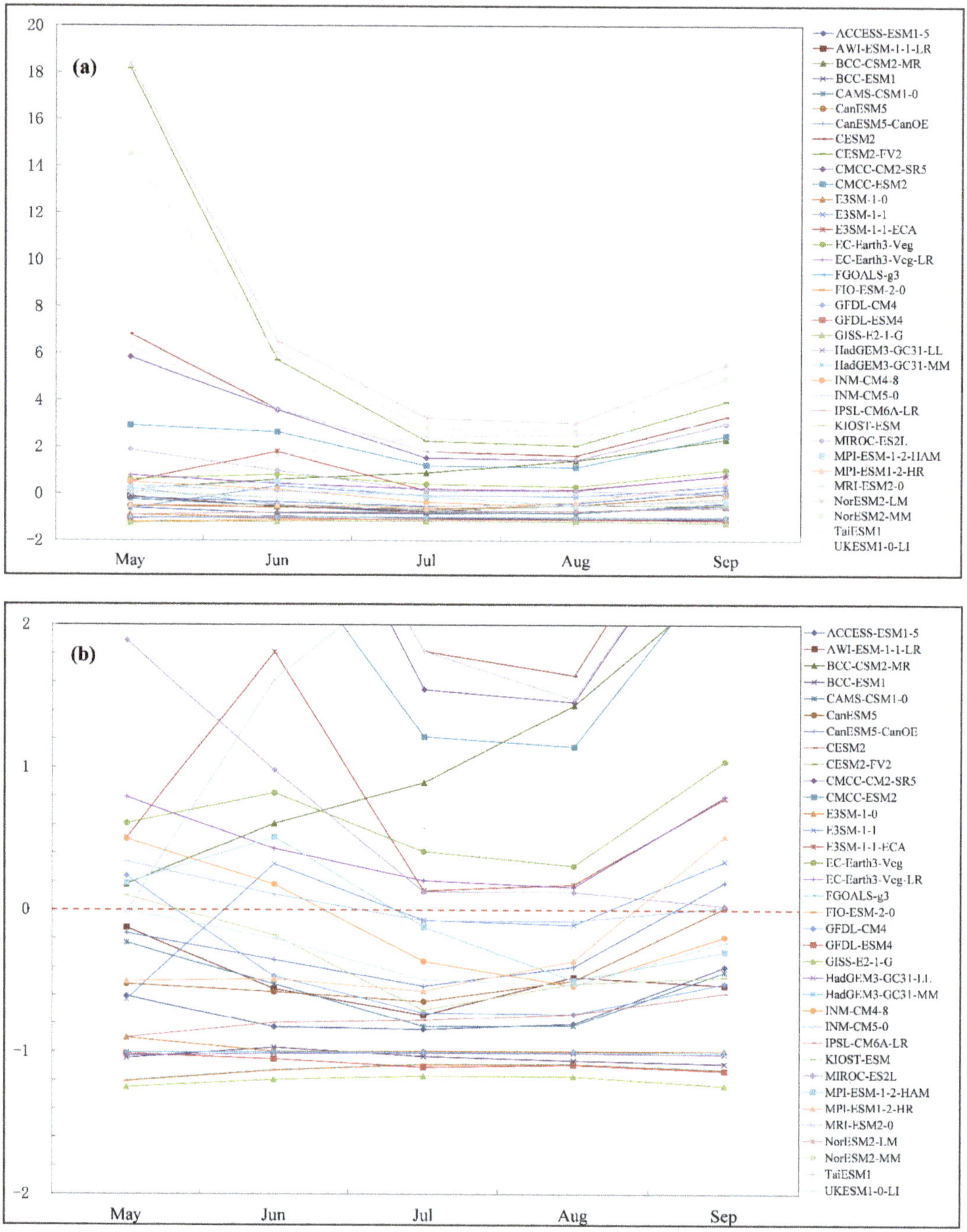

Figure 5. The ratio of the monthly LAI trend of the growing season between each CMIP6 model and the GLASS data. The y-axis is from -2 to 20 in (a), and from -2 to 2 in (b).

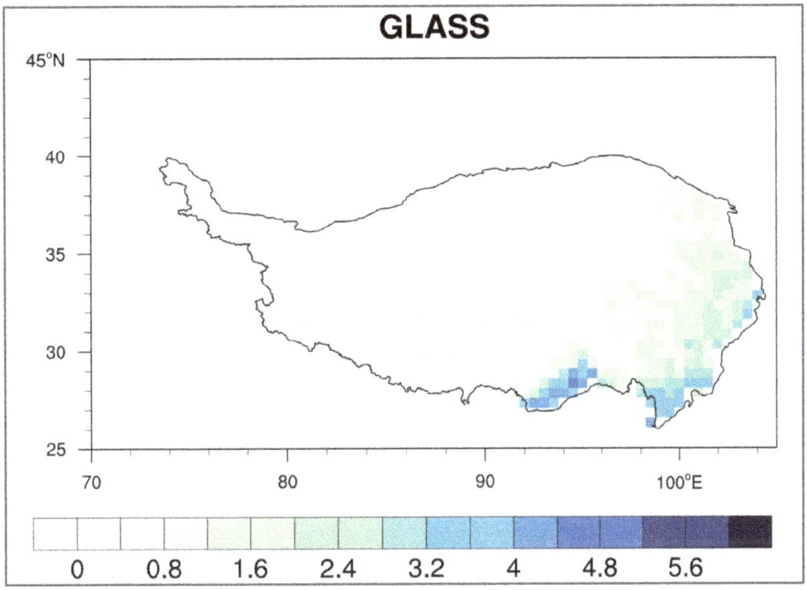

Figure 6. Spatial distribution of the GLASS LAI during the growing season.

Models that underestimated LAI did so mainly over meadows and alpine forest areas in southeast TP, while models that obviously overestimated LAI had great differences in their spatial bias (Figure 7). The obvious overestimation of BCC-CSM2-MR from the BCC family mainly came from the shrub area, while the overestimation of BCC-ESM1 was mainly from shrub areas, meadows, and part of the grassland, and there was high overestimation near river basins. The overestimation of GFDL-CM4 in the LM family came from the shrub areas, while the overestimation of GFDL-ESM4 was mainly distributed across all of Southeast Tibet and was significantly overestimated in the central part. The overestimation of the INM family mainly occurred in shrub areas, deserts, and grassland area, and the highest value of the overestimation bias was for the shrub areas. In addition, the CLM family (except for FIO-ESM-2-0) had abnormally high LAI values throughout the Tibetan Plateau region, and the overestimation was distributed throughout the region, especially for shrub areas and meadows in the southeast TP, with the bias values being 4.5–5.0 m^2 m^{-2}.

EC-Earth3-Veg and EC-Earth3-Veg-LR showed the best simulations for the average LAI (Figure 2), but none of them showed the exact spatial distribution of LAI (Figure 7). EC-Earth3-Veg and EC-Earth3-Veg-LR overestimated the southeastern edge of Tibet but underestimated the grassland and meadow regions of the TP; these positive and negative errors cancelled each other out.

Overall, the CMIP6 models had poor performance for the forest LAI simulation with the highest RMSE, and the bias of the CMIP6 models varied greatly with large overestimation and underestimation, but with the smallest relative bias (Figure S4). Although CMIP6 models had a small overestimation of forest average LAI generally (Figure S4), most models underestimated the forest LAI in the small areas where forests are concentrated on the southern edge of the TP (Figure 7). Similar to the forest LAI, the simulation of shrub was poor with large RMSE and bias, but the relative bias of the shrub was small. The performance of the CMIP6 models for simulating the grassland LAI was good among the different vegetation types with the smallest RMSE. The reason for the small absolute bias but large relative bias with grassland may be that the LAI value of grassland was small.

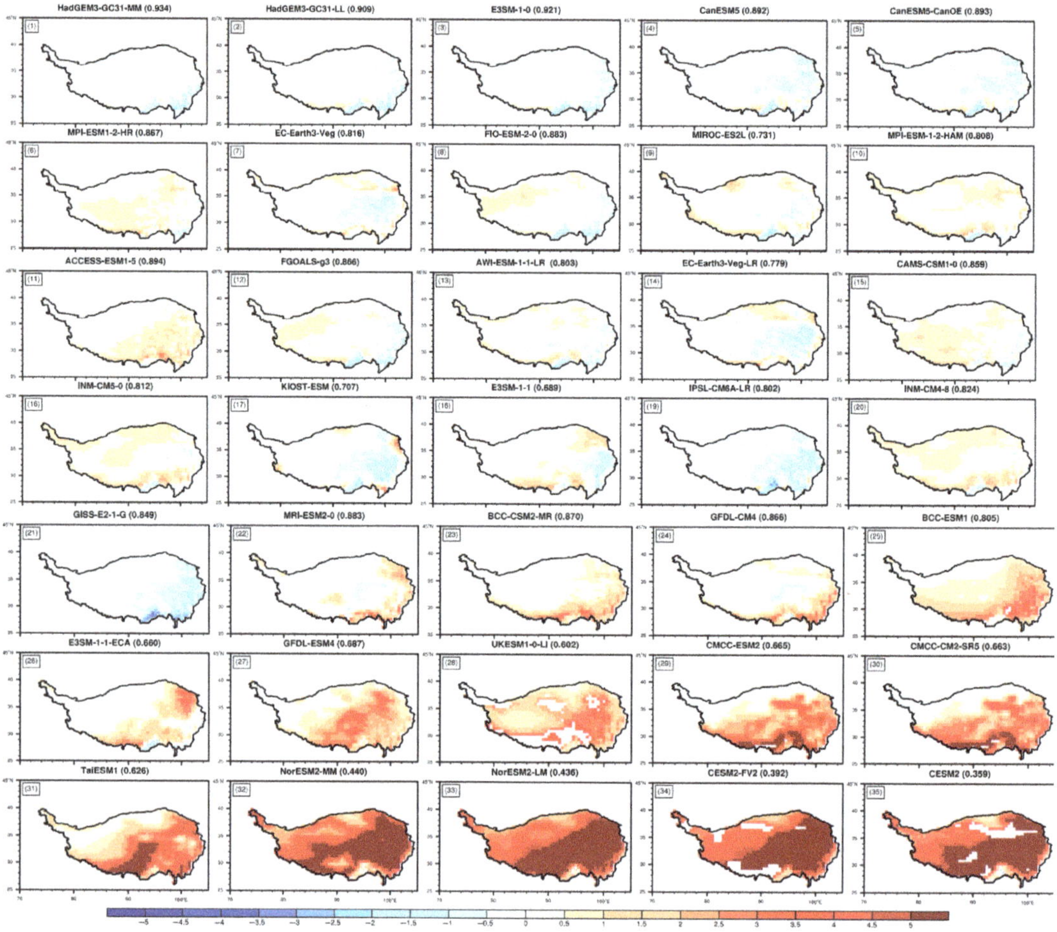

Figure 7. Spatial distribution of the bias of simulated and observed LAI during the growing season. The white part failed the significant difference test. The number in the top left corner is the ranking of each CMIP6 model for simulating the spatial distribution of the average LAI during the growing season in 1981–2014. The value in each title is the pattern correlation.

3.3.2. The Leaf Area Index Trend during 1981–2014

The GLASS LAI data showed a clear greening trend from 1981 to 2014 over the TP, except for some forest areas on the southern edge of the TP (Figure 8). The entire area had significant greening ($p < 0.05$) of 0.0047 m^2 m^{-2} yr^{-1} (Figure S4), especially in the river basins of the meadow area.

Similar to the analysis of the spatial distribution of the LAI, we ranked the performance in reproducing the LAI trend of the CMIP6 models (Table S2) and show the spatial distribution of the LAI trend from best to worst in Figure 9.

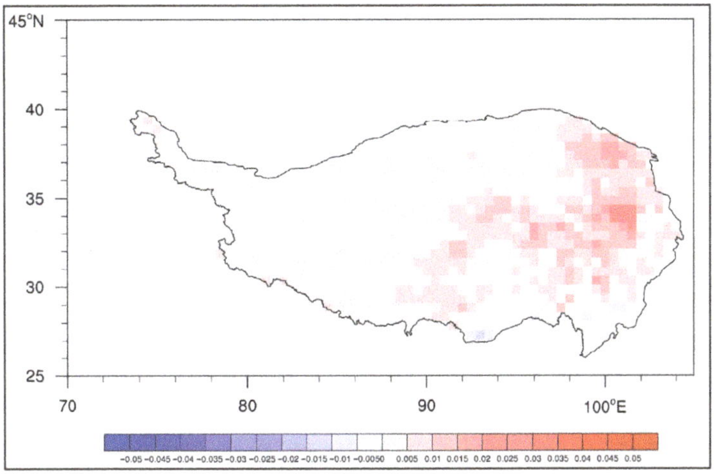

Figure 8. Spatial distributions of the linear trend of GLASS LAI during the growing season.

The CMIP6 models showed a poor ability to simulate the spatial distribution of the LAI trend across the whole Tibetan Plateau during 1981–2014, while most models could simulate the LAI trend in parts of the Tibetan Plateau (Figure 9). The pattern correlation of the LAI trend between all models and GLASS was less than 0.65, and a few models even had negative pattern correlations (Figure S6). There were five models (MPI-ESM-1-2-HAM, BCC-ESM1, BCC-CSM2-MR, EC-Earth-Veg, and EC-Earth-Veg-LR) that generally simulated the overall greening trend of the study area, and also captured the high value of the greening trend in the southeast region, where the spatial distribution of the greening trend was closer to the observation data, and five models (E3SM-1-1, AWI-ESM-1-1-LR, CESM2, GFDL-ESM4, and TaiESM1) simulated the obvious decline in the Southern TP better than other 30 models. E3SM-1-1 and MPI-ESM-1-2-HAM had the best performance in simulating the distribution of the LAI trend and could capture the greening and the decline as well.

Compared with other vegetation types, the simulation of forest LAI trend was poor with the highest RMSE, and the CMIP6 models generally overestimated the forest LAI trend. The simulation of the forest LAI trend showed great differences. Some models showed largely overestimations (NorESM2-MM with a bias of 0.026 m^2 m^{-2} $year^{-1}$) and some models showed large underestimations (GFDL-ESM4 with a bias of –0.017 m^2 m^{-2} $year^{-1}$), which resulted in a larger LAI bias range across all CMIP6 models than for other vegetation types (Figure S7). The alpine vegetation and grassland were also overestimated by CMIP6 models, but the meadow and shrub were underestimated (Figure S7).

In total, 70% of the models accurately simulated increases and decreases in the LAI trend of 80% of the area of the Tibetan Plateau, but the simulation of the value of the LAI trends on the grids was poor (Figure 9). Six models (FIO-ESM-2-0, HadGEM3-GC31-LL, FGOALS-g3, UKESM1-O-LI, GISS-E2-1-G, and GFDL-ESM4) all had obvious gray areas, which mean that the models showed a contrary trend to the GLASS data and had not captured the greening or the declining—especially for GISS-E2-1-G, the gray area was distributed across almost the entire area. Neither FIO-ESM-2-0 nor FGOALS-g3 captured the LAI trend in Northern Tibet, and neither UKESM1-O-LI nor GFDL-ESM4 captured the LAI trend in the southwestern region.

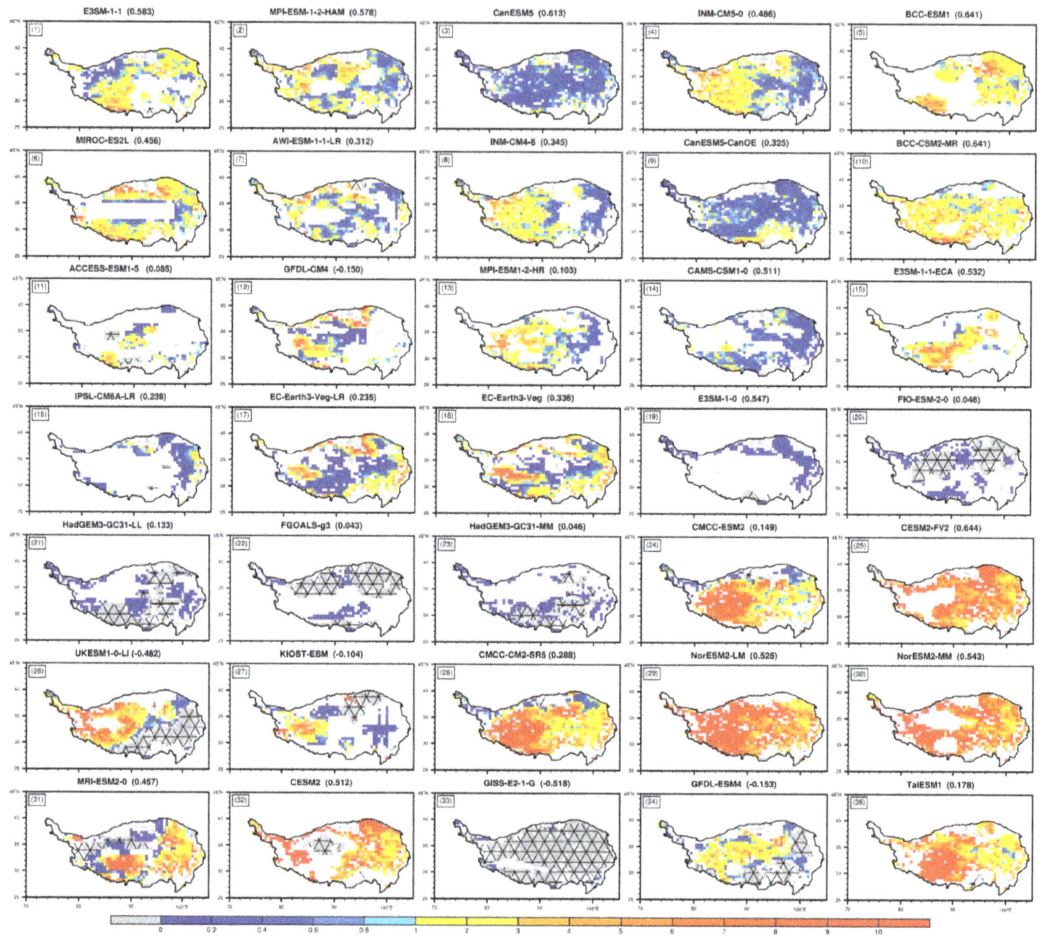

Figure 9. Spatial distributions of the ratio of simulated and observed linear trends in LAI during the growing season. The grid cells with colors all showed a statistically significant interannual change ($p < 0.05$). Gray areas mean the grid cells did not capture greening or a declining trend during 1981–2014 in the Tibetan Plateau, blue areas mean the grid cells captured the greening or the declining trend but underestimated them, and red areas indicated overestimations of the greening or the declining trend. Cross-hatched areas indicate that the LAI trend was negative. The number in the upper left corner is the ranking of each CMIP6 model for simulating LAI trends. The value in each title is the pattern correlation.

The remaining models all captured the greening in 1981–2014, while there were still underestimations and overestimations of the value of the LAI trend in grid cells (Figure 9). The underestimation of the LAI trend mainly came from the shrub, whole meadow area or part of the meadow area, and the greening of the shrub and meadows was underestimated. While the overestimation of the LAI trend came from the grasslands, the CLM family (except for FIO-ESM-2-0) overestimated the LAI trend in almost the whole area, especially the greening of the grassland, which was greatly overestimated. Similarly, 13 models (E3SM-1-1, INM-CM5-0, MIROC-ES2L, INM-CM4-8, MRI-ESM2-0, GFDL-CM4, MPI-ESM1-2-HR, E3SM-1-1-ECA, EC-Earth-Veg, EC-Earth-Veg-LR, UKESM1-O-LI, KIOST-ESM, and MRI-ESM2-0) all overestimated the greening of grasslands. Although the trend of forest LAI was generally overestimated by CMIP6 models (Figure S7), the decline trend of forest LAI was underestimated in parts of the southeast where alpine forests were concentrated (Figure 9).

4. Discussion

Our study chose GLASS LAI as our reference LAI because it is one of the leading data sources for studying long-term series vegetation changes with good representations of various surface LAI distributions. In an evaluation of the authenticity of GLASS LAI products in the grasslands of Xilinhot [84], it was found that the observational accuracy and consistency of GLASS LAI were better than those of MODIS LAI, making it more suitable for related research. When GLASS LAI data were used to analyze changes in the Amazon rainforest from 1982 to 2012 [53], it was demonstrated that the GLASS LAI data can be used for detecting changes in the large-scale surface vegetation status in long sequences. As early as 2014, Xiang et al. [85] compared LAI products (MODIS LAI, CYCLOPES LAI, and CCRS LAI) with ground measurement LAI data, and found that the accuracy of GLASS LAI data products was significantly higher than that of MODIS and CYCLOPES. At the same time, through a comparison of LAI products (MODIS LAI, CYCLOPES LAI, and CCRS LAI), it was found that, compared with other LAI products, GLASS LAI has the best temporal continuity and integrity, and smoother trajectories, and is an ideal data product for studying temporal changes in LAI. The spatial distribution of the GLASS LAI data is reasonable, and it also has good consistency with the global spatial distribution of MODIS LAI. It thus has great advantages in studies of the spatial distribution of LAI.

Although many studies using remote sensing products found an overall increasing trend of vegetation growth (greening) over the Tibetan Plateau, like the GLASS remote sensing products, controversy remains regarding how vegetation on the Tibetan Plateau has changed. Xu showed that spring warming advanced spring leaf-out time and increased the biomass [86]. However, Yu (2010) argued that the warm winter may also have led to delayed spring phases due to insufficient fulfillment of chilling requirements [87]. Zhang et al. [13] argued for the earlier start date of plant phenology and a longer growing season, but some still doubt this [88,89]. Regarding the change in the trend, a declined trend for the vegetation dynamics of the TP was found in some studies over the last 30 years (about 1980-2010) according to the Global Inventory of Modeling and Mapping and Studies (GIMMS) [90,91], but others found an increasing trend of vegetation growth in northeastern TP using other NDVI datasets for 1982–2011 [92]. These different understandings indicate that a combination of ground observations, remote sensing datasets, and land/vegetation models is necessary to fully understand past and future vegetation changes on the Tibetan Plateau.

In the CMIP6 models, some model groups showed consistency in simulating LAI and LAI trends using the same land surface model, but others showed great differences. In order to understand the possible reasons for these differences, we briefly summarized the differences among the model groups using the same land surface model (Table 2).

By combining the simulation results of the model for the average LAI and LAI trends in Figures 2 and 3 and the different characteristics of the models in Table 2, we found that the simulation results of the models using different land surface models were quite different on the whole; the simulation results of models using the same land surface model had overall consistency, whereas the simulation results of models using different versions of the same land surface model were different. There are many possible reasons for the large difference in the simulations of vegetation growth, such as the simplified parameterization, uncalibrated parameters, and the atmospheric forcing data that drive the model. The vegetation growth in the land surface model also subject to the simulations of other processes directly affecting vegetation growth, such as the simulation of soil temperature and moisture, surface radiation transfer, etc. Using the Community Land Model (CLM) as an example, Luo et al. [100] used the simulated data of Weather Research and Forecasting Model (WRF) to apply to the forcing data sets of the CLM model in the Tibetan Plateau, and found that there are deviations between simulated and observed surface temperatures with RMSE in the range of 2.0–4.2 °C. CLM4.0 simulated [101] lower soil temperature by −0.83 °C and higher sensible heat flux up to 60 W.m^{-2}, except in winter at Maqu Alpine Grassland. Xie et al. [102] found that the simulation of the winter radiation balance component and the surface energy balance component of CLM4.5 was poor, especially the simulation of the surface reflected radiation with the highest RMSE of 165.16 W.m^{-2} in January, and sensible heat flux in winter had a serious deviation with the highest RMSE of 145.15 W.m^{-2} in February. Song et al. [103] used CLM4.5, which underestimated soil temperature and latent heat flux in winter at the Naqu site, which indicated that the parameterization schemes of snow processes and surface albedos in CLM4.5 need to be improved. All these discrepancies in land surface simulations may lead to poor simulations of vegetation growth. Mao et al. [104] found that the GPP and LAI both had a positive correlation with precipitation and a strong negative correlation with incident shortwave radiation globally. Due to the special geography of the TP, especially the complex lower cushion surface characteristics, there is a particularity and complexity of the land–air interaction in the area, which has caused difficulties for CLM land surface simulation. How to improve and perfect the simulation performance of the CLM model on the vegetation of the TP requires more in-depth research in the future. However, there are factors that can be improved, such as continuing to optimize the parameter schemes of simulating the temperature, precipitation, radiation flux, and the coverage of snow on the TP in CLM. Although almost all CLM models overestimated LAI and the LAI trend, there were differences in the degree of overestimation. An obvious difference is that the FIO-ESM-2-0 model with the ocean wave model added to the coupling had better performance in simulating the area-averaged LAI of the Tibetan Plateau from 1981 to 2014 (Figure 2) than other CLM models. Other modules such as the ocean wave model in the coupled model might also have had a large impact on the CLM model.

Most models had a worse performance in simulating the forest LAI and LAI trend compared with other vegetation types. The reasons for this difference may be that, compared with grasslands and meadows, the vegetation growth mechanism in forest ecosystems is more complex, the species of forest ecosystems are more abundant, and it is more difficult to establish mathematical structures for simulations with different species. Changes in the long-term processes of different species within the forest system are more complex, and it is more difficult to establish mathematical structures with simulations.

From CMIP5 to CMIP6, the average LAI over the Tibetan Plateau still showed overestimation but of an even higher magnitude. Bao et al. [37] found that 10 out of 12 CMIP5 models overestimated LAI with bias of between 0.44 and 3.6 m^2 m^{-2} from 1986 to 2005. We found that 25 out of 35 CMIP6 overestimated LAI of TP, with bias ranging from 0.07 to 5.38 from 1981 to 2014. For the same model from CMIP5 to CMIP6, we found that some models had better performance: for example, HadGEM3-GC31 had the smallest bias of the CMIP6 models. Some models showed poor performance in CMIP6—for example, CESM2 in CMIP6 showed much a higher average LAI than its previous version, CCSM4 in CMIP5; additionally, INMCM4, with the lowest bias of 12 CMIP5 models [37], ranked 23rd in area-averaged bias among the 35 CMIP6 models. Both CanESM2 from CMIP5 and CanESM5 from CMIP6 maintained a better simulation of the average LAI on the TP with the smaller bias, the same as the MPI-ESM1-2-HR and the old version MPI-ESM-LR. There were also models, whether in the CMIP5 or in the CMIP6, where the simulation performance was relatively poor, such as bcc-csm1.1-m and the new version, BCC-CSM2-MR, in CMIP6, and NorESM1-ME and NorESM2-MM/LM from CMIP5 to CMIP6.

Song et al. [105] found that CMIP6 generally overestimated the global multiyear average LAI, and the overestimation of growing season length (GSL) contributed to the overestimated LAI in boreal and some temperate areas. We found that CLM family also overestimated the average LAI during the growing season in 1981–2014 on the TP. We analyzed the monthly average LAI of 35 models from 1981 to 2014 and found that most of the models had a longer growing season (Figure S8). CMIP6 LAI in April, October, and November were still large. Part of the reason for the global multi-year average LAI and the TP LAI overestimation was the same. Moreover, we found that LAI increased greatly during the leaf emerge stage in most CLM family models, which suggested too much carbon was being allocated to leaves. Improving the phenology and carbon allocation is crucial for improving LAI simulations over the Tibetan Plateau.

Table 2. Summary of the different models.

Land Surface Model	CMIP6 Models	The Difference of Models
BCC-AVIM2.0	BCC-CSM2-MR BCC-ESM1	BCC-CSM2-MR uses the carbon emissions provided by CMIP6 as the forcing, but BCC-ESM1.0 uses the chemical reaction gas and aerosol emission data provided by CMIP6 as the forcing [93]
CLASS3.6-CTEM1.2	CanESM5 CanESM5-CanOE	CanESM5-CanOE is exactly the same physical model as CanESM5, but it couples it with the CanOE ocean biogeochemical model [60]
CLM4.0	FIO-ESM-2-0 TaiESM1	The FIO-ESM-2-0 model adds an ocean surface wave model to the traditional atmosphere–land–ocean–sea ice coupled model of CPL7; TaiESM1 was developed on the basis of the Community Earth System Model version 1.2.2 by implementing several improvements to the parameterization schemes in the atmospheric component [94,95]
CLM4.5	CMCC-CM2-SR5 CMCC-ESM2	CMCC-CM2-SR5 does not include ocean biogeochemistry model in the model, but the BFM5.1 ocean biogeochemistry model was added to CMCC-ESM2
CLM5.0	CESM2 CESM2-FV2 NorESM2-LM NorESM2-MM	CESM2-FV2 reduces the horizontal resolution of the atmosphere and land on the basis of CESM2; NorESM2-LM and NorESM2-MM are similar to the CESM2 and CESM2-FV2 models in terms of the framework and model composition; the differences are that NorESM2 uses completely different oceans and oceano-biogeochemical model and uses a different ocean and oceano-biogeochemistry model and the atmosphere component of NorESM2-MM and CAM-Nor; the difference between NorESM2-LM and NorESM2-MM is the resolution [78,96]
ELM	E3SM-1-0 E3SM-1-1 E3SM-1-1-ECA	On the basis of E3SM-1-0, E3SM-1-1 has corrected several vulnerabilities and made improvements; on the basis of E3SM-1-1, E3SM-1-1-ECA uses the ECA plant and soil carbon and nutrient mechanisms, soil carbon and the effects of nutrients representing carbon, nitrogen and phosphorus, and it excludes the effect of coupled ocean and sea ice biogeochemistry [97]
HTESSEL	EC-Earth3-Veg EC-Earth3-Veg-LR	EC-Earth3-Veg-LR has a lower resolution than EC-Earth3-Veg
INM-LND1	INM-CM4-8 INM-CM5-0	On the basis of INM-CM4-8, the key improvements in INM-CM5-0 include an increase in the vertical resolution in the atmospheric module, a revision of the large-scale condensation and cloud formation parameterizations, the newly developed aerosol block, the horizontal resolution of the oceanic model, and a reworking of the INMCM5 program code for better performance on parallel computers [71]
JSBACH 3.2	AWI-ESM-1-1-LR MPI-ESM-1-2-HAM MPI-ESM1-2-HR	AWI-ESM-1-1-LR is based on AWI-ESM and adds a dynamic land change model to it; MPI-ESM1-2-HR and MPI-ESM-1-2-HAM both are based on MPIESM1.2, and the difference between the two is that MPI-ESM-1-2-HAM adds the Hamburg aerosol mode and MPI-ESM1-2-HR improves the resolution of MPIESM1.2, which has a higher resolution than MPI- ESM-1-2-HAM [76,96]
JULES	HadGEM3-GC31-LL HadGEM3-GC31-MM UKESM1-0-LI	HadGEM3-GC31 is a coupled atmosphere–land–ocean–sea ice model. Compared with HadGEM3-GC31-LL, HadGEM3-GC31-MM has a higher resolution. UKESM1 takes HadGEM3-GC31 as the core of the physical model and adds the carbon and nitrogen cycle and atmospheric chemical composition to it [98,99]
LM3.0	KIOST-ESM	Atmosphere–land–ocean–sea ice coupled model [68,96]
LM4.0	GFDL-CM4	A coupled ocean–atmosphere model [68,96]
LM4.1	GFDL-ESM4	A fully coupled chemistry–climate model [68,96]

Climate change has led to changes in vegetation on the TP in recent decades. From the 1980s to the beginning of the 21st century, the vegetation coverage rate of the TP showed an overall increasing trend [21], with large seasonal and spatial variations. The spring vegetation coverage of the Tibet Plateau showed the larger increasing rate [106] than other seasons. The humid areas in the Southeast TP showed increasing vegetation coverage while the Central and Northwest TP showed declined vegetation coverage [21,107]. The upper limit of the vertical natural zone of vegetation over the TP has changed significantly. The forest lines migrated to high altitudes [107]. The glacier retreat and permafrost ablation will aggravate the degradation of regional alpine grassland [108] on the TP. Due to changes in the permafrost environment, the soil moisture and nutrients in the root layer of vegetation are decreased, resulting in the drying out of swamp wetlands and the transformation into meadows in Zoige, according to the measured data on temperature precipitation [109], and shrub invasion of alpine meadows [107]. Species diversity in the native Kobresia humilis meadow community decreased in a simulation of a five-year temperature increase run a greenhouse in TP [110]. The degradation of permafrost, the drying out of some swamps, and the aggravation of surface salinization all exacerbated the desertification of permafrost area in the TP [111]. Meanwhile, many of the variables that cause changes in vegetation growth in the context of global change have also changed. Temperature and precipitation, which have a positive correlation with LAI [112], showed an overall increasing trend on the TP, with warming of 0.4 °C. 10 yr^{-1} over the last 30 years [12,13] and precipitation increasing by 1.96 mm.10 yr^{-1} in 1994–2015 [14]. Zhu et al. [113] found that in the past 50 years, the highest value of Photosynthetically Active Radiation (PAR) in China appeared in the southwest of the Tibetan Plateau (with an annual PAR of 35 $mol.m^{-2}d^{-1}$), while the PAR in the northwest of the Tibetan Plateau showed an upward trend in different seasons. By analyzing the daily temperature data provided by the National Meteorological Information Center, China Meteorological Administration, for the Tibetan Plateau stations from 1961 to 2007, Fan et al. [114] found that spring and summer are starting earlier while autumn and winter are starting later.

Some of these changes can be monitored by remote sensing, e.g., glacier retreat [115], widespread grassland variation [116] with grassland biomass dynamics [117], rising forest lines, shrub intrusion into alpine meadows, etc. However, it is difficult for vegetation growth models to simulate these complex processes. The phenology and allocation schemes were not designed to capture tree line migration or grassland transformation. Moreover, the land surface model also could not simulate the well permafrost thawing or the glacier retreat processes over the Tibetan Plateau.

Some researchers also found that the model had large errors in other simulation variables on the TP. Xiao et al. [118] evaluated the performance of the state-of-the-art global high-resolution models in simulating hourly precipitation and extreme precipitation in summer over the TP in 1950–2050 with eight CMIP6 high-resolution models (HighResMIP) and found that the CMIP6 HighResMIP overestimated the precipitation amount and frequency. Chen et al. [119] found that, although the CMIP6 models could simulate the spatial distribution characteristics of the average annual precipitation on the Tibetan Plateau, this was generally overestimated, with an average of more than 397.8 $mm.a^{-1}$. The simulations of temperature and precipitation, which have a greater impact on the LAI simulation of vegetation, showed a large error in the TP. The inaccuracy of the temperature and precipitation simulation may also be one of the reasons for the large error in vegetation simulations on the TP.

The acquisition of field data in TP was limited due to geographical, topographical, and environmental factors. However, continuous actual observation data from the plateau site are also very important for the accurate description of land–atmosphere interactions and the improvement of the parameterization of different physical processes [120–122].

Therefore, there are three pathways that may improve the performance of models in simulating LAI over the TP. The first is to incorporate missing physical mechanisms that directly or indirectly impact on vegetation growth, such as aerosol effects [123], elevated CO2 concentration, and the impact of volcanic eruptions on the climate [124]. Moreover, incorporating land surface processes such as permafrost thawing processes and the winter surface parameterization scheme [102] may be particularly important over the TP. The second is to calibrate and optimize the internal parameters [104] to better represent vegetation growth over the TP. Some of the parameters were not calibrated or validated over the TP, so using artificial intelligence to train models could improve the model simulations. The third is to further improve the observation system and obtain continuous and complete atmospheric observations, as site-observed vegetation growth is also very important for improving simulations of the vegetation on the TP.

As the temperature continues to rise, the impact of the climate on plant phenology becomes more complex [125] and the acquisition of the forcing data becomes harder due to the extreme weather problems caused by global warming, which will make simulation of the vegetation growth in the Tibetan Plateau more challenging in the future.

5. Conclusions

In this study, we evaluated the performance of CMIP6 models in simulating LAI and the LAI trend during the growing season of the Tibetan Plateau over the period 1981–2014, compared with the GLASS LAI. We found the following:

1. In total, 40% of the models overestimated the greening, 48% of the models underestimated the greening, and 11% of the models showed a declining LAI trend for 1981–2014 over the Tibetan Plateau. For the LAI, 70% of the models overestimated this, while about 17% of the models underestimated it.
2. Both the models underestimating greening, and the models underestimating LAI, showed the greatest underestimation bias in July and August. The biases and ratio of LAI (with the exception of the CLM family) and trend between the simulations and observations had the same change during the growing season.
3. CMIP6 models overestimated the LAI trend of alpine vegetation, forest, and grassland, but underestimated the meadow and shrub. The greening of grasslands was overestimated, and the greening of meadows was underestimated in CMIP6. Compared with other vegetation types, the performance of simulating the forest LAI trend was poor with the highest RMSE, and the declining trend in forest pixels showing a declining trend on the TP, was generally underestimated.
4. The performance in simulating the spatial distribution of LAI was better than the LAI trend. The underestimation of LAI was mainly in meadows and alpine forest areas in southeast TP. Similar to the forest LAI trend, the simulation performance of forest LAI was also poor, with the highest RMSE, and the forest LAI in parts of the southeast where alpine forests were concentrated on the TP was underestimated by 20 of 35 CMIP6 models.

Supplementary Materials: The following supporting information can be downloaded at: https://www.mdpi.com/article/10.3390/rs14184633/s1. Figure S1. The relative bias of the monthly mean LAI with simulations and observations; Figure S2. The bias of the monthly mean linear trend with simulations and observations; Figure S3. Spatial distributions of the simulated LAI during the growing season; Figure S4. The distribution of bias, relative bias and RMSE between the simulated and observed LAI with 35 CMIP6 models for different vegetation types; Figure S5. The area-averaged linear trend of simulated and observed LAI the during the growing season ($p < 0.05$); Figure S6. Spatial distributions of the simulated LAI linear trend during the growing season; Figure S7. The distribution of bias and RMSE between the simulated and observed LAI trend with 35 CMIP6 models for different vegetation types; Figure S8. The monthly LAI of 35 CMIP6 models from 1981 to 2014; Table S1. Summary of evaluation metrics and error ranking for models with the performance to simulate the LAI of Tibetan Plateau during growing season in1981–2014; Table S2. Summary of evaluation metrics and error ranking for models with the performance to simulate the LAI trend of Tibetan Plateau during growing season in1981–2014.

Author Contributions: Writing—original draft preparation, J.L. and Y.L.; writing—review and editing, J.L. and Y.L. All authors have read and agreed to the published version of the manuscript.

Funding: This research was funded by the Strategic Priority Research Program of Chinese Academy of Sciences grant number No. XDA20050102, XDA23060601, and the National Natural Science Foundation of China grant number No. 41975135.

Data Availability Statement: The study did not report any data.

Acknowledgments: Thank you to the anonymous reviewers who gave useful advice and made our article more organized. This study was supported by the Strategic Priority Research Program of Chinese Academy of Sciences (No. XDA20050102, XDA23060601) and the National Natural Science Foundation of China (No. 41975135). We are grateful for the data set of the Tibetan Plateau Boundary high frequency (HF) and the 1:1 million vegetation data set in China provided by the National Tibetan Plateau Data Center (http://data.tpdc.ac.cn, accessed on 9 December 2021). We thank the Beijing Normal University for providing the Global Products of Essential Land Variables (GLASS). We acknowledge the World Climate Research Programme, which, through its Working Group on Coupled Modelling, coordinated and promoted CMIP6. We thank the climate modeling groups for producing and making available their model output, the Earth System Grid Federation (ESGF), for archiving the data and providing access, and the multiple funding agencies that supported CMIP6 and ESGF.

Conflicts of Interest: There is no conflict of interest.

References

1. Bonan, G.B. Forests and climate change: Forcings, feedbacks, and the climate benefits of forests. *Science* **2008**, *320*, 1444–1449. [CrossRef] [PubMed]
2. Braswell, B.H.; Schimel, D.S.; Linder, E.; Moore, B. The response of global terrestrial ecosystems to interannual temperature variability. *Science* **1997**, *278*, 870–872. [CrossRef]
3. Nolan, C.; Overpeck, J.T.; Allen, J.R.M.; Anderson, P.M.; Betancourt, J.L.; Binney, H.A.; Brewer, S.; Bush, M.B.; Chase, B.M.; Cheddadi, R.; et al. Past and future global transformation of terrestrial ecosystems under climate change. *Science* **2018**, *361*, 920–923. [CrossRef]
4. Qin, D.; Stocker, T. Highlights of the IPCC Working Group I Fifth Assessment Report. *Clim. Change Res.* **2014**, *10*, 1.
5. Myneni, R.B.; Keeling, C.D.; Tucker, C.J.; Asrar, G.; Nemani, R.R. Increased plant growth in the northern high latitudes from 1981 to 1991. *Nature* **1997**, *386*, 698–702. [CrossRef]
6. Tucker, C.J.; Slayback, D.A.; Pinzon, J.E.; Los, S.O.; Myneni, R.B.; Taylor, M.G. Higher northern latitude normalized difference vegetation index and growing season trends from 1982 to 1999. *Int. J. Biometeorol.* **2001**, *45*, 184–190. [CrossRef] [PubMed]
7. Zhou, L.; Kaufmann, R.K.; Tian, Y.; Myneni, R.B.; Tucker, C.J. Relation between interannual variations in satellite measures of northern forest greenness and climate between 1982 and 1999. *J. Geophys. Res.-Atmos.* **2003**, *108*, ACL-3. [CrossRef]
8. Piao, S.L.; Ciais, P.; Huang, Y.; Shen, Z.H.; Peng, S.S.; Li, J.S.; Zhou, L.P.; Liu, H.Y.; Ma, Y.C.; Ding, Y.H.; et al. The impacts of climate change on water resources and agriculture in China. *Nature* **2010**, *467*, 43–51. [CrossRef]
9. Amagai, Y.; Kudo, G.; Sato, K. Changes in alpine plant communities under climate change: Dynamics of snow-meadow vegetation in northern Japan over the last 40 years. *Appl. Veg. Sci.* **2018**, *21*, 561–571. [CrossRef]
10. Kharuk, V.I.; Im, S.T.; Petrov, I.A. Alpine ecotone in the Siberian Mountains: Vegetation response to warming. *J. Mt. Sci.* **2021**, *18*, 3099–3108. [CrossRef]
11. Jin, Y.H.; Zhang, Y.J.; Xu, J.W.; Tao, Y.; He, H.S.; Guo, M.; Wang, A.L.; Liu, Y.X.; Niu, L.P. Comparative Assessment of Tundra Vegetation Changes Between North and Southwest Slopes of Changbai Mountains, China, in Response to Global Warming. *Chin. Geogr. Sci.* **2018**, *28*, 665–679. [CrossRef]
12. Chen, H.; Zhu, Q.A.; Peng, C.H.; Wu, N.; Wang, Y.F.; Fang, X.Q.; Gao, Y.H.; Zhu, D.; Yang, G.; Tian, J.Q.; et al. The impacts of climate change and human activities on biogeochemical cycles on the Qinghai-Tibetan Plateau. *Glob. Change Biol.* **2013**, *19*, 2940–2955. [CrossRef] [PubMed]
13. Zhang, G.L.; Zhang, Y.J.; Dong, J.W.; Xiao, X.M. Green-up dates in the Tibetan Plateau have continuously advanced from 1982 to 2011. *Proc. Natl. Acad. Sci. USA* **2013**, *110*, 4309–4314. [CrossRef] [PubMed]
14. Duan, A.; Xiao, Z.; Wu, G. Characteristics of Climate Change over the Tibetan Plateau Under the Global Warming During1979—2014. *Clim. Change Res.* **2016**, *12*, 374–381.
15. Gao, J.; Yao, T.D.; Masson-Delmotte, V.; Steen-Larsen, H.C.; Wang, W.C. Collapsing glaciers threaten Asia's water supplies. *Nature* **2019**, *565*, 19–21. [CrossRef]
16. Wang, S.; Niu, F.; Zhao, L.; Li, S. The thermal stability of roadbed in permafrost regions along Qinghai–Tibet Highway. *Cold Reg. Sci. Technol.* **2003**, *37*, 25–34. [CrossRef]
17. Gao, L.; Liao, J.J.; Shen, G.Z. Monitoring lake-level changes in the Qinghai-Tibetan Plateau using radar altimeter data (2002–2012). *J. Appl. Remote Sens.* **2013**, *7*, 073470. [CrossRef]
18. Liu, J.; Gao, J.; Wang, W. Variations of Vegetation Coverage and Its Relations to Global Climate Changes on the Tibetan Plateau during 1981—2005. *J. Mt. Sci.* **2013**, *31*, 234–242.
19. Wei, Y.; Lu, H.; Wang, J.; Sun, J.; Wang, X. Responses of vegetation zones, in the Qinghai-Tibetan Plateau, to climate change and anthropogenic influences over the last 35 years. *Pratac. Sci.* **2019**, *36*, 1163–1176.
20. Zhang, J.; Yuan, M.S.; Zhang, J.; Li, H.W.; Wang, J.Y.; Zhang, X.; Ju, P.J.; Jiang, H.B.; Chen, H.; Zhu, Q.A. Responses of the NDVI of alpine grasslands on the Qinghai-Tibetan Plateau to climate change and human activities over the last 30 years. *Acta Ecol. Sin.* **2020**, *40*, 6269–6281.

21. Xu, X.; Chen, H.; Levy, J.K. Temporal and spatial changes of vegetation cover characteristics in Qinghai-Tibet Plateau under the background of climate warming And its cause analysis. *Chin. Sci. Bull.* **2008**, *53*, 456–462.
22. Zhang, F.; Tiyip, T.; Ding, J.L.; Sawut, M.; Johnson, V.C.; Tashpolat, N.; Gui, D.W. Vegetation fractional coverage change in a typical oasis region in Tarim River Watershed based on remote sensing. *J. Arid Land* **2013**, *5*, 89–101. [CrossRef]
23. Xiao, Z.Q.; Liang, S.L.; Wang, J.D.; Xiang, Y.; Zhao, X.; Song, J.L. Long-Time-Series Global Land Surface Satellite Leaf Area Index Product Derived From MODIS and AVHRR Surface Reflectance. *Ieee Trans. Geosci. Remote Sens.* **2016**, *54*, 5301–5318. [CrossRef]
24. Liu, Y.; Liu, R.G.; Chen, J.M. Retrospective retrieval of long-term consistent global leaf area index (1981–2011) from combined AVHRR and MODIS data. *J. Geophys. Res.-Biogeosci.* **2012**, *117*. [CrossRef]
25. Zhu, Z.C.; Bi, J.; Pan, Y.Z.; Ganguly, S.; Anav, A.; Xu, L.; Samanta, A.; Piao, S.L.; Nemani, R.R.; Myneni, R.B. Global Data Sets of Vegetation Leaf Area Index (LAI)3g and Fraction of Photosynthetically Active Radiation (FPAR)3g Derived from Global Inventory Modeling and Mapping Studies (GIMMS) Normalized Difference Vegetation Index (NDVI3g) for the Period 1981 to 2011. *Remote Sens.* **2013**, *5*, 927–948.
26. Hua, T.; Wang, X.M. Temporal and Spatial Variations in the Climate Controls of Vegetation Dynamics on the Tibetan Plateau during 1982–2011. *Adv. Atmos. Sci.* **2018**, *35*, 1337–1346. [CrossRef]
27. Zhu, X.C.; Shao, M.G.; Tang, X.Z.; Liang, Y. Spatiotemporal variation and simulation of vegetation coverage in a typical degraded alpine meadow on the Tibetan Plateau. *Catena* **2020**, *190*, 104551. [CrossRef]
28. Lu, J.; Ji, J. A Simulation Study of Atmosphere-Vegetation Interaction over the Tibean Plateau Part II: Net Primary Productivity and Leaf Area Index. *Chin. J. Atmos. Sci.* **2002**, *26*, 255–262.
29. Tian, D.; Guo, Y.; Dong, W.J. Future Changes and Uncertainties in Temperature and Precipitation over China Based on CMIP5 Models. *Adv. Atmos. Sci.* **2015**, *32*, 487–496. [CrossRef]
30. Zhang, F.; Li, W.; Mann, M.E. Scale-dependent regional climate predictability over North America inferred from CMIP3 and CMIP5 ensemble simulations. *Adv. Atmos. Sci.* **2016**, *33*, 905–918. [CrossRef]
31. Xia, J.J.; Yan, Z.W.; Jia, G.S.; Zeng, H.Q.; Jones, P.D.; Zhou, W.; Zhang, A.Z. Projections of the advance in the start of the growing season during the 21st century based on CMIP5 simulations. *Adv. Atmos. Sci.* **2015**, *32*, 831–838. [CrossRef]
32. Anav, A.; Friedlingstein, P.; Kidston, M.; Bopp, L.; Ciais, P.; Cox, P.; Jones, C.; Jung, M.; Myneni, R.; Zhu, Z. Evaluating the Land and Ocean Components of the Global Carbon Cycle in the CMIP5 Earth System Models. *J. Clim.* **2013**, *26*, 6801–6843. [CrossRef]
33. Zhao, Q.; Zhu, Z.; Zeng, H.; Zhao, W.; Myneni, R.B.J.A.; Meteorology, F. Future greening of the Earth may not be as large as previously predicted. *Agric. For. Meteorol.* **2020**, *292*, 108111. [CrossRef]
34. Lawrence, D.M.; Fisher, R.A.; Koven, C.D.; Oleson, K.W.; Swenson, S.C.; Bonan, G.; Collier, N.; Ghimire, B.; van Kampenhout, L.; Kennedy, D.; et al. The Community Land Model Version 5: Description of New Features, Benchmarking, and Impact of Forcing Uncertainty. *J. Adv. Model. Earth Syst.* **2019**, *11*, 4245–4287. [CrossRef]
35. Brovkin, V.; Boysen, L.; Raddatz, T.; Gayler, V.; Loew, A.; Claussen, M. Evaluation of vegetation cover and land-surface albedo in MPI-ESM CMIP5 simulations. *J. Adv. Model. Earth Syst.* **2013**, *5*, 48–57. [CrossRef]
36. Sellar, A.A.; Jones, C.G.; Mulcahy, J.P.; Tang, Y.M.; Yool, A.; Wiltshire, A.; O'Connor, F.M.; Stringer, M.; Hill, R.; Palmieri, J.; et al. UKESM1: Description and Evaluation of the UK Earth System Model. *J. Adv. Model. Earth Syst.* **2019**, *11*, 4513–4558. [CrossRef]
37. Bao, Y.; Gao, Y.H.; Lu, S.H.; Wang, Q.X.; Zhang, S.B.; Xu, J.W.; Li, R.Q.; Li, S.S.; Ma, D.; Meng, X.H.; et al. Evaluation of CMIP5 Earth System Models in Reproducing Leaf Area Index and Vegetation Cover over the Tibetan Plateau. *J. Meteorol. Res.* **2014**, *28*, 1041–1060. [CrossRef]
38. Zhou, T.J.; Zou, L.W.; Chen, X.L. Commentary on the Coupled Model Intercomparison Project Phase 6(CMIP6). *Clim. Change Res.* **2019**, *15*, 445.
39. Watson, D.J. Comparative Physiological Studies on the Growth of Field Crops. 1. Variation in Net Assimilation Rate and Leaf Area between Species and Varieties, and within and between Years. *Ann. Bot.* **1947**, *11*, 41–76. [CrossRef]
40. Wang, L.W.; Niu, Z.; Wei, Y.X. Detecting the areas at risk of desertification in Xinjiang based on modis NDVI imagery. *J. Infrared Millim. Waves* **2007**, *26*, 456–460.
41. Wang, Z.; Liu, S.R.; Sun, P.S.; Guo, Z.H.; Zhou, L.D. The Variability of Vegetation Beginning Date of Greenness Period in Spring in the North-South Transect of Eastern China Based on NOAA NDVI. *Spectrosc. Spectr. Anal.* **2010**, *30*, 2758–2761.
42. Wang, Y.; Tian, Q.J.; Huang, Y.; Wei, H.W. NDVI Difference Rate Recognition Model of Deciduous Broad-Leaved Forest Based on HJ-CCD Remote Sensing Data. *Spectrosc. Spectr. Anal.* **2013**, *33*, 1018–1022.
43. Ewert, F. Modelling plant responses to elevated CO_2: How important is leaf area index? *Ann. Bot.* **2004**, *93*, 619–627. [CrossRef] [PubMed]
44. Sellers, P.J.; Dickinson, R.E.; Randall, D.A.; Betts, A.K.; Hall, F.G.; Berry, J.A.; Collatz, G.J.; Denning, A.S.; Mooney, H.A.; Nobre, C.A.; et al. Modeling the exchanges of energy, water, and carbon between continents and the atmosphere. *Science* **1997**, *275*, 502–509. [CrossRef] [PubMed]
45. Zhang, Y.; Liu, L.; Li, B.; Zheng, D. Redetermine the region and boundaries of Tibetan Plateau. *Acta Geogr. Res.* **2021**, *40*, 1543–1553.
46. Zhang, Y.; Liu, L.; Li, B.; Zheng, D. Comparison of boundary datasets covering Tibetan Plateau between 2021 and 2014 versions. *J. Glob. Change Data Discov.* **2021**, *5*, 322–332.
47. Lingzhiduojie, L. *Introduction to Environment and Development of Qinghai-Tibet Plateau*; China Tibetology Publishing House: Beijing, China, 1996.

48. Zhang, Y.; Ren, H.; Pan, X. *Integration Dataset of Tibet Plateau Boundary*; National Tibetan Plateau Data Center: Beijing, China, 2019.
49. Editorial Committee of Chinese Vegetation Map. Chinese Academy of Sciences. 1:1 Million Vegetation Data Set in China. National Cryosphere Desert Data Center. 2020. Available online: www.ncdc.ac.cn (accessed on 9 December 2021).
50. Li, L.; Liu, S.; Zhou, Z.; Zhou, G. Study on Drought Characteristics in Sichuan Province Based on Time Series GLASS LAI Data. *Sci. Technol. Assoc. Forum* **2013**, 123–124. [CrossRef]
51. Kim, K.; Wang, M.C.; Ranjitkar, S.; Liu, S.H.; Xu, J.C.; Zomer, R.J. Using leaf area index (LAI) to assess vegetation response to drought in Yunnan province of China. *J. Mt. Sci.* **2017**, *14*, 1863–1872. [CrossRef]
52. Li, M.; Du, J.; Li, W.; Li, R.; Wu, S.; Wang, S. Global Vegetation Change and Its Relationship with Precipitation and Temperature Based on GLASS-LAI in 1982–2015. *Sci. Geogr. Sin.* **2020**, *40*, 823–832.
53. Liang, B.; Liu, S. Changes in the Amazon rainforest from 1982 to 2012 using GLASS LAI data. *J. Remote Sens.* **2016**, *20*, 149–156. (In Chinese)
54. Li, X.; Qu, Y. Evaluation of Vegetation Responses to Climatic Factors and Global Vegetation Trends using GLASS LAI from 1982 to 2010. *Can. J. Remote Sens.* **2018**, *44*, 357–372. [CrossRef]
55. Xiao, Z.Q.; Wang, T.T.; Liang, S.L.; Sun, R. Estimating the Fractional Vegetation Cover from GLASS Leaf Area Index Product. *Remote Sens.* **2016**, *8*, 337. [CrossRef]
56. Shi, X.X.; Lohmann, G.; Sidorenko, D.; Yang, H. Early-Holocene simulations using different forcings and resolutions in AWI-ESM. *Holocene* **2020**, *30*, 996–1015. [CrossRef]
57. Ziehn, T.; Lenton, A.; Law, R.M.; Matear, R.J.; Chamberlain, M.A. The carbon cycle in the Australian Community Climate and Earth System Simulator (ACCESS-ESM1)—Part 2: Historical simulations. *Geosci. Model Dev.* **2017**, *10*, 2591–2614. [CrossRef]
58. Wu, T.W.; Lu, Y.X.; Fang, Y.J.; Xin, X.G.; Li, L.; Li, W.P.; Jie, W.H.; Zhang, J.; Liu, Y.M.; Zhang, L.; et al. The Beijing Climate Center Climate System Model (BCC-CSM): The main progress from CMIP5 to CMIP6. *Geosci. Model Dev.* **2019**, *12*, 1573–1600. [CrossRef]
59. Rong, X.Y.; Li, J.; Chen, H.M.; Su, J.Z.; Hua, L.J.; Zhang, Z.Q.; Xin, Y.F. The CMIP6 Historical Simulation Datasets Produced by the Climate System Model CAMS-CSM. *Adv. Atmos. Sci.* **2021**, *38*, 285–295. [CrossRef]
60. Swart, N.C.; Cole, J.N.S.; Kharin, V.V.; Lazare, M.; Scinocca, J.F.; Gillett, N.P.; Anstey, J.; Arora, V.; Christian, J.R.; Hanna, S.; et al. The Canadian Earth System Model version 5 (CanESM50.3). *Geosci. Model Dev.* **2019**, *12*, 4823–4873. [CrossRef]
61. Danabasoglu, G.; Lamarque, J.F.; Bacmeister, J.; Bailey, D.A.; DuVivier, A.K.; Edwards, J.; Emmons, L.K.; Fasullo, J.; Garcia, R.; Gettelman, A.; et al. The Community Earth System Model Version 2 (CESM2). *J. Adv. Model. Earth Syst.* **2020**, *12*, e2019MS001916. [CrossRef]
62. Cherchi, A.; Fogli, P.G.; Lovato, T.; Peano, D.; Iovino, D.; Gualdi, S.; Masina, S.; Scoccimarro, E.; Materia, S.; Bellucci, A.; et al. Global Mean Climate and Main Patterns of Variability in the CMCC-CM2 Coupled Model. *J. Adv. Model. Earth Syst.* **2019**, *11*, 185–209. [CrossRef]
63. Golaz, J.C.; Caldwell, P.M.; van Roekel, L.P.; Petersen, M.R.; Tang, Q.; Wolfe, J.D.; Abeshu, G.; Anantharaj, V.; Asay-Davis, X.S.; Bader, D.C.; et al. The DOE E3SM Coupled Model Version 1: Overview and Evaluation at Standard Resolution. *J. Adv. Model. Earth Syst.* **2019**, *11*, 2089–2129. [CrossRef]
64. Doscher, R.; Acosta, M.; Alessandri, A.; Anthoni, P.; Arsouze, T.; Bergman, T.; Bernardello, R.; Boussetta, S.; Caron, L.P.; Carver, G.; et al. The EC-Earth3 Earth system model for the Coupled Model Intercomparison Project 6. *Geosci. Model Dev.* **2022**, *15*, 2973–3020. [CrossRef]
65. Xie, Z.H.; Liu, S.; Zeng, Y.J.; Gao, J.Q.; Qin, P.H.; Jia, B.H.; Xie, J.B.; Liu, B.; Li, R.C.; Wang, Y.; et al. A High-Resolution Land Model With Groundwater Lateral Flow, Water Use, and Soil Freeze-Thaw Front Dynamics and its Application in an Endorheic Basin. *J. Geophys. Res.-Atmos.* **2018**, *123*, 7204–7222. [CrossRef]
66. Bao, Y.; Song, Z.Y.; Qiao, F.L. FIO-ESM Version 2.0: Model Description and Evaluation. *J. Geophys. Res.-Ocean.* **2020**, *125*, e2019JC016036. [CrossRef]
67. Held, I.M.; Guo, H.; Adcroft, A.; Dunne, J.P.; Horowitz, L.W.; Krasting, J.; Shevliakova, E.; Winton, M.; Zhao, M.; Bushuk, M.; et al. Structure and Performance of GFDL's CM4.0 Climate Model. *J. Adv. Model. Earth Syst.* **2019**, *11*, 3691–3727. [CrossRef]
68. Dunne, J.P.; Horowitz, L.W.; Adcroft, A.J.; Ginoux, P.; Held, I.M.; John, J.G.; Krasting, J.P.; Malyshev, S.; Naik, V.; Paulot, F.; et al. The GFDL Earth System Model Version 4.1 (GFDL-ESM 4.1): Overall Coupled Model Description and Simulation Characteristics. *J. Adv. Model. Earth Syst.* **2020**, *12*, e2019MS002015. [CrossRef]
69. NASA Goddard Institute for Space Studies (NASA/GISS). *NASA-GISS GISS-E2.1G Model Output Prepared for CMIP6 CMIP*; Earth System Grid Federation: Washington, DC, USA, 2018.
70. Walters, D.; Baran, A.J.; Boutle, I.; Brooks, M.; Earnshaw, P.; Edwards, J.; Furtado, K.; Hi, P.; Lock, A.; Manners, J.; et al. The Met Office Unified Model Global Atmosphere 7.0/7.1 and JULES Global Land 7.0 configurations. *Geosci. Model Dev.* **2019**, *12*, 1909–1963. [CrossRef]
71. Volodin, E.M.; Mortikov, E.V.; Kostrykin, S.V.; Galin, V.Y.; Lykossov, V.N.; Gritsun, A.S.; Diansky, N.A.; Gusev, A.V.; Iakovlev, N.G.; Shestakova, A.A.; et al. Simulation of the modern climate using the INM-CM48 climate model. *Russ. J. Numer. Anal. Math. Model.* **2018**, *33*, 367–374. [CrossRef]
72. Volodin, E.M.; Mortikov, E.V.; Kostrykin, S.V.; Galin, V.Y.; Lykossov, V.N.; Gritsun, A.S.; Diansky, N.A.; Gusev, A.V.; Iakovlev, N.G. Simulation of the present-day climate with the climate model INMCM5. *Clim. Dyn.* **2017**, *49*, 3715–3734.
73. Boucher, O.; Servonnat, J.; Albright, A.L.; Aumont, O.; Balkanski, Y.; Bastrikov, V.; Bekki, S.; Bonnet, R.; Bony, S.; Bopp, L.; et al. Presentation and Evaluation of the IPSL-CM6A-LR Climate Model. *J. Adv. Model. Earth Syst.* **2020**, *12*, e2019MS002010. [CrossRef]

74. Milly, P.C.D.; Malyshev, S.L.; Shevliakova, E.; Dunne, K.A.; Findell, K.L.; Gleeson, T.; Liang, Z.; Phillipps, P.; Stouffer, R.J.; Swenson, S. An Enhanced Model of Land Water and Energy for Global Hydrologic and Earth-System Studies. *J. Hydrometeorol.* **2014**, *15*, 1739–1761. [CrossRef]
75. Hajima, T.; Watanabe, M.; Yamamoto, A.; Tatebe, H.; Noguchi, M.A.; Abe, M.; Ohgaito, R.; Ito, A.; Yamazaki, D.; Okajima, H.; et al. Development of the MIROC-ES2L Earth system model and the evaluation of biogeochemical processes and feedbacks. *Geosci. Model Dev.* **2020**, *13*, 2197–2244. [CrossRef]
76. Mauritsen, T.; Bader, J.; Becker, T.; Behrens, J.; Bittner, M.; Brokopf, R.; Brovkin, V.; Claussen, M.; Crueger, T.; Esch, M.; et al. Developments in the MPI-M Earth System Model version 1.2 (MPI-ESM1.2) and Its Response to Increasing CO2. *J. Adv. Model. Earth Syst.* **2019**, *11*, 998–1038. [CrossRef]
77. Yukimoto, S.; Kawai, H.; Koshiro, T.; Oshima, N.; Yoshida, K.; Urakawa, S.; Tsujino, H.; Deushi, M.; Tanaka, T.; Hosaka, M.; et al. The Meteorological Research Institute Earth System Model Version 2.0, MRI-ESM2.0: Description and Basic Evaluation of the Physical Component. *J. Meteorol. Soc. Jpn.* **2019**, *97*, 931–965. [CrossRef]
78. Seland, O.; Bentsen, M.; Olivie, D.; Toniazzo, T.; Gjermundsen, A.; Graff, L.S.; Debernard, J.B.; Gupta, A.K.; He, Y.C.; Kirkevag, A.; et al. Overview of the Norwegian Earth System Model (NorESM2) and key climate response of CMIP6 DECK, historical, and scenario simulations. *Geosci. Model Dev.* **2020**, *13*, 6165–6200. [CrossRef]
79. Lee, W.L.; Wang, Y.C.; Shiu, C.J.; Tsai, I.C.; Tu, C.Y.; Lan, Y.Y.; Chen, J.P.; Pan, H.L.; Hsu, H.H. Taiwan Earth System Model Version 1: Description and evaluation of mean state. *Geosci. Model Dev.* **2020**, *13*, 3887–3904. [CrossRef]
80. Chang, Q.; Dou, J. Exploration on Spatial Correlation of Two Space Variables. *Stat. Appl.* **2016**, *5*, 397–403.
81. Brunke, M.A.; Fairall, C.W.; Zeng, X.B.; Eymard, L.; Curry, J.A. Which bulk aerodynamic algorithms are least problematic in computing ocean surface turbulent fluxes? *J. Clim.* **2003**, *16*, 619–635. [CrossRef]
82. Decker, M.; Brunke, M.A.; Wang, Z.; Sakaguchi, K.; Zeng, X.B.; Bosilovich, M.G. Evaluation of the Reanalysis Products from GSFC, NCEP, and ECMWF Using Flux Tower Observations. *J. Clim.* **2012**, *25*, 1916–1944. [CrossRef]
83. Wang, A.H.; Zeng, X.B. Evaluation of multireanalysis products with in situ observations over the Tibetan Plateau. *J. Geophys. Res.-Atmos.* **2012**, *117*. [CrossRef]
84. Sun, C.; Liu, L.; Guan, L. Validation of the GLASS LAI products in Xilinhot Grassland. *Remote Sens. Technol. Appl.* **2013**, *28*, 949–954.
85. Xiang, Y.; Xiao, Z.; Liang, S.; Wang, J.; Song, J. Validation of Global LAnd Surface Satellite (GLASS) leaf area index product. *J. Remote Sens.* **2014**, *18*, 573–596.
86. Xu, X.Y.; Riley, W.J.; Koven, C.D.; Jia, G.S.; Zhang, X.Y. Earlier leaf-out warms air in the north. *Nat. Clim. Change* **2020**, *10*, 370–375. [CrossRef]
87. Yu, H.Y.; Luedeling, E.; Xu, J.C. Winter and spring warming result in delayed spring phenology on the Tibetan Plateau. *Proc. Natl. Acad. Sci. USA* **2010**, *107*, 22151–22156. [CrossRef] [PubMed]
88. Shen, M.G.; Sun, Z.Z.; Wang, S.P.; Zhang, G.X.; Kong, W.D.; Chen, A.P.; Piao, S.L. No evidence of continuously advanced green-up dates in the Tibetan Plateau over the last decade. *Proc. Natl. Acad. Sci. USA* **2013**, *110*, E2329. [CrossRef] [PubMed]
89. Wang, T.; Peng, S.S.; Lin, X.; Chang, J.F. Declining snow cover may affect spring phenological trend on the Tibetan Plateau. *Proc. Natl. Acad. Sci. USA* **2013**, *110*, E2854–E2855. [CrossRef] [PubMed]
90. Chen, B.X.; Zhang, X.Z.; Tao, J.; Wu, J.S.; Wang, J.S.; Shi, P.L.; Zhang, Y.J.; Yu, C.Q. The impact of climate change and anthropogenic activities on alpine grassland over the Qinghai-Tibet Plateau. *Agric. For. Meteorol.* **2014**, *189*, 11–18. [CrossRef]
91. Piao, S.L.; Cui, M.D.; Chen, A.P.; Wang, X.H.; Ciais, P.; Liu, J.; Tang, Y.H. Altitude and temperature dependence of change in the spring vegetation green-up date from 1982 to 2006 in the Qinghai-Xizang Plateau. *Agric. For. Meteorol.* **2011**, *151*, 1599–1608. [CrossRef]
92. Shen, M.G.; Piao, S.L.; Jeong, S.J.; Zhou, L.M.; Zeng, Z.Z.; Ciais, P.; Chen, D.L.; Huang, M.T.; Jin, C.S.; Li, L.Z.X.; et al. Evaporative cooling over the Tibetan Plateau induced by vegetation growth. *Proc. Natl. Acad. Sci. USA* **2015**, *112*, 9299–9304. [CrossRef]
93. Xin, X.; Wu, T.; Zhang, J. Introduction of BCC models and its participation in CMIP6. *Clim. Change Res.* **2019**, *15*, 533–539.
94. Song, Z.Y.; Bao, Y.; Qiao, F.L. Introduction of FIO-ESM v2.0 and its participation plan in CMIP6 experiments. *Clim. Change Res.* **2019**, *15*, 558.
95. Wang, Y.C.; Hsu, H.H.; Chen, C.A.; Tseng, W.L.; Hsu, P.C.; Lin, C.W.; Chen, Y.L.; Jiang, L.C.; Lee, Y.C.; Liang, H.C.; et al. Performance of the Taiwan Earth System Model in Simulating Climate Variability Compared With Observations and CMIP6 Model Simulations. *J. Adv. Model. Earth Syst.* **2021**, *13*, e2020MS002353. [CrossRef]
96. Keeble, J.; Hassler, B.; Banerjee, A.; Checa-Garcia, R.; Chiodo, G.; Davis, S.; Eyring, V.; Griffiths, P.T.; Morgenstern, O.; Nowack, P.; et al. Evaluating stratospheric ozone and water vapour changes in CMIP6 models from 1850 to 2100. *Atmos. Chem. Phys.* **2021**, *21*, 5015–5061. [CrossRef]
97. Burrows, S.M.; Maltrud, M.; Yang, X.; Zhu, Q.; Jeffery, N.; Shi, X.; Ricciuto, D.; Wang, S.; Bisht, G.; Tang, J.; et al. The DOE E3SM v1.1 Biogeochemistry Configuration: Description and Simulated Ecosystem-Climate Responses to Historical Changes in Forcing. *J. Adv. Model. Earth Syst.* **2020**, *12*, e2019MS001766. [CrossRef]
98. Andrews, M.B.; Ridley, J.K.; Wood, R.A.; Andrews, T.; Blockley, E.W.; Booth, B.; Burke, E.; Dittus, A.J.; Florek, P.; Gray, L.J.; et al. Historical Simulations With HadGEM3-GC3.1 for CMIP6. *J. Adv. Model. Earth Syst.* **2020**, *12*, e2019MS001995. [CrossRef]

99. Sellar, A.A.; Walton, J.; Jones, C.G.; Wood, R.; Abraham, N.L.; Andrejczuk, M.; Andrews, M.B.; Andrews, T.; Archibald, A.T.; de Mora, L.; et al. Implementation of UK Earth System Models for CMIP6. *J. Adv. Model. Earth Syst.* **2020**, *12*, e2019MS001946. [CrossRef]
100. Luo, L.; Zhang, Y.; Zhou, J.; Pan, X.; Sun, W. Simulation and application of the land surface model CLM Drive by WRF in the Tibetan Plateau. *J. Glaciol. Geocryol.* **2013**, *35*, 553–564. (In Chinese)
101. Xiong, J.; Zhang, Y.; Wang, S. Influence of soil moisture transmission scheme improvement in CLM4.0 on simulation of land surface process in Qinghai-Xizang Plateau. *Plateau Meteorol.* **2014**, *33*, 323–336. (In Chinese)
102. Xie, Z.; Hu, Z.; Liu, H.; Sun, G.; Yang, Y.; Lin, Y.; Huang, F. Evaluation of the surface energy exchange simulations of land surface model CLM4.5 in alpine meadow over the Qinghai-Xizang Plateau. *Plateau Meteorol.* **2017**, *36*, 1–12. (In Chinese)
103. Song, Y.; Fan, Y.; Ma, T. Evaluation of simulation performance of land surface model NCAR_CLM4.5 at a degrated glassland station in semi-arid area. *Trans. Atmos. Sci.* **2014**, *37*, 794–803. (In Chinese)
104. Mao, J.F.; Thornton, P.E.; Shi, X.Y.; Zhao, M.S.; Post, W.M. Remote Sensing Evaluation of CLM4 GPP for the Period 2000–09. *J. Clim.* **2012**, *25*, 5327–5342. [CrossRef]
105. Song, X.; Wang, D.Y.; Li, F.; Zeng, X.D. Evaluating the performance of CMIP6 Earth system models in simulating global vegetation structure and distribution. *Adv. Clim. Change Res.* **2021**, *12*, 584–595. [CrossRef]
106. Yang, Y.; Piao, S. Variation in grassland vegetation cover in relation to climatic factors on the Tibetan Plateau. *Chin. J. Plant Ecol.* **2006**, *30*, 1. (In Chinese)
107. Yu, H.; Xu, J. Effects of climate change on vegetations on Qinghai-Tibet Plateau: A review. *Chin. J. Ecol.* **2009**, *28*, 747–754. (In Chinese)
108. Yang, J.; Ding, Y.; Shen, Y.; Liu, S.; Chen, R. Climatic Features of Eco-Environment Change in the Source Regions of the Yangtze and Yellow Rivers in Recent 40 Years. *J. Glaciol. Geocryol.* **2004**, *26*, 7–16. (In Chinese)
109. Wang, Y.; Zhao, Z.; Qiao, Y.; Li, C. Characteristics of the climatic variatation in zoig(e) in the past 45 years and its effects on the eco-environment in the area. *J. Geomech.* **2005**, *11*, 328–332, 340.
110. Li, Y.; Zhao, L.; Zhao, X.; Zhou, H. Effects of a 5-years mimic Temperature Increase to the structure and productivity of kobresia humilis meadow. *Acta Agrestia Sin.* **2004**, *12*, 236–239.
111. Wang, S.; Zhao, L.; Li, S. Interaction between Permafrost and Desertification on the Qinghai-Tibet Plateau. *J. Desert Res.* **2002**, *22*, 33–39.
112. Zhang, J.; Fu, C.; Yan, X.; Seita, E.; Hiroshi, K. Global respondence analysis of LAI versus surface air temperature and precipitation variation. *Chin. J. Geophys.* **2002**, *45*, 631–637. (In Chinese) [CrossRef]
113. Zhu, X.D.; He, H.L.; Liu, M.; Yu, G.R.; Sun, X.M.; Gao, Y.H. Spatio-temporal variation of photosynthetically active radiation in China in recent 50 years. *J. Geogr. Sci.* **2010**, *20*, 803–817. [CrossRef]
114. Fan, S.; Fan, G.; Lai, X. Analysis on variation trend of season start dates over the Tibetan Plateau during 1961–2007. *Clim. Environ. Res.* **2013**, *18*, 71–79. (In Chinese)
115. Wu, K.; Liu, S.; Bao, W.; Wang, R. Remote sensing monitoring of the glacier change in the Gangrigabu Range, southeast Tibetan Plateau from 1980 through 2015. *J. Glaciol. Geocryol.* **2017**, *39*, 24–34.
116. Xia, L.; Song, X.; Cai, S.; Hu, R.; Guo, D. Role of surface hydrothermal elements in grassland degradation over the Tibetan Plateau. *Acta Ecol. Sin.* **2021**, *41*, 4618–4631.
117. Feng, Q.; Gao, X.; Huang, X.; Yu, H.; Liang, T. Remote sensing dynamic monitoring of grass growth in Qinghai-Tibet plateau from 2001 to 2010. *J. Lanzhou Univ.* **2011**, *47*, 75–81, 90.
118. Xiao, Y.; Li, J.; Li, N. Evaluation of CMIP6 HighResMIP models in Simulating Precipitation over Tibetan Plateau. *Torrential Rain Disasters* **2022**, *41*, 215–223. (In Chinese)
119. Chen, W.; Jiang, D.; Wang, X. Evaluation and Projection of CMIP6 Models for Climate over the Qinghai-Xizang (Tibetan) Plateau. *Plateau Meteorol.* **2021**, *40*, 1455–1469. (In Chinese)
120. Luo, S.; Lv, S.; Zhang, Y. Simulation analysis on land surface process of BJ site of Central Tibetan Plateau using CoLM. *Plateau Meteorol.* **2008**, *27*, 259–271. (In Chinese)
121. Yang, Y.; Li, M.; Hu, Z. Influence of surface roughness on surface-air fluxes in alpine meadow over the Northern Qinghai-Xizang Plateau. *Plateau Meteorol.* **2014**, *33*, 626–636. (In Chinese)
122. Gao, Z.; Chae, N.; Kim, J. Modeling of surface energy partitioning, surace temperature, and soil wetness in the Tibetan prairie using the Simple Biosphere Model 2 (SiB2). *J. Geophys. Res. Atmos.* **2004**, *109*, D06102. [CrossRef]
123. Swingedouw, D.; Mignot, J.; Ortega, P.; Khodri, M.; Menegoz, M.; Cassou, C.; Hanquiez, V. Impact of explosive volcanic eruptions on the main climate variability modes. *Glob. Planet. Change* **2017**, *150*, 24–45. [CrossRef]
124. Singh, M.; Krishnan, R.; Goswami, B.; Choudhury, A.D.; Swapna, P.; Vellore, R.; Prajeesh, A.G.; Sandeep, N.; Venkataraman, C.; Donner, R.V.; et al. Fingerprint of volcanic forcing on the ENSO-Indian monsoon coupling. *Sci. Adv.* **2020**, *6*, eaba8164. [CrossRef]
125. Ma, X.; Chen, S.; Deng, J.; Feng, Q.; Huang, X. Vegetation Phenology dynamics and its response to climate change on the Tibetan Plateau. *Acta Ecol. Sin.* **2016**, *25*, 13–21.

remote sensing

Article

Retrieving Soil Moisture in the Permafrost Environment by Sentinel-1/2 Temporal Data on the Qinghai–Tibet Plateau

Zhibin Li [1], Lin Zhao [1,2,3,*], Lingxiao Wang [1], Defu Zou [3], Guangyue Liu [2,3], Guojie Hu [3], Erji Du [2,3], Yao Xiao [3], Shibo Liu [2,3], Huayun Zhou [2,3], Zanpin Xing [2,3], Chong Wang [1], Jianting Zhao [1], Yueli Chen [4], Yongping Qiao [3] and Jianzong Shi [3]

1 School of Geographical Sciences, Nanjing University of Information Science & Technology, Nanjing 210044, China
2 University of Chinese Academy of Sciences, Beijing 100049, China
3 Cryosphere Research Station on Qinghai-Tibet Plateau, State Key Laboratory of Cryospheric Sciences, Northwest Institute of Eco-Environment and Resources, Chinese Academy of Sciences, Lanzhou 730000, China
4 Department of Geography, Ludwig-Maximilians-Universität München, 80333 Munich, Germany
* Correspondence: lzhao@nuist.edu.cn

Citation: Li, Z.; Zhao, L.; Wang, L.; Zou, D.; Liu, G.; Hu, G.; Du, E.; Xiao, Y.; Liu, S.; Zhou, H.; et al. Retrieving Soil Moisture in the Permafrost Environment by Sentinel-1/2 Temporal Data on the Qinghai–Tibet Plateau. *Remote Sens.* **2022**, *14*, 5966. https://doi.org/10.3390/rs14235966

Academic Editor: Emanuele Santi

Received: 14 October 2022
Accepted: 21 November 2022
Published: 25 November 2022

Abstract: Soil moisture (SM) products presently available in permafrost regions, especially on the Qinghai–Tibet Plateau (QTP), hardly meet the demands of evaluating and modeling climatic, hydrological, and ecological processes, due to their significant bias and low spatial resolution. This study developed an algorithm to generate high-spatial-resolution SM during the thawing season using Sentinel-1 (S1) and Sentinel-2 (S2) temporal data in the permafrost environment. This algorithm utilizes the seasonal backscatter differences to reduce the effect of surface roughness and uses the normalized difference vegetation index (NDVI) and the normalized difference moisture index (NDMI) to characterize vegetation contribution. Then, the SM map with a grid spacing of 50 m $\times$ 50 m in the hinterland of the QTP with an area of 505 km $\times$ 246 km was generated. The results were independently validated based on in situ data from active layer monitoring sites. It shows that this algorithm can retrieve SM well in the study area. The coefficient of determination (R^2) and root-mean-square error (RMSE) are 0.82 and 0.06 m^3/m^3, respectively. This study analyzed the SM distribution of different vegetation types: the alpine swamp meadow had the largest SM of 0.26 m^3/m^3, followed by the alpine meadow (0.23), alpine steppe (0.2), and alpine desert (0.16), taking the Tuotuo River basin as an example. We also found a significantly negative correlation between the coefficient of variation (CV) and SM in the permafrost area, and the variability of SM is higher in drier environments and lower in wetter environments. The comparison with ERA5-Land, GLDAS, and ESA CCI showed that the proposed method can provide more spatial details and achieve better performance in permafrost areas on QTP. The results also indicated that the developed algorithm has the potential to be applied in the entire permafrost regions on the QTP.

Keywords: soil moisture; SAR; retrieval algorithm; high spatial resolution; permafrost

1. Introduction

Soil moisture (SM) is an essential component of the terrestrial hydrological cycle ecosystem [1,2]. In ecology, SM affects the growth and activity of vegetation and microorganisms by controlling the division of water [3], and can also mitigate changes in soil organic carbon content caused due to climate warming [4]. In hydrology, SM is not only an important parameter of the water cycle but is also used to infer surface- and ground-water exchanges [5]. In climatology, SM affects regional climate through changing surface albedo, evapotranspiration intensity, and sensible and latent heat fluxes [6]. In the permafrost environment of the Qinghai–Tibet Plateau (QTP), the SM in the active layer is significantly altered by the seasonal freezing and thawing processes and influences the energy exchange between permafrost terrain and the atmosphere [7,8]. The accurate information on the spatial and temporal distribution of SM helps advance hydrological, ecological, and climatological studies in permafrost areas [9].

In the permafrost area of 1.06×10^6 km^2 of the QTP [10], it is critical for various scientific studies to obtain accurate spatial distribution data of SM on a large scale. Due to the harsh environment and inconvenience of accessing the QTP, the traditional methods of obtaining SM by sampling measurement and monitoring are limited. Several meteorological and hydrological stations have been deployed over the past few decades, and data scarcity has been filled to a degree [11]. These restricted sites are unevenly distributed on the edge of permafrost areas or on seasonally frozen ground areas, which makes it challenging to study SM with high spatial heterogeneity [6,12]. Remote sensing technology has achieved significant advances in SM monitoring, with its unique large-area observation capabilities, and some remote sensing products and reanalysis data have been produced, such as the fifth generation of the land component of the European Centre for Medium-Range Weather Forecast atmospheric reanalysis (ERA5-Land) [13], the European Space Agency Climate Change Initiative (ESA CCI) [14], and the Noah land surface model driven by Global Land Data Assimilation System (GLDAS-Noah) [15]. Xing et al. evaluated seven SM data products (SMAP, SMOS-IC, ASCAT, ERA5-Land, ESA CCI, LPRM, AMSR2) over the permafrost region of the QTP based on in situ SM measurements and found that the SM data of ESA CCI had the highest accuracy [16]. Due to the lack of adequate measurement data, SM products generated by the model assimilation are significantly biased in the QTP [17]. In addition, the SM product data at a spatial resolution of tens of kilometers are affected by mixed pixels and do not accurately describe the SM distribution. The coefficient of variation (CV) analysis is often used to describe significant patterns in regional mean SM content, with the relationship between CV and SM often showing a hysteresis pattern in spatial variability [18–20]. The spatial–temporal SM variations have never been revealed over the permafrost area in the QTP, where the relationship between the CV and mean SM is similar to that found in other regions is unclear [21,22]. The lack of high-spatial-resolution SM data in the permafrost region of the QTP greatly limits the studies of the spatial–temporal distribution characteristics of SM. Few studies are able to analyze the spatial–temporal variations of SM in permafrost areas on a fine scale.

Synthetic Aperture Radar (SAR) has proven its high potential for retrieving high-spatial-resolution SM using the backscatter coefficient ($\sigma°$) [23,24]. Some empirical [25,26], semi-empirical [27], and physical scattering models [28] have been developed to relate backscattering with surface SM, roughness, and vegetation. Surface scattering models are commonly used for bare soil including the semi-empirical Dubios model [29], the Oh model [30], and the physically based Integrate Equation Model (IEM) [31]. These are usually integrated with vegetation scattering models, such as semi-empirical water cloud models (WCM), to predict scattering from vegetated areas [32,33]. Studies have developed algorithms for retrieving SM based on IEM or WCM inversion using minimization, LUT, and machine learning approaches. In 2014, He et al. estimated SM in the alpine meadow region by coupling the IEM and the WCM with R^2 and RMSE reaching 0.71 and 0.03 m^3/m^3, respectively [34]. In 2017, Bai et al. first estimated SM in the alpine steppe region of Magu using Sentinel-1 (S1) data with the WCM [35]. In 2021, Yang et al. coupled the improved Oh model in 2004 and WCM to estimate SM in the Nagqu region based on S1 data and MODIS optical data with the assumption of constant surface roughness [36]. Despite the significant advance in scattering modeling, SM inversion from these microwave scattering models are commonly ill-posed and complicated [37,38]. Besides utilizing microwave scattering, several algorithms have been successively developed and widely used to retrieve SM, such as change detection (CD) and neural network (NN) [39–41]. NN are mathematical models that are commonly trained by vegetation coefficients, backscatter coefficients, and other parameters in studies of SM retrieval using SAR data, which have high requirements for the data volume. However, the NN method requires a large amount of data for training and validation, which greatly limits its application in the permafrost region of the QTP.

The principle of the CD algorithm is based on the assumption that by differencing the backscattering coefficients in two periods, the effect of soil roughness and vegetation is reduced, and the backscatter difference is mainly due to the changes in SM [39]. Thus, it is more suitable than the NN algorithm for permafrost areas where a large amount of training data acquirement is difficult. Gao et al. mapped SM distributions at a 100 m spatial resolution located in Urgell using a CD algorithm by combining S1 and Sentinel-2 (S2) data [40]. Bauer-Marschallinger et al. used the CD algorithm to build the first global SM dataset with 1 km spatial resolution, which greatly advanced the progress of SM studies [42]. Zhu et al. proposed an unsupervised CD method as a pre-processing procedure for multi-temporal retrieval and improved the accuracy of the CD algorithm by reducing the uncertainty caused by changes in vegetation and roughness [43,44].

Over the QTP, studies of SM retrieval using SAR data have also been conducted in recent years. Yang et al. and Bai et al. estimated SM in alpine grassland environments in Nagqu and Magu, respectively, both with favorable results [35,36]. In the Beiluhe, Zhang et al. used estimated SM in alpine meadows and alpine deserts and improved the accuracy of estimating SM by the WCM and CD algorithms [45]. However, these SM retrieving studies on the QTP were conducted outside or in the margin/border of the permafrost area or within a very small permafrost region. The study area was very small, i.e., covering only a few square kilometers, and the effectiveness and accuracy of the retrieving algorithms or models were not tested in other areas. To our knowledge, no SM retrieving study has yet been conducted on the large-scale hinterland permafrost regions.

The freeze–thaw cycle of active layer soils and the water barrier of the frozen layer in the permafrost area play an active role in determining vegetation growth and SM retention [46,47]. It implies that in areas with high SM, the vegetation cover has significant interference with the radar signal. The vegetation canopy complicates the extraction of underlying soil water, as the canopy contains water and can also block or scatter radar signals [48]. Therefore, in the retrieval process, vegetation is another important factor that affects the radar signal in addition to SM. In the many ecological, hydrological, and agricultural studies, the vegetation canopy is usually expressed by the vegetation indices, such as Leaf Area Index (LAI), Normalized Difference Vegetation Index (NDVI), and the Enhanced Vegetation Index (EVI), biomass, vegetation height, etc. [35,45,49–52]. Several studies have shown that NDVI is easier to derive and has fewer errors than other vegetation indices and is widely used in SM retrieval studies [50,53]. In addition, in partially vegetated areas, Bao et al. found that Normalized Difference Moisture Index (NDMI) can also perform well in SM retrieval studies based on S1 data [54].

In the permafrost region on the QTP, the ground has a distinct freeze–thaw cycle process. The soil water is in a liquid state during summer and in a state of combination of ice and unfrozen water in other seasons [9]. The real implication of SM values obtained by in situ monitoring and sampling drying measurements could be different. In situ monitoring measures the unfrozen water (liquid water) content by monitoring the dielectric constant in the soil, such as the Hydra soil moisture sensor. The field-oven sampling acquires the total soil water content (unfrozen water and ice) by collecting in the field and then calculating the volumetric water content from the wet and dry weight of the soil. Meanwhile, σ° is sensitive to unfrozen soil water, and the frozen part is neglected in the retrieval process. The σ° could not represent the gross soil water content in all seasons except for the thawing season. Therefore, we need to be careful when choosing field "SM" data in developing and training SM retrieving algorithms [55].

In summary, a retrieval algorithm for SM is urgently needed to obtain SM spatial data which could promote hydrological, ecological, climatic, and engineering studies in the permafrost region of the QTP. In this study, the hinterland of the QTP was selected for SM retrieval, where a variety of surface types are included. The retrieval algorithm is trained and validated using multi-year in situ observations of different surface environments. We chose the months of July and August as the study period, which can reduce the errors caused by the freeze–thaw process of the soil. The SM retrieval during the thawing season could represent the gross soil water content. In addition, the CD algorithm is a promising method for SM retrieval in the permafrost region where a priori knowledge is scarce [39,40]. The liquid soil water is very small in the coldest winter season, by reference to the CD algorithm, the effect of surface roughness is minimized. Then, the vegetation effect is further represented and reduced by vegetation indexes (NDVI, NDMI) from optical data. Finally, the objectives of this study are: (1) to develop an SM retrieval algorithm suitable for permafrost environments on the QTP using high spatial-resolution SAR data to obtain spatial data of SM for the thawing season; and (2) to explore the spatial–temporal distribution characteristics of SM on the large extent of permafrost region on the QTP.

2. Materials and Methods

2.1. Study Area

In this study, an area of 505 km × 246 km in the hinterland of QTP along the Qinghai–Tibet Highway was selected as the study area, as shown in Figure 1. This area covers most of the stations of the SM and temperature monitoring of the permafrost networks [11]. The study area includes typical permafrost regions and seasonally frozen ground regions with an altitude between 4189 m and 6402 m a.s.l. The average annual temperature in this region is between −5.8 °C and −2.4 °C, and the trend of temperature increase is consistent, with an average rate of change of about 0.05 °C/a. The annual precipitation is in the range of approximately 210–580 mm, with sizeable interannual variation [56,57]. The precipitation is mainly concentrated between May and September, and there is an apparent upward fluctuation in annual precipitation, with an average variable rate of 7.49 mm/a from 2004 to 2016 [58].

The vegetation types in the study area are classified as alpine swamp meadows, alpine meadows, alpine steppes, and alpine deserts. The alpine meadows cover the largest areas, followed by alpine steppes and alpine swamp meadows at the least [59]. The degradation of the permafrost has affected the ecological situation in the QTP. The vegetation ecosystem degradation is significant, mainly manifested as the degradation of alpine swamp meadow to the alpine meadow and alpine meadow to alpine steppe [59].

The frozen ground undergoes seasonal freeze–thaw cycles [8]. On the QTP, the thawing process begins in mid-to-late May and lasts until late September to early November each year [58,60,61]. The effect of water transport during soil freezing and thawing on SM distribution is very significant. The measured data in the in situ show that the SM in the thawing season varies roughly between 0.1 m^3/m^3 and 0.5 m^3/m^3. When the soil is frozen in winter, the unfrozen water content is low.

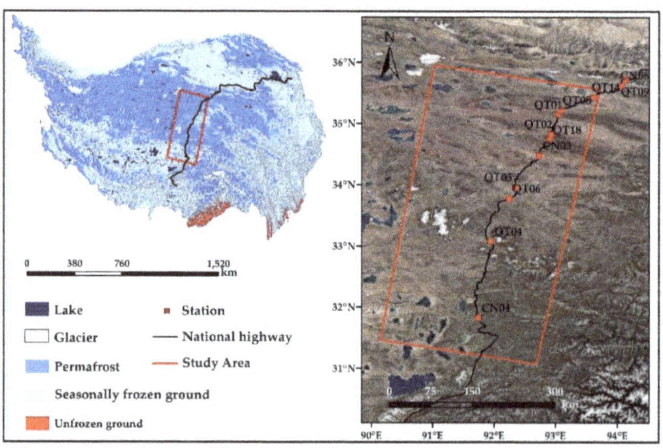

Figure 1. The overview of the study area. The base map is the map of the permafrost [10], glaciers [62], lakes [63], and topographic map from SRTM data [64].

2.2. Datasets

2.2.1. In Situ Observations

The in situ SM measurements are acquired from the SM and temperature observation network built by the Chinese Academy of Sciences (CAS) [11]. SM was measured by a Hydra soil moisture sensor and recorded by a CR10X/CR1000/CR3000 data logger, with an accuracy of ±2.5%. Table 1 shows detailed information on the stations used in the study. The elevation of these stations ranges from 4468 m to 5100 m a.s.l and spans about 420 km from north to south. The sites also contain several typical vegetation types of the QTP, including alpine swamp meadows, alpine meadows, and alpine steppes. They could represent the varied permafrost environment on the QTP to some extent. We collected 129 in situ SM data during the thawing season (months July, and August) from 2016 to 2019 at these sites, of which 100 were used for training the model and 29 for validation.

Table 1. Information of SM monitoring sites.

Sites	Lon. (°E)	Lat. (°N)	Location	Altitude(m)	Vegetation Types
CN03	92.727	34.47	Wuli	4625	Alpine steppe
CN04	91.737	31.81	Liangdaohe	4808	Alpine swamp meadow
CN06	94.063	35.62	Kunlun Pass	4746	Alpine meadow
QT01	93.043	35.14	Hoh Xil	4734	Alpine meadow
QT02	93.921	34.82	Beiluhe	4656	Alpine swamp meadow
QT04	91.941	33.07	Tanggula	5100	Alpine meadow
QT05	92.338	33.95	Kaixinling	4652	Alpine meadow
QT06	92.239	33.77	Tongtian	4650	Alpine steppe
QT08	93.084	35.22	Wudaoliang	4783	Alpine steppe
QT09	94.125	35.72	Xidatan	4538	Alpine steppe
QT14	93.600	35.43	Suonandaje	4468	Alpine meadow
QT18	92.892	34.73	Fenghuo	4773	Alpine swamp meadow

2.2.2. Sentinel-1

The Sentinel-1 (S1) satellites were launched by the European Space Agency (ESA) in the frame of Europe's Copernicus program, including Sentinel-1A (S1A) and Sentinel-1B (S1B). The orbital period of S1A is 98.6 min, the revisit period is 12 days, and the combined AB satellite is 6 days. The interferometric wide (IW) imaging model with a spatial resolution of 5 m × 20 m provides a more accurate σ° and better meets the needs of this study [50].

We use the ground range detection (GRD) products of S1 in IW acquisition mode with the VV and VH polarizations from Google Earth Engine (GEE). Compared to the VH polarization of S1, the VV polarization has greater potential for SM retrieval [40,65–67]. Some researchers have used both VV and VH in SM retrieval studies and characterized the effect of vegetation by the ratio of VV and VH [41,68]. However, in the study, only VV single polarized data is acquired before February 2017. Therefore, the optical vegetation index NDVI is used in this study for characterizing vegetation. The backscattered images of VV polarization in the completely thawed season (months July, and August) and the completely frozen season (January to February) from 2016 to 2019, in a total of 89 acquisitions, are employed for the SM retrieval algorithm development and validation.

2.2.3. Sentinel-2

Sentinel-2 (S2) is a high-resolution imaging satellite that carries a multi-spectral imager (MSI). The revisit period for one satellite is 10 days, and the revisit period for two satellites is 5 days [69,70]. In this study, cloud-free S2 TOA Level-1C data one week before and after the S1 acquisition is selected to calculate vegetation and water indices using the green band (B3: Green), red band (B4: Red), near-infrared band (B8: NIR), and short-wave infrared (B11: SWIR) [71].

2.2.4. SRTM DEM

Shuttle Radar Topography Mission (SRTM) data is mainly measured jointly by the National Aeronautics and Space Administration (NASA) and the National Imagery and Mapping Agency (NIMA). Interferometric radar data is captured using dual radar antennas and converted into digital terrain data. This study uses the "NASA SRTM Digital Elevation 30-m" elevation dataset provided by the GEE platform [64].

2.2.5. SM Data Products

In this study, we also compared the retrieval SM with three SM data products, such as the fifth generation of the land component of the European Centre for Medium-Range Weather Forecasts atmospheric reanalysis (ERA5-Land) [13], the Noah land surface model driven by Global Land Data Assimilation System (GLDAS-Noah) [15], and the European Space Agency Climate Change Initiative (ESA CCI) [14]. The specific information of the three datasets and the first layer of the depth range are shown in Table 2. In this study, the SM product data for 2 July and 19 August 2018 were selected for comparing the spatial distribution and average SM content with the retrieval results.

Table 2. Information on SM products.

Product type	Sensor	Period	Spatial Resolution	Temporal Resolution	Depth
Remote sensing products	ESA CCI	1978–2019	0.25° × 0.25°	Daily	~0–5 cm
Reanalysis products	ERA5-Land	2000–present	0.1° × 0.1°	3-Hourly	0–7 cm
	GLDAS-Noah	1948–present	0.25° × 0.25°	3-Hourly	0–10 cm

2.3. Methods

The workflow of the SM retrieval algorithm development is illustrated in Figure 2. The main steps are summarized as (1) S1 backscatter preprocessing; (2) Reducing the effect of surface roughness; (3) Reducing the effect of vegetation; (4) SM retrieval algorithm construction; (5) SM result post-processing; and (6) Retrieval result validation.

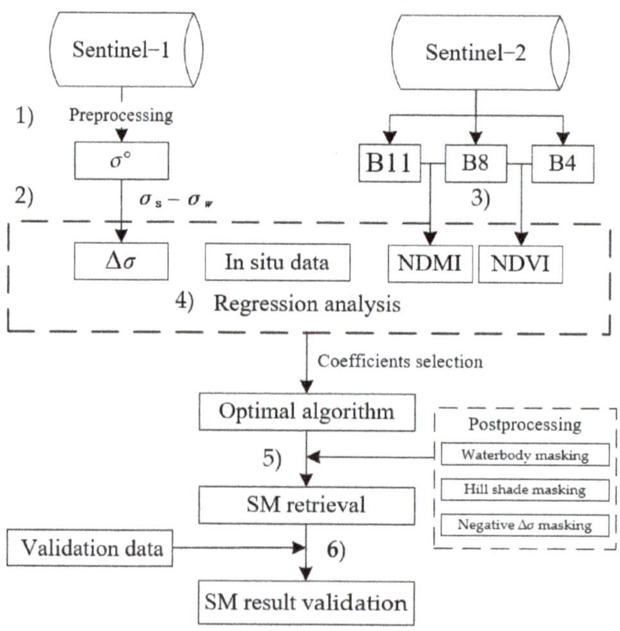

Figure 2. Workflow of SM retrieval and validation.

2.3.1. S1 Backscatter Preprocessing

Data preprocessing was performed on the GEE platform [72]. GEE has performed some preprocessing of the S1 data using the ESA S1 Toolbox (S1TBX), including applying orbit files, removing thermal noise, removing GRD border noise, radiometric calibration, and Range-Doppler terrain correction. Furthermore, the S1 incident angle normalization and spatial filtering are needed to make the data as correct as possible.

- S1 incident angle normalization

The σ° is affected by the incidence angle (θ^0) of S1 and has a slight deviation from the actual situation. There is a certain correlation between θ^0 and σ°, which can be expressed as a slope β [73,74]. This study chose the central incidence angle of the study area (38°) as the reference angle to reduce the overall error caused by extrapolation [42]. Therefore, as shown in Equation (1), we uniformly correct the σ to the value corresponding to the incident angle of 38° (σ° (38°)).

$$\sigma^{\circ}(38^{\circ}) = \sigma^{\circ}\,(\theta^0) - \beta\,(\theta^0 - 38^{\circ})\ [\text{dB}]. \tag{1}$$

The calculation of Equation (1) can be performed on the GEE.

- Refined Lee Filtering

In order to reduce the speckle noise in the image while preserving the image edge information, the Refined Lee filter with a window size of 7 × 7 is used in this study [75,76].

2.3.2. Sensitivity of Backscattering Coefficient to Soil Liquid Water

The correlation between SM and σ° in the permafrost region was analyzed at sites QT08 and QT09. As shown in Figure 3, the σ° and SM at both stations showed obvious seasonal variations, i.e., high in summer and low in winter. When the soil freezes, the soil's liquid water content decreases sharply, and therefore, the dielectric constant decreases, and consequently, the σ° drops significantly.

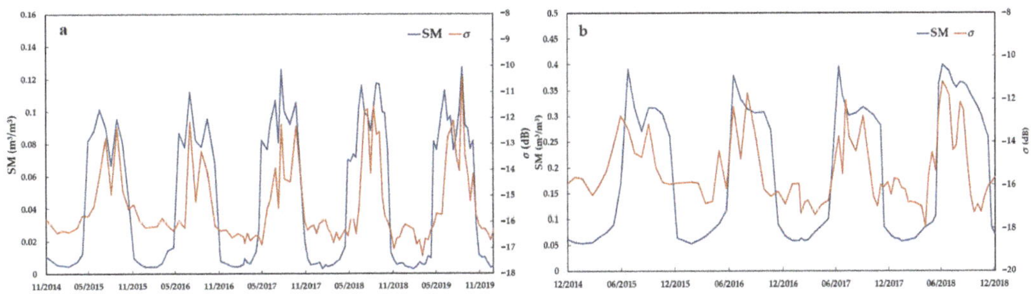

Figure 3. The time-series variation of SM and σ° at sites. (**a**) QT08 and (**b**) QT09.

Figure 3 also revealed that the same σ° corresponded to different SM values at the two sites. At site QT08, the σ° was in the range of −17 to −10 dB, corresponding to SM of about 0.003 to 0.13 m^3/m^3, while at site QT09, the σ° range was −17 to −11 dB, corresponding to the SM range of about 0.05 to 0.4 m^3/m^3. This shows that the range of σ° does not vary much between the two sites, but the corresponding SM ranges are dramatically different. Many studies have demonstrated that surface roughness and vegetation are the major factors affecting the correlation between surface backscatter intensity and SM [77]. Therefore, it is critical to reduce the effects of surface roughness and vegetation in the retrieval process to improve accuracy.

2.3.3. Reducing the Effect of Surface Roughness

The CD algorithm originally proposed by Wagner et al. determines the SM by linearly scaling the observed backscatter between that at the driest and wettest conditions [78].

The algorithm has been validated and used in many regions in SM retrieval studies in recent years, including semi-arid areas and mountainous areas [42,79]. Rignot EJM et al. conducted experiments in a mountainous region of the Katmai National Park and Preserve in Alaska and found that the CD algorithm is effective in removing the topographic influences [80]. Therefore, this study refers to the principle of the CD algorithm to eliminate the effect of topography in permafrost areas.

According to the characteristics of soil liquid water in the permafrost regions of QTP, the soil surface is mostly frozen in winter and holds little liquid water. Meanwhile, due to low precipitation during the period of January–February, the snowmelt or wet snowfall effect is limited, and the change in surface roughness in winter is not significant compared to the melting season. We assumed that the smallest σ° in winter of its temporal curves could represent the lowest liquid water content of the soil during this period. Therefore, we took the smallest value of σ° in the freezing season (January, February) as the reference value and subtracted it from the backscatter signal during the thawing season. The backscatter difference ($\Delta\sigma$) between the freezing and thawing seasons (months July, and August) represents the changes in SM and vegetation. It can be expressed as:

$$\Delta\sigma = \sigma_{\text{s}} - \sigma_{\text{w}} \tag{2}$$

where σ_{s} is the σ° of the thawing season, and σ_{w} is the σ° of the freezing season. Figure 4 shows the distribution and details of σ^{s}, σ^{w}, and $\Delta\sigma$ in the study area. In the case of Figure 4a,b, the topographic factor has a significant effect on the backscattered signal. Figure 4c shows that the method of calculating the σ° difference between summer and winter is effective in weakening this effect.

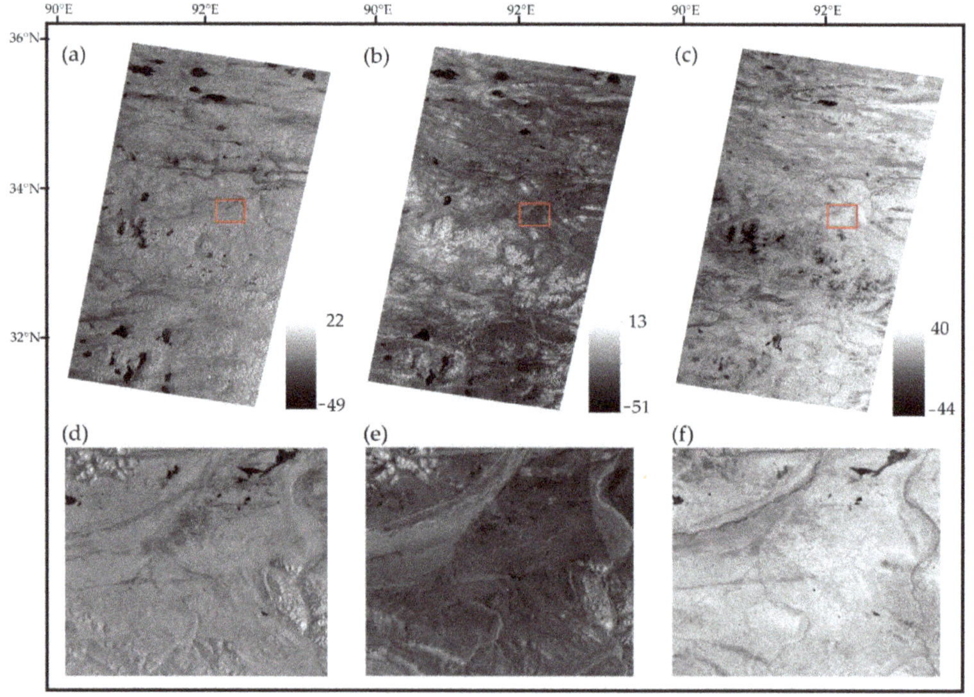

Figure 4. The distribution of σ_s, σ_w, and $\Delta\sigma$ and details in the study area. (**a**): The σ° on 2 July 2018. (**b**): The σ_{w} in 2018. (**c**): The $\Delta\sigma$. (**d–f**): The details after scaling up of σ°, σ_w, and $\Delta\sigma$ within the red box.

2.3.4. Reduce the Effect of Vegetation

We applied NDVI and NDMI to reflect vegetation characteristics and the water content in vegetation and proposed to use the combination of NDVI and NDMI to characterize the effect of vegetation on the σ. These vegetation indices were calculated as follows:

$$NDVI = (\rho_{nir} - \rho_{red})/(\rho_{nir} + \rho_{red}) \tag{3}$$

$$NDMI = (\rho_{nir} - \rho_{swir})/(\rho_{nir} + \rho_{swir}) \tag{4}$$

where ρ_{red}, ρ_{nir}, and ρ_{swir} are the reflection value in the red spectrum, the near-infrared spectrum, and the shortwave infrared spectrum, respectively. It is noteworthy that the contribution of soil to NDMI is primarily negative for some areas, while the contribution of green vegetation is mostly positive [81].

2.3.5. SM Retrieval Algorithm Construction

We collected 129 datasets (in situ data, and corresponding $\Delta\sigma$, NDVI, NDMI) during the thawing season (months July and August) from 2016 to 2019. A multiple linear regression model was constructed based on the linear relationship between SM and the $\Delta\sigma$, NDVI, and NDMI, and the SM retrieval algorithm can be expressed as follows:

$$SM = a \times \triangle\sigma + b \times NDVI + c \times NDMI + d \tag{5}$$

where a, b, and c are the coefficients of the three variables ($\triangle\sigma$, NDVI, NDMI), respectively, and d is a constant. In order to ensure the universality of the retrieval algorithm, we arrange the in situ data of different years together and then performed 10,000 random divisions with a ratio of nearly 8 to 2 for determining the optimal coefficient. One part is used to obtain model coefficients (a, b, c, d). The other is used to verify the accuracy of retrieved SM. Thus, we can obtain 10,000 sets of coefficients, training, and validation of R^2. Finally, we calculate the sum of R^2 for training and validation processes using their sample size as the weights. The coefficients corresponding to the maximum sum of R^2 are determined as the optimal coefficient.

2.3.6. SM Result Post-Processing

There are some outlier regions in the retrieval results, which are removed in the post-processing steps.

- Waterbody masking

The sensitivity of the σ° to soil liquid water is its advantage of SM retrieval, but σ° will present an anomaly and deviate from the normal range when the sensor scans water bodies such as rivers and lakes. The normalized difference water index (NDWI) is calculated using the green band, and the near-infrared band can effectively identify the water body information [82]. The NDWI is calculated as follows:

$$NDWI = (\rho_{green} - \rho_{nir})/(\rho_{green} + \rho_{nir}) \tag{6}$$

In Equation (6), ρ_{green} is the reflection in the green spectrum, corresponding to the B3 band of the S2. Then, a mask is created with 0 as the threshold to remove the water body part of the retrieved results.

- Shadow masking

It is found that the σ° in the hillside or foothills usually shows outliers in our study area due to the radar signal being obscured and distorted in these areas. The local incidence angle (θ) can be calculated by using the zenith angle and azimuth angle of S1 to represent the illumination condition of the radar signal, expressed in Equation (7),

$$\cos\theta = \cos\theta_{za} \times \cos S + \sin\theta_{za} \times \sin S \times \cos(\theta_{aa} - A) \tag{7}$$

where θ_{za} is the zenith angle, which is the same angle as θ^0, θ_{aa} is the azimuth angle, S is the slope, and A is the aspect. The pixels with θ smaller than a threshold to be defined are masked out. According to a visual comparison of shadow outliers and local incidence masks, the threshold of 15° is adopted in our study.

- Negative $\Delta\sigma$ masking

Theoretically, there is a positive correlation between the σ° and the SM, and the σ° in the thawing season should be higher than that in winter. Hence, the area where the $\Delta\sigma$ is less than zero is considered abnormal and masked out during the post-processing.

2.3.7. SM Retrieval Algorithm Validation

In all, 29 samples were used to evaluate the accuracy of the retrieved SM using the proposed algorithm. The root-mean-square error (RMSE) and coefficient of determination (R^2) are applied to indicate the accuracy of the SM retrieval result.

3. Results

3.1. Reduce the Effects of Surface Roughness and Vegetation

Figure 5 shows the correlation of σ° and $\Delta\sigma$ with the SM observations. There is no apparent correlation between the original σ° and SM, with a Pearson correlation coefficient (r) of 0.06. However, after subtracting the winter reference σ_w, the r between $\Delta\sigma$ and SM reached 0.76. In Figure 5a, the sensitivity of the radar signal to SM is weakened by the effect of surface roughness. The comparison of SM with σ° and $\Delta\sigma$ has proven that our method is able to reduce the effect of surface roughness and essentially improve the sensitivity of radar signals to SM.

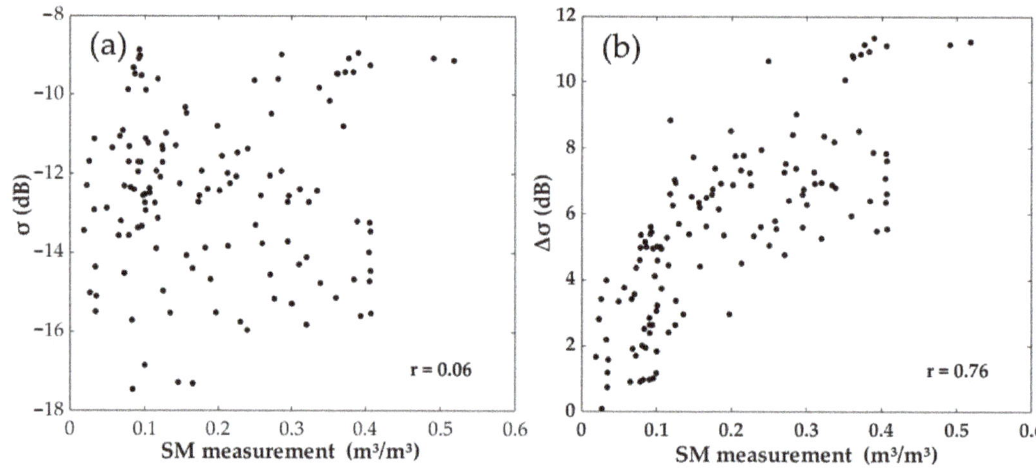

Figure 5. Relationship between the σ° and the SM measurements. (**a**): none-corrected $\sigma^\circ$and SM; (**b**): $\Delta\sigma$ and SM.

NDVI and NDMI jointly characterize the contribution of vegetation to the retrieval of SM. As shown in Figure 6, NDVI and NDMI have high correlations with SM, with r of 0.76 and 0.74, respectively. The results indicate that NDVI and NDMI are suitable for characterizing the vegetation contribution in σ° of the study area.

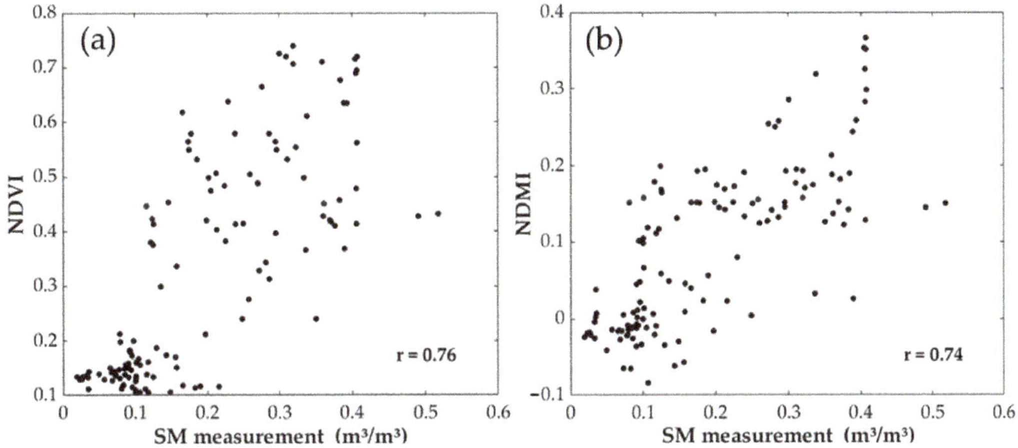

Figure 6. Relationship between the vegetation parameters and the SM measurements at the site. (**a**): NDVI and SM; (**b**): NDMI and SM.

3.2. SM Retrieval Algorithm and Validation

Table 3 lists the model coefficients and R^2 value of 10,000 regressions. The coefficients which generate the highest weighted R^2 were set as the optimal coefficients. The SM retrieval algorithm is expressed as follows:

$$SM = 0.02 \times \Delta\sigma + 0.24 \times NDVI + 0.28 \times NDMI + 0.003 \quad (8)$$

As illustrated in Table 3 and Figure 7, the retrieved result is satisfactory, with R^2 and RMSE reaching 0.82 and 0.07 m^3/m^3, respectively. As shown in Table 3, the mean values of the 10,000 sets of regression coefficients are very close to the optimal values, and the standard deviation is also relatively small. It indicates that the model coefficients are relatively stable, and not largely influenced by different divisions of training and validation samples, which demonstrates the robustness of the model.

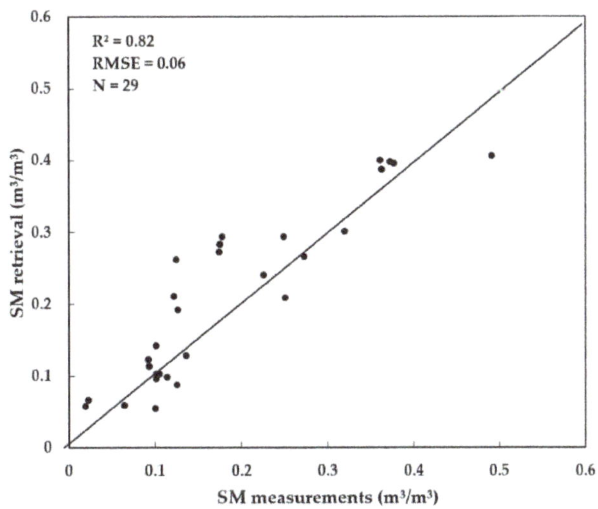

Figure 7. Comparison of SM-retrieved results with measurement data.

Table 3. The optimal coefficients are selected after the regression analysis. Mean is the mean value of each coefficient, STD is the standard deviation, and OPT is the optimal coefficient solution.

	a	b	c	d	R^2
MEAN	0.02	0.23	0.28	0.004	0.81
STD	0.0001	0.02	0.04	0.004	0.004
OPT	0.02	0.24	0.28	0.003	0.82

3.3. Map of Retrieved SM

Figure 8 shows retrieved SM on eight days of S1 acquisitions with the grid spacing of 50 m × 50 m. The white areas in the retrieval results are caused by post-processing, where water bodies, mountain shadows, and anomalous areas are masked. To better show the spatial distribution characteristics of SM, Figure 9 was made by overlaying the spatial distribution map of SM with the topographic map. In hill areas with undulating terrain, SM is usually higher. In order to show the complexity of the spatial distribution of SM in mountainous areas of the permafrost region, this study designed two transect lines to further show the SM variation in the hill areas and extracted the SM values corresponding to the two transect lines, as shown in Figure 10. The variability of SM in hill areas is well presented. It demonstrates the high variability of SM, which could not be revealed by coarse SM products.

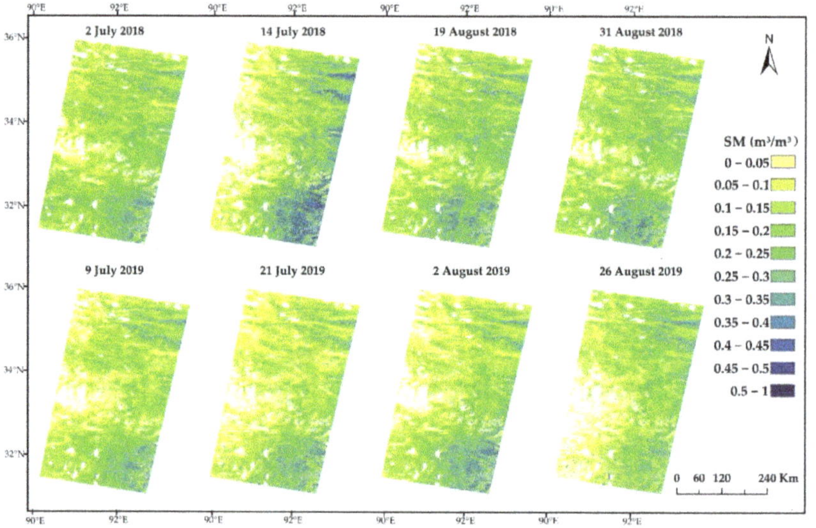

Figure 8. The map of SM retrieval results for thawing seasons 2018 and 2019 after post−processing.

Figure 11 shows the relationship between the spatial and temporal CV and the mean SM in this study area. The CV tends to decrease with increasing mean SM in both spatial and temporal dimensions, which shows that the variability of SM is higher in drier environments and lower in wetter environments. This pattern is related to the water-holding capacity of the soil and its spatial variability [83]. The large differences in the CV in different regions are related to the soil water content, bulk density, and soil texture [22,84,85]. In permafrost areas, the physicochemical properties of soils vary greatly in different areas of topography and vegetation cover, resulting in a high spatial heterogeneity of SM. In addition, the high CV in areas of low SM may also be explained by frequent precipitation and strong evapotranspiration during the thawing season.

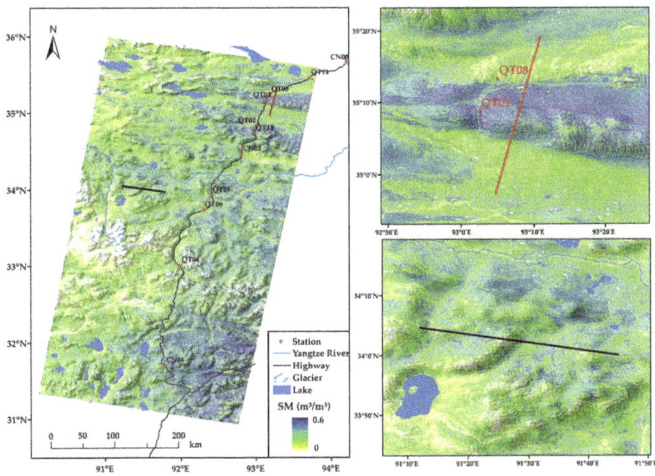

Figure 9. The spatial distribution of the SM and the location of the two transect lines.

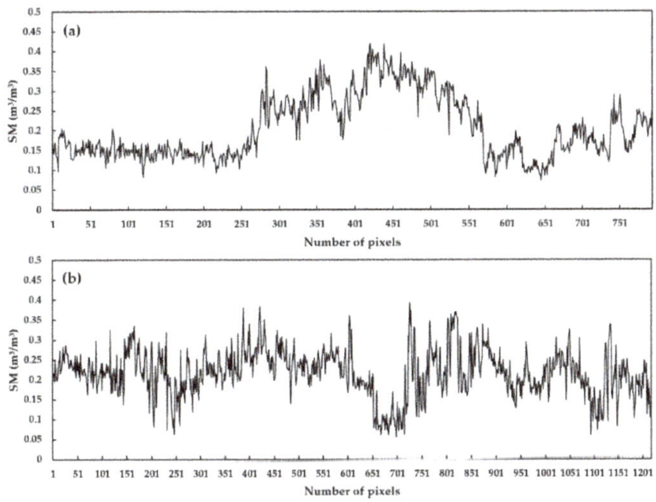

Figure 10. The SM values for the two transect lines. (**a**): Red transect line; (**b**): Black transect line.

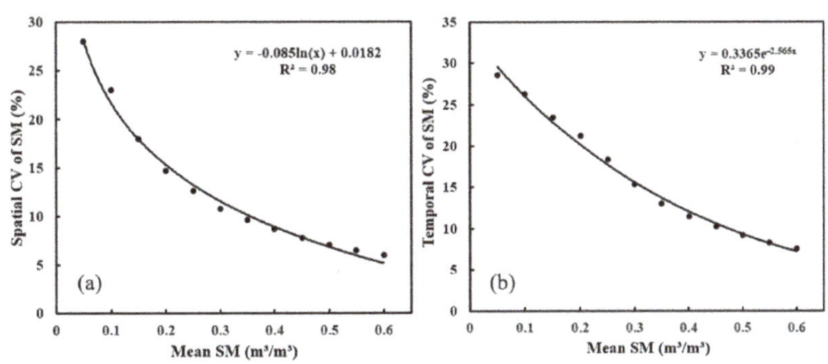

Figure 11. The corresponding CV for each mean SM interval. (**a**): Spatial; (**b**): Temporal.

4. Discussion

4.1. Comparison of S1-Retrieved SM with SM Products

We chose the Tuotuo River basin as an example, where the detailed investigation was conducted, to compare with widely used SM products, such as GLDAS-Noah, ESA CCI, and ERA5-Land (Figure 12). At the local scale, the retrieval results can present the distribution characteristics of SM in different surface environments. The widely used SM data products are unable to characterize the heterogeneity of SM spatial distribution in detail. For example, the GLDAS and CCI can only identify high SM in the southernmost glacial regions, while the distribution of SM in other areas is varied. They can only give a very rough description of the moisture distribution over tens of kilometers limited by its coarse resolution, and the details on SM distribution are lost. In addition, the distribution of the three SM products in this region is also different, which also confirms the demand for high-accuracy SM data in the QTP.

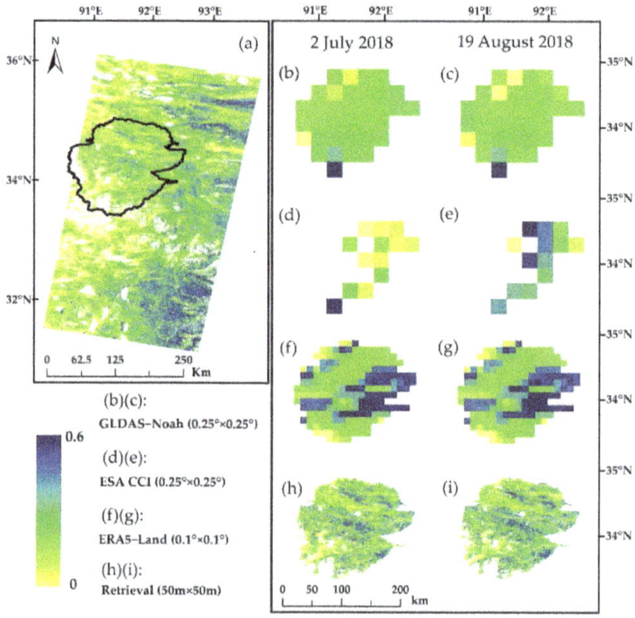

Figure 12. The comparison of SM distribution between retrieved SM and widely used SM products. (**a**) The place circled by the black line indicates the location of the Tuotuo River basin. (**b–i**): The SM distribution in this region with different data. The white areas in the retrieval results are caused by post-processing, where water bodies, mountain shadows, and regions with negative $\Delta\sigma$ are masked. The white areas of SM products are caused by the lack of effective data in QTP.

Figure 13 shows the statistics of SM distribution in the study area on 2 July 2018, from three SM products and S1 retrieval results. The average SM content of ESA CCI, ERA5-Land, GLDAS-Noah, and retrieval result is 0.34, 0.5, 0.29, and 0.19, respectively. The upper and lower quartiles of the in situ SM for the thawing season are 0.29 and 0.09, respectively. Compared with the in situ data, the SM values of the three products are significantly overestimated, while the retrieval results are in a reasonable range. In addition, previous research found that the SM data of the ESA CCI product has the best accuracy on the QTP compared with in situ observations, with an r of 0.63 [16]. In terms of accuracy, our SM retrieval results also showed substantially higher accuracy, and r reached 0.9.

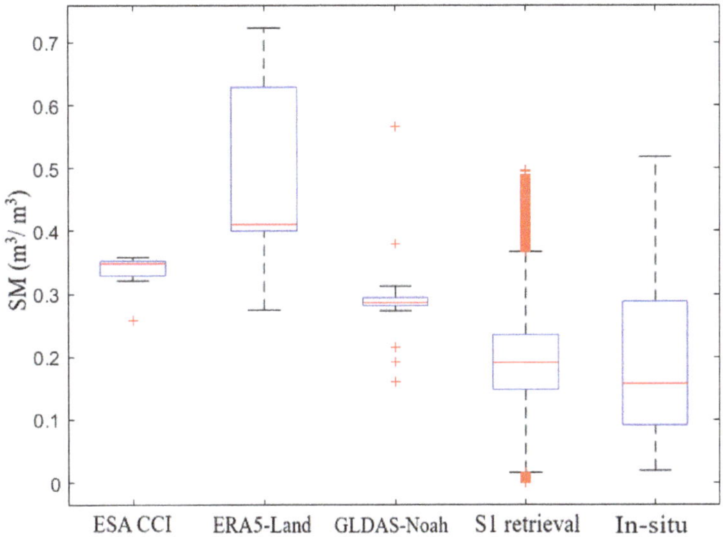

Figure 13. The box plot of SM for the three SM products and S1 retrieval results for the Tuotuo river basins on 2 July 2018. The box line diagram has six parts: lower edge, lower quartile, median, upper quartile, upper edge, and outliers beyond the upper and lower edges.

4.2. SM Distribution Characteristics at the Local Scale

Figure 14 shows the spatial distribution of different vegetation types [86] and SM in the Tuotuo River basin. We summarized the characteristics of SM over different vegetation types, as shown in Figure 15. In areas with high vegetation cover, such as alpine swamp meadows and alpine meadow areas, the SM content is significantly higher than in alpine steppe and alpine desert areas. The average SM content over different vegetation types from high to low is alpine swamp meadow (0.26), alpine meadow (0.23), alpine steppe (0.20), and alpine desert (0.16).

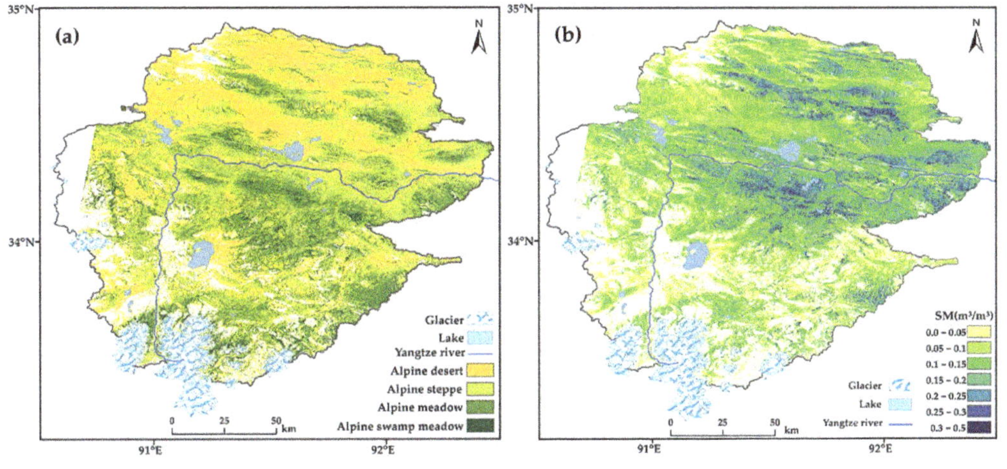

Figure 14. The vegetation types and spatial distribution of SM in the Tuotuo River basin. (**a**): The vegetation type. (**b**): S1retrieved SM.

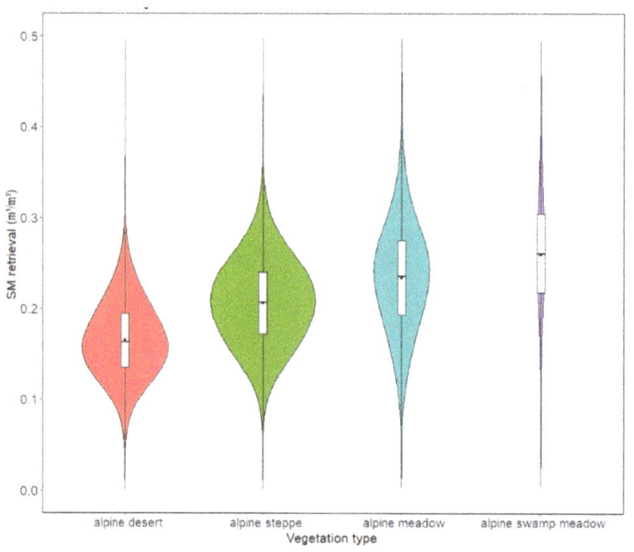

Figure 15. Violin plots of SM content of the four vegetation types. Each violin plot contains a box plot and a kernel density plot. A kernel density plot overlays each box plot. The black marks indicate the median, and the boxes indicate the quarter range of values. The number of samples defines the width of the violin in each subplot.

4.3. Regions with Very Low σ° in the Thawing Season

Normally, the value of σ° in the thawing season is higher than that in winter because the liquid water content in the thawing state is usually higher than in the frozen state [87]. However, during the SM retrieving process, we noticed that σ° during the thawing season in some regions is close to or even lower than σ° in winter. Therefore, we tried to find the reason by comparing the variation of the backscattering coefficients for long time series. As shown in Figure 16a, we consider the regions with significant seasonal variations in the σ° as normal areas, i.e., σ° is higher in summer than in winter, and the disordered areas as abnormal areas. To explain the potential reasons for this phenomenon, we further examined the precipitation, vegetation, and soil texture in these particular regions.

- Precipitation

The precipitation process during the thawing season is one of the main reasons for SM variations [88]. During precipitation events, the wet or flooded ground surface will cause the σ° to deviate from its normal range. We examined the precipitation conditions in three regions as in Figure 16a to test the possibility of this conjecture. The precipitation data is from ERA5-Land precipitation reanalysis data [13], and the temporal curves are shown in Figure 16b. The average annual precipitation of the three regions is 606 mm, 574 mm, and 624 mm, respectively, with little difference in precipitation. Therefore, precipitation is not the cause of low σ° in the thawing season, and the speculation that the wet ground surface causes low σ° is not true. The inference of the abnormal σ° caused by the accumulation of surface water is also not valid.

- Vegetation and soil texture

The impact of vegetation coverage and soil texture on SM content should not be neglected [89–91]. We found some differences between the two regions by examining the temporal changes in the NDVI values in the normal and abnormal regions. As shown in Figure 16c, the NDVI values in the abnormal areas are all relatively low (smaller than 0.1) and do not exhibit seasonal variations. Meanwhile, we referred to the soil texture dataset published by Liu et al. [92,93] and combined it with the field records in the anomaly areas.

We found that the soil in this abnormal area is composed of sand. There is a big chance that these abnormal areas are bare ground and are extremely dry during the particular period in the thawing season, therefore leading to low values of σ°.

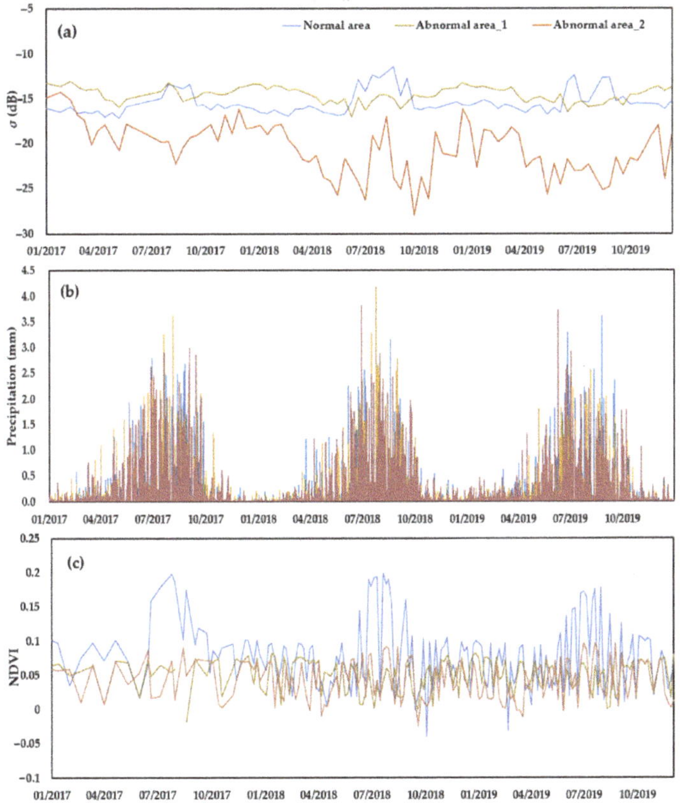

Figure 16. Comparison of normal and abnormal regions. (**a**): σ°; (**b**): Precipitation; (**c**): NDVI.

5. Conclusions

This study developed a concise and practical algorithm for SM estimation using Sentinel-1/2 temporal data in a permafrost environment on the QTP in the thawing season. The R^2 of this SM retrieval algorithm reached 0.82 with an RMSE of 0.06 m^3/m^3.

Our retrieved SM results were compared with current SM products (ERA5-Land, GLDAS-Noah, and ESA CCI) in the Tuotuo River basin and showed that our results have more strength and advantage in characterizing the spatial heterogeneity of SM distribution. By analyzing the SM distribution of different vegetation types, the alpine swamp meadow had the largest SM of 0.26 m^3/m^3, followed by the alpine meadow (0.23), alpine steppe (0.2), and alpine desert (0.16). We also found a significantly negative correlation between the CV and SM in the permafrost area that the variability of SM is higher in drier environments and lower in wetter environments.

The study also explored the reasons for abnormal SM retrievals in some places. The developed algorithm is not applicable in some extremely bare and dry ground with very low SM. Overall, the proposed algorithm shows great potential to derive the detailed SM distribution in the permafrost environment on the entire QTP, which has great significance in studying the SM characteristics in spatial detail and helps facilitate the studies of the response of permafrost to climate change.

Author Contributions: Z.L., L.Z. and L.W. conceived and designed the experiments; Z.L., L.W., D.Z., E.D., Y.X., Z.X., S.L., H.Z. and Y.C. performed the experiments; and Z.L., L.Z., L.W., G.L., G.H., J.Z., C.W., Y.Q. and J.S. analyzed the data and wrote the paper. All of the authors contributed to the editing of the manuscript. All authors have read and agreed to the published version of the manuscript.

Funding: This research was funded by the Second Tibetan Plateau Scientific Expedition and Research Program (2019QZKK0201), the National Natural Science Foundation of China (41931180, 42001054, 42001051), and the Youth Fund for Basic Research Program of Jiangsu Province (BK20200828).

Acknowledgments: We are grateful to the anonymous reviewers and editors for appraising our manuscript and offering instructive comments. We are very grateful to Lei Fan and Xiaojing Bai for their guidance on the experiments of this paper. We also appreciate the free access to data sets from the Chinese Academy of Sciences. In addition, we would like to thank the Google Earth Engine platform for providing the data and processing methods needed in this study.

Conflicts of Interest: The authors declare no conflict of interest.

References

1. Jin, R.; Li, X.; Liu, S. Understanding the heterogeneity of soil moisture and evapotranspiration using multiscale observations from satellites, airborne sensors, and a ground-based observation matrix. *IEEE Geosci. Remote Sens. Lett.* **2017**, *14*, 2132–2136. [CrossRef]
2. Seneviratne, S.I.; Corti, T.; Davin, E.L.; Hirschi, M.; Jaeger, E.B.; Lehner, I.; Orlowsky, B.; Teuling, A.J. Investigating soil moisture–climate interactions in a changing climate: A review. *Earth Sci. Rev.* **2010**, *99*, 125–161. [CrossRef]
3. Assouline, S. Infiltration into soils: Conceptual approaches and solutions. *Water Resour. Res.* **2013**, *49*, 1755–1772. [CrossRef]
4. Crowther, T.W.; Todd-Brown, K.E.; Rowe, C.W.; Wieder, W.R.; Carey, J.C.; Machmuller, M.B.; Snoek, B.; Fang, S.; Zhou, G.; Allison, S.D. Quantifying global soil carbon losses in response to warming. *Nature* **2016**, *540*, 104–108. [CrossRef] [PubMed]
5. Moghaddam, M.A.; Ferre, T.; Chen, X.; Chen, K.; Ehsani, M.R. Application of Machine Learning Methods in Inferring Surface Water Groundwater Exchanges using High Temporal Resolution Temperature Measurements. *arXiv* **2022**, arXiv:2201.00726.
6. Zeng, J.; Li, Z.; Chen, Q.; Bi, H.; Qiu, J.; Zou, P. Evaluation of remotely sensed and reanalysis soil moisture products over the Tibetan Plateau using in-situ observations. *Remote Sens. Environ.* **2015**, *163*, 91–110. [CrossRef]
7. Yi, S.; Wang, X.; Qin, Y.; Xiang, B.; Ding, Y. Responses of alpine grassland on Qinghai–Tibetan plateau to climate warming and permafrost degradation: A modeling perspective. *Environ. Res. Lett.* **2014**, *9*, 074014. [CrossRef]
8. Zhao, L.; Cheng, G.; Li, S.; Zhao, X.; Wang, S. Thawing and freezing processes of active layer in Wudaoliang region of Tibetan Plateau. *Chin. Sci. Bull.* **2000**, *45*, 2181–2187. [CrossRef]
9. Xu, L.; Abbaszadeh, P.; Moradkhani, H.; Chen, N.; Zhang, X. Continental drought monitoring using satellite soil moisture, data assimilation and an integrated drought index. *Remote Sens. Environ.* **2020**, *250*, 112028. [CrossRef]
10. Zou, D.; Zhao, L.; Sheng, Y.; Chen, J.; Hu, G.; Wu, T.; Wu, J.; Xie, C.; Wu, X.; Pang, Q. A new map of permafrost distribution on the Tibetan Plateau. *Cryosphere* **2017**, *11*, 2527–2542. [CrossRef]
11. Zhao, L.; Zou, D.; Hu, G.; Wu, T.; Du, E.; Liu, G.; Xiao, Y.; Li, R.; Pang, Q.; Qiao, Y. A synthesis dataset of permafrost thermal state for the Qinghai–Tibet (Xizang) Plateau, China. *Earth Syst. Sci. Data* **2021**, *13*, 4207–4218. [CrossRef]
12. Zhuang, R.; Zeng, Y.; Manfreda, S.; Su, Z. Quantifying Long-Term Land Surface and Root Zone Soil Moisture over Tibetan Plateau. *Remote Sens.* **2020**, *12*, 509. [CrossRef]
13. Muñoz-Sabater, J.; Dutra, E.; Balsamo, G.; Boussetta, S.; Zsoter, E.; Albergel, C.; Agusti-Panareda, A. ERA5-Land: An improved version of the ERA5 reanalysis land component. In Proceedings of the 8th Workshop-Joint ISWG and LSA-SAF Workshop, Lisbon, Portugal, 26–28 June 2018; pp. 26–28.
14. Dorigo, W.; Wagner, W.; Albergel, C.; Albrecht, F.; Balsamo, G.; Brocca, L.; Chung, D.; Ertl, M.; Forkel, M.; Gruber, A. ESA CCI Soil Moisture for improved Earth system understanding: State-of-the art and future directions. *Remote Sens. Environ.* **2017**, *203*, 185–215. [CrossRef]
15. Rodell, M.; Houser, P.; Jambor, U.; Gottschalck, J.; Mitchell, K.; Meng, C.-J.; Arsenault, K.; Cosgrove, B.; Radakovich, J.; Bosilovich, M. The global land data assimilation system. *Bull. Am. Meteorol. Soc.* **2004**, *85*, 381–394. [CrossRef]
16. Xing, Z.; Fan, L.; Zhao, L.; De Lannoy, G.; Frappart, F.; Peng, J.; Li, X.; Zeng, J.; Al-Yaari, A.; Yang, K. A first assessment of satellite and reanalysis estimates of surface and root-zone soil moisture over the permafrost region of Qinghai-Tibet Plateau. *Remote Sens. Environ.* **2021**, *265*, 112666. [CrossRef]
17. Liu, J.; Chai, L.; Lu, Z.; Liu, S.; Qu, Y.; Geng, D.; Song, Y.; Guan, Y.; Guo, Z.; Wang, J. Evaluation of SMAP, SMOS-IC, FY3B, JAXA, and LPRM Soil moisture products over the Qinghai-Tibet Plateau and Its surrounding areas. *Remote Sens.* **2019**, *11*, 792. [CrossRef]
18. Ivanov, V.Y.; Fatichi, S.; Jenerette, G.D.; Espeleta, J.F.; Troch, P.A.; Huxman, T.E. Hysteresis of soil moisture spatial heterogeneity and the "homogenizing" effect of vegetation. *Water Resour. Res.* **2010**, *46*, W09521. [CrossRef]
19. Srivastava, A.; Saco, P.M.; Rodriguez, J.F.; Kumari, N.; Chun, K.P.; Yetemen, O. The role of landscape morphology on soil moisture variability in semi-arid ecosystems. *Hydrol. Process.* **2021**, *35*, e13990. [CrossRef]

20. Fatichi, S.; Katul, G.G.; Ivanov, V.Y.; Pappas, C.; Paschalis, A.; Consolo, A.; Kim, J.; Burlando, P. Abiotic and biotic controls of soil moisture spatiotemporal variability and the occurrence of hysteresis. *Water Resour. Res.* **2015**, *51*, 3505–3524. [CrossRef]
21. Baroni, G.; Ortuani, B.; Facchi, A.; Gandolfi, C. The role of vegetation and soil properties on the spatio-temporal variability of the surface soil moisture in a maize-cropped field. *J. Hydrol.* **2013**, *489*, 148–159. [CrossRef]
22. Rosenbaum, U.; Bogena, H.R.; Herbst, M.; Huisman, J.A.; Peterson, T.J.; Weuthen, A.; Western, A.W.; Vereecken, H. Seasonal and event dynamics of spatial soil moisture patterns at the small catchment scale. *Water Resour. Res.* **2012**, *48*, W10544. [CrossRef]
23. Tomer, S.K.; Al Bitar, A.; Sekhar, M.; Zribi, M.; Bandyopadhyay, S.; Kerr, Y. MAPSM: A spatio-temporal algorithm for merging soil moisture from active and passive microwave remote sensing. *Remote Sens.* **2016**, *8*, 990. [CrossRef]
24. Torres, R.; Snoeij, P.; Geudtner, D.; Bibby, D.; Davidson, M.; Attema, E.; Potin, P.; Rommen, B.; Floury, N.; Brown, M. GMES Sentinel-1 mission. *Remote Sens. Environ.* **2012**, *120*, 9–24. [CrossRef]
25. Baghdadi, N.; Pedreros, R.; Lenotre, N.; Dewez, T.; Paganini, M. Impact of polarization and incidence of the ASAR sensor on coastline mapping: Example of Gabon. *Int. J. Remote Sens.* **2007**, *28*, 3841–3849. [CrossRef]
26. Zribi, M.; Baghdadi, N.; Holah, N.; Fafin, O. New methodology for soil surface moisture estimation and its application to ENVISAT-ASAR multi-incidence data inversion. *Remote Sens. Environ.* **2005**, *96*, 485–496. [CrossRef]
27. Ulaby, F.T.; Sarabandi, K.; Mcdonald, K.; Whitt, M.; Dobson, M.C. Michigan microwave canopy scattering model. *Int. J. Remote Sens.* **1990**, *11*, 1223–1253. [CrossRef]
28. Chen, K.; Yen, S.; Huang, W. A simple model for retrieving bare soil moisture from radar-scattering coefficients. *Remote Sens. Environ.* **1995**, *54*, 121–126. [CrossRef]
29. Dubois, P.C.; Van Zyl, J.; Engman, T. Measuring soil moisture with imaging radars. *IEEE Trans. Geosci. Remote Sens.* **1995**, *33*, 915–926. [CrossRef]
30. Oh, Y.; Sarabandi, K.; Ulaby, F.T. Semi-empirical model of the ensemble-averaged differential Mueller matrix for microwave backscattering from bare soil surfaces. *IEEE Trans. Geosci. Remote Sens.* **2002**, *40*, 1348–1355. [CrossRef]
31. Fung, A.K.; Shah, M.R.; Tjuatja, S. Numerical simulation of scattering from three-dimensional randomly rough surfaces. *IEEE Trans. Geosci. Remote Sens.* **1994**, *32*, 986–994. [CrossRef]
32. Attema, E.; Ulaby, F.T. Vegetation modeled as a water cloud. *Radio Sci.* **1978**, *13*, 357–364. [CrossRef]
33. Allen, C.; Ulaby, F. Modelling the polarization dependence of the attenuation in vegetation canopies. In Proceedings of the IGARSS, Strasbourg, France, 27–30 August 1984; pp. 119–124.
34. He, B.; Xing, M.; Bai, X. A Synergistic Methodology for Soil Moisture Estimation in an Alpine Prairie Using Radar and Optical Satellite Data. *Remote Sens.* **2014**, *6*, 10966–10985. [CrossRef]
35. Bai, X.; He, B.; Li, X.; Zeng, J.; Wang, X.; Wang, Z.; Zeng, Y.; Su, Z. First Assessment of Sentinel-1A Data for Surface Soil Moisture Estimations Using a Coupled Water Cloud Model and Advanced Integral Equation Model over the Tibetan Plateau. *Remote Sens.* **2017**, *9*, 714. [CrossRef]
36. Yang, M.; Wang, H.; Tong, C.; Zhu, L.; Deng, X.; Deng, J.; Wang, K. Soil moisture retrievals using multi-temporal sentinel-1 data over nagqu region of tibetan plateau. *Remote Sens.* **2021**, *13*, 1913. [CrossRef]
37. Kornelsen, K.C.; Coulibaly, P. Advances in soil moisture retrieval from synthetic aperture radar and hydrological applications. *J. Hydrol.* **2013**, *476*, 460–489. [CrossRef]
38. Zhu, L.; Walker, J.P.; Tsang, L.; Huang, H.; Ye, N.; Rüdiger, C. A multi-frequency framework for soil moisture retrieval from time series radar data. *Remote Sens. Environ.* **2019**, *235*, 111433. [CrossRef]
39. Wagner, W.; Lemoine, G.; Rott, H. A method for estimating soil moisture from ERS scatterometer and soil data. *Remote Sens. Environ.* **1999**, *70*, 191–207. [CrossRef]
40. Gao, Q.; Zribi, M.; Escorihuela, M.J.; Baghdadi, N. Synergetic use of Sentinel-1 and Sentinel-2 data for soil moisture mapping at 100 m resolution. *Sensors* **2017**, *17*, 1966. [CrossRef]
41. Paloscia, S.; Pettinato, S.; Santi, E.; Notarnicola, C.; Pasolli, L.; Reppucci, A. Soil moisture mapping using Sentinel-1 images: Algorithm and preliminary validation. *Remote Sens. Environ.* **2013**, *134*, 234–248. [CrossRef]
42. Bauer-Marschallinger, B.; Freeman, V.; Cao, S.; Paulik, C.; Schaufler, S.; Stachl, T.; Modanesi, S.; Massari, C.; Ciabatta, L.; Brocca, L.; et al. Toward Global Soil Moisture Monitoring With Sentinel-1: Harnessing Assets and Overcoming Obstacles. *IEEE Trans. Geosci. Remote Sens.* **2019**, *57*, 520–539. [CrossRef]
43. Zhu, L.; Si, R.; Shen, X.; Walker, J.P. An advanced change detection method for time-series soil moisture retrieval from Sentinel-1. *Remote Sens. Environ.* **2022**, *279*, 113137. [CrossRef]
44. Zhu, L.; Walker, J.P.; Ye, N.; Rüdiger, C. Roughness and vegetation change detection: A pre-processing for soil moisture retrieval from multi-temporal SAR imagery. *Remote Sens. Environ.* **2019**, *225*, 93–106. [CrossRef]
45. Zhang, X.; Zhang, H.; Wang, C.; Tang, Y.; Zhang, B.; Wu, F.; Wang, J.; Zhang, Z. Soil Moisture Estimation based on the Distributed Scatterers Adaptive Filter over the QTP Permafrost Region using Sentinel-1 and High-resolution TerraSAR-X Data. *Int. J. Remote Sens.* **2021**, *42*, 902–928. [CrossRef]
46. Li, Y.; Guan, D.; Zhao, L.; Gu, S.; Zhao, X. Seasonal frozen soil and its effect on vegetation production in Haibei alpine meadow. *J. Glaciol. Geocryol.* **2005**, *27*, 311–319.
47. Li, H.; Liu, F.; Zhang, S.; Zhang, C.; Zhang, C.; Ma, W.; Luo, J. Drying–Wetting Changes of Surface Soil Moisture and the Influencing Factors in Permafrost Regions of the Qinghai-Tibet Plateau, China. *Remote Sens.* **2022**, *14*, 2915. [CrossRef]

48. Bindlish, R.; Barros, A.P. Parameterization of vegetation backscatter in radar-based, soil moisture estimation. *Remote Sens. Environ.* **2001**, *76*, 130–137. [CrossRef]
49. Baghdadi, N.; El Hajj, M.; Zribi, M.; Bousbih, S. Calibration of the water cloud model at C-band for winter crop fields and grasslands. *Remote Sens.* **2017**, *9*, 969. [CrossRef]
50. Zeng, X.; Xing, Y.; Shan, W.; Zhang, Y.; Wang, C. Soil water content retrieval based on Sentinel-1A and Landsat 8 image for Bei'an-Heihe Expressway. *Zhongguo Shengtai Nongye Xuebao/Chin. J. Eco-Agric.* **2017**, *25*, 118–126.
51. Hajj, M.E.; Bégué, A.; Lafrance, B.; Hagolle, O.; Dedieu, G.; Rumeau, M. Relative radiometric normalization and atmospheric correction of a SPOT 5 time series. *Sensors* **2008**, *8*, 2774–2791. [CrossRef]
52. Kumari, N.; Srivastava, A.; Dumka, U.C. A long-term spatiotemporal analysis of vegetation greenness over the himalayan region using google earth engine. *Climate* **2021**, *9*, 109. [CrossRef]
53. El Hajj, M.; Baghdadi, N.; Zribi, M.; Belaud, G.; Cheviron, B.; Courault, D.; Charron, F. Soil moisture retrieval over irrigated grassland using X-band SAR data. *Remote Sens. Environ.* **2016**, *176*, 202–218. [CrossRef]
54. Bao, Y.; Lin, L.; Wu, S.; Deng, K.A.K.; Petropoulos, G.P. Surface soil moisture retrievals over partially vegetated areas from the synergy of Sentinel-1 and Landsat 8 data using a modified water-cloud model. *Int. J. Appl. Earth Obs. Geoinf.* **2018**, *72*, 76–85. [CrossRef]
55. Wigneron, J.P.; Kerr, Y.; Waldteufel, P.; Saleh, K.; Escorihuela, M.J.; Richaume, P.; Ferrazzoli, P.; Rosnay, P.D.; Gurney, R.; Calvet, J.C. L-band Microwave Emission of the Biosphere (L-MEB) Model: Description and calibration against experimental data sets over crop fields. *Remote Sens. Environ.* **2007**, *107*, 639–655. [CrossRef]
56. Zhao, L.; Zou, D.; Hu, G.; Du, E.; Pang, Q.; Xiao, Y.; Li, R.; Sheng, Y.; Wu, X.; Sun, Z. Changing climate and the permafrost environment on the Qinghai–Tibet (Xizang) plateau. *Permafr. Periglac. Process.* **2020**, *31*, 396–405. [CrossRef]
57. Lin, Z.; Guojie, H.; Defu, Z.; Xiaodong, W.; Lu, M.; Zhe, S.; Liming, Y.; Huayun, Z.; Shibo, L. Permafrost changes and its effects on hydrological processes on Qinghai-Tibet Plateau. *Bull. Chin. Acad. Sci.* **2019**, *34*, 1233–1246.
58. Cheng, G.; Zhao, L.; Li, R.; Wu, X.; Sheng, Y.; Hu, G.; Zou, D.; Jin, H.; Li, X.; Wu, Q. Characteristic, changes and impacts of permafrost on Qinghai-Tibet Plateau. *Chin. Sci. Bull.* **2019**, *64*, 2783–2795.
59. Wang, Z.W.; Wang, Q.; Zhao, L.; Wu, X.D.; Yue, G.Y.; Zou, D.F.; Nan, Z.T.; Liu, G.Y.; Pang, Q.Q.; Fang, H.B.; et al. Mapping the vegetation distribution of the permafrost zone on the Qinghai-Tibet Plateau. *J. Mt. Sci.* **2016**, *13*, 1035–1046. [CrossRef]
60. Lin, Z.; Burn, C.R.; Niu, F.; Luo, J.; Liu, M.; Yin, G. The thermal regime, including a reversed thermal offset, of arid permafrost sites with variations in vegetation cover density, Wudaoliang Basin, Qinghai-Tibet plateau. *Permafr. Periglac. Process.* **2015**, *26*, 142–159. [CrossRef]
61. Li, R.; Zhao, L.; Ding, Y.; Wu, T.; Xiao, Y.; Du, E.; Liu, G.; Qiao, Y. Temporal and spatial variations of the active layer along the Qinghai-Tibet Highway in a permafrost region. *Chin. Sci. Bull.* **2012**, *57*, 4609–4616. [CrossRef]
62. Shiyin, L.; Wanqin, G.; Junli, X. *The Second Glacier Inventory Dataset of China (Version 1.0) (2006–2011)*; National Tibetan Plateau Data Center: Beijing, China, 2012.
63. Messager, M.L.; Lehner, B.; Grill, G.; Nedeva, I.; Schmitt, O. Estimating the volume and age of water stored in global lakes using a geo-statistical approach. *Nat. Commun.* **2016**, *7*, 13603. [CrossRef]
64. Farr, T.G.; Rosen, P.A.; Ro, E.C.; Crippen, R.; Duren, R.; Hensley, S.; Kobrick, M.; Paller, M.; Rodriguez, E.; Roth, L. The Shuttle Radar Topography Mission. *Rev. Geophys.* **2007**, *45*, 361. [CrossRef]
65. Bai, X.; He, B. Potential of Dubois model for soil moisture retrieval in prairie areas using SAR and optical data. *Int. J. Remote Sens.* **2015**, *36*, 5737–5753. [CrossRef]
66. Amazirh, A.; Merlin, O.; Er-Raki, S.; Gao, Q.; Rivalland, V.; Malbeteau, Y.; Khabba, S.; Escorihuela, M. Retrieving surface soil moisture at high spatio-temporal resolution from a synergy between Sentinel-1 radar and Landsat thermal data: A study case over bare soil. *Remote Sens. Environ.* **2018**, *211*, 321–337. [CrossRef]
67. Bousbih, S.; Zribi, M.; Lili-Chabaane, Z.; Baghdadi, N.; Mougenot, B. Potential of Sentinel-1 Radar Data for the Assessment of Soil and Cereal Cover Parameters. *Sensors* **2017**, *17*, 2617. [CrossRef]
68. Dabrowska-Zielinska, K.; Musial, J.; Malinska, A.; Budzynska, M.; Gurdak, R.; Kiryla, W.; Bartold, M.; Grzybowski, P. Soil moisture in the Biebrza Wetlands retrieved from Sentinel-1 imagery. *Remote Sens.* **2018**, *10*, 1979. [CrossRef]
69. Li, J.; Roy, D.P. A global analysis of Sentinel-2A, Sentinel-2B and Landsat-8 data revisit intervals and implications for terrestrial monitoring. *Remote Sens.* **2017**, *9*, 902. [CrossRef]
70. Drusch, M.; Del Bello, U.; Carlier, S.; Colin, O.; Fernandez, V.; Gascon, F.; Hoersch, B.; Isola, C.; Laberinti, P.; Martimort, P. Sentinel-2: ESA's optical high-resolution mission for GMES operational services. *Remote Sens. Environ.* **2012**, *120*, 25–36. [CrossRef]
71. Srivastava, A.; Rodriguez, J.F.; Saco, P.M.; Kumari, N.; Yetemen, O. Global Analysis of Atmospheric Transmissivity Using Cloud Cover, Aridity and Flux Network Datasets. *Remote Sens.* **2021**, *13*, 1716. [CrossRef]
72. Gorelick, N.; Hancher, M.; Dixon, M.; Ilyushchenko, S.; Thau, D.; Moore, R. Google Earth Engine: Planetary-scale geospatial analysis for everyone. *Remote Sens. Environ.* **2017**, *202*, 18–27. [CrossRef]
73. Baghdadi, N.; Zribi, M.; Loumagne, C.; Ansart, P.; Anguela, T.P. Analysis of TerraSAR-X data and their sensitivity to soil surface parameters over bare agricultural fields. *Remote Sens. Environ.* **2008**, *112*, 4370–4379. [CrossRef]
74. Pathe, C.; Wagner, W.; Sabel, D.; Doubkova, M.; Basara, J.B. Using ENVISAT ASAR global mode data for surface soil moisture retrieval over Oklahoma, USA. *IEEE Trans. Geosci. Remote Sens.* **2009**, *47*, 468–480. [CrossRef]

75. He, L.; Qin, Q.; Ren, H.; Du, J.; Meng, J.; Du, C. Soil moisture retrieval using multi-temporal Sentinel-1 SAR data in agricultural areas. *Trans. Chin. Soc. Agric. Eng.* **2016**, *32*, 142–148.
76. Lee, J.-S.; Grunes, M.R.; De Grandi, G. Polarimetric SAR speckle filtering and its implication for classification. *IEEE Trans. Geosci. Remote Sens.* **1999**, *37*, 2363–2373.
77. Zhang, Y.; Gong, J.; Sun, K.; Yin, J.; Chen, X. Estimation of soil moisture index using multi-temporal Sentinel-1 images over Poyang Lake ungauged zone. *Remote Sens.* **2017**, *10*, 12. [CrossRef]
78. Wagner, W.; Lemoine, G. A study of vegetation cover effects on ERS scatterometer data. *IEEE Trans. Geosci. Remote Sens.* **1999**, *37*, 938–948. [CrossRef]
79. Pasolli, L.; Notarnicola, C.; Bertoldi, G.; Bruzzone, L.; Remelgado, R.; Greifeneder, F.; Niedrist, G.; Della Chiesa, S.; Tappeiner, U.; Zebisch, M. Estimation of soil moisture in mountain areas using SVR technique applied to multiscale active radar images at C-band. *IEEE J. Sel. Top. Appl. Earth Obs. Remote Sens.* **2015**, *8*, 262–283. [CrossRef]
80. Rignot, E.J.; Van Zyl, J.J. Change detection techniques for ERS-1 SAR data. *IEEE Trans. Geosci. Remote Sens.* **1993**, *31*, 896–906. [CrossRef]
81. Gao, B. NDWI-a normalized difference water index for remote sensing of vegetation liquid water from space. *Remote Sens. Environ.* **1996**, *58*, 257–266. [CrossRef]
82. McFeeters, S.K. The use of the Normalized Difference Water Index (NDWI) in the delineation of open water features. *Int. J. Remote Sens.* **1996**, *17*, 1425–1432. [CrossRef]
83. Brocca, L.; Morbidelli, R.; Melone, F.; Moramarco, T. Soil moisture spatial variability in experimental areas of central Italy. *J. Hydrol.* **2007**, *333*, 356–373. [CrossRef]
84. Li, T.; Hao, X.; Kang, S. Spatiotemporal variability of soil moisture as affected by soil properties during irrigation cycles. *Soil Sci. Soc. Am. J.* **2014**, *78*, 598–608. [CrossRef]
85. Tomer, M.; Anderson, J. Variation of soil water storage across a sand plain hillslope. *Soil Sci. Soc. Am. J.* **1995**, *59*, 1091–1100. [CrossRef]
86. Zou, D.; Zhao, L.; Liu, G.; Du, E.; Hu, G.; Li, Z.; Wu, T.; Wu, X.; Chen, J. Vegetation Mapping in the Permafrost Region: A Case Study on the Central Qinghai-Tibet Plateau. *Remote Sens.* **2022**, *14*, 232. [CrossRef]
87. Ulaby, F.T.; Kouyate, F.; Brisco, B.; Williams, T.L. Textural Infornation in SAR Images. *IEEE Trans. Geosci. Remote Sens.* **1986**, *24*, 235–245. [CrossRef]
88. Qiu, Y.; Fu, B.; Wang, J.; Chen, L. Spatiotemporal prediction of soil moisture content using multiple-linear regression in a small catchment of the Loess Plateau, China. *Catena* **2003**, *54*, 173–195. [CrossRef]
89. Hu, W.; Si, B.C. Estimating spatially distributed soil water content at small watershed scales based on decomposition of temporal anomaly and time stability analysis. *Hydrol. Earth Syst. Sci.* **2016**, *20*, 571–587. [CrossRef]
90. Francis, C.F.; Thornes, J.B.; Diaz, A.R.; Bermudez, F.L.; Fisher, G.C. Topographic control of soil moisture, vegetation cover and land degradation in a moisture stressed mediterranean environment. *Catena* **1986**, *13*, 211–225. [CrossRef]
91. Mohanty, B.P.; Skaggs, T.H. Spatio-temporal evolution and time-stable characteristics of soil moisture within remote sensing footprints with varying soil, slope, and vegetation–ScienceDirect. *Adv. Water Resour.* **2001**, *24*, 1051–1067. [CrossRef]
92. Liu, F.; Wu, H.; Zhao, Y.; Li, D.; Yang, J.-L.; Song, X.; Shi, Z.; Zhu, A.-X.; Zhang, G.-L. Mapping high resolution National Soil Information Grids of China. *Sci. Bull.* **2021**, *67*, 328–340. [CrossRef]
93. Liu, F.; Zhang, G.L.; Song, X.; Li, D.; Yang, F. High-resolution and three-dimensional mapping of soil texture of China. *Geoderma* **2019**, *361*, 114061. [CrossRef]

remote sensing

Article

Precipitable Water Vapor and Fractional Clear Sky Statistics within the Big Telescope Alt-Azimuthal Region

Artem Yu. Shikhovtsev [1,*], Pavel G. Kovadlo [1], Vladimir B. Khaikin [2] and Alexander V. Kiselev [1]

1 Institute of Solar-Terrestrial Physics, the Siberian Branch of the Russian Academy of Sciences, 664033 Irkutsk, Russia
2 Special Astrophysical Observatory of the Russian Academy of Sciences, Nizhnij Arkhyz, Zelenchukskiy Region, 369167 Karachai-Cherkessian, Russia
* Correspondence: ashikhovtsev@iszf.irk.ru; Tel.: +7-908-6464257

Abstract: The development of a network of ground-based telescopes requires detailed astroclimatic studies. This paper presents the spatial distributions of precipitable water vapor (PWV), total cloud cover (TCC) and cloud base height (CBH). With the aim of a representative description of the precipitable water vapor, a method for correcting this characteristic which takes into account the underlying surface is proposed. The method uses the exponential decrease in the water vapor content with the altitude and is based on the calculation of the averaged elevation of the grid nodes around the site. By applying this correction method, the seasonal changes in the median PWV values at the sites of Ali, Muztag-Ata and Suffa, as well as within the Chajnantor area are estimated. We show that the decrease of PWV with the altitude is exponential with a height scale of 1000 m for the sites in South America and Eurasia. The astroclimatic characteristics within the Big Telescope Alt-azimuthal (BTA) region (40°N–50°N; 35°E–55°E) are estimated. In this region, the sites suitable for the millimeter and submillimeter (mm/submm) observations are revealed. New sites are Mt. Horai and Mt. Kurapdag. In addition, we show that the Era-5 reanalysis data overestimate the PWV values by 1–2 mm and describe changes in the monthly medians of PWV. Comparison of the calculated medians with the measured PWV show that the correlation coefficient between these characteristics is 0.97.

Keywords: precipitable water vapor; clear sky; millimeter telescope; Ali; Muztag-Ata; BTA

Citation: Shikhovtsev, A.Y.; Kovadlo, P.G.; Khaikin, V.B.; Kiselev, A.V. Precipitable Water Vapor and Fractional Clear Sky Statistics within the Big Telescope Alt-Azimuthal Region. *Remote Sens.* **2022**, *14*, 6221. https://doi.org/10.3390/rs14246221

Academic Editors: Massimo Menenti, Yaoming Ma, Li Jia and Lei Zhong

Received: 11 October 2022
Accepted: 7 December 2022
Published: 8 December 2022

1. Introduction

Astronomical observations in the millimeter/submillimeter (mm/submm) range of the electromagnetic spectrum are limited by the Earth's atmosphere. In this spectral range, the Earth's atmosphere causes a significant attenuation of radiation. One of the main atmospheric gases that is responsible for attenuating radiation in the atmosphere is water vapor. In particular, precipitable water vapor (PWV) is often used for the site characterization. The amount of precipitable water vapor corresponds to the height of the water layer (in mm or cm), which would be formed by complete condensation of the water vapor from a column of atmospheric air with the area of 1 cm^2. With a decrease in the content of water vapor, which is distributed along the telescope's line of sight, transparency of the atmosphere increases.

A number of areas and sites with low water vapor content suitable for mm/submm astronomical observations are known in the world. For example, one of the areas with low water vapor content is the Chajnantor plateau in the Chilean Andes at ~5000 m elevation above sea level, where the Atacama Large Millimeter/Submillimeter Array (ALMA) is located [49]. Table 1 lists medians of precipitable water vapor at the sites with good astroclimatic conditions [2–6]. In particular, the measurement data show that the medians of PWV at the Chajnantor Plateau are 2.56 mm and 0.72 mm in January and August, respectively. At the Chajnantor summit, the medians of PWV decrease significantly, to 1.08 mm in January and 0.54 mm in August.

Table 1. Median values of precipitable water vapor at the sites with good astroclimatic conditions [2–6].

-			mm
Chajnantor Plateau	0.72 (August)	2.56 (January)	1.05
Cerro Chajnantor summit	0.54 (August)	1.08 (January)	0.67
Dome A	-	-	0.21
Dome C	-	-	0.28
South Pole	-	-	0.30
Cerro Macon	-	-	1.02
Mauna Kea	-	-	1.44
Mauna Loa	-	-	2.00
Karakaya Hills, Erzurum	2.7 (October–June)	4.0 (January–September)	2.7
Muztag -Ata	~1.0 (December–January)	~7.0 (July)	2.3
Hanle	~1.0 (December–January)	-	2.23
Merak	-	-	2.16
Ali	-	-	2.22
Salt	-	-	1.31
Yang	-	-	1.37

A pronounced minimum of PWV is observed in Antarctica. Year-round median values of the PWV at the sites of Dome A, Dome C and the South Pole are listed in Table 1. Due to low air temperatures, the sites of Dome A, Dome C [7,8] and the South Pole are characterized by very low content of water vapor. At these sites, annual medians of PWV do not exceed 0.30 mm.

For comparison, Table 1 lists the medians of PWV for sites, such as Cerro Macon and Mauna Kea [2], which are characterized by very high astroclimatic parameters. The medians at these sites are slightly higher than in Antarctica and equal to 1.02 mm and 1.44 mm at Cerro Macon and Mauna Kea, respectively.

In addition, one of the sites with low and moderate PWV is the Karakaya Hills in Erzurum city. An analysis of the radiosonde data at this site (located at 3170 m above sea level) showed that the median of PWV is 2.7 mm for the period 6 October 2016–15 January 2017. The median increased to 4.00 mm for 1 January 2016–26 September 2017 [6].

In the northern hemisphere, the Tibet Plateau is an area suitable for millimeter and submillimeter observations [9,10]. In particular, Zhao et al. [11] showed that the Lenghu site has very low PWV, ~1 mm. Amplitudes of intraday PWV fluctuations do not exceed 0.5 mm. In addition, Shantikumar S. Ningombam et al. studied the driest locations in Tibet and compared them with well-known astronomical sites in the world [12,13]. The authors highlighted three sites with high astroclimatic characteristics, including Hanle, Merak and Ali sites within the Himalayan region. The year-round medians of PWV at the Himalayan sites varied from 2.16 mm (Merak) to 2.23 mm (Hanle). Another promising site is Muztag -Ata. Xu et al. found that the annual median of PWV was 2.28 mm and 2.37 mm in 2018 and 2019, respectively [5]. At this site, the driest period with the median less than 2 mm is observed from October to March.

Although some sites with low water vapor content are known, further detailed astroclimatic studies are required for developing a ground-based network of mm/submm telescopes [14,15].

PWV is often estimated using radiosonde data [16–20]. However, the rather sparse network of stations, especially in mountainous regions, as well as the rare frequency of launching radiosondes, do not allow for detailed studies. Other possibilities for estimation of PWV are based on use of GPS data, GNSS data or MODIS observations [21,22].

The most optimal approach for preliminary estimation of PWV within a large region is use of reanalysis data [23–25]. The reanalysis data makes it possible to evaluate statistically representative atmospheric parameters including not only PWV, but also other characteristics relevant for astronomical observations (for example, total cloud cover, wind speed, air temperature, atmospheric boundary height, seeing, the Richardson parameter) [26–28]. Comparison between PWV changes estimated with the help of satellite and ground-based instruments and reanalysis-derived PWV changes shows a close relationship. For example, Kumar et al. found that at the Varanasi the correlation coefficient between the GPS-derived PWV and the PWV estimated from the Era-Interium reanalysis is close to 0.86 [29]. An analysis of the MODIS-PWV and the Era-5 derived PWV showed that the correlation coefficients are above 0.9 over most of China [22]. The area-averaged correlation coefficient is $\sim$0.98. The low correlation coefficients are observed near the southwestern border of China ($\sim$0.62).

Long-period changes in PWV show that the Mediterranean coasts, the Alps and northern and inland Europe are characterized by large correlation coefficients (more than 0.97) [30]. The root mean square deviations estimated between changes in the Era-5 PWV and the GPS-PWV range from 1.0 to 1.6 kg m^{-2} for the Mediterranean coasts and the Alps and from 0.5 to 1.0 kg m^{-2} for northern and inland Europe. In addition, the studies over Russia showed that MODIS data overestimates the atmospheric optical depth associated with PWV compared to the data of ground-based actinometric observations [31].

This paper is a continuation of our astroclimatic studies in the application of optical, millimeter and submillimeter astronomy [32–34]. We discuss the results of a quantitative analysis of atmospheric characteristics, including mean values, medians, mean absolute error, root mean square deviation and correlation coefficient [35–38]. The study is being carried out to determine the best sites for ground-based mm/submm telescopes as part of the ESMT Submillimeter Teselscopes (ESMT) project [34,39]. Within this project, it is planned to design three mm/submm telescopes in Russia, Uzbekistan and China. We should also note that our studies cover locations with low values of atmospheric diffuse light ranging from 8.3 to 13% (Figure 1). These locations are Baykal Astrophysical Observatory (BAO), Large Solar Telescope (LST-3), Big Telescope Alt-Azimuthal, Crimean Astrophysical Observatory and the Hapcheranga site [34,39].

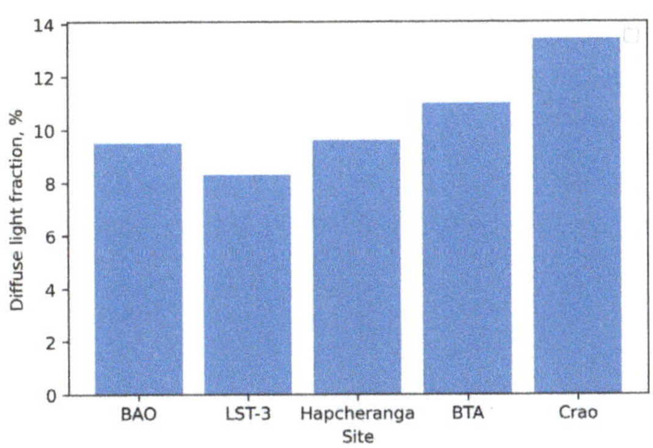

Figure 1. Diffuse light at astronomical observatories and sites suitable for observations.

2. Data Method to Correct Precipitable Water Vapor Values

In order to estimate spatial distributions of precipitable water vapor, we used the Era-5 reanalysis database [40]. The Era-5 reanalysis data are available with hourly resolution. The horizontal resolution is 0.25°. In our study, we used isobaric levels from 1000 hPa to 30 hPa. Each isobaric level corresponds to a certain altitude in the atmosphere.

One of the approaches used for estimating PWV is integration of specific air humidity vertical profiles. In particular, the PWV is calculated using the formula:

$$PWV = \frac{1}{\rho g} \int_{P_s}^{P_{hr}} q(P) dP, \tag{1}$$

where ρ is the water density (1000 kg/m^3), g is the gravitational acceleration (m/s^2), $q(P)$ is the air specific humidity (g/kg), P is the atmospheric pressure (hPa), P_s is the air pressure near the Earth's surface (hPa), which defines the lower integration limit, and P_{hr} is the air pressure at the altitude corresponding to the upper integration limit (hPa). The upper integration limit is chosen quite arbitrarily. The arbitrary choice of the upper limit is due to low content of water vapor in the upper layers of the atmosphere. At an altitude of 1.5–2 km, the water vapor pressure is half that at the Earth's surface, in average. The troposphere (atmospheric layer, from Earth's surface up to 10–12 km) contains 99% of the atmospheric water vapor. In the calculations, we assumed that the upper limit of integration is 30 hPa.

A significant problem is choice of the lower integration limit, as the lower part of the atmosphere (so-called dynamic atmospheric boundary layer) makes the largest contribution to PWV and, thereby, optical thickness. Radiosonde data and reanalysis make it possible to estimate PWV separately by integrating the vertical profiles in air humidity, as well as to estimate the lower integration limit at the reference radiosounding station. For choosing the lower integration limit, we use either surface air pressure measurements or the approximation of the standard atmosphere.

Another approach applied for estimation of atmospheric water vapor statistics is based on the hourly values of PWV, which are available in the Era-5 reanalysis database. Direct use of the reanalysis data leads to overestimated values of PWV, especially for mountainous areas. The input data in the reanalysis corresponds to a certain volume box limited by grid node. In this case, we can note that PWV values must be corrected. The Era-5 reanalysis data, which include PWV values, profiles of meteorological values and modeled relief, have a limited spatial resolution. PWV values are calculated for grid nodes that may not contain mountain peaks but include lower altitudes. A site suitable for millimeter and submillimeter observations can be remote from the nearest grid node. So, deviations in the PWV values may appear. To take into account the relief, we use a well-known Formula (2) containing an exponential term and a relative altitude difference δz:

$$PWV = K_{ref} PWV_0 exp\left(\frac{< C_{PWV} > \delta z}{h_l}\right) = K_0 PWV_0, \tag{2}$$

$$K_0 = K_{ref} exp\left(\frac{< C_{PWV} > \delta z}{h_l}\right) \sim K_{ref} exp\left(\frac{\delta z}{H_l}\right), \tag{3}$$

where $< C_{PWV} >$ is the non-dimensional proportionality coefficient, $H_l = h_l/0.439$ is the water vapor scale height, and h_l = 1000 m. A flowchart of the method for correcting PWV is shown in Figure 2.

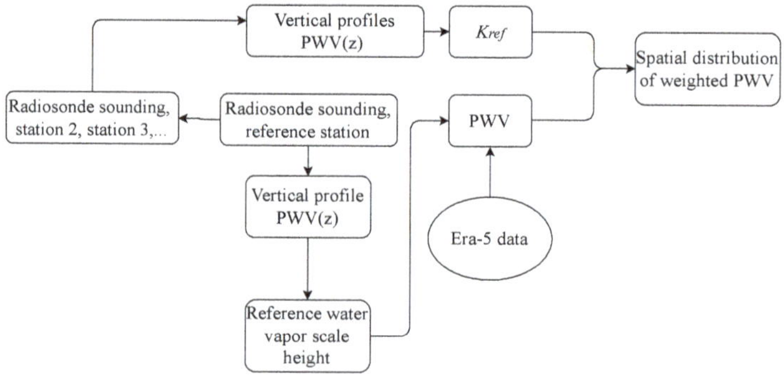

Figure 2. Flowchart of the method for correcting PWV. The values of PWV are computed using the radiosonde sounding at the nearest reference station. Additional radiosonde stations provide correction of PWV distribution within the region.

The water vapor scale height and the proportionality coefficient are calculated from the best exponential fit to the water vapor content profile obtained from radiosonde soundings launched from the nearest station (Mineralny Vody). At a given site, the coefficient $< C_{PWV} >$ and H_l vary over time. Under different atmospheric conditions, the scale height ranges from 1200 to 2700 m at the BTA. The mean value of H_l is 2280 m. The non-dimensional coefficients K_{ref} are calculated for grid nodes by interpolation of H_l estimated for a set of radiosonde stations within a given region. Adaptation of the method for the sites where the statistics of PWV are known shows that valid calculation of the δz value should take into account the limited area and depends on the relief model in the Era-5 reanalysis as well as the physical elevation of a given site. For a correct estimate of PWV at a given site, we calculate the relative altitude difference not for a single grid node but within the area around the site. In other words, PWV values are corrected using variable coefficients for different sites, taking into account the relief model and, in particular, the mean difference between the altitude of the mountain summit and the "surrounding terrain".

3. Precipitable Water Vapor Statistics within the Chajnantor Area

In order to determine how well the Era-5 reanalysis data reproduces real changes in precipitable water vapor, we used the data gathered from radiometer measurements at the Chajnantor Plateau [3]. Using Formula (2), we estimated the mean hourly PWV values within the Chajnantor Plateau. The Era-5-derived precipitable water vapor values are overestimated in comparison with the measurement data. We used the mean absolute error (MAE) for estimation of the PWV deviations assessed between the calculated and measured PWV:

$$MAE = \frac{1}{N}\sum_{i=1}^{N} \mid PWV_i(Era) - PWV_i(r) \mid, \tag{4}$$

where $PWV_i(Era)$ is the mean monthly PWV calculated from the Era-5 reanalysis data, and $PWV_i(r)$ is the mean monthly measured values of PWV. In average, the mean absolute error is 1.8 mm. The largest MAE falls in winter when water vapor content is the highest. The mean absolute error in winter is 3.1 mm. In summer, MAE decreases to 1.1 mm.

A correct estimation of statistical characteristics of precipitable water vapor, including hourly values, requires the use of a certain method, especially under conditions of low atmospheric moisture. Using Formula (2), we calculated the PWV variations within the Chajnantor area. Figure 3 shows the PWV variations for a period of five days within the Chajnantor area.

At the Chajnantor Plateau, measurements were performed with an APEX radiometer operating at 183 GHz [3]. At the Cerro Chajnantor summit, precipitable water vapor was measured by a water vapor radiometer (red line) at 183 GHz and the Tipper radiometer, which integrates radiation over a 103 GHz band, centered at 850 GHz. An analysis of intraday variations in the precipitable water vapor shows that the measured values of PWV significantly deviate from the hourly PWV values estimated from the Era-5 reanalysis data. The mean absolute error is 0.94 mm. We also estimated the root mean square deviation:

$$\sigma = \left(\frac{1}{N}\sum_{i=1}^{N}(PWV_i(Era) - PWV_i(r))^2\right)^{0.5} \tag{5}$$

and the correlation coefficient:

$$K = \frac{1}{N}\frac{\sum_{i=1}^{N}\left(PWV_i(Era) - \overline{PWV(Era)}\right)\left(PWV_i(r) - \overline{PWV(r)}\right)}{\sigma_{PWV(ERA)}\sigma_{PWV(r)}}, \tag{6}$$

where $\overline{PWV(Era)}$ and $\overline{PWV(r)}$ are the Era-5 derived PWV and measured PWV, respectively, and $\sigma_{PWV(ERA)}$ and $\sigma_{PWV(r)}$ are the root mean square deviations.

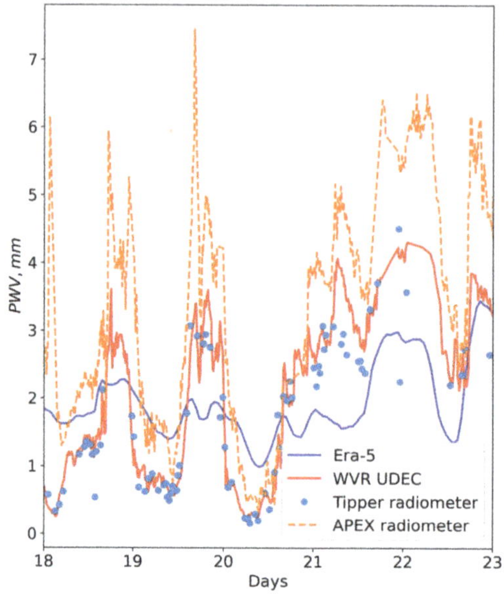

Figure 3. PWV variations for a period of five days in the Chajnantor area for the period 18 December 2011–23 December 2011. The blue line corresponds to the calculated Era-5 PWV variations. The red line and round markers correspond to the PWV variations at the Cerro Chajnantor summit. The orange line corresponds to the PWV variations at the Chajnantor Plateau.

The root mean square deviation of the PWV is 1.06 mm. The correlation coefficient between the PWV variations is 0.57. Thus, the Era-5 reanalysis data passably reflect atmospheric situations with low and high water vapor content in the atmosphere, which are of interest for mm/submm astronomical observations.

For longer averaging periods, consistency of PWV variations estimated from the Era-5 data and the data of the radiometric measurements improves. Figure 4 shows the changes in the mean monthly PWV values estimated from Era-5 data and the changes in the mean monthly measured radiometric PWV values. PWV values are averaged from 2011 to 2014. Statistics of PWV variations estimated from the Era-5 reanalysis data and radiometric data are shown in Table 2. In addition to the mean absolute error estimated for the hourly PWV values, we also estimated the RMS and the correlation coefficient between the PWV time series. Compared to intrahour variations, the mean absolute error of the mean monthly variations in PWV decreases to 0.23 mm, and the RMS decreases to 0.28 mm. The correlation coefficient is 0.97.

Table 2. Statistical characteristics of PWV variations estimated from Era-5 reanalysis data and radiometric measurement data.

Averaging	STD, mm	MAE, mm	K
Mean hourly	1.06	0.94	0.57
Mean monthly	0.28	0.23	0.97

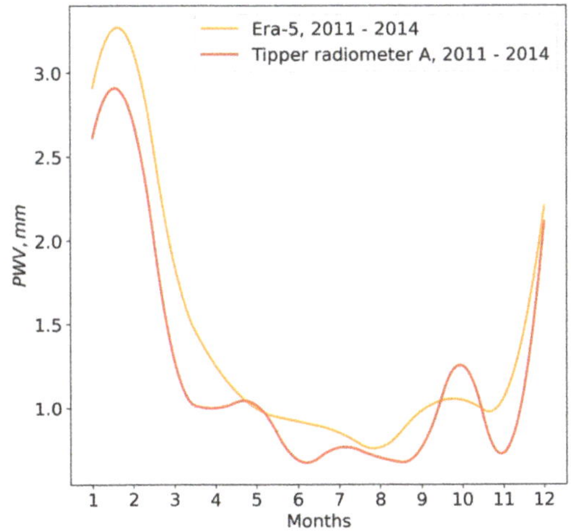

Figure 4. Changes in mean monthly PWV values estimated from Era-5 reanalysis data and changes in mean monthly measured PWV values within the Chajnantor area.

4. Atmospheric Parameter Statistics Relevant for the Millimeter/Submillimeter Observations within the BTA Region

Knowledge of statistics in air humidity and precipitable water vapor makes it possible to estimate absorption coefficients along the telescope's line of sight. In this section, we estimate the spatial changes and seasonal variability in water vapor content in the atmosphere within the BTA region [41]. Using the Era-5 reanalysis data, we obtained the statistically averaged spatial distributions of PWV for different seasons for the period 2010–2020.

4.1. Spatial Distributions of Precipitable Water Vapor within the BTA Region

With the aim of identifying the sites suitable for astronomical observations in the mm/submm wavelength range, we obtained spatial distributions of PWV within the BTA region. We used the Era-5 reanalysis data for mapping. The spatial distributions of PWV averaged for the period 2010–2020 for different seasons are shown in Figure 5. PWV averaging was performed using Formula (7):

$$PWV_0 = 1/N \sum_{i=1}^{N} PWV_{0i}, \tag{7}$$

where $PWV_0 i$ is the hourly mean PWV value, and N is the total number of hours for the period 2010–2020. The analysis of Figure 5 shows that there is a wide area of low PWV values within the BTA region. The average position of the area varies little during the year. However, the PWV values change significantly from season to season. The highest PWV values are observed in summer, while the lowest PWV values correspond to winter.

In our studies, in order to correct the PWV values, we compared the variations in precipitable water vapor estimated from the Era-5 data with the PWV obtained using the Global Navigation Satellite System in the BTA region [39]. The data comparison showed that the correlation coefficients varied from 0.71 to 0.94. The best relation between PWV variations was observed at the sites with altitudes from 727 m to 2095 m, with high repeatability of clear sky. In particular, at the Khunzakh with an elevation of 1627 m above sea level, the correlation coefficient was 0.94. At the Kislovodsk mountain astronomical station with an elevation of 2095 m above sea level, the correlation coefficient is also high (~0.91). The worst correlation coefficient was observed at the Peak Terskol, as the highest site of the analyzed (3121 m above sea level). At this site, the correlation coefficient was

0.71. It is also worth noting that, according to our estimates, the Chajnantor Plateau at an elevation of 5107 m has a correlation coefficient of 0.57.

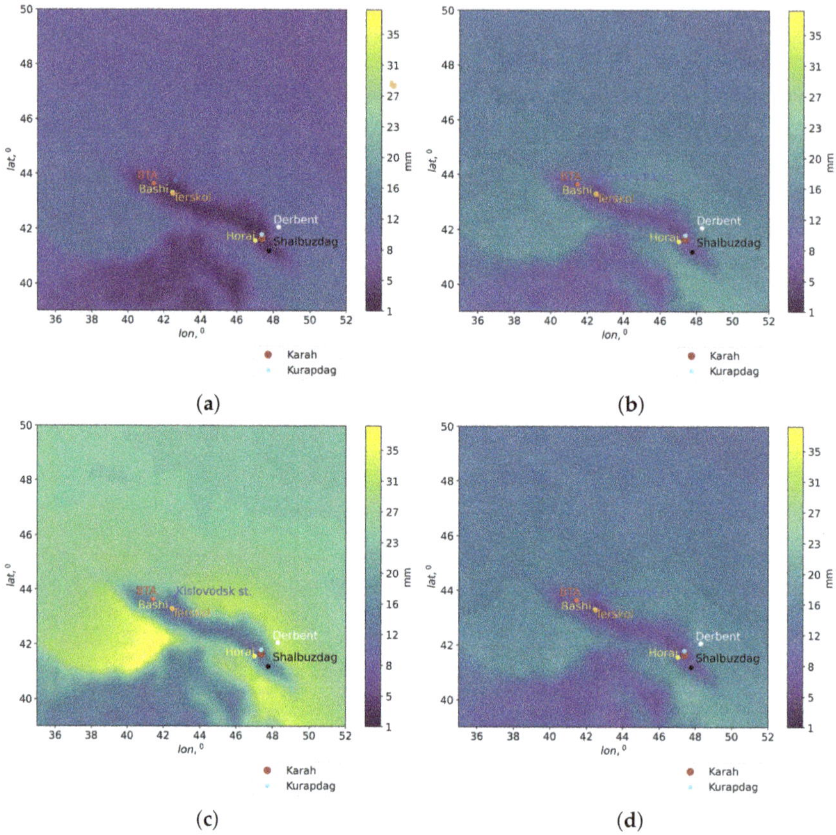

Figure 5. Spatial distributions of PWV_0 obtained from Era-5 reanalysis data for the period 2010–2020: (**a**) winter; (**b**) spring; (**c**) summer; (**d**) autumn.

Using Formula (2) and Era-5 reanalysis data, we estimated the spatial distributions of corrected PWV values for the period 2010–2020. Figure 6 shows the spatial distributions of the corrected PWV values. The correction of PWV values, taking into account the relief, was carried out using the relative altitude difference determined over a large area within the region. In this case, in low landforms, PWV values were not corrected; PWV values for mountain peaks were corrected using Formula (2). In the calculations, we used the Era-5 relief model.

The analysis of the corrected spatial distributions shows that PWV values for mountains decreased significantly. As a result of applying the correction method, PWV values at the summit and at the bottom of the mountain began to differ significantly. This is consistent with physical concepts and measurement data of precipitable water vapor.

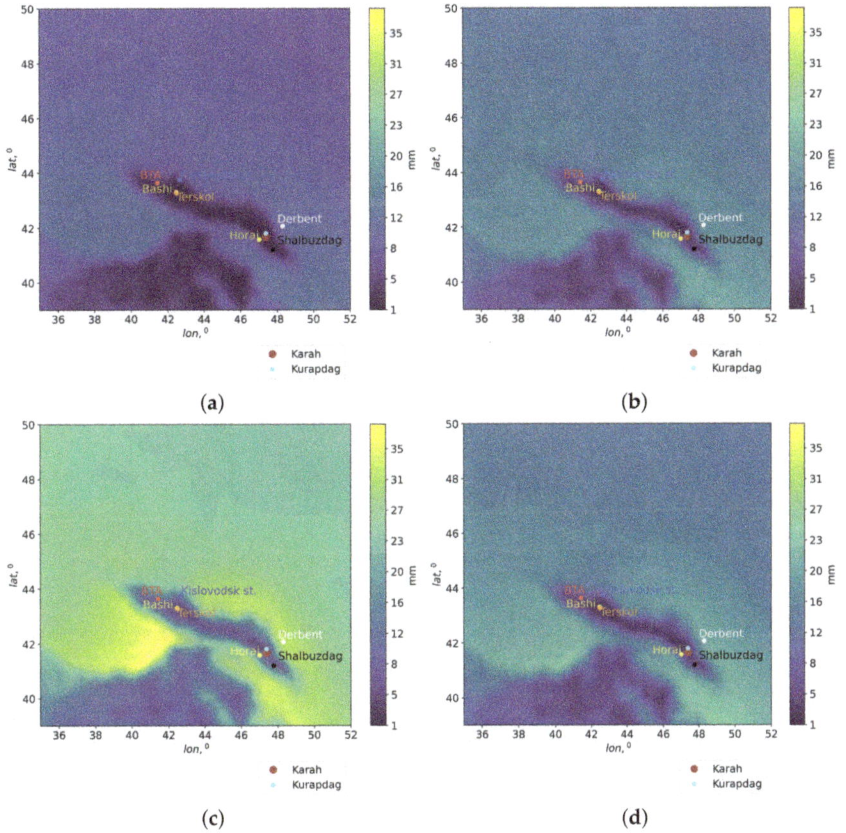

Figure 6. Era-5 PWV spatial distributions obtained for the period 2010–2020, adapted for mountain peaks: (**a**) winter; (**b**) spring; (**c**) summer; (**d**) autumn.

4.2. Spatial Distributions of Total Cloud Cover within the BTA Region

The duration of sunshine or total cloud cover determines the potential duration of astronomical observations. In order to reveal the sites with minimal total cloudiness, we obtained the spatial distributions of total cloud cover TCC (shown in Figure 7). The TCC values in the Era-5 reanalysis were averaged for the period 2010–2020 for different seasons using Formula (8):

$$TCC = 1/N \sum_{i=1}^{N} TCC_i, \tag{8}$$

where TCC_i is the hourly mean value of TCC, and N is the total number of hours from 2010 to 2020.

The analysis of Figure 7 shows that the TCC distributions are spatially inhomogeneous. Areas with low TCC are shown in dark blue. The best conditions in both TCC and PWV occur during winter. The sites with the best atmospheric conditions in TCC and PWV include the locations of mountainous Dagestan: Mt. Kurapdag and Mt. Horai. Table 3 shows mean values of TCC at the sites suitable for mm/submm observations. The best conditions in terms of TCC are noted at the location of Mt. Horai and Mt. Kurapdag.

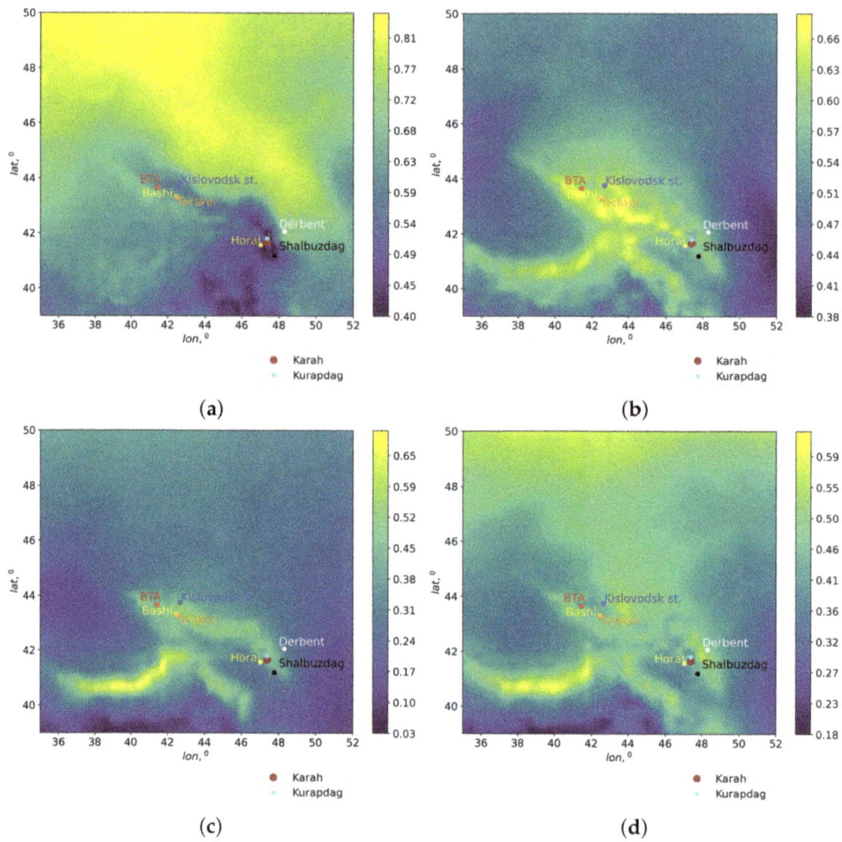

Figure 7. Spatial distributions of total cloud cover within the BTA region obtained for the period 2010–2020: (**a**) winter; (**b**) spring; (**c**) summer; (**d**) autumn.

Table 3. Mean values TCC at the sites suitable for mm/submm observations.

Site	Season	TCC
BTA	Winter	0.62
	Spring	0.65
	Summer	0.47
	Autumn	0.46
Terskol	Winter	0.63
	Spring	0.64
	Summer	0.49
	Autumn	0.47
Horai	Winter	0.44
	Spring	0.55
	Summer	0.33
	Autumn	0.41
Kislovodsk	Winter	0.57
	Spring	0.62
	Summer	0.47
	Autumn	0.46

Table 3. *Cont.*

Site	Season	TCC
Kurapdag	Winter	0.45
	Spring	0.57
	Summer	0.45
	Autumn	0.42

4.3. Spatial Distributions of Cloud Base Height within the BTA Region

For astroclimatic estimation of astronomical observing time within mountainous regions, an important parameter is the cloud base height (CBH). We attribute large values of cloud base height to a higher repeatability of high atmospheric mm/submm transparency. In this case, a high transparency of the atmosphere is due to low water content in the middle and upper troposphere compared to the lower layers of the atmosphere. Based on the Era-5 reanalysis data, we estimated CBH. Spatial distributions of cloud base height for different seasons are shown in Figure 8.

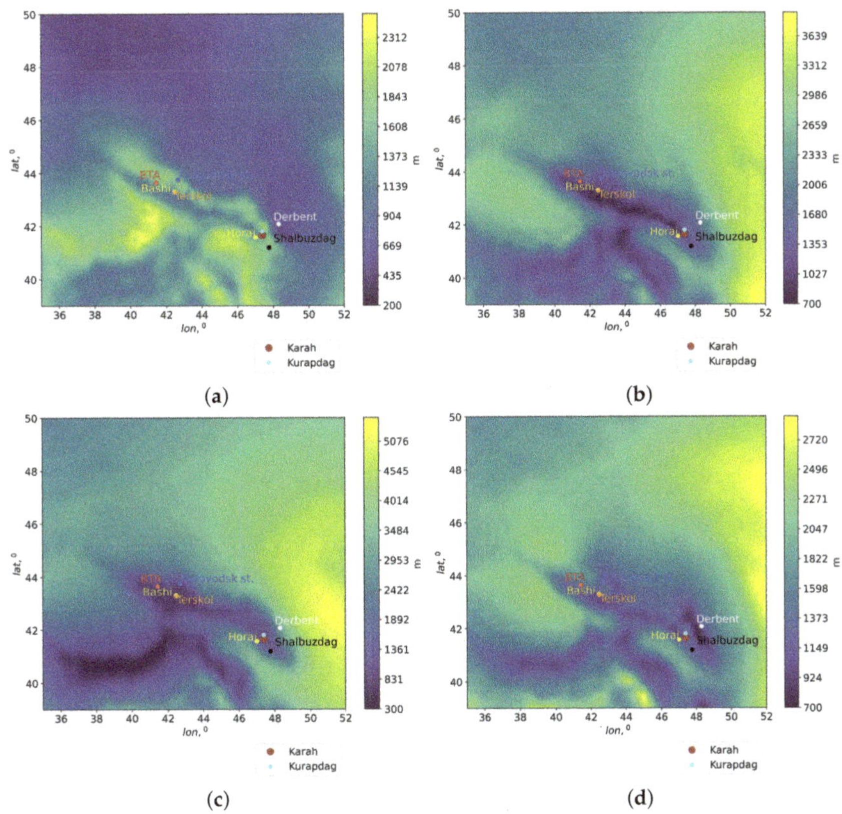

Figure 8. Spatial distributions of cloud base height for different seasons for the period 2010–2020: (**a**) winter; (**b**) spring; (**c**) summer; (**d**) autumn.

The analysis of spatial distributions of cloud base height indicates the following features of cloud formation:

(i) In all seasons, an area with low values of cloud base heights is observed over the Caucasus. The "depth" of this area in terms of CBH horizontal gradients varies throughout the year.

(ii) Excluding winter, an extended area with large CBH values is formed in the eastern part of the BTA region. In this region, maximum CBH occurs in summer. Cloud base heights range from ~3600 to ~5300 m. In winter, cloud base heights range from ~900 m to ~1500 m.

Table 4 shows mean values of CBH at the sites suitable for mm/submm observations. Typical cloud base heights at the BTA site vary from 800 m (summer) to 1500 m (winter). Similar conditions are also observed at the Peak Terskol. The minimum cloud base height values are observed in summer (750 m), and the maximum values correspond to the autumn–winter period (1100–1200 m).

Table 4. Mean values of CBH at the sites suitable for mm/submm observations.

Site	Season	CBH, m
BTA	Winter	1500
	Spring	1200
	Summer	800
	Autumn	1400
Terskol	Winter	1100
	Spring	950
	Summer	750
	Autumn	1200
Horai	Winter	1400
	Spring	1200
	Summer	1600
	Autumn	1200
Kislovodsk	Winter	1700
	Spring	1200
	Summer	900
	Autumn	1350
Kurupdag	Winter	1400
	Spring	1200
	Summer	1050
	Autumn	1100

One of the promising sites, namely Mt. Horai, has intermediate values in cloud base height: 1600 m in summer and 1200 m in autumn and spring. In comparison with Mt.Horai, Mt. Kurapdag has lower cloudiness in summer (1050 m).

In winter, the Kislovodsk Observatory has a high CBH value, ~1700 m. This value is close to the CBH at the BTA (1500 m).

Thus, Mt. Horai and Mt.Kurapdag are promising for mm/submm observations, not only in terms of PWV, but also in terms of total cloud cover and cloud base heights.

4.4. *Nighttime Cloud Fraction Fraction*

The estimation of the nighttime cloud fraction is one of the key parameters for evaluating the number of useful nights at an astronomical site [42–44]. In order to estimate the number of useful nights at the BTA, we used a similar approach to that described in paper [42]. We used the following methodology for classifying nights:

(i) Clear nights and partly clear nights: we supposed that cloud cover values range from 0 to 20% for the entire night. Clouds may be observed at night, but the total cloud cover ranges from 0 to 20% for 4 or more hours continuously.
(ii) Partly cloudy night: cloud coverage is less than or equal to 40% for 4 or more h.
(iii) Photometric night: a night is termed photometric if it satisfies (i).
(iv) Spectroscopic nights: a night is termed spectroscopic if it satisfies (i) and (ii).

In order to estimate the number of photometric and spectroscopic nights as well as nights with cloud coverage lower than or equal to 50% at the BTA site, we used hourly Era-5 data from 2010 to 2022. The duration of a night at a site is counted from an hour after sunset to an hour before sunrise. Figures 9 and 10 show box and whisker plots for photometric spectroscopic nights for 2010–2021. Table 5 summarizes the number of nights estimated with different thresholds for the cloud coverage at the BTA site. At the BTA site, the averaged number of spectroscopic nights per year estimated from long-term reanalysis data is 67.8 (71.2 for nights with cloud coverage lower than or equal to 50%). The number of photometric nights is less (it is equal to 25.3 per year).

Table 5. Number of nights for observations at the BTA.

Month	Nights with Cloud Cover ≤ 0.5	Spectroscopic Nights	Photometric Nights
January	4.2	4.1	1.3
February	4.2	4.1	0.9
March	7.0	6.2	1.3
April	6.9	6.2	1.7
May	8.2	8.0	3.7
June	8.1	8.0	4.0
July	7.8	7.1	4.0
August	7.1	7.0	3.0
September	5.7	5.1	2.1
October	4.0	4.0	1.1
November	4.0	4.0	1.1
December	4.0	4.0	1.1
Year	71.2	67.8	25.3

In addition, we compared the annual changes in amount of hours from visual observations at the BTA [45] with the hours estimated from Era-5 reanalysis data. Figure 11 shows annual changes in the amount of hours from visual observations at the BTA with the hours estimated from Era-5 reanalysis data. For spectral nights, the Era-5 data have the best agreement with visual observations at a threshold of 0.5. The total observing time at the BTA is 1453 h per year (Table 6). As Table 6 shows, one of the best sites in Dagestan, namely Mt. Kurapdag, is characterized by a large number of observation hours. The mean number of hours for the period from 2010 to 2021 was 1971 h.

Table 6. Estimations of observing time at the BTA. The value in December 2022 is the forecast variable.

BTA			Kurapdag	
Month	**Hours**	**Hours**	**Hours**	**Hours**
	2022	**2010–2021**	**2022**	**2010–2021**
January	66	108	189	214
February	109	97	200	173
March	63	91	170	139
April	102	88	127	104
May	96	62	123	84
June	50	72	142	101
July	105	106	87	116
August	180	154	185	163
September	139	155	193	180
October	200	178	233	209
November	148	188	250	231
December	142	154	305	257
Year	1400	1453	2204	1971

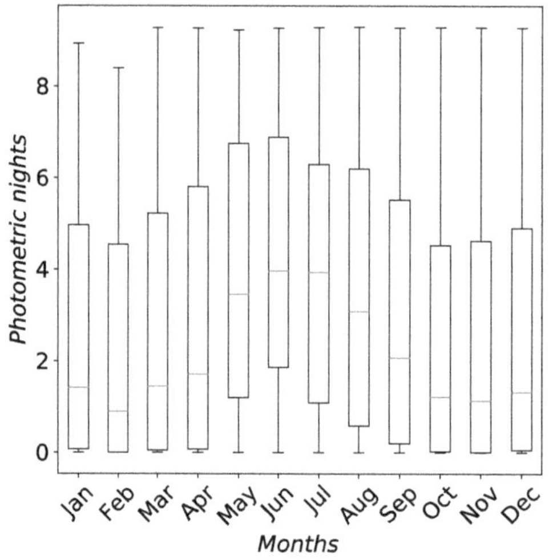

Figure 9. Box and whisker plots for photometric nights during 2010–2021 at the BTA site. The central horizontal lines inside the boxes are medians; bottoms and tops of the boxes are standard deviations; and ends of the whiskers are the minimum and maximum values.

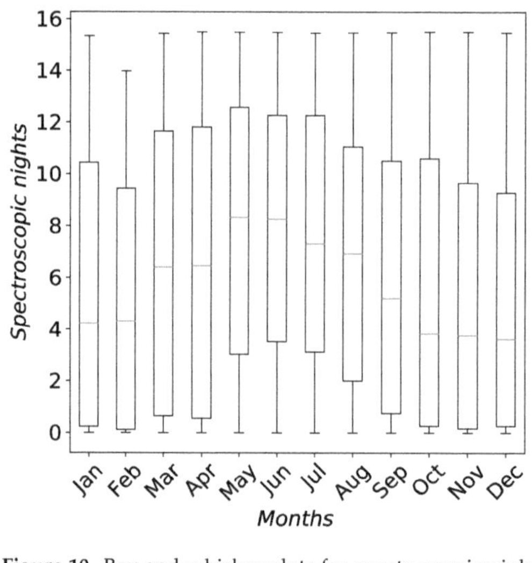

Figure 10. Box and whisker plots for spectroscopic nights during 2010–2021 at the BTA site. The central horizontal lines inside the boxes are medians; bottoms and tops of the boxes are ± standard deviation; and ends of the whiskers are the minimum and maximum values.

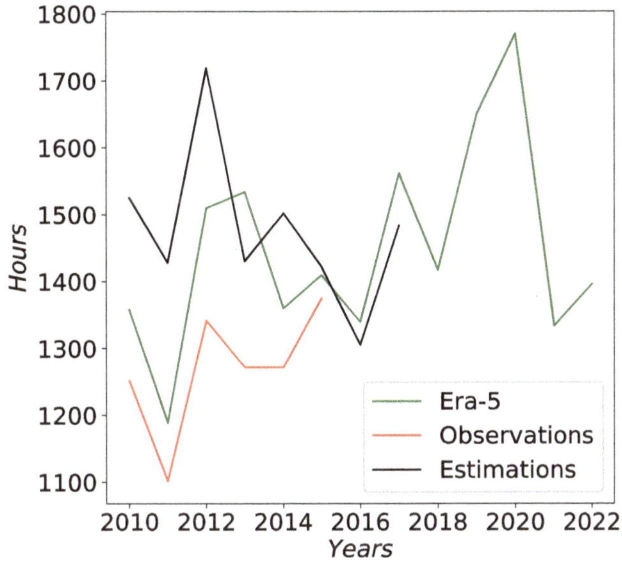

Figure 11. Annual changes in amount of hours from visual observations at the BTA with the hours estimated from Era-5 reanalysis data. The amount of observation time at the BTA according to the operation service is shown by the black line. The amount of observation time at the telescope according to astronomical observations is shown by the red line. The value in 2022 is the forecast variable (regression analysis).

5. Seasonal Variations of PWV at the Ali, Muztag-Ata, Suffa, Bta and Peak Terskol

For estimation of water vapor content within the BTA region, we corrected PWV values for a number of sites. Using Formula (2), we estimated the proportionality coefficients at different sites, taking into account ratios between the altitudes of the mountain peak and the "surrounding terrain".

Our method of PWV correction includes:

(i) The choice of area boundaries within which we should estimate the mean relative altitude difference. In particular, at the BTA, we used a limited area, which includes 4×5 grid nodes.

(ii) To calculate the relative altitude difference for an adjacent grid node, we shifted the selected area. Shifts are shown by the red box. The size of this area is fixed for each mesh node.

(iii) Knowing the relative altitude difference at a given site, we calculated the average ratios between precipitable water vapor at the mountain summit and the values corresponding to the "surrounding terrain".

Figure 12 shows seasonal changes in corrected median PWV values estimated from the Era-5 data at the sites of Ali, Muztag-Ata, Suffa, Bta and Peak Terskol.

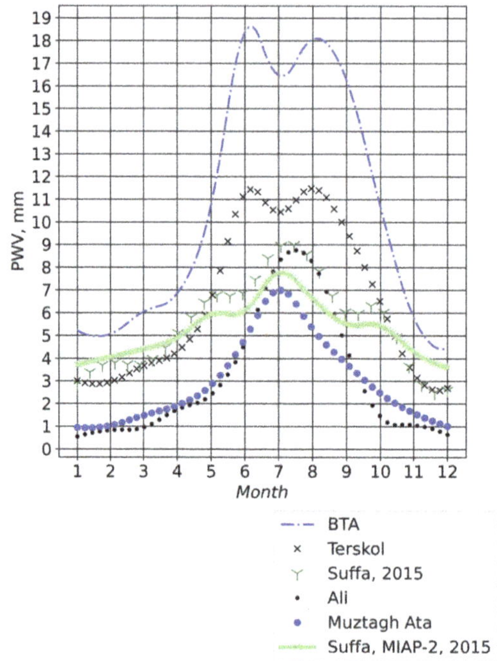

Figure 12. Seasonal changes in corrected median PWV values estimated from Era-5 reanalysis data at the sites of Ali, Muztag-Ata, Suffa, Bta and Peak Terskol.

Table 7 shows the mean ratios K_0 of the precipitable water vapor at the summit to the PWV values corresponding to the "surrounding terrain". Table 8 shows the mean values of PWV at the sites suitable for mm/submm observations.

Table 7. Mean ratios of precipitable water vapor at the summit to the PWV values corresponding to the "surrounding terrain" K_0 at the sites of Ali, Muztag-Ata, Suffa, Bta and Peak Terskol.

Site	Elevation, m	PWV/PWV_0
Ali	5050	0.72
Muztag -Ata	4536	0.75
BTA	2100	0.81
Terskol	3100	0.56
Suffa	2500	0.73

At the Ali site, the median value of PWV calculated from the Era-5 data is about 2.9 mm. This value exceeds the median value of PWV by a factor of 1.44 estimated in [10] (2.02 mm). The maximum PWV values are observed in summer. The median value is 7.0 mm. In winter, the median value is about 1 mm.

Similar astroclimatic conditions in terms of median PWV values are observed at Muztag-Ata. The median of PWV is 2.9 mm. In comparison with Ali, Muztag-Ata is characterized by a smaller intra-annual amplitude of PWV changes. In summer, the median is 5.7 mm; in winter, the median decreases to 1.0 mm.

Table 8. Mean values of PWV at the sites suitable for mm/submm observations.

Site	Season	PWV, mm
Ali	Winter	0.7
	Spring	1.7
	Summer	7.0
	Autumn	2.3
	Year	2.9
Muztag-Ata	Winter	1.0
	Spring	2.1
	Summer	5.7
	Autumn	2.6
	Year	2.9
Suffa	Winter	3.1
	Spring	5.2
	Summer	8.0
	Autumn	5.2
	Year	5.4
BTA	Winter	4.9
	Spring	6.5
	Summer	17.4
	Autumn	13.5
	Year	10.3
Terskol	Winter	2.9
	Spring	4.9
	Summer	11.1
	Autumn	6.5
	Year	6.3

At the Suffa International Observatory, medians in PWV are higher compared to Muztag-Ata and Ali. The annual median PWV derived from Era-5 reanalysis is 5.4 mm. The best conditions are observed in December–January when the median values are close to 2.6–2.9 mm. The worst conditions are in July with a median of 8.9 mm. Comparison of the calculated PWV with the PWV measured by MIAP-2 radiometer [46–48] shows that MAE is 0.79 mm, and the root mean square deviation is close to 0.87 mm. The correlation coefficient between the measured and calculated monthly PWV values is 0.98.

Following the proposed approach, the medians of PWV were estimated at the Peak Terskol and the site of BTA. At the BTA, the annual median is 10.3 mm. In winter, the median is minimal and equal to 4.9 mm; in summer, the median increases to 17.4 mm. In comparison with the BTA, at the Peak Terskol, the annual median of PWV decreases 1.63 times (the median is 6.3 mm). In summer, at the Peak Terskol, the median varies from 10.5 to 11.5 mm. In winter, the median is 2.9 mm (the minimum of PWV is observed in December).

6. Discussion

The study is aimed at the description of astroclimatic conditions within the BTA region (from 35°E to 55°E, from 40°N to 50°N). For the first time, spatial distributions of PWV within the BTA region were obtained. We show that a stable vast area with low water vapor content is formed within the BTA region. We associate this area with high transparency of the atmosphere for mm/submm radiation. In order to estimate water vapor content most closely matched to the measured values, we proposed a method for correcting PWV values, taking into account the relief. The method is based on averaging the elevation of grid notes within a certain area. In calculations, this area contains a certain number of grid nodes (4×5 nodes). For each grid node, the relative altitude differences between the mountain top and the surrounding terrain, as well as the corresponding proportionality coefficients, were calculated.

Using the correction method, we estimated the medians of PWV at the Chajnantor Plateau, Ali, Muztag-Ata, Suffa, BTA, Peak Terskol, Mt. Horai and Mt. Kurapdag. We showed that the Era-5 reanalysis reproduces changes in hourly PWV values with a correlation coefficient of 0.57. The consistency of the reanalysis data improves with the measured variations in terms of mean monthly PWV values. The correlation coefficient increases to 0.97.

In the calculations, we used the exponential dependence of PWV on the altitude [49]. Figure 13 shows the correspondence between medians of PWV and site elevation.

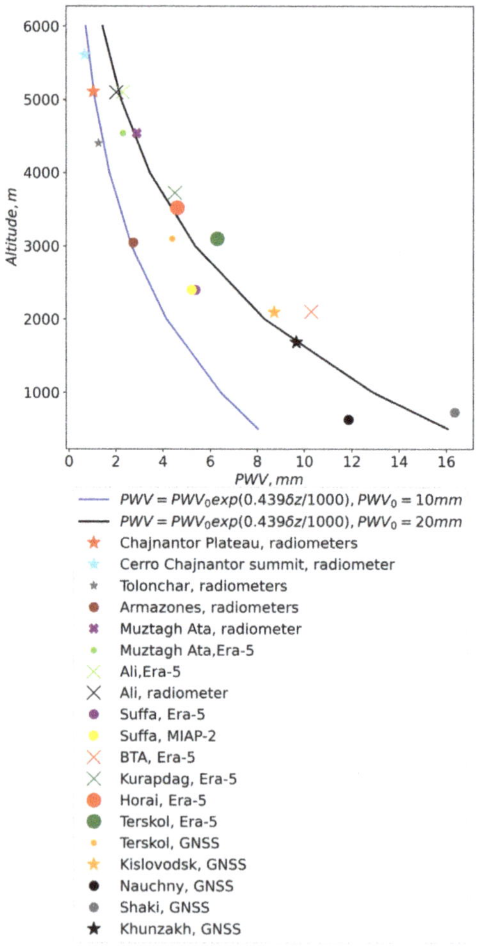

Figure 13. Dependencies of PWV medians on site elevation above sea level.

With the aim of comparing the PWV dependence, we estimated the medians of PWV at the Chajnantor Plateau, at the sites of Cerro Chajnantor Summit, Tolonchar and Armazones. It should be noted that the authors [4] considered a limited range of altitudes (from 3000 to 5640 m). In this range of altitudes, the authors obtained the dependence:

$$PWV = 2.75exp\left(-\frac{z-3000}{1820}\right). \quad (9)$$

However, if we consider a wider range of altitudes, we see that the Chajnantor Plateau, Cerro Chajnantor Summit, Tolonchar and Armazones are described well by the dependence:

$$PWV = 10exp\left(-\frac{0.439\delta z}{1000}\right) = 10exp\left(-\frac{\delta z}{2280}\right). \tag{10}$$

The analysis of the figure shows that in comparison with the sites of South America, Ali and Muztag-Ata have higher PWV and are described by the dependence:

$$PWV = 20exp\left(-\frac{0.439\delta z}{1000}\right) = 20exp\left(-\frac{\delta z}{2280}\right). \tag{11}$$

The PWV estimates obtained at these sites are in good agreement with theoretical dependencies of PWV on altitude. The mean absolute errors decrease with altitude and do not exceed 2.0 mm, on average. We can note that the reanalysis data somewhat overestimate the PWV medians by 1–2 mm compared to the measurement data. A small spread of PWV values indicates that using the reanalysis data and the proposed method, in general, we can estimate the medians of PWV reliably. Considering cases with low values of TCC as well as PWV, we can note that Mt. Kurupdag and Mt. Horai are located in an area suitable for mm/submm astronomical observations.

7. Conclusions

(i) In this article, we make use of the Era-5 reanalysis from 2010 to 2020 over the BTA region to summarize the empirical relation between the total amount of PWV in the atmospheric column and $PWV(z)$. Our results confirm that PWV in the surface layer of the atmosphere and the water vapor scale height affect the total amount of PWV. The functional relation between the total amount of PWV and $PWV(z)$ are similar among the discussed sites.

(ii) We proposed a method for correcting the PWV values which takes into account the water vapor scale height calculated for the nearest radiosounding station and underlying surface. The method is based on the calculation of the average elevation of the grid nodes around the site of interest. Within the BTA region, we calculated δz using 20 grid nodes for every site. We believe that taking into account the local orography makes it possible to more accurately parameterize the PWV. Based on the proposed method, the distributions of precipitable water vapor within the BTA region in different seasons were obtained. The analysis of PWV spatial distributions showed that the BTA is located in the belt with low water vapor content, extending southeastward. One of the main conclusions in this paper is that potential sites with low PWV are located east and southeast of BTA in the region (40.5°N–42.0°N; 46.2°E–48.7°E). In addition, using the method we estimated seasonal changes in corrected median PWV values at the sites of Ali, Muztag-Ata, Suffa, Bta and Peak Terskol. The statistics obtained are close to the measured PWV at these sites.

(iii) The Era-5 reanalysis passably describes the hourly fluctuations in PWV. The root mean square deviation between measured and calculated values of PVW within the Chajnantor area is 1.06 mm. The correlation coefficient is 0.57. For longer averaging periods, consistency of PWV variations estimated from the Era-5 data and radiometric measurements improves. The correlation coefficient increases to 0.97 for monthly PWV values.

(iv) We found that there are 68–71% spectroscopic nights per year at the BTA. These estimations are in a good agreement with the visual observations. At the BTA, the number of photometric nights derived from the Era-5 data is underestimated by 20% in comparison to that from the visual observations and equal to 25 per year. Using the reanalysis, we estimate that the observing time at the telescope is 1453 hours. At one of the best sites that we found, namely Mt. Kurapdag, the mean number of hours for the period from 2010 to 2021 was 1971 h.

Author Contributions: Investigation, visualization, writing—review and editing: A.Y.S.; methodology: P.G.K. and V.B.K.; formal analysis, investigation, visualization: A.Y.S. and A.V.K.; writing—review and editing: A.Y.S.; software, visualization: A.V.K. All authors have read and agreed to the published version of the manuscript.

Funding: This research was funded by RSF grant № 22-72-00049. The study of cloud cover was supported by the Ministry of Science and Higher Education of the Russian Federation.

Data Availability Statement: Data used are available on request from the corresponding author.

Acknowledgments: Measurements of diffuse light were carried out using the Unique Research Facility Large Solar Vacuum Telescope http://ckp-rf.ru/usu/200615/.

Conflicts of Interest: The authors declare no conflict of interest.

References

49. Wootten, A. The Atacama Large Millimeter Array (ALMA). *Proc. SPIE* **2002**, *4837*, 110–118. [CrossRef]
2. Tremblin, P.; Schneider, N.; Minier, V.; Durand, G.A.; Urban, J. Worldwide site comparison for submillimetre astronomy. *Astron. Astrophys.* **2012**, *548*, A65. [CrossRef]
3. Cortes, F.; Cortes, K.; Reeves, R.; Bustos, R.; Radford, S. Twenty years of PWV measurements in the Chajnantor Area. *Astron. Astrophys.* **2020**, *640*, A126. [CrossRef]
4. Otarola, A.; Breuck, C.D.; Travouillon, T.; Matsushita, S.; Nyman, L.A.; Wootten, A.; Radford, S.J.E.; Sarazin, M.; Kerber, F.; Perez-Beaupuits, J.P. Precipitable Water Vapor, Temperature, and Wind Statistics At Sites Suitable for mm and Submm Wavelength Astronomy in Northern Chile. *Publ. Astron. Soc. Pac.* **2019**, *131*, 045001. [CrossRef]
5. Xu, J.; Li, M.; Esamdin, A.; Wang, N.; Pu, G.; Wang, L.; Feng, G.; Zhang, X.; Ma, S.; Lv, J.; et al. Site-testing at Muztag-Ata site. IV. Precipitable Water Vapor. *PASP* **2022**, *134*, 015006. [CrossRef]
6. Ozdemir, S.; Yesilyaprak , C.; Aktug, B.; Ozturk, D.; Coker, D.; Balbay, R. Precipitable water vapor (PWV) estimations from the site of the Eastern Anatolia Observatory (DAG), a new astronomical observatory in Turkey. *Exp. Astron.* **2018**, *46*, 323–336. [CrossRef]
7. Valenziano, L.; Dall'Oglio, G. Millimetre Astronomy from the High Antarctic Plateau: Site Testing at Dome C. *Publ. Astron. Soc. Aust.* **1999**, *16*, 167–174. [CrossRef]
8. Kuo, C.-L. Assessments of Ali, Dome A, and Summit Camp for mm-wave Observations Using MERRA-2 Reanalysis. *Astrophys. J.* **2017**, *848*, 64. [CrossRef]
9. Deng, L.; Yang, F.; Chen, X.; He, F.; Liu, Q.; Zhang, B.; Zhang, C.; Wang, K.; Liu, N.; Ren, A.; et al. Lenghu on the Tibetan Plateau as an astronomical observing site. *Nature* **2021**, *596*, 353–356. [CrossRef]
10. Qian, X.; Yao, Y.; Wang, H.; Zou, L.; Li, Y.; Yin, J. Validation of the WRF Model for Estimating Precipitable Water Vapor at the Ali Observatory on the Tibetan Plateau. *Publ. Astron. Soc. Pac.* **2020**, *132*, 125003. [CrossRef]
11. Zhao, Y.; Yang, F.; Chen, X.; Zhang, X.; Ma, J.; Kong, X.; Fu, X.; Li, R.; Wei, Y.; Yao, Z.; et al. Long-term variations in precipitable water vapor and temperature at Lenghu Site. *Astron. Astrophys.* **2022**, *663*, A34. [CrossRef]
12. Ningombam, S.S.; Sethulakshmy, E.S.; Jade, S.; Shrungeshwara, T.S.; Vivek, S.G.; Angchuk, D.; Prabhu, T.P.; Mahay, T.T. Atmospheric opacity using 220 GHz (1.36 mm) radiometer data and water vapor trends over Indian Astronomical Observatory (IAO), Hanle. *J. Atmos.-Sol.-Terr. Phys.* **2020**, *208*, 4105404. [CrossRef]
13. Dumka, U.C.; Kaskaoutis, D.G.; Khatri, P.; Ningombam, S.S.; Sheoran, R.; Jade, S.; Shrungeshwara, T.S.; Rupakheti, M. Water vapour characteristics and radiative effects at high-altitude Himalayan sites. *Atmos. Pollut. Res.* **2022**, *13*, 101303. [CrossRef]
14. Anupama, G.C.; Maheswar, G.; Sriram, S.; Sivarani, T.; Parihar, P.S.; Nagabhushan, S.; Angchuk, D.; Barway, S.; Bhatt, B.C.; Banyal, R.; et al. A 10-m class national large optical-IR telescope. *J. Astrophys. Astron.* **2022**, *43*, 32. [CrossRef]
15. Erasmus, D.A.; Rooyen, R. A satellite survey of cloud cover and water vapor in northwest Africa and southern Spain. *Proc. SPIE* **2006**, *6267*, 62671O. [CrossRef]
16. Li, Z.; Muller, J.-P.; Cross, P. Comparison of precipitable water vapor derived from radiosonde, GPS, and Moderate-Resolution Imaging Spectroradiometer measurements. *J. Geophys. Res. Atmos.* **2003**, *108*, 4651. [CrossRef]
17. Dembelov, M.G.; Bashkuev, Y.B. Estimation of the Tropospheric Moisture Content Derived from GPS Observations, Radio Sounding Data, and Measurements with a Water Vapor Radiometer. *Atmos. Ocean. Opt.* **2022**, *35*, 359–365. [CrossRef]
18. Berezin, I.A.; Timofeyev, Y.M.; Virolainen, Y.A.; Volkova, K.A. Comparison of ground-based microwave measurements of precipitable water vapor with radiosounding data. *Atmos. Ocean. Opt.* **2016**, *29*, 274–281. [CrossRef]
19. Gong, Y.; Liu, Z. Evaluating the Accuracy of Jason-3 Water Vapor Product Using PWV Data from Global Radiosonde and GNSS Stations. *IEEE Trans. Geosci. Remote Sens.* **2021**, *59*, 4008–4017. [CrossRef]
20. Hu, H.; Yang, R.; Lee, W.-C.; Cao, Y.; Mao, J.; Gao, L. Multi-sensor study of precipitable water vapor and atmospheric profiling from microwave radiometer, GNSS/MET, radiosonde, and ECMWF reanalysis in Beijing. *J. Appl. Remote Sens.* **2020**, *14*, 044514. [CrossRef]
21. Gurbuz, G.; Akgul, V.; Gormus, K.S.; Kutoglu, S.H. Assessment of precipitable water vapor over Turkey using GLONASS and GPS. *J. Atmos.-Sol.-Terr. Phys.* **2021**, *222*, 105712. [CrossRef]

22. Zhu, D.; Zhang, K.; Yang, L.; Wu, S.; Li, L. Evaluation and calibration of MODIS near-infrared precipitable water vapor over China using GNSS observations and ERA-5 reanalysis dataset. *Remote Sens.* **2021**, *13*, 2761. [CrossRef]
23. Xu, K.; Zhong, L.; Ma, Y.; Zou, M.; Huang, Z. A study on the water vapor transport trend and water vapor source of the Tibetan Plateau. *Theor. Appl. Climatol.* **2020**, *140*, 1031–1042. [CrossRef]
24. Chen, B.; Yu, W.; Wang, W.; Zhang, Z.; Dai, W. A Global Assessment of Precipitable Water Vapor Derived From GNSS Zenith Tropospheric Delays With ERA5, NCEP FNL, and NCEP GFS Products. *Earth Space Sci.* **2021**, *8*, e2021EA001796. [CrossRef]
25. Khutorova, O.G.; Khutorov, V.E.; Teptin, G.M. Interannual Variability of Surface and Integrated Water Vapor and Atmospheric Circulation in Europe. *Atmos. Ocean. Opt.* **2018**, *31*, 486–491. [CrossRef]
26. Slättberg, N.; Lai, H.-W.; Chen, X.; Ma, Y.; Chen, D. Spatial and temporal patterns of planetary boundary layer height during 1979–2018 over the Tibetan Plateau using ERA5. *Int. J. Climatol.* **2022**, *42*, 3360–3377. [CrossRef]
27. Hellemeier, J.A.; Yang, R.; Sarazin, M.; Hickson, P. Weather at selected astronomical sites–an overview of five atmospheric parameters. *Mon. Not. R. Astron. Soc.* **2019**, *482*, 4941–4950. [CrossRef]
28. Han, Y.; Wu, X.; Luo, T.; Qing, C.; Yang, Q.; Jin, X.; Liu, N.; Wu, S.; Su, C. New C_n^2 statistical model based on first radiosonde turbulence observation over Lhasa. *J. Opt. Soc. Am.* **2020**, *37*, 995–1001. [CrossRef]
29. Kumar, A.; Kumar, S.; Pratap, V.; Singh, A.K. Performance of water vapour retrieval from MODIS and ECMWF and their validation with ground based GPS measurements over Varanasi. *J. Earth Syst. Sci.* **2021**, *130*, 41. [CrossRef]
30. Yuan, P.; Hunegnaw, A.; Alshawaf, F.; Awange, J.; Klos, A.; Teferle, F.N.; Kutterer, H. Feasibility of ERA5 integrated water vapor trends for climate change analysis in continental Europe: An evaluation with GPS (1994–2019) by considering statistical significance. *Remote Sens. Environ.* **2021**, *260*, 112416. [CrossRef]
31. Plahina, I.N.; Pankratova, N.V.; Makhotkina, E.L. Comparison of ground and satellite monitoring of aerosol optical thickness of the atmosphere in Russia. *Sovrem. Probl. Distantsionnogo Zondirovaniya Zemli Kosmosa* **2018**, *15*, 225–234. [CrossRef]
32. Shikhovtsev, A.Y.; Bolbasova, L.A.; Kovadlo, P.G.; Kiselev, A.V. Atmospheric parameters at the 6-m Big Telescope Alt-azimuthal site. *Mon. Not. R. Astron. Soc.* **2020**, *493*, 723–729. [CrossRef]
33. Nosov, V.V.; Lukin, V.P.; Nosov, E.V.; Torgaev, A.V.; Afanas'ev, V.L.; Balega, Y.Y.; Vlasyuk, V.V.; Panchuk, V.E.; Yakopov, G.V. Astroclimate Studies in the Special Astrophysical Observatory of the Russian Academy of Sciences. *Atmos. Ocean. Opt.* **2019**, *32*, 8–18. [CrossRef]
34. Khaikin, V.; Lebedev, M.; Shmagin, V.; Zinchenko, I.; Vdovin, V.; Bubnov, G.; Edelman, V.; Yakopov, G.; Shikhovtsev, A.; Marchiori, G.; et al. On the Eurasian SubMillimeter Telescopes Project (ESMT). In Proceedings of the 7th All-Russian Microwave Conference (RMC), Moscow, Russia, 25–27 November 2020; pp. 47–51. [CrossRef]
35. Burgan, H.I.; Vaheddoost, B.; Aksoy, H. Frequency Analysis of Monthly Runoff in Intermittent Rivers. In Proceedings of the World Environmental and Water Resources Congress 2017: Hydraulics and Waterways and Water Distribution Systems Analysis—Selected Papers from the World Environmental and Water Resources Congress 2017, Sacramento, CA, USA, 21–27 May 2017; pp. 327–334. [CrossRef]
36. Burgan, H.I.; Aksoy, H. Daily flow duration curve model for ungauged intermittent subbasins of gauged rivers. *J. Hydrol.* **2022**, *604*, 127249. [CrossRef]
37. Eris, E.; Cavus, Y.; Aksoy, H.; Burgan, H.I.; Aksu, H.; Boyacioglu, H. Spatiotemporal analysis of meteorological drought over Kucuk Menderes River Basin in the Aegean Region of Turkey. *Theor. Appl. Climatol.* **2020**, *142*, 1515–1530. [CrossRef]
38. Moriasi, D.N.; Arnold, J.G.; Van Liew, M.W.; Bingner, R.L.; Harmel, R.D.; Veith, T.L. Model evaluation guidelines for systematic quantification of accuracy in watershed simulations. *Trans. ASABE* **2007**, *50*, 885–900. [CrossRef]
39. Shikhovtsev, A.Y.; Khaikin, V.B.; Mironov, A.P.; Kovadlo, P.G. Statistical Analysis of the Water Vapor Content in North Caucasus and Crimea. *Atmos. Ocean. Opt.* **2022**, *35*, 165–175. [CrossRef]
40. Hersbach, H.; Bell, B.; Berrisford, P.; Hirahara, S.; Horányi, A.; Muñoz-Sabater, J.; Nicolas, J.; Peubey, C.; Radu, R.; Schepers, D.; et al. The ERA5 global reanalysis. *Q. J. R. Meteorol. Soc.* **2020**, *146*, 1999–2049. [CrossRef]
41. Kudryavtsev, D.O.; Vlasyuk, V.V. The Largest Russian Optical Telescope BTA: Current Status and Modernization Prospects. In Proceedings of the Ground-Based Astronomy in Russia—21st Century All-Russian Conference, Nizhny Arkhyz, Russia, 21–25 September 2020; pp. 21–31._2020_21. [CrossRef]
42. Ningombam, S.S.; Song, H.-J.; Mugil, S.K.; Dumka, U.C.; Larson, E.J.L.; Kumar, B.; Sagar, R. Evaluation of fractional clear sky over potential astronomical sites. *Mon. Not. R. Astron. Soc.* **2021**, *507*, 3745–3760. [CrossRef]
43. Zhang, J.-C.; Ge, L.; Lu, X.-M.; Cao, Z.-H.; Chen, X.; Mao, Y.-N.; Jiang, X.-J. Astronomical Observing Conditions at Xinglong Observatory from 2007 to 2014. *PASP* **2015**, *127*, 1292. [CrossRef]
44. Ningombam, S.S.; Narendra, A.; Bhatt, B.C.; Prabhu, T.P.; Anupama, G.C.; Angchuk, D.; Jorphail, S. Validation of estimated cloud fraction from MERRA-2 and AIRS data using ground based observation over IAO, Hanle. *Adv. Space Res.* **2020**, *66*, 826–843. [CrossRef]
45. Panchuk, A.V. Astronomical climate of the telescope installation site and observation time loss. *INASAN Sci. Rep.* **2020**, *5*, 344–350.
46. Bubnov, G.M.; Abashin, E.B.; Balega, Y.Y.; Bolshakov, O.S.; Dryagin, S.Y.; Dubrovich, V.K.; Marukhno, A.S.; Nosov, V.I.; Vdovin, V.F.; Zinchenko, I.I. Searching for New Sites for THz Observations in Eurasia. *IEEE Trans. Terahertz Sci. Technol.* **2015**, *5*, 64–72. [CrossRef]

47. Bubnov, G.; Vdovin, V.; Khaikin, V.; Tremblin, P.; Baron, P. Analysis of variations in factors of specific absorption of sub-terahertz waves in the Earth's atmosphere. In Proceedings of the 7th All-Russian Microwave Conference (RMC), Russia, Moscow, 25–27 November 2020; Volume 5, pp. 229–232. [CrossRef]
48. Balega, Y.; Bataev, D.K.-S.; Bubnov, G.M.; Vdovin, V.F.; Zemlyanukha, P.M.; Lolaev, A.B.; Lesnov, I.V.; Marukhno, A.S.; Marukhno, N.A.; Murtazaev, A.K.; et al. Direct Measurements of Atmospheric Absorption of Subterahertz Waves in the Northern Caucasus. *Dokl. Phys.* **2022**, *61*, 1–4. [CrossRef]
49. Khaikin, V.B.; Shikhovtsev, A.Y.; Mironov, A.P.; Qian, X. A study of the astroclimate in the Dagestan mountains Agul region and at the Ali Observatory in Tibet as possible locations for the Eurasian SubMM Telescopes (ESMT). *Proc. Sci.* **2002**, *425*, 72–79. [CrossRef]

MDPI
St. Alban-Anlage 66
4052 Basel
Switzerland
Tel. +41 61 683 77 34
Fax +41 61 302 89 18
www.mdpi.com

Remote Sensing Editorial Office
E-mail: remotesensing@mdpi.com
www.mdpi.com/journal/remotesensing

www.ingramcontent.com/pod-product-compliance
Lightning Source LLC
Chambersburg PA
CBHW041935260926
51850CB00018B/170
9783036565156